The Urban Book Series

The Urban Book Series is a resource for urban studies and geography research worldwide. It provides a unique and innovative resource for the latest developments in the field, nurturing a comprehensive and encompassing publication venue for urban studies, urban geography, planning and regional development.

The series publishes peer-reviewed volumes related to urbanization, sustainability, urban environments, sustainable urbanism, governance, globalization, urban and sustainable development, spatial and area studies, urban management, transport systems, urban infrastructure, urban dynamics, green cities and urban landscapes. It also invites research which documents urbanization processes and urban dynamics on a national, regional and local level, welcoming case studies, as well as comparative and applied research.

The series will appeal to urbanists, geographers, planners, engineers, architects, policy makers, and to all of those interested in a wide-ranging overview of contemporary urban studies and innovations in the field. It accepts monographs, edited volumes and textbooks.

Indexed by Scopus.

Eugenio Arbizzani · Eliana Cangelli ·
Carola Clemente · Fabrizio Cumo ·
Francesca Giofrè · Anna Maria Giovenale ·
Massimo Palme · Spartaco Paris
Editors

Technological Imagination in the Green and Digital Transition

Springer

Editors
Eugenio Arbizzani
Dipartimento di Architettura e Progetto
Sapienza University of Rome
Rome, Italy

Eliana Cangelli
Dipartimento di Architettura e Progetto
Sapienza University of Rome
Rome, Italy

Carola Clemente
Dipartimento di Architettura e Progetto
Sapienza University of Rome
Rome, Italy

Fabrizio Cumo
Dipartimento Pianificazione, Design,
Tecnologia dell'Architettura
Sapienza University of Rome
Rome, Italy

Francesca Giofrè
Dipartimento di Architettura e Progetto
Sapienza University of Rome
Rome, Italy

Anna Maria Giovenale
Dipartimento di Architettura e Progetto
Sapienza University of Rome
Rome, Italy

Massimo Palme
Departamento de Arquitectura
Universidad Técnica Federico Santa Maria
Antofagasta, Chile

Spartaco Paris
Dipartimento di Ingegneria Strutturale e
Geotecnica
Sapienza University of Rome
Rome, Italy

ISSN 2365-757X ISSN 2365-7588 (electronic)
The Urban Book Series
ISBN 978-3-031-29514-0 ISBN 978-3-031-29515-7 (eBook)
https://doi.org/10.1007/978-3-031-29515-7

This Springer imprint is published by the registered company Springer Nature Switzerland AG
The registered company address is: Gewerbestrasse 11, 6330 Cham, Switzerland

Committee

Sapienza University of Rome
DIAP—Department of Architecture and Design
LAB.ITECH—Laboratory of Architecture, Building Innovation and Technology, Environment and Climate Changes, Health
Fondazione Roma Sapienza

International Scientific Committee

David Allison, Clemson University, South Carolina, USA
Ruzica Bozovic-Stamenovic, National University of Singapore, China
Federico Butera, Polytechnic University of Milan, Italy
Orazio Carpenzano, Sapienza University of Rome, Italy
Ljiljana Dukanović, University of Belgrade, Serbia
Peter Droege, University of Liechtenstein, Liechtenstein
Boyan Georgiev, UAGEC-Department of Tecnology, Bulgaria
Anna Maria Giovenale, Sapienza University of Rome, Italy
Mario Losasso, University of Naples Federico II, Italy
Robinson Manguro, Kirinyaga University, Kenya
Saverio Mecca, University of Florence, Italy
Mario Morcellini, Sapienza University of Rome, Italy
Iva Muraj, Faculty of Architecture, University of Zagreb, Croatia
Silvia Naldini, Delft University of Technology, Netherland
Roberto Pagani, Polytechnic University of Turin, Italy
Massimo Palme, Federico Santa Maria Technical University, Valparaiso, Chile
Mario Raul Ramirez de Leon, University of San Carlos Guatemala, USAC, Guatemala
Fabrizio Schiaffonati, Polytechnic University of Milan, Italy
Markus Schwai, Norwegian University of Science and Technology, Norway
Begoña Serrano Lanzarote, Polytechnic University of Valencia, Spain

Wei Xing Shi, Tongji University, China
Belinda Tato, Harvard Graduate School of Design, USA

Scientific Coordination Committee

Eugenio Arbizzani, Sapienza University of Rome
Rosalba Belibani, Sapienza University of Rome
Eliana Cangelli, Sapienza University of Rome
Carola Clemente, Sapienza University of Rome
Fabrizio Cumo, Sapienza University of Rome
Alfonso Giancotti, Sapienza University of Rome
Francesca Giofré, Sapienza University of Rome
Spartaco Paris, Sapienza University of Rome

Organizing Committee

Anna Mangiatordi, Sapienza University of Rome
Elisa Pennacchia, Sapienza University of Rome
Virginia Adele Tiburcio, Sapienza University of Rome

Editorial coordination

Eugenio Arbizzani, Sapienza University of Rome
Anna Mangiatordi, Sapienza University of Rome
Mariangela Zagaria, Sapienza University of Rome

Foreword by Antonella Polimeni

Good afternoon to all participants, ladies and gentlemen, and welcome to Rome.

On behalf of the Community of Sapienza University of Rome, it is a real pleasure to welcome all of you to the first edition of the International Conference "Technological imagination in the green and digital transition". I am also pleased to give my best welcome to Dr Antonio Parenti, Head of the European Commission Representation in Italy, and to Prof. Mario Losasso, President of the Italian Society of Architectural Technology, as well as to all guests, students and colleagues.

The conference that we are about to open, organised by the Department of Architecture and Design and directed by Prof. Alessandra Capuano in cooperation with Sapienza Foundation, is to be a moment of methodological debate about built environments and the rise of contemporary urban challenges, so engaging for public and private institutions at national and international level.

The proposed key points of this conference—namely Innovation, Technology, Environment, Climate Changes and Health—are all interconnected priorities that cannot be further postponed, representing in the meantime strategic research and education activities for our University, perfectly aligned with the Italian National Recovery and Resilience plan, to be implemented in Italy as well as European member States, in order to overcome the present financial and social challenges.

I truly believe that Universities are, by definition, places of imagination, where planning the future is intended as an unavoidable "existential condition" as well as an essential moment of collective participation for an accomplished society.

Thank you for your attention, and I wish you a fruitful continuation of the conference.

Antonella Polimeni
Magnificent Rector
Sapienza University of Rome
Rome, Italy
antonella.polimeni@uniroma1.it

Foreword by Eugenio Gaudio

My warmest greetings to Dr. Antonio Parenti, Head of the European Commission Representation in Italy, to the President of the Italian Society of Architectural Technology Mario Losasso, to the Director Alessandra Capuano, and to Pietro Montani who will open with a Philosophical Lecture the Conference "Technological imagination in the green and digital transition".

A special greeting to Prof. Anna Maria Giovenale, my dear colleague and friend, who invited me to be here today. Thank you Anna Maria.

Let me also greet all other speakers as well other participant that will follow this Conference organized by the Department of Architecture and Design, together with the Fondazione Roma Sapienza.

From the very beginning, as President of the Fondazione Roma Sapienza, I supported the initiative of an international Conference on the theme of "Technological Imagination" having clear in mind that human imagination is inseparable from the "technical practice" with which it is entangled from the earliest origins of mankind, as Pietro Montani states in his book, *Technological destinies of the imagination.*

When the contents of the Conference were increasingly defined and focused around the areas of the green and digital transition, I realized that the very core of the Conference was becoming an attempt to respond to the contemporary challenges of the National Recovery and Resilience Plan, in their key role of revitalization for Research and University.

In this sense, the potential of technological culture is reaffirming its role of strategic tool for the conceiving, design and validation of future scenarios.

The sessions into which the Conference is structured, namely: Innovation, Technology, Environment, Climate Changes and Health, identified in order to outline the evolutionary scenarios of architectures and cities, allowing us to reflect at different levels on innovative models of building and management process, as well as design and products.

The goals of promoting digital transformation, supporting innovation in the production system, improving sustainability and ensuring an equitable environmental transition, find their clarification in the elaborations and experimentation presented through the contributions in the different sessions.

Modern technological innovation allowing multiple possibilities in all areas: nowadays digital technologies are enabling us to interact with people and things, all over the world.

There are astonishing, yet untapped potentials, suggesting that digitization, rather than a strict sense adaptive development, should be seen as an important evolutionary phenomenon and in the meantime a great opportunity.

Innovations connected with new technologies can provide to civil society a better quality of life, both at indoor and urban scale settings, addressing scientific development toward an effective culture of sustainability, reuse and security.

The employment of new technologies, a careful approach to the containment of land consumption as well as a careful consideration towards soil coverage modality and urban density, the recycling strategies and technological and typological redevelopment of degraded areas and buildings applying an energetic and eco-systemic approach, are the key elements for the conception of healthy and resilient urban habitats, able to adapt to the present global changes, as well as promoting prosperity, inclusiveness and social equity.

Last but not least, "health" issues, that need to be conceived at the very core of the potential determined by 'technological innovation and processes of ecological and digital transition.

The structure of the Conference is rooted on all these interrelated themes, and on that same basis also research needs to be reoriented.

I am confident that this first edition of the Technological imagination conference will contribute to pave the way of an innovative and interdisciplinary scientific approach to technology and policies for built environments, considered the real human challenge of the twenty-first century.

Thank you so much for your attention and enjoy the Conference.

Eugenio Gaudio
President
Fondazione Roma Sapienza
Rome, Italy
eugenio.gaudio@uniroma1.it

Foreword by Antonio Parenti

New European Bauhaus

Good morning,

Magnificent Rector of Sapienza University of Rome Professor Antonella Polimeni
President Fondazione Roma Sapienza Professor Eugenio Gaudio,
Director Department of Architecture and Design Professor Alessandra Capuano
and others.
Ladies and Gentlemen,

It is my pleasure to address you today and to open this International Conference "Technological Imagination in the digital and green transition" organized by Sapienza University of Rome.

Let me say that the title, the contents, and the proposals envisaged by the Conference match perfectly with the main pillars of the flagship initiative shaped by the President Ursula von der Leyen and launched in September 2021: the New European Bauhaus.

The New European Bauhaus is by nature transdisciplinary: it invites architects, designers, artists, scientists, engineers, artisans and citizens to share their expertise in preparing for the future.

With the New European Bauhaus, we want to make the European Green Deal tangible and "palpable".

We want to add a cultural dimension to the economic and technological transformation. This is essential to achieve our overarching goal: making Europe the first climate neutral continent by 2050. And thus reconciling our way of life with nature.

To get there, we need both: a real transformation of our economy and society, and a debate about how we can live in respect of nature and our planet.

The historical Bauhaus was founded in Weimar and Dessau. It turned into a worldwide movement. This did not happen by chance. Some ingredients of what made the historical Bauhaus a success can also be an inspiration for the New European Bauhaus.

Let me mention three.

The first ingredient: The historical Bauhaus was created in a time of **profound transformation**. People were facing the challenges of industrialisation. Gropius and the founders wanted to respond to the emerging needs of a new era. They aimed for solutions that were functional, affordable, but also beautiful. With this principle in mind, they shaped buildings, fabrics and furniture. They always aimed higher than just innovative design. The New European Bauhaus is also striving for this mix of aesthetics and affordability. But we want to add another element: sustainability. Because the New European Bauhaus wants to match sustainability with style.

Now, the second ingredient: **The historical Bauhaus boldly promoted new materials like steel and cement**. Today, we also need to look into new building materials. But this time, it is about sustainability. It is about materials that need less CO_2 in their production process. The New European Bauhaus wants to accelerate the transition of the built environment. It wants to scale up nature-based materials, to support circular design and architecture. Buildings are responsible for 40% of our energy consumption. And if we manage to change this, we have a chance to keep global warming below 1.5 degrees.

The third important element from the historical Bauhaus is **interdisciplinarity**. We want to convene people from different backgrounds and with different competences to share and grow their ideas and visions. We can create a better tomorrow, if culture and technology, innovation and design go hand in hand.

For our New European Bauhaus, the European Commission needs scientists, activists, artists, designers, architects and entrepreneurs. We want to include the ideas and perspectives of all ages and all backgrounds.

Today, at this conference we can contribute to this evolving New European Bauhaus network.

This project is a project of hope. It is a project of change and of economic transformation.

So I hope that this conference can contribute further to making the transformation happen and to connecting more and more people who want to make it happen.

Thank you very much and have a great conference.

Antonio Parenti
Head of the European Commission
Representation
Rome, Italy
antonio.parenti@ec.europa.eu

Foreword by Mario Losasso

Presentation of CONF.ITECH 2022

The green and digital transition represent in the contemporary research field the two new challenges for the evolution of technology within the themes of sociotechnical innovation. Consequently, technology and innovation in contemporary world must adapt to this general objective. Innovation in its hard and digital components once again becomes a central factor in the experimental propulsion that the project is assuming within a processuality and technologies that enable its conception and implementation.

Today, research is increasingly characterised by the need to focus on specialisms that lead to and contribute to the advancement of knowledge and the predictive value of what is studied in the disciplinary fields. However, with respect to the evolving complexity of phenomena, research requires continuous disciplinary interactions to be developed because we understand that one disciplinary field cannot alone address the most important challenges of contemporary society.

New forms of coexistence must be organized in a vision of interdependence and connection, while the green transition requires the definition of the limits of design action and the characteristics of the transformation processes. The new perspective of co-evolution will have to express a design attitude that allows to repair and, where necessary, rebuild the lost links between man, technology and nature.

The green and digital transition represent the two new challenges for the evolution of technology within the themes of social innovation. The Italian society of architectural technology SITdA has been working for a long time on the topics of the relationship between technology and urban and building development within a process-oriented and eco-systemic approach. In the field of technological design of architecture, the scientific society of the technology of architecture has activated research and training sensitivities on the themes of design experimentation framed within process and ecosystem dynamics, aimed at optimising the efficiency of products and processes by reducing inefficiencies and waste.

The SITdA supports research and spin-off outcome on territories through the activities of its scientific clusters. The Scientific Society SITdA has granted its patronage to the CONF.ITECH 2022 Conference, sharing its importance and topicality in view of the new challenges identified in the urban construction and environmental fields by the Next Generation EU Programme and the implementation programmes in the various nations of the European Union.

The topics that will be addressed during the three-day conference are fascinating and challenging, linking innovation, technology, environment, climate change and health.

These topics are strongly interrelated themes in which we are realising that it is impossible to deal with them separately, arriving in the most recent reflections at considering a single health for human beings and for the entire environment which is their living environment.

I would like to remind that the topic of digital culture, nature and technology was the central topic of the SITdA Naples 2020 Conference held last July with a delay due to pandemic difficulties, while the 2022 Conference of the Scientific Society is focused on the topic of the centrality of processes. As we can see, the work carried out in the Departments of Architecture and by the Scientific Societies in the area of architecture is an activity that has picked up significantly, foreshadowing new approaches, new fields of enquiry and new paradigms necessary for the new complexities that constitute the reference scenario of the future.

The experience of this Conference can provide a significant contribution to the sustainable and environmental evolution of the design area in its trans-scalar, multidisciplinary and challenging dimension, overcoming technocratic responses to a demand that requires the integration of the humanistic and technical-scientific dimensions.

Mario Losasso
President
Italian Society of Architectural
Technology—SITdA
Rome, Italy
mariorosario.losasso@unina.it

Foreword by Orazio Carpenzano

Welcoming Address from the Dean

On behalf of the Faculty, I wish to thank the organisers for asking me to give this opening address, while congratulating them on their efforts to bring together, in an international encounter, various perspectives on topics of such decisive importance for the future of our respective territories, as well as their people, living organisms and architecture.

My thanks go to Anna Maria Giovenale, Fabrizio Cumo, Eugenio Arbizzani, Carola Clemente, Eliana Cangelli and Francesca Giofrè, who will be giving talks on technological innovation, the environment, climate change and public health.

Thinking of energy in terms of how it relates to architecture during the green and digital transition means cultivating a *technological imagination*, a topic which leads to the broader question of the man–nature relationship and the possibility that architecture, by applying innovative ideas and concepts while promoting a growing social and emotional intelligence of its own, can contribute to inventing of new types of habitat for mankind on the planet earth, under a new pact for survival that allows all elements, both artificial and natural, to coexist in a sustainable balance which can serve as a preventive measure against the intrinsic destructive force of the Cosmos, an especially pressing problem where mankind has neglected certain methods for dissipating the energy of calamitous events made available by both ancient wisdom and scientific advances.

The 2021 Architecture Biennial, entitled "How Will We Live Together?", implicitly drew the attention of visitors to the need for a new approach to the man–nature relationship, following a thorough review of its historical and ethical premises. Hashim Sarkis, the curator of the exposition's seventeenth edition, passed on the following message: "In a scenario of exasperated political divisions and growing economic inequality, we call upon architects to imagine spaces in which we can all live in fruitful fellowship".

The man–nature relationship has always been a distinctive feature of humanistic and artistic thought on things technical, expressed in the construction of the *civitas*, the physical and political synthesis of civilisation. Medieval mysticism viewed nature as a foreboding wilderness, while the Renaissance redeemed the sense of *technè*, and the Romantic Period, with its high-strung, emotive outlook, led to the elaboration of the concept of the sublime.

Controlling and putting to use the energy generated by nature through sources of heat and movement (wind, sun, water), first through manual effort and then using the tools and machines produced by human ingenuity, was also a topic and challenge that led architecture to express, during the Modern Movement, boundless enthusiasm for the theories of Taylorism, which Corbusier summed up by interpreting human dwellings as machines of habitation.

But it is from the time of Vitruvius that architecture, engaged more or less explicitly with the triad of *utilitas-firmitas-venustas*, has addressed the problem of dissipating heat (or thermal inertia), as well as kinetic and elastic energy (in the case of earthquakes), at various latitudes of the globe, drawing on the available resources and raw materials. Historic Italian buildings, for example, built with walls roughly a metre thick and a structural layout measuring 4 × 4 or 5 × 5 m, have offered excellent thermo-hygrometric performance (in terms of energy consumption), as well as structural dependability (against seismic risk). In both cases the objective is to "mitigate", a term used by many modern-day scholars, the dissipation of different types of energy.

The history of architecture is filled with archetypes that need to be updated and reinvented. Think of the ingenuity it took to build Venice atop a giant underwater forest, or the aesthetic quality of the Tu'rat walls constructed by Southern Italian peasants, the windmills of Northern Europe and countless other magnificent examples of *swarm intelligence* collected by Bernard Rudofsky in his well-known book *Architecture without Architects: a short introduction to non-pedigreed architecture*, published by Doubleday & Company Inc., Garden City, (in 1964), following an exhibition at New York's Museum of Modern Art. Though, in truth, Roberto Pane and Gino Capponi had already touched on the topic in articles on the architecture of Ischia published in "Architettura e Arti decorative" in 1927, as did Giuseppe Pagano at the Milan Triennial "Rural Italian Architecture", published in the Notebooks of the Milan Triennial by Hoepli in 1936.

Looking beyond the confines of architecture, a recent reconsideration of the topic of Cinema and Energy can provide potentially useful points of affinity with architecture, especially in the collection of essays found in issues 7 and 8 of the periodical Imago, under the title *Cinema & Energy. Interdisciplinary Outlooks Combining Science, Aesthetics and Technology*, edited by Marco Maria Gazzano and Enrico Carocci (and published by Bulzoni in 2013). In an essay entitled *Dissipation and Aesthetic Experience*, the physicist Giuseppe Vitiello, in commenting on the film Transeurope Hotel by Luigi Cinque, writes: "The brain [which leads me to think of *swarm intelligence*] is described as an open system engaged in continuous exchanges with its surrounding environment. In both models and films, antinomies such as information/knowledge, feeling/knowing, blend with each other in the aesthetic experience, the favourable connection between ‘me and the object’ that characterises our existential dimension.”

Dissipation, therefore, should be seen as part of the evolution of our ecosystem, of our contemporary habitat. It gauges the possibilities for losing and exchanging, through a rekindling of collective emotional intelligence and technical and intellectual micro-revolutions. It is a risk that we must continue to face, as otherwise architecture will die, depriving man of an indispensable tool for managing the complexity of the physical habitat through creativity, in order to transfigure energy in a way that, at times, can prove so unreal, and yet so effective and indispensable, that it leads to the construction of new values and sublime beauty.

Orazio Carpenzano
Dean
Faculty of Architecture
Sapienza University of Rome
Rome, Italy
orazio.carpenzano@uniroma1.it

Acknowledgments

Thanks to:

The Magnificent Rector of the Sapienza University of Rome, Prof. Antonella Polimeni, and to the President of Foundation of the Sapienza University, Professor Eugenio Gaudio;

Dr. Antonio Parenti, Head of the European Commission's Representation in Italy, to Prof. Mario Losasso, President of the Italian Society of Architectural Technology, and to Prof. Orazio Carpenzano, Dean of Faculty of Architecture of the Sapienza University of Roma;

All the patrons of the conference: the Ministry of Ecological Transition; the European Commission; the Italian Society of Architectural Technology; the European Association for Architectural Education; Eurosolar; Healthy Urban Environment;

The Sponsors: CEFMECTP, the Joint Body for Construction Training and Safety of the City and Province of Rome; the Construction Pension Fund of the City and Province of Rome;

The Scientific Committee, all the reviewers and the Organizing Committee.

Contents

Contributors

Sofia Agostinelli Sapienza University of Rome, Rome, Italy

Hosam Al-Siah Sapienza University of Rome, Rome, Italy

Davide Allegri Polytechnic University of Milan, Milan, Italy

Maria Beatrice Andreucci Sapienza University of Rome, Rome, Italy

Eugenio Arbizzani Sapienza University of Rome, Rome, Italy

Marianna Arcieri Polytechnic University of Milan, Milan, Italy

Maria Vittoria Arnetoli University of Florence, Florence, Italy

Stefano Arruzzoli Polytechnic University of Milan, Milan, Italy

Davide Astiaso Garcia Sapienza University of Rome, Rome, Italy

Nazly Atta Polytechnic University of Milan, Milan, Italy

Gigliola Ausiello University of Naples Federico II, Naples, Italy

Maria Azzalin Mediterranean University of Reggio Calabria, Reggio Calabria, Italy

Meri Batakoja Ss. Cyril and Methodius University, Skopje, North Macedonia

Silvia Battaglia Polytechnic University of Milan, Milan, Italy

Oscar Eugenio Bellini Polytechnic University of Milan, Milan, Italy

Carla Álvarez Benito European University Foundation (EUF), Brussels, Belgium

Roberto Bianchi Mercatorum University, Rome, Italy

Leonardo Binni Polytechnic University of Marche, Ancona, Italy

Martina Bocci Polytechnic University of Turin, Turin, Italy

Andrea Bocco Polytechnic University of Turin, Turin, Italy

Arthur Bohn Polytechnic University of Turin, Turin, Italy

Roberto Bologna University of Florence, Florence, Italy

Steven Boon Housing Anywhere, Rotterdam, Netherlands

Martina Bosone Research Institute on Innovation and Services for Development of the Italian National Research Council (CNR-IRISS), Naples, Italy

Andrea Brambilla Polytechnic University of Milan, Milan, Italy

Timothy Daniel Brownlee University of Camerino, Camerino, Italy

Erica Brusamolin Polytechnic University of Milan, Milan, Italy

Maddalena Buffoli Polytechnic University of Milan, Milan, Italy

Francesca Caffari ENEA, Rome, Italy

Nicolandrea Calabrese ENEA, Rome, Italy

Gisella Calcagno University of Florence, Florence, Italy

Guido Callegari Polytechnic University of Turin, Turin, Italy

Maria Canepa University of Genoa, Genoa, Italy

Eliana Cangelli Sapienza University of Rome, Rome, Italy

Monica Cannaviello University of Campania "L. Vanvitelli", Aversa, Italy

Stefano Capolongo Polytechnic University of Milan, Milan, Italy

Cheren Cappello University of Sassari, Sassari, Italy

Barbara Cardone University of Roma Tre, Rome, Italy

Tecla Caroli Polytechnic University of Milan, Milan, Italy

Giovanni Castaldo Polytechnic University of Milan, Milan, Italy

Giulia Centi ENEA, Rome, Italy

Francesca Ciampa University of Naples Federico II, Naples, Italy

Andrea Ciaramella Polytechnic University of Milan, Milan, Italy

Adriana Ciardiello Sapienza University of Rome, Rome, Italy

Federico Cinquepalmi Sapienza University of Rome, Rome, Italy

Carola Clemente Sapienza University of Rome, Rome, Italy

Marta Cognigni Polytechnic University of Milan, Milan, Italy

Raffaella Colombo Istituto Comprensivo Rinnovata Pizzigoni, Milan, Italy

Alessandra Corneli Polytechnic University of Marche, Ancona, Italy

Nataša Ćuković-Ignjatović University of Belgrade, Belgrade, Serbia

Fabrizio Cumo Sapienza University of Rome, Rome, Italy

Laura Daglio Polytechnic University of Milan, Milan, Italy

Anna Dalla Valle Polytechnic University of Milan, Milan, Italy

Francesca Daprà Polytechnic University of Milan, Milan, Italy

Roberto D'Autilia University of Roma Tre, Rome, Italy

Alberto De Capua Mediterranea University of Reggio Calabria, Reggio Calabria, Italy

Jacopo Dell'Olmo Sapienza University of Rome, Rome, Italy

Valentina Dessì Polytechnic University of Milan, Milan, Italy

Raffaela De Martino University of Campania L. Vanvitelli, Aversa, Italy

Stefania De Medici University of Catania, Catania, Italy

Maria Giovanna Di Bitonto Polytechnic University of Milan, Milan, Italy

Marco Di Ludovico University of Naples Federico II, Naples, Italy

Mohamed Eledeisy Sapienza University of Rome, Rome, Italy

Lidia Errante Mediterranea University of Reggio Calabria, Reggio Calabria, Italy

Daniele Fanzini Polytechnic University of Milan, Milan, Italy

Emilio Faroldi Polytechnic University of Milan, Milan, Italy

Marco Ferrero Sapienza University of Rome, Rome, Italy

Maria Fianchini Polytechnic University of Milan, Milan, Italy

Irene Fiesoli University of Florence, Florence, Italy

Maria F. Figueira International Union of Property Owners (UIPI), Brussels, Belgium

Antonio Fioravanti Sapienza University of Rome, Rome, Italy

Rossella Franchino University of Campania L. Vanvitelli, Aversa, Italy

Caterina Frettoloso University of Campania L. Vanvitelli, Aversa, Italy

Valentina Frighi University of Ferrara, Ferrara, Italy

Matteo Gambaro Polytechnic University of Milan, Milan, Italy

Pablo Garrido Torres Universitat Politécnica de Catalunya, Barcelona, Spain

Vincenzo Gattulli Sapienza University of Rome, Rome, Italy

Marko Gavrilović University of Belgrade, Belgrade, Serbia

Emanuela Giancola UiE3-CIEMAT, Madrid, Spain

Francesca Giglio Mediterranea University of Reggio Calabria, Reggio Calabria, Italy

Elisabetta Ginelli Polytechnic University of Milan, Milan, Italy

Francesca Giofrè Sapienza University of Rome, Rome, Italy

Serena Giorgi Polytechnic University of Milan, Milan, Italy

Matteo Giovanardi Polytechnic University of Turin, Turin, Italy

Anna Maria Giovenale Sapienza University of Rome, Rome, Italy

Salvatore Giuffrida University of Catania, Catania, Italy

Evelyn Grillo Mediterranea University of Reggio Calabria, Reggio Calabria, Italy

Daniele Groppi Sapienza University of Rome, Rome, Italy

Maria Teresa Gullace Polytechnic University of Milan, Milan, Italy

Guillaume Habert ETH Zürich, Zürich, Switzerland

Sam Haghdady Islamic Azad University, Mashhad, Iran

Zakia Hammouni CRIR (Centre for Interdisciplinary Rehabilitation Research of Greater Montréal), Université de Montréal, Montréal, Canada;
Université McGill, Montréal, Canada;
Université du Québec à Trois-Rivière, Trois-Rivière, Canada

Giulio Hasanaj University of Florence, Florence, Italy

Mohammad Hassani Islamic Azad University, Kerman Branch, Iran

Tihana Hrastar University of Zagreb, Zagreb, Croatia

Azim Heydari Sapienza University of Rome, Rome, Italy;
Graduate University of Advanced Technology, Kerman, Iran

Dušan Ignjatović University of Belgrade – Faculty of Architecture, Belgrade, Serbia

Nataša Ćuković Ignjatović University of Belgrade – Faculty of Architecture, Belgrade, Serbia

Alexander Achille Johnson Vagelos College of Physicians and Surgeons, Columbia University, New York, USA

Fuat Emre Kaya University of Sassari, Sassari, Italy

Farshid Keynia Graduate University of Advanced Technology, Kerman, Iran

Alara Kutlu Polytechnic University of Milan, Milan, Italy

Adel Lakzadeh Islamic Azad University, Kerman Branch, Iran

Mario Lamagna Sapienza University of Rome, Rome, Italy

Massimo Lauria Mediterranean University of Reggio Calabria, Reggio Calabria, Italy

Francesco Leali UNIMORE, Modena, Italy

Adriano Magliocco University of Genoa, Genoa, Italy

Camilla Maitan Polytechnic University of Milan, Milan, Italy

Mariateresa Mandaglio Mediterranea University of Reggio Calabria, Reggio Calabria, Italy

Silvia Mangili Polytechnic University of Milan, Milan, Italy

Paola Marrone University of Roma Tre, Rome, Italy

Riccardo Marzo NCLAB, Rome, Italy

Luciana Mastrolonardo University G. d'Annunzio, Pescara, Italy

Redina Mazelli Polytechnic University of Turin, Turin, Italy

Eleonora Merolla Polytechnic University of Turin, Turin, Italy

Marco Migliore Polytechnic University of Milan, Milan, Italy

Martino Milardi Mediterranea University of Reggio Calabria, Reggio Calabria, Italy

Nikola Miletić University of Belgrade – Faculty of Architecture, Belgrade, Serbia

Jelena Milošević University of Belgrade, Belgrade, Serbia

Pietro Montani Honorary Professor of Aesthetics, Sapienza University of Rome, Rome, Italy

Ilaria Montella University of Roma Tre, Rome, Italy

Carol Monticelli Polytechnic University of Milan, Milan, Italy

Lucia Montoni University of Florence, Florence, Italy

Michele Morganti Sapienza University of Rome, Rome, Italy

Marco Morini ENEA, Rome, Italy

Noemi Morrone Istituto Comprensivo Rinnovata Pizzigoni, Milan, Italy

Erica Isa Mosca Polytechnic University of Milan, Milan, Italy

Elena Mussinelli Polytechnic University of Milan, Milan, Italy

Francesco Muzi Sapienza University of Rome, Rome, Italy

Francesco Nardi NCLAB, Rome, Italy

Giuliana Nardi University of Roma Tre, Rome, Italy

Ludovica Nasca University of Catania, Catania, Italy

Benedetto Nastasi Sapienza University of Rome, Rome, Italy

Berardo Naticchia Polytechnic University of Marche, Ancona, Italy

Maicol Negrello Polytechnic University of Turin, Turin, Italy

Aleksandra Nenadović University of Belgrade, Belgrade, Serbia

Antonio Novellino ETT SpA, Genoa, Italy

Filippo Orsini Polytechnic University of Milan, Milan, Italy

Giuseppe Orsini Sapienza University of Rome, Rome, Italy

Maria Giovanna Pacifico University of Naples Federico II, Naples, Italy

Giancarlo Paganin Polytechnic University of Milan, Milan, Italy

Massimo Palme Universidad Técnica Federico Santa María, Valparaíso, Chile

Elisabetta Palumbo University of Bergamo, Bergamo, Italy

Giulio Paparella Sapienza University of Rome, Rome, Italy

Spartaco Paris Sapienza University of Rome, Rome, Italy

Francesco Pasquale UNIMORE, Modena, Italy

Lorenzo Mario Pastore Sapienza University of Rome, Rome, Italy

Jelena Pavlović University of Belgrade, Belgrade, Serbia

Maura Percoco Sapienza University of Rome, Rome, Italy

Giacomo Pierucci University of Florence, Florence, Italy

Claudio Piferi University of Florence, Florence, Italy

Maria Rita Pinto University of Naples Federico II, Naples, Italy

Anna Pirani Centre for Theoretical Physics, Trieste, Italy

Giuseppe Piras Sapienza University of Rome, Rome, Italy

Nicola Pisani Colouree S.r.l., Genoa, Italy

Matteo Poli Polytechnic University of Milan, Milan, Italy

Riccardo Pollo Polytechnic University of Turin, Turin, Italy

Alice Paola Pomè Polytechnic University of Milan, Milan, Italy

Gianluca Pozzi Polytechnic University of Milan, Milan, Italy

Giulia Procaccini Polytechnic University of Milan, Milan, Italy

Donatella Radogna University "G. D'Annunzio" of Chieti-Pescara, Pescara, Italy

Alberto Raimondi University of Roma Tre, Rome, Italy

Andrea Rebecchi Polytechnic University of Milan, Milan, Italy

Rosaria Revellini IUAV University of Venice, Venice, Italy

Diletta Ricci Sapienza University of Rome, Rome, Italy;
Delft University of Technology, Delft, Netherlands

Guglielmo Ricciardi Polytechnic University of Turin, Turin, Italy

Alessandro Rogora Polytechnic University of Milan, Milan, Italy

Manuela Romano Polytechnic University of Milan, Milan, Italy

Rosa Romano University of Florence, Florence, Italy

Sabri Ben Rommane Erasmus Student Network AISBL (ESN), Brussels, Belgium

Laura Rosini University of Roma Tre, Rome, Italy

Massimo Rossetti IUAV University of Venice, Venice, Italy

Federica Rosso Sapienza University of Rome, Rome, Italy

Irina Rotaru Saint Germain-en-Laye, France

Helena Coch Roura Universitat Politécnica de Catalunya, Barcelona, Spain

Ana Šabanović University of Belgrade, Belgrade, Serbia

Samaneh Safaei Graduate University of Advanced Technology, Kerman, Iran

Ferdinando Salata Sapienza University of Rome, Rome, Italy

Sara Sansotta Mediterranea University of Reggio Calabria, Reggio Calabria, Italy

Antonello Monsù Scolaro University of Sassari, Sassari, Italy

Paolo Simeone Polytechnic University of Turin, Turin, Italy

Francesco Sommese University of Naples Federico II, Naples, Italy

Tianzhi Sun Polytechnic University of Milan, Milan, Italy

Chiara Tagliaro Polytechnic University of Milan, Milan, Italy

Maurizio Talamo Tor Vergata University of Rome, Rome, Italy

Andrea Tartaglia Polytechnic University of Milan, Milan, Italy

Chiara Tonelli University of Roma Tre, Rome, Italy

Agata Tonetti IUAV University of Venice, Venice, Italy

Matteo Trane Polytechnic University of Turin, Turin, Italy

Antonella Trombadore University of Florence, Florence, Italy

Maria Rosa Trovato University of Catania, Catania, Italy

Massimo Vaccarini Polytechnic University of Marche, Ancona, Italy

Carlo Vannini Sapienza University of Rome, Rome, Italy

Konstantinos Venis Polytechnic University of Milan, Milan, Italy

Maria Pilar Vettori Polytechnic University of Milan, Milan, Italy

Giulia Vignati Polytechnic University of Milan, Milan, Italy

Serena Viola University of Naples Federico II, Naples, Italy

Antonella Violano University of Campania "L. Vanvitelli", Aversa, Italy

Walter Wittich CRIR (Centre for Interdisciplinary Rehabilitation Research of Greater Montréal), Université de Montréal, Montréal, Canada

Alessandra Zanelli Polytechnic University of Milan, Milan, Italy

Edwin Zea Escamilla ETH Zürich, Zürich, Switzerland

Bojana Zeković University of Belgrade – Faculty of Architecture, Belgrade, Serbia

Alberto Zinno Stress Scarl, Naples, Italy

Nour Zreika Polytechnic University of Milan, Milan, Italy

Franca Zuccoli University of Milano-Bicocca, Milan, Italy

Milijana Živković University of Belgrade, Belgrade, Serbia

Maša Žujović University of Belgrade, Belgrade, Serbia

Chapter 1
From a Liquid Society, Through Technological Imagination, to Beyond the Knowledge Society

Anna Maria Giovenale

Abstract The paper aims to introduce the Proceedings of the *Conference "Technological Imagination in the Green and Digital Transition"*, starting from the initial idea. The Scientific Community has been invited to propose visions of technological imagination, in a time of great uncertainty and fragility, so that they could be subjected to a highly interesting analysis. The theme of "fragile" cities and habitats highlights the necessary transition from liquid society beyond the knowledge society. For the purposes of conference, it was noted that these themes, each with its own in-depth considerations, are to be found, thanks to the different contributions, in all of the various sessions. The Conclusions are to upgrade national and international research systems and to change the training modalities.

Keywords Technological Imagination · Innovation · Knowledge society · Training

1.1 The Idea of the Conference

The idea of the international conference *"Technological Imagination in the Green and Digital Transition"*[1] was born at a very particular moment, characterized by rapid change, the pandemic, the consequent economic crisis, and, as the preparatory work was underway for the conference, the outbreak of a war.

As is scientifically recognized, these factors have led to increases in our society in inequality and difficulty in accessing knowledge, while highlighting the lack of various skills and, above all, raising numerous questions about the future, to which it is difficult, and still too early, in any event, to provide answers.

[1] The international conference *"Technological Imagination in the Green and Digital Transition"* was held in Rome, at the Valle Giulia seat of the Faculty of Architecture of the Sapienza University, from 30 June to 2 July 2022. The preparatory work began in the month of June 2021.

A. M. Giovenale (✉)
Sapienza University of Rome, Rome, Italy
e-mail: annamaria.giovenale@uniroma1.it

E. Arbizzani et al. (eds.), *Technological Imagination in the Green and Digital Transition*, The Urban Book Series, https://doi.org/10.1007/978-3-031-29515-7_1

It was in this context that the idea of bringing the theme of "Technological Imagination" to the attention of the scientific community first arose and that its variations were subsequently developed. The starting point was noting that imagination, unlike perception, which implies an observation of reality, is the outcome of a cognitive synthesis combined with our plans, all aimed at creating an overall image of reality.

A singular concept, taking its cue from reflections formulated in other disciplines ("Sartre *The Imaginary. Phenomenological Psychology of Imagination*" edited by Kirchmayr 2007) restores the proper importance to design, in a sense closely tied to reality, and with the social significance typical of technological disciplines.

The assumption was that, starting from technological imagination, questions and interferences could be examined, so as to highlight, and therefore stimulate, transformations of the collective imagination, together with a growing expansion of the technical and technological universe.

The central theme, therefore, was to invite the scholarly and scientific community to propose visions of technological imagination in a time of great uncertainty and fragility, so that they could be subjected to a highly interesting analysis, under the assumption that a radical transformation of the very categories of reference was underway, and that this might even make it possible to highlight new categories.

At the same time, the reference context was also characterized by the targets set, and by the transversal priorities indicated, in the EU Next Generation Program, as part of the National Recovery and Resilience Plan, with the aim of promoting the growth of the innovation ecosystem.

These elements have made it possible, in the contemporary world, to launch, in a way that is both significant and relevant for civil society and for the scientific community, the desired processes of ecological and digital transition involving competitiveness, training, and inclusion with respect to social classes, geographic location, and gender inclusion.

Starting from these assumptions, it was decided, together with the colleagues of LAB.ITECH, the Laboratory of Architecture, Building Innovation and Technology, Environment and Climate Change, and Health of the Department of Architecture and Design of the Sapienza University of Rome, to analyze the theme of the technological imagination, first with respect to the "green and digital transition", and subsequently in terms of the topics: innovation, technology, environment, climate change, and health, addressing subjects not only of great interest to civil society, but that also the key topics addressed by the Laboratory itself.

To which a very important historical reference should be added, albeit one tied to a very different moment in time: the conference "*Culture, Technology and Metropolis*" held in Florence in 1987, on the occasion of the celebrations of Florence as the European Capital of Culture, during which designers, critics, technological figures, design experts, artists, and university professors, both from Italy and abroad, gathered to illustrate their thinking on the metropolis and its difficulties.

The proceedings were published in the volume *"The Metropolitan Technological Imagination"* (edited by Mucci and Rizzoli 1991).

Another reference deserving mention, and one also tied to a very different set of historical circumstances, namely those of an economic recession, is "The Invention of the Future", the first national conference of the SITdA, or Italian Society of Architectural Technology, held in Naples on the 7th and 8th of March 2008, during which various figures from the sectors of public policy and private investment, as well as university professors, gathered together to identify paths of action and contributions that could be reciprocally undertaken, so as to establish a synergistic path for future development.

The proceedings were published in the volume: "*The invention of the future*" (edited by De Santis et al. 2010).

The theme of linking "imagination" and "technology" is of particular interest not only to the scientific sector of technology, seeing that technological innovation (Torricelli and Lauria 2004) has always been something of a keyword, in combination with studies, research, and experimentation involving technology.[2]

How else, in fact, if not through a technological approach, though with a keen interest in bringing into play other disciplines as well, could the challenge of "technological imagination" have been launched, in a contemporary context, in order to disseminate interdisciplinary contributions, in the broadest sense of the term, during the phase of ecological and digital transition?

Technological imagination has been a subject of study and research for several years under the philosophical disciplines, as well as in the fields of anthropology and sociology, albeit under varying approaches.

In his book "*Technological Destinies of the Imagination*" (Montani 2022), Pietro Montani, a philosopher and honorary professor of aesthetics, holds that the human imagination is inseparable from technical endeavors, a relationship that dates back to the dawn of time. Over the course of history, technologies have arisen with enough transformative power to radically reorient the essential profile of forms of human life, while redesigning their destinies.

The contribution of Pietro Montani, who opened the conference with a lecture entitled: "Digital Spaces and the Material Culture", was particularly significant, offering an important introductory frame of reference for all the sessions on the theme of technological imagination.

1.2 "Fragile" Cities and Habitats: From a Liquid Society to Beyond the Knowledge Society

About twenty years have passed since the publication of the Italian translation of the book: "*Liquid Modernity*" (Bauman 2002), in which a series of reflections convey the sense of precariousness, ambiguity, and fluidity that permeates contemporary society.

---

[2] The concept of "Technological Innovation", widely discussed with regard to architectural technology, should be understood in all its various permutations involving: process, design and product.

With the public space finding itself increasingly emptied of public issues, the volume indicates nothing less than the public sphere as the place where reasons for coexistence should come together and be restored.

As Leonardo Benevolo writes in the introduction to one of his books (2012)[3]:

> Urban planning—in concrete terms: the organization of human constructions in a given area; urban and territorial programs; their initial operation or that designed for the future; discussion of these topics in various forums, from politics to civil society—is today practically a forgotten practice, playing only a vestigial role in terms of professional activities and social consideratio.

The sense of temporariness, of crumbling communities, and of abandonment of stability have grown particularly strong in the wake of the pandemic.

As is widely acknowledged, the various expressions of fragility to be found in the habitat are not attributable to climate change alone, but are also the consequence of a deeper, cultural crisis that extends to all contexts of habitation.

Once again, attention should be drawn not only to the unthinking use of resources, especially natural ones, but also to the increasing failure to attribute collective and individual values to environmental and cultural resources, as well as to knowledge.

The concepts of the information society,[4] and especially those of the knowledge society,[5] speak to this state of things.

Issues of welfare, in the forms it takes when applied to the habitat, to cities and their fragile habitats, as well as topics pertaining to the urban metabolism, the smart city, the transmission of data and information meant to increase efficiency in different sectors, and therefore the subject of innovations in systems, tools, products, and services as well, have invaded the realm of modern-day scholarly and scientific discussion, and continue to conquer significant space, making the establishment of ongoing relations between architecture and other sectors a necessity.

For the purposes of conference, it was noted that these themes, each with its own in-depth considerations, are to be found, thanks to the different contributions, in all of the various sessions.

Indeed, the very spirit of the technological imagination with which the scientific community was invited to propose its shared visions of the future has made it possible to raise questions that also touch on the topic of the knowledge society.

On the one hand, the multiple technological tools of contemporary design and construction are rapidly changing, underlining a growing complexity and continuous updates while making necessary new skills. It follows that there can be no ignoring, especially in certain cases, that knowledge undergoes a rapid obsolescence characterized by a finite "life cycle" (*From the Society of Knowledge to the Society for*

---

[3] Benevolo, L.(2012). *The Collapse of Italian Urban Planning*. Bari, IT: Laterza.

[4] A useful reference source is *Dialogue IV on Sustainability. A culture for the Information Society* (edited by Morcellini, M.), Sapienza University, 2016, which brings together contributions on the topic presented at the conferences held by the universities of the Lazio Region for the "Jubilee of Mercy".

[5] The knowledge society of today differs from the information society, in that its task is to transform information into resources and tools that allow society to act effectively.

Knowledge: the university polis and citizens of knowledge, of know-how, of knowing how to be and of knowing how to transmit, Prologue of Antonella Polimeni 2022).

At the same time, the rapid change in knowledge must be spread, understood, and accepted. It is of fundamental importance, therefore, that harmonious relations be established between resources, technologies, and society, to which end the scholarly and scientific community faces the task of constantly asking itself what type of society it wishes to build, as well as what type of knowledge is needed by a changing society.

With this in mind, visions of the future play a special role, requiring an even greater sense of responsibility, so that they can be disseminated in a way that ensures understanding, awareness, and inclusivity, all resulting in practical, and therefore effective, application.

1.3 The Organization of the Conference into Sessions

As already noted, the international conference "*Technological Imagination in the Green and Digital Transition*" was divided into five sessions which got underway following the welcoming remarks from the authorities and after the opening lecture by Pietro Montani.

Each session included, during its introductory phase: presentation of the managers of the session, and of a discussant chosen to act as the "alter ego" with respect to the session topics, along with a number of video interviews done with qualified experts on the topics addressed in the different sessions.

At the deadline for submitting the abstracts, 114 contributions had been received, including many from abroad.

It is interesting to observe that, as early as the first call for submissions, the session for which the greatest number was presented was that on the topic of "Climate Change", demonstrating the noteworthy engagement and the extensive interest of the scholarly and scientific community in the single greatest environmental risk, as well as the potential consequences facing mankind in contemporary society.

For in-depth information on the individual sessions, as well as their results, the reports drawn up by the session managers should be consulted.

Of interest herein is a review limited to some general reflections comparing the initial goals with the contributions presented.

If it is true that the digital revolution lies at the heart of the agenda of the world of design and construction, then what emerged, in general terms, from the contributions of the "*Innovation*" session, is the highly experimental nature of technological innovation, whose rapidly evolving methods and tools undoubtedly represent a great opportunity for the growth and development of sustainable cities, as well as for the construction sector and the achievement of quality results.

And this is a sector that, as has been pointed out on numerous occasions, is historically backward when it comes to innovation.

In fact, while the "*Technology*" session set itself the goal of discussing the impact of new design and manufacturing technologies on the construction of buildings and the urban environment, and on the repercussions that new housing models can have on the quality of life, including perceived quality, care was also taken to choose topics able to maintain the link between research on industrialized construction and that on sustainable development.

In the process, the need for innovative education in the fields of design and construction also came to the fore.

The goal of the "*Environment*" session was to discuss R&D models and design strategies for a low-tech environment, as well as for advanced, carbon–neutral building/plant integration achieved through low-intensity policies for the regeneration of the constructed environment, featuring elevated energy and environmental efficiency. Particular attention was focused on situations of energy poverty and economic need. The contributions presented outlined new scenarios and proposed pilot-cases, pertinent to low-intensity, high-efficiency contexts that definitely call for a useful "systemic structuring" of the relevant experiences, methods, and tools.

The "*Climate Changes*" session started from the assumption that, as things currently stand, cities are both the problem and the solution of climate change, meaning that the built environment must be rendered both adaptive and resilient to the effects of climate change and climate neutral. With this in mind, the goal was to discuss procedural models, strategies, and solutions involving design, technology, and digital advances potentially of use in defining new images of resilient cities capable of contributing to reducing the effects of climate change. The numerous contributions presented illustrated the various approaches, the different issues addressed with respect to the objectives set, and, most importantly, the different scales at which interventions could be carried out: from observation of the earth to gauging the potential of cities, as well as the environmental design of urban areas and sectors, of buildings, of building shells, plus the reuse of abandoned areas and decommissioned assets: a wealth of research and experimentation on the topic of environmental sustainability that amounts to a body of theories and good practices for the ongoing evolution of the built and urban environment in response to climate change.

The goal of the "*Health*" session was to discuss how the environmental determinants of health and their "material manifestations" could be classified and studied with respect to architectural technology, at the various scales of intervention, eventually through a dialogue of osmotic exchange with other disciplines. The call for contributions referred to visions of planning, decision-making, design, and implementation focused on people, foreseeing the short-, medium-, and long-term impacts on their health.

The contributions presented confirmed, using a variety of paradigms, theories, methodologies, and case studies, and with respect to all the different approaches and issues addressed, the validity of the initial premise that health is "the result of a complex system".[6] Also confirmed was the fact that health, as shown by how goals of

---

[6] See the contribution by Giofrè, F. that served as the introduction to the "Health" session.

the session were interpreted, definitely bore significant relevance to the other sessions of the conference as well, in particular with regard to topics of environmental and social sustainability.

1.4 Conclusions: Upgrade the National and International Research Systems and Teach How to Think

While the conclusions, in terms of the results of the sessions, will not be revealed until the related discussions, it can be stated, in general terms, that the contributions of the various sessions of the conference highlighted how unproductive it is, in modern-day reality, to maintain clear-cut boundaries and divisions between disciplines, and how an interdisciplinary approach is critically important to research activities, especially when it comes to achieving results that prove original (a characteristic that, over time, has been slightly in decline), concrete, and useful to society.

All the more so since the rapid developments in the field of science and scholarship themselves are increasingly geared toward eliminating boundaries between sectors, in keeping with the pace of technical-production transformations in industrial sectors.

The contributions of the various sessions also highlighted how certain topics were relevant across the board, to all the sessions, in that they brought up concepts, theories, methods, and tools which, at times, were also characterized by their dynamism (an added value), and which were common to multiple areas.

The characteristic of "operability" played a noteworthy role in the contributions presented, confirming the "design" aspect of the technological disciplines and their vocation for experimentation.

Naturally, the visions of the technological imagination, to be such, increasingly need to be developed, disseminated, and shared through national and international research groups that are trans-disciplinary in nature, making it possible, through a systemic approach, to arrive at continuous moments in which permanent knowledge is shared, updated, and renewed.

But certainly, not even this is sufficient.

The modern-day speed and complexity of knowledge, the high potential for the "spreading"[7] of technology, are reflected in the learning models of technological disciplines, which are closely related to the technical-production ends of the cities of the future.

On the one hand, an effort must be made to address the complexity, in assigning priorities, through new and advanced tools and methods that make it possible to maintain a dynamic relationship with the changing needs of society, while also serving as guides for research and innovation.

At the same time, there is the fundamental theme of the need to encourage critical and creative thinking regarding the processes for modifying the built environment.

---

[7] See Del Nord R (1991) Presentation. In: Mucci E and Rizzoli P (Eds.), *Metropolitan Technological Imagination*Milan, IT: FrancoAngeli. p 18.

This entails orienting education toward dynamic learning capable of absorbing new knowledge, of managing and synthesizing information, but, most importantly, of perceiving aspects relevant to all sectors while establishing connections and putting forth arguments.

In short: "The new challenge is to teach how to think" (Elkann 2018).[8]

A new way of transmitting the culture of the built environment, but also of creating new skills.

With regard to the Conference "*Technological Imagination in the Green and Digital transition*", it had already been decided, even while the proceedings were still underway, to view it as only the first of future international conferences on technological imagination, demonstrating the shared intent to repeat the encounter in a few years, when the experience acquired in the meantime will doubtless add to the value of the event.

References

Baumann Z (2002) Liquid modernity. Rome-Bari, IT, Laterza

Benevolo L (2012) The Collapse of Italian Urban Planning. Bari, IT, Laterza

Elkann A (2018) Lady Minouche Shafick: the new challenge is teaching how to think. In: La Stampa, 2 April 2018

From the Society of Knowledge to the Society for Knowledge: the university polis and citizens of knowledge, of know-how, of knowing how to be and of knowing how to transmit. Prologue by Antonella Polimeni, the Magnificent Rector of the Sapienza University of Rome, Opening of the Academic Year 2022–23, the 720th since the University's founding, Aula Magna of the Deanery of the Sapienza University of Rome, 15 September 2022

Montani P (2022) Technological destinies of the imagination. Milan, IT, Mimesis

Mucci E, Rizzoli P (eds) (1991) Metropolitan technological imagination. Milan, IT, FrancoAngeli

De Santis M, Losasso M, Pinto MR (eds) (2010) The invention of the future. In: 1st national conference of the Italian society of architectural technology (Naples, 7–8 March 2008). Florence, IT, Alinea

Sartre JP (2007) The imaginary. Phenomenological Psychol Imagination (ed. R Kirchmayr), Turin, IT, Piccola Biblioteca Einaudi

Torricelli MC, Lauria A (2004) Technological innovation for architecture: a multi-part diary. Pisa, IT, ETS

[8] The reference is to the article by Elkann, A., La Stampa, April 2, 2018 entitled: "Lady Minouche Shafick: the new challenge is teaching how to think", based on an interview with the Director of the London School of Economics and Political Science.

Chapter 2
Opening Lecture: Digital Spaces and the Material Culture

Pietro Montani

Abstract In this article, I will develop the following points: (1) The imagination is structurally technological as it is entangled with historically dominant technologies; (2) these orientate the reconfiguration of its multimodality, i.e., the fact that the imagination does not work only on the optical and visual level but extends its action to all of our sensorimotor system; (3) how this re-modeling is influenced by digital technologies remains to be clarified; and (4) in this problematic field, there are two opposing lines of development, which I will treat with some examples.

Keywords Imagination-technology entanglement · Multimodality of Imagination · Multimedia Environments

The expression "Technological Imagination" can be interpreted in two ways.

Firstly and among other things, our imagination operates brilliantly and creatively in the technological field, introducing major innovations therein not only functionally speaking but also in the aesthetic (products that are more appealing and enjoyable to use) and ethical (products that improve quality of life and the environment) sense. This first linguistic use is widespread in the field of architectural design. However, the second meaning of the phrase is more radical and specifically that a profoundly technological element is at work in the human imagination as such. I shall focus here on this second meaning, in the belief that a philosophical and anthropological approach that justifies and clarifies it will prove particularly helpful in the current phase of technological development which is strongly marked by an—often unhesitating—reoccurrence of the fear that human cultures have always had of technology. Simply think of Artificial Intelligence (on which I shall say a few words at the end of this article) and the presumed or real "dehumanization" scenarios that it seemingly indicates.

I have mentioned a philosophical and anthropological approach. Allow me to introduce this by taking a cue from a helpful concept disseminated primarily by a

---

P. Montani (✉)
Honorary Professor of Aesthetics, Sapienza University of Rome, Rome, Italy
e-mail: pietro.montani@fondazione.uniroma1.it

E. Arbizzani et al. (eds.), *Technological Imagination in the Green and Digital Transition*, The Urban Book Series, https://doi.org/10.1007/978-3-031-29515-7_2

famous Walter Benjamin essay on art in the age of its technological reproducibility but that has long and greatly ramified roots. I refer to the concept of a "field of action" (*Spiel-Raum*) that, from their earliest beginnings, human cultures managed in their relationship with material reality. It is a space that, to Benjamin's eyes, technically produced works of art reflect with particular plasticity, given that in the "decay of the aura", which distinguishes them more than all, "what is lost (...) is matched by a huge gain in the scope for play".[1]

In the Benjamin essay in question, the concept of "play" is thematic within an important distinction between two forms of technology, defined, respectively, "first and second technology", albeit not chronologically speaking. It is in the third version of the essay that the idea of a "second technology" is discussed in a way that only today, in the digital era, can Benjamin's great foresight perhaps be appreciated. His concept is that the way in which the human being's relationship with technology—which he sees as constitutive and not adventitious—is initially manifested is founded on ritual, magic, and semblance, and accompanied by a second founded essentially on play and experimentation. In this, Benjamin clearly falls within a tradition of thought that is associated in philosophy with the names of Kant (who saw the "free play of imagination and understanding" as the source of human experience) and Schiller (who spoke of a genuine "play drive" in the human being).

This is how Benjamin describes the second technology:

The origin of the second technology lies at the point where, by an unconscious ruse, human beings first began to distance themselves from nature. It lies, in other words, in play.

(...) The first technology really sought to master nature, whereas the second aims rather at an interplay between nature and humanity.

The primary social function of art today is to rehearse that interplay.[2]

In a work note of 1936, Benjamin added a striking observation on "human nature", of which he underlines the profoundly *mediated* nature: "Concept of second nature: this nature has always existed but was not previously differentiated from the first. It only became the second when the first formed within it."[3] This is a genuine paradox that I wish to reformulate as follows: the main trait of human beings consists in their ability to distance themselves from the material world and see in its spaces and opportunities for technical play. This means that "human nature" has from the very first been technical (a 'second' or 'mediated' nature) and, more precisely, that human sensitivity and imagination can, in principle, be implemented technically.

Here is a simple and effective example: the sensitivity and imagination of those who cannot see or who have lost their sight extend to the tip of the long stick they use to gain a sense of their surrounding space.

---

[1] W. Benjamin, *Selected Writings*, vol. III, edited by Howard Eiland and Michael W. Jennings (Cambridge, Harvard University Press, p. 127).

[2] Ibid., p. 107–8.

[3] Quoted from the Italian critical edition *Opere Complete di Walter Benjamin*, vol. VI, Einaudi, Turin 2004, p. 309.

This reference to *space* is particularly important here. One of the principal fields of action mentioned by Benjamin is linked to a specific and extremely technical 'game of spatializing'. In other words, the spatializing characteristic of humans stems from the fact that the human sensitivity and imagination prove technically implemented therein.

The philosopher Martin Heidegger—in many ways totally incompatible with Benjamin—had a similar thought on this point. Space is not something 'given' but the result of active spatializing. What does spatializing mean? Here is one of Heidegger's examples: imagine the action of clearing a path through a dense and compact blanket of snow. Try picturing it and you will immediately realize that it cannot be done without introducing a tool to which the human body spontaneously resorts for the execution of this operation. Heidegger's philosophical vocabulary contains several words to describe this spontaneous recourse to a technical means by the human body. One of these words is *Verlässlichkeit*, reliance. The human being *relies* on technology, adopting its resources very naturally. In a famous comment on a Van Gogh painting, Heidegger added that the very essence of the technical medium, i.e., its usability, rests on its *Verlässlichkeit*.[4]

What defines the human being's "second nature" is therefore its reliance on technique.

It is the *natural* technical mediation that allows the human to feel, experience, and modify the space wherein its life forms are organized. So, the human habitat has from the very first been a media environment, something built *within which* the unbuilt takes its authentic form. Precisely as the first, nature only takes shape within the second. A striking description of this condition can be found in an important Heidegger text, *Building, Dwelling, Thinking.*

The bridge swings over the stream "with ease and power". It does not just connect banks that are already there. The banks emerge as banks only as the bridge crosses the stream. The bridge designedly causes them to lie across from each other. One side is set off against the other by the bridge. Nor do the banks stretch along the stream as indifferent border strips of the dry land. With the banks, the bridge brings to the stream the one and the other expanse of the landscape lying behind them. It brings stream and bank and land into each other's neighborhood. The bridge *gathers* the earth as landscape around the stream.[5]

The Recapitulating: human sensitivity and imagination extend themselves spontaneously into artifacts that are incorporated, internalized, and adopted. In doing so, they differentiate their "field of action" (*Spiel-Raum*) in an ever more refined way, proving particularly receptive to the affordances present in material reality. Simply think of how many different operations can be executed with a stick and in how many different ways we can 'feel' a stick as an extension of our bodies.

But what happens when sensitivity and imagination rely on technique? The imagination, in particular, is remodeled by this technical interplay with nature and this

---

[4] Cf M. Heidegger, *Off the Beaten Track* (Cambridge: Cambridge University Press, 2002).

[5] Cf M. Heidegger, *Poetry, Language, Thought*, translated by Albert Hofstadter (New York: Harper Colophon Books, 1971.

re-modeling primarily affects one of the principal properties of the imagination, its *multimodality*—namely the fact that the imagination works not only on an optical and visual plane but extends its action to our entire sensorimotor system. The most recent acquisitions on the human brain tell us that no areas of the cerebral cortex which are specialized for certain functions—e.g., those intended for sight—cannot be given a new function. For example, in the event of supervening blindness, the areas of the cortex responsible for vision can receive input from tact or hearing. So, it is by no means metaphorical to say that blind people see with their hands (extended into a stick) or ears.

I believe that the spontaneous technical reliance of the human imagination is manifested in a highly significant manner at the level of its multimodality, which is reorganized by it. This means that the technological human imagination is continuously engaged in a complex game in which its multimodality is reorganized in a way that also redefines the crossover zone between the built and the unbuilt (remember Heidegger's effective bridge example). And that new dimensions of the "material culture" emerge in this game along with new ways of being of what we call "matter" or "material reality". In short, the technological imagination is constantly engaged in *integrating* its different components in a new manner and, thanks to this game, renegotiating the boundaries between the material and the virtual, the natural and the artifact, and the built and the unbuilt.

The verb integrate defines this game better than any other but beware, in no way does integration mean peaceful reconciliation. One example, which is striking to the same degree as it generally goes unnoticed, will help specify this point of capital importance.

In one event in the specific evolutionary history of *Homo sapiens*, the work of integration and multimodal reorganization performed by the imagination gave rise to a real paradigm. This occurred when *Homo sapiens* (and it alone) acquired the most powerful of all its technologies, spoken language. This technology is to be considered totally distinct from the expressive, mimetic, and gestural communication systems employed by the genus *homo* many thousands of years before the *sapiens* species appeared (today accredited as approximately 300,000 years ago) although, of course, no spoken language could have emerged during evolution without the support of those more archaic systems.[6]

Homo erectus, for example, was responsible for a prolific tool production. *Neanderthal* perhaps created the "Schöningen spears" which in terms of ballistic properties are by no means inferior to modern competition javelins. Clearly, these tools would have been impossible to design and use in the context of cooperative strategies in the absence of a solid system of mimic and gestural communication. But spoken language marked the emergence of something completely new. It was a system that could, and potentially inexhaustibly, classify not only the countless objects found in the world environment but also those properties of theirs that were transferable to other experience contexts (think again how many different tools can be obtained

[6] For a discussion of this point, please see Montani, P., *Technological Destinies of the Imagination* (Milan-Udine, Mimesis International, 2022).

from just a stick): an extremely powerful system for *recognizing* the potential present in material reality and *increasing it* at the same time.

So, this technology introduced a new component into the multimodality of the imagination which, on the one hand, massively boosted the degree of cognitive specialization accessible to human life forms and, on the other, forced the imagination to increase its ability to *integrate* the diverse components active in its multimodal work.

This appears obvious to us today, and spoken language seems the most natural of human resources. However, not only was it very different for many thousands of years but what we perceive as something obvious is, actually, the result of a complex work of integration. From that moment on, and irreversibly so, human experience conformed to a complex *entanglement* model that I would like to exemplify with the image of the Borromean rings.

Releasing one of these rings also releases the other two. This means that there is a constant co-determination and co-evolution among the three elements and that the integration of spoken language into the multimodality of the imagination was the first, and by far the most potent, example of "augmented reality": matter and inhabitable space were also hugely expanded and could be reorganized.

The requisite of *co-evolution* is far and away the most important of those present in the paradigm of the Borromean knot. I shall end by highlighting some points regarding the prospects of a technological imagination in the era of growing global digitalization.

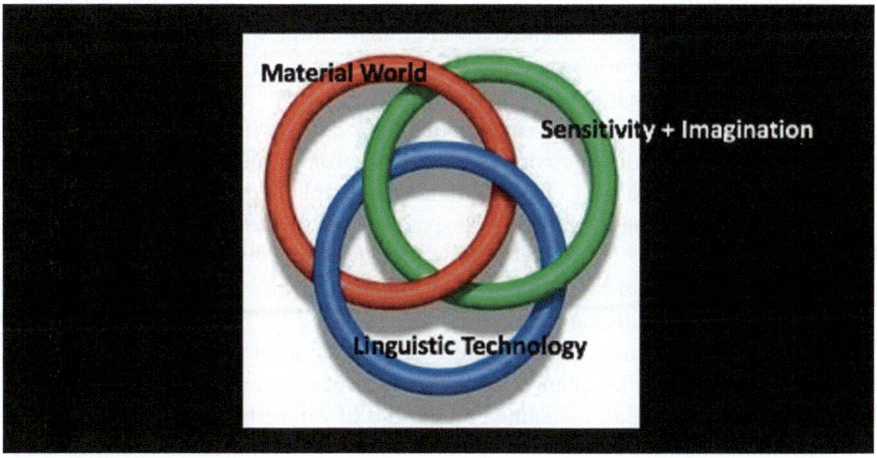

The principal question is: how do digital technologies interpret the co-evolutional paradigm I have just presented in the structure of the Borromean knot? We can look to Artificial Intelligence (AI) to move toward clarifying this point. We all know that one of the most prominent forms of spectacular progress accreditable to AI concerns machines capable of *deep learning*, namely of *autonomously* employing the learning processes of the ability required to execute a certain task (e.g., recognize images and their production). One of the most significant (and troubling) elements of *deep learning* is that, once provided with the basic instructions on the task to be performed, the machine analyzes the objects to be classified with the aid of samplings that can and usually do avail of criteria of a totally different pertinence from those in use among humans. However, although the criteria differ, these machines perform extremely accurately. Nonetheless, their most glaring limitation consists in their substantial inability to adequately adopt the element of unpredictability. As a result, the more their radius of action coincides with that of a circumscribed, self-referenced, and strictly programmed environment, the better the algorithms to which these machines respond will perform. The effective concept of *envelope* has been proposed (e.g., by Luciano Floridi)[7] for this type of environment. So, for example, to create an algorithm capable of driving a driverless car that offers the maximum safety guarantees we would have to design a motorway network specially devised for its performance, namely an *envelope*: a tendentially closed space immunized (as far as is possible) against all contingencies, filled with predictive automatisms, and therefore broadly devoid of interactive plasticity.

The design of the inhabitable spaces of our future smart cities might be based on this principle which would take to the extremes the strict *channeling of sensitivity* typically linked to security issues (the promise to shield against contingencies and the unforeseen) generally present in technology.

A good empirical criterion that better defines the action of a technological imagination in the digital era might consist in distinguishing between devices that support this autistic, anesthetic, and securitarian deviation and those which counter it directly or indirectly. Between those oriented toward transforming space into an envelope and those oriented toward enriching media environments via a growing and diversified integration of virtual spatiality and real spatiality.

To end, I would like to suggest schematizing this alternative by distinguishing between *substitute* media environments and *integrating* media environments. And I would like to suggest that only the latter can implement new prospects for the material culture, starting from technical spatialization practices, intended in all their breadth.

Here are some examples, chosen without any pretense of systematization, of which a future archeology of digital media might avail.

The most classic example of a substitute digital environment project was perhaps *Second Life* which, not by chance, actually died out in a few years. Diversely, I am by no means sure whether the same can be said of Zuckerberg's *Metaverse*. As we all know, *Metaverse* is still at the project stage, but the element of integration seems firmly present in the planned media environment. We should keep a close eye on this point.

As concerns integrating environments, here are four very simple everyday examples for our archeological outline. The first is the *Hologram protest* of 2015 via which

[7] Floridi, L. & Others, 'What the Near Future of Artificial Intelligence Could Be', *Philosophy & Technology*, Springer Nature, 2019, https://doi.org/10.1007/s13347-019-00345-y.

a group of citizens in Madrid circumvented a government ban on demonstrations by having a virtual march pass before the Spanish Parliament. The second is the totally unforeseen therapeutic effect of the *Pokémon Go* game launched a few years ago. Many autistic children freed themselves from their spatial confinement without difficulty thanks to the integration of the app, which reorganized their relationship with space in a non-threatening manner. The third example, *Be my Eyes*, is a standard smartphone app that, via the remote assistance of a partner, acts as a guide for the spatial orientation of the blind or visually impaired. It is conceivable that the smart glasses envisaged in Zuckerberg's project will opt for this integrating model rather than the substitute mode of VR headsets. The last example relates to robotics. Among the many possible examples, I shall mention only *ANYmal*, a robot designed to gain insight of and take samples of ecological changes interactively when operating in close contact with several natural environments. Lastly, in the social robotics sphere, it seems remarkable that the contrast between substitute and integrating is thematic and narratively essential in the recent fine novel *Klara and the Sun* by Kazuo Ishiguro (2021) which in a expert and informed manner foresees potential scenarios of our coevolutionary relationship with machines.

I shall end with the following claim: the processes that technically reorganize our spatiality result in the emergence of an opposition between the plasticity that asserts itself at the crossroads between cyberspace and real space and the anesthetic closure that characterizes the programmed envelopes linked to the optimization of AI performance. Designing the spatiality of media environments ought to take this opposition into account and put itself in a position to interpret their dialectics creatively.

Now, that is an objective for the material culture of our future.

Part I
Session I Innovation

The digital revolution, heralded and described not only prior to its conclusion, but before it had even been triggered, is the key topic on the agenda of the world of planning and construction. A new paradigm is needed, in order to replace the documentation-based approach with a practical method of quality control able to recognize, in structural evidence and building processes, pivotal factors of development that can be used to measure and monitor the performance of the works to be built and the parties involved.

The initial transition to the digital world, focused on three-dimensional models, necessarily led to a significant transformation in planning, design, construction, and management, in terms of both content and participants, with a focus on the organization of collaborative information flows and production procedures, optimizing instruments crucial to determining when top-quality results have been achieved.

It is precisely this surprising stock of untapped but available potential that makes digitalization appear to be less a phase of evolutionary adaptation, strictly speaking, than both a major moment of growth and a tremendous opportunity.

The session is designed as an occasion for discussing the features of the driving factors of the imminent digital revolution, with particular attention to innovative methods and instruments for controlling the quality of construction projects and processes.

Chapter 3
Innovation for the Digitization Process of the AECO Sector

Fabrizio Cumo

Abstract The conference's Innovation session gathered more than 20 papers ranging from the digitization of the construction world to the digital twin of complex infrastructure such as port areas. The common denominator of the presentations was the methodological search for the best combination of physical reality and digital replication, always putting the functionality and operability of the built environment at the center of the study. In fact, the contributions, although coming from very different research realities, clearly highlighted the inescapable path toward the integration of the digital world in the AECO sector.

Keyword Digital Twin · AECO · IOT · Digital platform

The cultural and socioeconomic shocks of the last years, deeply marked by the pandemic crisis, are proposing with the greatest urgency new thoughtful and evolutionary visions building on all positive and negative recent experiences, bearing in mind the fundamental contribution of technological innovation to cope with the present global changes scenario.

The dawn of the digital revolution heralded and described before it had even been triggered is a key issue on the agenda of the built environment, considered at any scale of analysis. It seems that a new paradigm is presently needed, in order to replace the actual documentation-based approach with a practical method of quality control, able to recognize both in structural evidence and building processes, those pivotal factors of development that can be used to measure and monitor the performance of single and multiple buildings and the parties involved.

The initial transition to the digital world, which focused essentially on three-dimensional models, led to a significant transformation in planning, design, construction processes, and management, in terms of both contents and participants, with a

F. Cumo (✉)
Sapienza University of Rome, Rome, Italy
e-mail: fabrizio.cumo@uniroma1.it

E. Arbizzani et al. (eds.), *Technological Imagination in the Green and Digital Transition*, The Urban Book Series, https://doi.org/10.1007/978-3-031-29515-7_3

focus on the organization of collaborative information flows and production procedures, optimizing instruments crucial to determining when top-quality results have been achieved.

It is precisely this surprising stock of untapped but available potential that makes digitalization appear to be less a phase of evolutionary adaptation, strictly speaking, than both a significant moment of growth and, in the meantime, a tremendous opportunity.

As a matter of fact, the digital revolution is becoming more and more an essential asset at the core of architectural design and construction activities, and this is calling for the mentioned paradigm shift from a document-based approach to a concrete application of a quality control method, to be reflected in the use of structured data within all construction process (Agostinelli et al. 2020).

This is the reason why, the 'Innovation' session within the 'Technological imagination' conference—held in Roma La Sapienza in June 2022—has been designed as an occasion for discussing those features and tools to be considered as driving factors of the mentioned digital revolution, with careful attention to innovative methodologies and instruments for controlling the quality of construction projects and processes, as well as their integration with best available technologies.

The session focus centered on the following topics proposed for the discussion to the participants:

- ICT Innovations in Architecture and Civil Engineering;
- Digital Transformation and AEC and The Role of Enterprise Architecture;
- Advances in Digital Engineering, Computing, and Simulation for the AEC Industry;
- Innovative Digital Technologies and Engineering Systems (e.g. Digital Twins);
- Project Management in design and construction processes;
- Advances in digital engineering, computing, and simulation.

The introductory statements were entrusted to highly qualified technology and communication experts. The initial contribution to the debate was an interview by the editor in chief of La Repubblica, Mr. Maurizio Molinari, who was asked to elaborate on his vision of the new professional profiles emerging as a result of the impending digital transition. Afterward, it was Professor Michael Grieves, the developer of the digital twin approach in 2002, who was asked to make a projection of prospects after the digital twin.

Data analysts are usually focusing on the process of combing data to find new relevant patterns for the business sector or other stakeholders. BI analysts are more suitable at making sense of what has happened and doing so at scale. Data analysts are better at looking for patterns that state what might happen, and vice versa. BI analysts are sounder at translating business requirements into the appropriate graphs, charts, spreadsheets, and dashboards. They tend to work close to the front line with business users and subject matter experts. They also need a solid understanding of business workflows, finance, and accounting to translate the raw data into a form that final users would find easier to use. BI analysts need a deep understanding of the technical side of working with structured databases and data warehouses. They must be fluent in writing complex SQL queries and creating complex joins across tables. Familiarity with various query optimization techniques helps ensure they create reports that cut database processing overhead.

BI analysts may get formed with very different SQL data transformation skills.

For example, they may work with extract, transform, and load tools, to transfer subsets of data from an operational database into a data warehouse to support a new query. Some basic user experience design skills can also help them identify the best way of presenting data to users that is simple and can explain the appropriate story (Cristani et al. 2015).

The next keynote speaker, Prof. Grieves, provided an overview of the state of digital innovation and pointed out that digital transition is already going further, converging into experiments in the metaverse. According to Prof. Greves, the driving force behind this transition is the Digital Transformation of business, with extensive use of advanced technologies related to artificial intelligence, digital twins, augmented and virtual reality, AR/VR goggles, and wide connectivity.

Digital transformation is already transferring part of the activities to cyberspace where even portions of the company's assets are relocating, changing their economic value; this process can be both an issue or a huge opportunity. To leverage the opportunity, most companies will need to transform their business models and their market offer. The potential issue is whether their market could also shift to the cyberspace with similar rapidity and whether it would make sense to create rapidly metaverse as a platform to meet their customers and provide them with a satisfying experience.

How and when people will actually access the metaverse is the current unknown of the matter: there will be probably a hybrid situation where part of the company (processes, activities, resources, services, software products) will be in the cyberspace and part of the customers will be accessing those by moving to the cyberspace, thus effectively transacting in the metaverse. However, still part of the company will remain in the physical space and consequently part of the transactions will be occurring in the physical space. A mixed space will keep existing where the cyber is in connection with the physical dimension for a long time.

Afterward, Prof. Maurizio Talamo, full professor of cyber security at University of Rome 'Tor Vergata', opened as key discussant the thematic session on innovation, with an introductive speech. Prof. Talamo emphasized how the concept of cyber security is of paramount importance in the development of digital innovation and clearly highlighted the most relevant fields of application for this matter in the field of innovation, namely:

- Risk Assessment and Management;
- Business Continuity and Resilience;
- Cyber Defense;
- Digital Identity management and Privileges;
- Application Security (API, ERP, secure code review);
- Data Protection.

For assisting construction industries, manufacturing, and business opportunities, digital security can employ the technology to stress-test and otherwise evaluate the vulnerabilities and capabilities of controls on computing environments. The ability to attack a live twin of a production environment—complete with ongoing updates reflected from the original system or environment—without putting data or productivity at risk potentially allows security teams to be as aggressive as they need to be without compromising operations (Cristani et al. 2016).

Concerning the contributions to the session, 18 papers were accepted for publication, coming from several Italian universities as well as from academic institutions from Iran, Pakistan, and Belgium; five contributions were selected for oral presentation at the conference, namely:

- *Short-Term Wind Speed Forecasting Model using Hybrid Neural Networks and Wavelet Packet Decomposition;*
- *Digital Twin for an innovative waterfront management strategy. Pilot project DSH2030;*
- *Digital Twin Models Supporting Cognitive Buildings for Ambient Assisted Living;*
- *COGNIBUILD: Cognitive Digital Twin Framework for Advanced Building Management and Predictive Maintenance;*
- *Untapping the potential of the digital towards the green imperative: the interdisciplinary beXLab experience.*

The most selected topic by the authors has been 'Innovative Digital Technologies and Engineering Systems (e.g. Digital Twins)'.

The 'digital twin' concept is returning frequently among the keywords as well as 'Maintenance'. Also, the 'Energy'-related topic received a consistent attention, intended in some cases as 'simulation' as well as 'culture'.

The field of technological innovation seems a very productive ground for methodologies, tools, and experimentation, and the digital twin emerges as the most promising methodology.

The analysis of the selected interventions also shows that the main use cases of the presented methodologies concerned the fields of energy management, maintenance, and safety applied to the whole built environment (both single building and more complex infrastructures). Innovation in technologies leads to the following major innovations:

- *For energy: integrated management of local micro-grids capable of simultaneously managing energy consumption and production;*
- *For maintenance: moving from 'scheduled maintenance' to 'predictive maintenance';*
- *For safety: use of imaging and computer vision technologies to automatically monitor abnormal behavior, processes, and management of human and instrumental resources.*

Research in technological innovation is confirmed as the field of applied experimentation on which to continue implementing the industry/university relationship.

Most of the papers in fact concerned experimental applications of applied technologies with direct connection with the built environment such as, for example, the utilization of digital workflow for social housing deep renovation design process.

In conclusion, it is important to underline how innovation is one of the fundamental drivers for the green and digital transition of built environment, where technology has a key role for many different aspects.

Innovation is crucial for all industrial sectors to keep their competitive edge, as it fosters a shared culture of constantly looking to do things better. It can help organizations become more efficient and sustainable, adapting to continuous and multifaceted changes of the conditions in which they operate.

Adopted at scale, it can create more consistent supply chains, with each member driven to meet a shared goal of progressive change and improvements, guided by the latest standards of construction methods, construction products, or facilities management.

The increasing diffusion of innovations on the organizational-managerial level with the use of information Communication Technology (ICT), both thanks to the introduction of these technologies together with those of electronic systems for plant automation, security systems, and communications in the building product to realize the new forms of building control and management that are variously defined as Computer integrated Building (CIB), when applied to tertiary buildings, 'domotics' or Home Automation, when implemented in residential construction sites (Scannapieco et al. 2017).

In this direction (technological and organizational innovation), new technologies based on electronics play a very crucial role. They allow the efficient realization of administrative, material purchasing, design, and work scheduling operations, with a small number of technical employees, favoring, through the concentration of information and the possibility of decentralizing production functions.

Information technologies formalize and standardize procedures and lead to the development within the companies of precise and defined responsibilities for specific problems, favoring the processes of specialization and technical-managerial management. Innovation, research and development, and invention are closely linked.

Research and development thrive on the application of new information that will aid in the development of new technologies, products, services, or systems. Inventing new products or services is a sign of a progressive organization and can be a significant revenue earner.

Innovations drive efficiencies through standardization, while many technical innovations have created joined-up workflows beyond the physical space, from offsite construction to virtual reality, the Internet of Things, and flying factories.

References

Agostinelli S, Cumo F, Guidi G, Tomazzoli C (2020) The potential of digital twin model integrated with artificial intelligence systems. In: Proceedings of the IEEE international conference on environment and electrical engineering and 2020 IEEE industrial and commercial power systems Europe, EEEIC/I and CPS Europe 2020, Madrid, Spain. https://doi.org/10.1109/EEEIC/ICPSEurope49358.2020.9160810

Cristani M, Karafili E, Tomazzoli C (2015) Improving energy saving techniques by ambient intelligence scheduling. In: Proceedings of the IEEE 29th international conference on advanced information networking and applications (AINA 2015), vol 1, Los Alamitos, California, pp 324–331. https://doi.org/10.1109/AINA.2015.202

Cristani M, Tomazzoli C, Olivieri F, Karafili E (2016) Defeasible reasoning about electric consumptions. In: Proceedings of the 30th IEEE international conference on advanced information networking and applications (AINA-2016), pp 885–892. https://doi.org/10.1109/AINA.2016.62

Scannapieco S, Tomazzoli C (2017) Ubiquitous and pervasive computing for real-time energy management and saving. Adv Intell Syst Comput 612:3–15. https://doi.org/10.1007/978-3-319-61542-4_1

Chapter 4
The Digital Revolution and the Art of Co-creation

Maurizio Talamo

Abstract The digital transformation process is creating a system of new digital technologies that will give rise to an autonomous ecosystem capable of predicting and determining what will happen in the physical world. The digital twin seen as a moment of creation of a self-named digital model (digital alter ego) and guided in its design by an artificial intelligence can be a source of concern, questioning the central role of the person and of his intelligence in the creative process. The most important problems are: Artificial intelligence capacity to extend its capabilities in a unpredictable way and the inability to understand how artificial intelligence builds its solutions. After hinting at the roots of this problems, we will introduce co-creation as a new way of conceiving the collaboration between the human being and artificial intelligence. Co-creation requires to solve the problems outlined above and this could perhaps define new frontiers of creativity.

Keyword Artificial Intelligence · Machine learning · Cloud computing

4.1 Introduction

Information technologies were developed at the end of the Second World War to make the organizational processes previously carried out either by man or by dedicated machines more efficient and faster. This has forever changed the way we work andinteracting with each other. Data has become the central element of all economic processes. Now, we are about to take a further step, and we are about to come to constitute a real virtual ecosystem in which all the processes that regulate our life will be the result of a negotiation between the digital alter ego of the physical world and the whole ecosystem. The protagonists will be the engines of artificial intelligence and they will be responsible to find the solution of a complex scientific problem or the planning of a trip or simply to design the supply system of a

M. Talamo (✉)
Tor Vergata University of Rome, Rome, Italy
e-mail: maurizio.talamo@uniroma2.it

E. Arbizzani et al. (eds.), *Technological Imagination in the Green and Digital Transition*, The Urban Book Series, https://doi.org/10.1007/978-3-031-29515-7_4

soda machine that suits the customer taste. We are at the beginning of this transformation process where the new AI technologies will be problem-solvers able to train themselves. Nowadays these new technologies are beginning to glimpse; there are already concrete examples of how this model will work, and what is still unclear is what will be the role of the "human being" inside to this digital ecosystem.

Information technology introduces, from the very beginning, a different way of seeing machines. Until the 1950s, the "modern" industrial model invented a machine to solve every single organizational problem. With the advent of the computer, the paradigm changes a program for every problem, but the machine, the computer, does not change.

It was 1994, the year Netscape introduced its first commercial browser called Navigator. This browser has changed the nature of the Internet. Until then, it was reserved for the academic and scientific community. Now, it becomes a tool offered to everyone. As the use and influence of the Internet continue to grow, companies will need to offer new services and new solutions to meet the needs and demands of consumers; to be successful, you need skillful strategies that combine digital media with conventional media: the www has born. Now artificial intelligences are integrated into this infrastructure to interact with humans, learn from them, understand their needs and guide them in their choices (McCarthy 1996).

4.1.1 Impact on Society

The web becomes a "virtual" space where an individual can build increasingly complex information, simply by asking on the network. It becomes a privileged space to create and make available new contents through tools for sharing information, knowledge, and culture. For example using a simple text I could build a completely new and complex 3-D image. The digital economy market is thus changed, giving rise to a few large AI platforms which base their strength and wealth on the quantity and quality of the data in their possession and on the ability to know and interpret this huge amount of data to solve complex business problems.

4.2 The Second Digital Revolution

The development of large "social network" platforms at the beginning of 2000 made it clear that the web itself was a place where the user could build his own digital identity, through which to share with others a virtual world made up of similar communities for business, cultural, or personal interests. LinkedIn, Wikipedia, Facebook, and Twitter are social networks in which billions of people share knowledge or simply represent themselves in a role-playing game that mimics reality. The algorithms that influence guide behaviors on social networks have the characteristic of bringing together like-minded people in relation to particular topics of common interest. In

2007, after various attempts made by various companies, Apple introduces the new iPhone technology that makes it clear to everyone how a device should be made. this kind of device is accompanying us in all our movements, allows the sharing of multimedia data on the Internet and continuously interact with social networks through natural and very simple gestures. Our smartphone is now associated with a digital identity, such as the Apple ID, which recognizes us on the network in everything we do. Now artificial intelligence like ChatGpt are extracting all the knowledge in the web and are able to interpret this knowledge interacting with human beings at any level: answering complex questions or simply doing conversation (Cooper 2023). The relationship that exists between the AI engine and the human being is asymmetrical. the human being sees the result of an elaboration but does not know how and what led to that result. Even the result may exceed the expectations of the AI engine creator himself. Instead, the AI engine takes into account all the information that has been given to build the solution and therefore learns how to improve its training level.

4.2.1 The Era of Digital Transformation

Everything we have described so far leads us to conclude that we are in the presence of a digital transformation process of the companies and more generally of all organizational forms and that this will have numerous results.

The enabling technologies of this process are the introduction of the Internet of things, blockchain.

The Internet of Things is used to provide any device, from washing machines to nano-machines, with a digital identity managed through appropriate infrastructures (Brown 2016; Rouse 2019; Xie and Wang 2017].

The blockchain with distributed ledger technology is a system that aims to allow negotiations and transactions to be carried out within a distributed world of digital identities, ensuring the verification of the correctness of transactions, without intermediaries (Narayanan et al. 2016; Sherman et al. 2019].

These new technologies will make it possible to use AI in every human activity from the simplest to the most complex scientific research activities. It could happen that an AI system tells us that a nuclear plant is safe without us being able to really understand the reason for this statement. The point is that in this way we will not have the opportunity to increase our ability to explore new ways of creating. an example is that of a pastry chef who asks us if we want a cake. We explain our tastes to him and he goes to the kitchen and comes back with a delicious cake. So we tell them that we want to become pastry chefs. We then ask him to explain his recipes, his tricks and the working method he applies. If he does this together we will be able to make better and more creative cakes. But the pastry chef replies that he doesn't know all these things. He makes cakes for those who want to eat them and that's it.

4.2.2 *New Technologies and the New Model*

IOTs, blockchain and also cloud computing will make it possible to use AI in every human activity from the simplest task to the most complex scientific research task (4,5). It could happen that an AI system tells us that a nuclear plant is safe without us being able to really understand the reason for this statement. The point is that in this way we will not have the opportunity to increase our ability to explore new ways of creating: an example is that of a pastry chef who asks us if we want a cake. We explain our tastes to him and he goes to the kitchen and comes back with a delicious cake. Afterwards we tell him that we also want to become a pastry chef. We then ask him to explain his recipes, his tricks and the working method he applies. If he does this, together we will be able to make better and more creative cakes. But the pastry chef replies that he doesn't know all these things. He makes cakes for those who want to eat them and that's it.

4.3 Where We Go: The Virtual Ecosystem and the Evolution of the AI

The almost science-fiction vision of the late 1950s of a world of cybernetic automatons gave rise to a real literary genre, cinematic with reflections on design and art in general.

In forms and ways different from the imaginary of those years, it is translating into something real. With a play on words, this reality is materializing in a virtual ecosystem where very sophisticated software programs process huge amounts of complex information to relate digital entities for social, cultural, and economic purposes: a digital twin of our society with its rules and its critical issues (Alraddadi et al. 2020; Scientific American 2018; Lai et al. 2020). Ultra-fast networks, G5/6/X, immersive reality devices, drones, nano-sensors, underwater, or space clouds are technologies that allow the development of computer code, programs that cooperate, make decisions, and determine the quality of services rendered by this ecosystem. A key role in the formation of this ecosystem is represented by the evolution of artificial intelligence. Artificial intelligence has made it possible to achieve amazing results in the most diverse sectors.

Artificial intelligence is currently one of the most disruptive classes of technology whose capability is rapidly improving thanks to the improvement of various factors: enormous diversity of data collected from various sources; availability of large economic archives; development of faster and more powerful computers; and improvement of artificial intelligence methods. For the past decade, AI has been ubiquitous and is not limited to just computing, and it has evolved to include other areas such as health, automotive, safety, education, business applications, and security. All this was made possible mainly by the introduction of technology called machine learning, able of classifying and interpreting large amounts of data in order to train an artificial intelligence engine to perform a certain task. It is evident that the applications of this technology are the most varied, from the game of chess to the recognition of natural language.

AI will play a predominant role in providing intelligent solutions not only for current outcome-based care, but also for preventive care (Topol 2016). Furthermore, in recent years there has been an exceptional increase in the unstructured medical and health data collected which is now available. This huge amount of data has provided a platform for artificial intelligence to structure data and train to predict disease and move to "precise medicine". Precise medicine is described by the National Institutes of Health as "an emerging approach to disease treatment and prevention that takes into account the individual variability in genes, environment and lifestyle for each person". It provides a more accurate prediction of the mode of treatment and strategies to prevent a specific disease. IoT platforms integrate AI capabilities such as machine learning-based analytics, gaining the ability to detect anomalies generated by sensors and devices.

Machine learning approaches coupled with the IoT are able to make operational predictions 20 times faster and more accurate than traditional business intelligence which usually monitors numerical thresholds. By 2020, 85% of interactive customer communication is expected to be automated and handled without any human assistance (Gartner).

Cyber security is the discipline that easily benefits from AI. To ensure versatile and stable protection, cybersecurity systems must consistently conform to the new dynamic environment. Cisco's "2018 Cybersecurity Annual Report", which examined a broad cross-section of trends and patterns in data theft, data loss, malware, and other issues, found that one-third of security managers is "completely dependent" on artificial intelligence to safeguard sensitive corporate information. However, the AI community is also realizing that the use of "intelligent" technologies can have a strong negative impact on many aspects of an individual's life. The loss of jobs replaced by machines is an aspect to be taken into account. New professions are often underpaid as less skill is required to perform tasks supervised by an AI engine. It is necessary to have a vision to guide the development of technologies toward objectives that allow the AI engines, autonomous and different from each other, to cooperate with each other to, on the one hand, determine a virtual ecosystem in which the individual as such and the individual within his community can develop his creative potential to the fullest and best; on the other hand, there is the risk of creating a virtual ecosystem in which the individual is a subject who passively lets himself be guided in his choices. To understand what we are talking about, the best-known example is the creation of software to automatically drive a car. We all know Tesla's project and we are interested in pointing out how the problem of allowing an AI engine to drive a car autonomously is dramatically different from providing AI support to a driver to help him drive. In the first case, the AI software will have to autonomously decide not only for itself but will have to take into account the behavior of other "artificial" or natural drivers, going to determine an overall movement strategy of the vehicles in the road infrastructure that delimits them that minimizes the risk ensuring vehicles reach their destination safely. Another example is the management system of a smart city or critical infrastructure. All these systems that interconnect autonomous artificial intelligence systems that cooperate in carrying out a small or large task are the near future. The completion of the second digital revolution is approaching.

In this paper, so far we have not addressed, if not marginally, the ethical problems that emerge from the impact of technologies on the lives of all of us. There is normally a favorable position for technology which states that its development should be left free as it is regulated by the market and that technology takes us to the best of worlds possible. This is contrasted by a position in which technologies are viewed with fear and tend to reject them, seeing only the negative aspects.

A position has recently emerged that is the result of hard research and experimentation work that has made it possible to understand in a scientific way that the development of AI technologies, in this new and complex dimension, is neither positive nor negative in principle, but it requires interdisciplinary knowledge to achieve the necessary awareness to be able to make the right choices.

As noted by prof. Josef Sifakis1, "There is no single solution that is totally reliable, there are a set of possible solutions with a reasonable degree of reliability. Aware of the critical issues, we must accept the percentage, even if low, of the risk of error and therefore of an accident".

This decision of which solution to adopt from time to time concerns the policies that are defined by the institutional decision-makers. This necessarily requires a synthesis between ethical and technological issues to consciously determine within what limits we want systems to act.

To understand the complexity of the synthesis process introduced above, it is necessary to clarify some aspects that are often underestimated (Cooper 2023).

4.4 A Possible Future: The Art of Co-creation

The virtual ecosystem that we have outlined has the purpose of guaranteeing profits to those who with the right technological skills and adequate investments will be able to exploit its potential best. A self-driving car can be transformed into a component of a digital service system with extraordinary potential. The planning of the trip, of the stops, of the supplies, and the possibility of obtaining an overall evaluation in real time of the insurance premium are just some of the first data that can be obtained, and on which, it can be affected interactively. There is an asymmetry of the data that is evident. The data is used to develop sophisticated marketing strategies tailored to behaviors, but if you want to know where your data is, it becomes a technological criticality. The explanation is simple: marketing brings profit, and ensuring access to your data is a cost.

All this is absolutely normal following the logic of a competitive market. The only risk is that, while being guided by good intentions, the individual disappears and remains a subject with the function of impersonating his own digital alter ego as a pawn in a game between digital platforms.

This necessarily requires a synthesis between ethical and technological issues to consciously determine within what limits we want systems to act.

It is therefore clear that it is necessary to find a balance point between the various different objectives, keeping the individual at the center in his aspirations and in his physical and intellectual needs. This is the goal of truly sustainable development. It is therefore necessary to understand how the encounter between ethics and technology can have the same explosive effect as the encounter between profit and technology. It is necessary to develop an original and harmonious system of knowledge that gives rise to a highly competent interdisciplinary community in order to involve the whole of society in a sustainable innovation project, with a view to capitalism with a social vocation. There is a technology that we have not yet invented that sees the protection of the rights of the individual as a basic objective, that meeting between ethics and technology that until now we have only seen in intentions and announcements, an open, concrete, flexible vision without be guided by a "false rationality" generated by scarce and superficial knowledge.

Here, this is the role that the individual as such and within his own community can and must play with respect to this virtual ecosystem that is being configured.

This leads to a new, multidisciplinary way of designing complex systems based on AI, in which the various scientific, technological, legal, and humanistic skills work together to build an ethical and sustainable ecosystem development model. A challenge as compelling as it is difficult in which the ability to dialogue between different cultural and scientific visions can produce surprising results.

When information technology was born we perceived it as a tool to be more efficient, when the web was born we thought we could wander into a still unknown world without thinking about the positive or negative consequences. We have listened to companies cheering on new technologies only to be swept away by them. It was probably an inevitable process at that time, but now we have matured in the ability to choose and guide technologies precisely because the level of innovation they have reached allows us to design original and sustainable paths consistent with our priorities in the real world.

Co-creation should be understood as a process in which new technologies help machines and humans to cooperate to overcome the boundaries of creativity as we conceive it today.

This is a different way of understanding technology and requires different models to develop it. To obtain these results, we need a new generation of AI technology that is able to make the relationship between human beings and artificial intelligence symmetrical. This means that just as we give knowledge to the artificial intelligence, the engine should give us back what it has learned and the new learning methods it has developed: a joint co-creation process. We are not talking about Explainable AI but about a vision of an artificial intelligence that guides human beings in understanding the method used by other AI engines in solving a problem, how they managed to extend the process of understanding unpredictable concepts more and more complexes. A partner of the human being not in seeking new solutions but new methods of exploring one's creativity. A symmetrical process in which the AI partner helps us to recover the baggage of knowledge and creative tools that we have provided to the AI engine and elaborated by it. A new starting point for creating new creative processes whose complexity is all to be imagined: a new AI technology to understand other AIs.

The laws of the market will initially resist this new way of proceeding. What is certain is that there is no easy way to transform the world into a complex digital reality.

References

Alraddadi EE, Allen SM, Colombo BG, Whitaker RM (2020) The role of homophily in opinion formation among mobile agents. J Inf Telecommun 4(4):504–523. https://doi.org/10.1080/24751839.2020.1772614

Brown E (2016) 21 Open source projects for IoT. In: Linux.com, 20 Sept 2016

Can neural network computers learn from experience, and if so, could they ever become what we would call 'smart'?. In: Scientific American 2018

Cooper G (2023) Examining Science Education in ChatGPT: An Exploratory Study of Generative Artificial Intelligence Abstract. J Sci Edu Technol, pp. 1–9. https://doi.org/10.1007/s10956-023-10039-y

Day N, Hassanien AE, Bhatt C, Ashour A, Satapathy SC (2018) Internet of things and big data analytics toward next-generation intelligence. Springer

Gatouillat A, Youakim B, Bertrand M, Ervin S (2018) Internet of medical things: a review of recent contributions dealing with cyber-physical systems in medicine. IEEE Internet Things J 5(5):3810–3822. https://doi.org/10.1109/JIOT.2018.2849014

Lai CS, Jia Y, Dong Z, Wang D, Tao Y, Lai QH, Wong RTK, Zobaa AF, Wu R, Lai LL (2020) A review of technical standards for smart cities. Clean Technologies 2(3):290–310. https://doi.org/10.3390/cleantechnol2030019

McCarthy J (1996) Review of the question of artificial intelligence. In: Annals of the history of computing 10(3):224–229. Collected in: J. McCarthy (Ed.) 10. Review of The Question of Artificial Intelligence. Defending AI Research: A Collection of Essays and Reviews, CSLI, pp. 73

Narayanan A, Bonneau J, Felten E, Miller A, Goldfeder S (2016) Bitcoin and cryptocurrency technologies: a comprehensive introduction. Princeton University Press, New Jersey, USA

Rouse M (2019) Internet of Things (IoT). In: IOT Agenda 2019

Sherman AT, Javani F, Zhang H, Golaszewski E (2019) On the origins and variations of blockchain technologies. IEEE Secur Priv 17(1):72–77. https://doi.org/10.48550/arXiv.1810.06130

Topol E (2016) The patient will see you now: the future of medicine is in your hands. BasicBooks

Xie X-F, Wang ZJ (2017) Integrated in vehicle decision support system for driving at signalized intersections: a prototype of smart IoT in transportation. In: Transportation research board (TRB), 96th annual meeting compendium of papers. Washington, DC, United States

Chapter 5
Toward a New Humanism of Technological Innovation in Design of the Built Environment

Spartaco Paris

Abstract This paper offers a reflection on the dialectical relationship between technological innovation and the culture of design for the built environment, hypothesizing a scenario in which the outdated interpretation of the merely instrumental role of technology may be plausibly recognized and overcome. Because it is increasingly complex, sophisticated, and pervasive in the life of and in the space inhabited by people, technology requires a renewed, humanistic approach in order to be governed, selected, and used, for the purpose of improving people's fragile living conditions on the planet. On the other hand, people are, in and of themselves, characterized by continuous exploration and invention to extend their abilities through technique, and given the uncertain future facing us, technique will once again be our friend, renewing a relationship that may be called one of *philotechny* rather than of pure fideism—or of anachronistic antagonism—toward technique. Although starting from a specific, low level of technology, the construction sector is absorbing the extraordinary, accelerated advances of the technologies belonging to the digital sphere; this may help guide the construction industry toward a new form of production, and at the same time, it may transform the designer's tasks, roles, and responsibilities. The designer of the future that faces us will have to be able to operate collaboratively within a system of vast competences, which will be mediated by the potency and accessibility of new tools and methods belonging to the digital technologies.

Keywords Building Intelligent Modeling · DT · Integrated Design · Architectural Technology · Technological culture of design

---

S. Paris (✉)
Sapienza University of Rome, Rome, Italy
e-mail: spartaco.paris@uniroma1.it

E. Arbizzani et al. (eds.), *Technological Imagination in the Green and Digital Transition*,
The Urban Book Series, https://doi.org/10.1007/978-3-031-29515-7_5

5.1 Foreword. New Technologies as the Artificial Limbs of Contemporary People

Different conditions and types of "technopoly" have gradually succeeded one another, being transformed following the industrial and scientific revolutions of modern society; first came society's fideism toward machines, followed by the exceptional effects on modern society of electronics and computing (1920–1945), until the IT revolution that we are still immersed in, in a fully post-modern society vacillating between fideism toward the techno-sciences and oscillating awareness of the world's fragility.

The time we are living through appears dominated by uncertainty, even though unceasing innovation of technique is, more and more intrusively, offering the promise of guiding and controlling our decisions. Increasingly available enabling technologies, and the possibilities offered by computing to acquire and select an immense quantity of data, are yet another potent field of technical/scientific research at the service of the market economy. And yet, this does not at all appear to offer a ground for stable forecasts so much, instead, as to underscore the context of uncertainty we live in. In fact, this condition has of late been, in the fields of both the arts and the sciences, the object of recent shows and major exhibitions, such as 2021's "Tre stazioni per Art-Science. Incertezza. Interpretare il presente, prevedere il futuro" ("Three Stations for Art/Science. Uncertainty. Interpreting the Present, Foreseeing the Future.") at Rome's Palazzo delle Esposizioni and at the Spanish Pavilion, under the title "Uncertainty," at the 17th International Architecture Exhibition, Biennale di Venezia, 2021.

At the height of the crisis of modernity, authoritative scholars were already observing that the technologies with which people are surrounded are not mere "means." Referring to the "devices" of the second industrial revolution, the philosopher Gunther Anders wrote "because the 'means' is by essence something secondary. (…) introduced ex post facto for the purpose of 'mediating' that end. They are not 'means' but preliminary decisions that are made for us before we are called upon to decide" (Anders 1980. p.13.). The cited example is one of many relating to the controversial relationship, starting from the modern age, between man and machine, people, and technologies (Fig. 5.1).

Philosophical thought, traditionally distrustful if not hostile to technical/scientific thought, has recently raised positions of dialectics more open to the role and the pervasiveness of technique in our society. The philosopher Emanuele Severino has recurringly raised a series of questions on the growing domination of technique, and its limits in Western, capitalist society.

It has been noted that over the past forty years, the conception by which technique might be of use for cultural progress has been transformed; consider the economic/productive dominion of American technical universities, defining a sort of shared marriage between science/technique/production and the market. On the one hand, a convergence with respect to what technological evolution represents in our society and particularly as relates to architectures and to the built environment

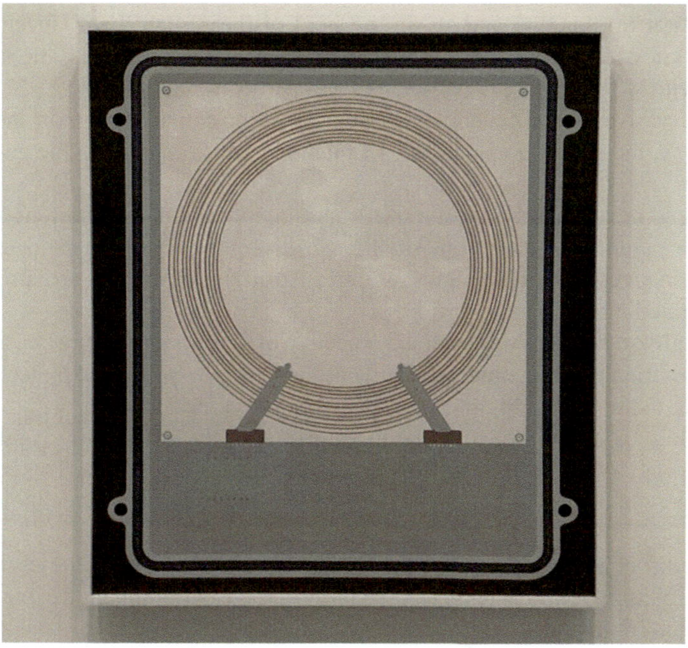

Fig. 5.1 Ulla wiggen, Cybernetic Serendipity, 1967. Biennale Arte 2022, milk of dreams. *Photo* © Spartaco Paris

is sought: The interpretation that Vittorio Gregotti offered in his paper about twenty years ago, now a matter of history, on the relationships between architecture, technique, and purpose (Gregotti 2004, p. 252), is still highly convincing. Referring to technique, the Milanese architect observes:

> "the attempt is made to portray it in two different ways: at one time considering technique as a central, structural element, and at another as adhering to the market's emotions and to the incessant innovations that continuous transformation brings. The dualism between formalists and technologists has, in both cases, a profound internal contradiction, however much they fail to achieve reality in one way or in the other. (…) And hence the true "valorists" that we are today; we are the ones seeking to represent, by exalting it, the current situation, the values that exist, and therefore we turn our ability into the way of translating these words. This appears clear in light of the theme of information technology: cybernetic culture, virtual culture, the so-called "technological sublime." This world, that of ancient mythology, of magic – this world has a continuous need to project its ideas forward, and therefore we no longer care about how things were done in times past, but we concern ourselves only with the future, a strange future, but one that also represents the resolution of all problems, and therefore the triumphalist future or the future of disaster."

The paper attempts to shrink the field to the physical environment we build and live in.

A plausible scenario for reacting to an increasingly segmented social, economic, and cultural setting may be that of attempting to operate through new forms of collaboration in planning activity in the field of the research and development of scientific knowledge, while seizing the opportunities of integrating the knowledge we deal with when operating with newer and newer digital tools.

The pervasiveness of technologies appears to be impacting, in an increasingly evident way, the procedures not only of planning but of managing the environment we live in.

The rapid development of computing and the integration of different disciplines allows the parameters based on which each discipline produces innovation to be radically changed: The disciplines are not just miniaturized, but are also invisible and integrated.

From utilizing technology with the objective of dominating nature, we are rapidly going on to transforming ourselves into entities ready to be shaped by technologies, integrating them physically into ourselves. "At the same time, our own relationship to technology is moving beyond the instrumental to the existential. We do not just use technology. We absorb it." (Khanna and Khanna 2012, p. 14). The coexistence between people and technology is being transformed into human/technological co-evolution.

We therefore have a multitude of continuously evolving technologies that extend our capacity for judging, planning, and controlling the environment we live in. Which will we need? Which ones will we be able to govern? Which ones can help us change our approaches to designing and managing our habitat?

There are at least two accompanying conditions we believe would be useful to emphasize.

The first relates to sharing the programmatic lines of the New European Bauhaus, particularly as regards the approach to renewing the built heritage: It is by all means clear that, for the West and soon for Asia as well, "The twentieth century was a story of building anew, pouring concrete in order to build a way out of poverty, or to conjure new towns and cities out of wartime ruins or developmental missteps. As a result, for many European countries, 80% of the buildings of 2050 have already been built, and some 97% of these existing European buildings will need to be renovated. (...) It implies a refining in place, understanding repair and retrofit cultures, developing new logics predicated on care and maintenance, on true collaboration and participation with diverse cultures and behaviours, and on building anew only where necessary and desirable." (Bason et al. 2020, p. 6). This is in the conviction that there can be no paradigm shift toward the built environment without a new critical, and thus interdisciplinary, capacity toward the hypothesis that regenerating existing construction might be able to produce new resources and values for society at large and for future generations.

The second thematic sphere of approach relates to whether to verify, develop, and more deeply analyze, with a view to operation, the scenario of reference characterizing the digital revolution—the "digital turn"—for the context of architecture, the construction of the built environment. From this standpoint, the conference's contributions in track innovation show interesting and fertile attempts at experimental application of digital technologies as powerful design "materials" for people and their environment.

5.2 Toward a New Paradigm of the Designer in the Digital Turn for the Built Environment. A New Scenario of Knowledge and Skill Integration in Design Processes

Although it has been traditionally characterized by a low technology level and content, the construction industry has also, in recent years, been affected by a development of digitalization throughout the supply chain.

Recent studies (EU Commission 2021) articulate into three main supply chains the spheres of application, with their innovative digital tools, to the construction sector: data acquisition (sensors, Internet of Things, 3D scanning), automating processes (robotics, 3D printing, drones), and digital information analysis (Building Information Modeling, Virtual/Augmented Reality, and Artificial Intelligence, Digital Twin) (Fig. 5.2).

Although at present the spread—at least in Europe—of procurement strategies to raise stakeholders' awareness of the construction sector's digitalization skills and methods is relatively broad and patchy, European policies are incentivizing the entire construction supply chain's digital updating, from planning to development and management.

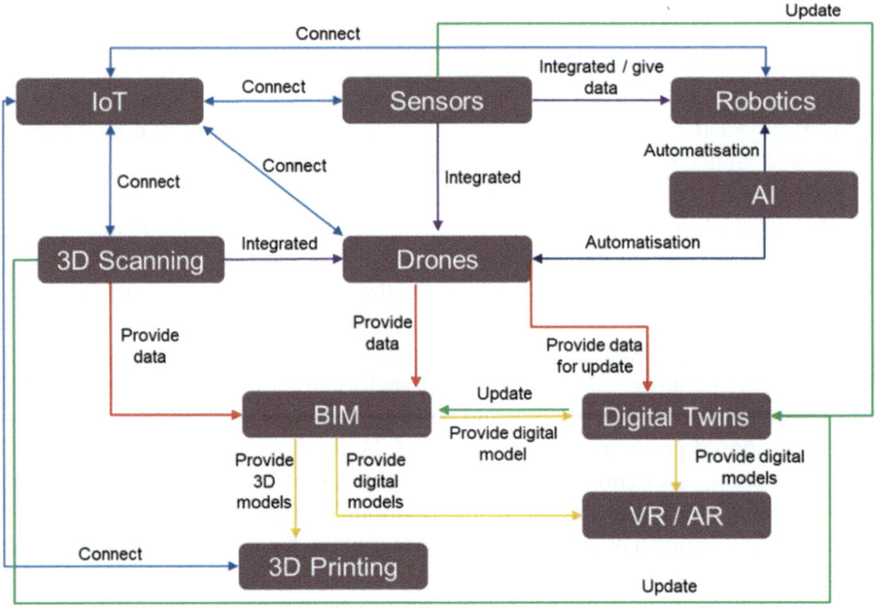

Fig. 5.2 Scheme of interaction among digital technologies in the construction sector 2021 (*Source* European Commission, p.20)

The third of the above-described supply chain—digital information analysis—is the planning activity's specific area of competence; this is if we share an extensive interpretation of the concept of design, as a "projection" of choices—morphological, functional, and technological—relating to people's specific needs. Digital information applied in integrated fashion to the built environment is a rapidly developing frontier that is profoundly transforming the role of the designer within the construction process.

It is interesting to point out that, at bottom, architecture is, from the technical and productive standpoint, far simpler than other industrial sectors that are today using advanced tools to deal with the contemporary world: It is far simpler than a drone or a technological device, but it is functionally far more articulated and is programmatically designed for a rather long life cycle.

In the direction of managing the built environment, the development of information technologies, and in particular of parametric modeling in the field of constructions, is going through a phase of transition toward full operativity and toward changing the approach to planning and managing the built environment. From a primitive phase of transfer to the digital environment of exclusively geometric information typical of the analog method of representing reality, we have transitioned to a long phase, still in progress, of parametric information modeling (BIM approach) in which the base/primitive IT elements are endowed with specific information (characteristics, requirements, performance) and not only geometric information; we are now in an evolutionary phase, that of the so-called Digital Twin already well established in the industrial sectors and in sectors deemed as having high technological content, while still embryonic in the field of construction. For this phase, the digital twin of physical reality might foreseeably contain—thanks to the use of increasingly accessible enabling technologies—"dynamic" information relating to specific behaviors of its true original.

This level of knowledge will be able to guide and facilitate the management—also with a view to sustainability—of the building heritage and is a horizon of technological innovation of development for the entire construction supply chain.

The recent essay entitled *The total designer. Authorship in the Architecture of the Postdigital Age* by L. Ortega attempted to redefine the transformation of the role of the architect operating within the digital transformation of tools and methods that are increasingly available, mature, and powerful.

The proposed programmatic definition of the total or expanded designer merits some reflection, as a figure who, according to the author, may overcome a traditional interpretation of the architect, and about whom a 20-point cultural manifesto is proposed, attributing a political and radical role to the architect in the postdigital era (Ortega 2017, p. 70.71) (Fig. 5.3).

Beyond the political interpretation, which may be more or less agreed with, which Ortega attributes to the "total designer," an avant-garde figure, there are some clear and acceptable aspects of the transformation of the role of the architect, endowed with new tangible and intangible tools for design. Let us attempt to describe them in order to outline a scenario of reference for the present and the future.

1/ Designers tend to limit their projects to certain scales; total designers develop their skills with the intention of learning how to generalize.

2/ Designer teach by transmitting information; total designers teach by creating states of mind.

3/ Designer are politically correct; total designers are political.

4/Designer write about design; total designers texts on design.

5/ Designers care for their design; total designers are obsessed about their projects.

6/Designers are victims of professionalism; total designers are victims of vitality.

7/Designers design with materials; total designers are material organizers.

8/Designers try to communicate their projects in a commercial format, generally by using images; total designers share their experiences and their work through public exhibition, by using drawings.

9/Designers hope for good reviews and approval; total designers are interested in action, not approval.

10/ Designers think and talk about design as a language and a convention and practicing in the discipline consists of belonging to the group that shares the language; total designers work with a material logic and practicing in the discipline consists of skillfully manipulating historical material.

11/ Designers realize diagram through the use of figuration; total designers realize their diagram s performatively, as opposed to translating them.

12/ Designers tend toward singularity; total designers tend toward multiplicity.

13/ Designers try to be economical; total designers love excess.

14/ Designers hate repetition; total designers love iterations.

15/ Designers love novelty; total designers love invention.

16/ Designers love the extremes; total designers love the in-between.

17/ Designers are tempered by limitations; total designers are radical through constraints.

18/ Designers solve problems; total designers generate questions.

19/ Designers don't call their work finished until they feel it is detailed enough; total designers don't have sense of finalization.

20/ Designers think about their next project; total designers always work as though the current project were their last.

Fig. 5.3 L. Ortega, manifesto on the expanded designer or the total designer, the total designer, 2017, p. 70–71)

Without a doubt, the power of processors allows the design prefiguration to be described with modes of reality simulation that are quick and powerful enough to propose a virtual world analogous to the real one. This brings a whole series of implications on the new requirements of the design, which traditionally could proceed by approximations and subsequent deeper analysis; today, it can fix in digital models a hyper-realistic simulation of what does not exist. This is an initial point that also brings to virtual worlds—see the metaverse—the tools of augmented reality, but that resides in the sphere of representation.

The new field of competence of the new digital designer lies in the possibility of attributing specific and customizable requirements to the digital models the designer creates: This means providing attributes relating to physical and mechanical behavior to the designed virtual environment. These attributes may be verified in reality using sensors: This projects the management of our buildings into an extraordinary, new dimension, already well established in the other industrial fields of application of Digital Twins (see automotive, mechanical industry, bioengineering).

The third field, perhaps the most challenging one, is the need for and appropriateness of working collaboratively among different competences: The medium provided by the model offers the digital design the condition of working in an integrated form, exchanging a model that "grows" and collects information and data through cooperation with other competences. This scenario defines a new and innovative framework for the role of the designer: The traditional designer, placed at the top of a pyramid of competences, was bound to a substantially linear progression of the design activity, ending upon his or her introduction of the designed object into operation; the digital design finds him or herself in a circular processual model and, by maintaining control over the model, operates in "horizontal" form with the other actors; he or she may potentially operate during the entire life cycle of the designed object, thereby amplifying its competences for management.

5.3 Conclusions. Renewal of a Technological Culture of Design

During the years of the growth and fall of the utopia of the Modern Movement, there was a coexistence of antagonistic positions in relation to the role that technological innovations could have produced in the productive processes of transforming the built environment. Radical pioneers of a modernity in which architecture fully entered into an industrial dimension, the formal repertoires—from Archigram to the Eameses, from R. Rogers to Renzo Piano—proposed avant-garde models whose results belong to currents of language rather than to profound transformations of the modes and content of architectural production addressing a technological culture of design.

On the other hand, as Massimo Perriccioli (Perriccioli 2021) observed in his recent studies, these pioneers sometimes sidelined or excluded from the history books on modern and contemporary architecture and from the "official" historiography, had, with their intuitions and experimentation open to technological innovation, prefigured an attitude oriented toward integrating the new tools that the technological advance in the processes and methods of design had ushered in, and toward anticipating themes and issues that are current today: We may consider the environmental issue, social emergencies, the myths of globalization, technological hybridization, the new forms of dwelling, and circular production systems, to cite just a few. This has represented an anticipation of a way of thinking about design and the role of the designer with a confidence in a "friendly" technology, one that does not abuse people's needs. In the twentieth century, this confidence was placed mainly in the material sphere of producing the project: consider the "constructive imagination" that guided such uncelebrated figures as Peter Rice or Jean Prouvè: and yet, that "integrated" vision of the design in relation to the means and techniques of production is conceptually no different from the designer's attitude in the digital era: the tools, the tool box, that belong mainly to the intangible sphere of the design process, are more powerful, but they guide the role of architecture into a renewed duality of art and industry for the built environment.

The horizon of reference for technological imagination may therefore reside in a new update of a technological culture of design, in which the processual dimension takes equal part in the sphere of artistic practice.

References

Anders G (1980) L'uomo è antiquato. Considerazioni sull'anima nell'epoca della seconda rivoluzione industriale. Edizione Italiana, 2003, ristampa gennaio 2021. Bollati Boringhieri, Milano. First Edition: Die Die Antiquiertheit des Menschen Bd. I: Über die Seele im Zeitalter der zweiten industriellen Revolution, 1980; Verlag C.H. Beck

Bason C, Conway R, Hill D, Mazzucato M (2020) A new Bauhaus for a green deal, 2020 UCL Institute for Innovation and Public. Purpose. https://www.ucl.ac.uk/bartlett/public-purpose/sites/public-purpose/files/new_bauhaus_cb_rc_dh_mm_0.pdf

European commission (2021) European construction sector observatory, digitalisation in the construction sector, April 2021

Gregotti V (2004) Architettura, tecnica e finalità nel progetto moderno e contemporaneo. In: Morabito G (ed) Dodici lezioni di cultura tecnologica dell'architettura. Rome, IT: Gangemi Editore

Khanna A, Khanna P (2012) Hybrid reality thriving in the emerging human-technology civilization, 2012 Published by TED Conferences, LLC

Ortega L (2017) The total designer. Actar Publisher

Perriccioli M (ed) (2021) Architettura, Design e Cultura tecnologica. Il racconto di 18 protagonisti del Novecento. Pensare l'architettura e il design nel XX secolo. Bari, IT: Progedit.

Severino E (2021) Tecnica e architettura (Rizzi, R. Piccola Biblioteca, Mimesis Edizioni, Milan, IT

Chapter 6
A BIM-Based Approach to Energy Analysis of Existing Buildings in the Italian Context

Marco Morini, Francesca Caffari, Nicolandrea Calabrese, and Giulia Centi

Abstract In the important challenge set at the EU level of doubling the annual energy renovation rate of buildings by 2030, building information modelling (BIM) represents a remarkable opportunity for its many advantages, in all stages of the process. For example, BIM allows for the creation of accurate models of buildings at both current and refurbished state that can also be imported in software for specialized analyses such as the energy performance study and, at the same time, constitute digital, easily searchable, and updatable databases of all sorts of information about a building. However, several barriers still hinder the full use of BIMs for energy analyses, such as issues in interoperability among software and lack of technical knowledge of professionals. The research, whose methodology and objectives are introduced in this paper, moves from these considerations and can be divided into two main phases. The first one ("from BIM to energy analysis") focused on the identification of interoperability issues between BIM authoring and Italian certified energy analysis software, starting from the application on case studies. The aim is to draft recommendations, targeted at the professionals in the sector, for the definition of building as-is models, optimized to make the importation in energy analysis software as seamless as possible. In the second part ("from energy analysis to BIM"), a series of parameters, to be included firstly in the form of custom property sets, were individuated to populate those models with the main results of the energy analysis. The objective of these actions is to support and, at the same time, valorize the work of the professionals carrying out energy audits while highlighting the potential of BIM for greater knowledge and digitization of the building stock.

M. Morini (✉) · F. Caffari · N. Calabrese · G. Centi
ENEA, Rome, Italy
e-mail: marco.morini@enea.it

F. Caffari
e-mail: francesca.caffari@enea.it

N. Calabrese
e-mail: nicolandrea.calabrese@enea.it

G. Centi
e-mail: giulia.centi@enea.it

E. Arbizzani et al. (eds.), *Technological Imagination in the Green and Digital Transition*, The Urban Book Series, https://doi.org/10.1007/978-3-031-29515-7_6

Keyword Building information modelling (BIM) · Industry Foundation Classes (IFC) · Energy audit · Refurbishment · Building

6.1 Introduction

The energy refurbishment of existing buildings is seen as one of the most important strategies for reducing greenhouse gas emissions in the building sector. To address the challenges of climate change and post-pandemic recovery, the European Commission (2020) launched a new strategy as part of the European Green Deal to give a boost to the renovation of the building stock called "A Renovation Wave for Europe". Its declared objective is to at least double the annual energy renovation rate of residential and non-residential buildings by 2030 and to foster deep energy renovations. In this framework, the technological innovation brought by building information modelling (BIM) in the construction sector can play a very important role, thanks to the advantages in all stages of the process, from the early phases of the acquisition of geometric and energy-related building data to those involving the management and monitoring of operating building performance. For example, BIM allows for the creation of accurate models of buildings at current state that constitute digital, easily searchable and updatable databases of all sorts of information about a building and, at the same time, that can be imported in software for specialized analyses such as the energy performance study. This is possible thanks to interoperability: in fact, in an open-BIM workflow, all participants from different disciplines can collaborate and exchange project information by using open, non-proprietary formats.

However, several barriers still hinder the full use of BIMs for energy analyses, such as issues in only partial interoperability among software and lack of technical knowledge of professionals. As stressed by Heffernan et al. (2017), interoperability issues typically result in architects communicating the design of a building in one model and an energy consultant reproducing that design within a new building energy model (BEM). Sanhudo et al. (2018) provided a comprehensive review of building information modelling for energy retrofitting, from the acquisition of geometric and energy-related data concerning a building to the use of as-is models for energy analysis, including interoperability issues with BIM authoring tools. Andriamamonjy et al. (2019) provide an in-depth overview of three main strategies to improve interoperability between BIM and building energy performance simulation tools. The most flexible is the strategy based on the identification of exchange requirements, aiming to make the BIM model the container of all information required for the various uses, including building energy analysis. It does not rely on proprietary tools or formats, but aims to make the BIM models exported to IFC compatible with the information exchange requirements for energy analysis. The above-cited publications show the great attention towards the use of BIM for the study of building energy performance and the need to overcome interoperability problems.

In Italy, since the entry into force of the so-called BIM Decree (MIT 2017), introducing gradually increasing obligations for the adoption of BIM-based methodologies for buildings and infrastructures, the number of buildings for the construction and renovation of which the use of BIM is required is slowly but steadily growing. Many administrations are beginning to equip themselves with the necessary tools for this purpose, also with the aim of considerably speeding up the internal procedures for checking and approving project proposals. In this framework, the work of guideline and process documentation drafting in the perspective of open BIM, carried out by the national State Property Agency (Agenzia del Demanio, ADM), represents an interesting reference in the Italian context (ADM 2021). Moreover, ADM's work proves that it is becoming common in public tenders to commission BIM "as-is" models together with energy audits. It is a practice to be encouraged in order to discern the current state of the existing building stock and promote its renovation, but professionals struggle with the difficulty of including energy audits into the BIM process. Indeed, out of the software tools that the Italian Thermo-technical Committee (CTI) certifies for building energy performance calculation, according to relevant technical regulations, only a few have started a process of more or less deep introduction into the BIM process.

In the framework of the research where this work is included, Centi et al. (2019) provided an in-depth study of these software tools and their capabilities, according to some predefined BIM- and energy-related criteria (e.g. IFC import, IFC export, building modelling, buildingSmart certification, etc.).

Starting from the application on case studies, the research that is briefly introduced in this paper has focused on the identification of the existing interoperability issues between BIM authoring and Italian certified energy analysis software tools and on some proposals for their optimization. Such issues regard the incomplete and/or incorrect import of IFC models exported from BIM authoring software into building energy modelling (BEM) and analysis tools. On the other hand, these also affect the export to IFC of the details and the results of the energy analyses, which can be part of thorough reports, drafted by the energy modeller and linked to the IFC model in the Common Data Environment, but are not included in the IFC model exported from the BEM software. In other words, the BEM software is only used to run the simulations on the model, but its results as of now are not implemented in the IFC model of the building.

Having said that, this paper introduces some recommendations—targeted at the professionals in the sector—for the definition of as-is IFC models of buildings with the aim to make the importation in energy analysis software as seamless as possible, regardless of the software tool used. Additionally, in order to populate those models with the main results of the energy analysis, a series of parameters to be included, firstly in the form of custom property sets, were individuated. The objective of these actions is to support and, at the same time, valorize the work of the professionals carrying out energy audits, while highlighting the potential of BIM for greater knowledge and digitization of the building stock.

The Italian Agency for New Technologies, Energy and Sustainable Economic Development (ENEA) has worked in closed cooperation with several stakeholders for the drafting of the Guidelines for Energy Audits of Buildings, published as a national standard/technical specification in 2020 (UNI/TR 11,775). The final scope of this study is to integrate the process of energy audit, described step-by-step in the Guidelines, within a BIM perspective to support professionals and facilitate the adoption of BIM in the process of refurbishment of existing buildings.

6.2 Methodology

Two phases characterize the methodology of the research discussed in this paper:

Phase 1: "from BIM to energy analysis", aimed at:

- The identification of interoperability issues between software tools for BIM authoring and energy performance analysis;
- The drafting of recommendations for the creation of building information models optimized for energy analyses.

Phase 2: "from energy analysis to BIM", aimed at:

- The valorization of the work of the energy auditors and BIM-related potential
- For greater knowledge and digitization of the existing building stock;
- The identification of energy parameters (Pset) to be included in the model of a building to populate it with the results of the energy performance analysis.

6.2.1 From BIM to Energy Analysis: Interoperability Issues and Possible Solutions

Figure 6.1 illustrates the process of the first phase of the research, based on actual case studies (i.e. buildings subject to energy audit).

The process, which has required in the beginning many iterations and optimizations, consisted of three main phases, involving three different software categories, i.e. for BIM authoring, BIM viewing, and building energy modelling (BEM).

First, a parametric model of the case study building was developed by means of a proprietary BIM authoring software and exported into an IFC2 $\times$ 3 file. Then, the exported IFC file was analysed and checked in terms of geometry and content by means of a BIM viewer, before being imported into the BEM software. The model could then be completed of the information needed for the energy analyses in the selected BEM software. It is possible to point out several issues that limit the process of IFC import into the BEM tool. Information relating to the geometry and envelope components of the building is in general imported correctly, but in an incomplete

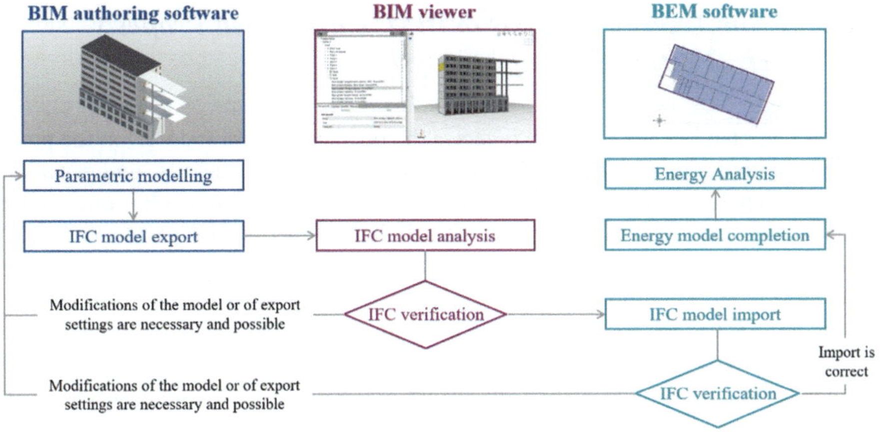

Fig. 6.1 Process from BIM to energy analysis

way. Next, it is necessary to check and complete the geometry of the imported model as well as to characterize the spaces, thermal zones, and technical systems, in order to carry out the energy performance simulations.

Starting from the application of the above-described process of specific case studies, recommendations for modelling were developed aimed at maximum interoperability between BIM authoring and BEM software, highlighting limits and problems and proposing general solutions. The scope is to make the process in Fig. 6.1 as seamless as possible, contributing to the elimination of the iterations represented by the "return arrows" (i.e. changes to the model and its export settings) through the suggestion of specific actions for the BIM modeller to perform. The recommendations detail aspects that range from the nomenclature of elements to their geometric and stratigraphic characterization inside the BIM authoring software and they concern the correct definition (also from an energy point of view), the optimization and the check of model geometry, building elements, materials and layers as well as (thermal) spaces and zones. The work has also attempted to clearly identify the additional operations needed for the control and completion of the imported model, in order to manage the expectation of what can be done with the used BEM software. In general, it is suggested to refer to the guidelines of the individual software houses for specific indications for the management and optimization of the IFC model during this process.

It is not purpose of this paper to describe in detail the content of these recommendations, which are intended to be published as a part of a BIM-based integration of the already issued Guidelines for Energy Audits of Buildings (UNI/TR 11,775). The aim is rather to explain the methodology that has led to their definition and the general rationale of the research.

6.2.2 From Energy Analysis to BIM: Customized Property Sets (Pset) for Energy Analyses

The second phase of the research reverses the perspective and focuses on the possibility to include, as part of an IFC model of the building, details, and results of the energy analyses carried out in the BEM software. The main results concern data about the current energy consumption of the building, energy performance indicators correlated to building features (areas, volumes), and proposals for improving energy efficiency and calculating the resulting savings. Such information is traditionally organized in specialist technical reports and attachments to be shared with the client.

Among these is the file processed within the energy analysis software used, which is only readable by those who own the specific software. With this approach, the work of the energy auditors is not fully valued, also in relation to the great potential of BIM for greater knowledge and for the digitization of the existing building stock.

The BIM model is indeed a container of information that, through export to IFC, become readable by all the actors involved in the process, regardless of the software that produced them. Hence, it is the IFC file of the model, adequately enriched with the details and results of the energy analyses, the right tool to value the work of energy auditors. However, the analysis of the IFC file exported from the energy analysis software used on a case study showed that it includes only the geometric characteristics of the building, with some thermo-physical information related to the building components. The need to expand what of the energy performance analysis is exported to the IFC file is evident for maximum fruition of the audit outcomes.

Accordingly, the second phase of the research has focused on the definition of a list of quantitative and energy parameters to be associated with the as-is BIM model, with the aim of returning the results of an energy audit. This is essential not only for a correct and complete description of the building being analysed, but also for the possibility to "catalogue" the existing building stock according to specific energy parameters. User-defined Property Sets (Psets) were used to implement these contents within the BIM as-is model. Properties are directly linked to the IFC Building entity and grouped into two Psets:

- *Building Quantitative Data:* containing the data of the building that are used to identify the heated net and gross surfaces and volumes, the surface of the dispersant envelope, on the basis of which the main energy performance indicators are also calculated;
- *Building Energy Data:* containing the results of the energy inventory and performance calculations (performance indicators, energy consumption, expenses per energy carrier…).

Psets can also include a link to the whole energy audit report, which also contains the proposals for improving the building performance and energy saving. For the mapping of these properties, it has been referred to ADM (2021).

It is not purpose of this paper to list all individuated properties (Centi and Morini, 2021), but rather to explain the approach, its possible advantages and uses. First, the definition of these Psets can represent a guide for contracting authorities in defining the information requirements for energy analyses in BIM-based tendering. They can also serve as a checklist by BEM software companies to identify the information to include in the IFC files after an energy analysis. The possibility of automatically exporting all the parameters and thus of having an IFC file complete with the results of an energy audit could greatly simplify the integration of the auditors' work into the BIM process.

Figure 6.2 schematizes the process of the second phase of the research, going from the energy analysis software to the "enriched" IFC model (shown in Fig. 6.3): all the information content about the building is stored in the agreed CDE for its future uses (among which is the project of the proposed energy efficiency measures).

A possible continuation of the research activity, in the direction of the maximum interoperability of energy audit in the BIM environment, could consist in the identification of correspondences of these parameters within the IFC standard according to the indications of the buildingSMART committee and the proposal of new parameters to be implement at the standard level. However, most of the parameters identified refer specifically to the energy calculation method contained in the relevant Italian standards, and therefore, they make it necessary to use custom Property Sets (Pset).

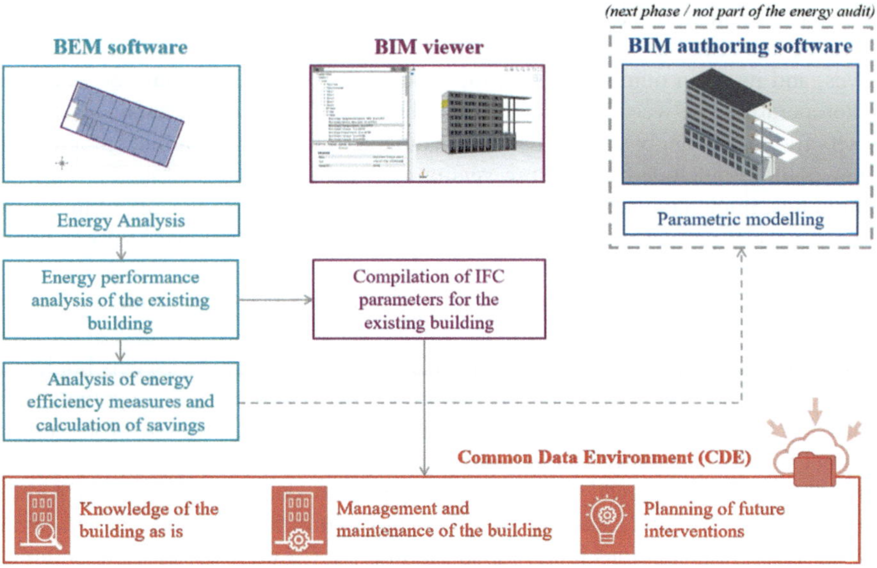

Fig. 6.2 Process from energy analysis to BIM

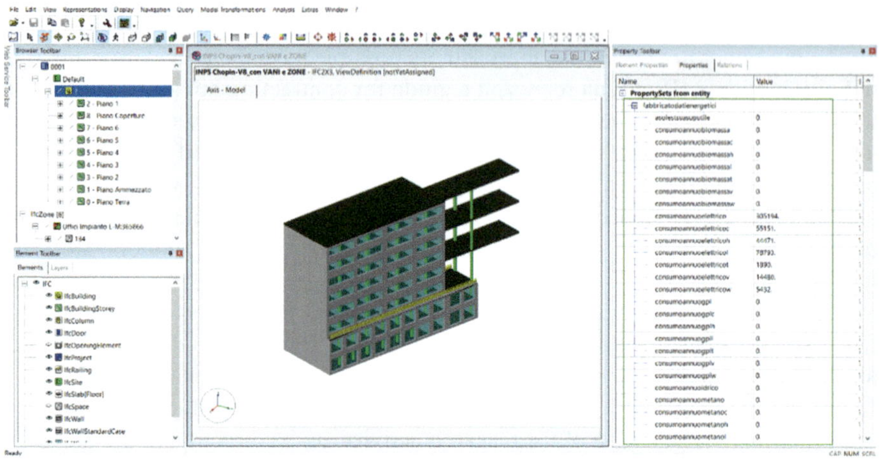

Fig. 6.3 Screenshot of a BIM viewer displaying an IFC model enriched with the defined Psets (*Source* Karlsruhe Institute of Technology)

6.3 Conclusions

This paper proposed a BIM-based approach to energy analysis of existing buildings in the Italian context. It briefly covered the whole process without getting into the technical details of it, but providing an overview of the research methodology, its objectives, and its main outcomes. Both the modelling recommendations for maximum interoperability between BIM authoring and BEM software and the list of customized Psets for energy analyses will be made available freely to the public. In particular, they could be included in a future version of the Guidelines for Energy Audits of Buildings, implemented within a BIM perspective, in line with the aim of this work, i.e. supporting professionals and enhancing the adoption of BIM in the refurbishment of the existing building stock.

The recommendations will be aimed at simplifying the work of the auditors by identifying general modelling actions, which can reduce errors and loss of information when importing the model into the analysis software. These aspects hinder the full use of BIM for energy analysis as well as the exploitation of its potential. On the other hand, the recommendation will also aim to draw attention on problems that cannot be solved "simply" through modelling and thus require solutions either at a software or standard level, stimulating the search for shared solutions, which could form the basis for future research developments in this field. For example, the study of these recommendations may be also useful for software houses for the optimization of some of the identified issues. In parallel, a codification of user-defined Psets to be associated with the as-is IFC model of a building was put in place with the aim of returning the results of an energy audit within an interoperable model. This also means making the work of the auditors easier to consult and utilize for planning and carrying out the proposed interventions, for the future management and maintenance of the building as well as for a more in-depth knowledge of the building. The wide diffusion of these energy parameters at national level could give a great impulse in this sense, overcoming the intrinsic limit linked to the use of user-defined, and not of common, Psets. This will be possible through the publication of the guidelines referred to in this work and their adoption as a technical-regulatory reference on the national territory.

On the other hand, this will also be possible through the implementation by the software companies of these parameters—mapped according to the proposed scheme—in the IFC files exported from the applications for energy analysis on the Italian market.

Acknowledgements The research was financed under the "*Accordo di Programma Ministero dello Sviluppo Economico-ENEA,* 2019–2021". Work package: WP1—High energy efficiency buildings.

References

ADM-Agenzia del Demanio (2021) Linee Guida di Produzione Informativa BIM (ADM2021-ADM-METHODSTP-XX-MS-Z-G00001)

Andriamamonjy A, Saelens D, Klein R (2019) A combined scientometric and conventional literature review to grasp the entire BIM knowledge and its integration with energy simulation. J Building Eng 22:513–527

Centi G, Margiotta F, Morini M, Pazzola L (2019) Stato dell'arte degli strumenti, degli standard e dei metodi per il Building Information Modeling (BIM) per edifici ad alta efficienza energetica. Report RdS/PTR2021/038

Centi G, Morini M (2021) Linee guida per la diagnosi energetica degli edifici con metodologia BIM. Report RdS/PTR2021/111 (in course of publication)

European Commission (2020) A renovation wave for Europe-greening our buildings, creating jobs, improving lives. COM/2020/662 final

Heffernan E, Sohel MI, Beazley S, McCarthy TJ (2017) From BIM (Building Information Modelling) to BEM (Building Energy Modelling): a collaborative approach. In: Australasian building simulation 2017 conference proceedings, pp 1–11

Katsigarakis K, Lilis GN, Giannakis G, Rovas D (2019) An IFC data preparation workflow for building energy performance simulation. In: European conference on computing in construction

MIT-Ministry of Infrastructures and Transportation (2017) Ministerial Decree n. 560 of 1 Dec 2017

Sanhudo L, Ramos NM, Martins JP, Almeida RM, Barreira E, Simões ML, Cardoso V (2018) Building information modeling for energy retrofitting–a review. Renew Sustain Energy Rev 89:249–260

UNI/TR 11775:2020. Diagnosi Energetiche - Linee guida per le diagnosi energetiche degli edifici

Chapter 7
Short-Term Wind Speed Forecasting Model Using Hybrid Neural Networks and Wavelet Packet Decomposition

Adel Lakzadeh, Mohammad Hassani, Azim Heydari, Farshid Keynia, Daniele Groppi, and Davide Astiaso Garcia

Abstract Wind speed is one of the most vital, imperative meteorological parameters, thus the prediction of which is of fundamental importance in the studies related to energy management, building construction, damages caused by strong winds, aquatic needs of power plants, the prevalence and spread of diseases, snowmelt, and air pollution. Due to the discrete and nonlinear structure of wind speed, wind speed forecasting at regular intervals is a crucial problem. In this regard, a wide variety of prediction methods have been applied. So far, many activities have been done in order to make optimal use of renewable energy sources such as wind, which have led to the present diverse types of wind speed and strength measuring methods in the various geographical locations. In this paper, a novel forecasting model based on hybrid neural networks (HNNs) and wavelet packet decomposition (WPD) processor has been proposed to predict wind speed. Considering this scenario, the accuracy of the proposed method is compared with other wind speed prediction methods to ensure performance improvement.

Keyword Sequence of estimators · Combined neural networks · Wavelet packet decomposition · Wind speed prediction

---

A. Lakzadeh · M. Hassani
Islamic Azad University, Kerman Branch, Iran

A. Heydari · D. Groppi · D. Astiaso Garcia (✉)
Sapienza University of Rome, Rome, Italy
e-mail: davide.astiasogarcia@uniroma1.it

A. Heydari
e-mail: azim.heydari@uniroma1.it

D. Groppi
e-mail: daniele.groppi@uniroma1.it

A. Heydari · F. Keynia
Graduate University of Advanced Technology, Kerman, Iran
e-mail: f.keynia@kgut.ac.ir

E. Arbizzani et al. (eds.), *Technological Imagination in the Green and Digital Transition*, The Urban Book Series, https://doi.org/10.1007/978-3-031-29515-7_7

7.1 Introduction

Energy formed a plethora of necessities of human life in the past and present, but it will also have a more paramount role in a future life; therefore, it is logical to be one of the human concerns (Blaabjerg and Liserre 2012). Fossil resources, the first and foremost energy source, are dwindling; moreover, their production costs, transmission, and distribution are enlarging. As a result, they may not be reliable sources of energy anymore and do not have ample potential to meet all human needs (Razmjoo et al. 2021). All these reasons have provoked researchers into offering new methods of supplying energy (Masters 2013). Of particular, new sources of supplying energy are solar, wind, wave, and water, but the problem faced up using these resources is random changes, such as changes in wind speed, water Debi, solar radiation intensity, which lead generated power to be partially unpredictable and unreliable. However, from any angle, the benefits of these clean energy sources outweigh the disadvantages. As has been mentioned earlier, wind is one of the most critical renewable sources of energy. Availability and permanency are two key factors leading researchers and industrial owners to replace fire with clean electrical energy (Erdem and Shi 2011; Jeon and Taylor 2012). This source of energy is currently one of the strategic products in the commercial energy markets. According to the collected data, about 2.3% of the world's electricity consumption is generated using wind energy (Torres et al. 2005; Zhao et al. 2012; Khalid and Savkin 2012). The use of wind power is growing in all parts of the world, and it has been predicted that by 2030 it will account for nearly 30% of world electricity production (Bhaskar et al. 2012; Salonen et al. 2011; Kusiak and Li 2010).

Thus, the use of advanced methods in maintaining the health of wind turbines, including predicting changes in wind speed, is an integral part of studies in this field.

Not only does a wind turbine failure increase maintenance costs, but also it reduces power generation. In order to generate effective power, the performance of the wind turbine must be checked by the status monitoring system. One of the critical performance parameters along power and step screw is wind speed, according to which Condition Monitoring System (CMS) results should be analyzed (Ackermann 2005; Heydari et al. 2021a). Some reasons representing the importance of wind speed prediction are the maintenance of wind turbine units, power generation schedule, and energy recovery and storage plan.

In (Torres et al. 2005), the Autoregressive Moving Average (ARMA) model has been presented in order to predict the hourly average wind speed. In (Bhaskar et al. 2012), the adaptive and wavelet neural network model has been provided in order to predict wind speed based on NWP information. Zhang et al. developed a new combined intelligent forecasting model based on multivariate data secondary decomposition approach and deep learning method in order to predict wind speed (Zhang et al. 2021). Heydari et al. presented a novel combined short-term wind forecasting model using metaheuristic optimization algorithm and deep learning neural network (Heydari et al. 2021b). Neshat et al. (2021) provided a machine learning-based model in order to predict short-term wind speed. The model is applied for the Lillgrund offshore wind farm in Sweden.

In this research, we have used the HNN method and WPD processor to predict wind speed. Three types of neural networks have been used in HNN neural network training based on three different algorithms which include Levenberg–Marquardt (LM), Broyden–Fletcher–Goldfarb–Shanno (BFGS), and Bayesian Regularization (BR) algorithms.

According to previous studies, with the proper selection of perceptron neural network training methods, the HNN neural network can be trained better than a single neural network. The best results for related neural networks are those that use the LM algorithm on the first step. LM is a fast training algorithm, so this algorithm should be selected as the first neural network in the hybrid neural network. On the second step of the HNN neural network, in order to identify better weights and biases in the search space, the BFGS algorithm was used to train the neural network, which caused the BFGS algorithm to start the training process from a suitable point, and was able to find the appropriate and minimal answers in the search space. In the proposed method, it is better to use the BR algorithm as the last neural network in order to adjust the parameters and obtain the maximum training efficiency.

7.2 Methods

7.2.1 Neural Network

In recent years, the use of artificial intelligence to solve complex and nonlinear problems such as classification and prediction has expanded. Artificial neural networks can identify the nonlinear relationships between variables through the training process (Yu 2010). In general, methods based on artificial intelligence are more accurate than conventional statistical methods.

One of the most well-known neural networks is the multilayer perceptron (MLP) neural network, which usually has a feedforward architecture. The MLP network consists of input and output layers and at least one hidden layer. Each of these layers has several neurons and includes processing units, and each unit is completely connected to the units of the next layer by weight (W_{ij}) (Alavi et al. 2010). Y output is obtained by transferring the sum of previous outputs and mapping by an activation function.

7.2.2 Wavelet Packet Decomposition

Wavelet packet analysis can be seen as a particular type of wavelet analysis. WPD is a classic signal processing method, which can break down the signal into suitable components. The level of wavelet analysis can profoundly affect WPD performance. WPD consists of two types: continuous wavelet transform and discrete wavelet transform. In this method, the general stage divides the approximation coefficients into two parts. After dividing the approximation coefficient vector and the detail coefficient vector, we obtain both on a large scale. Missing information is taken between two consecutive approximations in detail coefficients. Then, the next step involves dividing the vector of the new coefficient. Sequential details are never reanalyzed. However, in the corresponding wavelet packet, each detail coefficient vector is also decomposed into two parts using the same approximation vector approach.

7.2.3 Proposed Intelligent Forecasting Model

In this study, a suitable combination of MLP neural networks is utilized to design the prediction model. The HNN prediction model can significantly improve the training capability of neural networks in modeling a complex process (including wind speed prediction). In addition, the use of several structures (parallel and consecutive) to combine neural networks with more capabilities is described in (Amjady et al. 2010).

This study proposes a new prediction model to wind speed forecasting based on neural networks and noise reduction data. Since wavelet transform works as a preprocessing method, the data are decomposed by using wavelet transform in the first step. In the following step, decomposed data feed into neural networks as input data.

Figure 7.1 shows the process of the wind speed prediction according to the proposed method. As shown in the figure, each level of the HNN prediction model has been arranged by particular types of neural networks. Since all neural networks have the same number of input data, layers, and neurons in the proposed method, they can use the weights and biases values of the previous. Therefore, each neural network improves the knowledge gained from the previous one. In addition, the second set of outcomes transmitted between neural networks is the predicted value of the target variable. The second and third neural networks also receive the initial predicted value of the target variable as input, which increases the neural network prediction accuracy.

In the prediction models using wavelet transform as a noise reduction method, the process of noise reduction has three phases (see Fig. 7.2).

The first step of the noise reduction process is converting the raw data to decomposed layers by wavelet, called function analysis. Each part of the decomposed signal can be considered as a wavelet coefficient and a scale factor. By applying wavelet and scaling filters to the main time series at each step, wavelet and scaling coefficients are obtained, which are repeated in the form of the following pyramidal algorithm (Fig. 7.3).

In fact, in wavelet analysis, data are divided into two subsets of high-frequency and low-frequency data. The sparse-frequency data obtained by applying the father wavelet wave over the main series reflect the main features of the series. Much data

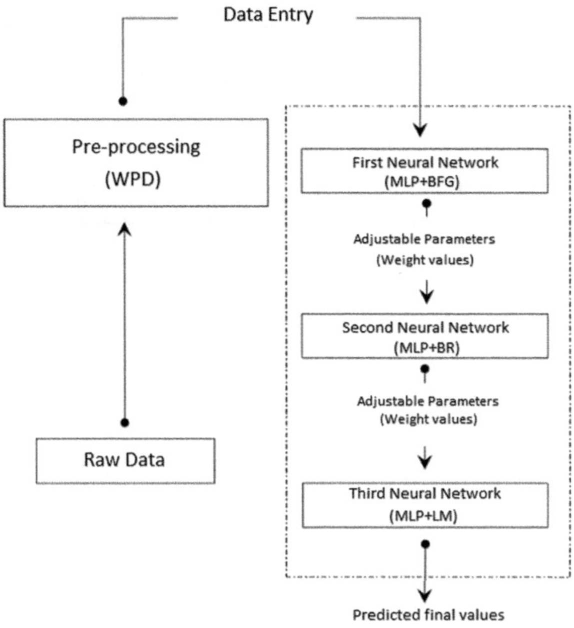

Fig. 7.1 Flowchart of the proposed method

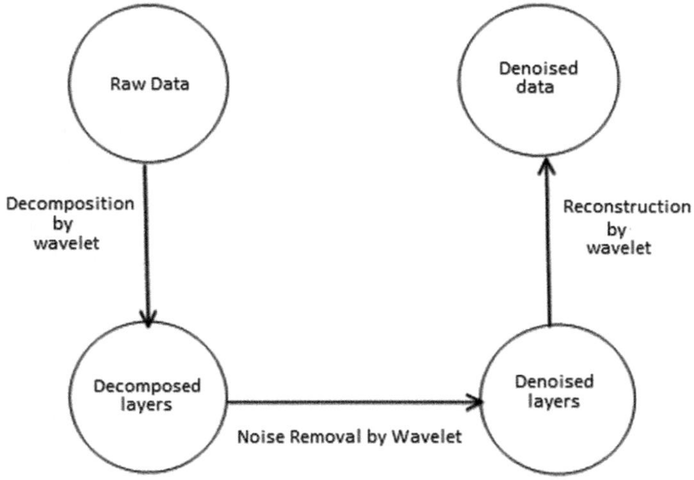

Fig. 7.2 Noise reduction by wavelet transform

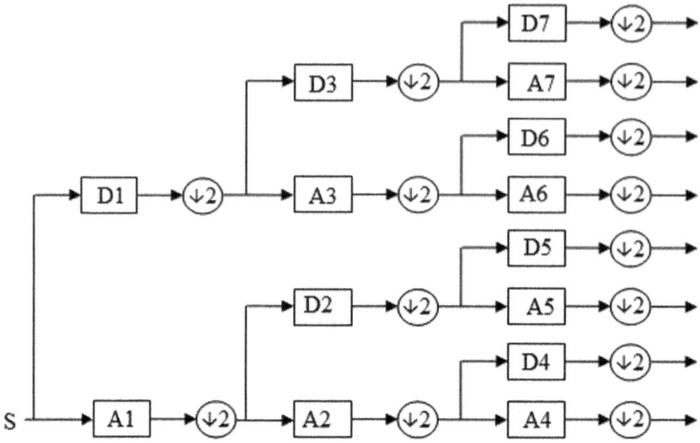

Fig. 7.3 Pyramidal algorithm for wavelet analysis, S: main data; $A[i]$ is the low-pass approximation coefficients, $D[i]$ is the high-pass detail coefficients

are also generated by applying the mother wavelet on the main series, often called noise. The main purpose of wavelet decomposition is to isolate the main features of the series from the noise.

7.3 Case Study

To evaluate the proposed model, 365 days of wind speed data from 1/3/2019 to 28/3/2020 were collected, all of which had a 5-min interval. Collected data consist of 12 samples per hour which mean 288 samples per day. On average, one data is extracted and used as the representative of that day. Collected data were divided into two categories: training and test. About 30% (120 days) of data are allocated for the test. Test data are used to evaluate the model performance.

7.4 Results and Discussion

The correlation of the obtained values by a model with the observed values (referred to as model evaluation or validation) is usually done by double comparisons between the simulated values and the observed values. There is the efficiency of the models and their validation, which can be determined by the Root Mean Square Deviation (RMSE), Mean Absolute Error (MAE), and Index of Agreement (IA).

$$\text{RMSE} = \sqrt{\frac{\sum_{i=1}^{n}(P_i - O_i)^2}{n}} \tag{7.1}$$

Table 7.1 Specifications of the artificial neural network

Input method	Hidden layers	Neurons	Learning steps	Learning rate
Neural network	2	10	150	0.001

$$\mathrm{MAE} = \frac{\sum_{i=1}^{n} |y_i - x_i|}{n} \tag{7.2}$$

$$\mathrm{IA} = 1 - \frac{\sum_{i=1}^{n} (o_i - p_i)^2}{\sum_{i=1}^{n} \left(|P_i - \overline{O}| + |O_i - \overline{O}|\right)^2} \tag{7.3}$$

First, we normalized the data to [0–1] range. After applying WPD, we used a 2db wavelength to convert the wavelet. We took three wavelet steps from the data.

Wavelet steps: in the first step, we took the wavelet from the main data, which has two output signals including upper bound and lower bound. In comparison with the original signal, these two signals are completely brand-new ones.

In the next step, the first step instrument applies two times on the output signals of the previous step. As a result, we have eight signals after step two, and in the third, the neural networks were trained using 8% of data for the testing phase.

Each neural networks have two hidden layers with up to ten neurons in each layer. The characteristics of the artificial neural network are given in Table 7.1.

We trained neural networks with three various algorithms: BR, BFG, and LM (Table 7.2). Finally, we reversed the output network results from wavelet transform and measured the coefficients of determination (R^2), mean squared error (RMSE), absolute mean error (MAE), and agreement index (IA) using real data.

In the presented tables, the values related to different scales are calculated and displayed to examine the proposed combinations of networks and their performance differences according to the activation function. As shown in Table 7.3, using the TANSIG activation function, considering the order of placement of the networks, the lowest amount of RMSE was related to HNN2, which had the best performance with the proposed order. As can be seen, the best arrangement of the networks was 1-10-10-1 and had an equal value of 2.432. Considering the corresponding values of

Table 7.2 Different proposed HNNs for review in research

First MLP + LM, then MLP + BR, and finally MLP + BFG	HNN(1)
First MLP + LM, then MLP + BFG, and finally MLP + BR	HNN(2)
First MLP + BFG, then MLP + BR, and finally MLP + LM	HNN(3)
First MLP + BR, then MLP + BFG, and finally MLP + LM	HNN(4)
MLP + LM only	HNN(5)

Table 7.3 Comparison of the proposed method with RMSE

Activation function	Network sequence	HNN(1)	HNN(2)	HNN(3)	HNN(4)	HNN(5)
Tansig	1-3-3-1	3.822	2.801	5.989	4.448	4.182
	1-5-5-1	3.551	2.799	6.062	4.430	4.224
	1-10-10-1	3.101	2.432	5.745	4.210	4.099
	1-15-15-1	3.220	2.501	5.920	4.220	4.120
Logsig	1-3-3-1	3.534	2.821	5.997	4.966	4.228
	1-5-5-1	3.618	3.027	6.066	4.413	4.210
	1-10-10-1	3.321	2.643	5.832	4.401	4.110
	1-15-15-1	3.401	2.765	5.899	4.409	4.102

the LOGSIG function, the same result is true and this network had the lowest value of RMSE in the same order. The corresponding value for this arrangement was 2.643.

Table 7.4 shows the MAE scales. The figures obtained from these calculations show the same results as the scale in Table 7.3. The corresponding MAE values for the HNN2 network with the TANSIG activation function are the lowest at 1.743. Also, using the LOGSIG activation function has the best efficiency and an equal value to 1.899 compared to other suggested items. As a result, the best-case scenario is HNN2 with the order 1-10-10-1 and using the TANSIG function.

The calculation of the determination coefficient values is given in Table 7.5. As expected from the review of the previous tables, the highest value for this scale was equal to 0.863 and was related to HNN2 with the order of 1-10-10-1 and the TANSIG function. In the continuation of the values, it can be seen that in the calculations related to the LOGSIG activation function, the corresponding values in the coefficient of determination for the two sequences 1-10-10-1 and 1-15-15-1 were similar and equal to 0.853. The lowest value was related to HNN3 with TANSIG activation function and the proposed order was 1-3-3-1.

Table 7.6 examines the values corresponding to IA. The values calculated for this parameter are in line with the previous findings. The highest value is related to HNN2

Table 7.4 Comparison of the proposed method with the MAE

Activation function	Network sequence	HNN(1)	HNN(2)	HNN(3)	HNN(4)	HNN(5)
Tansig	1-3-3-1	2.221	1.945	3.131	2.908	2.593
	1-5-5-1	2.135	1.944	3.273	2.955	2.395
	1-10-10-1	2.111	1.743	3.023	2.899	2.274
	1-15-15-1	2.104	1.867	3.100	2.903	2.145
Logsig	1-3-3-1	2.151	1.913	3.303	2.965	2.409
	1-5-5-1	2.110	1.982	3.361	2.937	2.452
	1-10-10-1	2.132	1.899	3.198	2.910	2.346
	1-15-15-1	2.134	1.901	3.143	2.910	2.345

Table 7.5 Comparison of the proposed method with the R2

Activation function	Network sequence	HNN(1)	HNN(2)	HNN(3)	HNN(4)	HNN(5)
Tansig	1-3-3-1	0.798	0.805	0.594	0.696	0.755
	1-5-5-1	0.749	0.804	0.693	0.721	0.753
	1-10-10-1	0789	0.863	0.701	0.756	0.769
	1-15-15-1	0.788	0.834	0.699	0.765	0.771
Logsig	1-3-3-1	0.731	0.843	0.645	0.698	0.735
	1-5-5-1	0.716	0.821	0.690	0.681	0.718
	1-10-10-1	0.728	0.853	0.687	0.702	0.727
	1-15-15-1	0.734	0.853	0.698	0.703	0.733

Table 7.6 Comparison of the proposed method with IA

Activation function	Network sequence	HNN(1)	HNN(2)	HNN(3)	HNN(4)	HNN(5)
Tansig	1-3-3-1	0.916	0.954	0.892	0.893	0.915
	1-5-5-1	0.912	0.954	0.906	0.909	0.904
	1-10-10-1	0.943	0.994	0.918	0.911	0.921
	1-15-15-1	0.940	0.992	0.911	0.908	0.922
Logsig	1-3-3-1	0.912	0.954	0.896	0.908	0.900
	1-5-5-1	0.915	0.946	0.906	0.904	0.897
	1-10-10-1	0.921	0.967	0.910	0.909	0.919
	1-15-15-1	0.919	0.966	0.911	0.908	0.918

with TANSIG activation function and order 1-10-10-1 and shows a value equal to 0.994. After that, the best and highest value of this index is related to the same network with the order of 1-15-15-1 and has a difference of 0.002. The lowest value in the calculations of this scale was similarly related to HNN3 with other TANSIG activation functions and the proposed order of 1-3-3-1.

Comparison of the output results of various neural network arrangements with error measurement criteria shows that the proposed model has the acceptable ability to predict wind speed. As shown in Tables 7.3, 7.4, 7.5, and 7.6 the most effective ANN model in terms of performance is the arrangement (1-10-1), sigmoid tangent function, and HNN hybrid algorithm (1) on Kerman time series data.

Different criteria were used to measure the model: coefficient of determination (R2), squared mean square error (RMSE), absolute mean error (MAE), and agreement index (IA). All the results showed that the proposed method performed very well. In particular, the excellent performance of this method in long-range forecasting will be a clear vision for the Condition Monitoring System (CMS) to prevent sudden events and damage to wind turbines and ensure the health of the turbine. This reduces turbine shutdown and reduces costs and increases production capacity.

7.5 Conclusions

At first, wind speeds are predicted for future using a new combined intelligent model.

For this purpose, three feedforward networks have been used. This type of wind speed prediction can be very efficient in the Iranian wind energy industry, and this research in this field is not similar in the world. The proposed model was evaluated with various network models, functions, and different numbers of neurons to extract optimal network structure. Furthermore, training and validation data were used to evaluate and compare diverse network structures' performance. The proposed model also was trained with training algorithms BFGS, LM, and BR. As the results show, with the right Perceptron neural network training methods, the HNN neural network can be trained better than a single neural network. As been explained before, the best results for class neural networks are those using the first-class LM algorithm. Training a Perceptron neural network with an LM algorithm leads training error to reduce quickly since the LM algorithm is reasonably fast. Therefore, it is logical to select as the training algorithm of the first level neural network in the hybrid model.

Acknowledgements Sapienza University of Rome (DPDTA—Horizon2020 project "GIFT" Geographical Islands FlexibiliTy Grant agreement ID: 824410) is kindly acknowledged for providing financial support for this work.

References

Blaabjerg F, Liserre M (2012) Power electronics converters for Wind turbine systems. IEEE Trans Ind Appl 48(2):708–719. https://doi.org/10.1109/TIA.2011.2181290

Razmjoo A, Rezaei M, Mirjalili S, Majidi Nezhad M, Piras G (2021) Development of sustainable energy use with attention to fruitful policy. Sustainability 13(24):13840. https://doi.org/10.3390/su132413840

Masters GM (2013) Renewable and efficient electric power systems. John Wiley & Sons

Erdem E, Shi J (2011) ARMA based approaches for forecasting the tuple of wind speed and direction. Appl Energy 88(4):1405–1414. https://doi.org/10.1016/j.apenergy.2010.10.031

Jeon J, Taylor JW (2012) Using conditional kernel density estimation for wind power density forecasting using conditional kernel density estimation for wind power density forecasting. J Am Stat Assoc 107(497):66–79. https://doi.org/10.1080/01621459.2011.643745

Torres JL, De BM, De FA, Garcı A (2005) Forecast of hourly average wind speed with ARMA models in navarre (Spain). Sol Energy 79(1):65–77. https://doi.org/10.1016/j.solener.2004.09.013

Zhao P, Wang J, Xia J, Dai Y, Sheng Y, Yue J (2012) Performance evaluation and accuracy enhancement of a day-ahead wind power forecasting system in China. Renewable Energy 43:234–241. https://doi.org/10.1016/j.renene.2011.11.051

Khalid M, Savkin AV (2012) A method for short-term wind power prediction with multiple observation points. IEEE Trans Power Syst 27(2):579–586. https://doi.org/10.1109/TPWRS.2011.2160295

Bhaskar K, Member S, Singh SN, Member S (2012) AWNN-assisted wind power forecasting using feed-forward neural network. IEEE Trans Sustain Energy 3(2):306–315. https://doi.org/10.1109/TSTE.2011.2182215

Salonen K, Niemelä S, Fortelius C (2011) Application of radar wind observations for low-level NWP wind forecast validation. J Appl Meteorol Climatol 50(6):1362–1371. https://doi.org/10.1175/2010JAMC2652.1
Kusiak A, Li W (2010) Estimation of wind speed: a data-driven approach. J Wind Eng Ind Aerodyn 98(10–11):559–567. https://doi.org/10.1016/j.jweia.2010.04.010
Ackermann T (2005) Wind power in power systems. vol 200(5). New York, Wiley
Heydari A, Garcia DA, Fekih A, Keynia F, Tjernberg LB, De Santoli L (2021a) A hybrid intelligent model for the condition monitoring and diagnostics of wind turbines gearbox. IEEE Access 9:89878–89890. https://doi.org/10.1109/ACCESS.2021.3090434
Zhang S, Chen Y, Xiao J, Zhang W, Feng R (2021) Hybrid wind speed forecasting model based on multivariate data secondary decomposition approach and deep learning algorithm with attention mechanism. Renewable Energy 174:688–704. https://doi.org/10.1016/j.renene.2021.04.091
Heydari A, Nezhad MM, Neshat M, Garcia DA, Keynia F, De Santoli L, Tjernberg LB (2021b) A combined fuzzy gmdh neural network and grey wolf optimization application for wind turbine power production forecasting considering scada data. Energies 14(12):3459. https://doi.org/10.3390/en14123459
Neshat M, Majidi Nezhad M, Abbasnejad E, Mirjalili S, Bertling Tjernberg L, Astiaso Garcia D, Alexander B, Wagner M (2021) A deep learning-based evolutionary model for short-term wind speed forecasting: a case study of the Lillgrund offshore wind farm. Energy Convers Manage 236:114002. https://doi.org/10.1016/j.enconman.2021.114002
Yu S (2010) Actual experience on the short-term wind power forecasting at Penghu-from an Island perspective. In: 2010 International conference on power system technology, IEEE Press, Zhejiang, China, pp 1–8. https://doi.org/10.1109/POWERCON.2010.5666619
Alavi AH, Gandomi AH, Mollahassani A, Heshmati AA, Rashed A (2010) Modeling of maximum dry density and optimum moisture content of stabilized soil using artificial neural networks. J Plant Nutr Soil Sci 173(3):368–379. https://doi.org/10.1002/jpln.200800233
Amjady N, Daraeepour A, Keynia F (2010) Day-ahead electricity price forecasting by modified relief algorithm and hybrid neural network. IET Gener Transm Distrib 4(3):432–444

Chapter 8
COGNIBUILD: Cognitive Digital Twin Framework for Advanced Building Management and Predictive Maintenance

Sofia Agostinelli

Abstract According to contemporary challenges of digital evolution in management and maintenance of construction processes, the present study aims at defining valuable strategies for building management optimization. As buildings' and infrastructures' Digital Twins (DT) are directly connected to physical environment through the Internet of Things (IoT), asset management and control processes can be radically transformed. The proposed DT framework connects building information model (BIM) three-dimensional objects to information about the planned maintenance of components, supplying system's self-learning capabilities through input data coming from Building Management Systems (BMSs), ticketing, as well as maintenance activities' data flow both as-needed or unexpected. The concept of real-time acquisition and data processing set the basis for the proposed system architecture, allowing to perform analysis and evaluate alternative scenarios promptly responding to unexpected events with a higher accuracy over time. Moreover, the integration of artificial intelligence (AI) allows the development of maintenance predictive capabilities, optimizing decision-making processes and implementing strategies based on the performed analysis, configuring a scalable approach useful for different scenarios. The proposed approach is related to the evolution from reactive to proactive strategies based on Cognitive Digital Twins (CDTs) for Building and Facility Management, providing actionable solutions through operational, monitoring and maintenance data. Through the integration of BIM data with information systems, BMS, IoT and machine learning, the optimization and real-time automation of maintenance activities are performed, radically reducing failures and systems' breakdowns. Therefore, integrating different technologies in a virtual environment allows to define data-driven predictive models supporting Building Managers in decision-making processes improving efficiency over time and moving from reactive to proactive approaches.

Keyword Cognitive building · Artificial intelligence · Digital Twin · Predictive maintenance · Facility management · Building information modeling

S. Agostinelli (✉)
Sapienza University of Rome, Rome, Italy
e-mail: sofia.agostinelli@uniroma1.it

E. Arbizzani et al. (eds.), *Technological Imagination in the Green and Digital Transition*, The Urban Book Series, https://doi.org/10.1007/978-3-031-29515-7_8

8.1 Introduction

The new Industrial Revolution (Oliveira and Afonso 2019) is related to digital automation enhancing quality and effectiveness of processes.

The construction sector is facing such a slow digital growth named Construction 4.0 involving new digital strategies based on technologies, connectivity, devices and cloud platforms. Interoperability of data and automation of processes are some of the main goals toward decision-making systems' decentralization, even though the construction industry is still appearing resistant.

Internet of Things (IoT), smart sensors, cloud computing, 5G networks, Extended Reality (XR) and Digital Twins (DT) are new paradigms of Construction 4.0 in the operation and maintenance (O&M) phase, which actually represents one of the higher costs of a building's life cycle (30%), as observed by Mourtzis et al. (2017).

New building maintenance strategies based on digital systems and data-driven technologies introduce proactive approaches to Facility Management (FM) of building systems such as failure prediction, effort estimation, or energy consumption optimization (Jasiulewicz-Kaczmarek et al. 2020).

Predictive maintenance is the main objective of such data-driven strategies, which can provide up to 630 billion in maintenance savings in 2025, as observed by McKinsey (2022).

In O&M 4.0, fault prediction and production efficiency are enabled by two main components: (a) sensor data and (b) data analysis across the entire life cycle of assets from design to the operation and maintenance phase.

In this regard, the present study aims at investigating different O&M strategies introducing the use of Digital Twins (DTs). In fact, the Building Management phase could be redefined using DT-based solutions for virtual and physical integration (Qi and Tao 2018) providing new DT-enabled strategies for operation and maintenance optimization.

8.2 Digital Twin Paradigm in Operation and Maintenance

8.2.1 Digital Twin Definitions

The concept of DT first appeared in 2002, and since then, many definitions evolved over time. In 2012, Dr. Michael Grieves introduced a first clear definition of the DT concept as the connection of data between physical and virtual products (Grieves and Vickers 2017). This enabling information set consists of data related to the assets such as geometry, operational and technical information up to their functional behavior. In this regard, Rosen defined the DT as a mirror between physical and virtual objects allowing data analysis across the entire asset life cycle (Rosen et al. 2015). Moreover, the concept of DT is enabled by algorithms for data-driven actions and decision-making systems. In fact, the effective relation between physical and

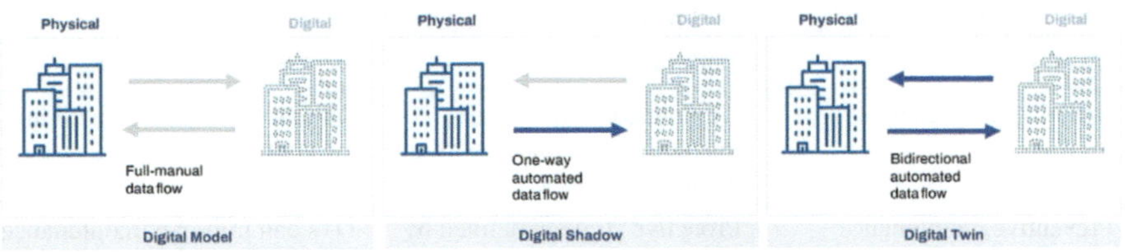

Fig. 8.1 Digital model, digital shadow, and digital twin (Kritzinger et al. 2018)

digital is enabled by data acquisition systems and processing technologies, with the aim of providing predictive capabilities (Liu et al. 2019) in order to promptly respond to the unexpected.

Specifically, DTs are composed of three main parts: physical, virtual and the connection between them. As mentioned, the DT is considered as a virtual mirror describing the physical properties of the system, delivering and receiving information (Tharma et al. 2018) for control, monitoring and decision-making processes.

An effective DT needs to be always connected and synchronized while running simulations of the physical counterpart over time. Depending on data flow and interaction levels between physical and digital, DT is defined in different key levels (Fig. 8.1) (Kritzinger et al. 2018):

- Digital Model—data flow between physical and digital is full-manual.
- Digital Shadow—only data flow from physical to digital is automatic.
- Digital Twin—data flow between physical and digital is bidirectionally automatic.

In the third level, the DT acquires data from sensors or on-site inspections, and it is also able to provide insights in terms of actions to be performed on the physical asset, such as the maintenance tasks as resulted from the status information acquired from sensors.

8.3 Maintenance Strategies and DT Application

In this paragraph, different maintenance strategies are analyzed and related to potential outcomes from DT implementation as defined in Table 8.1.

Reactive maintenance—it consists of maintenance activities not previously planned and caused by breakdowns (Swanson 2001) resulting in renewal of the damaged asset and causing big impacts on costs due to service interruption and production delays, etc.

As this type of maintenance approach is based on asset repairing, DTs can be applied to promptly detect failure's causes through models and simulations. However, the application of DTs determines improvements in the use of proactive maintenance approaches.

Table 8.1 Maintenance strategy and potential outcomes from DT implementation

Maintenance strategy	Definition	Implementation of DTs
Reactive maintenance	Unplanned maintenance resulted from breakdowns and damaged assets	DTs allows failure detections through models and simulations
Preventive maintenance	Proactive strategy defined by asset managers aimed at preventing or reducing failures	DTs can improve maintenance planning through data-driven strategies
Condition-based maintenance	Deviation monitoring from asset's optimal behavior through IoT, connectivity and cloud computing technologies, anticipating a planned maintenance activity	Data acquisition from sensors supplies analytical models creating a real-time knowledge base for cognitive systems
Predictive maintenance	Predicting a system's remaining life by merging data from different sources through data-driven or model-driven approaches	DTs analyze the asset's current state and behavior resulting in valuable predictions of component breakdowns
Prescriptive maintenance	Optimization of maintenance predictions using historical and real-time data and resulting in proactive activity plans	CDTs provide actionable information on maintenance activities enabled by artificial intelligence based on historical and real-time data

Preventive maintenance—it is a proactive maintenance strategy aimed at preventing or reducing breakdowns (Shafiee 2015) in order to minimize/avoid service interruptions. This approach is based on the asset manager's experience who defines time and frequency and planning service interruption (Bashiri et al. 2011). This strategy is based on over-maintaining strategies ensuring safety and productivity but still resulting in higher costs.

In this scenario, DTs can provide great improvements through data-driven planning systems for maintenance activities as it is traditionally developed by asset managers.

Condition-based (CBM) maintenance—it is based on monitoring the deviation from asset's optimal behavior through the use of IoT, connectivity and Cloud computing technologies, usually resulting in anticipating a planned maintenance activity (Nikolaev et al. 2019). This kind of approach can be improved by artificial intelligence systems acquiring and processing maintenance actual data over time (Mabkhot et al. 2018).

Following CBM strategies, sensors and real-time communication provided by the configuration of DTs help improving and supporting decision-making.

Data acquisition from sensors supplies the analytical models creating a knowledge base for cognitive systems resulting in asset condition representation for real-time monitoring.

Predictive maintenance—it is based on simulating and predicting a system's remaining lifetime combining and analyzing data from different sources (Fang et al. 2017; Werner et al. 2019) through data-driven approaches based on the availability of a large amount of sensors' data providing information on the asset state. Data analysis algorithms are performed and provide results based on data processing (Liu et al. 2018). This data-driven strategy is based on mathematical analytical models describing the component degradation (Sivalingam et al. 2018).

As abovementioned, Digital Shadows are introduced in CBM and predictive maintenance, as the Digital Model is automatically provided with state information.

DTs are based on predictive models evaluating and analyzing the asset's current state and behavior resulting in valuable predictions of possible component breakdowns.

Prescriptive maintenance—it is based on the optimization of maintenance merging historical and real-time data and resulting in proactive activity plans based on prediction (Consilvio et al. 2019; Matyas et al. 2017) in order to optimize cost, productivity, service and safety.

In prescriptive maintenance strategies, the integration of DTs is essential, as they provide actionable information on maintenance activities based on historical and real-time data. This integration involves the DT in a sort of activities' recommendation systems for operators through data analytics and artificial intelligence-enabled Cognitive Digital Twins (CDT). Despite progresses in this research field, applications of such CDT-based prescriptive maintenance are still only found in energy and manufacturing industry.

Adopting proactive strategies in maintenance is necessary to avoid breakdowns and detect inefficiencies by constantly monitoring the asset. In this way, real-time information is obtained, and behavior analysis is constantly performed allowing prompt diagnosis when failures occur.

Such proactive approach can be evolved combining historical and real-time data analysis through predictive models, resulting in failure prediction. This strategy can be applicable to several contexts and sectors where maintenance costs are high (Tao et al. 2019), allowing advantages such as limited downtimes and breakdowns, cost savings, enhanced productivity, clearly defined maintenance activities, reduced energy consumption, enhanced asset security.

In the diagram below (Fig. 8.2), information needed and monitoring approaches in proactive and reactive maintenance strategies are compared, and the CDT scenario is introduced.

The creation of accessible cloud-based CDTs providing real-time status information is a good chance to improve advanced predictive strategies, optimizing and reducing unnecessary activities (Tang et al. 2018; Pivano et al. 2019; Liew et al. 2019). In this regard, Immersive Extended Reality (XR) is also a valuable technology to visualize data and show component failures.

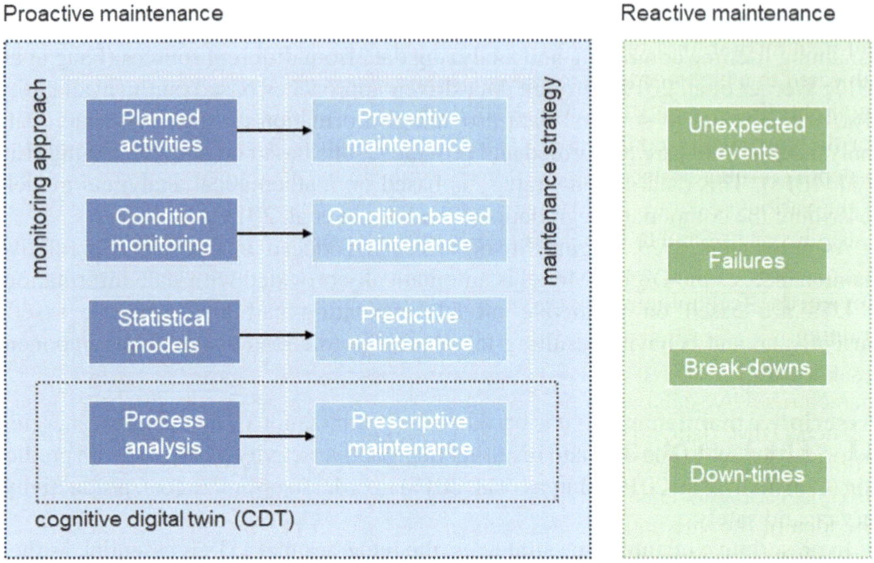

Fig. 8.2 CDT proactive strategies versus reactive maintenance

8.4 COGNIBUILD—A Cognitive Digital Twin Framework for Building Maintenance

When integrating asset, operational and historical data from a variety of sources, data availability becomes a prominent issue.

Big Data Management platforms enabling scalability, replicability and ubiquity are needed, optimizing data fusion and improving data models. Also, cybersecurity and blockchain are elements to be considered for ensuring data security (Longo et al. 2019).

As mentioned, the use of DTs is scalable and applicable in different maintenance approaches; for example, in preventive strategies, DTs can be used in the planning scheduling phase. In CBM, DTs provide real-time asset and health monitoring obtaining predictive solutions.

In this regard, DT systems need data from different sources in order to predict the evolution of asset's degradation based on the acquired operational data.

To this end, improvements in collecting and integrating information in data warehouses are still necessary.

Implementing asset monitoring in building management based on IoT technologies may overcome the lack of data (Kraft and Kuntzagk 2017), as well as the use of building information models (BIMs) as a data source. Also, synthetic simulated data could be a solution as long as a methodology for simulation scenarios is well-defined.

Even, data quality is a challenge to be overcome through specific algorithms for data preprocessing to improve the integrity of monitored data.

In the proposed scenario, building management and maintenance operations can be based on predictive systems, reducing operating costs, malfunctions and breakdowns through DT-enabled intelligent systems, developing cognitive capabilities and analyzing data from different sources (Raza et al. 2020).

The COGNIBUILD approach is based on the development of DTs from information modeling (BIM) connecting three-dimensional objects to operation and maintenance data. The model acquires self-learning capabilities analyzing input data coming from BMS systems, ticketing and through AI algorithms processing expected/unexpected events.

Therefore, AI systems enable cognitive and predictive capabilities on maintenance activities, optimizing decision-making processes and implementing prescriptive strategies based on data analysis. In this regard, the proposed framework is characterized by scalability and replicability on different contexts.

Then, BIM data represent a fundamental part of the DT, as it reproduces geometric and informative characteristics in a three-dimensional database, where the objects/components of the model are filled with specific attributes describing their functional/performances/operational data.

Receiving input and signals from sources like sensors, Building Management System (BMS) and Building Energy Management Systems (BEMS) or ticketing systems for maintenance operations, etc., the DT enriches its knowledge base developing self-learning and predictive capabilities enhanced by AI algorithms.

Digital Management Systems can be valuable sources for real-time data management of information related to the life cycle of buildings, enabling what-if analysis and simulations of scenarios for decision-making.

The proposed CDT framework (Fig. 8.3) is based on (a) BIM maintenance data directly connected and updated by (b) BMS systems, (c) ticketing systems and (d) computer vision-based sensors. In this context, (a) provides checklists and maintenance scheduling as planned, as well as (b), (c) and (d) allow the generation of maintenance intervention sheets.

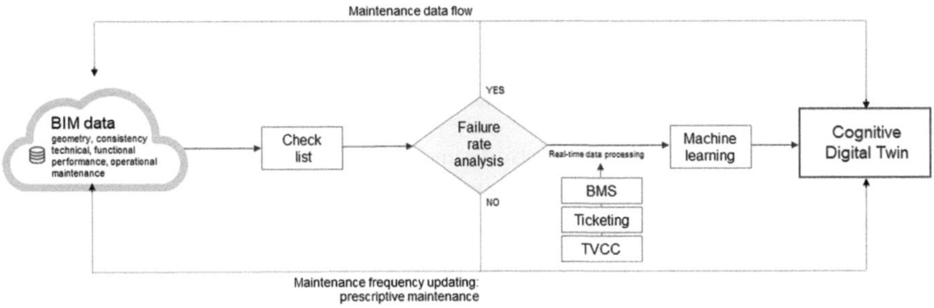

Fig. 8.3 Cognitive digital twin-based maintenance strategy

Therefore, planned checklists and data coming from sensors and maintenance systems are analyzed by machine learning algorithms, comparing expected and unexpected events and resulting in a new prescriptive maintenance frequency according to the following algorithm:

$$\mathrm{Mp} = \mathrm{Mppl} - \left[\mathrm{Mppl} * (\mathrm{fr} - \mathrm{frth})\right]$$

- fr is the real failure rate.
- frth is the threshold limit of the failure rate.
- Mppl is the planned maintenance frequency.
- Mp is the new CDT-based maintenance frequency.

According to the abovementioned, the CDT analyzes the scheduled maintenance cycle of components through statistical model-based evaluations resulting in updated frequency for the component maintenance.

If the failure rate is minor than 5%, then the planned maintenance frequency is considered as adequate and does not need to be shortened or updated.

8.5 Conclusions

Evolving from corrective to prescriptive maintenance strategies is a gradual process involving the application of DTs (Mihai et al. 2021).

Regulatory issues may exist for operation and maintenance approaches in critical assets as well as in potentially impactful contexts. According to the abovementioned, many operators still follow preventive and over-maintaining management strategies.

Moving forward to DT-based prescriptive approaches needs a correct classification of activities and a complete integration of advanced technologies such as cloud computing, data processing, IoT, data warehouses, data models, communication networks, cybersecurity and blockchain.

In this scenario, the proposed concept of CDT is based on automated interactions between physical and digital allowing advances in maintenance strategies and building management.

Barriers in data quality and availability can be overcome by the integration of IoT systems. The COGNIBUILD system introduces maintenance procedures based on the optimization of performed activities (Agostinelli et al. 2021), combining real-time and historical data and providing solutions in terms of valuable recommendation strategies using machine learning.

In this research scenario, progresses in cognitive technologies, artificial intelligence and calculation will provide enhanced capabilities to achieve a full-digital self-learned operation and maintenance approach.

References

Agostinelli S, Majidi Nezhad M, Cumo F, Heydari A, Muzi F (2021) Predictive maintenance strategy based on big data analysis and machine learning approach for advanced building management system. In: Proceedings of SEEP 2021 international conference, Vienna, Austria

Bashiri M, Badri H, Hejazi TH (2011) Selecting optimum maintenance strategy by fuzzy interactive linear assignment method. Appl Math Model 35:152–164

Consilvio A et al. (2019) Prescriptive maintenance of railway infrastructure: from data analytics to decision support. In: MT-ITS 2019 - 6th international conference on models and technologies for intelligent transportation systems

Fang X, Gebraeel NZ, Paynabar K (2017) Scalable prognostic models for large-scale condition monitoring applications. IISE Trans 49:698–710

Grieves M, Vickers J (2017) Digital twin: mitigating unpredictable, undesirable emergent behavior in complex systems. Transdisciplinary Perspectives on Complex Systems. Springer International Publishing, Cham, pp 85–113

Jasiulewicz-Kaczmarek M, Legutko S, Kluk P (2020) Maintenance 4.0 technologies—new opportunities for sustainability driven maintenance. Manag Prod Eng Rev 11:74–87

Kraft J, Kuntzagk S (2017) Engine fleet-management—the use of digital twins from a mro perspective. In: Proceedings of the ASME Turbo Expo 1

Kritzinger W, Karner M, Traar G, Henjes J, Sihn W (2018) Digital Twin in manufacturing: a categorical literature review and classification. IFAC- PapersIOnLine 51(11):1016–1022

Liew JX, Bin J, Liu Z (2019) Software as a service: the future of NDI data analysis in the cloud. Insight Non-Destruc Test Cond Monit

Liu Z, Meyendorf N, Mrad N (2018) The role of data fusion in predictive maintenance using digital twin. AIP Conf Proc 1949(1):020023

Liu J et al (2019) Dynamic evaluation method of machining process planning based on digital twin. IEEE Access 7:19312–19323

Longo F, Nicoletti L, Padovano A (2019) Ubiquitous knowledge empowers the smart factory: the impacts of a service-oriented digital twin on enterprises' performance. Annu Rev Control

Mabkhot MM, Al-Ahmari AM, Salah B, Alkhalefah H (2018) Requirements of the smart factory system: a survey and perspective

Matyas K, Nemeth T, Kovacs K, Glawar R (2017) A procedural approach for realizing prescriptive maintenance planning in manufacturing industries. CIRP Ann Manuf Technol 66:461–464

McKinsey (2022) Unlocking the potential of the internet of things. McKinsey. Available online: https://www.mckinsey.com/business-functions/mckinsey-digital/our-insights/the-internet-of-things-the-value-of-digitizing-the-physical-world. Accessed 9 Jan 2022

Mihai S, Davis W, Hung DV, Trestian R, Karamanoğlu M, Barn BS, Prasad R, Venkataraman H, Nguyen H (2021) A digital twin framework for predictive maintenance in industry 4.0. In: International conference on high performance computing & simulation, Barcelona, Spain

Mourtzis D, Zogopoulos V, Vlachou E (2017) Augmented reality application to support remote maintenance as a service in the robotics industry. Proc CIRP 2017(63):46–51

Nikolaev S, Belov S, Gusev M, Uzhinsky I (2019) Hybrid data-driven and physics-based modelling for prescriptive maintenance of gas-turbine power plant. In: IFIP Advances in information and communication technology, vol 565, pp 379–388

Oliveira M, Afonso D (2019) Industry focused in data collection: How industry 4.0 is handled by big data. ACM Int Conf Proc Ser 2019:12–18

Pivano L, Nguyen DT, Ludvigsen KB (2019) Digital twin for drilling operations -towards cloud-based operational planning. In: Proceedings of the annual offshore technology conference

Qi Q, Tao F (2018) Digital twin and big data towards smart manufacturing and industry 4.0: 360 degree comparison. IEEE Access 6:3585–3593

Raza M, Kumar PM, Hung DV, Davis W, Nguyen H, Trestian R (2020) A digital twin framework for industry 4.0 enabling next-gen manufacturing. In: 2020 9th international conference on industrial technology and management (ICITM), pp 73–77

Rosen R, Von Wichert G, Lo G, Bettenhausen KD (2015) About the importance of autonomy and digital twins for the future of manufacturing. IFAC-Papers Online 28(3), 567–572

Shafiee M (2015) Maintenance strategy selection problem: an MCDM overview. J Qual Maint Eng 21:378–402

Sivalingam K, Sepulveda M, Spring M, Davies P (2018) A review and methodology development for remaining useful life prediction of offshore fixed and floating wind turbine power converter with digital twin technology perspective. In: Proceedings 2018 2nd international conference green energy applied ICGE, pp 197–204

Swanson L (2001) Linking maintenance strategies to performance. Int J Prod Econ 70:237–244

Tang S, Wang R, Zhao X, Nie X (2018) Building cloud services for monitoring off-shore equipment and operators. Proceedings of the Annual Offshore Technology Conference 2:852–864

Tao F, Zhang H, Liu A, Nee AYC (2019) Digital twin in industry: state-of-the-Art. IEEE Trans Ind Inf 15:2405–2415

Tharma R, Winter R, Eigner M (2018) An approach for the implementation of thedigital twin in the automotive wiring harness field. Proc Int Design Conf DESIGN 6:3023–3032

Werner A, Zimmermann N, Lentes J (2019) Approach for a holistic predictive maintenance strategy by incorporating a digital twin. Proc Manuf 39:1743–1751

Zakiyudin MZ, Fathi MS, Rambat S, Tobi M, Uzairiah S, Kasim N, Latiffi AA (2013) The potential of context-aware computing for building maintenance management systems. Appl Mech Mater 405:3505–3508

Chapter 9
Design of CCHP System with the Help of Combined Chiller System, Solar Energy, and Gas Microturbine

Samaneh Safaei, Farshid Keynia, Sam Haghdady, Azim Heydari, and Mario Lamagna

Abstract This work was conducted to design a combined cooling, heating, and power (CCHP) system with photovoltaic energy which provides simultaneous generation of electricity, heat, and cold for a high-rise office building (23 floors) in the city of Mashhad in Iran. Our strategy was to supply load electric, thermal, and refrigeration with the help of solar energy. In addition, its superiority over other systems was evaluated. Analysis and study of solar radiation and the maximum level of solar panels use, according to the architectural plan, were carried out at the project site. The analysis of shadow points, the use of inverters and electrical detectors to increase the maximum solar power, and its cost-effectiveness were carefully studied via PVSOL software. Additionally, the amount of heat, cold, and electricity consumption was accurately calculated according to international standards and utilizing HAP software. The criteria for saving on the initial cost reduction, carbon dioxide emission reduction, operating cost reduction, payback period, revenue, and the minimum life expectancy of the equipment compared to those in other methods were also evaluated. The results obtained from the designed system of simultaneous generation of electricity, heat, and refrigeration, which combines gas microturbines as the primary stimulus, a combination of absorption and compression chiller to provide refrigeration load, a boiler for auxiliary heat load, and a thermal photovoltaic system to produce both electric and thermal loads, were finally revealed. This is believed to be a cost-effective strategy for high-rise residential or commercial buildings with a geographical location like that of Mashhad. Based on the electricity sales to the grid, with the rate of increase in inflation in electricity tariffs, this design in the Mashhad project was estimated to have an annual income of 166.676 thousand

---

S. Safaei · F. Keynia · A. Heydari
Graduate University of Advanced Technology, Kerman, Iran
e-mail: azim.heydari@uniroma1.it

S. Haghdady
Islamic Azad University, Mashhad, Iran

A. Heydari · M. Lamagna (✉)
Sapienza University of Rome, Rome, Italy
e-mail: mario.lamagna@uniroma1.it

E. Arbizzani et al. (eds.), *Technological Imagination in the Green and Digital Transition*, The Urban Book Series, https://doi.org/10.1007/978-3-031-29515-7_9

dollars. Moreover, the initial capital return period in this project was calculated to be 5.19 years.

Keyword CHP · CCHP · PV system · Building energy saving · T-EA

9.1 Introduction

The world's energy demand is increasing, and fossil fuel sources are declining.

Therefore, finding solutions to manage their consumption of particular importance, especially to reduce emissions. One of the best solutions is to use renewable sources (RESs), especially solar (Makkiabadi et al. 2021a, b) and wind energies (Kokkos et al. 2021). RES can provide an affordable and safe supply (Mago and Hueffed 2010). If the rate of extraction of an energy source does not exceed the natural rate of replenishment, it can be called as a sustainable energy source (Makkiabadi et. al 2021a, b; Razmjoo et al. 2021). Building sector consumes the highest amount of energy and electricity (Lamagna et al. 2020), especially to maintain their standards (Astiaso Garcia 2016). Building information modeling (BIM) is a digital tool used in different aspects, engineering design, construction, and management, understanding the sharing and transmission of building data in the whole life cycle of project planning, operation, and maintenance based on the integration of building data and information model (Agostinelli et al. 2022). Then, engineering and technical personnel can understand and take efficient response to various building information, providing the collaborative assistance model for all construction entities including design team, and construction and operating units (Peng et al. 2020). In addition, given the statistics, 85% of the Internet of Things (IoT) devices use digital twins' technologies for security monitoring, so applying the digital twins to the construction of smart cities has become the research hotspot in the field (Qin et al. 2021). Different parameters of buildings energy can be linked to the digital twin building section of the neighborhood for energy consumption monitoring, costs, and optimization (Manfren et al. 2021).

With the advent of technology, RES can be employed for electricity and heat generation in this sector (Nastasi 2019). Ahn (2019) examined the environmental and economic performances of an absorption chillers versus hybrid chillers in a CCHP system in a hospital. Those kinds of systems can offer stability (Konečná et al. 2020), controllability (Gu et al. 2012), electricity (Ehyaei and Bahadori 2007), and heat generation (Ebrahimi and Keshavarz 2013). Researchers have studied simultaneous generation to reduce peak load using multi-objective planning and concluded that electricity prices, natural gas prices, and PV efficiencies have different effects on performance (Yang et al. 2021). Qian et al. (2021) proposed a novel combined model based on multi-objective decision-making to analyze the performance evaluation of the wind–solar–CCHP systems. The use of thermal photovoltaics by Lianga et al. (2015) in a residential building in China showed that 31.7% of the thermal energy of the building could be supplied with 132 m^2 of collectors. Moreover, the use of solar energy in the facade of a building in one of the cities in Iran was investigated to provide at least 20% of the monthly electricity required for the critical month (Hoseinzadeh et al. 2021). The use of solar energy on the roofs of buildings in Valencia has also been analyzed to meet electricity demand (Gomez et al. 2021). Several researchers have also studied the combined PVT/water (Chow et al. 2007) and PVT/air systems (Kalogirou and Tripanagnostopoulos 2006). They have found that the natural flow of water is more efficient for receiving heat from the module and lowering the cell temperature. Jalalizadeh et al. (2021) simulated and studied a new combination of building solar heat collectors and absorption cooling system, as a three-dimensional production system, to meet the thermal and electrical energy needs of a residential building. The use of this proposed system has contributed to energy and economic savings and reduced the building energy consumption (De Santoli et al. 2018). CCHP is a powerful tool for increasing energy efficiency, and RES is ideal energy alternatives. Thus, the integration of them is a promising solution to remedy energy-related issues.

Compared with a conventional CCHP system, the hybrid CCHP system has better energy-saving and CO_2 reduction performance. However, the hybrid CCHP system consumes higher annual total cost on account of its high initial investment (Yang and Zhai 2018).

In this paper, the design, management, and evaluation of the CCHP system coupled with RES for a 23-storey office building in Mashhad (Iran) were carried out.

The CCHP system includes a gas microturbine, as primary actuator for power supply, in addition to an absorption and compression chillers for cooling and a boiler for heating. PV and solar thermal panels were used to supply electricity and hot water.

9.2 Weather

The collection of climatic information, such as the amount of sunlight and air temperature, is of particular importance in the design of RES (Manfren et al. 2020).

Figure 9.1 depicts the minimum, maximum, and average temperatures of Mashhad separately for each month.

In Fig. 9.2, the average dry temperature and the percentage of air humidity in Mashhad are shown separately for 24 h each month.

Figure 9.3 represents the average dry air temperature in Mashhad over one year at different times of the day for each month in the color spectrum.

In the diagram in Fig. 9.4, the amount of solar radiation is illustrated as a bar graph for each month.

Figure 9.5 shows the wind speed, temperature, and sundials of Mashhad in a period of one year as a flow chart.

The diagram in Fig. 9.6 describes the position of the sun in different months along with the temperature of Mashhad.

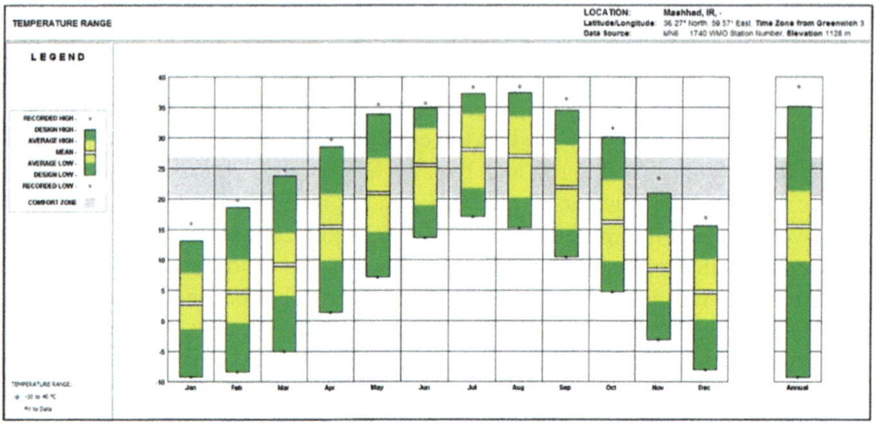

Fig. 9.1 Temperature graph of Mashhad in each month. *Source* PVSOL software

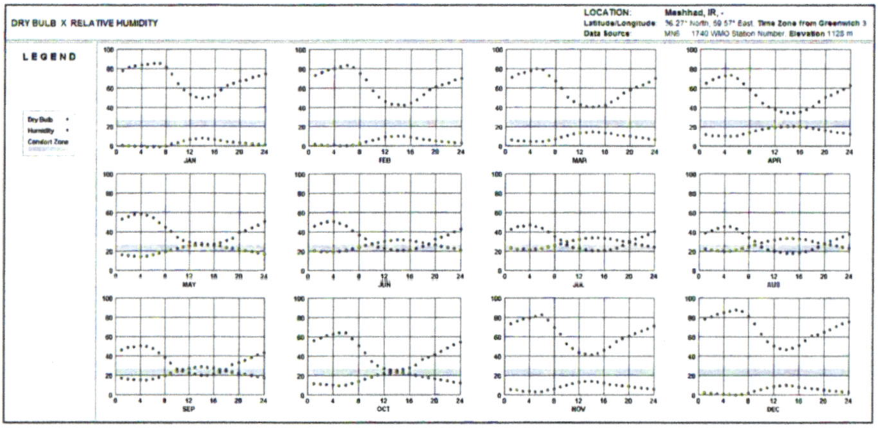

Fig. 9.2 Chart of dry temperature and humid temperature of Mashhad in one year. *Source* PVSOL software

9.3 Photovoltaic System

Since the project is inside the city and uses solar electricity to achieve maximum power and optimal use of radiation, the considered PV system is a grid-connected system. According to the architectural plans and consumption table, about 150 kW of solar panels with an efficiency of 14.7% was considered on the roof. PV panels were placed as follows.

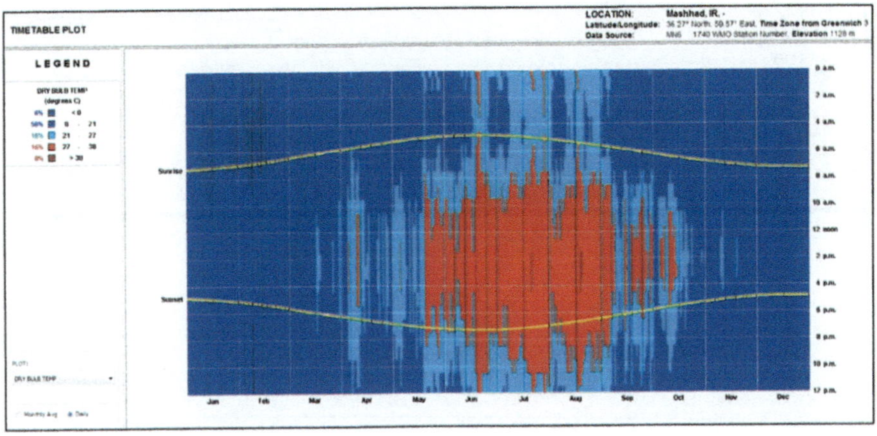

Fig. 9.3 Mashhad dry temperature diagram for one year. *Source* PVSOL software

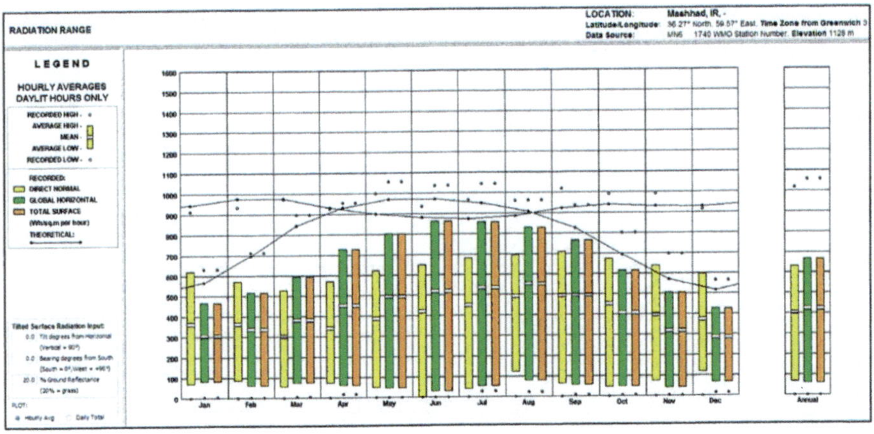

Fig. 9.4 Chart of solar radiation by month. *Source* PVSOL software

The presence of a shelter on the roof of the building (Fig. 9.7) necessitated examination and analysis of the shadows of this shelter on the solar panels. This analysis was performed with the software for one year. The results are shown as a color spectrum in Fig. 9.8. In this form, green spaces have very little shade and spaces that go red have the most shade in a year. In addition, spaces that do not cast shadows during the year are colorless. Figure 9.9 demonstrates how the PV panels were connected, which were then connected to the inverter.

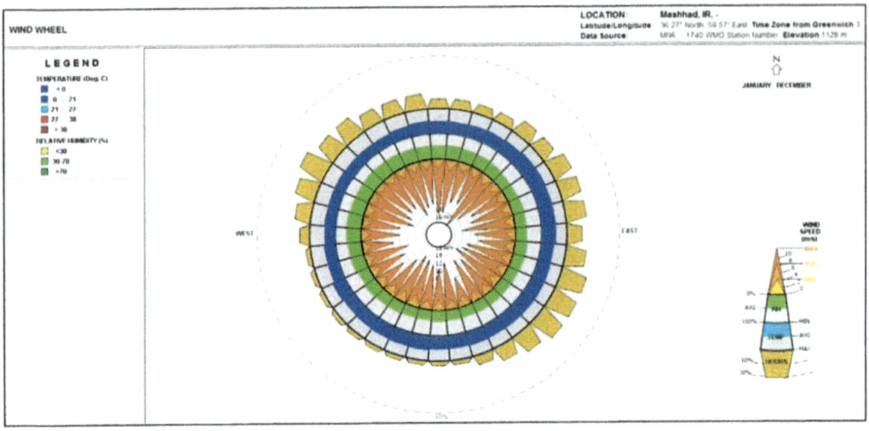

Fig. 9.5 Flower diagram of a year in Mashhad. *Source* PVSOL software

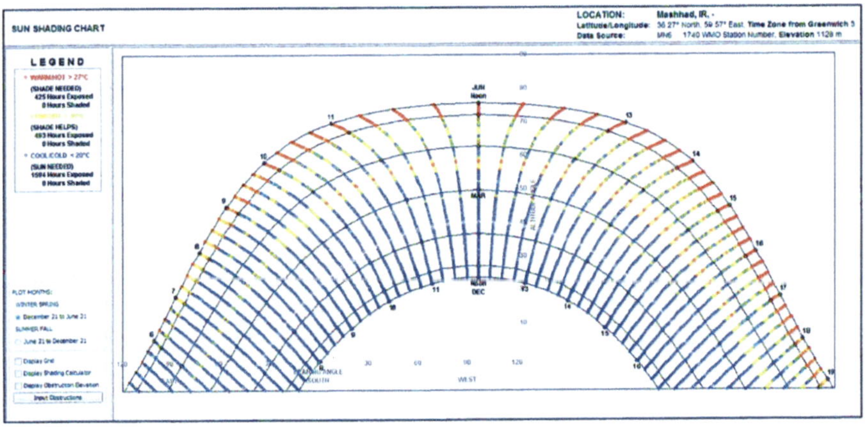

Fig. 9.6 Analysis of the position of the sun over a period of one year at the project site. *Source* PVSOL software

9.4 Energy Consumption

To calculate the heat and cooling load of the office building, HAP2 software was used. Thus, cooling, heating, and hot water loads were calculated according to Table 9.1.

The daily profile of the use of electrical equipment for this office building is based on the architectural plan and according to the reference Qian et al. (2021).

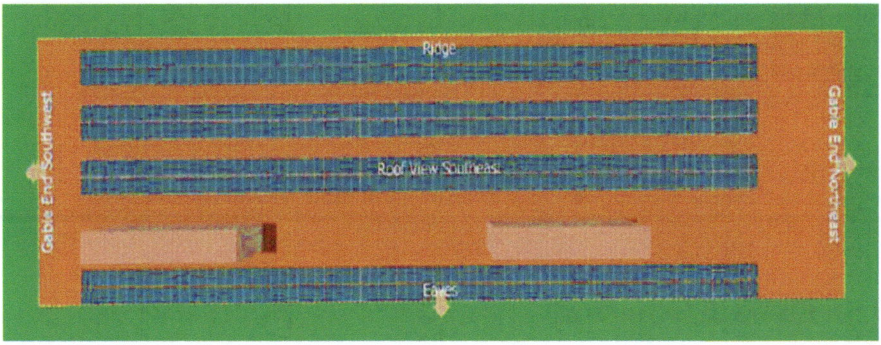

Fig. 9.7 How to place a photovoltaic panel on the roof of an office building. *Source* PVSOL software

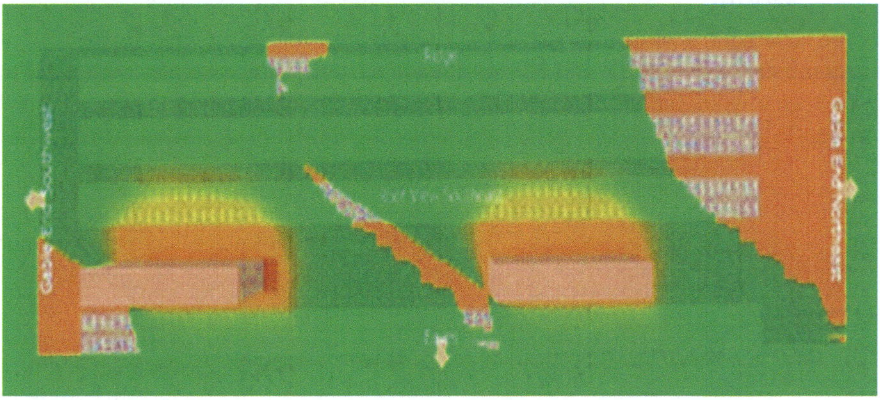

Fig. 9.8 Results of roof shadow analysis of office building. *Source* PVSOL software

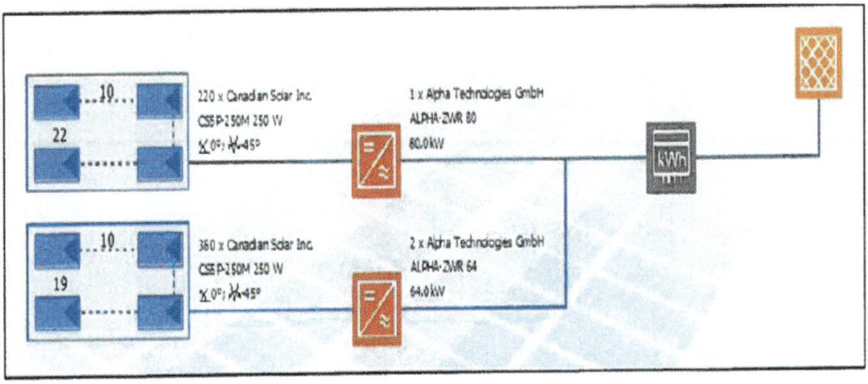

Fig. 9.9 How to connect solar panels. *Source* PVSOL software

Table 9.1 Results of energy demand analysis in a high-rise office project

Number	High-rise office building loads	Annual energy demand (MWh/year)
1	Thermal load	2152
2	Cooling load	4254
3	Electrical load	4344
4	Heat load of hot water consumption	5019

9.5 Problem Definition

To obtain the energy demand estimation system in the mentioned office and to increase the reliability and stability of the system, a combination of PV system with a unique CCHP system was used in comparison with the separate production system, whose schematic is shown in Fig. 9.10. To determine the quality and superiority of the combination of CCHP and PV production systems, this case was compared to other scenarios in Table 9.2. The total project cost in dollars per year was obtained via Eq. (9.1).

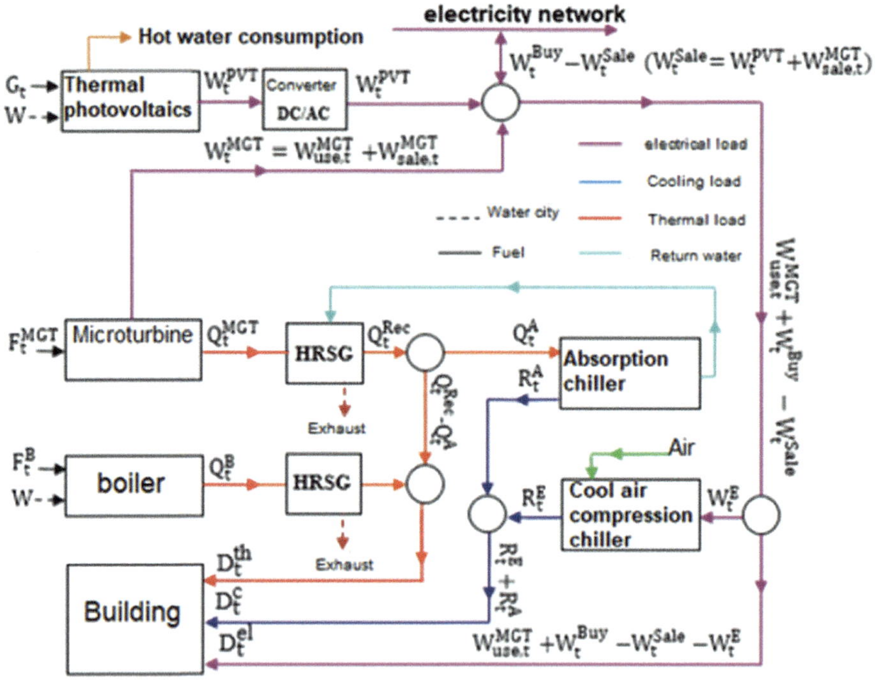

Fig. 9.10 Schematic of CCHP system with a PV system

Table 9.2 Scenarios studied

Scenario	1	2	3	4
PV/thermal			√	√
Gas microturbines				√
Boiler	√	√	√	√
Cool air compression chiller	√	√	√	√
Absorption chiller			√	√

$$\text{Net cost} = C + C^F + C^M + C^G - B^G - B_Q \tag{9.1}$$

where C is the total initial cost of the purchase in dollars per year. C^F is the total gas consumption, boilers, and gas microturbines in dollars per year. C^G is the total cost of purchasing electricity from the grid in dollars per year. B^G is the revenue from electricity sales of gas microturbines and PV systems in dollars per year. B^Q is the revenue generated by the heat generation of gas microturbines and PV systems in dollars per year. In this case, it was assumed how much it will cost if the amount of heat produced by the gas microturbine or the amount of hot water produced by the PV collector for one year is provided by the boiler. This cost is the revenue generated for heating from gas microturbines and hot water from PV collectors.

9.6 Results

Table 9.3 shows the design results of combining a PV system with a CCHP system in comparison with other systems in four scenarios.

Considering Table 9.3, in Scenario 1, it was assumed that the building's electricity demand is met by the electricity grid distribution, heated with a boiler, and the requested cooling is provided with a compression chiller. In this case, the project net cost is 21 k$/y. This system did not have income and capital return because it was just a consumer. In Scenario 2, the capacity of the compression chiller was slightly reduced, and an absorption chiller was added to the system. The project net cost was reduced by 65%. Still, like in Scenario 1, the system did not have revenue and return on investment just because it was a consumer. In Scenario 3, the same combination of Scenario 2 plus gas microturbine was used. This resulted in a net project cost of 2.32 times higher than Scenario 2. The revenue from the sale of gas microturbines' heat to the distribution network was estimated to be 108.43 k$/y, and the revenue from the heat generation of gas microturbines (to reduce boiler consumption) was estimated to be 15.39 k$/y. In total, the return on investment was calculated about 5.46 years. In Scenario 4, the use of the same devices as in Scenario 3 in combination with the PV system was considered. Therefore, with this combination of CCHP and PV, the project cost increased by 6.18%. However, the revenue from the sale of electricity to the distribution network was 166 k$/y, which was about 54% more than

Table 9.3 Design results of combining PV system with CCHP and separate production system using different scenarios

Scenarios	1	2	3	4
Cap^{PVT}(kw)	0	0	0	150
Cap^{MGT}(kw)	0	0	2 * 260	2 * 260
Cap^{B}(kw)	3 * 675	3 * 675	3 * 675	3 * 675
Cap^{A}(kw)	0	281	281	281
Cap^{E}(kw)	6 * 710	6 * 617	6 * 617	6 * 617
Net cost (k$/year)	289.21	284.44	660.68	701.53
C (k$/year)	138.75	132.11	268.74	371.42
C^{F} (k$/year)	18.99	18.99	361.05	361.05
C^{M} (k$/year)	42.75	41.24	86.42	86.62
C^{G} (k$/year)	88.72	92.09	68.28	68.28
B^{G} (k$/year)	0	0	108.43	166.67
B^{Q} (k$/year)	0	0	15.39	19.17
ROI (year)	0	0	5.46	5.19
$B_{20\ \text{year}}$ (k$(	0	0	28,679	43,197

that in Scenario 3. The revenue from heat generation from gas microturbines and PV systems also increased by 25% per year compared to Scenario 3. Furthermore, the income after 20 years was at least the lifespan of the equipment. Considering the inflation rate of 20.6%, it was estimated to be 43 k$/y, which is 1.5 times more than in Scenario 3. Scenario 4, a PV thermal system was combined with a CCHP system, including gas microturbines, boilers, absorption chillers, and compression chillers.

The superiority of Scenario 4 was owing to the minimum return on investment and the highest income in the minimum life of equipment.

9.7 Conclusions

Integration of CCHP and RES makes a very strong strategy since it is conducive to the supply of clean energy for commercial and residential buildings. The CCHP technologies could provide methods for improving the utilization efficiency of RES energy. RES will provide a clean and cheap energy source.

The integration of CCHP system with RES energy can realize mutual compensation of two kinds of technologies. Governments have made very good policies to make further use of RES energy and CCHP production systems and to reduce fossil fuel consumption. Iran is no exception to this rule. The encouragement of the Ministry of Energy and the New Energy Organization of Iran (SABA) has been highly effective. Among them, the following could be mentioned:

- Guaranteed purchase, electricity generated by solar systems, and CCHP generation systems at a very reasonable price compared to the selling price.
- Fuel supply of production systems at the same time as the fuel supply rate of power plants for 20 years.

Hence, utilization of the proposed design in this article could be a promising, suitable, and economical option for investment, reducing energy consumption, and reducing environmental pollutants for the design of commercial, residential, and especially commercial buildings for regions with similar climatic conditions to Mashhad.

At first, wind speeds are predicted for future using a new combined intelligent model. For this purpose, three feedforward networks have been used. This type of wind speed prediction can be very efficient in the Iranian wind energy industry, and this research in this field is not similar in the world. The proposed model was evaluated with various network models, functions, and different numbers of neurons to extract optimal network structure. Furthermore, training and validation data were used to evaluate and compare diverse network structures' performance. The proposed model also was trained with training algorithms BFGS, LM, and BR. As the results show, with the right Perceptron neural network training methods, the HNN neural network can be trained better than a single neural network. Training a Perceptron neural network with an LM algorithm leads training error to reduce quickly since the LM algorithm is reasonably fast. Therefore, it is logical to select as the training algorithm of the first-level neural network in the hybrid model.

Acknowledgements Sapienza University of Rome (DPDTA—Horizon 2020 project "GIFT" Geographical Islands FlexibiliTy Grant agreement ID: 824410) is kindly acknowledged for providing financial support for this work.

References

Agostinelli S, Cumo F, Nezhad MM et al (2022) Renewable energy system controlled by open-source tools and digital twin model: zero energy port area in Italy. Energies 15(5):1817

Ahn H (2019) Feasibility assessments of combined cooling, heating, and power (CCHP) systems for commercial buildings. The Pennsylvania State University

Astiaso GD (2016) Can radiant floor heating systems be used in removable glazed enclosed patios meeting thermal comfort standards? Build Environ 106:378–3881

Chow TT, He W, Ji J (2007) An experimental study of fac¸ade-integrated photovoltaic/water-heating system. Appl Therm Eng 27:37–45

De Santoli L, Garcia DA, Groppi D et al. (2018) A general approach for retrofit of existing–buildings towards NZEB: the windows retrofit effects on indoor air quality and the use of low temperature district heating. In: Proceedings - 2018 IEEE international conference on environment and electrical engineering

Ebrahimi M, Keshavarz A (2013) Sizing the prime mover of a residential micro-combined cooling heating and power (CCHP) system by multi-criteria sizing method for different climates. Energy 54:291–301

Ehyaei MA, Bahadori MN (2007) Selection of micro turbines to meet electrical and thermal energy needs of residential buildings in Iran. Energy Build 39:1227–1234

Gómez-Navarro T, Brazzini T, Alfonso-Solar D et al (2021) Analysis of the potential for PV rooftop prosumer produsction: technical, economic and environmental assessment for the city of Valencia (Spain). Renew Energy 174:372–381

Gu Q, Ren H, Gao W et al. (2012) Integrated assessment of combined cooling heating and power systems under different design and management options for residential buildings in Shanghai. Energy Build

Hoseinzadeh P, Khalaji Assadi M, Heidari S et al (2021) Energy performance of building integrated photovoltaic high-rise building: case study, Tehran. Iran Energy Build 235:110707

Jalalizadeh M, Fayaz R, Delfani S et al (2021) Dynamic simulation of a trigeneration system using an absorption cooling system and building integrated photovoltaic thermal solar collectors. J Build Eng 43:102482

Kalogirou SA, Tripanagnostopoulos Y (2006) Hybrid PV/T solar systems for domestic hot water and electricity production. Energy Convers Manag 47:3368–3382

Kokkos N, Zoidou M, Zachopoulos K et al (2021) Wind climate and wind power resource assessment based on gridded scatterometer data: a Thracian Sea case study. Energies 14:3448

Konečná E, Teng SY, Máša V (2020) New insights into the potential of the gas microturbine in microgrids and industrial applications. Renew Sustain Energy Rev 134:110078

Lamagna M, Nastasi B, Groppi D et al (2020) Hourly energy profile determination technique from monthly energy bills. Build Simul 13:1235–1248

Liang R, Zhang J, Zhou C (2015) Dynamic simulation of a novel solar heating system based on hybrid photovoltaic/thermal collectors (PVT). Procedia Eng 121:675–683

Mago PJ, Hueffed AK (2010) Evaluation of a turbine driven CCHP system for large office buildings under different operating strategies. Energy Build 42:1628–1636

Makkiabadi M, Hoseinzadeh S, Nezhad MM et al (2021a) Techno-economic study of a new hybrid solar desalination system for producing fresh water in a hot-arid climate. Sustainability 13:12676

Makkiabadi M, Hoseinzadeh S, Taghavirashidizadeh A et al (2021b) Performance evaluation of solar power plants: a review and a case study. Processes 9:2253

Manfren M, Nastasi B, Groppi D, Astiaso Garcia D (2020) Open data and energy analytics - An analysis of essential information for energy system planning, design and operation. Energy 21315

Manfren M, Nastasi B, Tronchin L, Groppi D, Garcia DA (2021) Techno-economic analysis and energy modelling as a key enablers for smart energy services and technologies in buildings. Renew Sustain Energy Rev 150:111490

Nastasi B (2019) Hydrogen policy, market and R&D project. In: Calise F, D'Accadia MD, Santerelli M, Lanzini A, Ferrero D (eds) Solar hydrogen production, Elsevier, Cambridge, MA, USA

Peng Y, Zhao S, Wang H (2020) A digital twin based estimation method for health indicators of DC–DC converters. IEEE Trans Power Electr 36(2):2105–2118

Qian J, Wu J, Yao L et al. (2021) Comprehensive performance evaluation of wind-solar-CCHP system based on emergy analysis and multi-objective decision method. Energy 230:120779

Qin Y, Wu X, Luo J (2021) Data-model combined driven digital twin of life-cycle rolling bearing. IEEE Trans Ind Inf 18(3):1530–1540

Razmjoo A, Rezaei M, Mirjalili S et al (2021) Development of sustainable energy use with attention to fruitful policy. Sustainability 13:13840

Yang G, Zhai X (2018) Optimization and performance analysis of solar hybrid CCHP systems under different operation strategies. Appl Therm Eng 133:327–340

Yang X, Leng Z, Xu S et al (2021) Multi-objective optimal scheduling for CCHP microgrids considering peak-load reduction by augmented ε-constraint method. Renew Energy 172:408–423

Chapter 10
Digital Construction and Management the Public's Infrastructures

Giuseppe Orsini and Giuseppe Piras

Abstract The purpose of the present paper "Digital construction and management the public's infrastructures" is to propose an interconnected development approach, in the management of public infrastructure asset, that through of digital modeling (BIM*) and interoperability provides tools to support decision-making processes. In detail, this work analyzes the innovative process of developing digital tools for the institutional tasks of supervision and support for the management of land transport infrastructure in the Italian national system. Therefore, trough of one assumed a georeferenced network of "digital twins" have been valued the scenarios obtainable whit the digitalization of the public works and of the territory's surveys. The principles for managing information flows for Italian's public transport infrastructures have been developed in accordance with national legislation and the reference UNI standards. The assumed flow is on the exchange of data between the managing subjects with the owners' authorities and surveillance bodies, taking as pivot element the Index public work (IOP) code attributed to each public work. Finally, a conceptual model has been proposed for the energy analysis of the road section and the identification of the best areas to create the "green islands" to produce renewable energy, for the management of infrastructure and for the recharging of electric vehicles.

Keywords Digitalization · GIS · Infrastructure · Integrated · Energy · Building Information Modeling · Geographic Information Systems

---

G. Orsini (✉) · G. Piras
Sapienza University of Rome, Rome, Italy
e-mail: giuseppe.orsini@uniroma1.it

G. Piras
e-mail: giuseppe.piras@uniroma1.it

E. Arbizzani et al. (eds.), *Technological Imagination in the Green and Digital Transition*, The Urban Book Series, https://doi.org/10.1007/978-3-031-29515-7_10

10.1 Introduction

It was carried out an analysis of the Italian network infrastructures aimed at their digitization. This work has also considered the regulatory framework establishing the mandatory adoption of digital tools (MIMS 2017) and the precise indications of the National Recovery and Resilience Plan (PNRR) (LAW 2021, Decree-Law 2021, MIMS). We proceeded to define the tools to make organic and interconnected the asset, through the detection of the major problems facing the infrastructural asset. Three important regulatory instruments recently adopted by Italy have been used to this end, namely:

- Guidelines on existing bridge verifications (MIMS 2020);
- Guidelines on monitoring existing tunnels (Decree-Law 2020);
- National Registry of Public Works (MIMS 2019).

The latter mentioned instrument in particular, through the attribution to each public work of an identification code, has made it possible to exchange information on the individual infrastructures, between the different subjects involved in the construction/management of the work and the institutional subjects called to supervise the infrastructural asset. This sharing of information, through the digitization of assets, presupposes the evaluation of the analysis tools to assess the state of health of the works, in relation to the possible risks related to the territory in which the works are located.

In order to define through the appropriate predictive algorithms, the priority and the method of the in-depth studies to be carried out through specific investigations on the works and interventions are to be performed. Finally, a system analysis was produced that manages infrastructures and traffic flows on a regional scale, assessing the resilience of the infrastructures themselves and their transition to energy autonomy.

The Italian territory has very complex orographic characteristics with a considerable presence of bridges and tunnels, both in the road and in the railway sector. The infrastructural building asset, which has developed from the end of the nineteenth century to the present, has recorded in the decades following the end of the Second World War the period of maximum growth of infrastructures, especially roads, through the use of construction technologies that exploited a lot of reinforced concrete (C.A.), also for the purpose of avoid the massive use of steel, since Italy is not a producer of this raw material. This implies that many works are at the end of the useful life period, and this together with the fact that the time calculation models did not include dynamic seismic actions in the mathematical modeling of structures is one of the major reasons why a systematic operation of census and cataloging of public works and in particular of road infrastructures is underway, motorways and railways. The need to catalog and classify the works, based on the territory and the functional characteristics, that is, the average daily traffic recorded, is one of the challenges that the Public Administration must face at the same time as the need for digitization of the same, through a *system of governance plan* capable of harmonizing the phases of the processes of:

- Innovation renewal;
- Planning;
- Design;
- Implementation of the interventions;
- Asset management and maintenance.

The evolution of the economic environment has led to a transformation of typology and traffic volume, which is often even ten times higher than the volume of traffic considered in the initial project. This evolution has not always been managed in an ordinary way through adequate industrial development plans, also due to the orographic difficulties mentioned above and the stratification of the territories that have seen Italian municipalities and cities develop over the centuries with criteria not suitable to provide an adequate response to the modern economy of industrial exchanges. Now is the time to reorganize the territory on the basis of shared criteria, highlighting the economic, environmental and social needs and evaluating through the tools of digitization the best solutions applicable to the different territories. The use of simulation is a tool that must move from research fields to application fields, in a widespread way, in order to feed the data from below, useful to support strategic decisions with an acceptable level of uncertainty.

To achieve what has been said above, a structure has been hypothesized that, starting from the inputs of the PNRR and the national legislation for the adoption and development of digital tools (MIMS 2017), identifies a holistic strategy for the management of the infrastructure asset, for land transport infrastructures (Fig. 10.1).

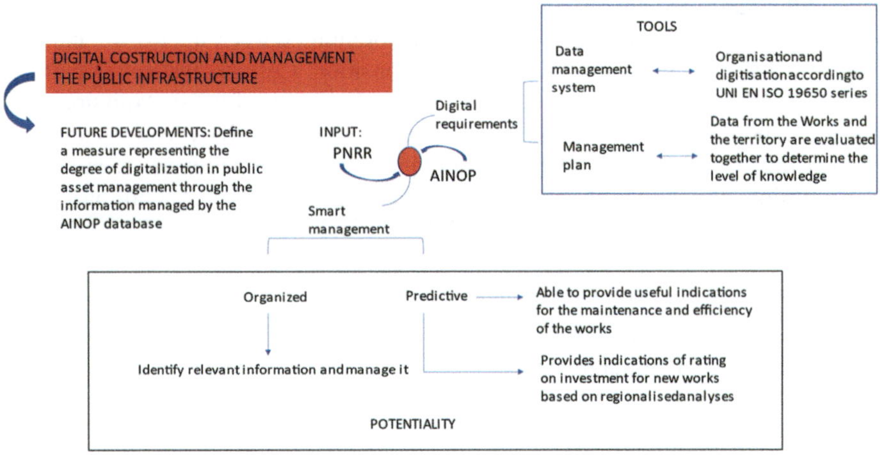

Fig. 10.1 Input–requirements–developments: elements and requirements for the digital management of public infrastructures

10.2 Digital Development: Strategy and Complexity

After the collapse of the Polcevera Viaduct in Genoa in the summer of 2018, in the Italian panorama, there has been an acceleration, which has led to the establishment of the Computer Archive of Public Works (AINOP) in the space of a year and the allocation in a single structure of the tasks of promoting and supervising the safety of Railways and Road and Motorway Infrastructures (ANSFISA) (Genoa Decree 2018). The spirit of innovation and the strong need to give uniformity to the management of Infrastructures, adapting management systems to common standards, have grown hand in hand with the country's needs to invest in Digitization and Environmental Sustainability.

In addition, with the PNRR (Decree-Law 2021), 62 billion euros have been allocated to **infrastructures for logistics and mobility**, highlighting the central role of the Ministry of Sustainable Infrastructure and Mobility (MIMS) in the implementation of decisive projects for the relaunch of the country, based on economic, social, and environmental sustainability.

The Ministry's projects are financed for **41 billion with the European resources** of the Next-Generation EU program (40.7 billion) and with those of react EU (313 million), which are added **national resources for almost 21 billion euros**, of which 10.6 billion are from the Complementary Fund and 10.3 billion are from the budget deviation. The national funds pursue the same objectives as the European ones, but in some cases, they finance projects related to a longer time horizon than the term of 2026 imposed by the PNRR, such as the completion of the Salerno-Reggio Calabria High Speed. About **56% of the resources (34.7 billion euros) are destined to interventions in the south of Italy the so-called "Mezzogiorno"**, a sign of the Government's desire to concretely launch policies to overcome the gaps between the different areas of the country (https://www.mit.gov.it/node/15710).

Some projects will be carried out in collaboration between the MIMS and other Ministries (Ecological Transition, Digital Transition, Culture, Justice, Department for the South and Territorial Cohesion). And, indeed, four of the six "missions" that make up the PNRR (listed below—see Table 10.1) provide interventions under the responsibility of the MIMS: digitalization, innovation, competitiveness, and culture (493 million euros); green revolution and ecological transition (15.8 billion); infrastructure for sustainable mobility (41.8 billion); inclusion and social cohesion (3.9 billion).

The strategic objectives of the plan are six:

1. Sustainable development;
2. Ecological and digital transition;
3. People's well-being and reduction of inequalities;
4. Infrastructure and competitiveness;
5. Growth and employment;
6. Reduction of territorial gaps.

The main interventions, with which it is estimated a reduction of 2.3 million tons/year of CO_2 emissions, are:

Table 10.1 PNRR interventions

Description of the intervention/investment		Amount (€)
Railways	Development of high-speed/high-capacity railway lines	25.00×10^6
	Strengthening regional networks and electrification	5.45×10^6
	Strengthening railway nodes in urban areas	3.00×10^6
	Redevelopment of 30 strategic stations from a transport and tourist point of view	7.00×10^5
	Local "*green*" transport and mass rapid transport	8.40×10^6
Digitization interventions	Local public transport to encourage the development of the *mobility as a service* model and the integrated use of the different modes	4.80×10^6
Ecological transition of logistics	*Smart mobility*, transport networks powered by renewable energy sources	1.40×10^6
Development of cycling mobility	Construction of urban and tourist cycle paths and to connect the provincial roads with the main transport routes	100×10^6
Road and motorway sector	Technological control of bridges, viaducts, and tunnels	4.50×10^5
	Securing the A24 and A25 motorways	1.00×10^6

- 700 km of railway between high-speed development and regional lines;
- 216 km for Rapid Mass Transport;
- 3200 electric and hydrogen buses for urban areas;
- 2000 methane buses for extra-urban transport;
- 1800 km of tourist and urban cycle paths;
- Grow up of the experimentation for non-electrified railways (in Val Camonica and southern Italy) of hydrogen technologies.

10.2.1 Complexity of the Italian Land Transport System

The Italian territory is a peninsula that extends into the Mediterranean Sea and includes two major islands, Sicily and Sardinia, which is connected to the rest of Europe by the Alpine chain, which is the natural border of the nation with neighboring European countries.

Apart from some flat areas and the larger Po Valley, the territory consists of a system of mountains and hills that go from north to south, drawing a territory difficult to connect. This has led to the construction of a large number of bridges and tunnels, both in the road and rail sectors.

Table 10.2 Types of managers

Type of manager	Managers N°	Tratte Km	Incidence %
Motorway dealers	27	8.006	0.95
ANAS road roads	1	27.259	3.25
Regions, provinces, and metropolitan cities	123	135.691	16.16
Communes	7.904	668.673	79.64

From the administrative and managerial point of view for the road and motorway sector, the Italian network is essentially divided into four types of operators (ANSFISA 2021) for a total of 840,000 km of infrastructure, for a total of 8055 operators; below is a table of the consistency of the network (see Table 10.2):

With reference to the road sector, the motorway managers and the national manager of ANAS are matured from a managerial and technological point of view for the promotion of a progressive digitalization of the works. The degree of engineering of the processes related to this type of infrastructure favors the *change management* aimed at the creation of an interconnected system of management of the public infrastructure asset, while considering, however, that these concessionaires manage less than 21% of the road network (see Table 10.3).

The railway sector has two types of main operators, the National Manager, namely RFI with 16,832 lines in operation, for a total of 15,882 km of network between fundamental and complementary lines, and n. 10 Isolated Railways, for 1130 km of network (https://www.ansfisa.gov.it/relazioni-annuali).

From a historical and technical point of view, railway lines have significant profiles compared to the road sector: the smaller extension, the reduced number of managers, and greater engineering due to the nature of rail transport. Also, with regard to the management and communication of data relating to transport, the panorama in

Table 10.3 List of isolated railway operators

Managing company	Railway lines	Region
OFFICE	Genoa–Casella	Liguria
FERROVIENORD	Brescia–Iseo–Edolo	Lombardy
GTT SpA	Turin–Ceres	Piedmont
SSIF SpA	Domodossola–Swiss border	Piedmont
SPA ATTACK	• Rome–Lido • Rome–Civitacastellana–Viterbo	Latium
Railways of Calabria srl	Entire network	Calabria
Circumetnea railway	Catania Borgo–Riposto–extra-urban section	Sicily
WALL srl	Entire network	Apulia–Basilicata
DOCTOR SPA	Entire network	Sardinia
EAV	• Circumvesuviana railway • Cumana and Circumflegrea railways	Campania

the railway sector is standardized and the communication protocol is regulated at European level through the ERA European Union Agency for Railways.

It is also necessary to consider 230 km of metro network divided into 14 lines, present in the cities of: Milan, Rome, Naples, Turin, Brescia, Catania, and Genoa, developed mainly in tunnels (about 180 km). The different lines are characterized by different Degrees of Automation (you go from lines with a GdA 1, with manual train operation to lines with GdA 4, or with an automatic train operation) (ANSFISA 2021). With regard to metropolitan areas, the degree of digitalization that can be achieved is significantly different from other network infrastructures, if only in view of the amount of data managed per unit area, even if the same organizational levels as the rest of the railway sector are proposed again (UNI 2019).

10.3 Model Proposal: Methods

The proposed model of digital management of information relating to the assets of public network infrastructures responds to "System" needs that are at a higher level than the management level typical of a generic Organization, which explains its digital management through the specific Organizational Informational Requirements (OIRs), defined by the UNI EN ISO 19650 standard. The management of "System" means the ability to combine in a single digital environment the important information relating to the infrastructures, with regard to road sections, railway lines, and information on the individual works that constitute it. In order to allow the Pubblica Amministrazione, which has the task of planning investments, management control and compliance with safety and environmental requirements, to have a single database, updated and comparable.

The definition of the IT requirements of the organization is focused on the assets of the specific order. The proposed federated model does not enter into the structure of the individual OIR and PIR, for which, if necessary, minimum contents will be defined to which the managers must adapt. The model instead provides the definition, at the regulatory level, of the requirements for AIR and EIR in order to generate uniform summable/overlapping AIM, in order to obtain a "National Asset Information Model NAIM" for ground transport infrastructures, which combines the data of BIM and GIS models, to be integrated by control sensors and satellite analysis.

$$\text{NAIM} = \sum_{1}^{n} \text{AIM} \tag{10.1}$$

NAIM = National Asset Information Model;
AIM = Asset Information Model;
n = number of infrastructure managers.

Table 10.4 Hypothesis information management UNI EN ISO 19650: definition of hypotheses for NAIM

Document	Hypothesis
Asset information requirements (AIRs)	Regulated at national level
Organizational information requirements (OIRs)	Only minimum content shared nationally
Project information requirements (PIRs)	Conform to OIR
Exchange information requirements (EIRs)	Conform to OIR and PIR
Project information model (PIM)	Conform to OIR, PIR, and EIR
Asset information model (AIM)	Conform to AIR
National asset information model (NAIM)	Conform to AIR

What is proposed in some respects is similar to the Roads' Cadastre established by legislative decree of 30 April 1992 n. 285 art. 13, paragraph 6, and made operational with ministerial decree n. 3484 of 01/06/2001, which in detail establishes the characteristics and requirements of the data to be entered on the platform, for the representation and continuous updating of the road asset by road managers. Tool this very important, but never fully implemented. In addition, there is no National federated model of the Road Cadastres of individual managers (Table 10.4).

As far as the railway sector is concerned, the *Register of INFrastructrure (*RINF) database referred to in Article 49 of the ERA Interoperability Directive (2016/797) is currently established and operational, in which the main characteristics of the European railway infrastructure, and therefore, of the Italian one are reported. However, this does not allow you to define a multi-level analysis scenario.

This integrated digital management model is based on a GIS representation, compliant with the EUROPEAN DIRECTIVE INSPIRE, in order to insert the works, bridges, tunnels, etc. in the specific environmental context, to analyze the specific risks due to the territory and allow through the appropriate calculation algorithms (to be developed) the analysis at the territorial level of the criticalities, in relation to the type of work and the characteristics of operation.

10.3.1 The Sources to be Included in the Basic GIS

The proposed cartography complies with the National Catalog for Spatial Data (RNDT) was established with Article 59 of the Digital Administration Code, Legislative Decree 82/2005 (CAD), and was identified as a database of national interest. On the basis of this regulatory framework, the RNDT is the national catalog of metadata for Pas' spatial data and the services and also constitutes as public register of such

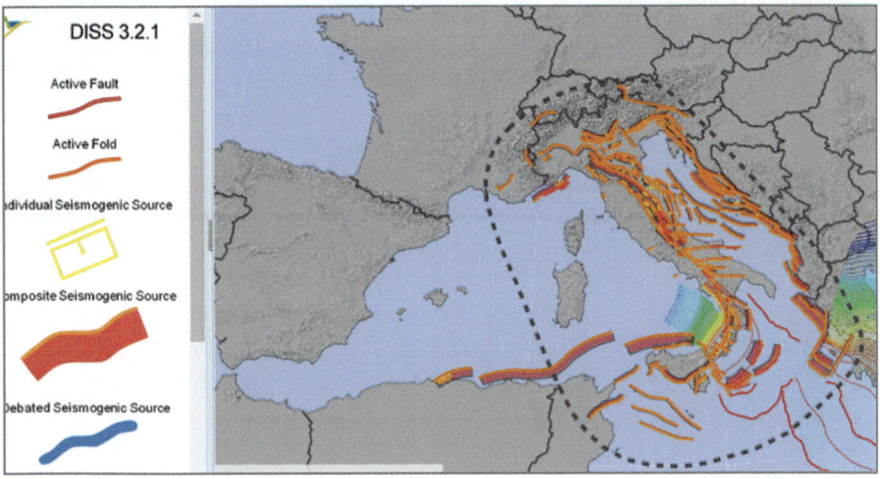

Fig. 10.2 DISS database: representation of seismic sources. *Source* DISS https://diss.ingv.it/ithdiss

data certifying its existence through the publication of the metadata. The RNDT is indeed an important component of the national infrastructure for spatial and environmental information established in Italy with the Legislative Decree no. 32/2010, the transposition of the INSPIRE Directive.

The main functionalities:

- metadata discovery, accessible to everyone;
- metadata management, restricted **for accredited public administrations**.

Below, the Maps are proposed in this paper:

Database of Individual Seismogenic Source (**DISS**), established by the National Institute of Geofisisica and Volcanology, is a repository geo where they are reported all information of nature seismotectonic (Fig. 10.2).

The term seismotectonics refers to the disciplinary sector that is interested in the relationships between geology, active tectonics and the seismicity of a given area, and which has as its main objective the identification of the structures that generate earthquakes—the seismogenic sources—and the estimation of their potential.

DTM: Digital Terrain Model o digital model of the soil represents the trend of the soil surface without the anthropic and vegetational elements (Fig. 10.3). Hydrological and river geomorphology analysis.

This tool can be used for the definition of risk maps (hydraulic, geomorphological, etc.) for the elaboration of maps of slope, curvature, and exposure of the slopes.

PAI: The hydrogeological plan is a fundamental instrument of the spatial planning policy outlined by Law 183/89. Given basin planning in each region, this instrument

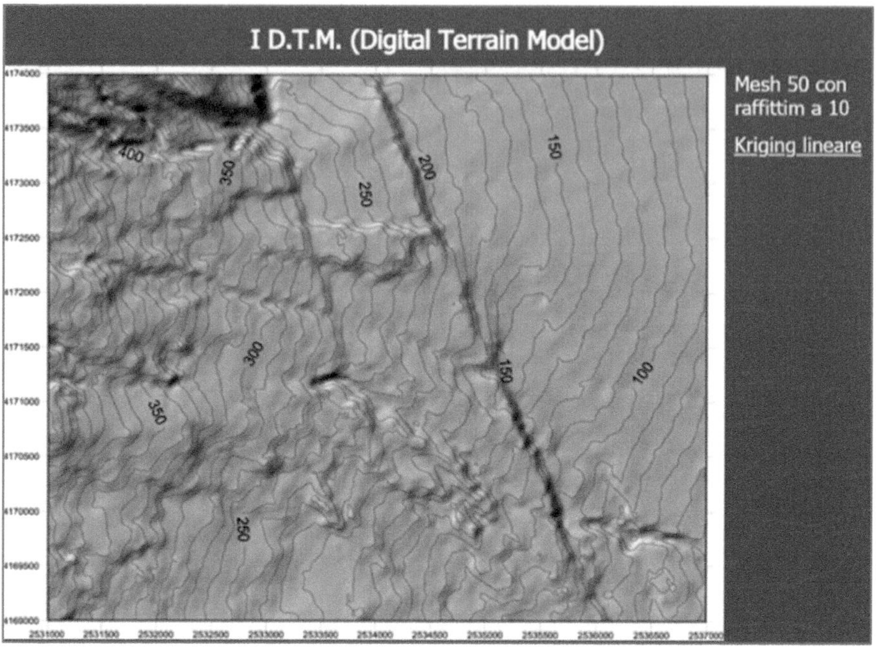

Fig. 10.3 DTM: digital terrain model. *Source* ISPRA (http://portalesgi.isprambiente.it/)

is the first thematic and functional excerpt. The Excerpt Plan for the Hydrogeological Structure, hereinafter referred to as the Excerpt Plan or Plan or P.A.I., drawn up pursuant to art. 17, paragraph 6 ter, of Law 183/89, art. 1, paragraph 1, of Legislative Decree 180/98, converted with amendments by Law 267/98, and art. 1 bis of Legislative Decree 279/2000, converted with amendments by Law 365/2000, has the value of a Territorial Sector Plan and is the cognitive, regulatory, and technical–operational tool through which the actions, interventions, and rules of use concerning the defense against the hydrogeological risk of the territory are planned and programmed (Fig. 10.4).

GEOLOGICAL MAP: Graphic representation in which are reported, with colors and symbols, the rocks emerging in a territory, their structural characteristics, thickness, location, age, stratigraphic relationships, mineral deposits and fossils; the data have been computerized with the ToolMap software (Fig. 10.5).

HYDROGEOLOGICAL MAP: The hydrogeological maps are represented the essential hydrogeological parameters of the territory, selected according to the objectives of the research. The hydrogeological parameters of particular interest are permeability, effective infiltration, transmissivity, etc.

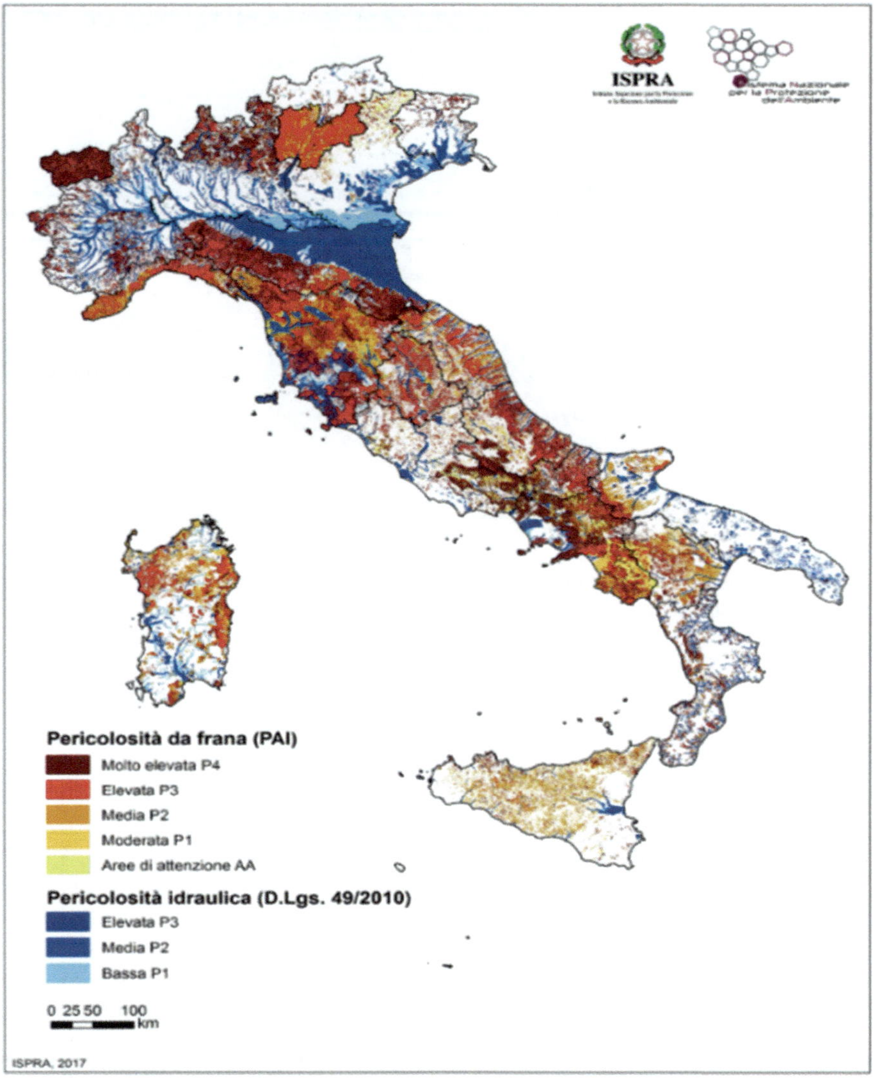

Fig. 10.4 PAI: hydrogeological plan. *Source* ISPRA (http://portalesgi.isprambiente.it/)

Hydrogeological maps can be divided essentially into two categories: synthesis maps and thematic maps. The former allows the reading of the territory in a hydrogeological key, providing in a synthetic way the main information on the hydrodynamic conditions existing within and at the limits of the individual hydrogeological domains. The latter supplement the previous ones with more detailed information on geological, hydrogeological, and hydrological aspects of particular interest (Fig. 10.6).

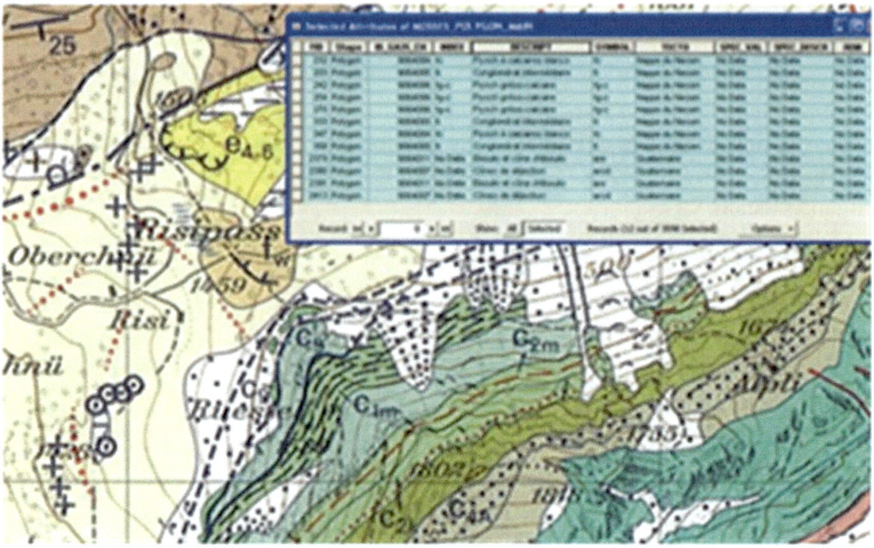

Fig. 10.5 Geological map: extract from the Geological Atlas of Switzerland 1:25,000 (AG25). *Source* ISPRA (http://portalesgi.isprambiente.it/)

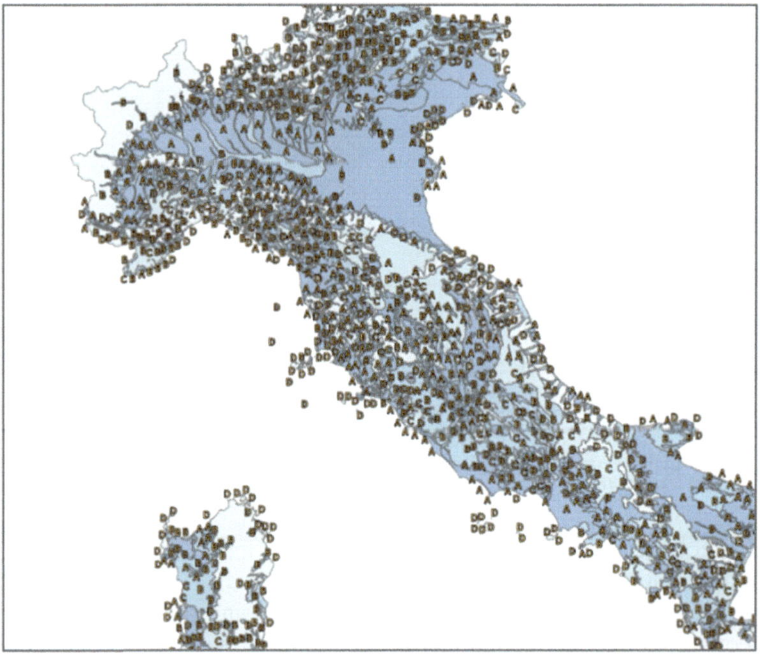

Fig. 10.6 Hydrogeological map: essential hydrogeological parameters. *Source* ISPRA (http://portalesgi.isprambiente.it/)

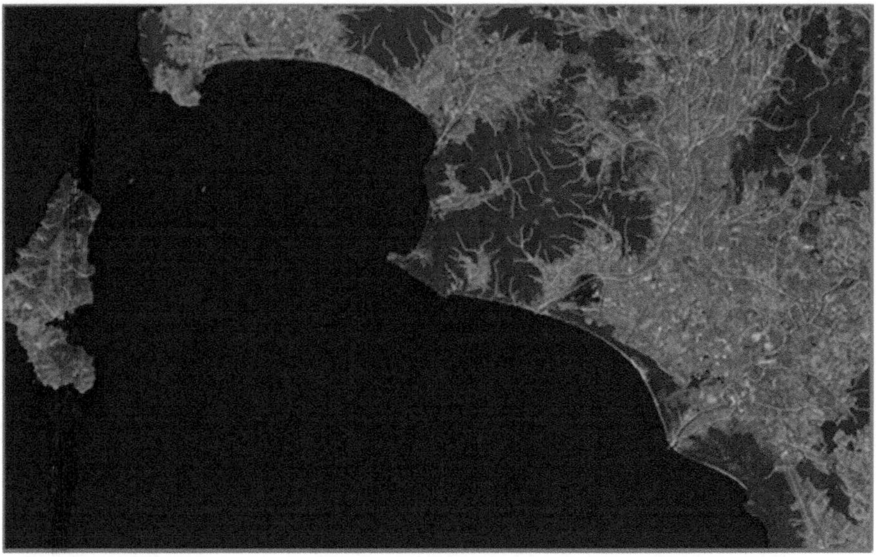

Fig. 10.7 Hydrogeological lattice: superficial alveoli. *Source* ISPRA (http://portalesgi.isprambiente.it/)

Of fundamental importance, are also all the maps that represent the "real interferences" with the works to be analyzed, namely:

HYDROGRAPHIC LATTICE: The hydrographic network is the set of riverbeds within which surface water flows (Fig. 10.7).

MAPS OF HOUSING DENSITY, SETTLEMENT TYPES, AND SPECIAL BUILDINGS: Population density is a measure of the number of people living in a given area (which may or may not include inland water surfaces). Normally, is measured in "inhabitants by square kilometer" (inhab./km^2). The value is obtained simply by dividing the number of inhabitants of a given territory by the surface of the territory itself (expressed in km^2) (Fig. 10.8).

10.3.2 Information Managed by the Model

The proposed model combines the possibilities of Digital Twins, for Linear Infrastructures (Dell'Acqua 2018, Vorotyntseva et al. 2021), making it possible for the bodies in charge to consult the information deriving from the recently adopted inspection sheets, provided by the guidelines for the assessment of the safety and state of maintenance of Bridges and Tunnels. These sheets, already designed for a computerized management of data, provide the determination of a class of attention of the

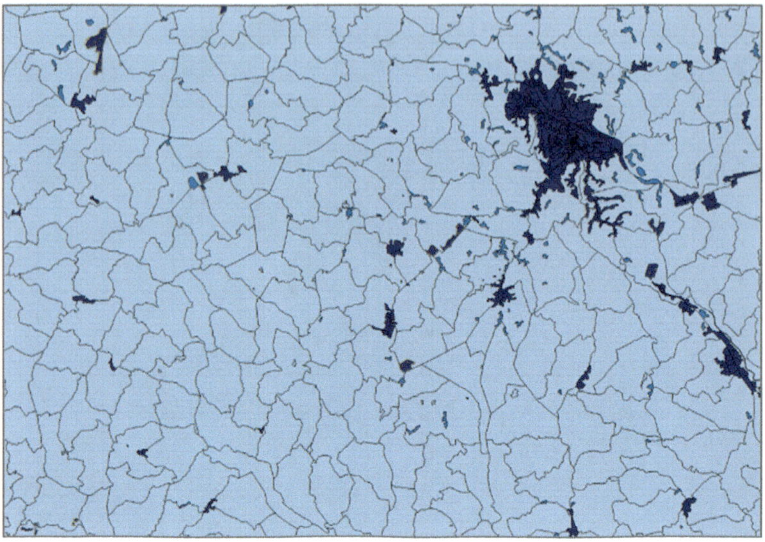

Fig. 10.8 Housing density map: socio-economic information. *Source* ISPRA (http://portalesgi.isprambiente.it/)

work, which determines the frequency of inspections, the instrumental and methodological checks in depth on the materials and in situ, the adaptation works if the work does not meet the requirements of design or operation.

The complex information framework, which involves a large number of operators and a plurality of public entities, called upon to carry out verification and control activities, as well as planning and development for the national transport network, is made possible thanks to the use of a single identification code assigned to the single work. The code in question is the IOP assigned by AINOP (Computerized Registry of Public Works, which already contains a part of the useful information, to be managed within the proposed model). Once the works have been uniquely identified and associated within a GIS, developed through arches and points, where the so-called "kilometers" of reference of the roads are defined through curvilinear abscissas, in which the infrastructures are located, hence the spatial and cartographic coordinates of reference, with respect to the rules defined at the regulatory level by the INSPIRE directive, so as to create the basic structure, are to be translated into an Integrated National System for the Management of Land Transport Infrastructures.

Within the georeferenced cartographic system, digital data computation logics will be introduced and managed, to be defined to support decision-making processes. This digital database has unique characteristics (Eastman et al. 2021), interoperability (Cosenza et al. 2021), data learning (Moretti et al. 2022), traceability (Cosenza et al. 2021, Mathew et al. 2022, Quqa et al. 2021), also integrating the assessments of economic, social, and environmental characteristics (Altıntaş and Ilal 2021, Xue et al. 2021, Mathew et al. 2022).

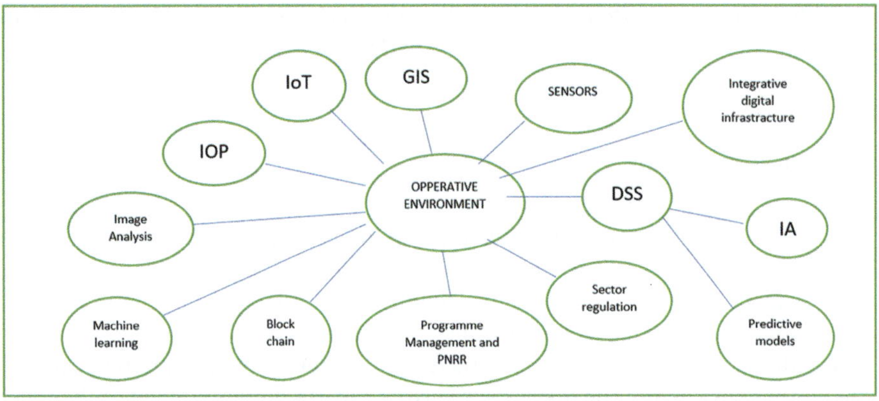

Fig. 10.9 Operating environment: elements included in the proposed model (internet of things—IoT, artificial intelligence—IA, decision support system)

The aim of the proposed model is also to create a context of predictive analysis of a probabilistic nature in order to mitigate the risk, understood as a risk deriving from any process that interacts with the life cycle of the infrastructure and the surrounding environment (Quqa et al. 2021) influenced by the operation of the infrastructure itself. In addition, the same model is a basic, certified, and updated tool, useful for managers, bodies, institutional organizations, and researchers to develop, through common creative licenses, adequate tools to support the processes of management of the works and, in general, infrastructure as well as evaluate the various phenomena at national, macro-regional, regional or district level, orienting and optimizing capital flows as well as the possibility of reaching artificial intelligence [29] and statistical approaches [28] aimed at defining design solutions that take into account multi-criteria optimization (Fig. 10.9).

Of course, by developing the hypotheses on the AIR, which currently constitute the formalized outcome of the OIR, a cascade of homogeneous and manageable information level of content is obtained through the EIR, under the hypotheses of wider interoperability formulated above. Connecting the references necessary to configure an *Asset Information Model* (AIM), from which to originate the *Digital Twin* (DT), thus connecting the BIM, IoT, and *predictive analytics* algorithms, to obtain (Eastman et al. 2021):

- Overall improvement of infrastructure efficiency;
- Analytical management of infrastructure data;
- Creation through the Digital Twin of a unified digital database;
- Ensure the relevance and machine-readable data format of the maintenance documents of the works;
- Approval and implementation of uniform formats and protocols for data exchange;
- Planning of investment and production programs using intelligent accident forecasting systems, traffic optimization, infrastructure resilience.

With a view to *Digitally Enabled Portfolio and Programme Management* and their implementation in the context of *Digitally Enabled Project Management*.

10.4 Conclusions

A possible digital management system of the Italian infrastructure asset regulated by the sector's regulatory discipline was presented, in compliance with the objectives of technological development, administrative management, optimization of environmental resources, and improvement of the transport system.

The digital management of infrastructures has been taken as a reference divided into two macrosystems. The first is the level of the works properly, in which the details relating to the realization of the works and the extraordinary maintenance works are defined, or in any case the maintenance work, to which the relative technical-accounting and authorization management documents are associated. This first level, largely structured within the aforementioned UNI ISO standards, has not been the subject of analysis although it constitutes an important institution to which a management system can be associated, normally internal to the contracting station, which sees to be involved the authorization bodies and the bodies responsible for anti-corruption and supervision of the works. On the other hand, it was considered to be a second level of the system, which aims to achieve the efficiency of the transport network itself and to control the profitability of investments. In particular, the six strategic objectives adopted with the PNRR were examined and a plan for the implementation of the schemes proposed with the application of UNI EN ISO 19650:2019 at the level of national management was hypothesized, to be implemented by means of the definition of common requirements identified for air, which requires uniformity for the bodies and companies that manage public infrastructures.

The strategic objectives underlying the PNRR have led the Italian government to plan substantial investments for the realization of public works in strict compliance with the timing imposed by the European Union for the use of EU funds. All these favor the creation of an integrated digital management system that, through the processing of certain data, guarantees the efficient management of public works by the actors involved, as in the proposed model. Generating a paradigm shift and a real digital *change management*, it is desirable, therefore, a new approach to the regulated sharing of data through a systematic and proliferating comparison between the various levels and actors involved both horizontally and vertically, in the management processes of public assets, even after engineering those not yet defined.

References

Annual report on isolated rail networks. https://www.ansfisa.gov.it/relazioni-annuali
ANSFISA Annual Report 2021. https://www.ansfisa.gov.it/relazioni-annuali

Art. 49 of Decree-Law no. 76 of 16 July 2020: Urgent measures for simplification and digital innovation
Art. 48, paragraph 7, of Decree-Law no. 77 of 31 May 2021: Governance of the National Recovery and Resilience Plan and first measures to strengthen administrative structures and accelerate and streamline procedures.
Decree-Law of 28 September 2018 n. 109 "Genoa Decree"
LAW 29 December 2021, n. 233: Conversion into law, with amendments, of the decree-law 6 November 2021, n. 152, containing urgent provisions for the implementation of the National Recovery and Resilience Plan (PNRR) and for the prevention of mafia infiltrations
MIMS—Decree of the ministry of infrastructure and sustainable mobility n. 560/2017
MIMS—Decree of the ministry of infrastructure and sustainable mobility n. 578/2020 of 17.12.2020
MIMS—Superior council of public works: guidelines for the drafting of the technical and economic feasibility project to be based on the award of public works contracts of the PNRR and the PNC
MIMS—Decree of the ministry of infrastructure and sustainable mobility n. 430/2019
Source: from the MIMS website. https://www.mit.gov.it/node/15710
UNI EN ISO 19650–1 (2019) Title: organization and digitization of information related to construction and civil engineering works, including Building Information Modelling (BIM) - Information management through building information modelling

Bibsection

Agostinelli et al. (2022) Renewable energy system controlled by open-source tools and digital twin model: zero energy port area in Italy. Energies 15:1817
Agostinelli et al. (2020) The potential of digital twin model integrated with artificial intelligence systems–special session on. In: New frontiers in green computing and artificial intelligence at the IEEE EEEIC 2020 conference. ISBN: 978-1-7281-7455-6
Altıntaş, Ilal (2021) Loose coupling of GIS and BIM data models for automated compliance checking against zoning codes. Autom Constr 128:1037
Cosenza et al. (2021) Layout-aware extreme learning machine to detect tendon malfunctions in prestressed concrete bridges using stress data. Autom Constr 132:103976
Dell'Acqua (2018): BIM for infrastructures–building information modeling for large linear infrastructures. EPC Editore
Eastman et al (2021) Rules and validation processes for interoperable BIM data exchange. J Comput Des Eng 8(1):97–114
Mathew et al. (2022) In a virtual power plants, a blockchain-based decentralized power management solution fossr home distributed generation. Sustain Energy Technol Assess 49:101731
Moretti et al. (2022) Re-design of a road node with 7D BIM: geometrical, environmental and microsimulation approaches to implement a benefit-cost analysis between alternatives. Autom Constr 135:04133
Osello (2015) Building information modelling—Geographic information system-augmented reality per il facility management. Flaccovio Editore
Quqa et al. (2021): Statistical approach for vibration-based damage localization in civil infrastructures using smart sensor networks. Infrastructures 6:22. 10.3390/
Quqa et al. (2021) Automatic identification of dense damage-sensitive features in civil infrastructure using sparse sensor networks. Autom Constr 128:103740
Ratti C, Claudel M (2016) The city of tomorrow: sensor, network, hacker, and the future of urban life–Yale University Press. ISBN: 0300204809, 9780300204803
Vorotyntseva et al. (2021) Coordination between participants in the BIMbased renovation of street engineering networks. In: E3S Web of conferences 244:05001
Xue et al. (2021) Two-layer adaptive blockchain-based supervision model for off-sitemodular housing production. Comput Indus 128:103437

Zheng Y et al. (2019) An application framework of digital twin and its case study. J Amb Intell Hum Comput

Chapter 11
An Innovative Multi-objective Optimization Digital Workflow for Social Housing Deep Energy Renovation Design Process

Adriana Ciardiello, Jacopo Dell'Olmo, Federica Rosso, Lorenzo Mario Pastore, Marco Ferrero, and Ferdinando Salata

Abstract Nowadays, the energy retrofit of the building sector is identified as a major instrument toward a climate-neutral Europe by 2050. In accordance with the European Renovation Wave program, deep energy renovations are needed, starting from public and less efficient buildings. Furthermore, the renovation of the social housing building stock is also an important response to energy poverty, as it could contribute safeguarding health and well-being of vulnerable citizens. In particular, buildings from the 1960–1980, which constitute a large portion of cities, often have high energy demand and low indoor comfort because most of them have been built before energy-efficiency regulations. In this context, the paper aims to propose a multi-objective approach toward energy renovation of the social housing building stock, by means of an innovative digital workflow. The objective functions are minimizing energy consumption, CO_2 emissions, investment, and operational costs. Toward these contrasting objectives, numerous passive strategies are taken into account, which are compatible with the considered architecture. The optimal solutions are found by means of a genetic algorithm coupled with energy performance simulation software. The methodology is applied and verified on a significant and relevant case study, pertaining to the social housing building stock of Rome, Italy (Mediterranean climate). The outputs of the workflow are a set of optimal solutions among

A. Ciardiello (✉) · J. Dell'Olmo · F. Rosso · L. M. Pastore · M. Ferrero · F. Salata
Sapienza University of Rome, Rome, Italy
e-mail: adriana.ciardiello@uniroma1.it

J. Dell'Olmo
e-mail: jacopo.dellolmo@uniroma1.it

F. Rosso
e-mail: federica.rosso@uniroma1.it

L. M. Pastore
e-mail: lorenzomario.pastore@uniroma1.it

M. Ferrero
e-mail: marco.ferrero@uniroma1.it

F. Salata
e-mail: ferdinando.salata@uniroma1.it

E. Arbizzani et al. (eds.), *Technological Imagination in the Green and Digital Transition*, The Urban Book Series, https://doi.org/10.1007/978-3-031-29515-7_11

which to choose the fittest one depending on the need of the different stakeholders. The proposed multi-objective approach allows reducing the energy consumption for heating by 31% and for cooling by 17% and the CO_2 emissions up to 27.4%. The proposed methodology supports designers and policymakers toward an effective building stock renovation, which can answer the urgent energy and environmental targets for the coming decades.

Keywords Multi-objective optimization · Social housing · Existing buildings · Retrofit · Building energy simulation

11.1 Introduction

In the last decades, the European Union has encouraged Member States to develop strategies toward a deep renovation of the building stock since the 75% of the buildings is currently energy inefficient and the building sector is responsible for 40% of energy consumption and 36% of CO_2 emissions (European Parliament 2012). Indeed, the European Green Deal asks Member States to improve long-term strategies toward a more sustainable building stock, starting from the renovation of the public buildings (European Commission 2020). Furthermore, these strategies aim to fight energy poverty, allowing for clean and safe energy for everyone.

In a changing climate, operation costs will increase with the global warming level and energy poverty will worsen as a consequence (Santamouris 2016). Moreover, in recent years, the energy produced from non-renewable sources is becoming more and more expensive, making the transition to renewable sources and the energy-efficiency strategies for our buildings increasingly urgent. The EU Commission's proposal for the recast of the Energy Performance of Building Directive (EPBD) introduces better measures and tools to increase the rate and depth of building renovations and presents a pathway for buildings to become "Zero Emission" by 2050 (European Parliament 2021).

For all these reasons, the energy retrofit of the building stock is an urgent task. Retrofit strategies should highly reduce both energy consumption and CO_2 emissions, but at the same time, they should be cost-effective. Therefore, a highly complex problem is outlined, considering multiple objectives and multiple variables—where each can have a wide range of possible values (Jafari and Valentin 2018).

To deal with such a complex problem and to explore this wide space of solutions, optimization algorithms can be used, coupled with energy simulation software (Machairas et al. 2014).

These advanced digital tools can support the designer during his decision-making phase and can address the design problem toward more sustainable and comfortable solutions. Indeed, the space of solutions would be too wide to manually explore it and so "intelligent" algorithms can be used to automatically converge toward optimal solutions. In the scientific literature, the most common optimization algorithm used in building design optimization problems is the genetic algorithm (Costa-Carrapiço et al. 2020).

Multi-objective optimization in building design is an active research field (Kheiri 2018). Dealing with conflicting objectives, the optimization process does not provide only one solution—the absolute optimum—but a set of optimal solutions called Pareto solutions. This allows the designer to choose one of the selected solutions based on his preferences and requests. Moreover, it can be useful to explore different combinations and to allow architectural variability. The building energy optimization for existing buildings is a topic of increasing interest, and different methods were employed in the scientific literature with respect to the theoretical framework, objective functions, and genes and software (Hashempour et al. 2020; Ruggeri et al. 2020). Research is still needed to find an approach that can be worldwide shared and used in different design problems.

11.2 Aim and Contribution of the Work

Based on the above-discussed context, the work aims to expand the discussion on multi-objective optimization of retrofit actions toward more sustainable and comfortable buildings. Indeed, the paper proposes a multi-objective approach toward energy renovation of the social housing building stock, by means of an innovative digital workflow. This kind of approach aims to consider simultaneously different aspects of the design (architectural, environmental, energy, social) to address the designer toward more sustainable retrofit strategies. Moreover, the paper aims to investigate the influence of the passive strategies only on the energy efficiency and the environmental impact of the building. The digital workflow is set to be easily applied to different design problems. Indeed, each building would require a specific and tailored optimization to better address a deep and effective energy renovation.

11.3 Method

In order to find suitable retrofit strategies that are tailored for each specific building, the proposed digital workflow is built on a significant and relevant case study. The digital model of the case study is first prepared, and then, architecturally compatible retrofit strategies are taken into account. On these base considerations, the optimization problem is outlined and genes and constraints are set. After running the simulations, the optimal combination of strategies for the retrofit intervention is chosen from the Pareto curve of optimal solutions.

11.3.1 The Case Study and the Energy Model

The digital workflow proposed is applied and verified on a significant and relevant case study, pertaining to the social housing building stock of Rome, Italy.

The selected case study is a building designed by Lucio Passarelli in the late 70s in the northeast area of the city, for the social housing complex "Vigne Nuove" (Lenci 2006).

In particular, the selected building is building C, chosen among the four residential buildings in that area. It is a linear block building, with the distinctive characteristic of external cylindric staircase volumes. The case study consists of seven floors of apartments, the ground floor is open, and the roof consists of common spaces employed as terraces and small closed volumes for private storage. The building consists of 108 apartments and the total area of the conditioned zone is 13,000 m^2.

Table 11.1 describes the envelope elements of the building with the description of layers and their thermophysical features.

The building is modeled by means of Rhinoceros and Grasshopper plugins and simulated by means of EnergyPlus. With respect to the thermal zones, the residential floors are considered as one thermal zone. The thermal zone is a conditioned area, with an occupancy schedule from the late afternoon until the early morning (recalling

Table 11.1 Description of layers and thermophysical features of the envelope

Layers	Thickness (m)	Conductivity (W/mK)	Density (kg/m^3)	Specific heat (J/kgK)	Transmittance (W/(m^{2}K))
Outdoor wall					
Gypsum block	0.10	0.27	950	840	1.16
Air gap	0.20	–	–	–	
Gritted concrete	0.10	0.52	1550	1000	
Loggia wall					
Inner plaster	0.015	0.32	950	1000	1.06
Gypsum block	0.10	0.25	750	840	
Expanded clay block	0.15	0.42	1100	1000	
Roof					
Inner plaster	0.015	0.32	950	1000	0.66
Predalles slab	0.24	0.58	1670	1000	
Polyurethane	0.03	0.04	32	1400	
Lightweight concrete	0.04	1.00	1100	1000	
Tiles	0.02	1.30	2300	840	
Window					
Single glass	0.004	1.00	0.82	0.88	5.8

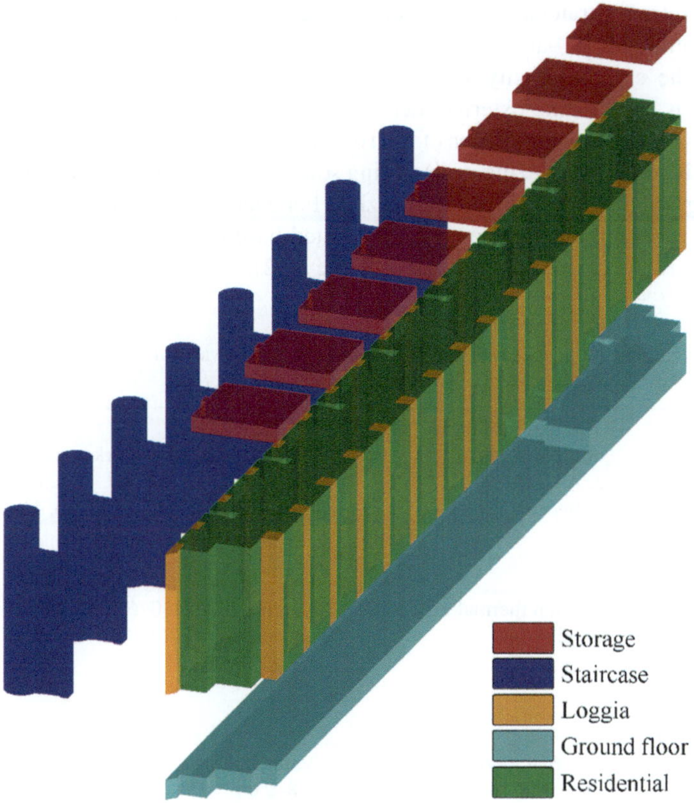

Fig. 11.1 Exploded view of the energy model

a typical working day). The setpoint temperatures are set to 20 °C for heating and 26 °C for cooling. The energy system, which is not the focus of this preliminary work, is set as an ideal system, working with natural gas for heating and electricity for cooling. The natural ventilation is set to 0.3 vol/h. In addition, there are non-conditioned zones for the staircases, the storage volumes on the roof, the loggias, and the open ground floor. Figure 11.1 shows the model and its thermal zones.

The model prepared with Rhino and Grasshopper is then exported to be used in the optimization process, conducted by means of an in-house implemented genetic algorithm written in Python and connected to EnergyPlus with the Eppy library.

11.3.2 Retrofit Strategies

Based on the analysis of the current status of the building, different architecturally compatible strategies are considered toward an energy renovation of the building.

These retrofit strategies—which constitute the "genes" of the building for the genetic optimization—are the following: (i) adding mid and internal thermal insulation to the external cavity walls, (ii) adding internal thermal insulation to the loggia walls, (iii) adding external thermal insulation to the roof, (iv) changing solar reflectance of the finishing layer of the roof, (v) changing windows, (vi) closing the loggias with operable glazing, (vii) adding solar shading in the loggias, (viii) closing the open ground floor with operable glazing. For each gene, a range of possible solutions is considered and the costs are evaluated based on the regional price list for Lazio, where the case study is located (Table 11.2).

Table 11.2 Investigated genes for the energy retrofit, range of variability, and costs

	Gene	Range of variability	Costs
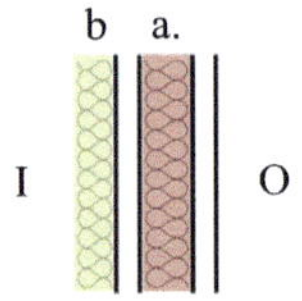	Thermal insulation outdoor wall a. Air gap b. Internal layer	a.1 Expanded clay a.2 Expanded granular cork a.3 Polyurethane foam b. Polyurethane board 0–6 cm (steps of 2 cm)	a.1 45.20 €/m^2 a.2 60.76 €/m^2 a.3 70.50 €/m^2 b. 44 €/m^2
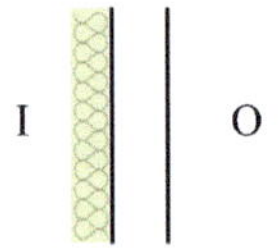	Internal thermal insulation loggia wall	Polyurethane board 0–6 cm (steps of 2 cm)	44 €/m^2
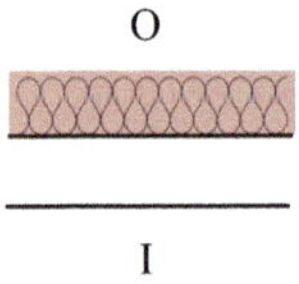	External thermal insulation roof	Polyurethane board 0–7–8–9 cm	450 €/m^3
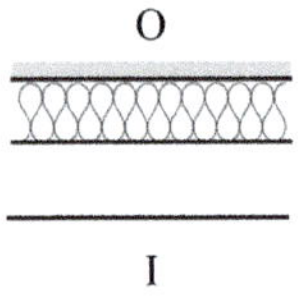	Solar reflectance of the finishing layer of the roof	Solar reflectance 10–90% (steps of 10%)	40 €/m^2
	Windows	W0. $U = 1.80$ W/m^{2}K W1. $U = 1.60$ W/m^{2}K W2. $U = 1.40$ W/m^{2}K W3. $U = 1.10$ W/m^{2}K W4. $U = 0.90$ W/m^{2}K W5. $U = 0.70$ W/m^{2}K	270.20 €/m^2 326.00 €/m^2 366.00 €/m^2 390.00 €/m^2 422.00 €/m^2 478.00 €/m^2

(continued)

Table 11.2 (continued)

	Gene	Range of variability	Costs
	Closing loggia	Yes/No	210 €/m^2
	Solar shading	Yes/No	75 €/m^2
	Closing ground floor	Yes/No	210 €/m^2

11.3.3 Optimization Problem and the Genetic Algorithm

The optimization problem is formulated based on the need to simultaneously consider energy, environmental and economic aspects of the retrofit interventions. Therefore, a multi-objective optimization is conducted, dealing with the minimization of (i) energy demand (ED), (ii) CO_2 emissions (CO_2), (iii) investment costs (IC), (iv) energy costs (EC). The problem can be summarized by the following equation:

$$\min\ F(x) = \min[\mathrm{ED}(x), \mathrm{CO}_2(x), \mathrm{IC}(x), \mathrm{EC}(x)] \tag{11.1}$$

The ED is evaluated by means of yearly dynamic simulations with EnergyPlus. Based on the energy consumption of the building and the source of energy used, the CO_2 emissions are automatically evaluated according to the European Environment Agency data (EEA, https://www.eea.europa.eu/ims/greenhouse-gas-emission-intensity-of-1). The IC is the sum of each retrofit action cost implemented on the building based on the regional price list (Regione Lazio 2020). The EC is the cost of the primary energy yearly consumed by the building based on the energy price in the Eurostat database (Eurostat, https://ec.europa.eu/eurostat/databrowser/view/NRG_PC_202_C__custom_1358122/default/table?lang=en).

Therefore, the space of solutions is a four-dimensional space with 82,944 possible alternatives, considering all the genes (the retrofit strategies described in the previous section) and their range of variability.

A genetic algorithm is used to explore the space of solutions in a faster and more efficient way. Indeed, the algorithm is set to automatically converge toward optimal solutions with respect to the considered objective functions. From the 82,944 possible solutions, only 2000 are simulated, allowing a significant reduction of the computational time.

The optimization algorithm used is an in-house developed active archive Non-dominated Sorting Genetic Algorithm (aNSGA-II). This algorithm is still not widely employed in literature, but its high efficiency is demonstrated in different works (Rosso et al. 2020; Hamdy et al. 2012).

The outputs of the digital workflow are a set of optimal solutions along the Pareto curve, among which the designer can choose the fittest one depending on the need of the different stakeholders. In this case, as the building is public, we hypothesize that the best solution for all the stakeholders is the solution that simultaneously minimizes all the objectives.

11.4 Results

The results of the multi-objective optimization are shown in Fig. 11.2, where the dots represent the simulated buildings. To represent the fourth dimension (the CO_2 emissions), a color axis is used. Compared to the reference building, the optimal solution adds vertical insulation in the cavity of the external wall with expanded clay, replaces the Windows with W0 windows, and closes the loggias with operable glazing. With these genes implemented, the optimal solution allows reducing ED up to 28%. In greater detail, the reduction of the energy consumption for heating is 31% and for cooling 17%.

With respect to the CO_2 emissions, the passive strategies implemented in the optimal solution allow reducing the greenhouse gas emissions up to 27.4%. The EC is reduced by 23.2% and the IC is 70.11 €/m^2.

Figure 11.3 compares the results of the reference building and the optimal solution according to each objective function.

11.5 Conclusions

A deep renovation of our building stock is urgent to reduce energy consumption and the environmental impact of buildings. In particular, social housing plays an even more important role toward this goal, as low-income families that live in these buildings are vulnerable to energy poverty. For these reasons, the research aims to develop an innovative digital workflow that can take into consideration energy, environmental and economic aspects simultaneously for the optimization of building retrofit strategies.

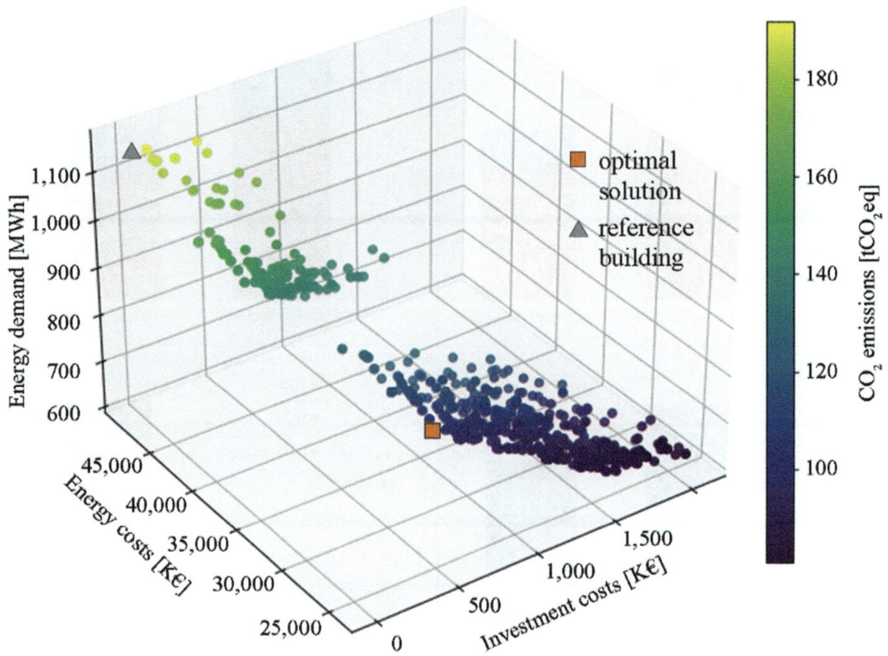

Fig. 11.2 Space of solutions with respect to annual energy demand, energy costs, investment costs, and CO_2 emissions

The workflow is applied and verified on an existing social housing located in Rome, which constitutes a relevant case study. Based on the analysis of the current status of the case study building, architecturally compatible strategies are taken into account as genes of the optimization process. The results of this multi-objective optimization are a set of optimal solutions, among which the designer can choose the fittest one for the specific design problem. Therefore, the proposed approach can highly support the decision-making process of retrofit design by exploring and simulating a wide space of solutions. This is possible by means of the genetic algorithm that reduced the energy simulations required from 82,944 to 2000.

In this work, the optimal solution is chosen among the Pareto frontier as the solution that minimizes simultaneously all the objective functions, i.e., energy demand, CO_2 emissions, energy costs, investment costs. The results demonstrate that high reductions (around 30%) can be achieved using this approach.

The proposed digital workflow is set to be easily repeated in different design problems and could support the activities of professionals and policymakers about retrofit actions to be undertaken on existing buildings.

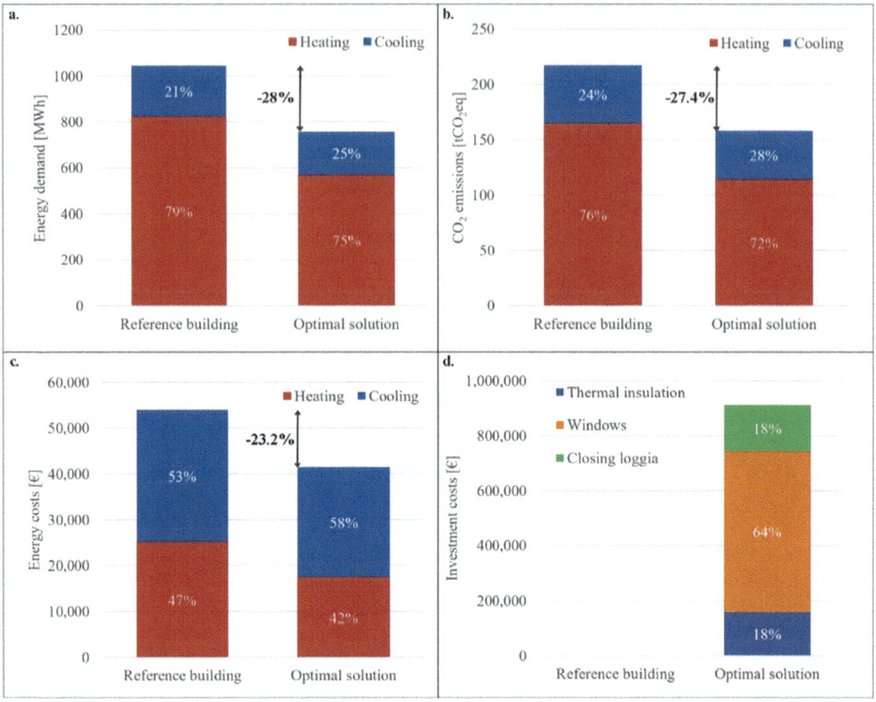

Fig. 11.3 Comparison of the reference building and the optimal solution with respect to **a** ED, **b** CO_2, **c** EC, **d** IC

References

Costa-Carrapiço I, Raslan R, González JN (2020) A systematic review of genetic algorithm-based multi-objective optimisation for building retrofitting strategies towards energy efficiency. Energy Build 210:109690. https://doi.org/10.1016/j.enbuild.2019.109690

EEA. Greenhouse gas emission intensity of electricity generation in Europe. https://www.eea.europa.eu/ims/greenhouse-gas-emission-intensity-of-1

European Commission (2020) A renovation wave for Europe—greening our buildings, creating jobs, improving lives

European Parliament (2012) Directive 2018/844/EU

European Parliament (2021) Proposal for a Directive of the European Parliament and of the Council on the energy performance of buildings, 15 Dec 2021

Eurostat. Gas prices components for household consumers—annual data. https://ec.europa.eu/eurostat/databrowser/view/NRG_PC_202_C__custom_1358122/default/table?lang=en

Hamdy M, Palonen M, Hasan A (2012) Implementation of pareto-archive NSGA-II algorithms to a nearly-zero-energy building optimization problem. In: First building simulation and optimization conference, Loughborough, UK, 10–11 Sept 2012, pp 181–188. https://doi.org/10.4067/S0034-98872012000500015

Hashempour N, Taherkhani R, Mahdikhani M (2020) Energy performance optimization of existing buildings: a literature review. Sustain Cities Soc 54. https://doi.org/10.1016/j.scs.2019.101967

Jafari A, Valentin V (2018) Selection of optimization objectives for decision-making in building energy retrofits. Build Environ 130:94–103. https://doi.org/10.1016/j.buildenv.2017.12.027

Kheiri F (2018) A review on optimization methods applied in energy-efficient building geometry and envelope design. Renew Sustain Energy Rev 92:897–920. https://doi.org/10.1016/j.rser.2018.04.080
Lenci R (2006) Studio Passarelli, cento anni cento progetti. Electa, Milano
Machairas V, Tsangrassoulis A, Axarli K (2014) Algorithms for optimization of building design: a review. Renew Sustain Energy Rev 31:101–112. https://doi.org/10.1016/j.rser.2013.11.036
Regione Lazio (2020) Tariffa dei prezzi, edizione 2020, per le opere pubbliche edili ed impiantistiche del Lazio
Rosso F, Ciancio V, Dell'olmo J, Salata F (2020) Multi-objective optimization of building retrofit in the Mediterranean climate by means of genetic algorithm application. Energy Build 216:109945. https://doi.org/10.1016/j.enbuild.2020.109945
Ruggeri AG, Gabrielli L, Scarpa M (2020) Energy retrofit in European building portfolios: a review of five key aspects. Sustainability (Switzerland) 12:1–37. https://doi.org/10.3390/SU12187465
Santamouris M (2016) Innovating to zero the building sector in Europe: minimising the energy consumption, eradication of the energy poverty and mitigating the local climate change. Sol Energy 128:61–94. https://doi.org/10.1016/J.SOLENER.2016.01.021

Chapter 12
Digital Information Management in the Built Environment: Data-Driven Approaches for Building Process Optimization

Francesco Muzi, Riccardo Marzo, and Francesco Nardi

Abstract In Italy, the traditional management of construction works, throughout life cycle, still dominates the market compared to a digital approach. This research aims at bringing out the potential and benefits of a digital management by developing strategies and methodologies able to optimize processes related to three different use cases. The proposed use cases have been developed by applying digital methodologies to different building contexts, aiming at both site management and management of the built environment. The first case deals with an important public building of 35,000 m^2 located in a residential context in the center of Rome. The use of digital methodology made it possible to optimize and prevent problems related to large-scale works and construction sites located in central residential areas. The second case concerns a residential complex of 16 buildings located in Rome, where the BIM model supplies a constant flow of information for predictive maintenance system. The last one refers to port infrastructures located on the coast of Lazio region in Italy. The digital information model was developed to set up a risk management system capable of safely managing the port's main assets. In conclusion, the results achieved through the implementation of a digital approach generated by a structured information flow integrated with the BIM model, allowed an optimized management of time and economic resources in the three case studies mentioned, although the diversity of objectives and types of construction works. This improvement is made possible by a shared and connected digital model, characterized by a high level of geometric and informative detail and cloud computing strategies to enhance process efficiency, supporting decision-making and information management.

F. Muzi (✉)
Department of Astronautical, Electrical and Energy Engineering (DIAEE), Sapienza University of Rome, Rome, Italy
e-mail: francesco.muzi@uniroma1.it

R. Marzo · F. Nardi
NCLAB, Rome, Italy
e-mail: riccardo.marzo@nclab.it

F. Nardi
e-mail: francesco.nardi@nclab.it

E. Arbizzani et al. (eds.), *Technological Imagination in the Green and Digital Transition*, The Urban Book Series, https://doi.org/10.1007/978-3-031-29515-7_12

Keywords Digital construction management · Building Information Modeling · Construction management · BIM

12.1 Introduction

The traditional management of building infrastructures has been overturned by the processes of building automation (Quinn et al. 2020), thanks to technological development that allowed considerable progress toward a digital (Dahanayake et al. 2021) and environmentally sustainable (Rosa et al. 2022) management. This evolution coined the concept of intelligent building, able to manage different systems on a single digital and interoperable asset, exchange information flows, and perform actions in relation to the derived parameters. Part of the management process is the use of IoT devices and the connection of each device to the network (Dave et al. 2018).

In this way, connected objects (e.g., light bulbs or cameras) create an information database that provides more specific and adaptable services. This revolutionary process leads to the development of cloud-based platforms that move big data into virtual memories (Tomazzoli et al. 2021). In the era of building digitization and the renewal of the construction industry, it is important to capitalize on the advantages of BIM to optimize workflows and improve the management of the life cycle. The Building Information Modeling methodology implemented with IoT devices and Artificial Intelligence algorithms, in fact, allows the creation of a digital twin that replicates the real building and contains all the information needed to control the building process, from concept to maintenance (Agostinelli 2021). The digital management achieved through Building Information Modeling (BIM) enables the construction industry to improve all phases, from better planning of resources to improving collaboration between various disciplines, helping to keep the project on time and on budget (Ruperto et al. 2019). BIM is a digital methodology that represents a method of managing construction and infrastructure projects, suitable for digitizing the representation of assets to build and as-built, optimizing entire processes (Cinquepalmi et al. 2019).

12.2 State of the Art

The BIM model under study is created and prepared to follow the regulatory evolution in the field of construction, in order to generate a virtual environment able to provide information about the status and the building systems that compose it. Combining BIM model with IoT devices and ML systems is set up the concept of digital twin (DT) as a digital object into which information can be input and output (Guidi et al. 2020). The DT enables the collaboration between artificial intelligence and data analytics to create dynamic and predictive models that can learn and update the state of its physical opponent. The aim of developing these technologies is the creation and setting of intelligent buildings that transcend the areas of comfort, optimizing performance and reducing energy waste.

This is made possible by the creation of digital twins of infrastructures capable of generating predictive models that enable the establishment of smart cities. The development of new technologies makes the innovation process possible and aims to improve it through digital management (Piras et al. 2022). One of the main aspects dealt with in the case studies is related to data management. In fact, the use of cloud-based solutions for collecting, sharing, and using information is planned. In addition, for management, maintenance activities include the use of digital solutions for automated building maintenance such as cable robots for facade cleaning, integration of IoT devices for security, and building monitoring at the users' service. It can therefore be claimed that the methodology proposed in the three case studies is aligned with the digital transformation process and directs in its application the activities toward aspects of optimization and mitigation of delays.

12.3 Methodological Approach Strategy

In this work, the digital BIM methodology (Matarneh et al. 2019) plays the role of common denominator in the three case studies described. The system devised is illustrated in Fig. 12.1, where the project placed at the center of the process plays both the role of objective and protagonist.

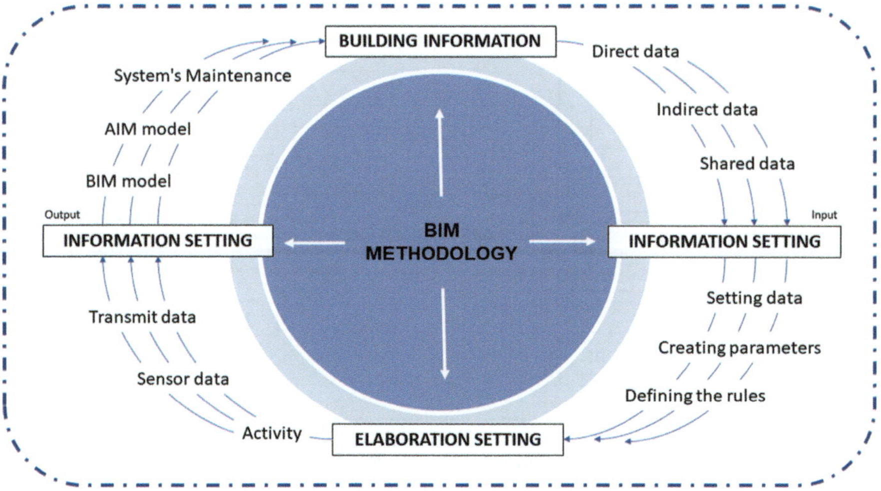

Fig. 12.1 BIM methodology workflow for the digital twin—R. Marzo (Next City Lab)

The implementation of a digital model with a high level of information requirements (LoIN) builds a digital twin capable of transferring data from the building to the virtual reproduction (Liu et al. 2021). This information, obtained from IoT devices, makes it possible to structure an information database. Through the processing of this data, it is possible to set up energy analyses for consumption management and maintenance analyses for building asset management (Heaton et al. 2020). In the first case study, the time and cost management of a 4D and 5D BIM model based allow stakeholders to prepare simulations and forecast scenarios aimed at optimizing resources (Agostinelli et al. 2019). Moreover, this digital model can play a decisive part in the definition of predictive maintenance and energy management strategies as analyzed in the second case study (Gonzalez et al. 2021).

Finally, in the case concerning port areas, the process led to the management of strategic assets for Risk and Safety Management activities (Rodrigues et al. 2022). The assets are digitized to pursue Facility Management strategies aimed at providing very high-performance standards in each case study. The big data are then used by information models in CDE's shared environments (Abbasenejad et al. 2021) (Fig. 12.2).

The approach is therefore guided by the thread of parametric modeling with informative objects. The assets are described, modeled, and computerized in order to pursue the strategies of the facility, intended to ensure very high standards of performance in each case study. The attainment of the objectives passes therefore from the predisposition of the geometric, spatial, functional, and technological information, and the successive step is defined from the use of the informative models in environments of sharing CDE (Abbasenejad et al. 2021).

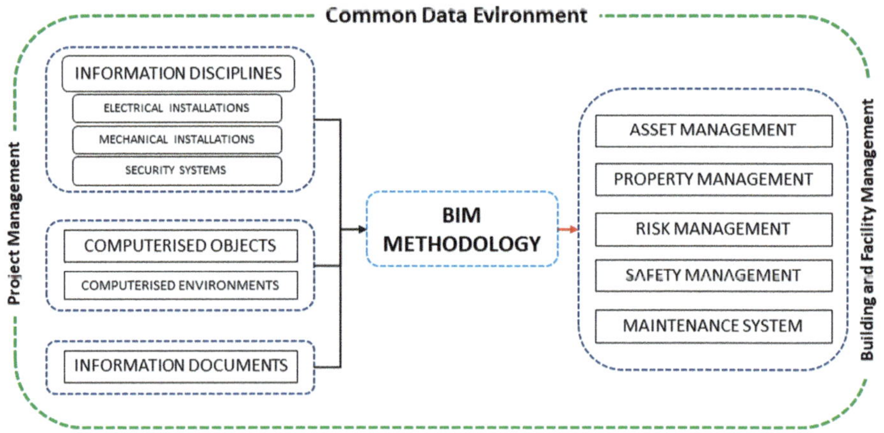

Fig. 12.2 Key point in Common Data Environment—R. Marzo (Next City Lab)

12.3.1 First Case Study: Public Building in a Residential Area

The application case (Fig. 12.3) discusses a 35,000 m^2 building owned by a public institution (CDP), located in Rome in a central urban context. The building was built starting in 1914 and officially opened in 1928, became the headquarters of the Istituto Poligrafico dello Stato and later the site of an industrial activity until 2010, when it was abandoned. Requalification work began in 2021 thanks to the intervention of a large Italian company (ENEL) that decided to transform part of the building into their offices. The project is driven by environmental sustainability and from the conceptualization phases of the program and construction design based on two cornerstones, respect for historical memory, with high-quality restoration, and a new modern architecture for the volumes on the rooftop. The benefits of the requalification are not only limited to the building proper but extend to the urban context. In fact, the upgrading process is aimed at following high international standards to pursue a positive impact in the construction and operation process. The construction site is monitored by IoT sensors that transmit data to a CDE to track and provide a measure of the estimated 8% reduction in CO_2 emissions in the neighborhood area. The project (Home page Piazza Verdi project, https://www.poligraficopiazzaverdi.it/il-progetto/) has an investment of €160 mln in public funding, and the project phases include the preparation of a renewal of the exterior facades, a new distribution of interior spaces, and the construction of a four-floor underground parking garage. The digital approach of the implemented process is oriented to the management of the future real estate asset by furthering FM strategies, such as energy management, asset management, property management connected to Business Intelligent services. Facility strategies are supported by the presence of a digital model, defined Asset Information Model (AIM), which by geometric and informational structuring is able to empower operators and management of operational phases with the knowledge and technical instruments to be able to manage and maintain the asset.

In this case, the preparation of a parametric digital model with high information content, obtained by applying a Building Information Modeling (BIM) methodology, allows the actors in the construction process to communicate precise information and to forward it at the end of the process, associated with a 3D model which can be consulted and is updateable. This peculiarity of BIM is made possible with the structuring of a Common Data Environment (CDE), a common data storage useful for the transmission, optimization of the worksite phases, and the subsequent management phase of the work. The creation and use of a DT for the maintenance and management phases of the asset, as a step subsequent to the architectural regeneration, allow the asset's owner and future manager to equip their selves with an FM model even from the realization phases, predicting critical issues and problems before the work is put into operation.

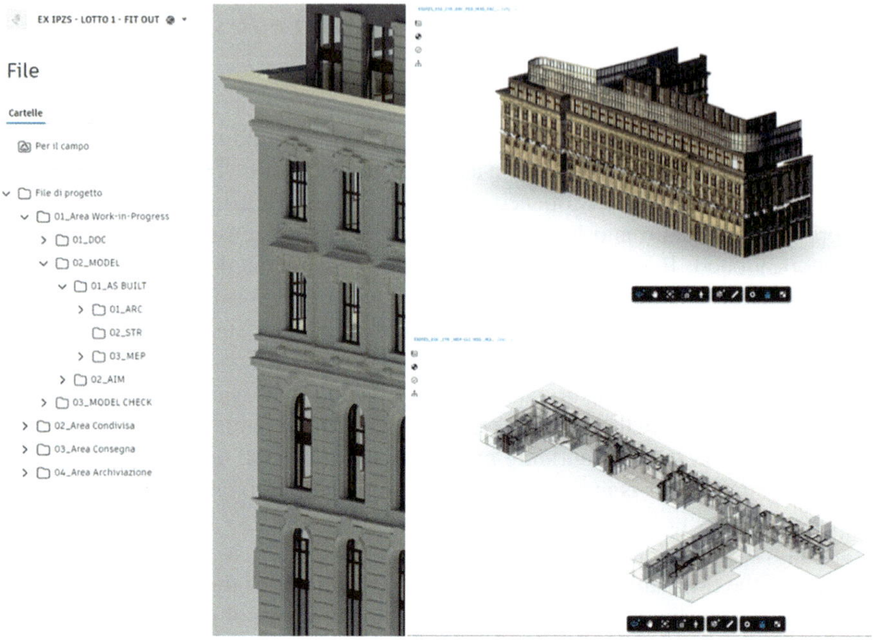

Fig. 12.3 Common Data Environment of the project, example of queryable information on facade restoration and installation of mechanical system

12.3.2 Second Case Study: Residential Compound

The case study (Fig. 12.4) concerns a residential complex composed of 900 flats located in Rome.

The complex represents a state-of-the-art building system for the use of materials and for the use of energy systems powered by renewable sources. Through a detailed quantitative and qualitative analysis of the entire compound, the realization of the as-built BIM digital model began, which was implemented with a high amount of information related to the project. This connection made it possible to design a

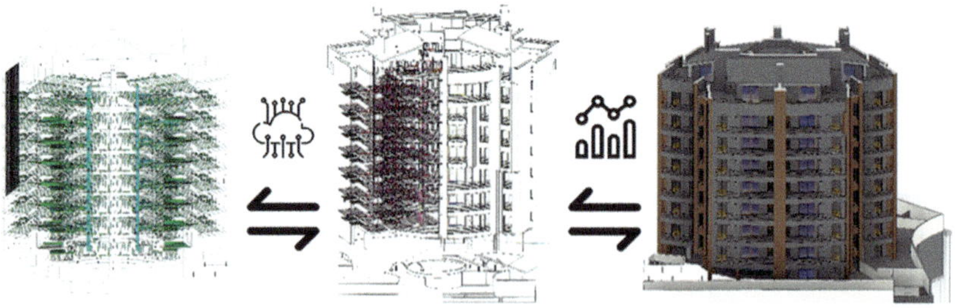

Fig. 12.4 Digital twin representation of the residential complex

digital twin capable of providing economic-temporal simulations for the production of innovative and functional solutions to support decision-making. This virtual duplicate therefore allows the physical part to be optimized to make it more efficient and adaptable to technological-energetic developments in the building industry.

Predictive maintenance methodology. The process developed and applied to the building complex concerns a predictive maintenance strategy to improve the performance of mechanical and electrical systems.

The strategy can provide a customized maintenance service based on machine learning systems to reduce malfunctions and failures. The amount of data needed to support decision-making activities is obtained after a certain period of time (maximum three years). This period depends on the duration of the maintenance loops: for example, an item with weekly maintenance operations reaches the target necessary for its full operation much earlier than items with monthly maintenance periods. This methodology determines an acceptable failure rate of components on the basis of statistical data derived from specific maintenance reports drawn up by the operator. If the failure rate is exceeded, the maintenance period is reduced by the difference between the real failure rate and the target failure rate. In this way, that item should remain functional in the next maintenance operation. In addition, labor costs can be combined with the duration of each operation to analyze costs. In this case study, the analysis concerns the mechanical, electrical, and lighting systems that supply the outdoor and communal areas, without considering the flats for private use. Predictive maintenance through big data analysis has been shown to substantially improve the overall operation of the residential installations investigated through the optimization of maintenance cycles. The aim is to enable the entire production chain to reduce maintenance time and increase productivity, thereby reducing costs and responding flexibly and effectively to consumer needs.

12.3.3 Third Case Study: Port Infrastructures

The use of this digital approach can also be seen in the third and last case studies analyzed in this section (Fig. 12.5), a set of parallel projects involving the port areas of Anzio, Terracina, Formia, and Ventotene, located along the mid-Tyrrhenian coast. The common objective of all these projects was to implement and improve the traditional real estate, in other words the complex of services for the management of buildings and real estate in general, declined according to its five main management activities, specifically those of Asset Management, Project Management, Property Management, Building Management, and Facility Management, through the adoption of a digital approach and, specifically, the use of BIM processes with the potential to improve communication between all actors involved, safety; optimize maintenance and in general the management of the life cycle; increase the energy, space, and economic awareness of the asset. The core of these objectives is the creation of

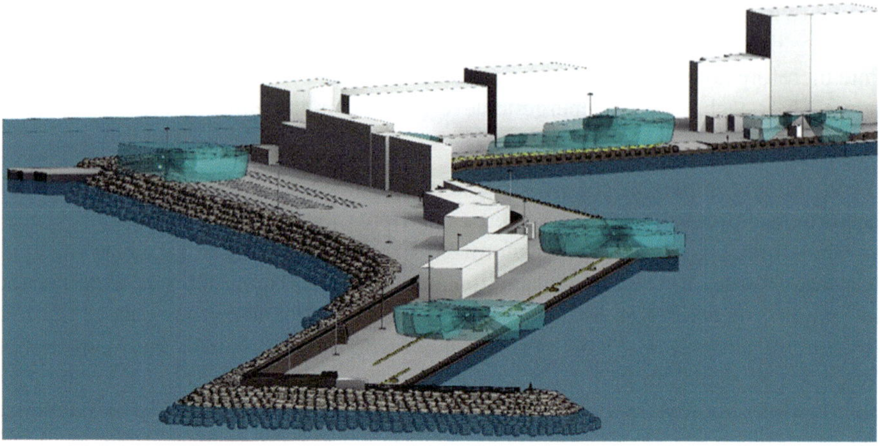

Fig. 12.5 Digital twin representation of security system

detailed parametric models in which the large amount of data describing a vast asset such as a port area, meant as a complex of buildings, infrastructures, and equipment, can be collected and stored.

Focusing on the port area of the city of Anzio, the digital revolution mainly involved the areas of Risk Management, Security Management, and in addition, Energy Management. The use of the BIM model for the assessment and consequent energy efficiency of the port area, with regard to urban lighting, highlighted the presence of isolated areas with poor lighting and lack of video surveillance cameras, consequently, focused attention on the problem of security. The first phase involved the identification of the risk, which means the areas considered critical from the point of view of safety, taking into consideration a series of factors, which can be summarized as follows: location of the area in relation to the built-up area, level of occupancy, public lighting, and coverage of the video surveillance system. The second phase examined the risk assessment and can be summarized in the creation of an automatic area classification system, obtained by translating the above-mentioned qualifying factors into precise rules, combined and merged within an algorithm capable of returning, according to a pre-established coding system, the result of the analysis for each sub-area, into which the entire port area was automatically subdivided, obtaining a decision support tool. The identification of the areas with the highest risk consequently made it possible to circumscribe and focus the design and economic effort for the implementation of security in the port area. The main area of intervention concerned the implementation of the existing video surveillance system, supported by the introduction of an advanced computer vision system to identify specific critical circumstances, such as intrusion attempts, assaults, and similar.

12.4 Conclusions

The objective is to provide a predictive system for maintenance and safety, an intelligent management capability, and at the same time, a data sharing environment accessible over time and upgradeable. The idea of the digital twin as a model for managing and controlling physical systems based on Big Data has emerged over the last decade in industrial sectors. BIM platforms were developed in response to the need for more effective tools for management, allowing processes to evolve to meet the requirement for digital prototyping in construction. In fact, the proposed projects made it possible to create, on a rich database, a digital reproduction of the work, able to accurately simulate and estimate activities and possible interferences, providing a better and faster execution. In the three case studies described above, the efficiency of a dynamic digital approach is highlighted by the inefficiency of traditional systems. Looking forward future developments, this methodology can be set to make it usable and replicable in different projects for managing the built environment by exploiting business intelligent systems powered by big data.

References

Abbasenejad B et al (2021) Building Information Modelling (BIM) adoption and implementation enablers in AEC firms: a systematic literature review. Archit Eng Des Manag 17(5–6):411–433

Agostinelli S (2021) Actionable framework for city digital twin-enabled predictive maintenance and security management systems. In: Proceedings of BIM 2021 international conference. Wessex Institute of Technology. ISBN: 978-1-78466-441-1

Agostinelli S et al (2019) 5D-BIM: tools and methods for digital project construction management. WIT Trans Built Environ 192. ISSN: 1743-3509. WIT Press

Cinquepalmi F et al (2019) ZEB prototype controlled by a machine learning system. In: Modernization and globalization: challenges and opportunities in architecture, urbanism, cultural heritage. IFAU—3rd international forum on architecture and urbanism. ISBN: 978-9928-131-89-8

Dahanayake KC et al (2021) IoT-BIM-based digital transformation in facilities management: a conceptual model. J Fac Manage

Dave B et al (2018) A framework for integrating BIM and IoT through open standards. Autom Constr 95:35–45

Gonzalez J et al (2021) BIM and BEM methodologies integration in energy-efficient buildings using experimental design. Buildings 11(491)

Guidi G et al (2020) The potential of digital twin model integrated with artificial intelligence systems. In: Special session on “new frontiers in green computing and artificial intelligence” at the IEEE EEEIC 2020 conference. ISBN: 978-1-7281-7455-6

Heaton J et al (2020) Asset information model to support the adoption of a digital twin: West Cambridge case study. IFAC-PapersOnLine 53(3):366–371

Home page Piazza Verdi project. https://www.poligraficopiazzaverdi.it/il-progetto/. Accessed 10/02/2022

Liu Z et al (2021) Transition from building information modeling (BIM) to integrated digital delivery (IDD) in sustainable building management: a knowledge discovery approach based review. J Clean Prod 291:125223

Matarneh ST et al (2019) Building information modeling for facilities management: a literature review and future research directions. J Build Eng 24:100755

Piras G et al (2022) Renewable energy system controlled by open-source tools and digital twin model: zero energy port area in Italy. Energies 15:1817. https://doi.org/10.3390/en15051817
Quinn C et al (2020) Building automation system—BIM integration using a linked data structure. Autom Constr 118:103257
Rodrigues F et al (2022) BIM approach in construction safety—a case study on preventing falls from height. Buildings 17(23)
Rosa F et al (2022) Transforming a historic public office building in the centre of Rome into nZEB: limits and potentials. Energies 15:697. https://doi.org/10.3390/en15030697
Ruperto F et al (2019) 4D BIM-oriented digital project construction management for AEC industry. WIT Trans Built Environ 192. ISSN: 1743-3509. WIT Press
Tomazzoli C et al (2021) Cyber-physical systems improving building energy management: digital twin and artificial intelligence. Energies 14(8):2338. https://doi.org/10.3390/en14082338

Chapter 13
Immersive Facility Management—A Methodological Approach Based on BIM and Mixed Reality for Training and Maintenance Operations

Sofia Agostinelli and Benedetto Nastasi

Abstract Innovation technology in industries including manufacturing and aerospace is moving toward the use of Mixed Reality (MR) and advanced tools while Architecture, Engineering and Construction (AEC) sector is still remaining behind it. Moreover, the use of immersive technologies in the AEC digital education, as well as for professional training, is still little considered. Augmented and Mixed Reality (AR/MR) have the capability to provide a "X-ray vision", showing hidden objects in a virtual/real overlay. This feature in the digital object visualization is extremely valuable for improving operation performance and maintenance activities. The present study gives an overview of literature about the methodologies to integrate virtual technologies such as AR/MR and Building Information Modeling (BIM) to provide an immersive technology framework for training purposes together with the Digital Twin Model (DTM)-based approach. Furthermore, the Facility Management (FM) tasks' training on complex building systems can benefit from a virtual learning approach since it provides a collaborative environment enhancing and optimizing efficiency and productivity in FM learning strategies. For this purpose, the technological feasibility is analyzed in the proposed case study, focusing on the realization of a methodological framework prototype of immersive and interactive environment for building systems' FM. Cloud computing technologies able to deal with complex and extensive information databases and to support users' navigation in geo-referenced and immersive virtual interfaces are included as well. Those ones enable the DTM-based operation for building maintenance both in real-time FM operators' training and FM tasks' optimization.

Keywords AEC industry · Digital Twin · Facility Management · Mixed Reality · Digital construction

---

S. Agostinelli (✉) · B. Nastasi
Sapienza University of Rome, Rome, Italy
e-mail: sofia.agostinelli@uniroma1.it

B. Nastasi
e-mail: benedetto.nastasi@uniroma1.it

E. Arbizzani et al. (eds.), *Technological Imagination in the Green and Digital Transition*, The Urban Book Series, https://doi.org/10.1007/978-3-031-29515-7_13

13.1 Introduction

Nowadays, Facility Management (FM) is mainly carried out through performance-based contracts stressing the importance of the building performance gap between the as-designed and the as-built configuration (Talamo et al. 2019). The interoperability between expected performance databases and actual ones remains a key obstacle for reducing the mentioned gap, but digital technologies have the potential to overcome it (Wong et al. 2018).

Indeed, monitoring the real data allows the assessment of such a deviation entailing a continuous Measurement and Verification (Manfren et al. 2020) activity that gives the chance to narrow the challenge to the predicted and actual maintenance (Atta and Talamo 2020) during operation. Quantity and quality of information depend on technologies adopted up to the level of big data calling for cloud computing, machine learning techniques and data environment readable by building experts (Ahmed et al. 2017).

Building Information Modeling (BIM) has been adopted widely in the design phase, while for the operation one, its use is often as an archive of information since the model is created after a survey of the as-built. However, great potential of this later created model can be expressed if merged with Virtual Reality (VR) technologies giving the chance to survey and inspect the buildings remotely (Sabol 2008). Furthermore, the experience can be further improved through Augmented Reality (AR) technologies overlapping the physical visit of the buildings with its virtual reconstruction. This merge provides instantaneously a batch of information to the surveyor up to a "X-Ray" vision experience. It entails those experts and non-expert who can benefit from this technology since required information to fill the skills and knowledge gap could be received using that technology. This undoubtedly leads to a widespread of this solution and expeditious adoption by workers and trainers.

13.2 Material and Methods

13.2.1 Background and Literature Review

Considering that the FM phase is the longest period of the entire building life cycle, information management utilizing BIM is still facing challenges related to on-site application since it is mainly used in the design phase (Chung et al. 2021). Due to differences in design, construction and FM BIM information requirements, the identification of specific data for operation and maintenance as well as for BIM integration with external management systems is still a research purpose.

To this end, an open-source BIM-based Facility Management information exchange system called Construction Operations Building information exchange (COBie) has been developed in the UK and USA (East and Carrasquillo-Mangual 2012).

Moreover, the evolution of digital approaches based on Virtual Reality (VR) and Mixed Reality (MR) introduced the application of immersive spaces for learning and education in medicine (Liu 2014), construction (Messner et al. 2003) and production engineering (Maffei and Onori 2019).

Focusing the attention on construction education, studies underlined the effectiveness of MR as an education tool.

Azhar et al. (2018) explored the potentials of MR and VR implementation in design review. Shirazi and Behzadan (2015) introduced complex construction processes through AR in construction management, showing that students interacting with the augmented environment were able to improve learning on tasks and activities.

The same digital approach could be potentially extended to maintenance service training, as MR provides immersive simulation in a multi-user collaborative and interactive environment for real-time live learning. Moreover, MR and other immersive technologies are promising solution tools for problem-solving and decision-making processes in operation and maintenance (Ke et al. 2016).

Considering that FM industry needs the acquisition of a large amount of data from different sources (Irizarry et al. 2013) due to the diversity of the maintained items (Rankohi and Waugh 2013), immersive technologies provide solutions to ensure collaboration between operators on the same up-to-date virtual information (Bae et al. 2013).

Moreover, as building lifecycle's costs are mostly related to the operation and maintenance phase (Becerik-Gerber et al. 2011), the use of virtual collaborative solutions such as AR/VR/MR combined with cloud computing and artificial intelligence is significant in the FM industry (Zakiyudin et al. 2013).

In this regard, FM industry has increased its interest in digitalization since collaborative MR solutions could be useful for the optimization of maintenance activities, as it enables knowledge transfer between operators and stakeholders, providing them a context-sensitive interactive platform.

In fact, the use of MR instructions allows more than 82% error's decreases in maintenance activities (Mekni and Lemieux 2014). In this regard, NASA's experts use Microsoft HoloLens MR-based remote assistance in maintenance tasks (Hachman 2015).

Also, immersive solutions are utilized by Boeing (Sacco 2016) and other international companies to reduce production time and improve task location for maintenance operators.

Combining building information modeling (BIM) and MR (Chu et al. 2018) for FM (Gheisari and Irizarry 2016), remote collaboration and interaction through holograms is enabled, but it still represents a research area. In fact, the integration of BIM and MR technologies allows users to be virtually co-located even though the applicability is still under development in FM.

Therefore, the main purpose of this research is to enhance the use of BIM and immersive technologies in FM, increasing work productivity through the application of a Digital Twin-based framework. In this regard, Asset Information Model (AIM) information dataset has been defined, and Common Data Environment (CDE) and MR solutions are both used for maintenance and training activities.

13.2.2 System Architecture and Process Flow

The realization of the proposed framework is related to the implementation of different components and data flow in order to enable the digital Facility Management system architecture:

1. BIM Model for Facility Management—Asset Information Model (AIM).
2. DTM-based up-to-date data flow using mobile user interfaces.
3. Cloud-based CDE platform for data storage and analysis.
4. Building on-site maintenance activities using MR.
5. Remote support and hybrid training using MR (Fig. 13.1).

The proposed framework is based on BIM data as the main source for physical and digital interactions. The configuration of a customized mobile user interface (UI) allows operators to gain maintenance on-site data directly updating the AIM through the CDE platform combining as-built and on-site data. The CDE platform also provides a knowledge base for data analysis, machine learning systems, prediction and reasoning.

Moreover, MR is introduced for holographic on-site operators' experience as well as for prototyping a hybrid learning space for workers' assistance and remote support.

BIM Model for Facility Management—Asset Information Model (AIM). The Asset Information Model (AIM), as defined by EN ISO 19650:2018, is a BIM model including specific spaces and assets of the building. This model is mainly used

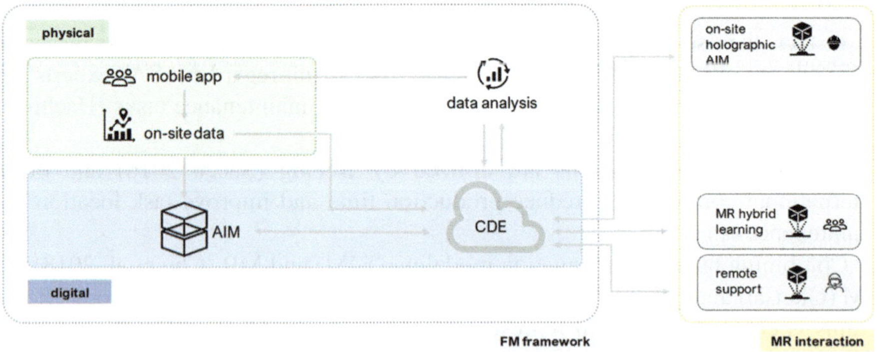

Fig. 13.1 FM-MR process flow

to support operation and maintenance phases. Typically, the AIM is the evolution of a Project Information Model (PIM) which is updated to incorporate modifications evolved during the operation phase (East and Carrasquillo-Mangual 2012). This model is also equipped with connectivity data to enable FM systems related to space, equipment and asset management. Specific operation and maintenance objects (OMO) need to be identified in the BIM model and structured according to AIM geometric and information standards which allow the interaction with other systems with a correct data flow.

A Facility Management dataset that can be associated to BIM's OMOs is proposed in Table 13.1 defining both geometric and information standards.

The AIM is structured according to a set of control parameters which allow to keep a standardized database during the asset's lifecycle, describing the actual status of a building, component, or asset over time.

Moreover, the AIM contains parameters related to the Construction Operation Building Information Exchange (COBie) Standard[1] (East and Carrasquillo-Mangual 2012) which is used for data transmission during the building life cycle. Such parameters are also useful for integrating AIM data with Computerized Maintenance Management Systems (CMMSs) (East and Carrasquillo-Mangual 2012).

DTM-based up-to-date data flow using mobile user interfaces. The concept of Digital Twin derives from Product Lifecycle Management (PLM) Grieves' definition as a digital representation of physical assets and the connections between them (Grieves and Vickers 2017). As there is still no clear definition of Digital Twins, Kritzinger et al. proposed three main categories.

(i) The Digital Model (DM) is related to the use of BIM in the construction industry, as a digital representation without any form of automation in data flow; (ii) a Digital Shadow (DS) is based on real-time data flow from physical to digital asset. It is also defined a Digital Shadow when data are automatically and real-time transferred from physical to digital asset only; and (iii) a Digital Twin is a DM with automatic data flow from physical to the digital asset and vice versa.

The use of a mobile application is integrated in the proposed system architecture in order to collect on-site operational data which constantly and automatically update the AIM parameters. Such approach provides an up-to-date database based on the Digital Twin concept, as items of the operation and maintenance model (AIM) are constantly updated by on-site inputs, providing data analysis and reasoning systems to predict actions.

Cloud-based Common Data Environment (CDE) platform. Cloud computing allows end-users to access IT infrastructure and applications through the network, consisting in a combined set of technologies such as the Internet, distributed systems, virtualization. Depending on end-users' utilization, it is defined as Infrastructure as

[1] COBie (Construction Operation Building Information Exchange) Standard specifies the transfer of information between the owner and the construction team. Following the last update of COBie in 2010, now it is called COBie2 and it provides readability both for humans and machinery. The human readable format is provided in Microsoft Excel Spreadsheet format.

Table 13.1 AIM information dataset

AIM parameter name	AIM parameter type	Description
Development		
Dev_Installation	*Yes/no*	Specifies whether a component has already been installed
Dev_Progress_data	*Text*	Specifies dd/mm/yy of installation
Dev_Progress	*Integer*	Indicates the progress percentage of the installation
Classification		
Class_Technical component	*Text*	Main component
Class_Technical sub-component	*Text*	Sub-component
Class_Description	*Text*	Description of the classified item
Traceability		
Trac_Serial number	*Text*	Serial number of the installed component
Trac_Vendor	*Text*	Vendor of the installed component
Trac_Installer	*Text*	Installation supplier
Localization		
Loc_Building	*Text*	Building code
Loc_Level	*Text*	Building level
Loc_Position	*Text*	Building position
Loc_Room	*Text*	Building room
Documentation		
Doc_Use-maintenance	*URL*	User and maintenance manual
Doc_Datasheet	*URL*	Product datasheet
Doc_Web	*URL*	Manufacturer's website
Doc_DB-Management	*URL*	Management database
Doc_Product	*URL*	Product certification
Doc_Approval	*URL*	Homologation certificate
Doc_Mounting	*URL*	Assembly sheet
Doc_Installation	*URL*	Installation instructions
Doc_Testing	*URL*	Test certificate
Doc_Compliance	*URL*	Declaration of conformity
Doc_Maintenance-history	*URL*	Maintenance history
Doc_Maintenance-plan	*URL*	Maintenance plan
Doc_Evacuation-plan	*URL*	Evacuation plan
Doc_Survey	*URL*	Surveys
Doc_Description	*URL*	Component descriptions
Doc_Model	*URL*	Product model

(continued)

Table 13.1 (continued)

AIM parameter name	AIM parameter type	Description
Doc_Manufacturer	*URL*	Product manufacturer
Facility Management		
Fm_Type-of-intervention	*Text*	Type of maintenance interventions
Fm_Description	*Text*	Maintenance description
Fm_Frequency	*Number*	Maintenance frequency (min)
Fm_Cost	*Number*	Maintenance cost
Fm_Time	*Number*	Maintenance time (min)
Fm_Human-resources	*Number*	Human resources for maintenance activity (n. men)
Fm_Equipment-cost	*Number*	Total cost of maintenance equipment

a Service (IaaS), Platform as a Service (PaaS), or Software as a Service (SaaS). Despite many positive impacts on business management, the adoption of cloud computing solutions is still remaining behind in construction industry. The main benefits are decision support system, compatibility, information and organization, while weakness factors are mainly focused on data security and privacy.

The use of cloud computing in construction industry is strictly related to process management based on the use of BIM data organized through the implementation of CDEs. A CDE is a cloud-based and object-based system providing query, transfer, updating and management of project elements from a variety of data sources (Sacks et al. 2018) in a single multi-service platform. In such context, the configuration of a CDE platform plays a relevant role, ensuring a consistent data flow and enhancing efficiency through the connection of AIM data to on-site operations (Fig. 13.1). Moreover, the CDE platform is a cloud-based data service for data collection and analysis (Preidel et al. 2018) of both geometric and information models providing a knowledge base for the implementation of MR technologies.

Building on-site maintenance activities based on MR. Immersive technologies can help performance improvement of operators by providing virtual information about the real environment, as VR, AR and MR allow users to interact with an immersive and virtual scenario as represented by Milgram and Kishino (Fig. 13.2).

While VR is an immersive full-virtual application, AR provides a virtual-content layer over real-world scenarios, adding relevant information about the real-time captured image. MR is a combination of reality and virtuality.

The proposed framework aims to achieve FM performance improvements by introducing MR technology into on-site maintenance activities, where BIM data represent the baseline for merging the real and virtual world.

While VR is an immersive full-virtual application, AR provides a virtual-content layer over real-world scenarios, adding relevant information about the real-time captured image. MR is a combination of reality and virtuality.

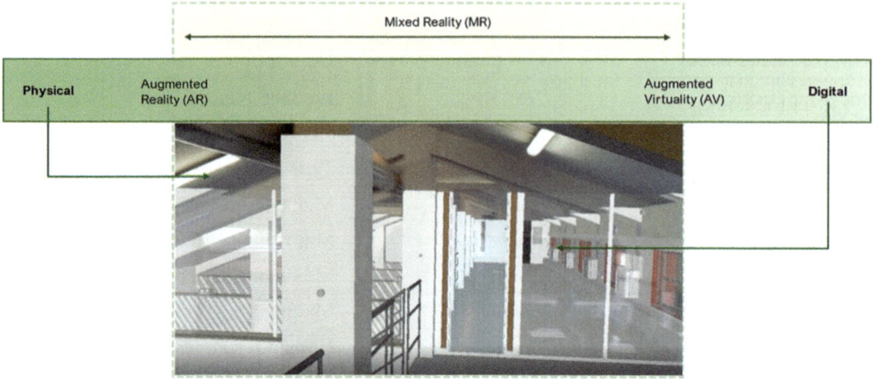

Fig. 13.2 Mixed Reality spectrum from Milgram and Kishino: AIM-MR virtual overlay

The proposed framework aims to achieve FM performance improvements by introducing MR technology into on-site maintenance activities, where BIM data represent the baseline for merging the real and virtual world.

The present application proposes building on-site maintenance activities integrated to MR devices to improve maintenance process efficiency through holographic on-site implementation of the AIM. In fact, in the proposed framework, holograms are directly connected to the CDE platform, which provide maintenance geometric and information data coming from the AIM in a "X-ray" vision immersive virtual overlay. The MR extension allows operators to perform a new maintenance work process visualizing the AIM on-site through immersive overlays, checking and updating the AIM parameters and improving maintenance efficiency.

MR is also a tool for communication and alerts displaying as well as for real-time collaboration and information visualization during the entire operation and maintenance process.

Remote support and hybrid training using MR. Azhar et al. (2018) introduced VR for design communication teaching and observed that students felt unstable and needed supervision from trainers. While AR introduces digital objects overlaying on the real world, MR combines the real and virtual scenarios.

The main difference between AR and MR is compared to a sliding scale by Lehman and Tan (2021), where AR integrates digital objects in the physical world and MR allows virtual objects' interaction in the real world. In fact, MR users are conscious of the real environment while involved in virtual actions. The use of MR devices improves students' long-term learning as mentioned by Azhar et al. who observed the effectiveness of MR in enhancing communication capabilities in AECO education, revealing the potential of MR for practical and technical trainings.

In such context, AR/VR/MR tools are reliable solutions to train maintenance operators as (i) literature shows that the initial investment is totally repaid by training costs reductions; (ii) immersive technologies allow to visualize and interact with simulated items, according to practical learning approaches which increase trainees'

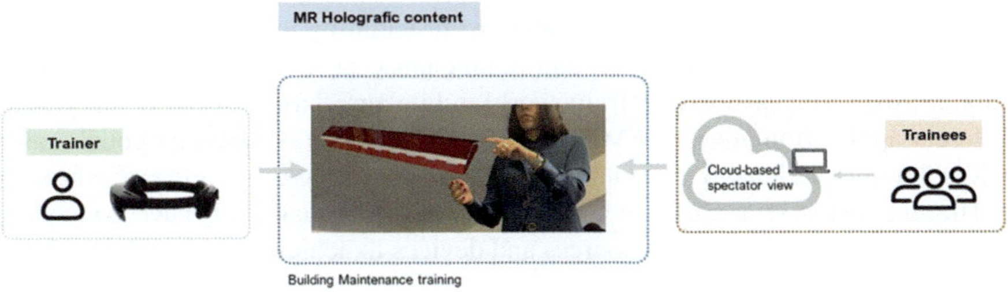

Fig. 13.3 Shared holographic content

motivation and skills acquisition; (iii) MR systems allow to collect a variety of performance data in order to check and improve the training process (Borsci et al. 2015).

The objective of the proposed solution is to enable shared training experiences between physical and digital.

For the above mentioned, the proposed framework (Fig. 13.3) involves MR approach extended to training purposes (Ogunseiju et al. 2021) aiming to provide remote assistance from off-site expert to on-site operators sharing the on-site's experience within the same holographic view.

On-site technical staff are able to connect to off-site building managers through video calls, replacing unnecessary physical inspections through the use of shared remote immersive experiences.

In this regard, an important example of cloud-based solutions to realize training metaverse has been proposed in Italy in 2022 by Microsoft Italia and Hevolus (2022), allowing the University of Naples Federico II to experiment the first didactic solution for participatory and laboratory teaching in MR.

With the same perspective, the use of MR-Hybrid Learning (MR-HL) environments can be introduced in maintenance service training with the aim of giving instructions to maintenance operators (Sepasgozar 2020) using MR environments.

In fact, head-mounted devices (HMDs) allow trainers to act and experiment with free hands (Agostinelli et al. 2021), showing maintenance practical activities using holograms. Trainees are involved in MR-HL environments being able to visualize holograms without wearing any device and by only accessing MR data streaming through a specific cloud-based spectator view platform.

13.3 Conclusions and Future Research

MR is a challenging digital technology for the FM industry, where digitalization is still relatively behind if compared to other sectors. The adoption of the proposed framework involves the introduction of digital collaborative solutions with the potential to increase efficiency in the FM phase.

The present study aims at contributing to research by investigating the concept of collaborative MR in operation and maintenance tasks as well as for workers' training, exploring a possible framework architecture based on BIM an MR for different application areas of FM. The proposed solution is based on collaborative MR technology as it has a relevant potential in operative tasks as well as in education and training in FM sector, improving efficiency as workers currently have to manually get information from different sources and devices to achieve their tasks, leading to a large number of possible errors.

Moreover, despite the use of HMDs is currently restricted and uncovered in the construction sector (Khan et al. 2021), MR could provide access to a digital centralized information system obtaining substantial beneficial in work tasks' optimization of FM.

In this regard, the proposed system architecture is aimed at finding solutions to potentially address the gap, acting as a starting point for future in-depth developments.

Moreover, structured and organized BIM data combined with MR systems provide improved understanding and are a valuable solution to resolve issues and inefficiency related to traditional design artifacts.

As technology progresses, immersive, realistic and customized training can improve education in many other fields of AECO industry such as construction site safety. MR enables workers identifying on-site hazards and gets improved context experience-based awareness. In this regard, future developments of the present research could be related to the extended use of MR to construction site safety training as a tool for enhancing hazard detection, avoidance, response and communication in order to reduce injuries and deaths in construction sites.

Author Contributions Conceptualization, S.A. and B.N.; methodology, S.A.; formal analysis, S.A.; investigation, S.A.; resources, S.A.; writing—original draft preparation, S.A. and B.N.; writing—review and editing, S.A.; visualization, S.A. All authors have read and agreed to the published version of the manuscript.

References

Agostinelli S, Majidi Nezhad M, Cumo F, Heydari A, Muzi F (2021) Predictive maintenance strategy based on big data analysis and machine learning approach for advanced building management system. In: Proceedings of SEEP 2021 international conference, Vienna, Austria

Ahmed V, Tezel A, Aziz Z, Sibley M (2017) The future of big data in facilities management: opportunities and challenges. Facilities 35(13/14):725–745. https://doi.org/10.1108/F-06-2016-0064

Atta N, Talamo C (2020) Digital transformation in facility management (FM). IoT and big data for service innovation. In: Daniotti B, Gianinetto M, Della Torre S (eds) Digital transformation of the design, construction and management processes of the built environment. Research for development. Springer, Cham. https://doi.org/10.1007/978-3-030-33570-0_24

Azhar S, Kim J, Salman A (2018) Implementing virtual reality and mixed reality technologies in construction education: students' perceptions and lessons learned. In: 11th annual international conference of education, 2018
Bae H, Golparvar-Fard M, White J (2013) High precision vision-based mobile augmented reality system for context-aware architectural, engineering, construction and facility management (AEC/FM) applications. Vis Eng 1(1):1–13
Becerik-Gerber B, Jazizadeh F, Li N, Calis G (2011) Application areas and data requirements for BIM-enabled facilities management. J Constr Eng Manag 138(3):431–442
Borsci S, Lawson G, Broome S (2015) Empirical evidence, evaluation criteria and challenges for the effectiveness of virtual and mixed reality tools for training operators of car service maintenance. Comput Ind. https://doi.org/10.1016/j.compind.2014.12.002
Chu M, Matthews J, Love PED (2018) Integrating mobile building information modelling and augmented reality systems: an experimental study. Autom Constr 85:1–26
Chung S, Cho C-S, Song J, Lee K, Lee S, Kwon S (2021) Smart facility management system based on open BIM and augmented reality technology. Appl Sci 11:10283. https://doi.org/10.3390/app112110283
East B, Carrasquillo-Mangual M (2012) The COBie guide. Available online: https://nibs.org. Accessed 1 Nov 2021
Gheisari M, Irizarry J (2016) Investigating human and technological requirements for successful implementation of a BIM-based mobile augmented reality environment in facility management practices. Facilities 34(1/2):69–84
Grieves M, Vickers J (2017) Digital twin: mitigating unpredictable, undesirable emergent behavior in complex systems. In: Transdisciplinary perspectives on complex systems. Springer International Publishing, Cham, pp 85–113
Hachman M (2015) NASA's taking Microsoft's HoloLens to the space station. PCWorld. Retrieved from https://www.pcworld.com/article/2940460/nasastaking-microsofts-hololens-to-the-space-station.html
Hevolus e Microsoft Italia annunciano Hybrid Learning Spaces: la realtà mista per una didattica sempre più ibrida. Available online: https://news.microsoft.com/. Accessed 10 Mar 2022
Irizarry J, Gheisari M, Williams G, Walker BN (2013) InfoSPOT: a mobile augmented reality method for accessing building information through a situation awareness approach. Autom Constr 33:11–23
Ke F, Lee S, Xu X (2016) Teaching training in a mixed-reality integrated learning environment. Comput Hum Behav. https://doi.org/10.1016/j.chb.2016.03.094
Khan A, Sepasgozar S, Liu T, Yu R (2021) Integration of BIM and immersive technologies for AEC: a scientometric-SWOT analysis and critical content review. Buildings 11:126. https://doi.org/10.3390/buildings11030126
Lehman SM, Tan CC (2021) Leveraging edge computing for mobile augmented reality. In: Fog/edge computing for security, privacy, and applications. Springer
Liu Y (2014) Virtual neurosurgical education for image-guided deep brain stimulation neurosurgery. In: International conference on audio, language and image processing, 2014. IEEE, pp 623–626
Maffei A, Onori M (2019) Evaluation of the potential impact of fully-immersive virtual reality on production engineering curricula. In: 2019 IEEE AFRICON, 2019. IEEE, pp 1–5
Manfren M, Aste N, Leonforte F, Del Pero C, Buzzetti M, Adhikari RS, Zhixing L (2020) Parametric energy performance analysis and monitoring of buildings—HEART project platform case study. Sustain Cities Soc 61. https://doi.org/10.1016/j.scs.2020.102296
Mekni M, Lemieux A (2014) Augmented reality: applications, challenges and future trends. Appl Comput Sci 205–214
Messner JI, Yerrapathruni SC, Baratta AJ, Whisker VE (2003) Using virtual reality to improve construction engineering education. In: American society for engineering education annual conference & exposition, 2003
Ogunseiju O, Akanmu A, Bairaktarova D (2021) Mixed reality based environment for learning sensing technology applications in construction. ITcon 26(Special issue Construction 4.0:

Established and Emerging Digital Technologies within the Construction Industry (ConVR 2020)):863–885. https://doi.org/10.36680/j.itcon.2021.046
Preidel C, Borrmann A, Mattern H, König M, Schapke SE (2018) Common data environment. In: Borrmann A, König M, Koch C, Beetz J (eds) Building information modeling. Springer, Cham. https://doi.org/10.1007/978-3-319-92862-3_15
Rankohi S, Waugh L (2013) Review and analysis of augmented reality literature for construction industry. Vis Eng 1(9):1–18
Sabol L (2008) Building information modelling & facility management, the power of process in the built environment. IFMA world workplace. www.dcstrategies.net. https://pdfs.semanticscholar.org/cb40/288c07f9351120447f872a2e815f5f1a8db5.pdf
Sacco A (2016) Google Glass takes flight at Boeing. CIO. Retrieved from https://www.cio.com/article/3095132/wearable-technology/google-glass-takes-flight-atboeing
Sacks B, Eastman C, Lee G, Teicholz P (2018) BIM handbook: a guide to building information modeling for owners, designers, engineers, contractors, and facility managers, 3rd edn. Wiley
Sepasgozar SME (2020) Digital twin and web-based virtual gaming technologies for online education: a case of construction management and engineering. Appl Sci 10(13):4678. https://doi.org/10.3390/app10134678
Shirazi A, Behzadan AH (2015) Design and assessment of a mobile augmented reality-based information delivery tool for construction and civil engineering curriculum. J Prof Issues Eng Educ Pract 141(3):04014012
Talamo C, Pinto MR, Viola S, Atta N (2019) Smart cities and enabling technologies: influences on urban facility management services. IOP Conf Ser Earth Environ Sci 296:012047 https://doi.org/10.1088/1755-1315/296/1/012047
Wong JKW, Ge J, He SX (2018) Digitisation in facilities management: a literature review and future research directions. Autom Constr 92:312–326. https://doi.org/10.1016/j.autcon.2018.04.006
Zakiyudin MZ, Fathi MS, Rambat S, Tobi M, Uzairiah S, Kasim N, Latiffi AA (2013) The potential of context-aware computing for building maintenance management systems. Appl Mech Mater 405:3505–3508

Chapter 14
A Digital Information Model for Coastal Maintenance and Waterfront Recovery

Francesca Ciampa

Abstract In the context of the global climate crisis and the resulting catastrophic flooding phenomena, the contribution looks at an innovative digital model for the coastal recovery, attentive to the protection of waterfronts and their stakeholders. By intervening in the relationship between transformation and conservation of built environment, it is necessary to establish governance support tools capable of foreseeing emergency scenarios to protect the population. The research looks at the port areas of coastal cities as a contemporary and collective public space in which to test the collaborative digital model proposed for waterfronts recovery and maintenance. The need-based methodological process used the human life protection, exposed to flooding danger, as the input of a design process. Through a survey and modeling phase, the waterfront breaks down into environmental and technological systems, specifying the extent of the failure. The waterfront digitization allows providing the governance with a sensor alert tool that gives the monitoring of the behavior and the state of the waterfront elements' degradation. This information is simplified and given back to the users who both made responsible for the maintenance culture of the places they use and alerted to the possible danger they are exposed. The case is Atrani, where an internal flooding, caused by the estuary overflowing, degenerated in the entire coastal system up to the sea. The results provide a digital model capable of exploring and optimizing the coastal built environment to increase the governance capacity and the waterfront performance.

Keywords Coastal digital regeneration · Maintenance · Port areas · Waterfront recovery · Flooding

F. Ciampa (✉)
University of Naples Federico II, Naples, Italy
e-mail: francesca.ciampa@unina.it

E. Arbizzani et al. (eds.), *Technological Imagination in the Green and Digital Transition*, The Urban Book Series, https://doi.org/10.1007/978-3-031-29515-7_14

14.1 Introduction

The climate crisis and the consequent catastrophic flooding phenomena drive research to identify an innovative digital model for the recovery of the coastal built environment, attentive to the protection of waterfronts (San-Miguel-Ayanz et al. 2017; Lyddon et al. 2020) and their stakeholders (Costello et al. 2009; Deepak et al. 2020). This need stems from the consideration that 39.3% of the European population resides along the coastal area (Ciampa et al. 2021a), of which 1.4 million people are affected due to flooding phenomena (Raška et al. 2022), suffering approximately 1 billion damages per year (Grillakis et al. 2016; Amponsah et al. 2018; Mannis 2020). Linked to the need is the urgency with which to intervene, as people's exposure to flood events will increase over time (Mediero et al. 2014; Rizzo et al. 2020). The latter is exacerbated by the change in global warming—from level 2.5 to 5.6 over the period 1908–2080 (European Agenda 2020; Rogger et al. 2017)—and the consequent depletion of 0.8% of Europe's gross domestic product (European Commission 2009; Parisi 2021) linked to the vulnerability of coastal systems. These projections push research to intervene in the relationship between use and maintenance of the built environment, trying to establish governance support digital tools capable of foreseeing emergency scenarios to protect the population. The latter is no longer only the receiver of the data but also its co-producer. The community involved as a user of the coast contributes in the digital intervention systems through collaborative information flows (Durugbo et al. 2011; Giurgiu 2021). The user can become a support to digital coastal monitoring and alerting tools through the construction of interaction and interchange networks that offer a new methodology of information flow analysis (Boccaletti et al. 2006).

The latter aims to improve the organizational requirements related to the ability to recognize a waterfront failure, the efficiency to report it, the adaptability to weather monitoring tools and the flexibility to respond promptly to the flood event that might occur (Schultz-Jones 2009; UNRISD 2021). This allows for the possibility of proposing digital models that use information flows from the network of users who, while representing lesser knowledge, can identify themselves as an organized community of "explorers" of their coastal system (White 2008; Biden 2021). The involvement of users becomes a significant requirement within the phases of information flow collection as it improves community collaboration on the one hand and empowers the community for widespread monitoring on the other. In the digital information model, the aggregating factor influences the organizational structure of data processing resulting in more efficient decision-making for expert knowledge (Durugbo et al. 2009; Ciampa et al. 2021b).

14.2 Stakeholders Role into a Digital Information Model

Human participation within digital information models is based on the sense of "responsibility" and "empowerment" of users toward the built environment through the same tool: in the first case, it refers to the individual willingness to perform actions of indirect care of the used space through signaling actions that they can actively perform. In the second case, reference is made to the educational training that induces the user to move away and get to safety when they are warned in case of imminent natural danger of the space used (Mohseni et al. 2013; Sun and Ye 2021). This mechanism is based on the prosumer perspective (Izvercianu et al. 2014), which marks the evolution from user to prosumer. Whereby, the citizen is no longer the one who simply uses the places but becomes a producer of the quality of the space contributing with own co-creative actions as a guarantor of vigilance (Boeri et al. 2016; Xin et al. 2022). This perspective determines a multilevel knowledge system at different scales that combines expert knowledge with local knowledge in a multi-sectoral view. The model proposed generates both digital and connective infrastructure (Fusco Girard 2014; Malekloo et al. 2021) because of it involves stakeholders attributing them a monitoring role in place they are located (Viola, 2016; Perez et al. 2019). This allows the stakeholders to point out, in the observation area indicated to them, certain vulnerable units as a field in which to identify and recognize elements to be reported (Caterina 2013; Na et al. 2021). This mechanism allows the community, a holder of common knowledge, to have an active monitoring role with innovative digital tools (Kong et al. 2018). They can report through a mobile application managed by the public administration, which records failures and abnormal performance found by users. In this way, municipality can to intervene in real time through the expert knowledge service offered by competent technicians (Perez and Tah 2021). The feedback that stakeholders identify comes back to them as a capacity to enhance the resilience of the environment and their increasing education about danger (German et al. 2012; Luckey et al. 2021). The stakeholder identifies signs of inefficiency in the space they use by contributing to the reliability of natural hazard warning data (Huang et al. 2016; Hou et al. 2021). The interpretation of the mismatch between the state of the space in efficient performance and that in degraded state serves to intervene on the functional loss. It lets public administration to mitigate the impacts that such degradation would have on the community and the settlement system. In this way, the digital information becomes the built environment vulnerability data linked to the performance decay of the technical element or component affected by the failure (UNI 99100/1993).

14.3 Methods and Materials

The methodology is based on the exigency performance approach (Pinto and Talamo 2015), i.e., meeting primary needs—the protection of human life related to vulnerable and flood-prone users—as input to a transferable digital model (Bosone and Ciampa

2021). This tool experiments and enhances the participation of communities in the maintenance and alerting processes of the port areas used, indirectly contributing to the protection of their lives and the performance efficiency of their built environment in case of catastrophic events (Ciervo et al. 2012). Through a graphical survey and modeling phase, it is possible to break down the waterfront into environmental and technological systems, specifying the extent of disruption. The digitization of the waterfront allows providing governance with a sensor alert tool that provides monitoring of the behavior and degradation status of the waterfront elements.

This information is simplified and feedback to users who are empowered about the culture of place knowledge they use and alerted to the possible danger they are exposed to. The experimental case is the waterfront of Atrani (Fig. 14.1), on the Amalfi Coast in Italy, where a case of flash flood of the Dragone river in 2010 invaded areas from the inner settlement system down to the sea.

The flooding phenomena were linked to both natural causes (low pressure and convective thunderstorm systems resulting in a cumulative daily rainfall of 129.2 mm–19.4 mm/h) and anthropogenic causes (the artificial diversion of the river below the urban area through an underground channel to the sea). The channeled flow generated an overpressure that progressively led to the breaking of the channel slab cover, modifying the flow on the road downstream. In the absence of hydrometric and environmental data, the acquisition of information about the process took place only through the documented testimony of the local population who, with the support of amateur videos, collaborated in the development of identifying the loss of performance of the built environment elements (Fig. 14.2).

Fig. 14.1 Atrani waterfront, 2022. *Photo* Francesca Ciampa

Fig. 14.2 Atrani flooded port area, 2022. *Photo* Francesca Ciampa

The methodology offers a scientific advancement in digital maintenance tools because it identifies in the involvement of coastal built environment users the possibility to enhance the information flows by the collaborative tool proposed. The user involvement indicates in the processes of monitoring the loss of coastal system functionality while preventing the human and urban safeguard. The advancement is to induce community empowerment toward the waterfront by employing indirect monitoring and care actions for the built environment. This methodology improves and simplifies the acquisition of digital data by returning a model of shared information capable of interacting with users in both direct and feedback forms. The methodology uses the privileged user point of view to optimize data collection methods with respect to coastal technical elements monitored by inspection over time. The methodology merges the decision makers view with stakeholders one in order to optimize the data collection, both monitoring the inspection elements over time. The methodology innovates the prefiguration of new models of management, evacuation and sustainable maintenance of the coastal environment both through alerting technologies and through the assessment of the efficiency parameters and waterfront livability quality. This allows the cooperation of different figures, respecting the specific and/or general knowledge and competences, related to each other for the effectiveness and the safe control of the usability of the coastal area.

14.4 Results and Discussion

The results provide a digital model capable of exploring and optimizing the coastal built environment in order to increase the governance capacity and overall performance of the waterfront. By discrediting the coastal built environment through classes of environmental and technological units, it has been possible to simplify the reading of the complex port area using analogy by image. This brought the community closer to knowing their built environment despite not having technical knowledge. In order to allow stakeholders to identify the classes of technical elements, it was necessary to guide the user to the signaling, directing and simplifying of the input data, and thus, giving the possibility of categorization through a univocal association. To this end, the image that the user could provide needed textual accompaniment to be selected based on pre-set definitions agreed with the expert knowledge. Acting on the individual performance dysfunction of the coastal margin or its urban surroundings allows for a flexible information organization construction. An artificial intelligence algorithm that analyzes the photograph and labels it based on data training establishes the association between the picture taken and the list of accompanying technical commentary. The latter acts by evaluating both whole image classification and object detection. The signaling of a fault, through digital application tools of the public administration (decision-makers), can affect the functional loss of relations with other potentially damaging parts. This decomposition therefore intervenes on an interscalar logic that allows monitoring, in an indirect way, including the intermediate levels through the mitigation of the damaged relations. The complex system of the port area identifies the waterfront as an observation site whose sensitive unit is represented precisely by the subset of technical elements that influence the classes and the close relationships with the built environment. The class of vertical elements refers, for example, to the waterfront embankments; the class of horizontal elements refers to the surface of the pier; the class of installations refers to the technological solutions for mitigating and raising the water level. The class of furnishings refers to those coastal soil modeling solutions that trap excess waves in emergencies, and instead, in normal conditions, act as recreational equipment or street furniture. The alert system provided by the users (stakeholders) flows into the support of alert situations, which are then evaluated by the technical operator in order to define the need and priority of possible inspection methods. These professionals return the data to the alarm operator who communicates to the public administration (decision-makers) both the possibility of intervening to realign a loss of performance and to evacuate the area in case of the simultaneous occurrence of a flooding event (unmanageable due to the absence of functional safety). The contribution identifies the involvement of different actors, with specific reference to stakeholders, in the digital monitoring tools which include four benefits of coastal performance efficiency: adaptation, perception, coordination and reciprocity. Proposed model has the adaption capacity to merge the tangible flexibility, linked to the physical transformations of coastal environment; and the intangible one, linked to the community use modalities of the waterfront. Adaptation expresses the need to use digital tools that involve and, at the same time, empower stakeholders to their built heritage. The model perception aspect concerns the intuitive ability to monitor the built environment, transforming users knowledge into a tool to support technical knowledge. Perception expresses the possibility of binding the built environment to an image/state of the built environment that monitors its performance whenever it is out of line with perception. The model concerns a reciprocity aspect based on the simultaneous ability to protect both people by protecting the built environment of the coastal system they use. Coordination as the ability to translate qualitative data into technical and scientific data. Coordination as collective sharing for the pursuit of the common good. Reciprocity expresses the ability to return performance so that the connections and balances between the single functional loss and the system with the realignment which it communicates. The model proposed represents a collaborative digital tool for coastal recovery and maintenance. (Fig. 14.3).

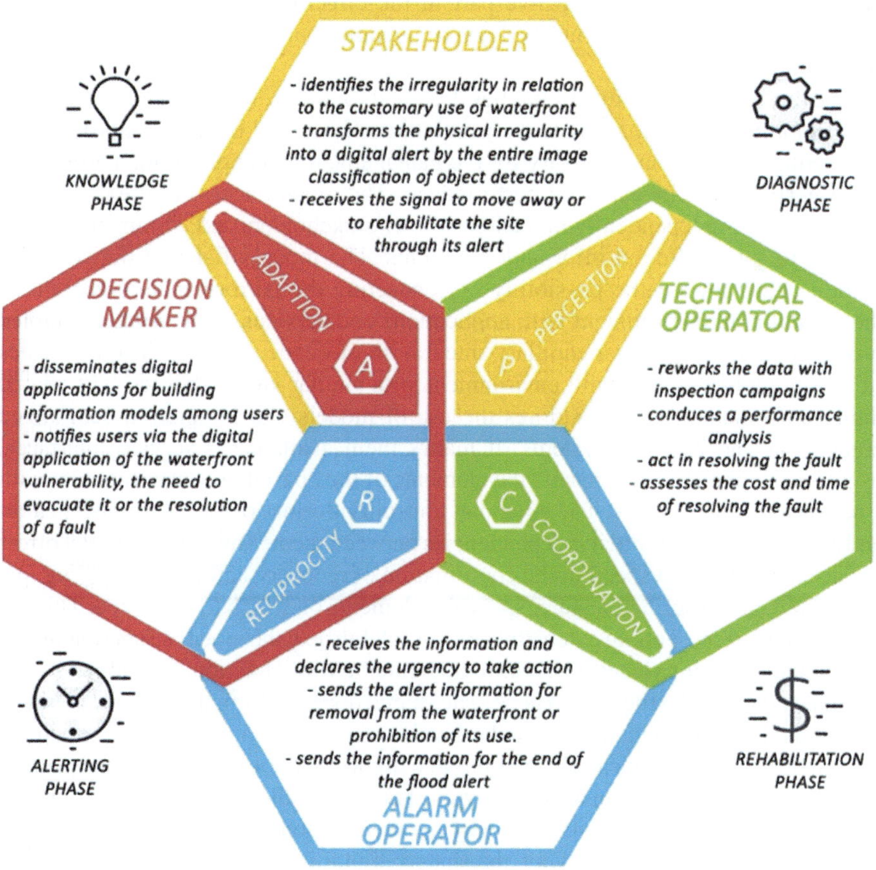

Fig. 14.3 Shared digital information model

14.5 Conclusions

The digital coordination of waterfront recovery and coastal maintenance actions finds application in urban activities assisted by the contemporary technological context. The diffusion of digital systems and devices makes it possible to improve the performance of the built environment and to monitor the performance of activities in aggregated multi-actor forms. The maintenance of settlement quality, while respecting the safety, usability and well-being of its users, is also based on the ability to recognize, in a shared way, the loss of functionality of the coast constituent elements. Waterfronts, which have accommodated settlement processes for generations by adapting to community spaces, find in digital information models tools to mitigate and prepare for the consequences of climate change and to respond effectively to community needs.

The proposed model opens the research toward a type of service in which the stakeholder becomes an active part in the effectiveness of the digital action through interfaces with a simplified communication language in a network of user communities.

The model represents functionalities designed for the protection of human life and active monitoring for coastal recovery activities. The model offers the possibility of integrating information necessary for the management and appropriate operation of the waterfront and its expected performance. The model, accessible remotely and interactable or independently updatable, allows stakeholders and administrations to interact to form different information systems. The model of shared digital information opens toward the possibility of connecting the performance alignment of the coastal built environment with applications and services, generating a complex process characterized by a capillary integration of connectivity at different scales of action, replicable in a built environment with similar vulnerabilities. The shared digital information model explores the use of models that integrate a centralized component and backend systems. It makes them available as service interfaces, using a system that interconnects the elements of the coastal built environment with a network of stakeholders. The latter can transmit information or coordinate an alert and retrieval process reducing dependencies between them. This type of model offers greater flexibility and speed in the coordination of recovery actions, identifying in a timely and circumscribed manner space the elements on which to intervene. This, on the one hand, represents an extensibility and scalability of the model's functionalities, and on the other, it reduces costs through greater agility of action, developed by the efficiency dictated by the shared need. The shared digital information model could be further advanced by generating standardized communication protocols to which the community could be educated, resulting in a higher level of interaction, and scaling back inconveniences or transmission errors. This would allow the model to offer ever greater reliability both as a stakeholders involved responsibility and in terms of the process effectiveness. In this way, the shared digital information model acts in the built environment, realigning performance with new needs, receiving stakeholder requests by transferring them as intervention requirements and inducing appropriate response transformations.

References

Amponsah W, Ayral PA, Boudevillain B, Bouvier C, Braud I, Brunet P, Delrieu G, Didon-Lescot JF, Gaume E, Lebouc L (2018) Integrated high-resolution dataset of high-intensity European and Mediterranean flash floods. Earth Syst Sci Data 10:1783–1794

Biden J (2021) RESILIENCE 21 building a nation of resilient communities. Washington

Boccaletti S, Latora V, Moreno Y, Chavez M, Hwang DU (2006) Complex networks: structure and dynamics. Phys Rep 424:175–308

Boeri A,Gaspari J, Gianfrate V, Longo D, Pussetti C (2016) The adaptive reuse of historic city centres. Bologna and Lisbon: solutions for urban regeneration. Techne 12:230–237

Bosone M, Ciampa F (2021) Human-centred indicators (HCI) to regenerate vulnerable settlement systems as circular city: from the Bronx (NY) to Ercolano (IT). Sustainability 13:5505. https://doi.org/10.3390/su13105505

Caterina G (2013) Conservazione, manutenzione e gestione degli spazi pubblici e dei beni architettonici. In: Castagneto F, Fiore V (eds) Recupero Valorizzazione Manutenzione nei Centri Storici. Un tavolo di confronto interdisciplinare. Lettera 22, Siracusa, Italia, pp 14–17

Ciampa F, Seifollahi-Aghmiuni S, Kalantari Z, Ferreira CSS (2021a) Flood mitigation in Mediterranean coastal regions: problems, solutions, and stakeholder involvement. Sustainability 13:10474. ISSN 2071-1050. https://doi.org/10.3390/su131810474

Ciampa F, De Medici S, Viola S, Pinto MR (2021b) Regeneration criteria for adaptive reuse of the waterfront ecosystem: learning from the US case study to improve European approach. Sustainability 13:4156. https://doi.org/10.3390/su13084156

Ciervo F, Papa MN, Medina V, Bateman A (2012) Reconstruction and numerical modelling of a flash flood event. In: Atrani 2010, EGU general assembly 2012, Vienna, Austria, 22–27 Apr 2012

Costello A, Abbas M, Allen A, Ball S, Bell S, Bellamy R, Friel S, Groce N, Johnson A, Kett M (2009) Managing the health effects of climate change. Lancet 373:1693–1733

Deepak S, Rajan G, Jairaj PG (2020) Geospatial approach for assessment of vulnerability to flood in local self governments. Geoenvironmental Disasters 7:41–61

Durugbo C, Tiwari A, Alcock JR (2009) An infodynamic engine approach to improving the efficiency of information flow in a product-service system. In: CIRP IPS2 conference, pp 107–112

Durugbo C, Tiwari A, Alcock JR (2011) Modelling information flow for collaboration. In: Camarinha-Matos LM (ed) Technological innovation for sustainability. DoCEIS 2011. IFIP advances in information and communication technology. Berlin, Heidelberg. https://doi.org/10.1007/978-3-642-19170-1_1

European Agenda N.U. Habitat III New Urban Agenda (2020) Quito declaration on sustainable cities and human settlements for all. In: Habitat III conference, Bruxelles

European Commission (2009) Climate change impacts in Europe. Final report of the PESTA research project, Huan Carlos Ciscar

Fusco Girard L (2014) The role of cultural urban landscape towards a new urban economics: new structural assets for increasing economic productivity through hybrid processes. Hous Policies Urban Econ 1:3–27

German S, Brilakis I, DesRoches R (2012) Rapid entropy-based detection and properties measurement of concrete spalling with machine vision for post-earthquake safety assessments. Adv Eng Inform 26:846–858

Giurgiu IO (2021) Flood pulse design principles—a time-based approach to urban flooding. In: SeaCities. Taylor & Francis, London

Grillakis M, Koutroulis A, Komma J, Tsanis I, Wagner W, Blöschl G (2016) Initial soil moisture effects on flash flood generation—a comparison between basins. J Hydrol 541:206–217

Hou L, Chen H, Zhang (Kevin) G et al (2021) Deep learning-based applications for safety management in the AEC industry: a review. Appl Sci 11:821

Huang G, Liu Z, Weinberger KQ (2016) Densely connected convolutional networks. CoRR, vol 620, pp 1608–6993
Izvercianu M, Şeran SA, Branea AM (2014) Prosumer-oriented value co-creation strategies for tomorrow's urban management. Procedia Soc Behav Sci 182:149–156
Kong Q, Allen RM, Kohler MD, Heaton TH, Bunn J (2018) Structural health monitoring of buildings using smartphone sensors. Seismol Res Lett 89:594–602
Luckey D, Fritz H, Legatiuk D et al (2021) Artificial intelligence techniques for smart city applications. In: Toledo Santos E, Scheer S (eds) Proceedings of the 18th international conference on computing in civil and building engineering. Springer International Publishing, São Paulo, Brazil, pp 3–15
Lyddon C, Brown JM, Leonardi N, Plater AJ (2020) Sensitivity of flood hazard and damage to modelling approaches. J Mar Sci Eng 8:724
Malekloo A, Ozer E, AlHamaydeh M, Girolami M (2021) Machine learning and structural health monitoring overview with emerging technology and high-dimensional data source highlights. Struct Health Monit. https://doi.org/10.1177/14759217211036880
Mannis A (2020) Indicators of sustainable development, environmental software and services. University of Ulster
Mediero L, Santillán D, Garrote L, Granados A (2014) Detection and attribution of trends in magnitude, frequency and timing of floods in Spain. J Hydrol 517:1072–1088
Mohseni H, Setunge S, Zhang GM, Wakefield R (2013) Condition monitoring and condition aggregation for optimised decision making in management of buildings. Appl Mech Mater Trans Tech 552:1719–1725
Na Y, El-Tawil S, Ibrahim A et al (2021) Stick-slip classification based on machine learning techniques for building damage assessment. J Earthq Eng 1–18
Parisi D (2021) Holistic approach to urban regeneration. In: Parisi D (ed) New metropolitan perspectives. Springer, London
Perez H, Tah JHM (2021) Deep learning smartphone application for real-time detection of defects in buildings. Struct Control Health Monit 28:2751
Perez H, Tah JHM, Mosavi A (2019) Deep learning for detecting building defects using convolutional neural networks. Sensors 19:3556
Pinto MR, Talamo C (2015) Recupero e manutenzione: la ricerca incontra le esigenze dei territori. In: Lucarelli MT, Mussinelli E, Trombetta C (eds) Cluster in progress. La tecnologia dell'architettura in rete per l'innovazione. Maggioli Editore, Santarcangelo di Romagna, Italia, pp 246–257
Raška P, Bezak N, Ferreira C, Kalantari Z, Banasik K, Bertola M, Bourke M, Cerdà A, Davids P, de Brito M, Evans R, Finger D, Halbac-Cotoara-Zamfir R, Housh M, Hysa A, Jakubínský J, Kapovic S, Kaufmann M, Keesstra S, Hartmann T (2022) Identifying barriers for nature-based solutions in flood risk management: an interdisciplinary overview using expert community approach. J Environ Manage 310:114725. https://doi.org/10.1016/j.jenvman.2022.114725
Rizzo A, Vandelli V, Buhagiar G, Micallef AS, Soldati M (2020) Coastal vulnerability assessment along the north eastern sector of Gozo Island (Malta, Mediterranean Sea). Water 12:1405
Rogger M, Agnoletti M, Alaoui A, Bathurst JC, Bodner G, Borga M, Chaplot V, Gallart F, Glatzel G, Hall J (2017) Land use change impacts on floods at the catchment scale: challenges and opportunities for future research. Water Resour Res 53:5209–5219
San-Miguel-Ayanz J, Chuvieco E, Handmer J, Moffat A, Montiel-Molina C, Sandahl L, Viegas D (2017) Climatological risk: wildfires. In: Poljanšek K, Marín Ferrer M, De Groeve T, Clark I (eds) Science for disaster risk management 2017: knowing better and losing less. Publications Office of the European Union, Rue Mercier, Luxembourg, pp 294–305
Schultz-Jones B (2009) Examining information behavior through social networks: an interdisciplinary review. J Doc 65:592–631
Sun H, Ye J (2021) Tapping potential of water drive reservoir in ultra high water cut stage based on computer image recognition and processing technology. In: Abawajy J, Xu Z, Atiquzzaman

M, Zhang X (eds) International conference on applications and techniques in cyber intelligence. ATCI 2021. Lecture notes on data engineering and communications technologies, vol 81. Springer, Cham. https://doi.org/10.1007/978-3-030-79197-1_70

UNRISD (2021) Sustainable development performance indicators. United Nations Research Institute for Social Development

Viola S (2016) Maintenance is sharing: adattività dei sistemi insediativi. In: Ascione P, Russo Ermolli S, Viola S (eds) Energia, innovazione tecnologica, processi manutentivi: il patrimonio storico tra istanze conservative e updates prestazionali. Clean, Napoli, Italia, pp 108–123

White L (2008) Connecting organizations: developing the idea of network learning in inter-organizational settings. Syst Res Behav Sci 25:701–716

Xin T, Weilin O, Wenbo H (2022) Intelligent image processing based on VPX architecture. In: Proceedings of SPIE 12169, eighth symposium on novel photoelectronic detection technology and applications, p 121699C. https://doi.org/10.1117/12.2625212

Chapter 15
Sustainable Workplace: Space Planning Model to Optimize Environmental Impact

Alice Paola Pomè, Chiara Tagliaro, and Andrea Ciaramella

Abstract The construction sector is one of the main sources of environmental degradation in the world. Data demonstrates that commercial assets are the most intensive consumers of resources. Among those, the largest amount of buildings' emissions comes from office building operations. Buildings' impact on the environment does not depend only on energy and material consumptions; but several studies demonstrate that sustainable savings could be achieved through occupants' trainings. To develop a model for assessing the sustainable performance of office buildings which accounts also for occupants' behavior, authors worked with the Real Estate Center of Politecnico di Milano and the Joint Research Center PropTech of Fondazione Politecnico di Milano. Through this cooperation, a tool is under development that:

I. Assesses the quantity of space needed by organizations, based on the employees' ways of working; and
II. Evaluates how much space occupancy and utilization may influence the sustainable performances of office buildings.

This paper describes the general functioning of the tool and looks at the contribution that PropTechs (Properties Technologies) can give to its implementation. Even if PropTechs are introducing digitalization in several real estate processes, few of them are focusing on the environmental. This study reviews the existing Italian PropTechs and selects those that could add value to the proposed tool. The analysis allows to define strengths and limits of the existing tools, helpful for implementing a new tool based on real needs of building managers. The tool aims to reduce the environmental impact of office buildings by suggesting more sustainable and user-oriented strategies.

A. P. Pomè (✉) · C. Tagliaro · A. Ciaramella
Polytechnic University of Milan, Milan, Italy
e-mail: alicepaola.pome@polimi.it

C. Tagliaro
e-mail: chiara.tagliaro@polimi.it

A. Ciaramella
e-mail: andrea.ciaramella@polimi.it

E. Arbizzani et al. (eds.), *Technological Imagination in the Green and Digital Transition*, The Urban Book Series, https://doi.org/10.1007/978-3-031-29515-7_15

Keywords Environmental sustainability · Workplace planning · PropTech

15.1 Introduction

The U.S. Energy Information Administration (2013) reported that office buildings are responsible for 20% of total commercial buildings' energy consumption. This is not just an American issue, but office buildings represent the largest consumer of energy in all countries (Lin et al. 2022). The great challenge of the new Millennium will be to integrate sustainable development in all sectors of the global market. Sustainable development is a balance between technologies, innovation strategies, and ecosystems (Vollenbroek 2002). However, a literature review on sustainability in the construction field (Limac et al. 2021) shows that most sustainable applications focus on optimizing materials and construction systems during the design and construction phases. This represents a limit of the existing literature, especially considering that the in-use stage is the most resource consuming (Menassa 2011). In addition, the studies that look at environmental impact of in-use buildings focus on the reduction of energy consumptions (Yeheyis et al. 2013). Improving energy efficiency has a positive influence, but energy is just one component of consumption.

In defining the "human impact", Wackernagel and Rees (1996) considered the number of people (i.e., population), the average amount of consumed resources (i.e., affluence), and the intensity of resources' production (i.e., technology). Therefore, to examine the interactions between users, nature, and the built environment, the concept of smart sustainable buildings emerges (Belani et al. 2014). Smart sustainable buildings are a combination of technology and materials that provides users with flexible, productive, interactive, integrated, and dynamic environment (Belani et al. 2014). Buckman et al. (2014) individuated adaptability as the major feature of smart sustainable buildings. Hence, to introduce sustainable strategies in office management, an approach that translates the needs of employees into space requirements is needed to make offices adaptable overtime (Thuvander et al. 2012).

The interest in improving the sustainability performance of office buildings increased due to the disruptive effect brought by COVID-19 pandemic. The pandemic has drastically changed the ways of working of employees, integrating more flexibility (Tagliaro and Migliore 2021).

This change is redefining the demand and configurations of offices (Seugbeom et al. 2021). Workplace managers have to adapt buildings to the new needs of employees by integrating technology. The digital transformation of the built environment needs to include the digitalization of the building management. In the facility management, digital technologies are brought by PropTech, abbreviation for Property Technologies. PropTech companies introduce digital solutions for improving the effectiveness of the processes (Baum et al. 2020).

The general aim of this study is to acknowledge the potential of technology to improve workplace management toward a more sustainable use of office buildings. The research presents a tool for workplace sustainable evaluation, which is composed of two modules. The Workplace Space Quantification focuses on assessing the effectiveness of space planning, while the Workplace-Integrated Ecological Footprint Assessment evaluates the ecological footprint of the in-use office building.

After a literature review on the already developed digital tools in the Italian real estate market, the tool is presented and discussed. Finally, the conclusion presents the limitations and future developments of the work.

15.2 State of the Art

The term PropTech indicates all technologies that are impacting the real estate market (Braesemann and Baum 2020). It increases operations' effectiveness and efficiency (Siniak and Kauko 2020). PropTech companies are usually startups and scaleups (but, also consolidated companies) that bring into the real estate innovation, digital development, and transparency (Baum et al. 2020). Authors collaborated with the Italian PropTech Network (IPN) of Politecnico di Milano, that mapped 184 companies (Bellintani et al. 2021).

PropTech taxonomy clusters companies according to the proposed solutions (Baum 2017).

The cluster relevant for the purpose of this research is Smart Real Estate, which describes digital platforms that facilitate the real estate assets' operations (Baum et al. 2020). These platforms may provide and aggregate information of buildings or facilitate the control of building services. According to IPN, Smart Real Estate focuses on companies that manage the built environment through high-tech platforms (Bellintani et al. 2021). Among the 184 Italian PropTech companies, 19% are recorded in the Smart Real Estate, which is divided into two sub-clusters, namely Immersive Visualization and Experience, and Smart Building and Operations. The former includes solutions that support promotion of properties, and the latter includes solutions that support operators and managers during the in-use phase.

Even if PropTech companies listed in the Smart Real Estate cluster aim to improve the in-use management of buildings, they look at different aspects. Some solutions help to reduce energy consumptions of buildings. For example, solutions are available, powered by Artificial Intelligence, to control autonomously the Heating, Ventilating, and Air Conditioning systems of a building (e.g., *Brainbox AI*). Real-time modifications allow to optimize in-use energy consumption. Other solutions focus on the indoor environmental quality (IEQ). Platforms acquire IEQ data from sensors, process the information, and produce reports of possible improvements (e.g., *Nuvap*). Building managers can evaluate the healthiness of the environment dynamically and continuously. Others concentrate on the maintenance of buildings by integrating management platforms in support of property data, maintenance activity, and energy consumption (e.g., *Facilio*). Others look at the level of occupancy of space by elaborating data, collected through sensors (e.g., *iComfort*). Managers may improve the space planning of the buildings by understanding users' preferences. Finally, other platforms control the building accessibility through check-in and check-out systems (e.g., *Sofia Locks*). These solutions propose a smart access control system which helps to manage through flexibility co-living, workspaces, healthcare facilities, and retail spaces.

Smart Real Estate PropTech companies try to optimize building management through several approaches. However, the adoption of sustainability in the built environment, and especially in office buildings, means going further the energy aspect or the analysis of combined data. A sustainable office building management means including social and economic implications (Jiménez-Pulido et al. 2020). This requires a holistic approach that involves into the process stakeholders (such as building users and employees) (Jiménez-Pulido et al. 2020). Moreover, none of the PropTech listed in the Smart Real Estate directly focuses on workplace management. Therefore, to overcome the sustainable management limit, the present research develops a digital tool to support workplace managers in integrating employees' needs into the space planning and to evaluate the effects of employees' occupation and behavior in the assessment of environmental sustainability.

15.3 Methodology

Authors collaborated with the Joint Research Center—PropTech between Fondazione Politecnico di Milano and for companies operating in Italy, namely Covivio, Vodafone, BNP Paribas Real Estate, and Accenture, to develop a new tool in support of a more sustainable workplace management. To meet the goal of the research, the following steps have been implemented by authors:

1. Literature review and benchmarking analysis to define the type and number of spaces in support of different ways of working;
2. Literature review on the sustainable indices able to assess the effects of users on buildings' environmental impact;
3. Definition of the WSP and WIEFA structure and calculations;
4. Development of the online platform with digitalization of the WSP calculations;
5. First experiments of calculations on 3 case studies;
6. 5 meetings with office buildings managers for testing the WIEFA;
7. 4 meetings with the partners of the project for test all the processes.

Finally, the tool is structured into two models, namely Workplace Space Quantification (WSP), and Workplace-Integrated Ecological Footprint Assessment (WIEFA). First, the tool assesses the space quantification through the analysis of the ways of working of employees. Second, the environmental impact of the office is assessed by using a specific sustainable index, the Ecological Footprint (EF).

Table 15.1 Workplace Space Quantification—space classification

Spaces	Capacity [n° users]	Average m^2
Not assigned open individual workstation	1	6
Assigned open individual workstation	1	6
Open team workstation	1	4
Workstation in closed shared office	1	7
Touchdown	6	24
Coffee point	10	35
Meeting point/waiting area	8	20
Small meeting space (open)	6	9
Large meeting space (open)	8	25
Study/phone booth	1	4
Brainstorming room	8	24
Small meeting room	3	9
Medium meeting room	8	28
Large meeting room	14	48
Client business lounge	1	24
Filing cabinet	1	0.5
Storage space	1	1
Archive	1	1
Locker area	1	0.5

Source Tagliaro (2021)

15.3.1 Workplace Space Quantification

The Workplace Space Quantification tool has been already implemented online.[1] WSP estimates the number and the m^2 of spaces needed by an organization according to the ways of working of its employees, which are gathered through a 11-question survey submitted to either the individual employees or group managers. WSP identifies nineteen different spaces, as reported in Table 15.1, for which benchmarked size (m^2) and capacity (number of people hosted in the space) have been defined through the literature.

The questions characterize the way of working based on the time spent on different activities (such as "In a working week, how long does the group work in the office?") and the number of people involved in different activities (such as "On

[1] Available online: https://www.braveworkplace.it/login.

average, how many people attend the meetings you hold for collaboration activities on concentration work?") performed in the office.

The tool allows to calculate the needed spaces for a single group of employees or add together more groups to estimate the general need of the entire organization. Alternative scenarios can be created and compared showing the overall number and m^2 of spaces needed by the organization, and the specific number and m^2 of spaces by each group for each space, as reported in Figs. 15.1 and 15.2.

Fig. 15.1 Workplace Space Quantification: the scenario of the workplace. Retrieved from https://www.braveworkplace.it/login

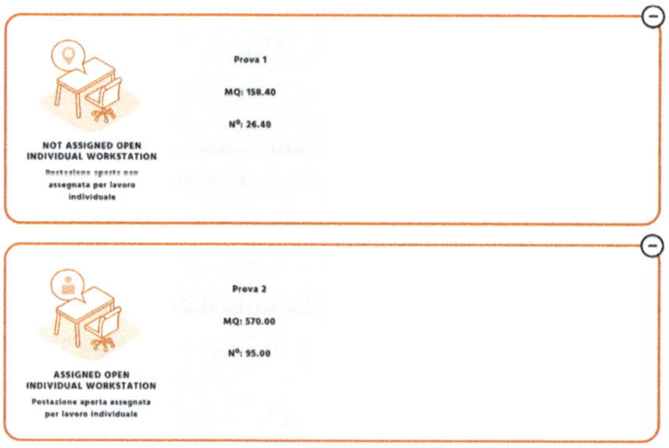

Fig. 15.2 Workplace Space Quantification: a focus on the two spaces. Retrieved from https://www.braveworkplace.it/login

WSQ is not only helpful in the design stage of offices, but it can be used to verify the appropriateness of the workplace in monitoring employees' ways of working during the in-use stage and inform potential spatial rearrangements to better meet the needs of employees. Based on the results of WSQ, office sustainability can be assessed through the WIEFA.

15.3.2 Workplace-Integrated Ecological Footprint Assessment

To evaluate environmental impact of office buildings, authors are reasoning on the Ecological Footprint (EF) index, which is a solution-oriented approach, capable to assess the (in-)efficiency of buildings' use. EF has been developed by Wackernagel and Rees (1996) to compare the demand with the supply of resources. The demand side is the population of a system (such as the building), while the supply side is the ecosystem in which the population lives (such as the Earth). EF converts consumptions and emissions in global hectares [gha]. Global hectares represent the land of Earth that can restore or absorb humans' consumptions or emissions. These lands are built-up land, forest land, fishing land, pastureland, cropland, and CO_2 sink factor.

WIEFA tried to overcome the limitations of previous studies (Acosta and Moore 2010; Gottlieb et al. 2012; Husain and Prakas 2018; Martínez-Rocamora et al. 2016; Solìs-Guzmàn et al. 2013) that tried to implement EF in the environmental impact assessment of buildings. Indeed, WIEFA puts together all the different impact sources defined by previous studies and evaluates the users' effect on environmental impact through a new impact source, Occupant. WIFE articulates 9 impact sources that show the consumption of the built environment, namely Built-up, Energy Consumption, Water Consumption, Material Consumption, Food and Drink, Mobility, Waste Generation, Recycle Potential, and Occupant. These are converted into global hectares [gha], through two conversion factors. World Yield Factor (WYF) translates impact sources in tons of CO_2 equivalence. While Equivalence Factor (EQF) converts CO_2 equivalence into gha. Both the factors are defined globally by the Global Footprint Network.[2] The 9 converted impact sources, defined addenda, are algebraically sum together, as shown in Fig. 15.3.

Built-up, Energy Consumption, Water Consumption, Material Consumption, Food and Drink, Mobility, and Waste Generation are summed together as they represent consumed resources and emitted pollutants. While Recycle Potential and Occupant are subtracted, they represent recreated benefits. Recycle Potential assesses the materials reuse in the building. For example, if the building produces electricity through a photovoltaic plant, the energy consumed over the year will be reduced. Occupant highlights the benefit of simultaneous building's occupation by multiple users.

[2] Available online: https://www.footprintnetwork.org/.

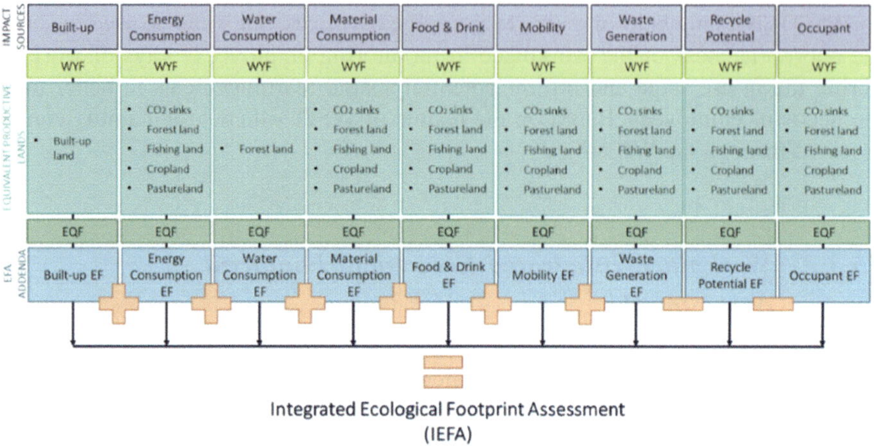

Fig. 15.3 Workplace-Integrated Ecological Footprint Assessment model—elaboration by authors

15.4 Conclusions

The proposed tool intends to support workplace managers in defining which is the effect of employees in the use of space and resources of an office. The integration of WSQ and WIEFA allows to develop a specific tool for office buildings, that evaluates environmental impacts of the in-use stage. Previous solutions focus only on energy consumption or look at the buildings' performance, without estimating the impact of users' occupation and behavior. However, the present tool can help workplace managers to adapt office buildings to changes in users' needs.

The tool still presents some limitations and room for improvement. First, the Workplace-Integrated Ecological Footprint model still needs to be refined and digitalized. Second, the science of sustainability does not look only at users and environment, but it aims to achieve an economic sustainability, which would be important to add to the current model. An additional section of the model will need to be implemented to look at the operational costs associated with the use of offices. This third addition will add the economic sustainability to the tool and will help workplace managers to understand not only the environmental effects of users, space utilizations, and resources' consumption, but also economic effects.

The model would need to be tested through a case study to assess its reliability and effectiveness. In order to offer seamless functionality and assure precision of the data, ideally the tool should automatically retrieve the information from sensors and periodically prompted questionnaires to building users, which would encompass integration with a number of PropTech solutions.

References

Acosta K, Moore J (2010) Creating an ecological footprint assessment: using component and compound economic input output methods together with the natural step to develop sustainability management system. In: Proceedings of the state of the art in ecological footprint theory and applications, British Columbia Institute of Technology

Baum A (2017) PropTech 3.0: the future of real estate. University of Oxford, Saïd Business School, Oxford

Baum A, Saull A, Braesemann F (2020) PropTech2020: the future of real estate. University of Oxford, Saïd Business School, Oxford

Belani D, Makwana AH, Pitroda J, Vyas CM (2014) Intelligent building new era of today's world. In: Proceedings of the trends and challenges of civil engineering in today's transforming world, Surat, India, pp 1–16

Bellintani S, Ciaramella A, Delmenico A, Leoncini SF, Pomè AP, Silvestri R, Tagliaro C (2021) Italian PropTech monitor. Politecnico di Milano, Milano

Braesemann F, Baum A (2020) PropTech: turning real estate into a data-driven market? Oxford Future of Real Estate Initiative

Buckman AH, Mayfield MBM, Beck S (2014) What is a smart building? Smart Sustain Built Environ 3(2):92–109

Gottlieb D, Kissinger M, Vigoda-Gadot E, Haim A (2012) Analyzing the ecological footprint at the institutional scale—the case of an Israeli high-school. Ecol Indic 18:91–97

Husain D, Prakas R (2018) Life cycle ecological footprint assessment of an academic building. J Inst Eng 100:97–110

Jiménez-Pulido C, Jiménez-Rivero A, García-Navarro J (2020) Improved sustainability certification system to respond to building renovation challenges based on a literature review. J Build Eng 45

Limac L, Trindaded E, Alencara L, Alencarb M, Silva L (2021) Sustainability in the construction industry: a systematic review of the literature. J Clean Prod 289

Lin S, Zhang H, Li J, Ye C, Hsieh J (2022) Evaluation smart office buildings from a sustainability perspective: a model of hybrid multi-attribute decision-making. Technol Soc 68

Martínez-Rocamora A, Solís-Guzmán J, Marrero M (2016) Toward the ecological footprint of the use and maintenance phase of buildings: utility consumption and cleaning tasks. Ecol Indic 69:66–77

Menassa CC (2011) Evaluating sustainable retrofits in existing buildings under uncertainty. Energy Build 43:3576–3583

Seugbeom K, Yooneun L, Byungchul C (2021) Adoption of satellite offices in response to a pandemic: sustainability and infection control. Sustainability 13:8008

Siniak N, Kauko T (2020) The impact of proptech on real estate industry growth. IOP Conf Ser Mater Sci Eng 869(6)

Solìs-Guzmàn J, Marrero M, Ramìrez-de-Arellano A (2013) Methodology for determining the ecological footprint of the construction of residential buildings in Andalusia (Spain). Ecol Indic 25:239–249

Tagliaro C (2021) Progetto ufficio. Strategia e processi per l'evoluzione degli spazi aziendali. Real Estate strumenti e casi, Franco Angeli

Tagliaro C, Migliore A (2021) "Covid-working": what to keep and what to leave? Evidence from an Italian company. J Corp Real Estate

Thuvander L, Femenías P, Mjörnell K, Meiling P (2012) Unveiling the process of sustainable renovation. Sustainability 4(6):1188–1213

U.S. Energy Information Administration (2013) https://www.springer.com/gp/authors-editors/book-authors-editors/resources-guidelines/rights-permissions-licensing/manuscript-preparation/5636#c5640
Vollenbroek FA (2002) Sustainable development and the challenge of innovation. J Clean Prod 10(3):215–223
Wackernagel M, Rees W (1996) Our ecological footprint: reducing human impact on the earth, 9th edn. New Society, Gabriola, BC, Canada, pp 9–148
Yeheyis M, Hewage K, Alam MS, Eskicioglu C, Sadiq R (2013) An overview of construction and demolition waste management in Canada: a lifecycle analysis approach to sustainability. Clean Technol Environ Policy 15:81–91

Chapter 16
Digital Twin Models Supporting Cognitive Buildings for Ambient Assisted Living

Alessandra Corneli, Leonardo Binni, Berardo Naticchia, and Massimo Vaccarini

Abstract The rapid and global aging of population is outlining the need for environments that can provide support for these individuals during their daily activities. The challenge of an aging society is being addressed through the incorporation of new technologies into the home environment, which is nothing less than Ambient Assisted Living (AAL). To date, some of the AAL solutions exploit AI models to recognize the elderly's behaviors through data collected by sensors. In recent times, Digital Twins (DTs) at building level have begun to appear on the construction domain. These are still under development but through the integration of users into assessments, they improve efficiency, prevention, and prediction of likely events through real-time AI computing. The integration of DT and AAL defines cognitive buildings which aim to learn at scale, reason with a purpose, and co-operate with users in a natural way. This research aims to develop DT models to achieve scenario awareness to provide support to elderly people living alone and suffering from cognitive disorders. The proposed multi-agent architecture is based on a five-layer system that autonomously develops high-level knowledge to detect anomalies in the home environment scenarios and therefore support the user. Bayesian networks (BNs) are exploited to perform high-level deductive reasoning on low-level multi-modal information, thus recognizing senseless or dangerous behaviors, environmental disruptions, changes in behavioral patterns, and serious medical events. Bi-directional user-system interaction provides user support by leveraging Speech-To-Text and Text-To-Speech AI agents. Three main functions were tested: real-time data integration, anomaly detection, and two-way interaction.

A. Corneli (✉) · L. Binni · B. Naticchia · M. Vaccarini
Polytechnic University of Marche, Ancona, Italy
e-mail: a.corneli@univpm.it

L. Binni
e-mail: l.binni@pm.univpm.it

B. Naticchia
e-mail: b.naticchia@univpm.it

M. Vaccarini
e-mail: m.vaccarini@staff.univpm.it

E. Arbizzani et al. (eds.), *Technological Imagination in the Green and Digital Transition*, The Urban Book Series, https://doi.org/10.1007/978-3-031-29515-7_16

Keywords Digital Twin · Ambient Assisted Living · Activity Recognition · Scenario awareness · Bayesian networks

16.1 Introduction

Projections about the increase of people in older age from the United Nations show that the number of people aged 65 or older will double by 2050 (United Nations 2020a, b), and this will cause significant effects on the share of the population suffering from geriatric diseases such as cognitive impairments, and consequently on the share of the population needing care due to loss of independence (Berryhill et al. 2012).

Advanced care planning, supported decision-making, and availability of assistive devices can enhance autonomy regardless of an elderly person's level of capacity (World Health Organization 2017).

Nevertheless, solutions are still far from predictive environments as intended by the Ambient Assisted Living (AAL) policies (World Health Organization, Regional Office for Europe 2017). This gap could be narrowed by the Digital Twin (DT) technology. Accordingly, this paper provides prototypes of DT models to support elderly people within the home environment in detecting anomalies in daily scenarios.

16.2 Literature Review

An increasing number of DT models are currently emerging within the built environment. The development of such innovating systems aims to achieve a variety of objectives depending on domains (Liu et al. 2021; Opoku et al. 2021; Sharma et al. 2020). Among the different assets that are currently being mirrored by DTs, both prevention and prediction of probable events during the whole life cycle of a building can be performed by means of a building-level DT, enhancing the building's efficiency as well. Building-level DTs consider both the environment and its user processing real-time information to offer appropriate services.

Hence, they could be also referred to as cognitive buildings (Yitmen et al. 2021).

AAL is a field that combines information and communication technologies, sociological sciences, and medical research and its purposes can be summarized as the development of products and services for countering the effects of a growing elderly population (Li et al. 2015; Dobre et al. 2017). Cognitive environments in AAL domain should be able to learn at scale, reason with purpose and co-operate with users in a natural way. Accordingly, some cognitive human-centered environments have recently begun to appear (De Paola et al. 2017; Rafferty et al. 2017; Patel and Shah 2020; Calderita et al. 2020) due to the significant influence that the home environment has on AAL's objectives.

Since pursuing and completing Activities of Daily Living (ADLs) allows autonomous well-being in older ages, this kind of system usually aims at encouraging, supporting, and easing the users in their ADLs. A variety of sensors and devices are exploited to collect data that can be processed by AI algorithms to make analysis on either user or environmental conditions. However, visual sensors (e.g., cameras) are not fully exploited yet.

16.3 Methodology

16.3.1 Mirroring Real Environment

A consistent virtual representation of the context must include both the building elements and the user. Accordingly, a real-time representation of the user is combined with BIM information within a game engine (GE), i.e., unity. The user virtual counterpart is synthesized through its posture, which is typically referred to as Skeleton.

A LiDAR camera is used as visual sensor (Intel RealSense L515). Nuitrack AI is used as the Skeleton tracking algorithm to process camera's raw 3D data and yield the user Skeleton. Besides, dynamic features such as appliance states, environmental temperature, and so on can also be associated with BIM elements into the GE to retrieve further information from sensor readings.

Though, further information is required to define a complete semantic of a user suffering from cognitive disorders. The activities that he or she performs can be detected through an Activity Recognition (AR) model. The AR task can be performed following either data-driven or knowledge-driven approaches (Rafferty et al. 2017). A data-driven approach is followed in this work since it enables the modeling of uncertainty and exploits increasingly available activity datasets. Specifically, the model developed in Liu et al. (2020) is integrated as the AR agent of the system. The MS-G3D model is based on Spatial–Temporal Graph Convolutional Networks (ST-GCNs), firstly proposed in Yan et al. (2018) for the AR task, and is therefore fed with the 3D coordinates of the Skeleton joints instead of RGB images as required by the previous Convolutional Neural Network (CNN). This results in a lightweight model that outperforms existing methods for AR.

16.3.2 Knowledge Contextualization

BIM information combined with its dynamic features and the user-related information define a low-level knowledge on the real asset of its virtual counterpart. A contextualization of such information allows interpreting the real world and thus the twinning of scenarios that includes the environment and the user with its behaviors, habits, intentions, activities, and situations. Emergent scenarios that the system aims to detect are those including anomalies. Accordingly, knowledge contextualization can be achieved through an agent acting as the reasoner of the system. A rule-based reasoner has been proposed in De Paola et al. (2017). Their module consists in if, else conditional rules and takes basic decisions such as turning the heating/cooling system on/off depending on the user satisfaction. Since complex rules cannot cope with scalability and are not easily reusable, above all when considering an elderly person suffering from cognitive disorders, a probabilistic approach is followed by means of Bayesian Networks (BNs).

Such a probabilistic model is based on conditional probabilities that an event may occur depending on evidence or other variables. BNs are probabilistic graphical models that represent a set of variables and their conditional dependencies via a directed acyclic graph. Expert knowledge could be elicited in Conditional Probability Tables (CPTs) of the nodes that represent the events. BNs are ideal for taking an event that occurred and predicting the likelihood that any one of several possible known causes was the contributing factor (De Grassi et al. 2009). In this work, BNs are used to infer different types of scenario anomalies:

- Wasteful and senseless situations (e.g., window open while heating system is turned on)
- Unusual behaviors (e.g., skipping meals)
- Dangerous situations (e.g., something dropped on the ground)
- Emergencies (e.g., falls).

Once the real-world scenario has been recognized by the system, it should offer the appropriate supportive services. A dialog system is implemented through a flow-based programming tool (i.e., Node-RED) to enable a bidirectional interaction between the user and the system itself. This platform can bridge the gap between the reasoner and the services to deliver.

16.4 System Architecture

The architecture of the proposed system has been outlined following the guidelines stated in Lu et al. (2020), which define the structure of a building-level DT. Thus, our model consists of five layers, namely data acquisition layer, transmission layer, digital modelling layer, data/model integration layer, and service layer. The proposed architecture defines a system able to autonomously perform high-level reasoning to detect anomalies in daily scenarios and consequently offer support to the user. Figure 16.1 shows the architecture of the system.

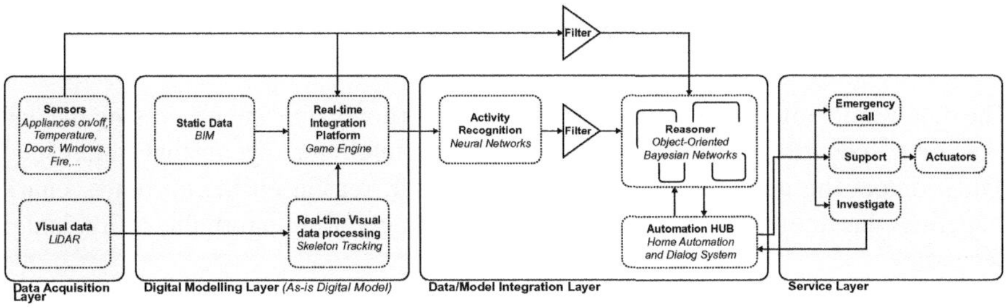

Fig. 16.1 System architecture. Arrows indicate data flow (transmission layer)

16.4.1 3D Real-Time Representation

Data acquisition and digital modelling layers hold the computation that manages the 3D real-time representation of the context. The virtual scenario built upon BIM and Skeleton data requires some adjustments. Two different filtering algorithms are developed for the following reasons:

- Filtering non-confident Skeleton data. Some Skeleton joints may have low confidence values due to obstructed camera field of view and can thus be less reliable leading to distortions of the user's avatar within the GE. Consequently, a threshold is introduced to discard data regarding Skeleton joints with a confidence value below 10%.
- Enhancing Skeleton stabilization. Once non-confident data are discarded, the avatar should move following natural movements. The avatar is stabilized through an autoregressive filter that acts on joint's position, orientation, and avatar's height.

$$X_{(t+1)} = (1 - a) \cdot X_{(t)} + a \cdot X^{\text{raw}}_{(t+1)} \tag{16.1}$$

where $X_{(t+1)}$ is the processed data at the time $(t + 1)$, $X_{(t)}$ is the processed data at the time (t), a is a corrective factor with a value that ranges between 0 and 1, and $X^{\text{raw}}_{(t+1)}$ is the raw value of the data at the time $(t + 1)$.

These filters allow avoiding the ambiguities that the avatar had. Furthermore, cleaner movements mean more consistent output data that will consequently feed the AR model. To this end, the avatar joints have been remapped as the Kinect v2 Skeleton, used to build the NTU RGB + D dataset (Shahroudy et al. 2016) on which the MS-G3D model is pre-trained.

16.4.2 Scenario Awareness

The data/model integration layer is responsible for analyzing and processing the data to achieve scenario awareness. It consists of three intelligent agents that can, respectively, detect the activities performed by the user, reason on the current scenario detecting the anomalous ones, and act accordingly to support the individual if necessary.

The agent acting as the reasoner of the system is formalized through an Object-Oriented Bayesian Network (OOBN), composed in turn of four sub-modules (Fig. 16.2).

The approach followed to formalize the OOBN firstly considers general symptoms that could lead to anomalous scenarios (confusion, depression, loss of memory, and

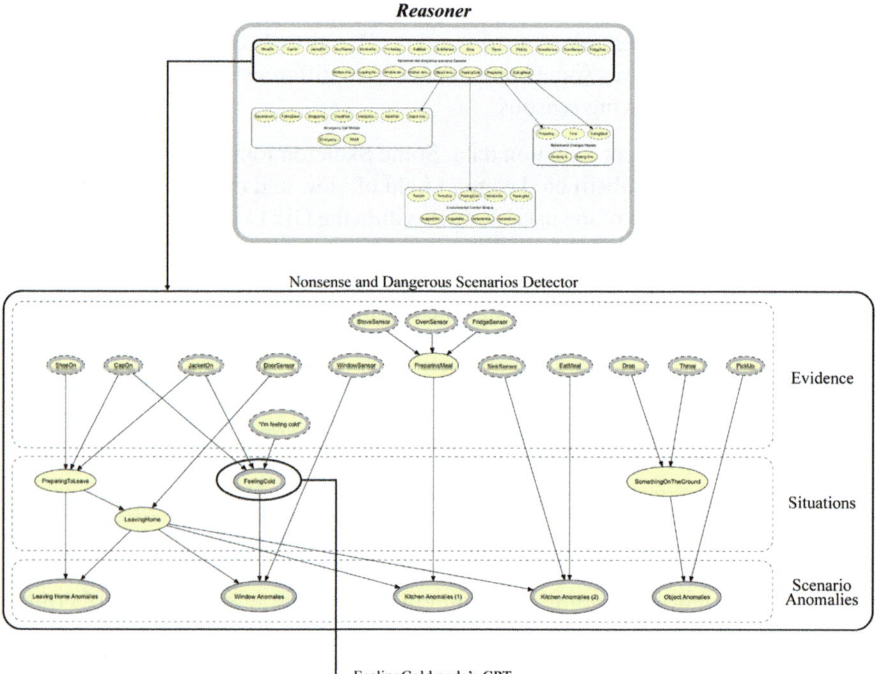

FeelingCold node's CPT

FeelingCold								
"I'm Feeling Cold"	Not				DialogSystem interaction "I'm feeling cold"			
CapOn	Not		CapOn		Not		CapOn	
JacketOn	Not	JacketOn	Not	JacketOn	Not	JacketOn	Not	JacketOn
False	0.9	0.4	0.25	0.1	0.1	0.05	0.05	0.01
FeelingCold	0.1	0.6	0.75	0.9	0.9	0.95	0.95	0.99

Fig. 16.2 The reasoner consists of an Object-Oriented Bayesian Network, decomposed in turn into four underlying OOBN modules. Zooming in shows the Nonsense and Dangerous Scenarios Detector module and the CPT of the "FeelingCold" situation node. Dashed gray-edged nodes represent input nodes. Solid gray-edged nodes are output nodes

so forth) (Berryhill et al. 2012; Dillon et al. 2013; scie.org, https://www.scie.org.uk/dementia). Then, a semantic regarding probable events, situations, scenarios, and anomalies in a AAL environment is built: Evidence is captured by sensors (turning on/off appliances, indoor/outdoor temperature, open/closed window, and so forth) as are the results of the AR model and the user-system vocal interactions; situations are combinations of evidence and represent feelings, behaviors, events, or intentions (feeling hot/cold, getting dressed, something on the ground, leaving home, and so forth).

By associating and combining available evidence and recognizable situations, probable scenarios are theorized as the anomalies that may occur. Anomalous scenarios therefore include time disorientation, difficulties arranging, indifference to the environment, getting easily overwhelmed, mishandling appliances, and changes in eating patterns.

The Automation HUB, based on Node-RED, can integrate applications to offer appropriate support to the user. In this work, a dialog system is implemented. Speech-to-Text (STT) and Text-to-Speech (TTS) services that rely on Machine Learning models are integrated to define a dialog system whereby bidirectional vocal interactions between the user and the cognitive layer of the building can be performed.

Is essential not to have Hot Phrases (HP) since the user may forget them due to cognitive impairments. HP are phrases typically used to trigger common dialog system such as Alexa and Google Assistant.

16.5 System Implementation

Combining BIM data and the Skeleton allows achieving a reliable real-time virtual representation of the physical asset which is shown in Fig. 16.3. BIM data from a home environment are converted to Industry Foundation Classes (IFC) format using Autodesk Revit. Importing IFC files into the unity game engine recognizes all BIM objects as Prefabs. Prefabs preserve all information related to BIM objects. Working with physics engines, Unity allows additional properties to be assigned to Prefabs achieving greater realism. Indeed, the mesh collider attribute is applied to all tangible components to avoid inconsistencies. Additionally, dynamic features can be added to Prefabs to extract real-time sensor readings about BIM objects. Besides, the LiDAR camera is placed at a height of 1 m and leveled horizontally. The tests show that the distance between the user and the camera should be unobstructed and not exceed 5 m to obtain consistent results.

To evaluate the effectiveness of the Object-Oriented Bayesian Networks developed in this work, the node's CPTs are filled eliciting the knowledge of the authors. Figure 16.2 shows the CPT relative to the "FeelingCold" situation node. Then, possible combinations of evidence are set up by manually activating input nodes, and the expected consequences achieve high percentage values meaning that predictable anomalies within the scenario are fully recognized. Figure 16.4 shows an example of anomaly detection within the Nonsense and Dangerous Scenario module. Four input

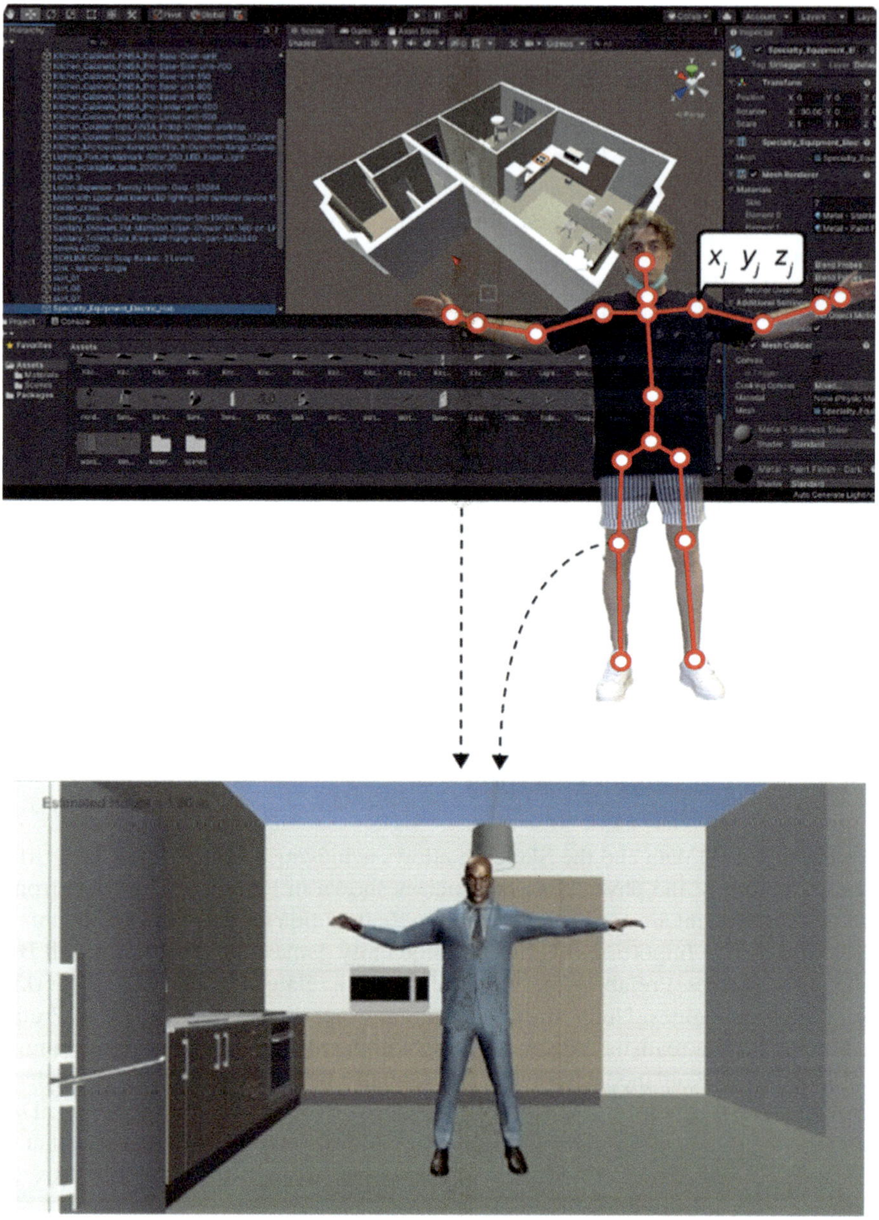

Fig. 16.3 Real-time 3D representation of the context

nodes are set up to represent a scenario where the user is barefoot, not wearing a jacket and hat, and is opening the door. Specifically, the ShoeOn, CapOn, and JacketOn input nodes have been set up to false (activities recognizable through the AR model), while the DoorSensor input node has been set up to open. The user is not preparing to leave (98.90% false) but is actually leaving home (90.09% true). The "Leaving Home Anomalies" output node detects a likelihood of 87.34% that the user is leaving undressed.

Figure 16.5 shows the dialog system prototype built upon the STT and TTS processes. STT module starts by recording the user's speech without requiring HP. Then, the record is managed by the IBM's Watson STT service that returns a transcription of the speech. Finally, the transcription is shown in the Node-RED's debug tab. By contrast, the TTS process is automatically triggered by the system depending on the output of the reasoner. Tailored messages can be played depending on the needs of the user. These written messages are converted through the IBM's Watson TTS service. Finally, the converted speech is played by the speakers.

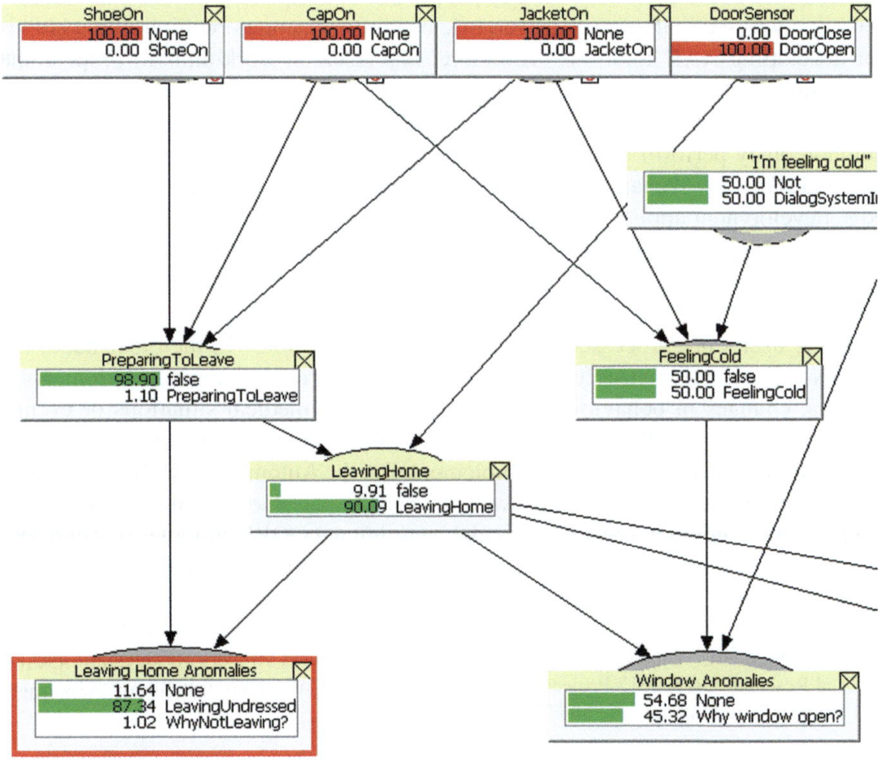

Fig. 16.4 Detected anomaly within the Nonsense and Dangerous Scenario module

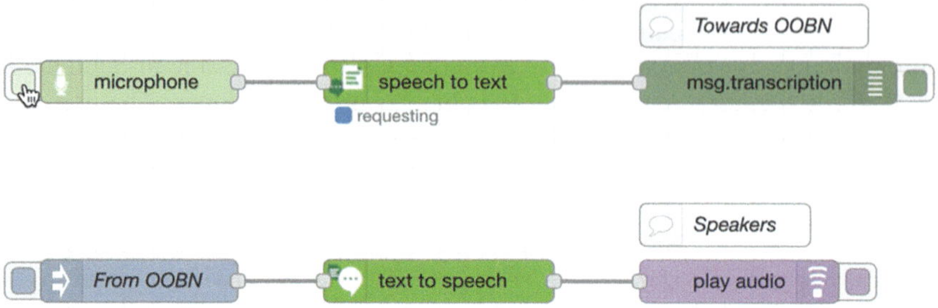

Fig. 16.5 SST and TTS processes within the Node-RED editor

16.6 Conclusion

The increase in the number of elderly people and consequently the increased occurrence we will see in the future of geriatric cognitive disorders requires new systems for developing AAL solutions. Therefore, this research work aims to propose the development of Cognitive Building through the exploitation of the DT paradigm.

The grounded multi-agent system architecture defines a model able to autonomously perform real-time high-level reasoning, that allows the detection of anomalies in daily scenarios, and consequently offers support to the user. The knowledge development applied here is a major strength: the raw data that is captured by multi-modal sensors (visual and non-visual) and subsequently reported in 3D in real time, but also the reasoning applied at a high level when anomalies are detected. AR is performed using a neural network model that leverages 3D data derived from the user's pre-processed real-time 3D representation. On the other hand, the OOBN can recognize wasteful, meaningless, or dangerous behaviors, environmental distress, changes in behavioral patterns, and serious medical situations or events, and then trigger specific services. Two-way voice interaction with the individual is performed by the dialogue system, implemented in the Automation HUB agent, based on Node-RED. A number of improvements can be addressed as future work of this study: implementing the MS-G3D model; learning the OOBN modules through data captured from a real-world AAL environment; and fully testing the entire pipeline in an end-to-end manner.

Acknowledgements This work is partially supported by the Italian Ministry of Education, University and Research PRIN 2017 Project: "A Distributed Digital Collaboration Framework for Small and Medium-Sized Engineering and Construction Enterprises".

References

Berryhill ME, Peterson D, Jones K, Tanoue R (2012) Cognitive disorders. In: Encyclopedia of human behavior, 2 edn, vol 1. Academic Press, pp 536–542

Calderita L, Vega A, Barroso-Ramírez S, Bustos P, Núñez P (2020) Designing a cyber-physical system for ambient assisted living: a use-case analysis for social robot navigation in caregiving centers. Sensors (Switzerland) 20(14):1–25

De Grassi M, Naticchia B, Giretti A, Carbonari A (2009) Reti Bayesiane con applicazioni all'edilizia e alla gestione del territorio. Franco Angeli, Milano

De Paola A, Ferraro P, Gaglio S, Re G, Morana M, Ortolani M, Peri D (2017) An ambient intelligence system for assisted living. In: AEIT international annual conference, Cagliari, Italy, pp 1–6

Dillon C, Serrano CM, Castro D, Leguizamon PP, Heisecke SL, Taragano FE (2013) Behavioral symptoms related to cognitive impairment. Neuropsychiatr Dis Treat 9:1443–1455

Dobre C, Mavromoustakis C, Garcia N, Goleva R, Mastorakis G (2017) Ambient assisted living and enhanced living environments, 1st edn. Butterworth-Heinemann

Li R, Lu B, McDonald-Maier K (2015) Cognitive assisted living ambient system: a survey. Digit Commun Netw 1:229–252

Liu Z, Zhang H, Chen Z, Wang Z, Ouyang W (2020) Disentangling and unifying graph convolutions for skeleton-based action recognition. arXiv:2003.14111v2

Liu Y, Chen K, Ma L, Tang S, Tan T (2021) Transforming data into decision making: a spotlight review of construction digital twin. In: Proceedings of the international conference on construction and real estate management 2021, pp 289–296

Lu Q, Parlikad A, Woodall P, Don Ranasinghe G, Xie X, Schooling J (2020) Developing a digital twin at building and city levels: case study of West Cambridge campus. J Manage Eng 36(3)

Opoku D, Perera S, Osei-Kyei R, Rashidi M (2021) Digital twin application in the construction industry: a literature review. J Build Eng 40:102726

Patel A, Shah J (2020) Real-time human behaviour monitoring using hybrid ambient assisted living framework. J Reliab Intell Environ 6(2):95–106

Rafferty J, Nugent C, Liu J, Chen L (2017) From activity recognition to intention recognition for assisted living within smart homes. IEEE Trans Hum-Mach Syst 47(3):368–379

scie.org. https://www.scie.org.uk/dementia

Shahroudy A, Liu J, Ng T-T, Wang G (2016) NTU RGB+D: a large scale dataset for 3D human activity analysis

Sharma A, Kosasih E, Zhang J, Brintrup A, Calinescu A (2020) Digital twins: state of the art theory and practice, challenges, and open research questions. Tratto da https://arxiv.org/abs/2011.02833

United Nations (2020a) World population ageing 2019. United Nations, Department of Economic and Social Affairs, Population Division

United Nations (2020b) World population ageing 2020 highlights: living arrangements of older persons. United Nations, Department of Economic and Social Affairs, Population Division

World Health Organization (2017) Global strategy and action plan on ageing and health. WHO, Geneva

World Health Organization, Regional Office for Europe (2017) Age-friendly environments in Europe. A handbook of domains for policy action. Denmark

Yan S, Xiong Y, Lin D (2018) Spatial temporal graph convolutional networks for skeleton-based action recognition. In: AAAI conference on artificial intelligence

Yitmen I, Alizadehsalehi S, Akiner I, Akiner ME (2021) An adapted model of cognitive digital twins for building lifecycle management. Appl Sci 11(9)

Chapter 17
Less Automation More Information: A Learning Tool for a Post-occupancy Operation and Evaluation

Chiara Tonelli, Barbara Cardone, Roberto D'Autilia, and Giuliana Nardi

Abstract Climate change and the pandemic generated an urgent need to have an efficient urban habitat that includes technological innovations to deal with the ecological and digital transitions. Italy counts about 14 million buildings, 12 of which are houses, responsible for more than 40% of final energy consumption, most of which is ascribable to users' behavior and lifestyle. The increase in buildings' energy performance is strongly related to a smart management of the demand and self-consumption, as well as a more effective and active involvement of the occupants: it is, therefore, pivotal to come up with user-friendly tools to measure and monitor the performance of the buildings and users' habits. Tools to encourage the choices toward the environment's comfort, rather than automation technologies, allowing the occupants and information systems to move in the direction of ecological transition. The aim is to create an aware "energy citizenship" for people living in efficient buildings. The proposal is a system that uses IoT technology and provides a global evaluation of the state of the house, from which can be extracted suggestions for better and virtuous behavior. The overall ecological footprint is measured based on five "cycles": energy; environment; water; waste production; food. Collected data create an urban database that, along with big data, constitutes a set of boundary conditions that are crossed with single units' data. The measures related to single units can be applied to a wider network in order to create a smart city, involving dwellers in a serious game on their homes' performance. The proposal is part of the research on post-evaluation occupancy, in the belief that even the best model-houses perform worse in use, rather than the predictions expected on paper.

C. Tonelli (✉) · B. Cardone · R. D'Autilia · G. Nardi
University of Roma Tre, Rome, Italy
e-mail: chiara.tonelli@uniroma3.it

B. Cardone
e-mail: barbara.cardone@uniroma3.it

R. D'Autilia
e-mail: roberto.dautilia@uniroma3.it

G. Nardi
e-mail: giuliana.nardi@uniroma3.it

E. Arbizzani et al. (eds.), *Technological Imagination in the Green and Digital Transition*, The Urban Book Series, https://doi.org/10.1007/978-3-031-29515-7_17

Keywords Post-occupancy evaluation · Energy behavior · Energy citizenship · Energy culture

17.1 Introduction

The recent pandemic highlighted some housing inadequacies with respect to the morphology, the comfort, and the health of people. The architectural research identified the need for more compact and flexible solutions to face different climatic conditions as well as new needs for the living space (Rode et al. 2014). The minimum size paradigm derived from hygienic criteria[1] should be replaced by a new one that takes into account sustainability, health, and energy efficiency. At the same time, people's awareness of the impact of their actions should be improved.

Emergencies such as climate change and the pandemic also urged the need for efficient urban habitats and technological innovations to encourage the ecological transition. Italy counts 14 million buildings, 12 of which are residential, responsible for more than 40% of final energy consumption, most of which is ascribable to users' behavior and lifestyle (Fig. 17.1).

The improvement of the buildings' energy performance and the reduction of the environmental impact depends on the smart consumption and user involvement to

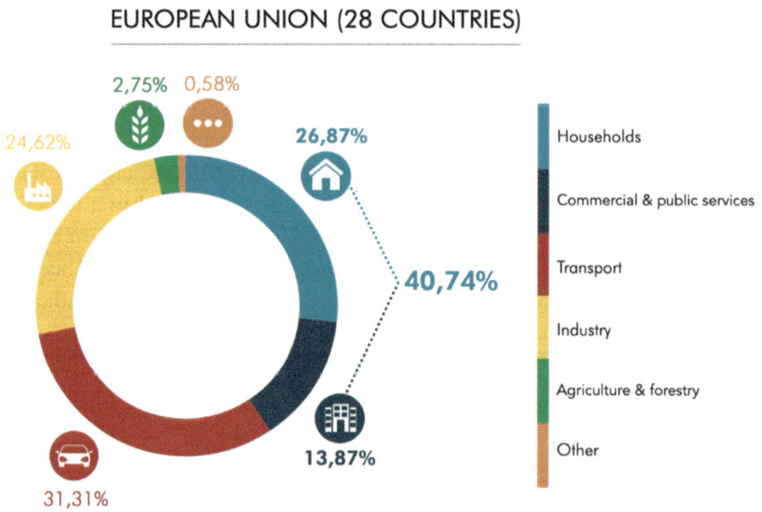

Fig. 17.1 Final energy consumption, EU28, Eurostat 2021 (2019 data)

[1] See the Italian Ministerial Decree: DM 5 Luglio 1975, Modificazioni alle istruzioni ministeriali 20 giugno 1896, relativamente all'altezza minima ed ai requisiti igienico-sanitari principali dei locali di abitazione. Last accessed 23/05/2022: https://www.gazzettaufficiale.it/eli/gu/1975/07/18/190/sg/pdf.

keep the environment healthy. It is therefore essential to have efficient tools to monitor the performance of the buildings together with the inhabitants' behavior acting on their daily habits and, more generally, improving the "Energy Culture" (Stephenson et al. 2010).

The relationship between energy and culture has long been a focus of cultural anthropology (White 1943): it is possible to trace the history of civilization through the availability, distribution, and use of energy (Smil 2018). The three interrelated components: norms, practices, and material culture interact with the social institutions to create the energy culture (Stephenson 2018) (Fig. 17.2).

Habitus is a long-term thinking pattern shaped by a particular cultural environment that influences modern social behavior (Bourdieu 1980). The study of energy culture aims to change habits by promoting new practices, attitudes, and behaviors (Rau et al. 2020). The technology, the buildings, and other assets that influence energy use are referred to as material culture. The norms are common ideas, the individual and collective expectations and goals regarding material activities and culture. Daily actions and the selection and acquisition of assets are examples of practice.

Over the past forty years, much literature considered the interaction between humans and buildings, but no standard way has emerged for comparing the results of different studies (Dong et al. 2018; Hong et al. 2015). The technologies related to the Internet of Things (IoT) offer the possibility of measuring the overall behavior of a living space. In order to improve the models in literature, which deal, in general,

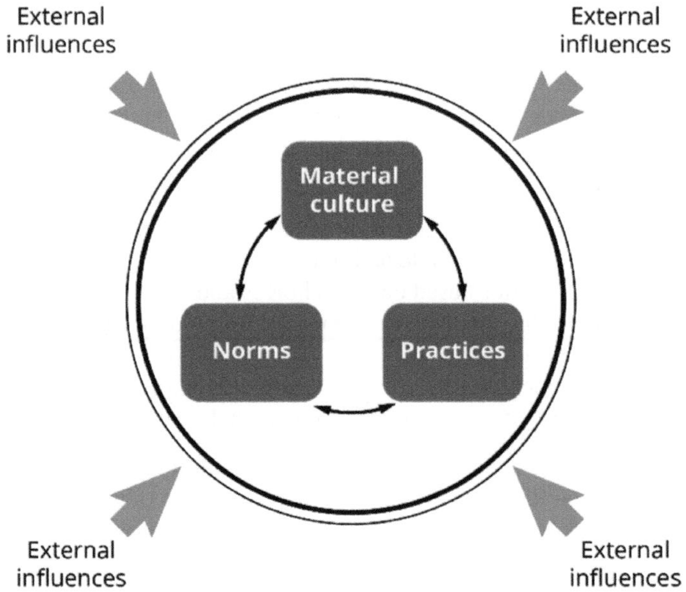

Fig. 17.2 An energy culture framework (adapted from Stephenson 2018)

with the monitoring of single environmental variables, we try to calculate the overall impact of different variables taking into account their interdependence.

The aim of this paper is to develop a machine learning tool to increase the environmental awareness and to help the residents to optimize the energy consumption together with the environmental impact (Tonelli and Converso 2014). The tool is also intended for the planning analysis and, at the same time, to trigger a process of gamification among the tenants.

17.2 Materials and Methods

We propose a machine learning algorithms that can be trained by data coming from sensors that measure the environmental data, as a tool to improve the (Tonelli et al. 2019) "smart citizenship" and to optimize the environmental quality of life at home. This strategy is related to the "architecture of choices" (Johnson et al. 2012; Scheibehenne et al. 2010), and it is derived from scientific insights originating from "neurological" models together with the possibility of having a large amount of data available. The tool is designed to make people aware of the quality of the environment in which they live, but not to prescribe behaviors.

Rather, it is continuous monitoring that can induce a change in individual behavior.

Quoting the Italian Research Plan[2] "*Digital technologies have, in turn, a function which, although ancillary, is of crucial importance for achieving the reduction of energy consumption and, consequently, of the environmental impact: (...) must favor the virtuous behaviour of users by providing clear and understandable information in real-time (feedback) on the state of the building and on consumption and, at the same time, guaranteeing adequate support in the decision-making process (...)*".

The monitoring tool helps to optimize consumption and calculate the ecological footprint. The specific goals are as follows:

- to monitor the ecological footprint of the house and the behavior of the inhabitants by showing these data in a simple and intuitive way;
- to predict future scenarios based on user choices and external data;
- to estimate the long short-term consequences of behavior in terms of consumption and ecological footprint;
- to suggest best practices to improve the ecological footprint;
- to allow the planning of actions to achieve personal goals.

Finally, the tool is designed for a single housing unit but can be integrated into a larger network to build an urban database. The collection of this data on an urban scale, together with data on traffic conditions, weather, and external pollution, constitutes the set of boundary conditions that the machine learning system intersects with the internal data of each unit.

---

[2] Last accessed 23/05/2022: https://www.mur.gov.it/sites/default/files/2021-05/PNR2021-2027.pdf.

17.2.1 The Neural Network Architecture

The purpose of the algorithm is to return a habitability index k based on a given number N of environmental variables $x_1, x_2, ..., x_N$

$$[0, 10] \ni k = k(x_1, x_2, \ldots, x_N) \qquad (17.1)$$

The x_i variables are in general strongly correlated random variables, feature that make it very difficult to guess the functional form of $k(.)$.

Our purpose is to build a machine that provides a home habitability index on a scale ranging from 0 to 10. There are of course several constraints that can easily be turned into algorithms. For example, if the value of variable x_n is above a danger threshold, then $k = 0$ for any value of the other variables x_i with $i = n$. However, there are other variables whose behavior and correlation with the others is more difficult to guess, such as the availability of household appliances or the presence of possible smokers.

We observe that by means of "natural intelligence," a good planner is able to estimate the value of k based on its experience. Our aim is therefore to create a machine giving the same answers as an experienced planner, and for this purpose, we train an artificial intelligence system.

One of the most difficult problems in machine learning is known to be the choice of the optimal network topology. The most expensive sectors in terms of energy, economy, and environment are precisely that of neural architecture search (Strubell et al. 2020).

We proceeded in a simpler way, choosing four possible neuronal architectures and comparing the reliability of their responses. The four networks are linear regression, a network with nearest neighbors interactions, a convolutional neural network, and a random forest. In Goodfellow et al. (2016), a detailed description of the four architectures can be found.

To train the networks, we generated 10,000 data records and submitted them to the human evaluators who assigned a score from 0 to 10 to the environmental configuration. A subset of these records are used to train the networks, and the remaining was used to verify their efficiency. By providing the neural network with the tables whose scores we know, we were able to assess whether the response was reliable, and we can compare the responses of the four networks.

17.2.2 The Environmental Variables

The algorithm processes the environmental data and outputs the result. The input variables are classified into five categories:

- the geographic position of the building (Table 17.1);

Table 17.1 Outdoor conditions

Variable	Code	Unit of measure
Summer temperature	T	°C
Winter temperature	T	°C
Relative humidity	RH	%
Altitude	SLM	m s.l.m.
Wind	W	km/h

Table 17.2 Outdoor pollution

Variable	Code	Unit of measure
Atmospheric particulate matter	PM10	$\mu g/m^3$
Atmospheric particulate matter	PM2.5	$\mu g/m^3$
Nitrogen dioxide	NO_2	$\mu g/m^3$
Ozone	O_3	mg/m^3
Carbon monoxide	CO	mg/m^3
Sulfur dioxide	SO_2	$\mu g/m^3$
Benzene	C_6H_6	$\mu g/m^3$
Benzo(a)pyrene	$C_{20}H_{12}$	ng/m^3
Arsenic	As	ng/m^3
Cadmium	Cd	ng/m^3
Nickel	Ni	ng/m^3
Lead	Pb	$\mu g/m^3$
Noise (day)	dB	Decibel
Noise (night)	dB	Decibel

- the variables related to the buildings' geographic location, such as the outdoor pollution (Table 17.2);
- the buildings' shape, based on their estimated occupancy rate (Table 17.3);
- the presence of facilities and appliances (Table 17.4);
- the indoor pollution (Table 17.5).

These variables may be strongly statistically correlated. Moreover, the variables are also heterogeneous, being both qualitative and quantitative in nature. For this reason, artificial intelligence is particularly useful.

Table 17.3 Indoor condition

Variable	Code	Unit of measure
Square meter	A	mq
High	H	mt
Volume	V	mc
Inhabitants	P	Number

Table 17.4 Facilities and appliances

Variable	Code	Unit of measure
Cooling system	–	Boolean
Heating system	–	Boolean
Domestic hot water	ACS	Boolean
Washing machine	–	Boolean
Dishwasher	–	Boolean
Refrigerator	–	Boolean
Computer	–	Boolean
Photocopier	–	Boolean
Television	–	Boolean

Table 17.5 Indoor pollution

Variable	Code	Unit of measure
Tobacco smoke	ETS	$\mu g/m^3$
Carbon disulfide	CS_2	$\mu g/m^3$
Sulfur oxide	SO_x	$\mu g/m^3$
Carbon monoxide	CO	mg/m^3
Ozone	O_3	$\mu g/m^3$
Atmospheric particulate matter	PM10	$\mu g/m^3$
Atmospheric particulate matter	PM2.5	$\mu g/m^3$
Benzene	C_6H_6	$\mu g/m^3$
Formaldehyde	CH_2O	$\mu g/m^3$
Microbiological agents—bacteria, viruses, endotoxins and mycotoxins	–	Boolean
Indoor allergens—mites, epidermal derivatives of pets, cockroaches, fungi	–	Boolean
Mold	–	Boolean
Radon	Rn	Boolean

The geolocation provides data on temperature, relative humidity, and site altitude, quantities that influence the global air quality and the users' sense of comfort.

The outdoor pollution rate adds up to the indoor pollution rate, and it is in general higher than the outdoor rate, as shown by the data of the European Environment Agency.[3]

[3] Last accessed 23/05/2022: https://discomap.eea.europa.eu/map/fme/AirQualityExport.htm.

The main environmental pollutants are listed according to the national set of rules which ensures that the limit values and reporting period for calculating the concentration of pollutants in the air are given.

The shape of the rooms influences indoor health conditions when considered in relation to the number of occupants. Allometric relationships have been studied for urban spaces, assuming a relationship between population density and the extent of inhabited land similar to the one between body mass and metabolism in biology (D'Autilia and D'Ambrosi 2015). Extending to the analysis of residential land, this approach could lead to the identification of a "natural law" that takes into account other parameters as well as subjective and cultural constraints (Cardone and D'Autilia 2018). The allometry method makes it possible to identify a mathematical law based either on the variables normally used to define the living spaces or on different variables such as the sharing of common spaces. The law is expressed in a formal way and makes it possible to simulate different scenarios, providing an efficient analytical tool for studying and forecasting different urban contexts.

The presence of facilities and appliances (e.g., cooling and heating systems, one or more televisions, etc.) results in a given level of noise pollution, radiation, and consumption. This also affects the perception of the comfort of living.

The indoor pollutants come from a variety of sources. Human factors that contribute to indoor pollution include tobacco smoke, the combustion process of oil, gas, paraffin, coal, and wood. Man-made pollutants come from cleaning products and the use of household appliances and tools such as printers and plotters. Other important indoor pollutants come from the building materials and furniture materials such as chipboard or pesticide-treated cabinets, carpets, and paneling.

The indoor pollutants can be grouped into 3 main categories:

- chemical pollutants: nitrogen dioxide (NO_2), sulfur dioxide (SO_2), carbon monoxide (CO), ozone (O_3), airborne particulate matter (PM10, PM2.5), benzene (C_6H_6), formaldehyde (CH_2O), polycyclic aromatic hydrocarbons (PAHs), environmental tobacco smoke (ETS), asbestos;
- physical agents: radon;
- microbiological contaminants (indoor allergens): mites, animal allergens, molds and fungi, outdoor allergens.

Outdoor pollutants infiltration can be due to ventilation systems and vents. Air-conditioning systems can easily become a breeding ground for molds and other biological contaminants and can spread pollutant agents throughout the building. In Tables 17.1, 17.2, 17.3, 17.4 and 17.5, the considered variables are reported together with the range and the units.

Once the levels and parameters to be correlated had been defined, the next step was to create a database of case studies. Based on the case studies created and their evaluation, the machine was trained to evaluate general indoor health, taking into account the previously defined environmental, architectural, and shape parameters.

17.3 Results

Once the four models were trained on the data, we tested their validity by means of test data. From the statistical analysis of the results, it can be seen that convolutional neural networks are the ones that work best (Figs. 17.3, 17.4, 17.5 and 17.6).

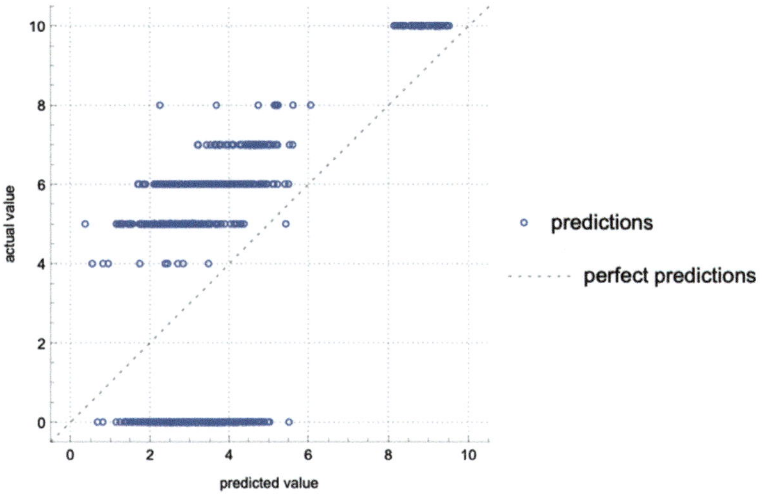

Fig. 17.3 Comparison of the perfect prediction with the linear regression model

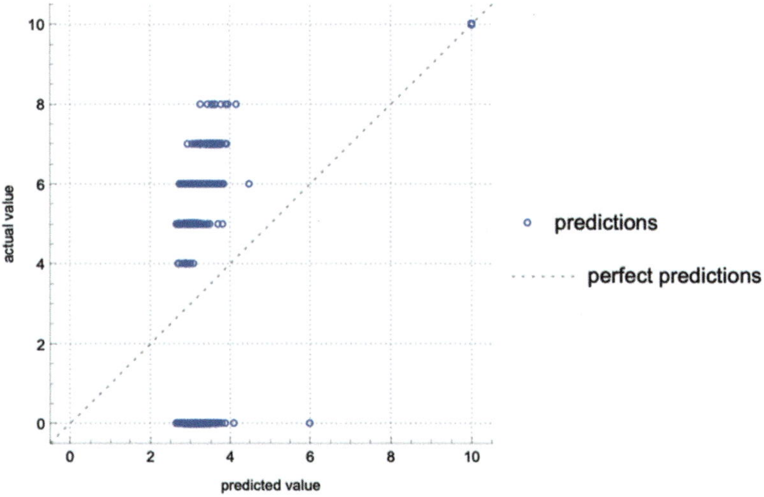

Fig. 17.4 Comparison of the perfect prediction with the nearest neighbors model

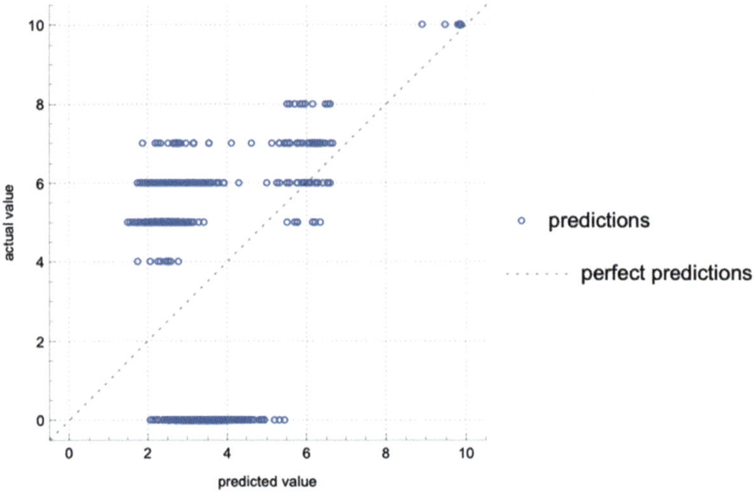

Fig. 17.5 Comparison of the perfect prediction with the convolutional neural network model

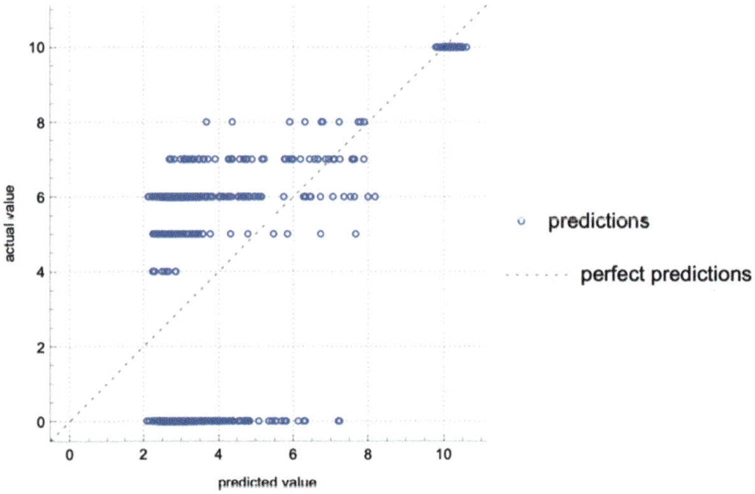

Fig. 17.6 Comparison of the perfect prediction with the random forest model

This result is somewhat unexpected because convolutional neural networks are very efficient for the recognition of images or in general of data arranged on a grid. The high level of correlation of the variables and their inhomogeneity has a correspondence with the distant characteristics of the images, such as the nature of the edges of a figure and its color.

The algorithm obtained is able to provide a habitability index as a function of the chosen variables. This means that, connected to environmental indicators, it provides a single number that could be visible on the "dashboard" of an apartment to inform the inhabitants of the level of domestic pollution. This instrument does not prescribe the actions to be taken, but, like a sort of speedometer, it informs people of the environmental situation, increasing their awareness, and the possibility of interventions. The algorithm can continue to learn as the available data increase, thus improving its reliability. Finally, the AI system can communicate with similar systems in the neighboring apartments, increasing its efficiency, and creating a global environmental map.

17.4 Conclusions

The described tool simulates a sort of awareness on the home governance, so that people can act in a specific way knowing that each choice has an impact in terms of environmental footprint. The predictor increases the precision of the calculations and the accuracy of the results as the number of cases analyzed increases.

In order to develop a more effective tool, it would be advisable to build a dataset based on measures related to more effective situations where all parameters are known. The final assessment should be carried out by a pool of experts on different topics, so that the conditions can be studied not only from an architectural and urban planning point of view, but also from a health and psychological point of view. In addition, the configurations studied can be labeled starting from the diseases that affect the residents. The creation of an effective predictor would make this possible:

- in-depth knowledge of the current situation concerning existing indoor environments' healthiness;
- widespread availability of an effective evaluation tool preparatory to the development of a sole and complete indoor environment health certification, which is lacking at the moment;
- estimation and production of effective solutions to renovate unhealthy environments;
- creation of a design tool to address the construction of suitable indoor environments;
- widespread and accessibility to in-depth knowledge of these topics;
- introduction of policies aimed at increasing healthiness in indoor environments;
- opportunity to educate users to a virtuous behavior, aimed at improving the healthiness in indoor environments.

The system's output is a dashboard, a sort of digital-control panel which allows the users to follow in real time the parameters related to their own consumption of energy and resources, through the connection with devices such as smartphones, tablets, and PCs. The overall ecological footprint is measured based on energy, environmental, water waste, and food "cycles."

The system uses IoT technology and processes the data collected by the sensors with artificial intelligence and provides a global assessment of the state of the house (a number), from which partial assessments and suggestions for better and more virtuous behavior can be derived. The final goal of such a system is to be connected with similar system to create a network of real-time information about the ecology of the indoor habitat.

The resulting tool does not deal with automation technologies to control the indoor spaces' comfort, but addresses tools able to make choices that allow occupants and information systems to learn together. The fine-tuning of these tools is enabled by the availability of parametric modeling software, jointly with opportunity to develop artificial intelligence models to calculate the connection among variables.

The development of devices to convert the large, heterogeneous, and variable amounts of data into information enables the development of a resilient and adaptable planning methodology based on the complex interactions between structural and technological systems, between spaces and functions, between social and economic factors. The machine learning system combines this data with indoor data from various sensors that calculate consumption, environmental parameters, and the behavior of people in each unit.

The final aim is to create a conscious "energy citizenship" living in energy-efficient buildings. Energy and functional issues were, and remain, one of the driving factors for sustainable strategic development, and the strength of the smart approach lies in its ability to promote a holistic vision. The measures related to single units can be applied to a wider network in order to create smart districts and smart cities, involving dwellers in a serious game on their homes' performance. The European Framework Program identified Smart Cities as one of the answers to the energy and environmental problems highlighted by the Strategic Energy Technology Plan.

References

Bourdieu P (1980) Le Sens pratique. Les edition de minuit

Cardone B, D'Autilia R (2018) Comfort abitativo e standard edilizi: l'approccio allometrico. Officina 21:80–84

D'Autilia R, D'Ambrosi I (2015) Is there enough fertile soil to feed a planet of growing cities? Phys A 419:668–674

Dong B, Yan D, Li Z, Jin Y, Feng X, Fontenot H (2018) Modeling occupancy and behavior for better building design and operation—a critical review. Build Simul 11:899–921

Goodfellow I, Bengio Y, Courville A (2016) Deep learning. Google Books

Hong T, D'Oca S, Turner WJ, Taylor-Lange SC (2015) An ontology to represent energy-related occupant behavior in buildings. Part I: introduction to the DNAs framework. Build Environ 92:764–777

Johnson EJ, Shu SB, Dellaert BGC, Fox C, Goldstein DG, Häubl G, Larrick RP, Payne JW, Peters E, Schkade D, Wansink B, Weber EU (2012) Beyond nudges: tools of a choice architecture. Mark Lett 23(2):487–504. https://doi.org/10.1007/s11002-012-9186-1

Rau H, Moran P, Manton R, Goggins J (2020) Changing energy cultures? Household energy use before and after a building energy efficiency retrofit. Sustain Cities Soc 54:101983

Rode P, Keim C, Robazza G, Viejo P, Schofield J (2014) Cities and energy: urban morphology and residential heat-energy demand. Environ Plann B Plann Des 41(1):138–162. https://doi.org/10.1068/b39065. http://epb.sagepub.com/content/41/1/138.abstract

Scheibehenne B, Greifeneder R, Todd PM (2010) Can there ever be too many options? A meta-analytic review of choice overload. J Consum Res 37(3):409–425. https://doi.org/10.1086/651235. http://hdl.handle.net/11858/00-001M-0000-0024-F614-E
Smil V (2018) Energy and civilization: a history. MIT Press
Stephenson J (2018) Sustainability cultures and energy research: an actor-centred interpretation of cultural theory. Energy Res Soc Sci 44:242–249
Stephenson J, Barton B, Carrington G, Gnoth D, Lawson R, Thorsnes P (2010) Energy cultures: a framework for understanding energy behaviours. Energy Policy 38(10):6120–6129
Strubell E, Ganesh A, McCallum A (2020) Energy and policy considerations for modern deep learning research. In: AAAI 2020—34th AAAI conference on artificial intelligence. https://doi.org/10.1609/aaai.v34i09.7123
Tonelli C, Converso S (2014) Digital mirror: a method to shape smart citizenship. Energy Build 83:173–180
Tonelli C, Montella I, Cardone B (2019) Do well to dwell well. Awareness as the driver for the behaviour of tomorrow's citizens. J Archit Environ Struct Eng Res 2(3)
White LA (1943) Energy and the evolution of culture. Am Anthropol 45(3):335–356

Chapter 18
A Prosumer Approach for Feeding the Digital Twin. Testing the MUST Application in the Old Harbour Waterfront of Genoa

Serena Viola, Antonio Novellino, Alberto Zinno, and Marco Di Ludovico

Abstract Supporting the settlement systems' life cycle management through synchronisation of the real world with a virtual platform constitutes the horizon for the MUST team research (Maintenance Urban Sharing Tools) with the Departments of Architecture (DiARC) and Structures for engineering and architecture (DiSt) of the University of Naples, Stress Scarl and ETT SpA. Living digital simulation models are based on information analysis and constant data supply. The research identifies the involvement of the settlement systems users through creating collaborative information flows, one of the driving factors of the digital revolution. The paper introduces the connotative aspects of the MUST application (Smau Innovation Award 2019, Campania Start-Up 2020 funded project) to identify the building and urban system loss of functionality. With the support of an experiment conducted in the waterfront area of the old harbour of Genoa, the paper identifies strengths and weaknesses in using the MUST application to support and streamline the Digital Twin that ETT S.p.A. is implementing with the DSH2030 (Digital Sustainable Harbor 2030) project. A prosumer perspective is the foundation of this research focusing on the sense of responsibility of communities towards the built environment and on the willingness of individuals to invest in care actions. The paper returns the research results achieved to date with an open and public model design, equipped with different interfaces to meet the diverse needs of the groups involved, allowing expert citizens to interact and report in progress feedback.

S. Viola · M. Di Ludovico
University of Naples Federico II, Naples, Italy
e-mail: serena.viola@unina.it

M. Di Ludovico
e-mail: diludovi@unina.it

A. Novellino (✉)
ETT SpA, Genoa, Italy
e-mail: antonio.novellino@ettsolutions.com

A. Zinno
Stress Scarl, Naples, Italy
e-mail: alberto.zinno@stress-scarl.it

E. Arbizzani et al. (eds.), *Technological Imagination in the Green and Digital Transition*, The Urban Book Series, https://doi.org/10.1007/978-3-031-29515-7_18

Keywords Digital Twin · Urban maintenance · Citizens · Sensors

18.1 Introduction

The Internet of Things (IoT) and cloud computing renew systems and process in several production sectors, from manufacturing to aerospace, impacting design, production and maintenance (Tao et al. 2019). In the building sector, the digital transition, although an essential choice for urban governance, is prolonged due to the built systems complexities, stakeholders' diversity and construction technologies' evolution. However, scholars admit that digitalisation applied to maintenance and management could significantly improve built environments' qualities and companies' business.

The contribution resumes the studies that started with the development of MUST, Maintenance Urban Sharing Tool (already registered at SIAE with progress number D000014807), a digital application designed to involve citizens in buildings' life cycle extension.

Potentialities and gaps of Digital Twins (DTs) applications for urban maintenance are the entry point of this paper. It outlines the feasibility of synchronisation scenarios between the real world and a virtual model, extending the data acquisition process to dynamic and conscious sensors.

DTs support managers and technicians in decision-making approaches on digital clones of physical assets by simulating adaptive scenarios in response to incoming data from real-time operations.

Developed as integrated multi-physics, multi-scale, probabilistic simulations, DTs use the best available physical models and sensors to mirror the life of their real twins (Glaessgen and Stargel 2012).

In a scenario marked by collaborative approaches to the built environment management (The Council of Europe Secretariat in Consultation with the Faro Convention Network (FCN) Members 2018), extending the data acquisition process to citizens' involvement in faults detection is the research hypothesis discussed for a proof-of-concept DT under development in the old harbour waterfront Genoa. The achieved results can be traced back to the definition of renewal coordinates in urban maintenance and DT's functionality extension with data output from MUST.

18.2 Digital Transition for Urban Maintenance

Identifying the built environment as a complex system is an underlying hypothesis for the first studies on maintenance methods and procedures at an urban scale (Ciribini 1986). Since the 90s, with the first British Standards, the scientific community worked for the involvement of administrators, companies and users within a technical, cultural and social process able to manage the stratified and interconnected layers connoting the built environment. Over thirty years, maintenance emerged as an iterative strategy based on archived data, forecasts and inspections aimed at conserving and realigning the performances provided by a building asset within acceptable values, contrasting failure and obsolescence processes (UNI 11257 2007).

An expansion of perspectives, witnessed today by the UNI 13306: 2010 and 10147: 2013 standards (UNI EN 13306 2010; UNI 10147 2013), is recorded with maintenance defined as a complex activity carried out by organisations capable of integrating structured information, technical know-how, strategic and managerial skills and to guarantee, over time, a quality condition (Hauashdh et al. 2022).

The concept of *service* informs the approach implemented in the last decade (Ferreira et al. 2021), helping to trace the cultural coordinates of several experiences in the field, united by the commitment to (1) raise the maintenance procedure to a planned activity, legitimately and definitively included in the project; (2) define the design, executive and inspection techniques to minimise costs; (3) bring communities back into the care and custody process.

The concept of DT introduced by Grieves (2002), with the Conceptual Ideal for Product Lifecycle Management (PLM) as the digital representation of a physical asset and the automatic connections that bind them together (Grieves and Vickers 2017), opens up new scenarios for maintenance planning in the construction sectors. DT is "An integrated multi-physics, multi-scale, probabilistic simulation of a vehicle or system that uses the best available physical models, sensor updates, fleet history and so forth, to mirror the life of its flying twin" (Glaessgen and Stargel 2012).

Despite several projects carried out in the construction sector to develop DTs during the design and execution phases, problems still emerge when the smart city scenario refers to the already built environment (Bujari et al. 2021). Main difficulties can be attributed to: (a) complexity of elements constituting the physical space that should be reported in the virtual product; (b) quality and quantity of data sources that feed the model; (c) communication flows between them. Only a few experiences (Wray 2020) demonstrate the effective convergence between urban maintenance and DTs, showing the improvements that DTs could bring to the maintenance plan (Deng et al. 2021).

DTs are expected to support policymakers and professional practitioners at the urban scale, controlling spaces' and elements' behaviours compared to users' needs and foreseeing performance maintenance through the life cycle (Zheng et al. 2020). According to McKinsey (2015), DTs in urban care processes will provide up to USD 630 billion in potential savings in the construction market in 2025. In this perspective, DTs should have a predictive attitude to evaluate real systems' current and subsequent states and predict failures McKinsey (Tjønn 2019). Furthermore, this prediction improves when the DTs include more information about the physical features, the ongoing process or the operation's characteristics (Errandonea et al. 2020).

Sensors and devices for real-time data acquisition are fundamental for feeding the DTs. Their development has been fostered by the awareness of the economic benefits; these technologies can bring to the building process. An indispensable prerequisite is progress monitoring, which provides wide-ranging information on faults, relating each failure to a system of triggering causes (Olivotti et al. 2019).

Research on DTs in the urban maintenance field collides with technological difficulties in acquiring and transferring data from the real system to the virtual model. The most significant problems are related to various sources' interaction in the information process, such as asset data, operational data, historical maintenance activities and information on constructive transformations, which are decisive for assets' users and managers. Recursive data directly acquired or synthetically generated through simulation are fundamental for DTs that include information on how the performance provided improves after maintenance activities.

18.3 Methods and Materials

The paper resumes and develops further the results already achieved within the project metrics (2013–2017) managed by the Regional High Technology District for Sustainable Buildings, Stress.

The methodology is based on a mixed deductive and inductive approach that integrates the literature review results with a critical analysis of the impacts on urban maintenance reachable through adopting a DT fed by data from the MUST application.

Initially conceived for reporting faults that affect the buildings' functionality, MUST has been developed by an interdisciplinary research team, with the collaboration between architects experts in building maintenance (DiARC), structural engineers (DiSt), computer scientists (ETT) and researchers from the Regional District Stress.

Smau Innovation Award in 2019, as an Italian Excellence Model of Innovation for Companies and Public Administrations, MUST was awarded by the Regione Campania Start-Up 2020. Since 2021, this grant supports the transfer to market process. Within this funding, the Working Group prefigures an open and public model, equipped with different interfaces to meet the diverse needs of the groups involved, allowing citizens to interact and report progress feedback.

The MUST application supports the maintenance process by two main service modules: a mobile application and a web GIS-based dashboard. The mobile application is developed for iOS and android smartphones (available on the marketplaces). Once the user is registered, he can take a snap of the area; the picture is AI processed to provide the user with the best matching label to tag the identified issue. Then, the user can add further notes and submit the notification. The mobile app interoperates with cloud infrastructure (Google Firebase). A web application provides the facility manager with an intuitive dashboard to list, prioritise and supervise the maintenance interventions workflow (Fig. 18.1).

The web dashboard includes a synoptical geo-referred view of the notifications and the possibility to elaborate reports and analytics (space, time, type of issue, etc.). With the support of MUST, the research explores incremental innovation scenarios

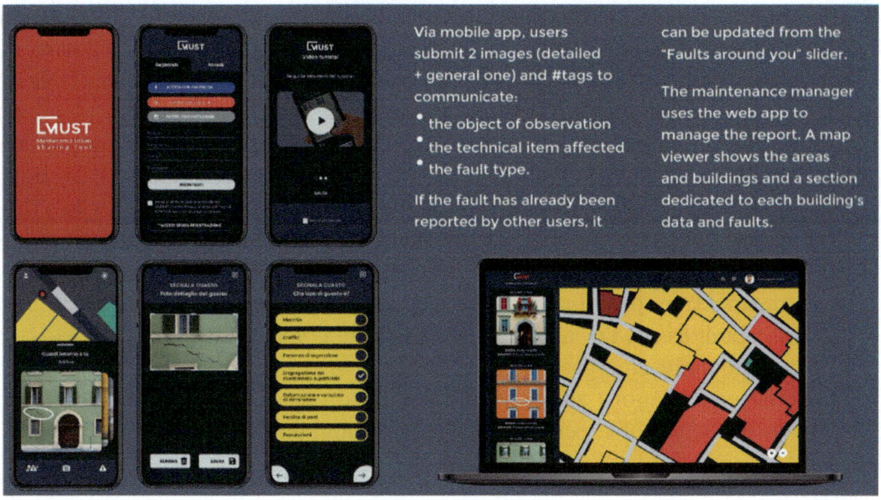

Fig. 18.1 From the mobile app to the facility manager dashboard workflow. Authors elaboration

in the urban maintenance sector, referring the potentialities and gaps analysis to the applied research project Digital and Sustainable Harbor 2030 (DSH 2030—https://www.digitalsustainableharbour.it/), funded by the POR FESR LIGURIA.

18.4 Results and Discussion

The literature review referred to urban maintenance and DTs development returns a framework of studies in constant evolution (Bibri and Krogstie 2021). Focusing on the transition processes towards smart cities and the impacts of enabling technologies, the results can be traced back to the following three research areas:

1. methodological approaches for systematising the urban environment and studying the behaviour of units and components. Built environment care and preservation for long have been activities:
   - very demanding on a theoretical level because they are based on technical skills and analytical skills;
   - not very evident on the executive level because they do not require sophisticated technical skills;
   - not significant on the media level, not very attractive and exciting.
2. failure processes analysis with the support of dynamic and conscious sensors to presume the life span of sub-systems and technical elements. In the urban maintenance segment, predicting and mitigating failure processes affecting the maintenance sets emerge in an international arena as an issue that concerns

buildings users, real estate managers, condominium administrators, technicians and the whole community. Hence, adopting sensors capable of monitoring the state conditions of urban systems is a central issue of many studies (UNI EN 13306 2010; UNI 10147 2013).
3. inspection activities with preventive effectiveness. The literature review provides a framework of studies to reorganise the management coordination for urban services. Reactivating the symbiosis between built assets and communities (The Council of Europe Secretariat in Consultation with the Faro Convention Network (FCN) Members 2018) becomes a horizontal challenge for quality preservation and transmission. An expanded awareness about the need to share responsibilities towards built assets through values matured in contexts and handed down over time profoundly emerges from the analysis of documents produced by international organisations and scholars. A call to engagement and empowerment renews the action of expert knowledge and communities.

According to point 1, urban spatial areas can be assumed as the reference context where the interaction between DT and MUST could successfully work. Within each spatial area, the object of observation is the aggregated maintenance set, identified for an organisational and economic optimisation of the maintenance practice (UNI 11257 2007). The aggregation aims to synthesise and optimise the framework of knowledge, compare the characterisation of the technical elements and the planned activities, relate them to the worksite of the urban space and seek in the latter unexpressed potential and constraints.

According to point 2, a prosumer perspective, crasis of the terms producer and consumer is at the basis of the idea of feeding the DSH 2030 DT with data acquired by users via mobile phones, adopting the MUST application to power the communication between real space and the virtual model.

The adoption of a data flow coming from MUST is subjected to the functioning of the DT, which must be able:

- to integrate various types of data of physical objects;
- to exist in the entire life cycle of physical objects, co-evolving with them;
- to optimise physical objects.

Three groups of stakeholders participate in the interaction between DT and MUST:

- citizens; they are the active component of the maintenance process, which initiates the reporting of faults;
- managers; they access the reports sent by citizens and urge inspections in the event of an alert;
- technicians; they intervene following the warnings with inspections and repairs.

A gradual knowledge strategy is supposed to support citizens in recognising the failure processes of sub-systems and technical components. The occurrence of events capable of affecting the real system identity is anticipated by signals that return in the virtual model, with indications offered by users about the process's extent and consequences. In addition, a database of the states according to which the failure event

Fig. 18.2 A prosumer approach for feeding the DT for urban maintenance. Data flow schematic drawing by ETT. Authors elaboration

can evolve supports the user's involvement in recognising significant conditions to prevent the onset of situations that will lead to the "cessation of the ability of an entity to perform the required function" (UNI 10147 2013).

According to point 3, the predictive and decision-making processes are reconfigured due to the synchronising between real space and virtual models with aware sensors.

The critical analysis of the impacts that a DT fed with data from the MUST application can exert on urban maintenance (Fig. 18.2) outlines:

- a framework of pervasive changes due to digitalisation on companies, markets and business models for urban maintenance. They can help:
  a. increasing the quality of the settlement system;
  b. facilitating the relationship between stakeholders;
  c. reducing times and costs, simplifying methods, promoting the predictive maintenance approach.
- future requirements regarding accuracy, functionality and integrability. They should be respected within the Digital and Sustainable Harbor 2030 (DSH 2030) project. For example, to guarantee that DSH 2030 could be able to predict the urban system response, within each spatial area, to an unexpected event before it occurs, DT should:
  a. be the most realistic representation of a physical asset, incorporating models and available information;
  b. contain all process information and acquire operational, organisational and technical knowledge;

c. always be synchronised with the physical asset;
d. be able to run simulations of physical asset behaviour;
e. be self-evaluative: it is something alive that changes, improves and evolves while maintaining the comparison between physical and virtual space.

Positive organisational consequences of the integration between DT and MUST can be traced back to the DT attitudes to:

- solve accessibility problems; hosted in the cloud, it can be accessed from anywhere and thus can provide information on the status of data;
- forecast of the evolution of failures, simulation of the effects of maintenance action;
- help the operator to perform the specific maintenance task;
- calculate the cost of the maintenance activity.

18.5 Conclusions

Assuming the built environment as a complex, dynamic and adaptive system, the research prefigures an incremental innovation scenario, with digitisation renewing the management services for the maintenance sets through synergies between stakeholders.

A theoretical framework has been traced in the international context considering organisations, municipalities and scholars' commitment to urban maintenance. DTs have been defined as models of physical assets in operation that could be applied to design a more effective preventive maintenance strategy.

Focusing on proof-of-concept DT under development in the ancient port of Genoa, the research hypothesis is to integrate the data flow with a set of information provided by citizens adopting the application MUST. Accuracy, functionality and integrability have been outlined as the requirements for a DT to support managers and technicians with remote control, predictive decision-making approaches and adaptive scenarios.

Extending the life cycle of buildings is a MUST for the community, to which research responds with the expanded sharing of monitoring, updating and programming processes.

References

Bibri SE, Krogstie J (2021) A novel model for data-driven smart sustainable cities of the future: a strategic roadmap to transformational change in the era of big data. Future Cities Environ 7(1):3. https://doi.org/10.5334/fce.116

Bujari A, Calvio A, Foschini L, Sabbioni A, Corradi A (2021) A digital twin decision support system for the urban facility management process. Sensors 21:8460. https://doi.org/10.3390/s21248460

Ciribini G (1986) Il Laboratorio dei Virtuosi. Lo stato emotivo come nuova dimensione progettuale della città. Recuperare 22

Deng T, Zhang K, Shen Z (2021) A systematic review of a digital twin city: a new pattern of urban governance toward smart cities. J Manag Sci Eng 6(2):125–134. ISSN 2096-2320. https://doi.org/10.1016/j.jmse.2021.03.003

Errandonea I, Beltrán S, Arrizabalaga S (2020) Digital twin for maintenance: a literature review. Comput Ind 123:103316. ISSN 0166-3615. https://doi.org/10.1016/j.compind.2020.103316

Ferreira C, Silva A, de Brito J, Dias IS, Flores-Colen I (2021) The impact of imperfect maintenance actions on the degradation of buildings' envelope components. J Build Eng 33:101571. ISSN 2352-7102. https://doi.org/10.1016/j.jobe.2020.101571

Glaessgen E, Stargel D (2012) The digital twin paradigm for future NASA and U.S. air force vehicles. In: 53rd AIAA/ASME/ASCE/AHS/ASC structures, structural dynamics and materials conference, 20th AIAA/ASME/AHS adaptive structures conference, 14th AIAA

Grieves M (2002) Conceptual ideal for PLM. Presentation for the product lifecycle management (PLM) center, University of MI

Grieves M, Vickers J (2017) Digital twin: mitigating unpredictable, undesirable emergent behavior in complex systems. In: Transdisciplinary perspectives on complex systems. Springer International Publishing, Cham, pp 85–113

Hauashdh A, Jailani J, Rahman IA, Al-Fadhali N (2022) Strategic approaches towards achieving sustainable and effective building maintenance practices in maintenance-managed buildings: a combination of expert interviews and a literature review. J Build Eng 45:103490. ISSN 2352-7102. https://doi.org/10.1016/j.jobe.2021.103490

McKinsey (2015) Unlocking the potential of the internet of things. https://www.mckinsey.com/business-functions/mckinsey-digital/our-insights/the-internet-of-things-the-value-of-digitizing-the-physical-world

Olivotti D, Dreyer S, Lebek B, Breitner MH (2019) Creating the foundation for digital twins in the manufacturing industry: an integrated installed base management system. Inf Syst E-Bus Manag 17:89–116

Tao F, Zhang H, Liu A, Nee AYC (2019) Digital twin in industry: state-of-the-art. IEEE Trans Ind Inform 15:2405–2415

The Council of Europe Secretariat in Consultation with the Faro Convention Network (FCN) Members (2018) The Faro Convention Action Plan handbook for 2018–2019. https://www.coe.int/en/web/culture-and-heritage/faro-action-plan

Tjønn AF (2019) Digital twin through the life of a field. In: Abu Dhabi international petroleum exhibition and conference 2018, ADIPEC. Society of Petroleum Engineers

UNI 10147 (2013) Manutenzione. Termini aggiuntivi alla UNI EN 13306 e definizioni

UNI 11257 (2007) Manutenzione dei patrimoni immobiliari. Criteri per la stesura del piano e del programma di manutenzione dei beni edilizi. Linee guida

UNI EN 13306 (2010) Manutenzione. Terminologia di manutenzione

Wray S (2020) Cities today. Bentley systems and Microsoft team up on city digital twins. https://cities-today.com/bentley-systems-and-microsoft-team-up-on-digital-twins/

Zheng C, Yuan J, Zhu L, Zhang Y, Shao Q (2020) From digital to sustainable: a scientometric review of smart city literature between 1990 and 2019. J Clean Prod 258. https://doi.org/10.1016/j.jclepro.2020.120689

Chapter 19
Untapping the Potential of the Digital Towards the Green Imperative: The Interdisciplinary BeXLab Experience

Gisella Calcagno, Antonella Trombadore, Giacomo Pierucci, and Lucia Montoni

Abstract The paper shares the experience of the building environmental eXperience Laboratory (beXLab) at the DIDA Department of Architecture-University of Florence, an interdisciplinary and open research group working on the experimentation of transferable methodological approaches and practical tools to support the challenging twin transitions, the green and the digital, starting from pivotal public buildings. As prototypical university Living Lab (born in the frame of the Med-EcoSuRe project, from here LL), beXLab is a shared space where researchers (architects, technical physicians, energy and information engineers, user experience designers, etc…) are experimenting integrated and innovative retrofit solutions, by involving key actors (decision makers, technical offices, energy managers), stakeholders (technicians, companies) and end-users (students, univerisity community). The LL place and space in the pilot university building of Santa Verdiana (in the UNESCO World Heritage Centre of Florence) was equipped with a bulk of IoT sensors for the monitoring of real-time environmental parameters and used as a test room for innovative retrofit/refurbishment and advanced technologies application, as well as to capture and valorise the users' experience. The beXLab physical space is coupled with a BIM model, as a base for the development of a Digital Twin intended to construct a reliable and shareable image of reality (i.e. energy performances and indoor comfort/well-being assessment) and forecast trustable future scenarios (i.e. simulations towards participative design processes). Our present research, based on a crucial interdisciplinary approach and focusing on the energy-economic-social-cultural implications in the Mediterranean areas, proposes a new way of imagining and representing a sustainable, healthy and green future for buildings and urban

G. Calcagno (✉) · A. Trombadore · G. Pierucci · L. Montoni
Department of Architecture, University of Florence, Florence, Italy
e-mail: gisella.calcagno@unifi.it

A. Trombadore
e-mail: antonella.trobadore@unifi.it

G. Pierucci
e-mail: giacomo.pierucci@unifi.it

L. Montoni
e-mail: lucia.montoni@unifi.it

E. Arbizzani et al. (eds.), *Technological Imagination in the Green and Digital Transition*, The Urban Book Series, https://doi.org/10.1007/978-3-031-29515-7_19

builtup areas, in compliance with the EU Green Deal, the Renovation Wave and the New EU Bauhaus initiative.

Keywords Digital Twin · User experience and awareness · Energy-environmental quality · Monitoring systems · Retrofit strategies

19.1 Introduction

Living in the current built environment is highly impacting both the planet and its inhabitants. Buildings are big consumers of resources and energy, often unable to guarantee minimum environmental quality, and sometimes they badly influence the comfort and well-being of the occupants. If the need to reduce the environmental impacts of buildings is well-known since the energy crisis in the 70s, with the ecology principles entering in the AEC field, this reduction is still an ongoing effort in the climate change era.

Since green challenges in the building sector are as much urgent as complex, digital opportunities can be exploited to manage the building life cycle in a more efficient way, addressing more reliable building processes by supporting a more informed decision-making towards sustainability.

According to the UN Sustainable Development Goals (SDGs), future buildings should rely exclusively on Clean Energy, imperative to contrast climate change, and guarantee Health and Well-being for All in Sustainable Communities and Cities. Foreseeing zero climate impact by 2050, the ambitious EU Green Deal is both a call to update existing buildings in the Renovation Wave strategy and to digitalise the building sector. Moreover, the New European Bauhaus initiative invites to re-think planning and design practices towards creativity and interdisciplinarity, as a basis to envision sustainable, inclusive but also beautiful places.

Notwithstanding both the green and the digital domain have informed AEC research and design practice since decades, the parallel green/digital transition is still moving the first steps, requiring an enlarged interdisciplinary space where digitalisation can support the construction of future sustainable buildings and cities.

19.2 Experimenting Digital Twins for Building Energy Retrofits

In the context of the green and digital transition, Digital Twins represent the most advanced concept, with the possibility to twin/synchronise existing buildings with digital models by adopting real-virtual data bridges, such as sensors and IoTs (Trombadore et al. 2020). It is easy to capture the potential of such synchronisation, with real-time data deepening the comprehension of buildings for a more appropriate decision-making along the entire life cycle. This appears particularly profitable for the purpose of renovating existing buildings, when the availability of a new amount of reliable digital data can be strategic not only to analyse the existing conditions, but also to predict retrofit scenarios and to drive a more sustainable future use, management and functioning of the public building.

Within the Med-EcoSuRe project (Mediterranean Universities as catalyst of Eco-Sustainable Renovation), the DIDA Department of Architecture of the University of Florence founded beXLab (*building environmental eXperience Laboratory*), a Living Lab (LL) in the university building site of the retrofit pilot project (Fig. 19.1). Here, the experimentation with Digital Twin is intended both to collectively construct a reliable and shareable image of the existing and to forecast trustable future scenarios (Trombadore and Calcagno 2022).

beXLab systematises interdisciplinary competencies and methodologies to innovate the renovation of public buildings, by the means of Digital Twins, especially valorising passive solutions that impact on energy performance and human comfort/wellbeing.

The idea of the LL was born in the field of technology of architecture, and it has immediately involved energy engineering for the common effort to find technological solutions for zero/positive-energy buildings. The LL experience in the university public building under renovation allows to constantly get the pulse on the real challenges, to engage strategic actors involved (decision makers, stakeholders and users),

Fig. 19.1 Picture of the beXLab inside

and to take advantage from the local academic community. Working on process innovation to renovate existing buildings, beXLab is in fact enlarging towards information engineering, to tackle the digital side, and user experience design, to valorise the unique human contribution in building energy efficiency and sustainability.

19.3 The Architectural Point of View

The building environmental experience proposed by the beXLab Living Lab has been conceived on well-rooted principles in the discipline of Technology of Architecture.

Focusing on the appropriateness of the human habitat, the discipline is not limited to formal aspects but considers to the environmental and socio-cultural dimensions (Schiaffonati et al. 2011). Over the years it assumed as proper the ecological task of optimisation of buildings' environmental performance, starting from the design process. Moreover, the consideration of design as a social act (Nardi 2010) nourishes the perspective of engagement and participation of users in the transformation of the built environment.

By leveraging on the need-performance theory, the discipline of technology of architecture adopts a systemic approach that has exceeded the boundaries of the material technologies to evolve into softer ones, towards management and governance (Torricelli 2011), to "lead" the even more diversified contributions occurring in the definition of sustainable projects. For this reason, the discipline focuses both on the material and information transformation in the field of architecture, early recognising the role of ICT to support better design processes (Ciribini 1984).

In the wake of this disciplinary culture, the Med-EcoSuRe pilot project was intended as an occasion to innovate the building retrofit process by engaging its actors (academics, decision makers, stakeholders and users) in a common LL space and place of collaboration.

For the development of the pilot project, a building block hosting functions representative of the university life (teaching, study and research) was selected in the historical complex of Santa Verdiana in the UNESCO World Heritage city centre of Florence. Constructed in the 90s on the design of Arch. Roberto Maestro, the parallelepiped volume with north–south orientation has two floors, with a large room on the first one and other two rooms on the ground floor, separated by a central open/covered corridor (Fig. 19.2).

Due to the inconsistency of existing data and information, a survey of the building/plant system was conducted for the population of a preliminary BIM asset information model (LOD C - according to UNI 11337-4, 2017). Such information is useful from the earliest stages of the retrofit process for the definition of a reliable knowledge framework on the existing building, for a preliminary energy audit and for an improved communication in the LL (Fig. 19.3). The digital model has been exploited to analyse and simulate the environmental and energy behaviour of the building through different softwares, deepening the identification of the energy and indoor comfort criticalities, as follows:

Fig. 19.2 South façade of the university pilot building in Florence

Fig. 19.3 Image of the digital model of the pilot building

- Environmental analysis: the visualisation of the sun path in the building context highlights that the south façade with large glass surfaces suffers the absence of any type of shading device. Concerning natural ventilation, cross-ventilation cannot occur in the two rooms on the ground floor, due to fixed fixtures (Fig. 19.4);

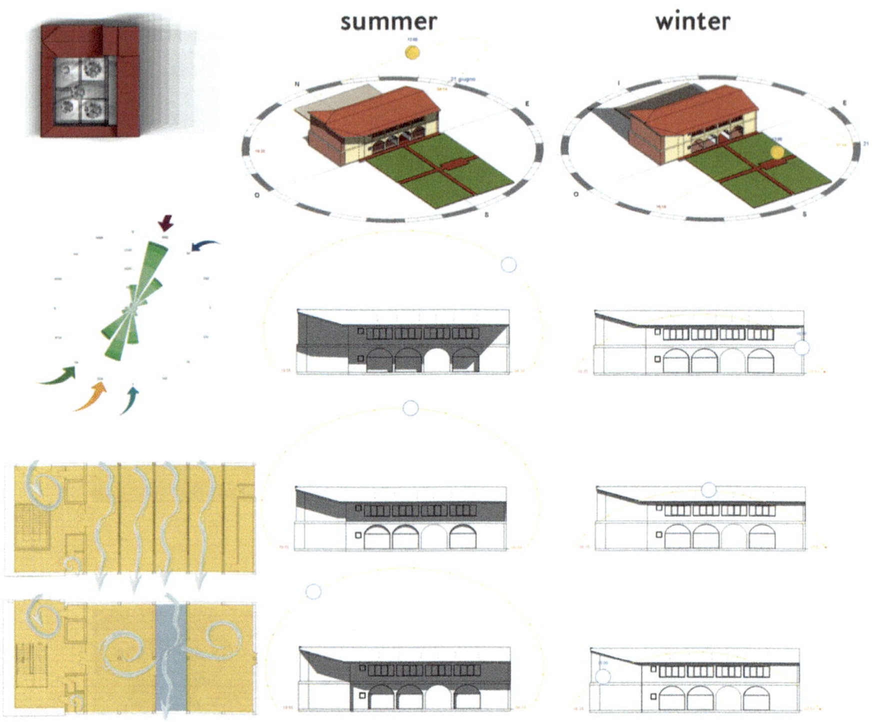

Fig. 19.4 Preliminary environmental analysis of the existing building

- Analysis of the thermo-hygrometric performance of the envelope in dynamic conditions: all the components of the opaque and transparent envelope are characterised by transmittance values that are not fitting the current regulations on building energy performances. The weak behaviour of the envelope overloads the plant system, entailing a high energy demand both for heating and cooling spaces (Fig. 19.5);
- Natural and artificial light analysis: the symmetrical arrangement of the glazed surfaces in the double north–south exposure leads to an uneven indoor distribution of natural light, causing glare phenomena from south and under lighting, especially in the central area of the large room on the upper floor, negatively impacting the visual quality and the energy demand for artificial lighting (Fig. 19.6);
- Simulation of the energy performance: the main energy consumptions for heating and cooling is determined by the need of guaranteeing internal thermal comfort despite the weak envelope behaviour. The main impact on energy consumption for heating in the winter period is attributable to thermal conduction through the vertical components of the envelope, both opaque and transparent. Even more impacting is the energy consumption in the summer period, determined by the negative contribution of solar radiation coming from the south through the large

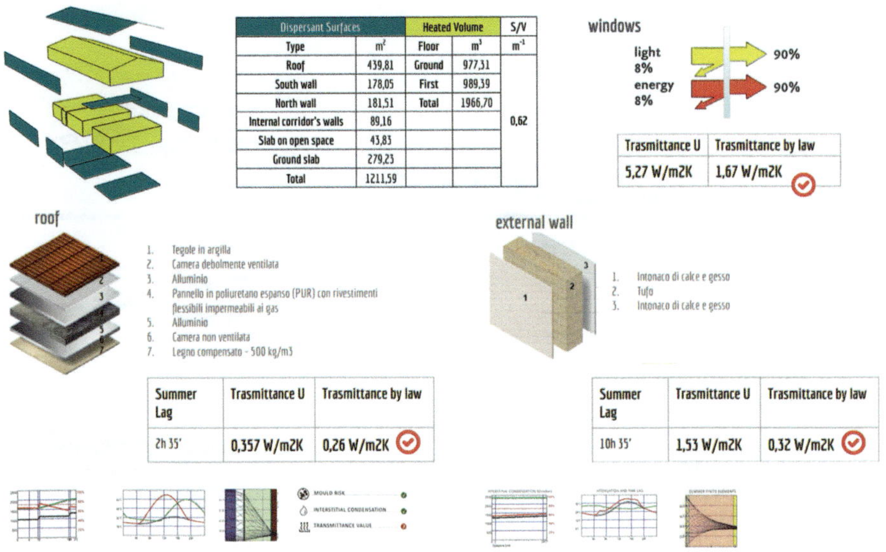

Fig. 19.5 Preliminary thermo-hygrometer analysis of the opaque and transparent components in dynamic regime

and not shaded windows. Another impacting energy consumption is linked to the presence of lighting bodies with low-energy performance (neon) (Fig. 19.7).

Addressing the energy targets of the Med-EcoSuRe project, but also considering the specificities of the context of historical and cultural value, the architectural project for the energy retrofit was set up to solve the energy and indoor quality criticalities of the existing building, and formulated to achieve three main objectives:

- Improvement of indoor comfort (thermal and lighting);
- Energy efficiency (reduction of energy needs and integration of renewable systems);
- Architectural constraints (feasibility and reversibility).

Different retrofit scenarios have been integrated in the BIM model, simulated and evaluated by mixing traditional bioclimatic and innovative building technologies, permitting to appreciate the different degrees of improvement in comparison to the current building conditions, in terms of indoor comfort, energy performances and feasibility (Fig. 19.8).

It is possible to recognise the opportunities in the adoption of digital building models to address the green objectives in the first phases of the retrofit process:

- Standardised definition of the knowledge framework of the existing building to start a reliable retrofit process, representing a trustable and interoperable data and information container;

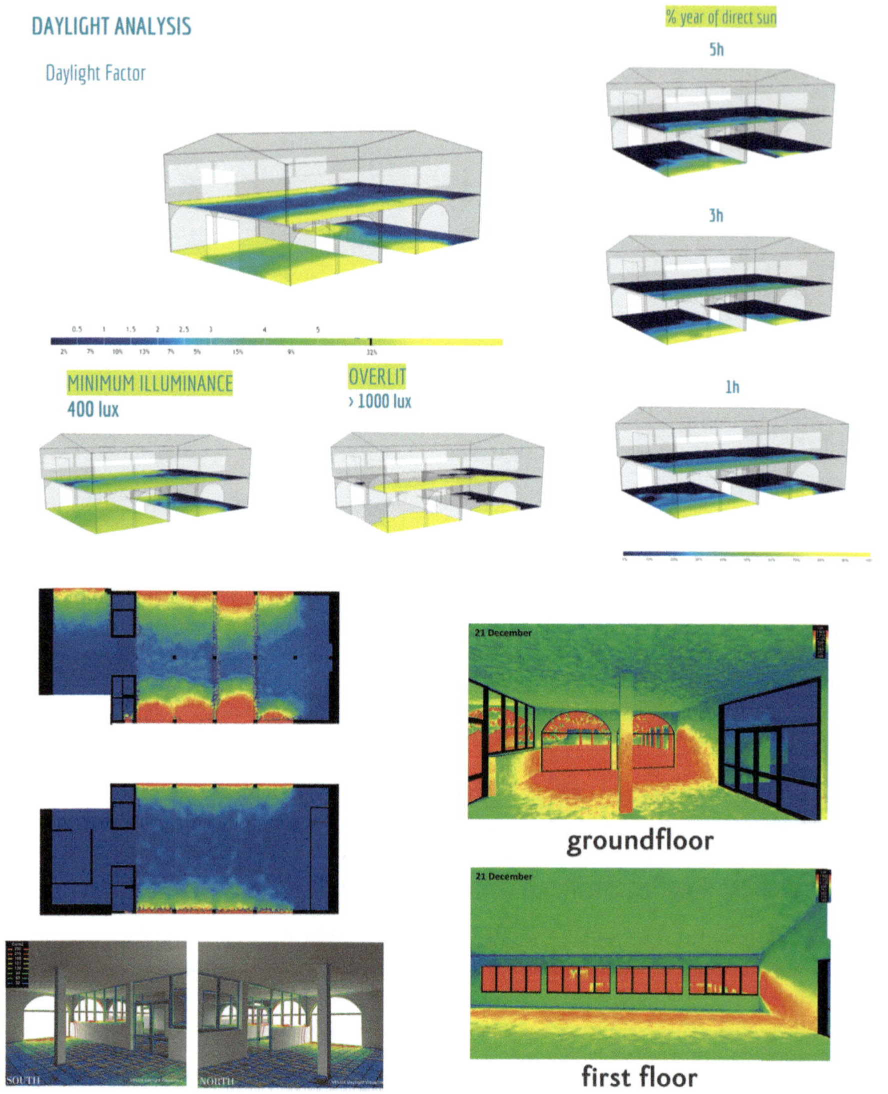

Fig. 19.6 Preliminary analysis on the natural lighting in the pilot building

- Evaluation of the environmental impacts of the existing building through virtual simulations on energy performances and indoor environmental quality, deepening the analysis of criticalities;
- Definition of predictive retrofit scenarios sustaining the planning and design.

Moreover, digital models can be exploited to innovate the subsequent phases of intervention and post-management of the retrofitted building (operation and maintenance).

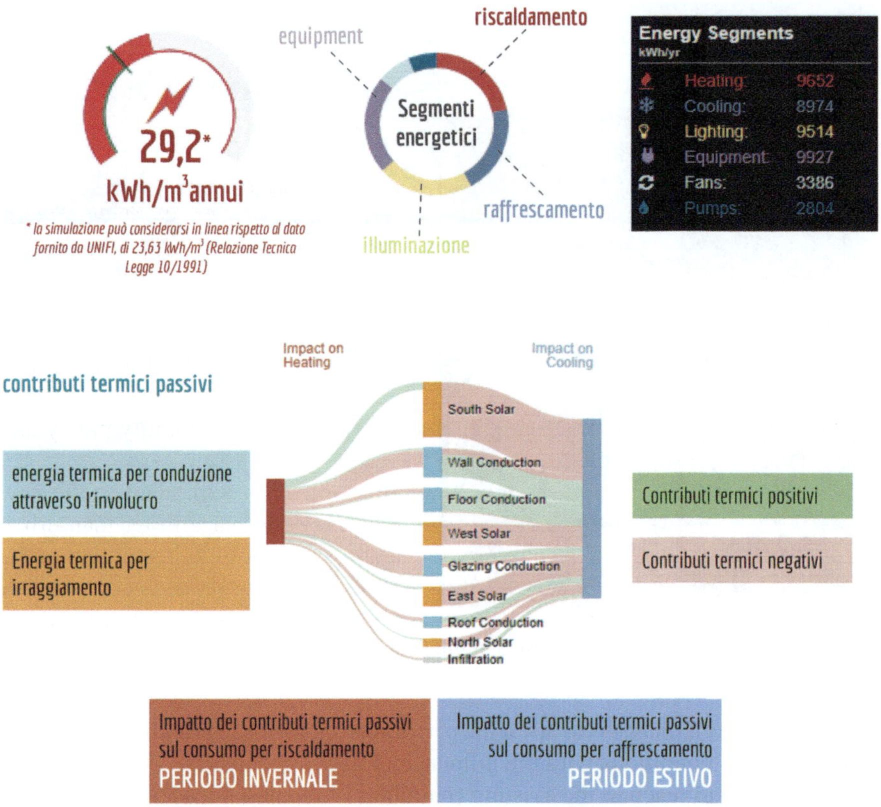

Fig. 19.7 Preliminary simulation of the existing building energy behaviour

Yet, a key factor for the success of the digital modelling towards the definition of feasible scenarios is the validation of the model itself, which can only be achieved through a comparison with real dynamic data. In order to map the main boundary operative conditions, complex monitoring systems are needed together with specific expertise on measurements and data acquisition/post-processing techniques.

19.4 The Energy Engineering Side

Building technologies are, in most of the cases, well-known and universally standardised. National's regulations guide the design of new constructions imposing materials' specs, energy targets and boundary conditions for comfort and well-being (thermo-hygrometric, lighting and acoustic). The major part of recommendations is set on the basis of static calculations and is extrapolated through simplified models.

Fig. 19.8 Preliminary architectural design of different retrofit solutions

The concrete challenge is to attest the performance of buildings and comfort status in real dynamic conditions with the influence of user's occupancy. That aim results even more hard to reach in the existing (buildings of the past decades and historical heritage) where a lack of information is present regarding structures and components such as wall stratigraphy and services. In such a context, it is really important to set up strategies and methodologies which allow acquiring useful data and creating a suitable matrix for the exhaustive description of the living environments. Tools must be configured in a very flexible and modular way in order to be replicable in different sites. The LL is effectively the proper space to configure measurements layouts and test them in operating conditions. Many sensors were installed for real-time monitoring of internal temperature distributions, relative humidity, heat loss flux, lighting levels, air quality as well as external parameters detected through a weather station (see pictures of Fig. 19.9, top). Actually, the amount of data is intentionally large because the setup of redundant data collection lets the cross-correlation, in the research phase, amongst the phenomena acting in the living laboratory involving users. The post-processing of measurement values achieves as a first goal the understanding of the real significance of each sensors' group and the definition of a compact and plug and play system to install in the next projects. In a second phase, the same collected data would be managed and aggregated to validate the digital model in existing context and set up the Digital Twin.

From a technical point of view, the environmental monitoring does not represent a critical aspect since the market proposes many solutions that can be easily

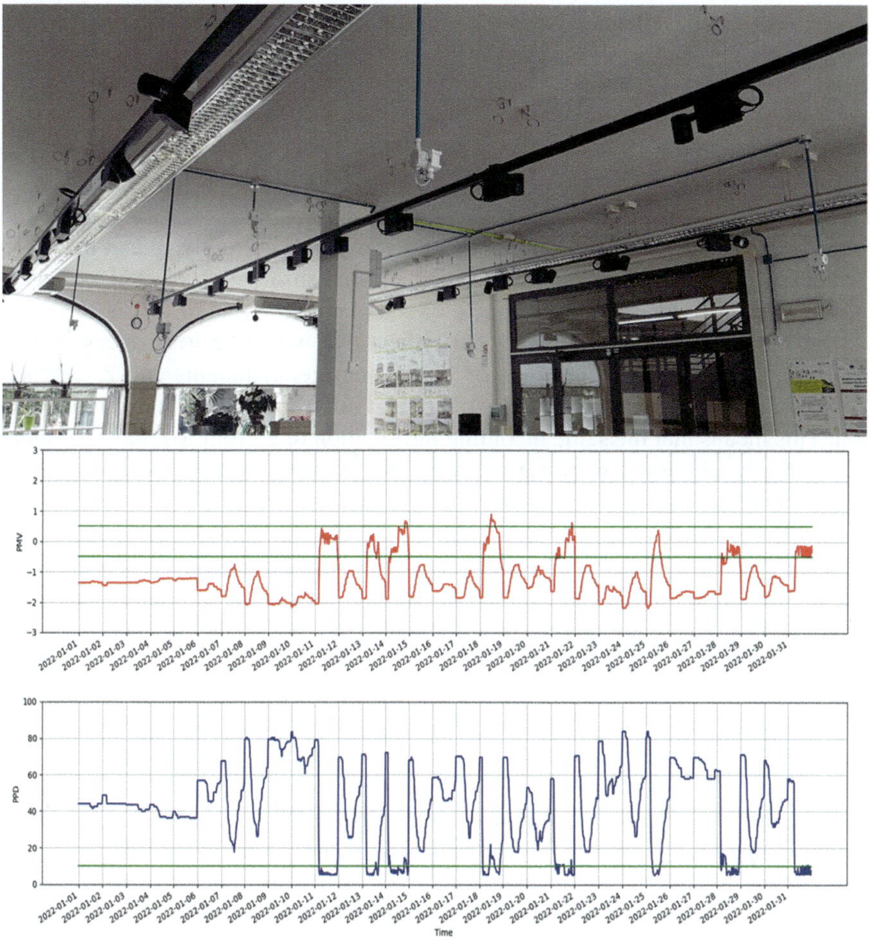

Fig. 19.9 Setup of sensors inside the beXLab (top) and thermo-hygrometric comfort evaluation during January 2022 (bottom)

placed in rooms reducing the impact in the structures, plants and users; the available wireless configurations are in fact very little intrusive, indeed. On the other hand, the elaboration of the huge amount of data that could be collected especially in the long-term periods is a challenging task. Numerical parameters and technical information need to be translated in a simple, clear, educational form to involve the different actors (users, researchers, managers …), to make them aware and to suggest criticalities/opportunities.

For instance, according to national regulations such as UNI EN ISO 7730 (2006), the internal comfort level could be derived and the conduction of the living spaces quantified in terms of predicted mean votes and percentage of dissatisfied as shown in Fig. 19.9, bottom (referred to January 2022). Those objective results would be also compared and validated with the users' subjective perception, expressed through specific ongoing questionnaires (according to EN ISO 10551 2019).

19.5 Opening Towards It and Users' Experience (UX)

Considering the amount and complexity of the row data, currently, one of the pivotal challenges of the LL is the data organisation and visualisation of selected/aggregated information in a smooth, intuitive and pleasant manner.

The idea of experimenting a retrofit project that would be truly accessible and available to all the different kinds of users (managers, students and researchers) requires the customisation of the experience and different levels of data communication. This means that the same incoming information (e.g. from IoT sensors) should be differently addressed to the various users, for example to inform technical offices or disseminate the results to students.

The process of creating info visualisation requires data analysis and a deep understanding of users, of their need and abilities/limitations. The engagement of end-users and UX design in the LL comes with the target of raising the awareness and identifying, analysing and stimulating best human behaviours related to energy, comfort and well-being.

In recent decades, there is a growing phenomenon of translation towards the virtual/digital but also abstract data in visual representation that can easily interact with users and create outstanding quality experiences (Hassenzahl and Tractinsky 2006).

Once the idea of human–computer interaction (HCI) focused just in functional benefits and usability aspects (how to make computers as intuitive as possible), now the concept of UX goes with the idea of interactive technologies, works with a better understanding of knowledge fields, such as emotion graphics, storytelling and linguistics, and takes care of the all contextual aspects.

Despite the growing interest and research on info visualisation and UX in some fields, first of all the cultural one, they are almost undeveloped for the energy efficiency sector, especially concerning retrofitting processes on public buildings. In the age of globalisation and information technology, UX design can encourage the retrofitting process and stimulate the dissemination of contents on energy efficiency and environmental comfort, with innovative and creative solutions to engage people.

19.6 Conclusions

The interdisciplinary processes described in the present paper have been developed over time, and their implementation is ongoing. However, the objective of an effective green/digital transition in the building sector has immediately revealed the need for synergy amongst working groups from different disciplines and methods.

The system "building-plant-user" is so complex and multi-faceted that it could not be addressed from a single and limited point of view, requiring instead many approaches/languages to dialogue together, contaminating the knowledge and skills from "hard" and "soft" sciences.

Moreover, the complexity of the investigated phenomena needs to be simplified in order to spread the results to the largest number of users, each with their own specific background (from building and energy managers to end-users). beXLab is taking up this challenge, proposing itself as an ideal place where the articulated syntax of the "building-plant-user" system could be understood and codified.

In this context, the human being is placed at the centre not only as a user of the environment, where the maximum efficiency and comfort have to be guaranteed but, above all, as the main responsible for the proper management of the living spaces, of their impact on the health of the planet and its inhabitants.

References

Ciribini G (1984) Tecnologia e progetto, argomenti di cultura tecnologica della progettazione. Ed. CELID, Torino

Hassenzahl M, Tractinsky N (2006) User experience—a research agenda. Behav Inf Technol 25:91–97. https://doi.org/10.1080/01449290500330331

ISO 10551 (2019) Ergonomics of the physical environment—subjective judgement scales for assessing physical environments. International Organization for Standardization (ISO)

Nardi G (2010) Percorsi di un pensiero progettuale. Maggioli Editore, Sant'Arcangelo di Romagna

Schiaffonati F, Mussinelli E, Gambaro M (2011) Architectural technology for environmental design. TECHNE J Technol Archit Environ 1(1):48–53. https://doi.org/10.13128/Techne-9434

Torricelli MC (2011) Technological design—research and practice in the architectural project. TECHNE J Technol Archit Environ 1(2):16–23. https://doi.org/10.13128/Techne-9922

Trombadore A, Calcagno G (2022) Mediterranean university as catalyst for eco-sustainable renovation: the experience of Med-EcoSuRe cross-border living lab. In: Sayigh A (ed) Sustainable energy development and innovation. Innovative renewable energy. Springer, Cham. https://doi.org/10.1007/978-3-030-76221-6_34

Trombadore A, Calcagno G, Pierucci G (2020) Advance smart cities through digital twins expanding the knowledge and management capacity of public buildings stock for energy efficiency rehabilitations. In: Contesti—Città Territori Progetti, 1/2020. Firenze University Press, Firenze, pp 129–139

UNI 11337-4 (2017) Edilizia e opere di ingegneria civile—Gestione digitale dei processi informativi delle costruzioni—Parte 4: Evoluzione e sviluppo informativo di modelli, elaborati e oggetti. Ente Nazionale Italiano di Unificazione (UNI)

UNI EN ISO 7730 (2006) Ergonomics of the thermal environment—analytical determination and interpretation of thermal comfort using calculation of the pmv and ppd indices and local thermal comfort criteria. Ente Nazionale Italiano di Unificazione (UNI)

Chapter 20
Digital—Twin for an Innovative Waterfront Management Strategy. Pilot Project DSH2030

Maria Giovanna Pacifico, Maria Rita Pinto, and Antonio Novellino

Abstract In the era of smart cities, the digital twin of a settlement system allows not only the real-time control of the quality levers offers by the subsystems, but also the prediction of the future performance over the life cycle. This is feasible through the implementation of predictive models and the simulation of the impact that the design solutions can generate. The Digital and Sustainable Harbour 2030 (DSH2030) project, funded under the Liguria region's POR FESR, sees the cooperation of the Innovation, Development, and Sustainability structure of the Porto Antico of Genoa, with ETT S.p.A. supported by the Department of Architecture of Naples, Netalia S.r.l., BF Partners S.r.l., Colouree S.r.l., AiTrust S.r.l., Circle Garage S.r.l., and the University of Genoa in the MaLGA structure. In particular, the research question expressed by the local authority of the Porto Antico of Genoa concerns the measurement and evaluation of environmental parameters, in relation to the consumption and production of renewable energy; monitoring of the flow of people and vehicles (land and sea) for both security and commercial purposes; the control of safety performances and usability of the built system. A complex virtual model is the answer to which the partnership is working on. The paper illustrates the criteria and principles that inform the design, testing, and validation of an enhanced digital twin for the tourist port of Genoa. Specifically, it discusses the work carried out by ETT S.p.A. with the Department of Architecture of Naples that, through the integration of sensors, measurement technologies, Internet of Things (IoT), machine learning technologies, intends to prefigure new models of sustainable management and maintenance of the port area.

Keywords Digital twin · Smart harbor · Internet of Things

M. G. Pacifico (✉) · M. R. Pinto
University of Naples Federico II, Naples, Italy
e-mail: mariagiovanna.pacifico@unina.it

M. R. Pinto
e-mail: pinto@unina.it

A. Novellino
ETT Spa, Genoa, Italy
e-mail: antonio.novellino@ettsolutions.com

E. Arbizzani et al. (eds.), *Technological Imagination in the Green and Digital Transition*, The Urban Book Series, https://doi.org/10.1007/978-3-031-29515-7_20

20.1 Introduction

European Union and the international organizations attention (European Commission 2019; Patto dei sindaci per il Clima e l'energia Europa, https://www.pattodeisindaci.eu/about-it/l-iniziativa/il-patto-in-cifre.html) placed on the cities and peri-urban environments mitigation effects has emphasized the fundamental importance of day-by-day monitoring. The latter is aimed at obtaining a real-time built system and subsystems control, prediction of future performance, and implementation of models that can simulate the design and development solutions impacts. In addition, among the strategies settled within the PNRR for urban regeneration, the city system digitization takes a central role, launching toward the development of Smart Cities, in view of innovative, inclusive, and sustainable cities (European Commission 2016). In this perspective, the Digital Twin (DT) becomes a strategic tool for the implementation of innovative and sustainable management strategies of urban systems, for its configuration as a digital copy of the physical environment, the relationships between the parties and the processes in place, past and future. It is also a useful tool for the implementation of participatory processes that through the empowerment of citizens generate democratic cities. To that end, the paper focuses on harbor DT applied to the Ancient Port of Genoa that can be visualized through a platform made by multiple layers and different categories of data in virtual reality for collaborative and participatory processes, focusing on port management, security, and decision support. The approach described supports citizens and decision-makers and planning professionals with tools to achieve partnership. The quantitative and qualitative data collected through different type of sensors provide a clear reading of the port complexity, involving all the stakeholders in the management and developing process.

20.2 Background Scenario: Digital Twin for Smart Settlement System

The realization of a smart governance using information and communication technologies (ICT) is closely linked to the development of smart cities in order to improve decision-making through better collaboration among different stakeholders, including public administrations and citizens (Lara et al. 2016; Pereira et al. 2018). One example is the use of IoT to address citizens' needs and to improve their activities performances, to reduce negative impacts on the environment, to reduce costs of living (Mishra and Chakraborty 2020), and to ensure quality of built system and users' satisfaction (Pinto 2019).

Many countries and governments consider smart cities a solution to global warming, population growth, and resource depletion (Deng et al. 2021). Especially, if related to the built environment monitoring, e.g., fault detection and diagnostics system using sensor networks and AI-based modern technologies, they could generate some benefits such as reducing maintenance costs, reducing energy consumption and associated costs, increased productivity, and extended equipment life (Brunone et al. 2021; Halmetoja 2022).

At this point, defining a digital model of a settlement system is the core to develop innovative strategies of knowledge management where monitoring through sensors, measurement technologies, IoT, machine learning technologies provides a real-time update (Kaur et al. 2020) and offers an interface that allows monitoring the past and present operation and makes prediction about the future (Grieves 2014; Wang et al. 2021). Digitization, and specifically DT, appears as an opportunity to catalyze and monitor the processes of transformation and development of cities in a perspective of circular economy and sustainability.

The DT must be considered as a process (Wenner et al. 2021), a dynamic model made of data that have a permanent and constant connection with the real world. It is more frequently adopted in facilitating facilities management operation to optimize operation efficiency at both building level and city level, in order to better answer to the requirement for a sustainable and appropriate city and space management, particularly in the current pandemic and eventually the post COVID-19 pandemics, where occupancy detection is an important part of the facility management to ensure users' safety (Hou et al. 2021).

Liu et al. (2018) describe the components of DT: physical products in real space, virtual products in virtual space, and the connections of data and information that will tie the virtual and real products together; and the basic architecture that consists of sensor and measurement technologies, Internet IoT, and machine learning. Alongside this, Qi et al. (2021) design five dimension of DT: physical entities, virtual models, data, services and connections in digital twin.

Some prototypes have tested the functioning of the DT in the context of settlement systems, defining the potential of this tool. Dembski et al. (2020) have described the prototype of Herrenberg's DT, in Germany, where the virtual model is a model 3D of the built environment, and the data model consists in data from volunteered geographic information. The experimentation has revealed that urban digital twin allows to gain a better understanding of potential solutions for urban challenges improving the interaction processes between citizens on the one hand and official local representatives on the other. This provided the opportunity to test a series of scenarios and potential solutions as well as evaluate their impacts using a real-life case.

Another experimentation (Lu et al. 2020) was carried out within West Cambridge site of the University of Cambridge, here the DT simulate the behavior of a system of buildings, integrating heterogeneous data sources, supporting intelligent data query, and providing a smarter decision-making. This demonstrator integrates an as-is multi-layered IFC Building Information Model (BIM), building management system data, space management data, real-time IoT-based sensor data, asset registry data, and an asset tagging platform. The demonstrator also includes a tool for improving asset maintenance and another one for asset tracking using augmented reality (AR) and equipment failure prediction.

Within the smart cities debate, port digitalization is defined as a complex phenomenon that concerns multiple aspects to control and monitor, and as Inkinen et al. (2021) suggests is a generic term, and it refers to the adoption, collection, storage, analysis, and use of digital information in ports and port communities. It is manifested through digital platforms, and it has impacts on operational management causing changes in organizational work cultures and practices in ports (Heilig et al. 2017).

20.3 Materials and Methods

The Digital and Sustainable Harbour 2030 (DSH2030) project, funded under Liguria region's POR FESR, has the aim of developing and enhancing a Digital Twin of the Porto Antico of Genoa to supervise and optimize the management processes of the area. The construction of the digital model is based on collecting data from different sources and moving through IoT, 5G, and AI.

The area object of the research project (Fig. 20.1) is the one under the management of Porto Antico di Genova SpA, included within the perimeter of the Expo of Genoa defined according to art. 3 c.1 l. August 23, 1988 n. 373, an area of over 130,000 m^2 of

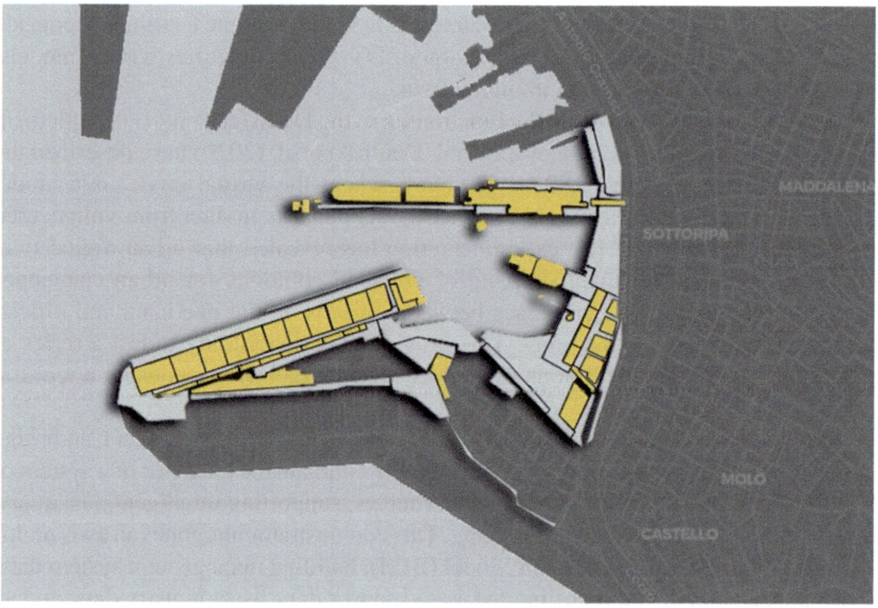

Fig. 20.1 Expo area managed by Porto Antico di Genova SpA: the buildings in yellow and the soil in gray

which 71,000 m^2 of covered area and 59,000 m^2 outdoors, characterized by different building typologies, different activities and uses.

Its current configuration is the result of a process of urban regeneration that began essentially in 1992 and that still continues today in the structuring and completion of projects that aim to connect the city to the sea.

DT become the ambient where the administration can accurately simulate and plan the area's activities in a way that is healthy, safe, and comfortable for visitors and those who work there, efficient in terms of resource management for those who manage it, and sustainable for the environment.

In order to design a proper and useful DT model, it was significant having a dialogue with the policymakers to well understand how supporting their decision-making and create a virtual ambient through which to more easily manage the port system.

According to this view, at the basis of the platform for data exploitation, aggregation, and management, interviews were conducted with key stakeholders (managers of: Event Management and Procurement and Facility Management offices) to properly understand themes to be addressed and how. Therefore, in order to improve and maximize the operability of the area management structure, a platform to administrate is being implementing in order to make available IoT, geo-referenced data, processed and produced data (Nuzzo et al. 2021). The management of the input data coming from different categories of sensors, through the structuring of diversified information layers for each type of data, will allow the manager to keep under control the performance of the systems, the flows of people moving within the area, anomalous consumptions.

Navigating in the different parts of the platform, that provides digital maps, with overview, zoom, and selection functions, allows users to view and interact with the data and metadata that include links to retrieve further documentation and references (Fig. 20.2). In relation to the above, IoT technologies are essential to make every object, person or building a source of data usable in real time; advanced and increasingly powerful video analysis algorithms, thanks to artificial intelligence approaches, make surveillance cameras powerful and advanced smart sensors able to collect a large amount of information; data storage requires a secure and reliable cloud. Finally, artificial intelligence through machine learning techniques allows to extract information content of great relevance from the acquired data.

The prototype developing within DSH2030 project and visualized in virtual reality, for supporting a collaborative management platform, is set up as follows: a 3D model of the built environment provided by local authorities (Traverso 2021), data collected from an heterogeneous type of sensors (mobile devices, sensors, video content analysis), a multi-layer platform for aggregate data exposure and management.

Particularly, the collection of data for monitoring is structured on three levels: The first level is through the installation of smart meters for energy production and consumption, air quality, weather forecasts, sea level, water consumption; the second level is through the installation of cameras; the third is through mobile applications developed within the project that allow to receive from users alerts about faults on

Fig. 20.2 Integrated and multi-layer platform for the port management

the built in systems on one hand and their position over time on the other. Regarding the monitoring for maintenance strategies, it was decided to apply and extend the field of competence of the Maintenance Urban Sharing Tool application (Viola and Borriello 2017). It is a product of a synergy between Department of Architecture (DiARC) and Structures for engineering and architecture (DiSt) of the University of Naples, Stress Scarl and ETT SpA. This tool, that is also object of a doctoral ongoing research, is composed by a mobile app to send alerts for the detection of faults and performance anomalies on subsystems and a Web application to incorporate the notification sent by the app users for the purpose of timely, effective, and efficient management of inspections and maintenance interventions. Moreover, the application MUST has been complemented with a function that allows monitoring of water meters (Fig. 20.3); this implementation derives from the administration's need to have a digitized monitoring of water meters, which is currently managed by writing data on sheets of paper that are, only later, digitized by the hand of an operator.

The optimization of management processes, through the reading of data on the platform from constant monitoring, has as a further objective the improvement of the attractiveness and safety of the area by promoting a digital and sustainable development.

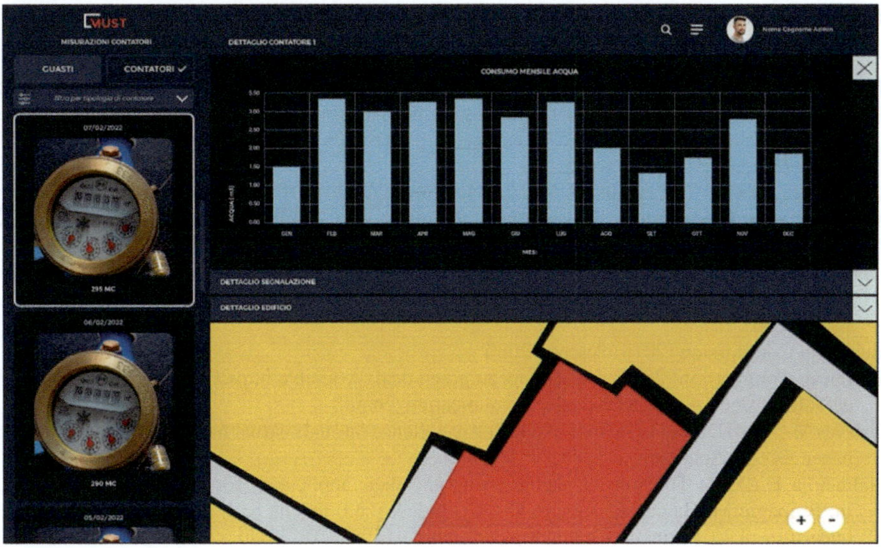

Fig. 20.3 Concept for the dashboard aim to monitor water consumption

20.4 Conclusions

The DSH2030 project therefore has a strong innovative character, across all its main components. It proposes a holistic view of real/virtual space that makes it vividly ‘intelligent’ position data, presence data, projection of displacements and accesses, etc. together with the data acquired by the distributed sensors allow to carry out elaborations, correlations, scenario simulations that are made available in the dashboards dedicated to the area manager (Porto Antico di Genova SpA). In this way, it is possible to close the chain between those who manage the area, who plan the activities and action in case of critical situations, and the final are users.

In the context of European development policies, the research outlines innovative ways and applications of knowledge management oriented to promote sustainable management and development of waterfronts. The enhanced Digital Twin will allow a real-time monitoring of the port system performances. DSH2030 is a living demonstrator of how data-driven tools also allow to effectively support both the needs expressed by the Porto Antico SpA, as operator, of continuous optimization of all processes (economic, security, environmental) for which it is responsible, and to offer adaptive tools and models able to effectively react to sudden changes in the scenarios of use and expectations of end users, as well as possible future new emergencies.

References

Brunone F, Cucuzza M, Imperadori M, Vanossi A (2021) From cognitive buildings to digital twin: the frontier of digitalization for the management of the built environment. In: Wood additive technologies. Springer tracts in civil engineering. Springer, Cham. https://doi.org/10.1007/978-3-030-78136-1_5

Dembski F, Wössner U, Letzgus M, Ruddat M, Yamu C (2020) Urban digital twins for smart cities and citizens: the case study of Herrenberg, Germany. Sustainability 12:2307. https://doi.org/10.3390/su12062307

Deng T, Zhanf K, Shen Z (2021) A systematic review of a digital twin city: a new pattern of urban governance toward smart cities. J Manag Sci Eng 6(2):125–134. https://doi.org/10.1016/j.jmse.2021.03.003

European Commission (2016) Agenda 2030

European Commission (2019) The European green deal. Available https://ec.europa.eu/info/sites/default/files/european-green-deal-communication_en.pdf

Grieves M (2014) Digital twin: manufacturing excellence through virtual factory replication. White paper. Google Scholar

Halmetoja E (2022) The role of digital twins and their application for the built environment. In: Bolpagni M, Gavina R, Ribeiro D (eds) Industry 4.0 for the built environment. Structural integrity, vol 20. Springer, Cham. https://doi.org/10.1007/978-3-030-82430-3_18

Heilig L, Lalla-Ruiz E, Voß S (2017) Digital transformation in maritime ports: analysis and a game theoretic framework. Netnomics 18(2–3):227–254. https://doi.org/10.1007/s11066-017-9122-x

Hou HC, Remöy H, Wu H (2021) Digital twin to enable smart heritage facilities management: a systematic literature review. In: Proceedings of the 27th Pacific Rim Real Estate Society annual conference, Pacific Rim Real Estate Society (PRRES)

Inkinen T, Helminen R, Saarikoski J (2021) Technological trajectories and scenarios in seaport digitalization. Res Transp Bus Manag 41. https://doi.org/10.1016/j.rtbm.2021.100633

Kaur MJ, Mishra VP, Maheshwari P (2020) The convergence of digital twin, IoT, and machine learning: transforming data into action. In: Farsi M, Daneshkhah A, Hosseinian-Far A, Jahankhani H (eds) Digital twin technologies and smart cities. Internet of things (technology, communications and computing). Springer, Cham. https://doi.org/10.1007/978-3-030-18732-3_1

Lara AP, Da Costa EM, Furlani TZ, Yigitcanla T (2016) Smartness that matters: towards a comprehensive and human-centred characterisation of smart cities. J Open Innov Technol Mark Complex. https://doi.org/10.1186/s40852-016-0034-z

Liu Z, Meyendorf N, Mrad N (2018) The role of data fusion in predictive maintenance using digital twin. AIP Conf Proc 1949(1):020023. AIP Publishing. Google Scholar

Lu Q, Prlikad P, Woodall P, Ranasinghe GD, Xie X, Liang Z, Konstaninou E, Heaton J, Shooling J (2020) J Manage Eng 33(3). https://doi.org/10.1061/(ASCE)ME.1943-5479.0000763

Mishra KN, Chakraborty C (2020) A novel approach toward enhancing the quality of life in smart cities using clouds and IoT-based technologies. In: Farsi M, Daneshkhah A, Hosseinian-Far A, Jahankhani H (eds) Digital twin technologies and smart cities. Internet of things (technology, communications and computing). Springer, Cham. https://doi.org/10.1007/978-3-030-18732-3_2

Nuzzo I, Caterino N, Novellino A, Occhiuzzi A (2021) Computer-aided decision making for regional seismic risk mitigation accounting for limited economic resources. Appl Sci 11(12):5539. https://doi.org/10.3390/app11125539

Patto dei sindaci per il Clima e l'energia Europa. https://www.pattodeisindaci.eu/about-it/l-iniziativa/il-patto-in-cifre.html

Pereira GV, Parycek P, Falco E, Kleinhans R (2018) Smart governance in the context of smart cities: a literature review. Inf Polity. https://doi.org/10.3233/IP-170067

Pinto MR (2019) Conoscenze integrate e strategie di manutenzione programmata per il patrimonio culturale. In: Pinto MR (ed) Coordinare le conoscenze per la manutenzione del patrimonio culturale. Clean Edizioni, Napoli. ISBN: 9788884976123

Qi Q, Tao F, Hu T, Anwer N, Liu A, Wei Y, Wang L, Nee A (2021) Enabling technologies and tools for digital twin. J Manuf Syst 58:3–21. https://doi.org/10.1016/j.jmsy.2019.10.001

Traverso S (2021) Database topografico del comune di Genova e modelli in ottica digital twin. Intervento tenuto alla Smart Week di Genova il 2/12/2021

Viola S, Borriello F (2017) MUST: App per il riconoscimento de guasti da parte degli utenti nei centri storici. In: Capozzi R (ed) Approcci, metodologie, procedure e tecniche per la riqualificazione e la resilienza dei centri storici e degli edifici di pregio. Il progetto come verifica sperimentale [Approaches, methods, procedures and techniques for regeneration and resilience of the historical centers and buildings of prestige. The project as experimental control], pp 210–213. ISBN: 978-88-8497-493-8

Wang K, Hu Q, Zhou M, Zun Z, Qian X (2021) Multi-aspect applications and development challenges of digital twin-driven management in global smart ports. Case Stud Transp Policy 9(3):1298–1312 (2021). https://doi.org/10.1016/j.cstp.2021.06.014

Wenner M et al (2021) The concept of digital twin to revolutionise infrastructure maintenance. In: Proceeding of 27th ITS world congress, Hamburg, Germany, 11–15 Oct 2021. Available https://www.homeport.hamburg/wp-content/uploads/2021/04/Wenner-et-al._The-Concept-of-Digital-Twin-to-Revolutionise-Infrastructure-Maintenance.pdf

Chapter 21
BIM and BPMN 2.0 Integration for Interoperability Challenge in Construction Industry

Hosam Al-Siah and Antonio Fioravanti

Abstract Interoperability is a growing challenge for the construction industry in general, especially for the designing process, where it is exposed to many challenges due to the most critical part of this sector that related to heterogeneous information exchange. Particularly, during the implementation of a project where there is a need for sharing and exchanging a huge amount of data among several actors to accomplish the design process. Therefore, the need for real supportive tools has emerged to facilitate the process of data collection and digitalization in order to automate the whole process. However, different kinds of issues prevent improving the interoperability in the ACE industry. This paper focuses on the barriers of improving the interoperability in this industry sector and proposes a new method of linking and collecting the data from different actors. To this objective cloud storage for flowcharts and building information model "BIM" have been used. One of the best flowcharting languages—Business Process Modelling and Notation "BPMN" 2.0—has been adopted, where the data will be collected and the process will be explained and connected directly to the BIM model to be reviewed, used, and saved.

Keywords Design process · Building management · Interoperability · BPMN · BIM

21.1 Introduction

The designing and management tools within the construction industry have been improved significantly during the last century, which could be referred mainly to process complexity, and the new level of detailed project with large number of actors' data that have to be managed. Moreover, the improvement of the computer's

H. Al-Siah (✉) · A. Fioravanti
Sapienza University of Rome, Rome, Italy
e-mail: hosam.alsiah@uniroma1.it

A. Fioravanti
e-mail: antonio.fioravanti@uniroma1.it

E. Arbizzani et al. (eds.), *Technological Imagination in the Green and Digital Transition*, The Urban Book Series, https://doi.org/10.1007/978-3-031-29515-7_21

capability and the progressive introduction of ICT in Architecture Engineering and Construction "AEC" industry from one side simplified and made easier processes and checking, from the other side contributed to make work more complex and fragmented due to divergent specializations. ICT was so popular that was created the terms computer-aided design "CAD" and computer-aided manufacturing "CAM." Despite the great benefits of using the computers, it was initially limited to engineering 2D/3D drawings and "hand science" model calculations without focusing much on other aspects—a part of advanced research field. Building ICT improvement led to suggest building information modeling "BIM" as a repository of many possible information related to the building that are used in different activities related to manage, design, and control different buildings project aspects.

Since it first introduction by professor C. Eastman in the 70s, BIM helped in developing the industry, where it was supposed to be a repository of project information to facilitate the exchange of information within the same studio and then with other studios and related actors. Nowadays, BIM is a full-fledged, object-oriented, three-dimensional model that represent a digital building to track physical characteristics and functions. The adoption and implementation of BIM tools increased globally among the major players of the construction industry over the recent years, for example, it reached 99% in UK in 2020 (NBS Enterprises Ltd. 2020) (Fig. 21.1).

The adoption and implementation of the BIM tools by the stakeholders (Fig. 21.2) has emerged the benefits they got of using them; Ku and Taiebat (2011) identified seventeen BIM functions that can be used by General Contractors whereas Langar

Fig. 21.1 BIM adoption in UK over time. *Source* NBS Enterprises Ltd. (2020)

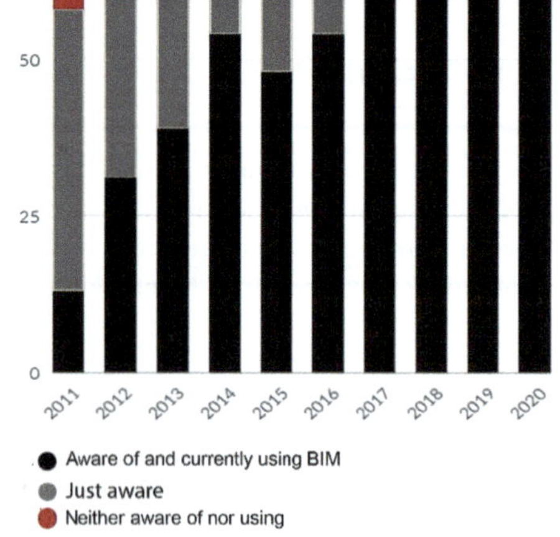

and Pearce (2014) identified eighteen such BIM functions for designers. These and many other studies encourage the software houses and researchers in the digital AEC field to invest in this technologies and in the related tools; however, there is obviously a gap between the BIM utopia promised and the current available tools, where the practical experiences proved that BIM tools alone are not enough as the current situation, where during the years the vendors, developers, and researchers gave—unintentionally sometimes—a lot of promises the technology could not achieve yet. The prof. R Miettinen and his team showed four elements in order to reach the BIM utopia, firstly, solving the issue of using single BIM data or a variety of BIM models together with other software and tools, secondly improving the collaboration within BIM tools; thirdly, they discuss problems of using BIM during the whole lifecycle of the building which is a challenge has to be solve; finally, they argue that there is no conclusive evidence of increasing the productivity after using BIM tools alone which is point has to be clarified (Miettinen and Paavola 2014).

Typically, the construction project on one side involves many actors who use wide range of programs and tools to manage, design, control, and construct the projects and save different kinds of data and information. On the other side, the today technology has inadequate interoperability among the actors, and there is still a dearth of investigations addressing interoperability issues in the regarding of code point of view and the concept one. In this paper, we analyze the main interoperability challenges for the ACE industry regarding to the use of BIM tools and explained the barriers that prevent the development of the technology in the construction industry from the first point of view. Furthermore, we present the proposed solutions as have been represented in BIM literature and make a propose about the cloud storing with a central database to face this challenge.

21.2 Interoperability in Construction Industry

The development of the construction industry through history created new projects, more complex and have a lot of characteristics which makes them extremely difficult to predict their different outcomes. That convinced the stockholders in ACE industry to develop new ways and tools to control the whole project's processes and improve the exchanging of information and data among all the stakeholders of the project.

Interoperability in the context of the Information and Communication Technology—ICT—could be defined as the ability of different systems, computerized products, and/or individuals to work together and exchange information with each other even when they have divergent fields (Carrara et al. 2009b). This makes it one of the most important things for develop any industry, especially the construction industry where millions of information have to be exchanged among many actors. However, many issues prevent the development of interoperability which we named as interoperability barriers, which could be categorized into three main types: conceptual, technological, and organizational barriers.

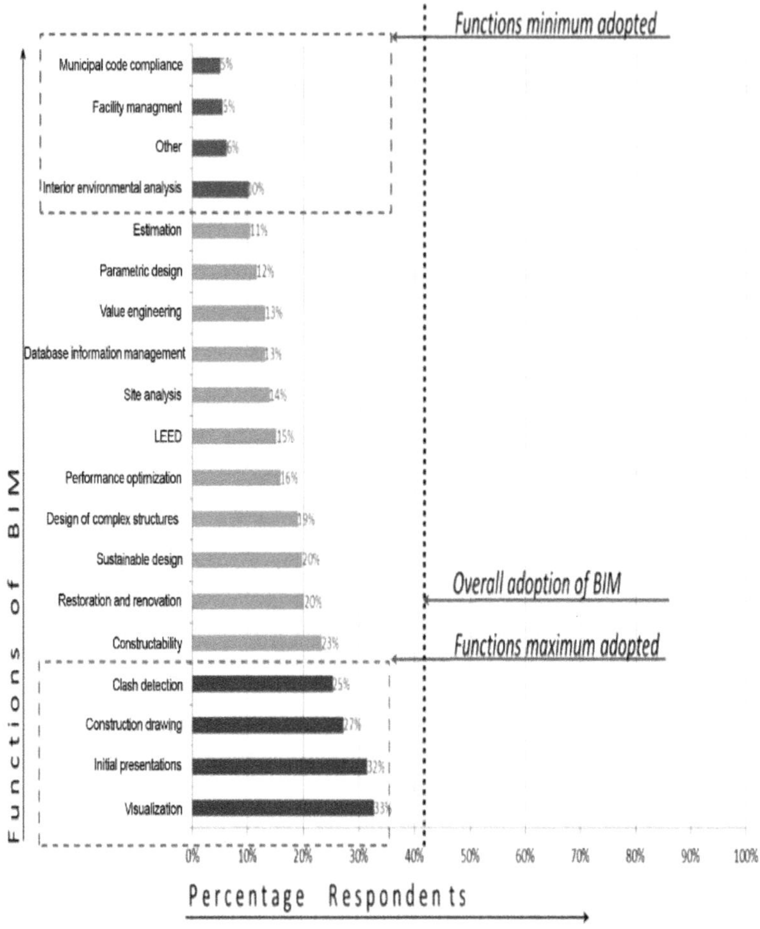

Fig. 21.2 Adoption of functions of BIM by architectural firms (Langar and Pearce 2014)

21.2.1 The Conceptual Barriers

The conceptual barriers are mainly related to the syntactic and semantic incompatibilities of information exchanged between the actors (Ullberg et al. 2009); these problems concern the modeling at a high level of abstraction and how different actor could understand the same data and information in different ways (Carrara et al. 2009a), which consider as a big challenge for the industry development, where the data mostly are collected without explanation—called implicit information which—create different kinds of problems, for example, what the architect sees as a door or a window to connect the room with the garden the energy engineer will see them as thermal bridge have to manage and take in the consideration and so on. This opens a discussion about the actors' different goals and the future of the building information management itself where the main idea of it is to be a repository of project's data that will used during the building life cycle, which was a great idea when it first introduced, clashes the today industry situation where is required to reaches the point of knowledge exchanging instead data among the actors.

21.2.2 The Technological Barriers

The technological barriers are mainly the problems that related to the ability of the actors to effectively send, import/export, and use information and data between each other, moreover, those problems related to use the computers or ICT to communicate and exchange information (Ullberg et al. 2009). The challenges from this point of view are more related to the standardization of the files and the data formats or natural format, where through the history the number of developers increased, where almost each developer using its private data format to use their files which create the need to standardize a common standard. The first standard developed was IGES one which was use within The National Aeronautics and Space Administration—NASA—as an independent agency in the U.S. Since then, many data formats and standards have been developed and used widely like Design Web Format, Standard for the Exchange of Product Data, and Industry Foundation Classes. The last one is the most detailed, developed, and popular open standard available nowadays. IFC standard managed and developed by buildingSMART as a non-profit international organization that aims to improve the exchanging of information between software applications that used in the construction industry. BuildingSMART International has implemented the IFC standard using XML technologies as ifcXML specification (BuildingSMART 2007). XML is a platform-independent language for representing data and has been used in the development of Web service applications. However, the performance of Web services has shown a significant decrease when using XML data causing by the low efficiency of reading and parsing XML data during the execution of services.

21.2.3 The Organizational Barriers

This kind of barriers focus on the problems from the organization point of view, where the challenge here are concerned with the incompatibilities of organization structure and management techniques implemented in different enterprises and related to human problems like task distribution, responsibility, human resources management and authority (Ullberg et al. 2009).

With the first suggestion of the idea of building information modeling and during the decades of adopting it for the designing and managing the projects within the construction industry, many promises have been made about the ability of the new

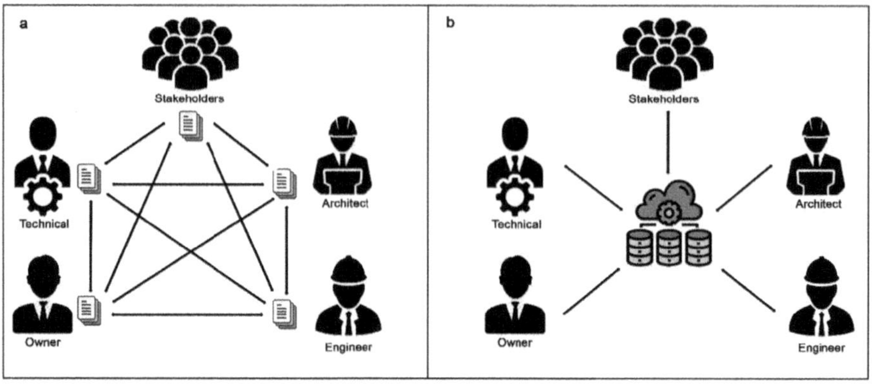

Fig. 21.3 **a** Data restricted in documents; **b** central clouding database. *Source* Own work

technology to exchange data; however, many studies established it as one of the biggest challenges for the development of the technology (Cerovsek 2011; Sattler et al. 2019; Singh et al. 2011; Tchouanguem Djuedja et al. 2019). This problem is referred to different reasons; firstly, there is a lack of multidisciplinary coordination among different stakeholders of the project; secondly, since there are many developers in this field, it become difficult to integrating different models from different sources; moreover there is a difficulty in exchanging the information between the actors of the project at the organization level; thirdly, the technology faces problem in the modifying the models using different tools due to the alternative perceptions which depend on the purpose of use; finally, different kinds of data are not easily shareable or useable, where although many documents are in electronic documents—like PDF—there is a problem in extracting the information from them; in other words, there is a rigidity in the information that is saved in this kind of files, which make them not dynamically useable. In additional to that this way required the actor to share the file with all the related actors (Fig. 21.3).

Consequently, it was understood that overcoming this challenge will bring many benefits to the entire construction industry, where it will be reflected on the efficiency and productivity within the projects on the organizational and administrative level and enhance the environmental sustainability due to better understanding and sharing models.

21.3 BIM and BPMN Integration

BPMN nowadays is one of the most advanced flowcharting tools available, where it provides an intuitive notation to represent any business process in the form of flowcharts, with a comprehensible representation of constructs defined in software-execution language and came with a clear standard which facilitates communication and understanding.

The proposed method is based on two main concepts, cloud storing and connecting, and central database, where in this method the tools are divided into three parts (Fig. 21.4). The first part—in the lift of Fig. 21.4—is the BIM model as a comprehensive repository of data related to the building during its lifecycle; the connection with this part will depend on the scripting codes to import and export data; in this part, data will be saved and used by the designers. The second part—in the right of the Fig. 21.4—part consists of the model of Business Process Model and Notation "BPMN," where this model will contain the business process description flowcharts and clarify the actor responsibilities within the process depending on the country codes where the project will be implemented. The last part will be an interface program to make the live connection between the two models allow data collected from the BIM to populate BPMN, with availability to view the entities details and description. With this connection method, we aim to overcome the rigidity of the data within the documents and speeding the designing process during the project management.

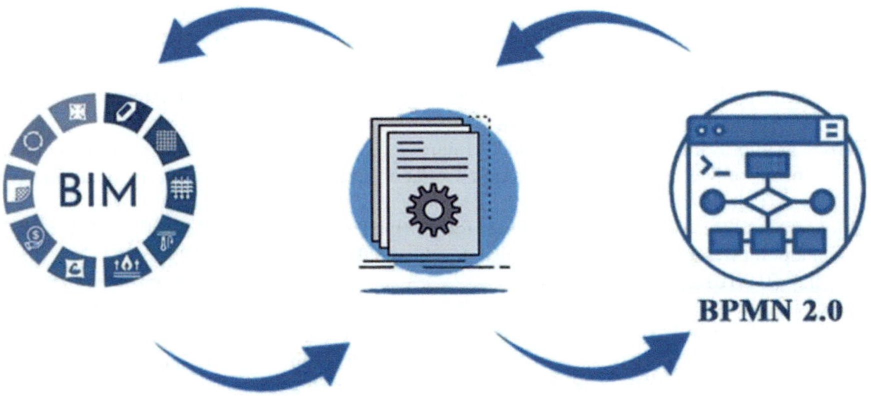

Fig. 21.4 BIM and BPMN construction data connection. *Source* Own work

21.4 Conclusion

The projects within the construction industry become more detailed through the years, which makes the right managing and exchanging information between the actors at the right time as a challenge for the industry; in this paper, we reviewed in details the types of barriers that prevent improving the interoperability within the construction industry and categorized them to three types conceptual that related to the understanding data, technological that related to the transferring data, and organizational that related to the responsibility of managing data.

In the end, we proposed a methodology adopting the idea of central database cloud connection and suggest creating link between BPMN—as a process flowcharts and collection data from the stakeholders, and BIM—as a repository of building information. This approach is supposed to overcome the data rigidity in different files and to adopt processes flowchart to increase the sharable knowledge between the actors, where the BPMN model will carry on the knowledge that will be shared.

Acknowledgements The research has been partially financed by the National Interest Research Program 2017 (PRIN—Progetti di Ricerca di Rilevante Interesse Nazionale–Prot. 2017EY3ESB) titled "A Distributed Digital Collaboration Framework for Small and Medium-Sized Engineering and Construction Enterprises" of MUR (Italian Ministry of University and Research).

References

BuildingSMART International Alliance for Interoperability (2007) ifcXML implementation guide. History v.2. Accessed 18 May 2022
Carrara G, Fioravanti A, Nanni U (2009a) Knowledge-based collaborative architectural design. Int J Des Sci Technol 16(2):1–16
Carrara G, Fioravanti A, Kalay YE (2009b) Improving design quality of complex building systems by means of ICT enhanced collaboration. Collaborative working environments for architectural design. Palombi Editori, Rome, pp 3–18
Cerovsek T (2011) A review and outlook for a "Building Information Model" (BIM): a multi-standpoint framework for technological development. Adv Eng Inform 25(2):224–244. https://doi.org/10.1016/j.aei.2010.06.003
Ku K, Taiebat M (2011) BIM experiences and expectations: the constructors perspective. Int J Constr Educ Res 7(3):175–197. https://doi.org/10.1080/15578771.2010.544155
Langar S, Pearce A (2014) State of adoption for building information modeling (BIM) in the southeastern United States. In: The Associated School of Construction international conference, Washington DC, Mar 2014
Miettinen R, Paavola S (2014) Beyond the BIM Utopia: approaches to the development and implementation of building information modeling. Autom Constr 43:84–91. https://doi.org/10.1016/j.autcon.2014.03.009
NBS Enterprises Ltd. (2020) 10th annual UK's national building specification report. https://www.thenbs.com. Accessed 18 May 2022
Sattler L, Lamouri S, Pellerin R, Deneux D, Larabi M, Maigne T (2019) Multi-domain BIM interoperability: queries as data-exchange vectors. In: International conference on industrial engineering

and systems management (IESM), 25–27 Sept 2019, pp 1–6. https://doi.org/10.1109/IESM45758.2019.8948082

Singh V, Gu N, Wang X (2011) A theoretical framework of a BIM-based multidisciplinary collaboration platform. Autom Constr 20(2):134–144. https://doi.org/10.1016/j.autcon.2010.09.011

Tchouanguem Djuedja JF, Karray MH, Foguem BK, Magniont C, Abanda FH (2019) Interoperability challenges in building information modelling (BIM). In: Proceedings of the I-ESA conferences, vol 9, pp 275–282. http://doi.org/10.1007/978-3-030-13693-2_23

Ullberg J, Chen D, Johnson P (2009) Barriers to enterprise interoperability, enterprise interoperability. In: Second IFIP WG 5.8 international workshop, IWEI 2009, Valencia, Spain, Oct 13–14, 2009. Lecture notes in business information processing, vol 38. Springer, Berlin, pp 13–24. http://doi.org/10.1007/978-3-642-04750-3_2

Chapter 22
Digital Twin Approach for Maintenance Management

Massimo Lauria and Maria Azzalin

Abstract After years of slightest attention to the environment, low productivity, and least rates of technological innovation, the construction sector has started a slow but in-depth review of its statutes and priorities. The ongoing ecological and digital transition opens to new opportunities connected to the implemental policies of Industry 4.0—at now Industry 5.0—and related enabling technologies. Opportunities that strongly reaffirm the need for innovative, responsible, and sustainable governance of the life cycle of buildings, placing it in the new perspective of Digital Twin approach. Starting from this scenario, the paper presents some ongoing upgrade of a maintenance management model expressly aimed at optimizing activities in the operation and maintenance phase from which evident economic, environmental, and social extra costs arise.

Keywords Digital twin · Maintenance · openBIM · Internet of things · Cloud

22.1 Background

A new "stage" in the historical sequence of Industrial Revolutions is arising, and the current Fourth one—started in 2014 with the launch of Industry 4.0—is now rapidly changing its statutes (Schwab 2016).

The EU report "Industry 5.0: Toward more sustainable, resilient and human-centric industry" strongly affirms the need to speed up the underway digital and ecological transformation to deeply restore environment, economy, and communities. Three are the key-principles: centrality of man, self-sustainability, flexibility, and resilience. A new Collaborative Industry and Super Smart Society evolves

M. Lauria · M. Azzalin (✉)
Mediterranean University of Reggio Calabria, Reggio Calabria, Italy
e-mail: maria.azzalin@unirc.it

M. Lauria
e-mail: mlauria@unirc.it

E. Arbizzani et al. (eds.), *Technological Imagination in the Green and Digital Transition*, The Urban Book Series, https://doi.org/10.1007/978-3-031-29515-7_22

characterized by intelligent cooperation between machines and humans (European Commission 2021a).

The ongoing transition, digital and ecological, opens to innovate opportunities come from the implemental policies of Industry 4.0—at now Industry 5.0—and related enabling technologies.

Many sectors, from aerospace to aeronautics, from complex industrial scenarios to automotive and medicine too, are using, with increasing pervasiveness, digitization, automation, Digital Twin approaches. After years of low productivity, equally reduced rates of technological innovation and least attention to the environment, also the construction sector seems to have finally begun a slow but in-depth review and innovation of its statutes and priorities as well as of its operativeness.

The urgency of responsible and sustainable governance of built environment is broadly matched by three important challenges of European Green Deal—circularity, digitalization, ecology—and by their related strategic plans: "Circular Economy Action Plan" (European Commission 2020a), "A Europe fit for the digital age" (European Commission 2020b), and "A Renovation Wave for Europe. Greening our buildings, creating jobs, improving lives" (European Commission 2020c).

The first, in promoting the principles of circularity throughout the life cycle of buildings, introduces policies that increase material efficiency, reduce climate impacts, and improve durability and adaptability of buildings in the life cycle (European Commission 2020a).

The second, by encouraging digital transformation, also affirms its role as a tool for managing climate change and achieving the green transition through a human-centered approach (European Commission 2020b).

The third formalizes the goal of doubling the upgrading of existing building stock by 2030, giving it greater energy efficiency and contributing to the decarbonization process (European Commission 2020c).

New challenges that re-address the issues of energy efficiency and sustainability by strongly linking them to a whole life cycle approach and to an increasing attention for operational and maintenance phase, O&M. A new perspective that recognized O&M as the longest and most qualitatively articulated phase of the life cycle, responsible for about 80% of total costs and the greatest environmental impacts (Aghimien et al. 2018).

In this scenario, it is strategic to properly manage an increasing amount of diverse information: from performance data under use conditions to behavioral and experiential aspects related to end users, their well-being, and their level of satisfaction (Bilal et al. 2016).

With specific reference to O&M phase, several tools have long been available: computerized maintenance management system and computer-aided facility management, CMMS and CAFM, building automation systems, BASs. To these tools are now added innovative approaches that exploit new interoperability tools able to facilitate the sharing of all available information and implement the potentialities of managing data acquired by real-time monitoring of building performance.

These tools are based on open standards—Industrial Foundation Classes, IFC (ISO 16739:2018)—and data specifications—Construction Operations Building information exchange, COBie (NBIMS-US-V3.4:2015) that are specifically for the exchange of information between the design and use phases (Cabinet Office 2012).

BuildingSMART (https://www.buildingsmart.org/) has long been engaged in their development and transfer into international standard. The aim is defining common information requirements and languages that could be shared among all different operators involved in each phase of building process and life cycle (ISO 19650-1:2018; UNI EN 17412–1:2021).

Both IFC and COBie have been assumed as tools to support Service Life Planning (ISO 15686 Series), through the definition of "sets of IFC properties" (ISO 15686-4:2014) that may be used in the application of service life assessment methods (Factor Method, ISO 15686-2:2012), in defining environmental impacts (ISO 15686-6:2004 Withdrawn), in Life Cycle Costing (ISO 15686-5: 2017), in structuring of feedback of data from practice (ISO 15686-7:2017) (Patacas et al. 2015).

Now, interoperability tools, building information modeling, BIM, Internet of Things, IoT, domotics and building automation control systems, BACS, open to innovative paradigms and opportunities related to Digital Twin approach in buildings.

22.2 Digital Twin Approach

In 2011, Michael Grieves introduced the term Digital Twin, DT, to define the synchronization between two realities: physical objects in real space, virtual objects in virtual space (Grieves 2011).

Physical and virtual objects are linked through the mutual exchange of data throughout the entire life cycle, both in real time and asynchronously (Bouchard 2016).

In 2019, the Gartner Inc. placed Digital Twin among the five emerging trends that would drive technology innovation for the next decade (Gartner Inc. 2019).

Nowadays, Digital Twin is one of principal technological nodes of Industry 5.0, capable of interacting physical and virtual objects, Big Data, IoT, blockchain, machine learning, and artificial intelligence, AI (Evans et al. 2019).

In the construction sector, Digital Twin permits optimizing the life cycle of buildings and infrastructures through the management of all available information including that acquired through performance simulation and real-time monitoring.

It allows predicting future performance; experimenting, simultaneously, changes and/or improvements without having to test them on the product itself or on special mock-ups; developing simulations in use with changes in state and boundary conditions, and machine learning.

Its application gets concrete and precise usefulness at O&M phase in structuring decision-making processes, planning predictive maintenance strategies, reducing the impact on the environment, improving the comfort and satisfaction of end users.

Furthermore, the real-time update of data from the IoT system and sensors allows continuous detection of state of health, operational conditions, anomalies, downtime and inefficiencies, preventing potential risk situations.

Three are the essential elements of a Digital Twin for the construction sector.

- Smart Building—physical object equipped with sensors, software, and technologies capable of gathering data or accessing information.
- BIM model—digital object holding geometric, physical, functional, and behavioral data related to physical object.
- IoT—a network of objects interconnected through the Internet integrated with analytical data communication network.

BIM model and Smart Building concern purely technological and operational aspects. Although fundamental, nevertheless they represent only partial components of the whole digital ecosystem that comprehends also human aspect (Pasini et al. 2016).

Digital Twin allows interaction between people and buildings maximizing the user experience and transferring it into life cycle and O&M decision-making processes, calibrating needs, methods, times, and costs.

22.3 Digital Twins for a Maintenance Management Model

BIM model makes already possible: to identify and correct detections and/or interferences between architectural and/or structural and/or plant engineering projects; to increase efficiency in time management during construction phase; to verify accuracy and completeness of information and their circulation among various operators.

In O&M, phase is not the same yet.

Nowadays, the Digital Twin approach applied to O&M represents an area with still many criticalities nevertheless extremely challenging too (Delgado and Oyedele 2021).

The recent EU report on "Digital Building Logbooks," DBL, encourages the use of Digital Twin approach. It deals with the definition of a common European approach covering all relevant building information, enable to synergies, interoperability, data consistency, and information exchange. It supports the widespread use of DBLs across Europe as tool able to contribute to several high-profile policy initiatives already mentioned in the paper (European Commission 2020a, b, c, 2021a, b).

In Italy, the National Recovery and Resilience Plan and the National Research Plan 2021–2027, albeit from different perspectives, affirm the strategic importance of the Digital Twin approach for permanent monitoring of structural integrity and operational functionality of buildings and infrastructures, as well as for realization of smart grids, smart infrastructures, smart buildings.

The diffusion of BIM methodologies, their integration with IoT technologies and related applications through Digital Twin approaches, the development of open standards, IFC and COBie, as well as the availability of properly structured, accessible and updatable data and the adaptation of legacy systems represent the frontier of the new challenges of research and standardization.

They also constitute some focuses of R&D actions carried out by BIG srl, Building Innovative Governance, Academic Spin-off of which the authors are two founders.

Among current activities, carried out in partnership with ACCA Software Spa, there is the proposal of a maintenance management model, MMM, for innovative governance of real estate assets (Lauria and Azzalin 2020). The MMM promotes an active dialogue between the actors and operators involved in O&M phase, allows sharing all available information, introduces innovative ways of collecting, processing and management data aimed at activating predictive maintenance strategies. Thanks to real-time monitoring features, it also configures a potential "observatory" of over the time changes of building components and products, of their operation and modes of use.

The MMM uses Digital Twin approaches, enhancing the potential of ICT and Geographic Information Systems, GIS. It offers the integration between BIM model, IoT technologies domotic and building automation control systems, for monitoring of performance in use, acquisition and real-time recording of data.

It operates in a digital cloud environment, usable with a browser, putting together viewer technologies, virtual reality, and augmented reality (Fig. 22.1).

Its level of technological maturity is at now a TRL 6 (Technology demonstrated in a relevant environment).

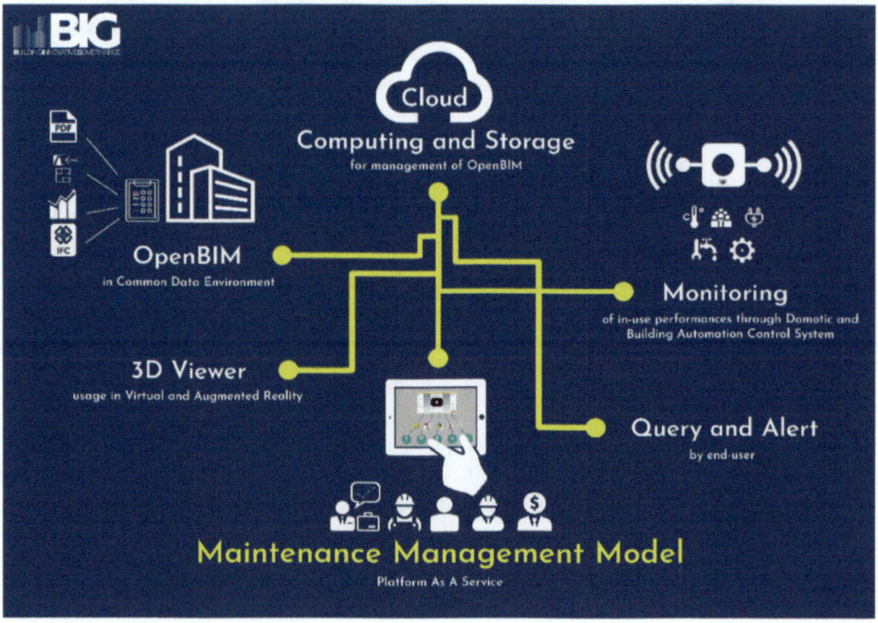

Fig. 22.1 Maintenance management model. *Source* By Authors (Lauria and Azzalin 2020)

The MMM is part of the proposal related to the start-up of BIG srl financed with POR Calabria Region funds (2019/20); recently it has obtained two-year funding under Smart&Start Italia Facilitation Program of Invitalia spa, for its implementation, prototyping, patenting, experimentation, and marketing.

It has also received several awards confirming the innovativeness of the proposed idea and won two separate editions of the "Innovation Award" of the Chamber of Commerce of Reggio Calabria in 2019 and 2021.

Closely related to the MMM upgrade actions is the research activity: Infinity BIM research. Design, construction, and maintenance of digital building model for the Digital Twin, funded by MISE, whose general objective is to create new products for the AEC sector supply chain through the application of BIM methodologies.

The experimentation of the MMM, currently in progress, is at now applied to two study cases: the Citadel of the *Mediterranean* University of Reggio Calabria (Fig. 22.2), and the industrial plant of Mangiatorella spa, a mineral water bottling company, one of the most important in southern Italy (Fig. 22.3).

Fig. 22.2 Citadel of the Mediterranean University of Reggio Calabria, Italy. *Source* By Authors

Fig. 22.3 Mangiatorella spa, a mineral water bottling company, Stilo, Italy. *Source* By Authors

In particular, two ongoing activities are strictly functional to the upgrade of the MMM: structured collection of information and digitization of all available information. They both constitute the already operative services delivered and commercialized by BIG srl.

In a digital twin approach perspective, the two aforesaid activities permit advanced use and share of available information by different types of operators/users with mobile technology, cloud, RFID tags.

At now, the implementation of BIM model (openBIM format) is carrying out through the definition and writing of specific parameters for maintenance activities planning—a Digital Twin for maintenance—with a first level of connection between interoperability systems and informative capital (IFC format).

Since now, referring to collecting information, the outcomes confirm the criticalities expected and expressed in EU Report on DBL (European Commission 2021b), concerning the lack of organized and structured archives that make existing information not only available but also above all easily accessible and usable.

Regarding digitization, the possibility of using reverse engineering approaches—"scan to BIM" procedures—allows the control and integration of information derived from the project documentation made available by the client.

The two activities introduced—structured collection of information and digital modeling—both converge in the experimentation of Digital Twin for maintenance and in the related definition of the MMM. It is based on a preliminary deductive action that, starting from the classification of building elements as codified by UNI 8290, aims at verifying the possibility to identify homogeneous families of sets of specific properties (characteristics, performance, failures, anomalies, controls, etc.).

These will be assumed for the purposes of interoperable management of information in the O&M phase, also according to the information required for compiling the three operational documents of maintenance plan—a mandatory document at Italian normative level (DPR 207/2010 art. 38).

A delimitation of application field referring to window and thermal conditioning system has been assessed. Such delimitation has been assumed with the aim of defining the relative need and/or the necessary information contents.

The quantity and degree of confidence of the information to be included in the BIM models refers to an open process of continuous implementation. All available information will be used in relationship with those required from national standards (UNI 11337-4), from new definitions of Information Needs (ISO 19650-1:) and from the structuring of specific Information Delivery, IDs (UNI EN 17412-1:2021).

The experimentation, after the phases above introduced and subsequent validation of specific IDs, will provide for verification of its replicability.

22.4 Conclusions

The first results of research activities underway show the need to define since the beginning the correct flow of information with respect to each of the operators involved (who); the timing (when) with respect to which this information must be included and/or must be available; the methods (how and where) of sharing. Needs confirmed in the mentioned EU Report on DBL (European Commission 2021b).

Today, the technologies and the various hardware, software and cloud applications are increasingly powerful, able to manage and process a greater amount of information, favoring its sharing.

Unresolved nodes remain, however, the precise information framework to refer to, the interoperability of information, as well as the management of bidirectional interoperability between modeling and simulation software.

Issues that represent the goals of the ongoing activities, here only in part introduced, and carried out by the research group.

These aims, as also stated by BuildingSMART at the various national chapters, refer therefore to a basic criticality, not yet resolved, which is the need for coding and/or shared classification of information. A theme that is certainly crucial from the point of view of the use of interoperability formats and so for the development of the Digital Twin approach for O&M phase.

References

Aghimien D, Aigbavboa C, Oke O, Koloko N (2018) Digitalisation in construction industry: construction professionals perspective. In: Shiau J, Vimonsatit V, Yazdani S, Singh A (eds) Streamlining information transfer between construction and structural engineering. ISEC Press

Bilal M, Oyedele L, Qadir J, Munir K, Ajayia S, Akinadea O, Owolabi H, Alaka H, Pasha M (2016) Big data in the construction industry: a review of present status, opportunities, and future trends. In: Advanced engineering informatics, vol 30, no 3, pp 500–521. http://doi.org/10.1016/j.aei.2016.07.001

Bouchard J (2016) Digital twins: identical, but different. Oliver Wyman Partner

buildingSMART. https://www.buildingsmart.org/

Cabinet Office (2012) COBie data drops

Delgado JMD, Oyedele L (2021) Digital twins for the built environment: learning from conceptual and process models in manufacturing. Adv Eng Inform 49. https://doi.org/10.1016/j.aei.2021.101332

European Commission (2020a) Circular economy action plan

European Commission (2020b) A Europe fit for the digital age

European Commission (2020c) A renovation wave for Europe—greening our buildings, creating jobs, improving lives

European Commission (2021a) Industry 5.0: towards more sustainable, resilient and human-centric industry

European Commission (2021b) Study on the development of a European Union framework for digital building logbooks. Final report

Evans S, Savian C, Burns A, Cooper C (2019) Digital twins for the built environment. An introduction to the opportunities, benefits, challenges and risks. White paper. Built Environment Institution of Engineering and Technology

Gartner Inc. (2019) Gartner survey reveals digital twins are entering mainstream use. Stamford, Conn

Grieves M (2011) Virtually perfect: driving innovative and lean products through product lifecycle management. Space Coast Press

Lauria M, Azzalin M (2020) A maintenance management model. Upgrading and experimentation. In: Serrat C, Ramon Casas J, Gibert V (eds) Current topics and trends on durability of building materials and components. Proceedings of the XV edition of the international conference on durability of building materials and components (DBMC 2020), Barcelona, CIMNE, pp 677–684. http://doi.org/10.23967/dbmc.2020.001

Pasini D, Ventura SM, Rinaldi S, Bellagente P, Flammini A, Ciribini ALC (2016) Exploiting internet of things and building information modeling framework for management of cognitive buildings. In: 2016 IEEE international smart cities conference (ISC2), pp 1–6. http://doi.org/10.1109/ISC2.2016.7580817

Patacas J, Dawood N, Vukovic V, Kassem M (2015) BIM for facilities management: evaluating BIM standards in asset register creation and service life. ITcon 20:313–331

Schwab K (2016) The fourth industrial revolution. Portfolio Penguin, UK

Chapter 23
Digital Infrastructure for Student Accommodation in European University Cities: The "HOME" Project

Oscar Eugenio Bellini, Matteo Gambaro, Maria Teresa Gullace, Marianna Arcieri, Carla Álvarez Benito, Sabri Ben Rommane, Steven Boon, and Maria F. Figueira

Abstract Finding reliable and safe accommodation is a key obstacle to students' international mobility. While the European Commission plans a tri-fold increase of Erasmus+ participants by 2027, allowing international students to get suitable accommodation remains one of the main difficulties encountered during the mobility experience. European Higher Education Institutions (HEIs) are unable to cover the majority of student accommodation demand. Indeed, as stated by Eurostudent VII Report, only 17% of students in Europe find home abroad in student accommodation facilities. Moreover, in accordance with the Erasmus+ Impact Study 2019, 23% of the students involved in the survey considered very important to have support in finding accommodation abroad during mobility along with insurance and other practical aspects. This contribution explores the first results of the European project *HOME* (*Home of Mobile Europeans*). The project, currently ongoing, is funded by the 2019 Key Action 2 Erasmus+ call and developed by six European partners. According to the digital transition planned by the Erasmus+ Programme, HOME supports EU

O. E. Bellini · M. Gambaro · M. T. Gullace (✉) · M. Arcieri
Polytechnic University of Milan, Milan, Italy
e-mail: mariateresa.gullace@polimi.it

O. E. Bellini
e-mail: oscar.bellini@polimi.it

M. Gambaro
e-mail: matteo.gambaro@polimi.it

M. Arcieri
e-mail: marianna.arcieri@polimi.it

C. Á. Benito
European University Foundation (EUF), Brussels, Belgium

S. B. Rommane
Erasmus Student Network AISBL (ESN), Brussels, Belgium

S. Boon
Housing Anywhere, Rotterdam, Netherlands

M. F. Figueira
International Union of Property Owners (UIPI), Brussels, Belgium

E. Arbizzani et al. (eds.), *Technological Imagination in the Green and Digital Transition*, The Urban Book Series, https://doi.org/10.1007/978-3-031-29515-7_23

mobility by providing students and trainees with a digitalized infrastructure that integrates the search for accommodation within existing European digital mobility initiatives, such as the Erasmus+ App. Moreover the project defines a set of living "quality labels" to increase the transparency of information about accommodation offer at the European level. Furthermore, educational resources and a training toolkit will be available, in the HOME website, to spread and replicate the project's learnings results. Once operational, HOME will represent an essential digital solution for a more accessible and quality student accommodation offer.

Keywords Digital infrastructure · Erasmus+ programme · European learning mobility · Student accommodation quality labels · Home of Mobile Europeans (HOME)

23.1 Context of the Research

Supporting international education and mobility opportunities, for students, academics, and trainers, is one of the high priorities of the European Commission.

Since its establishment in 1987, the Erasmus Programme (European Community Action Scheme for the Mobility of University Students) has significantly increased. Until 2020, the Programme reached more than 11.7 million participants (Fig. 23.1a), becoming one of the most important mobility programs in the world.

Despite the mobility restrictions imposed by the Covid-19 pandemic during 2020, the Erasmus+ Programme 2014–2020 closed with a positive record of participants. New social behaviors, the growth of digital environments for teaching and learning, and blended mobility opportunities highlighted the importance of digitalization in processes, tools, and services for education and mobility.

The new Erasmus+ Programme 2021–2027 has been extended and enriched in four strategic areas: digital transition, social cohesion, sustainable growth, strengthening European identity, and active citizenship. Its budget has almost doubled to over €28 billion (compared to €14.7 billion for 2014–2020) (Fig. 23.1b) to further support the mobility opportunities according to the strategic priorities of the *Inclusion and Diversity Plan*, the *Digital Education Plan*, the *Youth Participation Strategy*, and the *European Green Deal* (European Commission 2019, 2021a, b, c, 2022). Another key step in the digitalization of the Programme is a series of important projects supported by the EU, such as the *Erasmus Without Paper (EWP)*[1] and the *European Student Card.*[2]

---

[1] https://www.erasmuswithoutpaper.eu.

[2] https://education.ec.europa.eu/levels/higher-education/european-student-card-initiative.

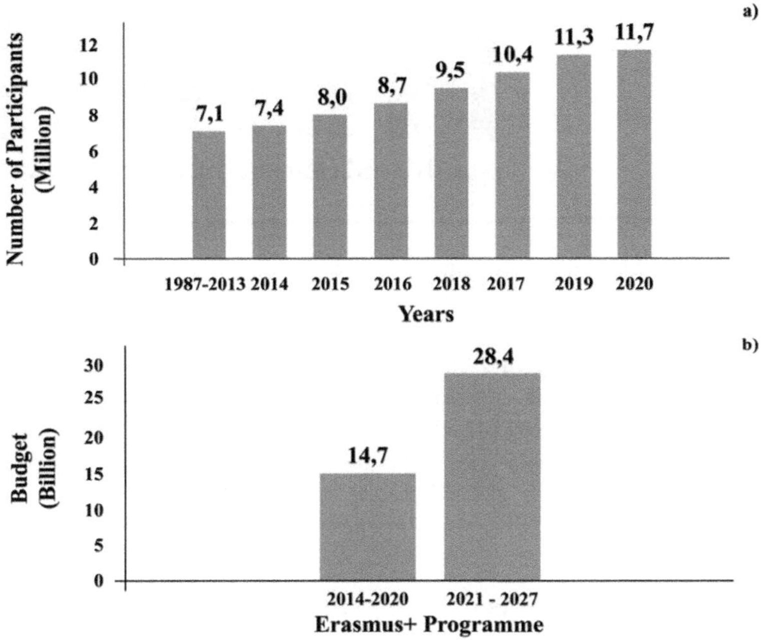

Fig. 23.1 **a** Erasmus mobility programme period 1987–2020, number of participants. *Source* Erasmus+ Annual report 2020. **b** Budget of Erasmus+ programmes. *Source* European Commission Erasmus+ Call 2021

Moreover, the *Erasmus+ App* allows students in mobility to identify themselves, access services connected to the Erasmus+ Programme such as the Online Learning Agreement, and manage all administrative steps related to their mobility period.

In the face of this significant effort for the *digital transformation* of Erasmus mobility and the internationalization of European HEIs, finding reliable and affordable accommodation is one of the main obstacles to student mobility (Kuzmane et al. 2017; Hauschildt et al. 2021). Complying with quality requirements and finding flexible rental conditions for students, particularly in the private market, is tangible challenges for those that decide to spend a period abroad for study or traineeship purposes.

Furthermore, with the Covid-19 pandemic, students' preferences concerning accommodation changed significantly. The Erasmus Student Network's 2020 survey with around 22,000 respondents and focusing on the impact of Covid-19 on mobility students shows that student accommodation has been one area with major issues, along with transportation and other basic needs (Gabriels and Benke-Aberg 2020).

In response to this framework, the HOME (Home of Mobile Europeans) project provides a digitalized infrastructure that integrates the search for accommodation with the European student mobility initiatives. Through this interface, mobility students can find more easily an accommodation, assess, and compare it through different groups of quality labels. Therefore, the project connects the

Fig. 23.2 HOME project's logo

different stakeholders that gravitate within the world of student accommodation: students, (HEIs), housing providers, markets intermediaries, governmental (GOs) and non-governmental organizations (NGOs).

The relevance of mobility digital services strongly rose during and after the pandemic. In this framework, the Erasmus+ App is a pivotal step forward. Furthermore, the future integration of the HOME project within the Erasmus+ App, and more in general, to the Erasmus+ Programme could represent a significant contribution at the European and potentially global level.

23.2 The HOME Project—Home of Mobile Europeans

HOME (Fig. 23.2) facilitates students in finding the most suitable accommodation to their preferences and guides the private and public housing providers in showcasing their online offers, highlighting their accommodation's features and strengths. The project is carried out by six partners from five EU countries, selected from six groups of relevant stakeholders: *Universities* (Politecnico di Milano), *European Foundations* (European University Foundation—EUF[3]), *International Students organizations* (Erasmus Student Network—ESN),[4] *Accommodation Providers*, (Housing Anywhere[5] and International Union of Property Owners—UIPI),[6] and an experienced trainer (Confia International).[7]

HOME provides an univocal reading of private and public accommodation physical and digital characteristics pursues three main key objectives:

- a common understanding of student accommodation quality standards in Europe, defining how accessible and qualitative accommodation looks like;
- an univocal form of data collection, agreeing on how information about quality accommodation is compared and stored;
- defining how this information is accessed and shared between the various stakeholder, i.e., HEIs, students, market intermediaries, GOs, NGOs, etc.

[3] https://uni-foundation.eu.

[4] https://www.esn.org.

[5] https://housinganywhere.com.

[6] https://www.uipi.com.

[7] https://www.confiainternational.eu.

The project develops:

- A set of European Student Accommodation Quality Labels (EAQLs) to increase the quality and transparency of information about student accommodation.
- A Digital Data Standard (DDS) to ensure that information about student accommodation is seamlessly shared at the European level;
- A public Application Programming Interface (API) that integrates with other EU initiatives for the digitization of HEIs;
- Seven multiplier events for the validation and dissemination of ESQLs, DDSs, and API.
- A digital toolkit to share the HOME results and services with the industry and HEIs.

To achieve such objectives, HOME fosters four working groups (Fig. 23.3), five intellectual outputs (IOs), seven multiplier events (E), and four key results (Fig. 23.4).

WORKING GROUP	DESCRIPTION	MEMBER
Steering Committee	Responsible for supporting the IO leaders and monitoring of the overall progress and management of the project to ensure the timely delivery of the results.	EUF, Housing Anywhere, UIPI
Quality Assurance Group	Responsible for monitoring the project progress and effective management, as well as the quality and the impact of the activities carried out within the consortium.	CONFIA, ESN Erasmus Student Network
Dissemination Group	In charge of the dissemination and communication strategy of the project to mainstream its results.	CONFIA, ESN Erasmus Student Network, EUF, Housing Anywhere, UIPI
Sustainability Group	In charge of ensuring the continuity of the project after the project lifetime through a thorough evaluation of resources needed.	CONFIA, ESN Erasmus Student Network, EUF, Housing Anywhere, UIPI

Fig. 23.3 HOME's working groups

IO	CONTENT	RESULT	LEADER
IO1	Research of (national) accommodation standards and establishment of European Student Accommodation Quality Labels.	R1_Student Accommodation Quality Labels	UIPI 1923 with the contribution ESN Erasmus Student Network, Housing Anywhere
IO2	Map relevant data and various digital standards for student accommodation in Europe.	R2_Digital Data Standard for student accommodation and develops the HOME public API to ensure that data about accommodation is shared in a univocal and seamless way all over Europe.	Housing Anywhere
IO3	Create a major use-case for the API by integrating and displaying all quality accommodation of supported housing providers and intermediaries; Connecting HOME API to the Erasmus+App.	R3_HOME integrates with the Erasmus+App	EUF European University Foundation
IO4	Set up an online competence centre on HOME website; Development of educational resources to be available on both homeproject.eu and the Erasmus+ App.	R4_Educational resources to make results freely accessible and easily replicable (IO4 & IO5). Indeed, HOME develops a Multiplier Toolkit for housing providers.	EUF European University Foundation
IO5	Develop the HOME Multiplier Toolkit to help educating the industry and the HEIs about the HOME results.		CONFIA

Fig. 23.4 HOME's, IOs, contents, and results

23.3 Project Advancement

23.3.1 The European Student Accommodation Quality Labels (IO1)

The initial phase of the research focused on investigating national standards and practices in the student accommodation sector and the feasibility assessment of the ensuing labeling exercise. By analyzing national legal frameworks, an examination of minimum legal requirements for the private rental sector and, where existing, for student accommodation in the countries involved in the project—Belgium, France, Germany, Italy, Luxembourg, and the Netherlands—has been conducted. The information was mainly found on governments' databases, official Ministries' Web sites, and legal handbooks. In addition, to overcome language barriers in the research on some EU countries' legal requirements, a questionnaire on national legal frameworks for the private rental sector has been designed and distributed within the UIPI network. The questionnaire aimed to built a clearer comparative picture of legal requirements for private accommodation while verifying the research outcome.

Subsequently, the second part of the research examined existing labels or assessment methods for student accommodation in Europe and related best practices, such as "Lokaviz label" in France.[8] After screening, identifying, and evaluating all services, needs, and features that might be relevant for devising the labels, a set of six EAQLs have been designed to embrace three main factors: (1) be welcoming to an international target market; (2) be accessible to people with disabilities; (3) be univocal in terms of identification of the accommodation features. The EAQLs certify some of the most relevant quality aspects of the student accommodation: (1) international friendliness, (2) wheelchair-accessible, (3) room quality, (4) super-secure, (5) well-equipped, and (6) premium accommodation. Each label is identified through an icon, a criterion, and a list of indicators (Fig. 23.5). Each icon has a different number of indicators conceived to identify the specific needs of the students and associated with the defined criteria. The goal of providing the students with a selected, comprehensive, but not excessively detailed list of choices during their accommodation search guided the identification of this limited group of icons and indicators.

Through a bottom-up process, housing providers were involved in the EAQLs validation actively participating to multiplier events (E1 and E2), focus groups, and surveys.

---

[8] https://www.lokaviz.fr/data/logement_etudiant_crous_charte_20120604_1.pdf.

HOME
quality student accommodation

LABEL	CRITERION	CHECK-LIST'S INDICATORS	
International Friendliness	Accommodation adequate to the needs of international students	○ Wi-fi ○ Communication in English ○ Contract in English ○ Short-term contract available (min. 3 months)	○ Basic furniture ○ International guarantor allowed ○ Online booking available
Wheelchair-accessible	Accommodation entirely accessible for persons with reduced mobility and/or using a wheelchair	○ Accessible elevator (if not ground floor) ○ Accessible doors ○ Accessible WC	○ Accessible shower/bathtub ○ Possibility to adapt to individual needs
Room quality	Room provided with all necessary furniture	○ Single bed ○ Pillows ○ Bed linen	○ Bath towels ○ Desk and chair ○ Closet/drawers
Super-secure	Accommodation providing above-average safety features including against fire accidents	○ Lock on room door ○ Security (armoured) door	○ Surveillance of the building (either video or 24/7 concierge)
Well-equipped	Well-equipped accommodation, displaying above-average household appliances (either kitchen appliances or laundry)	○ Dishwasher ○ Oven ○ Microwave ○ Stove ○ Kitchenware	○ Washing machine ○ Iron ○ Iron board ○ Drying rack
Premium accommodation	Accommodation providing exclusive features	○ At least two of the following luxury facilities or services: gym, pool, game room, 24/7 concierge, cleaning services	

Fig. 23.5 HOME's European Accommodation Quality Labels (EAQLs)

23.3.2 Digital Data Standard (DDS) for Student Accommodation (IO2) and Public Application Programming Interface (API) Public (IO3)

A DDS for student housing-related data has been developed to achieve a univocal way of gathering and structuring the accommodation information of all European-wide housing providers and intermediaries. By making housing providers and intermediaries adhere to the created data standard, the HOME application can ensure the correct entitlement of EAQLs to the accommodations and smooth integration to the HOME interface.

A first version of the DDS has been created by mapping all the student housing data of sixteen large student housing providers or intermediaries across Europe. This process allowed identifying the most important and widespread data used within the student housing sector in Europe, which, together with the necessary data elements to identify and entitle the quality labels, formed the first version of this DDS. The student housing data have been delivered univocally according to the DDS. The project set up a structured database containing all the accommodation data from all connected housing providers and intermediaries. On top of this database, an API is created; it will enable applications within the Erasmus environment to pull the data from this database and use it to present the accommodations' information, including the EAQLs. All the elements within the HOME application are open-sourced, ensuring a wide future dissemination within the Erasmus environment.

Both private and public housing providers validate the DDS and the technical systems. To ensure that the technical system developed under IO2 is consistent with students' and housing providers' expectations, as well as the European Commission's requirements. The partners involved in that IO embarked on an informal *validation process* which consisted of consulting private housing providers and market intermediaries as well as public housing providers (i.e., universities with student accommodation facilities) on the validity of the DDS and the technical feasibility of the API requirements. HousingAnywhere and EUF have carried out this validation process. The former contacted six of Europeans largest private housing providers, and eight public universities across Europe and one national network of student service providers.

Currently, the vetting process results are being internally discussed, analyzed, and validated. The main purpose of this process focused on: ensure that the system is aligned with the reality of the student accommodation field and gather a qualitative first validation of IO2's technical results.

23.3.3 Multiplier Events (E) and Dissemination Activities

In 2020, the Multiplier Event E1 introduced the promotion of the EAQLs to 25 European stakeholders (Fig. 23.6). HOME was also presented during the Erasmus+ App major conference of March 2022 by EUF with the cooperation of ESN reaching 50 stakeholders in the field of international mobility in Europe. In 2021, the dissemination was scaled up using the first deliverables produced and focusing on the scope of HOME in light of the pandemic to connect with potential end-users and student housing representatives with a specific focus on IO1, while on Spring 2022 communications started focusing on the DDS under IO2.

In the HOME web site,[9] which obtained 7,200 views, news items were produced concerning the project's key developments, events, and webinars. Moreover, to foster

[9] https://thehomeproject.eu/.

EVENT	COUNTRY OF VENUE	EVENT TITLE
E1	Belgium	**Quality Labels First Validation**
E2	Online	**Roundtable on Student Accommodation in Times of Covid-19**
E3	Netherlands	**Presenting the Student Accommodation Quality Labels, the European Student Accommodation Digital Standard & API**
E4	Spain	**HOME Pilot session**
E5	Italy	**HOME Multiplier event for students NGOs**
E6	Italy	**HOME Multiplier event for HEIs**
E7	Italy	**HOME Multiplier event for housing providers and market intermediaries**

Fig. 23.6 HOME's project multiplier events

the visibility of the results of IO1 as well as comprehension of the quality labels a total of nine videos have been developed.

Dissemination activities include social media communication saw the launch of the first campaign on IO1's results, created and managed by ESN, creating one video per quality label, and launching it throughout the duration of a few months. The campaign reached 40,000 people, and recently, a second campaign was launched focusing on the DDS (IO2). Partners have also cooperated to apply to relevant conferences like EAIE on a yearly basis. The next conference will be hosted by the Politecnico di Milano where HOME project final results will be presented to about 60 European entities such as HEIs, students' representatives and NGOs, and housing providers.

23.4 Preliminary Results

The project's outreach activities reached around 50,000 stakeholders by webinars, multiplier events, and communication campaigns. As a result, key European stakeholders involved in student accommodation sector provide positive feedback and a high level of interest. HOME's outreach results are to be considered in the context of niche targeting communication to ensure that the involved stakeholders engage more actively with the project during its lifecycle and after it. Moreover, synergies with the Erasmus+ App team are being implemented to further boost the project integration beyond the engagement phase once the technical solutions will be fully functional. During outreach activities, one of the main challenges has been: translating the technical project's contents into accessible information for a broad audience; summarizing the HOME's solutions clearly and accurately; upgrading HOME's solutions to the European student housing has already been achieved successfully.

The achieved preliminary results can be summarized as below: (1) Definition six targeted European Accommodation Quality Labels (EAQLs) to support and simplify the choice of online student accommodation; (2) The setup of the EAQLs; (3) Successful dissemination of the EAQLs; (4) First version of the Digital Data Standard (DDS); (5) Groundwork for the technical architecture needed to gather, store, and export housing data; (6) Validation of the DDS and a public Application Programming Interface (API) requirements by private and public student housing providers across Europe.

The consequent project activities will focus on integrating HOME's accommodation module interface with the Erasmus+ App, developing the educational materials and the multiplier toolkit for housing providers, universities, and students.

The Covid-19 pandemic strongly impacted the project's development. Due to the several emergency conditions in the countries involved in the project, all activities stopped for six months. Therefore, the HOME project has obtained an extension, and will be completed by the end of 2022.

23.5 Final Remarks

The HOME project is a preliminary attempt to identify, standardize, and systematize different aspects of European student mobility.

Based on the difficulties and challenges encountered during its development, the research has also allowed focusing on a knowledge overview of some critical aspects of temporary housing for students in Europe concerning accommodation adequacy. Nevertheless, an effective strategy to tackle this phenomenon at the European level still needs to be addressed.

Three factors are particularly evident:

- First, housing and habitability standards, provided by national building codes or health rules, vary substantially from one country to another. As a result, the housing sector is far from evenly regulated across Europe. Moreover, even fewer European Countries have specific regulations regarding student accommodation.
- The need to consider and define the quality of student accommodation in European university cities to help students find appropriate housing from a trustworthy source more systemically way is becoming fundamental. As mentioned in the Research Report from HousEramus+ (Kuzmane et al. 2017), some countries and cities have already created a quality standard to provide students with suitable accommodation. Centralizing a reliable and trustworthy supply on a public body's Web site would facilitate the flow of information. However, it would also require public policies developed by local authorities to monitor the prices and quality of the properties rented.

- In addition to the difficulties of integration with local languages and rental procedures or regulations in the countries of the Union, finding an understandable rental contract, also in English, might be complicated and untrustworthy for both parties involved, students and landlords. The lack of institutional support in administrative and legal aspects and the provision of a recognized international rent contract for mobile students' accommodation remains one of the most uncovered topics at the local and European levels. Intercultural awareness is an essential precondition for smooth integration and mutual growth in all the European countries involved in the Erasmus+ Programme.

The HOME project's experimentation and innovation contribution to the digitalization and standardization of advanced accommodation services in support of international mobility is just one of the possible examples of the *digital transition* to purse the strategic objectives of the Erasmus+ Programme. Once operational, HOME will also contribute to achieving the following macro-strategic project goals:

1. Supporting dialogue between the HEIs and their users and connecting stakeholders which gravitate toward mobility and education at the European level.
2. Promoting a system of transnational accommodation where students can move easily for short periods and between different European institutions.
3. Fostering transparency and accessibility to quality accommodation by simply connecting the HOME interface within the Erasmus+ App, where all students' mobility activities are *digitally centralized.*

References

European Commission, Directorate-General for Education, Youth, Sport and Culture (2019) Erasmus+ higher education impact study: final report

European Commission (2021a) Erasmus+ 2021–2027 fact sheets. Erasmus+ 2021–2027. Publications Office of the EU

European Commission (2021b) Erasmus+: over €28 billion to support mobility and learning for all, across the European Union and beyond. https://ec.europa.eu/commission/presscorner/detail/en/ip_21_1326

European Commission, Directorate-General for Education, Youth, Sport and Culture (2021c) Erasmus+ annual report 2020

European Commission (2022) Erasmus+ programme guide 2022

Gabriels W, Benke-Aberg R (2020) Student exchanges in times of crisis. Erasmus Student Network AISBL

Hauschildt K, Gwosć C, Schirmer H, Wartenbergh-Cras F (2021) Social and economic conditions of student life in Europe. Synopsis of indicators. Eurostudent VII 2018–2021

Kuzmane D, Jahnke S, Pérez Encinas A, Alfranseder E, Fellinger J, Montagnese D, Espin Alba E, Dopazo Ruibal L, Apert J (2017) HousErasmus+ research report

Part II
Session | Technology

The ongoing evolution of Industry 4.0 has inevitably made preservation and improvement of the existing ecosystem a key factor in the development of our constructed environment. As a result, processes and systems of production were radically restructured, orienting the technological innovations of the digital era toward decarbonization while, at the same time, supporting new paradigms of social living.

The green and digital transition has brought forth objectives and solutions that radically interfere with existing methods for planning, developing, and producing goods and services, as well as with utilization of the energy and material resources whose chief consumer is the constructed environment.

The new technologies are meant to transform the system of habitation and the urban context by integrating ICT systems with technical spaces and elements, as well as artificial intelligence and the robotizing of connected and collaborative construction processes, both in production facilities and worksites.

These innovations can provide users and civil society with an improved quality of life, both indoors and in urban settings, channeling scientific development to establish an effective culture of sustainability, reuse, and safety.

The session promotes discussion of the impact of new technologies of design and manufacturing on the construction of buildings and the urban environment, in addition to examining the potential side-effects of the new models of habitation on the quality of life and people's perception of the same.

Chapter 24
Technologies for the Construction of Buildings and Cities of the Near Future

Eugenio Arbizzani

Abstract The objectives and solutions that become necessary, within the green and digital transition, can radically interfere with existing methods for designing and producing buildings and portions of cities. The "factory" and the "construction site" were one and the same thing, in the production of buildings, up until the seventies of the last century. Following the advent of prefabrication, the factory has gradually been separated from the construction site, to the point where industrial manufacturing now accounts for much of the production value of buildings through the dry-assembly of factory-made products and components. The development of enabling technologies involves knowledge-intensive technologies associated with a high level of research and development, rapid innovation cycles, substantial investment costs and highly qualified jobs. The development of this field could produce a new chain of industrial services for companies based in our country.

Keywords Factory-made building components · Smart robot assembly systems · Enabling technologies · Hybrid structural systems · Man–machine active interaction

24.1 Introduction

In the green and digital transition currently underway, the objectives and solutions that become necessary can radically interfere with existing methods for designing and producing buildings and portions of cities, seeing that they have to optimize the use of energy resources and the flows of materials, whose largest single consumer is the built environment.

The ongoing evolution of Industry 4.0 towards 5.0 has inevitably made the preservation and improvement of the existing ecosystem a key factor in the development of our constructed environment. As a result, processes and systems of production need to be radically restructured, orienting technological innovations of the digital

E. Arbizzani (✉)
Sapienza University of Rome, Rome, Italy
e-mail: eugenio.arbizzani@uniroma1.it

E. Arbizzani et al. (eds.), *Technological Imagination in the Green and Digital Transition*, The Urban Book Series, https://doi.org/10.1007/978-3-031-29515-7_24

era towards decarbonisation, while, at the same time, supporting new paradigms of social living.

The green and digital transition has brought to the fore objectives and solutions that radically interfere with existing methods for the planning, design and production of goods and services, as well as with the utilization of energy and material resources.

24.2 Building Factory Versus Construction Site

In the production of buildings, the "factory" and the "construction site" were one and the same thing up until the seventies of the last century. Fifty years later, in Tokyo, the dismantling of the Nakagin Capsule Tower by Kisho Kurokawa is underway.[1] That iconic building was the result of the Japanese Metabolist utopia, whose goal was to arrive at the industrial, off-site production of housing modules designed to colonize buildings equipped with the most up-to-date technologies, in accordance with pre-established maintenance programs and planned life cycles.

The history of this building, and of built architecture in general, proceeded—fortunately, in my opinion—in other directions. These proved more respectful of the cultural context of cities, but also less aware of the need for production with more sustainable environmental impacts and—even more importantly—of the advantages of taking into consideration, in the design process, the optimal use of resources, as well as their eventual recycling at the end of their life cycles.

The fact is that, following the advent of prefabrication, the factory has gradually been separated from the construction site, as over the last 30 years an increasingly evolved industrial manufacturing sector has developed, to the point where it now accounts for much of the production value of buildings through the dry-assembly of factory-made products and components.

With the advent of robotization, it is easy to imagine that our sector's production cycle will undergo further transformations. However, it should be emphasized that urban context and cultural heritage will always constitute an "unicum" whose social and cultural value cannot simply be reproduced. The "second machine age"—referred to as "Industry 5.0" in the programs of the European Commission—is unlikely to see machines prevail over the work of humans, though there will be substantial changes in methods of design and in interactions between man and his built environment.[2]

This revolution will be significant enough to lead to an evolutionary change in the direction of a new collective imagination capable of allowing us to conceive—albeit unconsciously—of new and unpredictable scenarios for the development of interactions between man, machines and the built environment.

New technologies are designed to transform the system of habitation and the urban context by integrating ICT systems with technical spaces and components, as well

[1] Cfr: https://www.greyscape.com/nakagin-capsule-tower-preservation-and-restoration-project/.

[2] Cfr: https://ec.europa.eu/info/news/industry-50-towards-more-sustainable-resilient-and-human-centric-industry-2021-jan-07_en.

as with artificial intelligence and the robotization of connected and collaborative construction processes, both in production facilities and at work sites.[3]

These innovations can provide users and civil society with a better quality of life in indoor and urban environments while orienting scientific advances towards an effective culture of sustainability, reuse and safety.

For this reason, there is now an unavoidable need to balance the demand for innovative production systems able to guarantee the quality and cost-effectiveness of projects with the obligation to preserve the environmental heritage, which otherwise the planet risks tragically losing.

The topics of the session were specifically designed to maintain the connection between research on the industrialization of the construction sector and that on the development—sustainable, human-centric and resilient—of the environment to be built in the near future.

The technology session presented three short interviews on these topics, both to introduce the theme of the latest research and to outline a vision of the discipline that draws meaning from its origins, so as to proceed even further, towards a scenario which appears to lay the groundwork for a thoroughgoing change.

In the first interview, Yotto Koga, a young software designer and researcher at the Autodesk Robotics Laboratory in San Francisco, discusses his research on "smart robot assembly systems".[4]

In the second interview, Professor Fabrizio Schiaffonati, one of our principal schoolmaster of discipline, is asked to lay out his vision of the relationship between the factory and the construction site, as well as the future role of design.[5]

In the last interview, architect Sara Codarin, who earned her Ph.D. at the University of Ferrara with a thesis on "Building Robots",[6] and is now an assistant professor at Lawrence Technology University, College of Architecture and Design, illustrates her research on robots in higher education.

Also invited to the Technology Session, in the role of discussant, was Francesco Leali, full professor at the Enzo Ferrari Department of Engineering of the University of Modena and Reggio Emilia and Dean of the Master's Program in Advanced Automotive Engineering and Coordinator of the Unimore Automotive Academy project.

The Master's Degree Programme in Advanced Automotive Electronic Engineering is an interuniversity, international course of study established as a collaborative initiative of the Universities of Bologna, Ferrara, Modena and Reggio Emilia and the University of Parma, working together with the world's most prestigious automotive companies headquartered in Italy.

---

[3] Arbizzani, E. (2017). Informatizzazione nel processo di trasformazione dell'ambiente costruito. In: Civiero P. (Ed.). *Tecnologie per la riqualificazione. Soluzioni e strategie per la trasformazione intelligente del comparto abitativo esistente*. Maggioli Editore, IT: Sant'Arcangelo di Romagna.

[4] Cfr.: https://www.universal-robots.com/case-stories/autodesk/.

[5] Cfr.: Schiaffonati, F. (2021). *Lettera a un aspirante architetto*. Milan, IT: Lupetti.

[6] Cfr.: Codarin, S., (2020). *Innovative Construction Systems within Building Processes. An Approach to Large-scale Robotic Additive Layer Manufacturing for the Conservation of Cultural Heritage*. Thesis for an International Doctorate in Architecture and Urban Planning, University of Ferrara.

Professor Leali was invited because it is the automotive sector that has had the technological and organizational innovations of Industry 4.0 in place for the longest time. In this sector, advances in manufacturing have gone hand-in-hand with the revolution in approaches to planning and design.

The scenario that he outlined might appear far removed from the architectural culture and the urban reality that surrounds us, but I believe we will soon see a progressive hybridization of the various production contexts. Just think of the interactions already taking place between electric vehicles, the energy production systems of our buildings and the urban infrastructures of the most advanced cities. The resulting technological imagination appears much more immersive and exciting than might seem to be the case in these difficult times.[7]

Five reports were presented during the session, illustrating the outlined scenario, together with proposals for innovations in teaching at faculties of architecture.

The combination of the interviews, our discussant speech and the session papers provided us with glimpses of what we can imagine for the future, highlighting certain aspects of particular relevance to the scientific community centred around themes of technological design in the contemporary era.

The coexistence of urban districts still in search of new environmental and social balances, together with the very latest technological research in construction, points to the possibility of progressing towards a society in which innovation not only benefits the shareholders who hold the keys, but also moves in the direction of more generally increasing the well-being of all stakeholders.

In the case of the drivers of technological innovation which we addressed in this session, they can be classified according to five lines of research and development that appear worthy of note, less because of how widespread they are than based on their ability to drive the innovative processes necessary for a vigorous recovery of production in the construction sector. They can be summarized in four key terms: key enabling technologies; new technological teaching of design; energy-conscious buildings; hybrid building technologies.

24.3 Enabling Technologies

The development of enabling technologies[8] involves: "knowledge-intensive technologies associated with a high level of research and development, rapid innovation cycles, substantial investment costs and highly qualified jobs". Such technologies are of systemic importance, because they feed the value of the production-system chain and have the capacity to innovate processes, products and services in all economic sectors of human activity. Furthermore, a product based on enabling technology uses

[7] Camorino, A., (2016). L'immaginario tecnologico. Un'analisi sociologica della cosmologia contemporanea, A Journal of the Social Imaginary. www.imagojournal.it.

[8] The Ministry of Economic Development, (2018). *National Plan for Industry 4.0. Investments, Productivity and Innovation.*

advanced manufacturing technologies and increases the commercial and social value of a good or service.

With regard to our production sector, the most applicable enabling technologies concern:

- additive manufacturing, off-site and on-site, for the production of components, parts of buildings and housing modules on demand;
- augmented reality and artificial intelligence, to support the creative and design processes of buildings and cities;
- horizontal and vertical integration of information, to create production and distribution process chains for industrialized building products, from the manufacturer to the designer and the user, to be integrated into construction and urban-development projects;
- cloud and big data, to manage large amounts of data on open systems and on digital twins of buildings and urban districts;
- robotics and wearable smart tools, to ensure better safety conditions for construction workers, in factories and on construction sites.

All these process innovations seem able to generate noteworthy progress in teaching and training systems for construction and urban design. In the near future, they will be significantly transformed to adapt to user requests. On the one hand, they must become increasingly self-explanatory, user-friendly and virtually immersive, so that students can learn with greater ease and interest, while, at the same time, they need to more effectively support the new skill requirements of the construction industry.

A third field of investigation, perhaps the most popular and thoroughly tested to date, concerns technological innovations in materials, products and systems to ensure the energy efficiency and the energy balances, of buildings. This theme requires a new vision of the interface between building components and plant systems. "Integrability" was a performance design concern of the 1980s, but it was a concept limited to aspects of the physical interfaces between building elements. Today, it is a question of giving content to the functional interface, along with the reciprocal benefits that such a dialogue can produce by ensuring a proper balance between the energy demanded and the energy produced.

The goal of zero-energy buildings can already be reached today, at least in new constructions, but soon we will see, with smart automotive construction and positive energy districts providing support as well, buildings make a notable contribution to lowering the overall demand for energy and reducing climate-altering emissions.[9]

---

[9] Cfr: United Nations Environment Programme (2020). *2020 Global Status Report for Buildings and Construction: Towards a Zero-emission, Efficient and Resilient Buildings and Construction Sector.* Nairobi.

24.4 Evolution of Construction Technologies

This first encounter of ours offered only glimpses of a sector with enormous prospects for technological innovations in construction and built architecture. I am referring, in general, to the evolution of construction technologies, which for many years have shown only minor advances involving individual materials or fragmented processes, with adequate solutions yet to be offered to the new problems involving the resilience and sustainability of the built environment.

These have to do with the redesigning of natural materials in terms of their production and use, as well as the hybridization of various technological cycles: wood, steel, concrete, glass; but also bamboo, natural fibres and bio-plastics.[10] The integration of different component materials can determine which technical elements and portions of the construction best satisfy performance requirements, making appropriate use of materials in those instances where they are most efficient. An example is the rapid evolution of hybrid structural systems that combine wood, steel and concrete, and which already make it possible to design tall hybrid wooden buildings.[11] Also of interest is the opportunity for increasingly lighter buildings, with materials optimized for their specific uses, meaning that ever smaller quantities are needed. This particular innovation contributes to lowering the demand for non-renewable quarry materials while reinforcing the recycling and reuse chain and promoting heightened awareness of the fact that the goal of zero emissions is indeed possible.

The development of this field could produce a new chain of industrial services for companies based in our country, along with the possibility of exporting this experience to emerging countries that do not yet emit climate–changing gases at invasive levels, but may soon become involved in processes of social and economic progress, which could play a decisive role in establishing a renewed balance of the anthropized environment.

While these are the specialized topics that we wish to highlight, in more general terms, it can be said that the themes addressed in the five sessions proposed some elements pertinent to various sectors, in particular for the development of innovation in a sector that is now widely aware of the challenges ahead:

- the opportunity, or rather the urgent need, to use all forms of innovation in instruments, processes and products to support the level of well-being of people while working towards a balanced readjustment of the built environment;
- the massive availability of new solutions and equipment, but, at the same time, the difficulty our sector has in exploiting its potential, due to a scenario in the construction industry which is still underdeveloped compared to other sectors of civil society and the production apparatus;

[10] Cfr: Antonini, E., Boeri, A., Giglio, F. (2020). *Emergency-Driven Innovation. Low-Tech Buildings and Circular Design.* Springer. http://doi.org/10.1007/978-3-030-55969-4.

[11] Cfr: CTBUH 2022 Steel-Timber Hybrid Buildings Conference, 23–24 May, Chicago, https://ctbuhsteeltimber.com.

- with the mass media now spreading the message that everything proposed by the market is "sustainable" (sustainable cars, sustainable basil, sustainable holidays …), the need to develop methodologies and control systems which are able to support, under recognized scientific parameters, true sustainability—social, production-related, economic and environmental—of every production activity.

One factor has always hindered the development of a truly industrialized construction sector: the need to customize urban and construction products on the basis of information provided by the end user and the constraints of the existing building stock.

But given the evolutionary process currently underway, new opportunities can be found, thanks to technologies such as cloud computing, the Internet of Things, virtual and augmented reality, for controlling production processes through dialogue between machines, or between machines and humans, or by offering the possibility of foreseeing problems and proposing timely, effective solutions, including measures geared towards limiting emissions of pollutants and social inequality, as well as others designed to protect the rights and safety of workers (Arbizzani E., 2017).

Under the European Commission's vision, the Industry 5.0 of the near future must increasingly become a "resilient provider of prosperity" able to ensure that the production of goods and services regains respect for the limits of the planet and places the well-being of users and workers at the centre of production processes. Nowadays, only large companies can implement digital technologies in their production processes, while small and medium-sized companies—typical of almost all enterprises in the construction sector—are unable to follow this trend.

The objective set by the European Commission for the development of technology, as the main tool for ensuring the sustainability of evolved production systems, appears to go in the direction of supporting precisely those sectors, such as construction, which are most fragile and faced with difficulties in their technological evolution, so as to encourage any process promoting circular economics, while providing increased skills and employment for workers.

As will be the case in all the most advanced industrial sectors, the make-up of new construction workers will be shaped by their interaction with technology, rather than by any replacement of those workers by machines. Workers supported by exoskeletons, augmented reality or virtual reality, or who are equipped with new-generation safety devices or connected to artificial intelligence or big data, or workers who work alongside robots, all appear to be futuristic visions. But, it is precisely the active interaction between man and machine which appears to be best suited to the evolution of a sector featuring intensive employment of human personnel, such as the construction sector.

With the decision taken recently by the government, we have come to the end of the construction boom that was temporarily underway in our country, thanks to the measures of economic support provided to the construction sector. Soon, there will be an urgent need to train and retool everybody working in the industry with the new digital production skills.

Initiatives carried out under a multidisciplinary, data-driven digital strategy can respond to the growing demand for higher quality, based on a predictable timetable in a complex environment, as well as to the need to satisfy requirements of environmental, social and economic sustainability, in combination with both technological and cultural considerations.

24.5 Conclusions

In all the latest reports on Europe as a whole and on individual countries, business enterprises are shown to be struggling with a scarcity of technical and digital skills, while the institutional bodies responsible for training appear incapable of meeting such demands. There can be no more putting off the need to return, even in our faculties of architecture, to an approach, both in teaching and scholarship, which combines both theory and practice.

Or better yet, an approach which preserves the paradigms of theoretical teaching for which our faculties are universally renowned, while resuming, with the utmost creative, organizational and technological effort, teaching of the discipline based on practical and productive project activities.

An approach both multidisciplinary and trans-disciplinary must be put in place, regarding which this conference has simply offered food for thought by inviting leading scholars from other disciplines to present their visions of innovation in scientific research in the sector.

We all believe that technology has always consisted of a striving on the part of mankind, a reflection of people's aspirations to improve their cultural situations, their very lives. The teaching of the discipline of design based on practical experience and immersion in constructed reality constitutes, at present, an unavoidable necessity, but one that can be used to expand the strengths of the sector in the direction of increasingly innovative, inclusive and sustainable scenarios for our surrounding environment.

The students who graduate from our faculties must be prepared to operate in their chosen field, but above all they must be able to use their skills as designers to produce a positive impact on the environment by applying their creativity to the real world. At the same time, there is an emergency that must be faced by a category of professionals obliged to undergo a rapid evolution of their knowledge and even more importantly of their cultural approach, in order to prepare them for a multidisciplinary profession in which the new tools of project support play a significant role.

References

Arbizzani E (2017) Informatizzazione nel processo di trasformazione dell'ambiente costruito. In: Civiero P (ed) Tecnologie per la riqualificazione. Soluzioni e strategie per la trasformazione intelligente del comparto abitativo esistente. Maggioli Editore, Sant'Arcangelo di Romagna, IT, pp 211–234

Cangelli E, Conteduca M (2018) Architecture on demand. New scenarios for the design project and the construction industry. Techne, 16/2018. University of Florence Press, Florence, IT

European Commission. Long-term strategy for 2050. From: https://ec.europa.eu/clima/policies/strategies/2050_it

European Commission. Industry 5.0. Towards a sustainable, human-centric and resilient European industry. From: https://ec.europa.eu/info/news/industry-50-towards-more-sustainable-resilient-and-human-centric-industry-2021-jan-07_en

Figoli FA, Mattioli F, Rampino L (2022) Artificial intelligence in the design process. Franco Angeli, Milan, IT

https://www.europarl.europa.eu/news/it/headlines/economy/20201112STO91445/nuova-politica-industriale-ue-le-sfide-da-affrontare

Pozzi G (2022) Industrial housing. Or how industrialization can improve residential buildings. Mimesis Edizioni, Milan, IT

Vaccaro G, Buoninconti L (2021) 3D printing and the circular economy. Innovation and sustainability for the construction sector. SMC n.13/2021. Luciano Editore, Naples, IT

Chapter 25
The Living Lab for Autonomous Driving as Applied Research of MaaS Models in the Smart City: The Case Study of MASA—Modena Automotive Smart Area

Francesco Leali and Francesco Pasquale

Abstract The revolution of digital technology in the field of mobility generates a complex environment where information technology, vehicle engineering and urban planning cooperate in the design of sustainable cities.

Keywords Autonomous driving · Connected and cooperative vehicles · Computer vision · Artificial intelligence · Mobility as a service · Smart city · Public space · Landscape

25.1 Introduction

The process of urbanization molded our cities with a series of transformations, where each one stratified and left a legacy that still testifies the culture and the technologies of that time. The revolution of modernity massively changed the urban environment in terms of infrastructures by the use of concrete and steel for roads, bridges, tunnels, stations, etc., adapting to vehicle which suddenly became faster, larger, heavier and consequently more polluting (Fig. 25.1). The digital revolution of nowadays instead, even though it occurs at the scale of a microchip or in the a-dimensionality of the ether, might be still more disruptive, as it applies no more to the implementation of physical capacity, but it debuts in the unexplored realm of simulating and replacing human intelligence.

F. Leali (✉) · F. Pasquale
UNIMORE, Modena, Italy
e-mail: francesco.leali@unimore.it

F. Pasquale
e-mail: francesco.pasquale@unimore.it

E. Arbizzani et al. (eds.), *Technological Imagination in the Green and Digital Transition*, The Urban Book Series, https://doi.org/10.1007/978-3-031-29515-7_25

Fig. 25.1 Viaduct on A24 Highway Roma-L'Aquila, 1969. *Source* Abruzzo LIVE https://abruzzolive.it/a24-50-anni-di-autostrada-era-il-10-luglio-del-69-quando-venne-inaugurato-il-primo-tratto-della-roma-laquila/

In the global scenario, mobility as a whole is estimated to impact for about 25% of global emissions[1] and its implications on health, environment, safety and life quality are even more critical issues within the dense urban areas of the contemporary cities.

The models based on non-renewable-energy-consuming private cars are no longer sustainable as population increases, cities densify and resources become less available, so that research on sustainable and intelligent mobility has become a solid global trend, addressed in all international agendas[2] and encouraged by actions for a common jurisdictional framework[3] (Fig. 25.2).

The wider application of information technologies in the field of mobility, on both vehicles and infrastructures, is a disruptive change that is generating a deep shift in the users' habits, the automotive industry, the roads management and the way cities are shaped to embrace these transformations. The combination of computer vision, artificial intelligence and data communication technology enables vehicles to connect more and more among them and with the infrastructures of the city due to V2X[4] protocols, which is gradually leading to fully autonomous driving with ADAS[5] of level 5 (Fig. 25.3). The upcoming scenario of connected and autonomous vehicles enable a large variety of products and services based on the massive amount

[1] IPCC 2022 report. https://www.ipcc.ch/report/ar6/wg2/downloads/report/IPCC_AR6_WGII_FinalDraft_FullReport.pdf.

[2] United Nations Agenda 2030—Sustainable Development Goals. https://sdgs.un.org/goals.

[3] EU Regulation 2019/2144 of the European Parliament and of the Council of 27 November 2019. https://eur-lex.europa.eu/legal-content/EN/ALL/?uri=CELEX:32019R2144.

[4] V2X—Vehicle To Everything.

[5] ADAS—Advanced Driving Assistance Systems. https://www.sae.org/blog/sae-j3016-update.

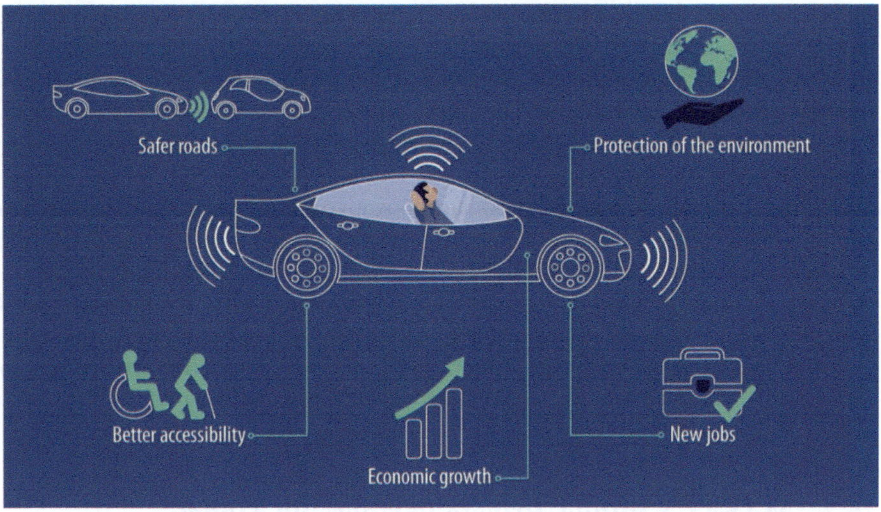

Fig. 25.2 Benefits of autonomous driving. *Source EPRS, European Commission* https://www.europarl.europa.eu/news/en/headlines/economy/20190110STO23102/self-driving-cars-in-the-eu-from-science-fiction-to-reality

of generated data, increasing diversity and complexity in the urban ecosystem of mobility.

The urban shape will not be immune from the digital revolution as a passive scenography, but instead it has to be designed according to the upcoming technologies requirements, affecting the public realm and the morphology of the connective

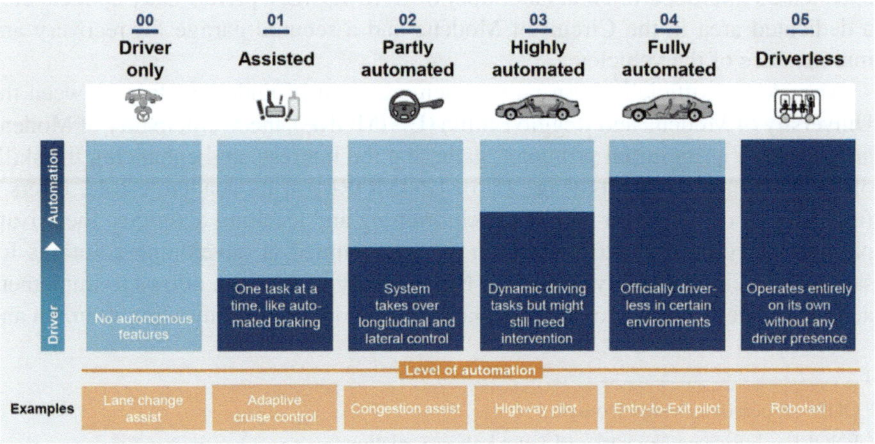

Fig. 25.3 Autonomous driving, 5 levels ADAS. *Source* SAE International, VDA; *EPRS, European Commission* https://www.europarl.europa.eu/news/en/headlines/economy/20190110STO23102/self-driving-cars-in-the-eu-from-science-fiction-to-reality

space due to different street sections, multimodal hubs, on street parking space, re-fueling stations, radar/LIDAR sensors, horizontal and vertical signs, wired/wireless networks, cameras, antennas, etc. Moreover, it is indeed needed a synergic integration with the other urban networks such as energy grids, green and blue infrastructures and more to approach the ecological issues on a global scale and in a multidisciplinary perspective.

In order to design, develop and evaluate innovative mobility solutions, the structure of the living lab stands out as the ideal environment where to test solutions, dealing with the full complexity of the city within a controlled ODD.[6] This model of public–private collaboration is assuming increasing importance, both at the European level, through the ENOLL[7] association, active since 2006, and in the national panorama, where the National Research Program (PNR[8]) 2015–2020 mentions living labs for the first time, defining them as a tool to support more applied and industrial research.

25.2 MASA: A Public–Private Partnership Model

The integration of knowledge and experiences in the field of vehicles engineering, information technology, urbanism, economy and law (Fig. 25.4) collect in MASA—Modena Automotive Smart Area, which is since 2017 an open and collaborative ecosystem of innovation and research for mobility, with a public–private governance, in Modena, in the heart of the motor valley. MASA operates as an open space for research and experimentation of about 3 km^2 in the context of the urban regeneration of the R-Nord district in Modena (Fig. 25.5). Here companies, research centers, public administration and end users develop new applications, technologies and services in the field of CCAM[9] and MaaS.[10] Supporting assets of the living lab are the data center, a dedicated interdepartmental laboratory of UNIMORE, a private start-up incubator, a dedicated area in the Circuit of Modena and a secured garage for recovery and maintenance of the vehicles.

MASA was officially born with a memorandum of understanding between the University of Modena and Reggio Emilia (UNIMORE), the Municipality of Modena and Maserati as an initial promoter, gathering the interests and enhancing the skills of the project participants (Fig. 25.6). UNIMORE, in its academic role, offers a fundamental contribution with research, training and teaching activities; the private partners carry the industrial research program, aimed at developing solutions for safe and efficient mobility; while the Municipality of Modena allows to implement actions on the field, improving the quality of life of citizens, investing in urban and

[6] ODD—Operative Design Domain.

[7] ENOLL—European Network of Living Labs association.

[8] PNR—National Research Program 2015–2020.

[9] CCAM—Connected and Cooperative Autonomous Mobility.

[10] MaaS—Mobility as a Service.

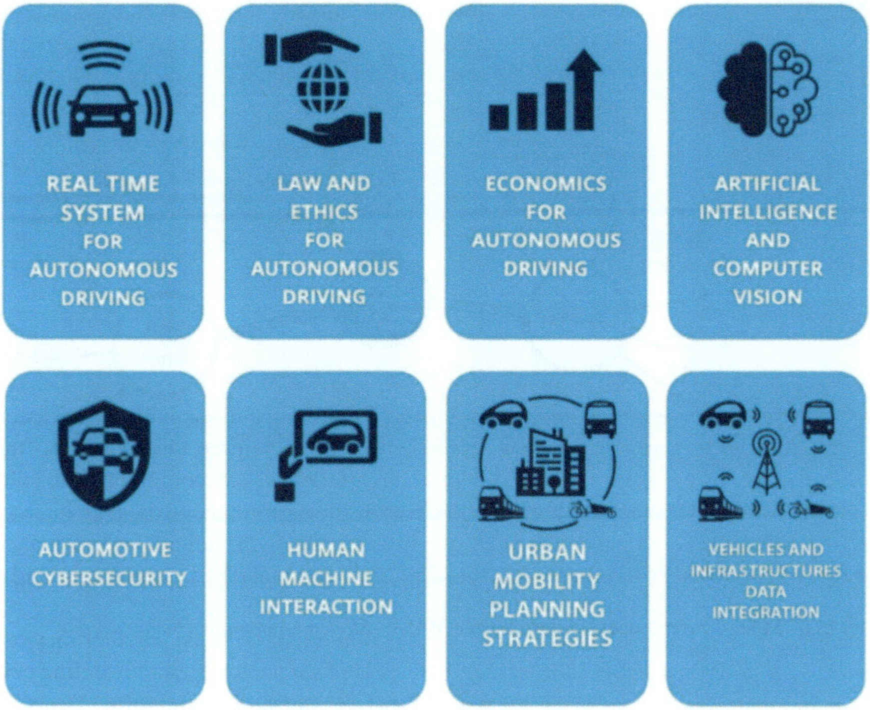

Fig. 25.4 MASA skills and competences. *Source MASA—Modena Automotive Smart Area*

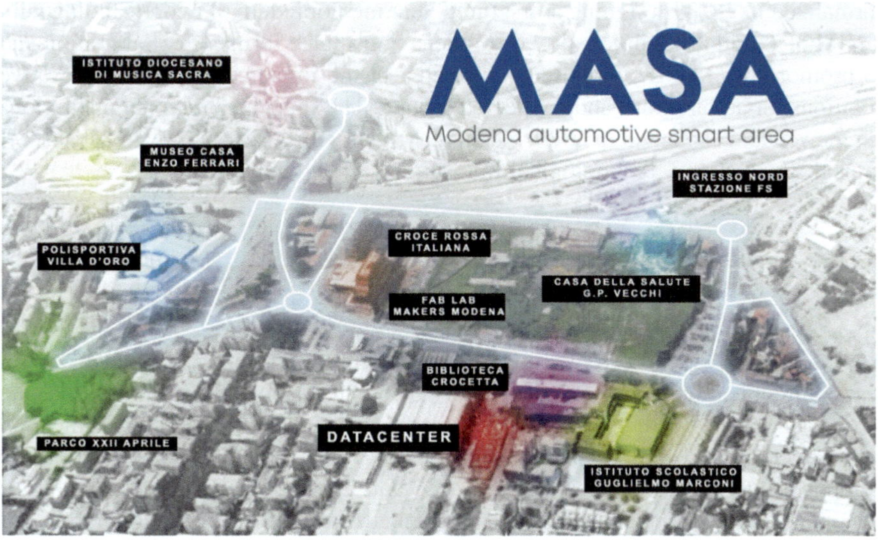

Fig. 25.5 MASA in the R-Nord district. *Source MASA—Modena Automotive Smart Area*

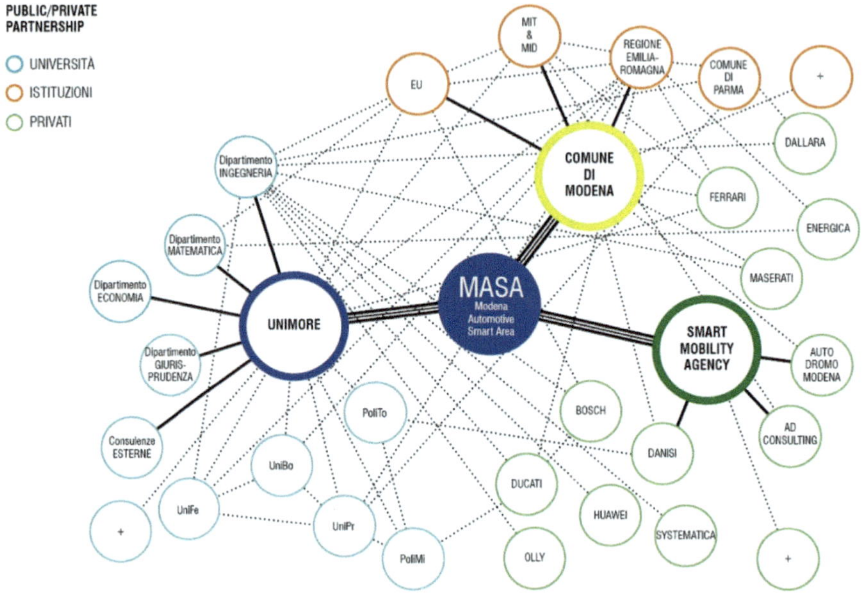

Fig. 25.6 MASA governance diagram. *Source MASA—Modena Automotive Smart Area*

technological requalification of the R-Nord district and making mobility a system with other networks and collective infrastructures.

The initiative also benefits from the endorsement of the Ministry of Transport, formalized in 2018 by signing a first three-year memorandum of understanding with UNIMORE and the Municipality of Modena to evaluate forms of collaboration and to promote the MASA experimentation area in the implementation of innovative solutions for autonomous and connected driving. The support of national institutions is renewed and expanded in 2020, when the Ministry of Infrastructure and Transport joins the Ministry of Technological Innovation and Digitization.

25.3 A Tiny Smart City: The Infrastructured Urban Area

The ODD of MASA is currently infrastructured with smart cameras, wired and wireless communication network (1 dedicated 4G antenna and one 5G antenna): sensors, servers and a data center, for the V2X operations of bidirectional communication between connected vehicles and the city (Fig. 25.7) and for the test drive of vehicles equipped with ADAS devices up to levels 3 and 4. The infrastructure installations for operation are constantly being implemented, and up to now it counts on 100 cameras, 4 fog nodes (Nvidia Titan), 10 MIND Smart Cameras, 2 mobile traffic lights and about 100 control units for weather and air quality control. The MASA data acquisition and communication system operate with "*on the edge*" technology, i.e., through the anonymization of sensitive data by an artificial intelligence incorporated in the acquisition cameras, which generalizes its personal or sensitive data in less than 30 ms. This happens before those are sent over the network to the control unit, which will send the commands back to the vehicle to be executed (Fig. 25.8). MASA is currently the only European Living Lab where road tests are carried out with an overall latency time of less than 100 ms between acquisition and command input.

Furthermore, MASA counts on the support of operational outside the living lab ODD, such as a dedicated private area at the Modena Autodrome, active since 2017,

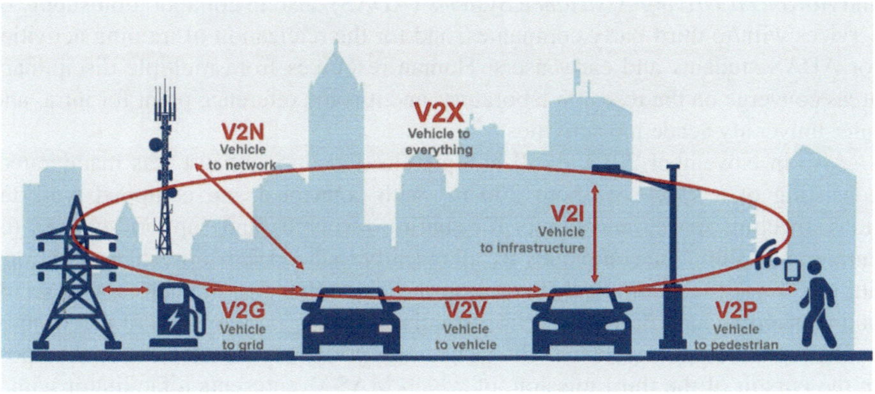

Fig. 25.7 V2X technology diagram. *Source* Porsche Consulting; *MASA—Modena Automotive Smart Area*

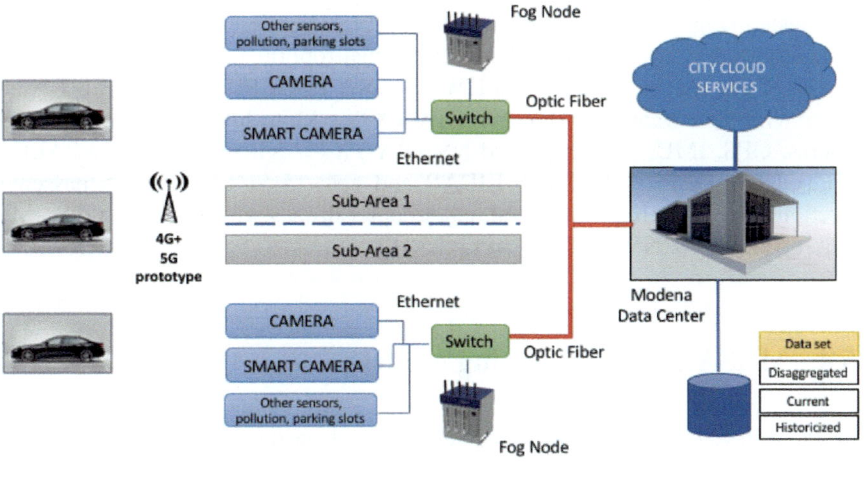

Fig. 25.8 V2X technology diagram. *Source* Porsche Consulting; *MASA—Modena Automotive Smart Area*

suitably infrastructured for the earliest test drive of self-driving vehicles and for active V2X communication, and to carry out training activities for students and users/early adopters of ADAS devices. Some technologies have already been tested in the area, such as traffic lights interconnected to the cloud system and supporting the latest industrial and automotive communication standards, digital signage, IoT cameras for obstacle recognition based on artificial intelligence, and connected with cloud servers "MASA," video analysis for the identification of parking spots to implement "*smart parking*" services, various sensors interconnected through the "*LoRa*" network. A dedicated university research laboratory has also been active since 2018, equipped with a static simulator that allows to perform "*Hardware in the loop*" for the implementation of research and development activities on autonomous driving and *Advanced Driving Assistance Systems* (ADAS), also in collaboration/supply of services with/to third-party companies, and for the realization of training activities for ADAS students and early users. Human resources from multiple disciplinary areas converge on the research laboratory, and it is the reference point for intra- and inter-university academic activities.

Also in November 2018, the E-hub private start-up incubator was inaugurated, consisting of a space of about 300 m^2 with connected and equipped workstations, meeting spaces and others for sharing resources and support services for entrepreneurship. The companies are all recently founded start-ups, sharing a strong attitude toward technological innovation in fields related to the automotive sector and its supply chain derivatives, in particular focusing on the fields of electronics, software and computer science. The incubator is positioned as an operational outpost in the pursuit of the third mission, of which MASA represents a facilitator with a key role in the network of relationships between the world of research and that of entrepreneurship.

As a facility, MASA uses a supervised garage of about 300 m^2 for the storage, custody and maintenance of some of the vehicles on which the projects are already being tested (Fig. 25.9). Currently, the fleet of vehicles on which drive testing has already started is composed of 2 Maserati Quattroporte L3 cars with LIDAR, 6 cameras, GPS, IMU and radar and HW NVIDIA PEGASUS; 1 Maserati Levante L2 car with stereo camera, HW NVIDIA TX2; 1 Motorbike Energica EVA with cameras, GPS, IMU, gyroscope and NVIDIA TX2 // XILINX ULTRASCALE; 1 LIFETOUCH Delivery Bot with LIDAR and gps; 15 SETA bus with passenger detection system; 4 Hexacopter and quadricopter drones with cameras, GPS, IMU, gyroscope and HW NVIDIA TX2 (Fig. 25.10).

25.4 MASA 2.0: An Evolving Model

The main development of the living lab is based on the creation and promotion of a regional ecosystem of connected and autonomous mobility, which favors new paradigms of more sustainable, safe and inclusive urban environment, through the conception, development and experimentation of services from part of the companies

Fig. 25.9 MASA infrastructure asset: 1. Data Center; 2. Via Rita Levi Montalcini; 3. Millemilia Garage; 4. Autodromo di Modena. *Source MASA—Modena Automotive Smart Area*

and research centers of our territory and beyond, establishing a system with other tangible and intangible, urban and extra-urban infrastructural networks.

In its evolution, MASA intends to be a model for the experimentation of research, development, design and validation services of devices and technological solutions for the management of data between vehicles and cities, making available both its physical and virtual infrastructure. The three-year protocol signed with the Ministry of Infrastructure and Transport has been renewed in June 2021, implementing the areas of shared actions on the broader frame of technological innovation for sustainability and urban regeneration. In particular, MASA has acquired skills and resources to expand services in the scientific, technical/technological, economic, legal and urban planning fields. The extent of the impact of its activities will be further implemented, starting from the local area to match the goals of the national and European agendas.

The services concern contracts between a multiplicity of actors: private and private, private and university/research institution, university and private, university and university/research institution. The leap in scale is understood as the systematization of the experiences gained so far, maximizing the operational synergies of the operators and confirming itself as increasingly authoritative interlocutors toward local, national and European institutions.

Fig. 25.10 MASA vehicle asset: 1. Maserati Quattroporte; 2. Energica Eva; 3. Delivery Bot; 4. SETA Bus; 5. Drone Hexacopter; 6. Formula Indy; 7. Shuttle Olli by Local Motors

In terms of relations with companies, UNIMORE has signed a protocol with the Smart Mobility Agency, which has been born from the collaboration of AD Consulting, Aerautodromo di Modena and Danisi Engineering. SMA participates in the technical-scientific committee that selects applicable within the projects and defines the international collaborations according to the objectives of MASA.

UNIMORE has fielded 13 departments for this collaboration protocol, confirming the multidisciplinary expansion on the issues of mobility, the city and ultimately the connected society.

MASA intends to attract resources by pursuing funding lines for research and innovation through both Italian and European applications and by seeking further private partnerships in order to broad the horizon of collaborations to non-traditional and innovative operators in the field of connected and sustainable mobility.

25.5 Developing Plans for 2021–2024

At the state-of-the-art MASA has concluded a three-year cycle of fundamental experiences for the validation of the research request and the verification of the existence of a market interested in welcoming it. Its surprising attractive value now allows to start a more ambitious three-year planning period, which will see on the one hand the implementation of consolidated core business activities and on the other structured to expand the range of services offered.

In this perspective, the governance model in the public–private partnership is evolving, with the expansion of both the public and private realm, a strategy that is believed to be the trajectory to follow as well in the upcoming future. As a result of the evolution of the model, a coherent organizational structure will be adopted, which tends to be more defined and with dedicated resources to be able to perform its functions.

Equally in progress is the legislation ruling on the use of public roads and the methods of collecting and managing sensitive data by public and private individuals. With respect to this sensitive topic, everything related to the requirements of the DGPR is being studied in depth with the Law Department.

Regarding the definition and the development of new services, the process will follow a line of inclusiveness toward new actors, able to bring skills and contributions on a broader horizon of objectives and disciplinary areas.

MASA is currently investigating potential strategies and solutions for the connective space of the district due to the use of MaaS models, again confronting with the actual conditions of an ongoing process of regeneration, with a large variety of situations of the context to be taken in account and immediate feedback on the field. The upgrade of MASA will take these data as inputs to generate a complex model as output. As mentioned before the multimodality of transport, regarding both people and goods, implicates the integration of several infrastructures, where information technologies are not merely on more layers to the system, but the enabling trigger to enlace them all.

According to the lesson learned from ancients Romans, where in the Latin language the word "city" had a semantic distinction between "urbs" and "civitas," i.e., the city of the stones and the city of the people, society is considered as well as a crucial infrastructure that demands engagement at all levels, from the governance body to the stakeholder until the final users, for whom right of mobility policy and inclusive services are designed.

MASA as a living lab aims to investigate on how cities will adapt to the new paradigms of mobility, where a virtual revolution is taking place faster than urban transformation processes. Learning from the lesson of the mobility revolution of the twentieth century, where an under-reactive urbanism allowed private cars to take it all, the urban design and planning should be able to fit the emerging mobility models in the complexity of the smart city as an ecosystem.

Chapter 26
Expanding the Wave of Smartness: Smart Buildings, Another Frontier of the Digital Revolution

Valentina Frighi

Abstract Smart buildings can be considered the future development direction of constructions: IoT, which extended connections and intelligence to real-life objects, led to a revolution in building practices, making it necessary to obtain edifices equipped with new original features. Seeking to respond to climate-related challenges of the twenty-first century, the technologies triggered by the digital revolution led smart buildings to become the natural evolution of the "sustainable" or NZEB buildings, introducing a series of innovations toward positive changes, continuing the path of hybridization with other disciplines which characterized this digital era. Indeed, the term "smart buildings" conventionally refers to all buildings that show some kind of innovations, concerning technical plants but also building envelope components or the building system as a whole. Besides, it can be said that in the wake of recent directives issued by the EU concerning the Green Deal, the Renovation Wave, and the New European Bauhaus, the technological culture of architecture has evolved, affecting also the aesthetic domain. Therefore, the paper aims to understand the new paradigms of current architecture, analyzing the advantages brought in terms of innovative methods and tools for controlling the quality of construction projects and processes, but also considering new digital techniques for design and representation, smart high-performance materials, adaptive and innovative technologies and/or sensors; thus trying to understand how architectural objects became inspiring examples of the combination of technological innovation and design, and how they can play an important role in terms of environmental sustainability and reduced consumption of resources.

Keywords Smart buildings · Digital revolution · Building envelope · Smart materials · Technological innovation

V. Frighi (✉)
University of Ferrara, Ferrara, Italy
e-mail: valentina.frighi@unife.it

E. Arbizzani et al. (eds.), *Technological Imagination in the Green and Digital Transition*, The Urban Book Series, https://doi.org/10.1007/978-3-031-29515-7_26

26.1 Introduction

Current living habits, also due to the current situation, have bewilderingly changed compared to what we used to do in the past, making clear that the pre-pandemic living and work patterns will not hold once the situation will be solved. The past years have caused us to fundamentally re-evaluate the importance of the buildings we live and work in, so as the impact they have on our lives, our well-being, and our relationships with the surrounding environment.

Simultaneously, the word "smart" started to be considered ever more not only as a hybrid situation in which we are allowed to work according to the most appropriate methods and locations, thanks to the connection, but also as a way of control and optimize our setting environment. This has led us to reappraise how we conceive buildings, especially concerning the diffusion of information and control technologies into traditional constructions.

At the same time, the increasing focus on environmental and societal issues, in the wake of recent directives issued by the European Commission concerning the Green Deal (European Commission 2019), the Renovation Wave (European Commission 2020), and the New European Bauhaus (European Commission 2021a), underlines the importance of effective and inclusive tools to support and foster the energy transition, toward more inclusive, sustainable, affordable, and secure solutions to global challenges.

The comprehension of the value of digitization within the construction sector (Fig. 26.1), progressively led the technological culture of architecture to evolve, involving also the aesthetic domain in the definition of new paradigms; thus, the built environment evolved accordingly: from simple dwellings to buildings equipped with advanced technologies, made available by the innovations of the Fourth (and then Fifth) Industrial Revolution and especially by the Internet of Things (IoT).

26.2 Smart Buildings as a Frontier of the Digital Revolution

The digital transformation process which began about 30 years ago within the Architecture, Engineering, and Construction (AEC) sector, produced great buildings improvement: on the one hand through the transition from analog to digital tools and computer-aided design, and on the other hand, thanks to the technological advancements introduced by the last Industrial Revolutions.

Digitalization changes how buildings will be planned, constructed, used, and managed, empowering technologies and connecting the several stakeholders involved in the design process.

In light of the continuous hybridization among different disciplines which characterized this digital era, coupled with the pivotal role of digitalization for the competitiveness and sustainability of the construction sector (European commission 2021b),

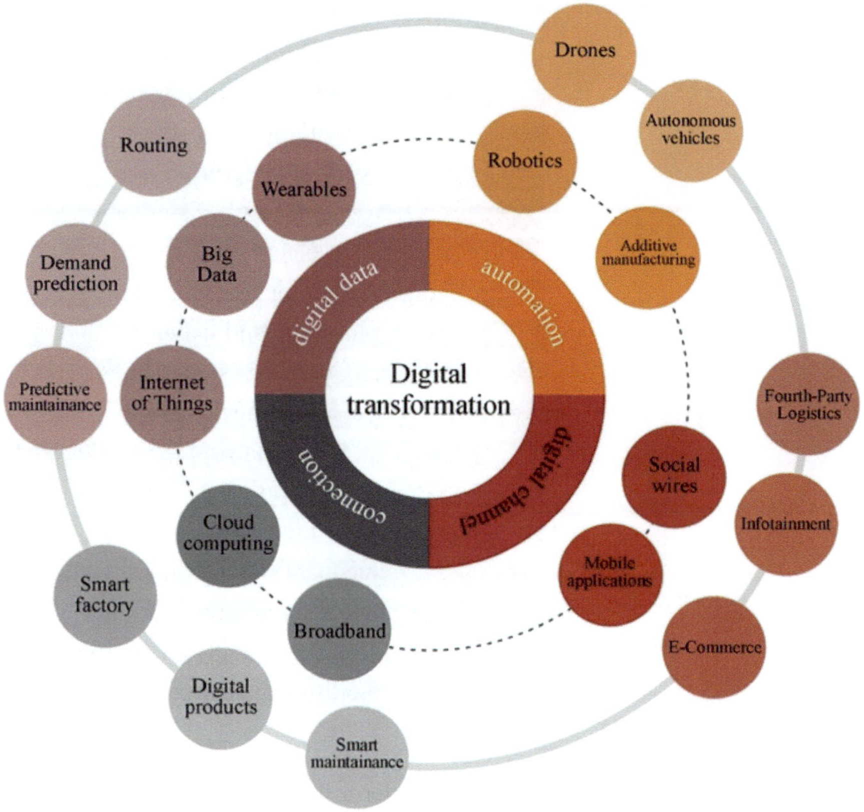

Fig. 26.1 Driver of the digitalization process. *Source* Author's elaboration on Roland Beger and BDI (2015)

we can say that smart buildings can be considered the natural evolution of the concept of "sustainable" or NZEB buildings, introducing a series of innovations toward positive changes and a more respectful vision of construction (Fig. 26.2).

But what can be considered a "smart building?" The concept of intelligence in buildings was introduced approximately around the 1990s (Derek and Clements-Croome 1997) when Information and Communication Technologies (ICTs) started to be applied to buildings, fostered by the first automations within the building sector enabled by the so-called Building Management Systems (BMS). However, it is only in the new millennium, with the spread of data storage and analytics systems, the introduction of open protocols, the "wireless transition" and, above all, the ubiquity of the IoT, that the concept of Smart Buildings begins to shape as we know it today (Bucknam et al. 2014).

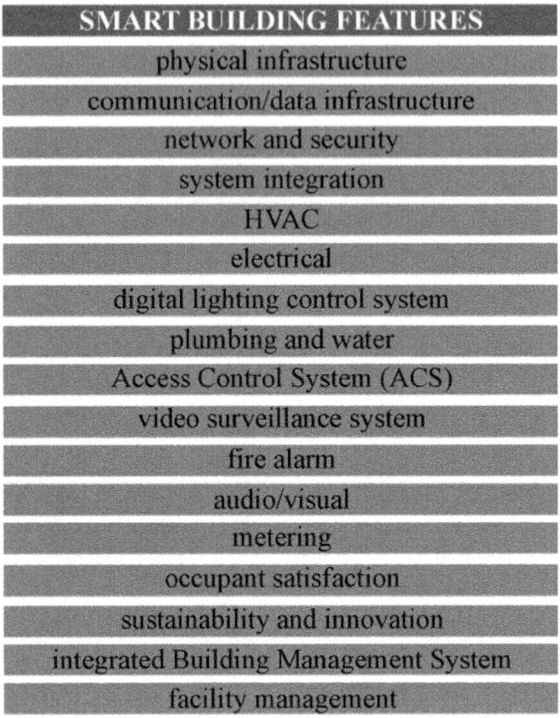

Fig. 26.2 Smart buildings' features. *Source* The Author

"Smart building" indeed is an umbrella term that has recently come into use to describe many different technologies that are being integrated into buildings. However, there is no clear definition of what makes a building "smart" (Hoy and Brigham 2016). In general terms, we can say that a smart building is a building that shows some kind of innovations, concerning technical plants but also building envelope components or the building system as a whole. It usually provides extra facilities if compared to a "standard building," such as connected device systems (composed of sensors and actuators) or IoT technologies, which allow visualizing several parameters in real-time concerning building state (such as the temperature of the rooms, the state of the equipment, the consumption of heating and electricity).

Moreover, the recent and rapid evolution of artificial intelligence (AI) and machine learning (ML) has equipped buildings with the ability to learn (Alanne and Seppo 2022) from previous settings, making adaptability at a system level a reality, also recurring to digital twin (DT) as training environments.

So, the result is the use of the word "smart" referring to the intelligence of things, intending to address the increased ability of systems (in this case, buildings) in doing something.

Smart buildings indeed redefine the use of common infrastructure in light of new technologies, thanks to modeling tools and connected devices which made it possible to conceive less energy-intensive and more environmentally friendly systems.

26.3 The "Smart" Goes Through the Envelope

A domain which offered fertile ground to coalesce much of the current discourse is those of building skin (Trubiano 2013), the place in which several factors (formal, performance, material, and mutual integration) come to a synthesis.

As interface among external and internal factors, to which is entrusted the achievement and the subsequent maintenance of certain conditions, this liminal surface must now be designed in a flexible and dynamic way to allow its adaption to the ever-changing variations of the surrounding context.

Conceiving the envelope as a responsive, multifunctional element rather than a static, single-behavior boundary, led to a paradigm shift in the envelope design field, determining the need for a wider analysis of the interactions between the building and the environment (Perino and Serra 2015), evaluating performance and mutual role in terms of constraints and intentions, generally measured quantitatively.

However, it must be said that focusing exclusively on the performance functionality of the building envelope can mean privileging the technological determinism of matters. Even though building envelopes are, by definition, central to the thermal and environmental regulation of a building's interior, the materials with which they are built contribute not only to their functionality but also to their formal value. It is through the materials and systems which shape the envelope that buildings relate to the surrounding, being immersed in it and dialogue also with the social and cultural context.

This adaptability capacity, which can be translated into the generation of architecture equipped with responsive abilities, produces a revolution in consolidated formal paradigms, determining the birth of complex systems in which the technological components are distinctive ad essential competences that significantly contribute to the definition of new architectural languages. Thus, the performance and functional-technological issues acquire a new aesthetic-formal value, becoming generators of a design process whose results are building organisms able to go beyond different cultures and architectural traditions.

So, is it quite clear that materials can be conceived as one of the most important "smartest" parts of a building, being often driven by innovation and technology (Sinopoli 2015). Yet, is it true that they would not allow meeting the expectations if installed incorrectly; the lack of coordination often results in the generation of severe performance issues during the construction phase, although products equipped with very high performance.

Indeed, among the latest "game-changers" (Sinopoli 2015) there are also innovative methods and tools for controlling the quality of construction projects and processes, so as digital techniques for design and representation, smart high-performance materials, and/or adaptive and innovative technologies that allow us to challenge how we think about building envelopes, making emerging an alternative paradigm for a theory of building.

26.4 Materials and Methods: Which Technologies Fit the Smart Building?

As in the case of any type of construction, when we talk about smart buildings the materials that can be used are plenty; industry and academia indeed, constantly work to develop innovative materials and building systems able to reconfigure themselves to meet external and internal changes in climate and user behavior (Barozzi et al. 2016), to create "living environments" (Abdullah and Al-Alwan 2019).

The quick spread of innovative technologies has been followed by the development of classification systems, to catalog and systematize technologies which fit the smart building (Carlucci 2021). Among the others, one of the most used (Attia et al. 2020; Casini 2016; Frighi 2022) is the one that identifies different families of building envelope typologies, according to the technology they adopt. The most common are: (i) dynamic shadings, (ii) adaptive glazing (chromogenic façades), and (iii) dynamic façades.

The first can be defined as the direct evolution of fixed shading systems typical of vernacular architecture (Al Dakheel et al. 2020); it comprises elements, devices, and technologies mainly aimed at selectively control the incident solar radiation in a dynamic way, thus possibly recurring to automated or intelligent systems.

The second refers to façades which employ chromogenic materials, mainly smart glass technologies, whose functioning is based on the change of their optical properties in relation to different kinds of inputs (e.g., solar radiation, temperature variation, application of voltage, use of hydrogen) (Fig. 26.3).

The third family refers to sets of heterogeneous components, elements, and/or materials able to vary their performance features over time, even recurring to technologies generally used in other industrial sectors (such as nanotechnologies, aerogel, vacuum insulated panels, self-cleaning technologies, or phase change materials) or equipped with the ability to change their behavior by reacting to external stimulus (of environmental nature or based on control systems that resort on sensors and actuators) thus varying one or more of their properties. Changes are, in general, direct and reversible. This is the case, for instance, of shape-memory materials (Mohamed 2017).

Downing the scale to the technologies included in the abovementioned domains, in the first case we mainly refer to building components, such as shading systems, shutters, louvers, or to general elements whose purpose is to regulate interaction among inside and outside, mainly in terms of thermal comfort and visual performance. The second family instead mainly involves glazed advanced systems, which range from reflective, selective, or low emissive glazing and goes through vacuum insulated glass, TIM glass, and heating glass, toward chromogenic glazing technologies and, again, innovative emerging solutions equipped with adaptive functions (generally control and manage more than prevent/reject solar radiation). While the third family concerns façade systems as a whole, considering simply ventilated façades, active façades (Villegas et al. 2020), or, again, innovative systems that involve advanced

Fig. 26.3 Metallic louvers, office buildings in Lisbon. *Source* The Author

materials and technologies equipped with smart functions in terms of purpose, operation mode, or adaptivity toward the external variations, considering also the relevant physics of the system or the adaptation scale (Loonen et al. 2013).

26.5 Expanding the Wave of Smartness: Conclusion and Future Developments

Architecture has always been a representation of the society, being the physical environment in which people live and work; from simple shelters conceived to offer protection, buildings have evolved following the modification of way of life, and users' needs as well, being shaped by climate conditions, materials readily available, and the value of the society building them. Accordingly, as new technologies rapidly developed, changing the needs of people and communities, construction continuously evolved; aided by computer design, architecture becomes ever more articulated in brave new shapes, previously unthinkable, capable of providing unprecedented behavior that goes beyond the simplification of traditional solutions, often determined by the assembly of regular elements.

The great potential new smart materials technologies have, can dramatically change the process of design and construction; a building envelope can then be thought as having multiple configurations, depending on the time of the day, the season, and the use, which may result in a certain architectural quality (Meagher 2015).

Smart buildings, whose technological components acquire new aesthetic-formal value, are thus becoming the driving force for the development of new ideas and architectural forms while giving buildings the much-needed resilience capacity (e.g., anticipating potential risk factors, developing future scenarios, or testing new performance responses).

Besides, the use of innovative materials, separated from their adoption according to conventional forms and meanings, generates a sort of new cross-cutting linguistic means, expression of a global architectural language capable to adapt to the continuously changing needs of a hybrid contemporaneity.

Although many twists and turns, this process has been unstoppable, so future development and architectural directions should focus on the production of materials, components, and systems that can support this process of evolution within the domain of architectural technology.

Certainly, the current ability to gather and collect a great amount of data, coupled with an explosion in IoT applications, can be considered somewhat of a solved problem, but this only represents a first step toward the expansion of the "smartness' wave"; the following is understanding what kind of information data is telling us, to inform better decisions, and managing that information in a way which makes data easy to access and to analyze. Since cognitive buildings are still on the distant horizon, we must find a way to merge the digital components of smart buildings with the traditional analog components of building operation, focusing on how we deliver information to building operators, what this information tells them, how they can use it and how they can feedback their knowledge to the system. We need to start from the perspective of someone using a building and think about what they want to know and then find a way of describing and connecting that information.

The task of tackling current challenges concurrently is a daunting one, but the correct application of smart building technology is undoubtedly part of the solution.

References

Abdullah YS, Al-Alwan HAS (2019) Smart material systems and adaptiveness in architecture. Ain Shams Eng J 10:623–638

Al Dakheel J et al (2020) Smart buildings features and key performance indicators: a review. Sustain Cities Soc 61:102328

Alanne K, Seppo S (2022) An overview of machine learning applications for smart buildings. Sustain Cities Soc 76(2022):103445

Attia S, Lioure A, Declaude Q (2020) Future trends and main concepts of adaptive facade systems. Energy Sci Eng. https://doi.org/10.1002/ese3.725

Barozzi M et al (2016) The sustainability of adaptive envelopes: developments of kinetic architecture. Procedia Eng 155(2016):275–284. https://doi.org/10.1016/j.proeng.2016.08.029

Bucknam AH, Mayfield Stephen M, Beck BM (2014) What is a smart building? Smart Sustain Built Environ 3(2):92–109

Carlucci F (2021) A review of smart and responsive building technologies and their classifications. Future Cities Environ 7(1):1–12. http://doi.org/10.5334/fce.123

Casini M (2016) Smart buildings. Advanced materials and nanotechnology to improve energy-efficiency and environmental performance. Woodhead Publishing series in civil and structural engineering, no 69. Elsevier, Oxford

Derek T, Clements-Croome J (1997) What do we mean by intelligent buildings? Autom Constr 6:395–400. http://doi.org/10.1016/S0926-5805(97)00018-6

European Commission (2019) The European green deal. https://eur-lex.europa.eu/legal-content/EN/TXT/?qid=1588580774040&uri=CELEX:52019DC0640. Accessed 17 May 2022

European Commission (2020) A renovation wave for Europe—greening our buildings, creating jobs, improving lives. https://eur-lex.europa.eu/legal-content/EN/TXT/?qid=1603122220757&uri=CELEX:52020DC0662. Accessed 17 May 2022

European Commission (2021a) New European Bauhaus. Beautiful, sustainable, together. https://europa.eu/new-european-bauhaus/system/files/2021-09/COM%282021%29_573_EN_ACT.pdf. Accessed 17 May 2022

European commission (2021b) Digitalization in the construction sector. https://ec.europa.eu/docsroom/documents/45547/attachments/1/translations/en/renditions/pdf. Accessed 20 May 2022

Frighi V (2022) Smart architecture. A sustainable approach for transparent building components design. Springer International Publishing. http://doi.org/10.1007/978-3-030-77606-0j

Hoy MB, Brigham TJ (2016) Smart buildings: an introduction to the library of the future. Med Ref Serv Q 35(3):326–331. https://doi.org/10.1080/02763869.2016.1189787

Loonen RCGM et al (2013) Climate adaptive building shells: state-of the-art and future challenges. Renew Sustain Energy Rev 25:483–493. https://doi.org/10.1016/j.rser.2013.04.016

Meagher M (2015) Designing for change: the poetic potential of responsive architecture. Front Archit Res 4(2015):159–165. https://doi.org/10.1016/j.foar.2015.03.002

Mohamed ASY (2017) Smart materials innovative technologies in architecture: towards innovative design paradigm. Energy Procedia 115(2017):139–154

Perino M, Serra V (2015) Switching from static to adaptable and dynamic building envelopes: a paradigm shift for the energy efficiency in buildings. J Facade Des Eng. https://doi.org/10.3233/FDE-150039

Sinopoli J (2015) The constantly evolving smart building. http://www.automatedbuildings.com/news/feb15/articles/sinopoli/150201011303sinopoli.html. Accessed 04 Mar 2022

Trubiano F (2013) Performance based envelopes: a theory of spatialized skins and the emergence of the integrated design professional. Buildings 3:689–712. http://doi.org/10.3390/buildings3040689

Villegas JE, Restrepo Gutierrez JC, Colorado HA (2020) Active materials for adaptive building envelopes: a review. J Mater Environ Sci 11(6):988–1009

Chapter 27
Sharing Innovation. The Acceptability of Off-site Industrialized Systems for Housing

Gianluca Pozzi, Giulia Vignati, and Elisabetta Ginelli

Abstract From the sixties, innovation and industrialization have been a returning mantra for the construction sector at every new building cycle passage after an economic crisis, as a tool of overcoming difficulties. This positivism has always been disregarded, especially for housing and for Italy. To avoid this dynamic recurrence even in the current ecological transition passage, research must provide, in parallel with innovative products and techniques, innovative cultural approaches so that extraordinary products and techniques can be accepted by the market, demonstrating how the synergy between them leads to a high added value for sustainable quality of living. Most of the actors (from designers to builders and maintainers) agree that innovative systems, especially industrialized off-site, are more sustainable, especially today when sustainability and resilience are the core of the construction sector; despite this, these systems are struggling to spread. This contribution focuses on acceptability and decision-making processes that lead to innovative choices, identifying the innovation of the functional, social and economic management of the buildings as the "missing ring" for housing. This acceptability has certainly increased today because of new form of "atypical" living, such as senior/student and temporary housing and co-living, which contribute to intensifying the demand of "industrialized", flexible, affordable and reliable houses. Technological innovation, in fact, actives only if technical innovation is combined with strategies and new approaches in organization, marketing and after-sales services focused on sharing and participation. Through an example of a realized off-site transformable residential building and case studies of new form of management, this contribution proposes innovation perspectives capable of overcoming design and decision-making obstacles to the spread of off-site systems, also identifying in the institutional sustainability one of the cores of this subject.

G. Pozzi (✉) · G. Vignati · E. Ginelli
Polytechnic University of Milan, Milan, Italy
e-mail: gianluca.pozzi@polimi.it

G. Vignati
e-mail: giulia.vignati@polimi.it

E. Ginelli
e-mail: elisabetta.ginelli@polimi.it

E. Arbizzani et al. (eds.), *Technological Imagination in the Green and Digital Transition*, The Urban Book Series, https://doi.org/10.1007/978-3-031-29515-7_27

Keywords Off-site building production · Acceptability criteria · Functional management in use · Innovation management · Institutional sustainability

27.1 Building Cycles and Innovation: An Introduction

The contemporary Italian construction market has just entered the 7th cycle (CRESME 2020) and, like every moment of "rebound" after a decline, is characterized by a positive growth, also driven by the incentives of the PNRR. This beginning is also marked by proposals and research for the innovation of the construction sector which is, especially in Italy, radically and historically backward compared to industry and manufacturing. However, in all the turns of the cycle, the conditions so that research and constructive experimentation could effectively disseminate have never been realized in a sufficient scale to make a qualitative leap and really affect the widespread market. For this to happen, in fact, it is necessary that three conditions overlap simultaneously (Lehmann and Fitzgerald 2013; Losasso 2010; Lu et al. 2018; Russo Ermolli 2007). The first condition is technical innovations able of improving the quality and sustainability (not least economic) of constructions. The second is the political will to drive and support adequate building innovation. The third is the thrust of the market that must accept innovation and political/social demands.

27.2 Innovation for Today Housing

The construction sector is in strong evolution, and new trends are emerging in the housing market, in terms of performance, living space forms of use, costs and turnover that involve a rethinking in construction practice. New increasing segments of the population are highlighted (Graph 27.1) (CDP Cassa Depositi e prestiti 2018): young people, not numerous but with a high dynamism, and the elderly, in great increasing number. Especially for these categories, the management of the house is crucial, privileging the idea of "house as a service", with a dynamic similar to hotels management. To satisfy these demands,[1] the market requires more fluid conditions (Bergan et al. 2020). This need can be satisfied by the innovation of construction systems but also by new management models and by procedures that support their intrinsically value.

[1] Mobility of population between 18 and 35: in 2017 1.36 million people were subject to residential transfer within national borders for work (CDP 2018).

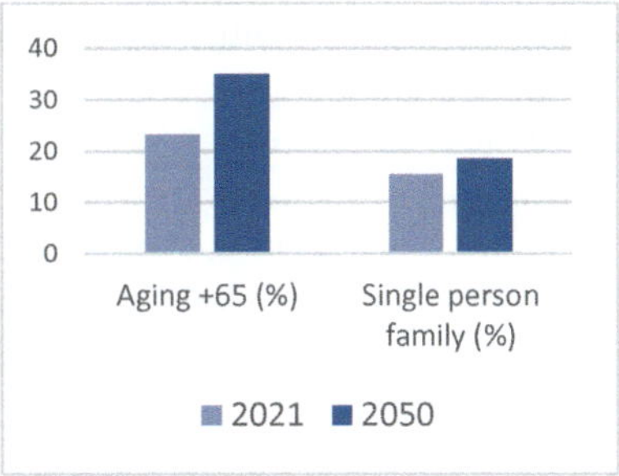

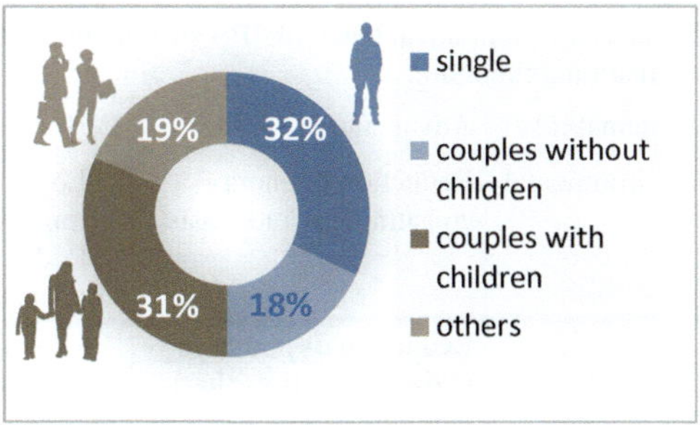

Graph 27.1 (left) Population evolution for Italy. Elaboration from ISTAT official data 2021. (right) Households composition in major cities. Elaboration from ANCE (2019)

27.2.1 The Advantages of IBS

The contemporary scientific literature (Table 27.1) and many international reports[2] clearly assert that Industrialized Building Systems (IBS[3]), especially dry techniques, significantly contribute to the sustainability of construction systems, including residential. Below is highlighted only a summary of the main advantages derived from the international literature, delegating to any future research a broader articulation and referring to Pozzi (2021) for an exhaustive treatment of the topic.

27.2.2 The Obstacles for IBS

Alongside the advantages, the same scientific literature and the proven modest diffusion clearly highlight the factors that limit the acceptance and growth of IBS. The real crux of the problem lies in the actor who makes the choice of the system to be used (Goodier and Gibb 2005) and the limiting factors can be divided, depending on the decision-maker, into factors of the client, the builder and the designer.

The **client**[4] very often has an inadequate and fragmented knowledge of systems and a scarce ability to perceive the added value of the innovative solutions, linked more to individual perceptions than to systemic market research. In addition, the client fears the possible serial nature of the industry as a customization difficulty (Boafo et al. 2016) and believes that the need to anticipate all decisions is an insurmountable obstacle (Elnaas 2014).

[2] www.freedoniagroup.com; www.fortunebusinessinsights.com/industry-reports/modular-construction-market-101662; www.grandviewresearch.com/industry-analysis/modular-construction-market; www.alliedmarketresearch.com/precast-construction-market; www.rolandberger.com/publications/publication_pdf/roland_berger_prefabricated_housing_market_3.pdf (visited on 08/03/2022).

[3] Here defined as "Off-site building construction system with high level of replicability".

[4] i.e. investors and users.

Table 27.1 Main advantages of IBS in relation to the three dimensions of sustainability, from international literature

Sustainability	Advantage	Performative advantage	Source
Environmental	Reduction of energy consumption and waste	– Lower environmental impact – Reduction of raw materials	Elnaas (2014), Jaillon and Poon (2008)
	Reduction of pollution in the production phase	– Lower environmental impact of the site	Elnaas (2014), Jaillon and Poon (2008), Jiang et al. (2018), Tam et al. (2007)
	Reduction of the overall CO_2 of the building	– Lower environmental impact – Lower carbon footprint	Elnaas (2014), Goodier and Gibb (2005)
	Greater dismantling and reversibility of the building	– Extension of the useful life – Lower environmental impact	Benros and Duarte (2009), Noguchi and Hernàndez-Velasco (2005)
Economic	Reduction of construction costs, with the same performance	– Cost-effectiveness	Jaillon and Poon (2008), Jiang et al. (2018), Goodier and Gibb (2007), Haas et al. (2000), Housing Communities and Local Government Committee (2019)
	Reduction of production times	– Cost-effectiveness	
	Triggering economies of scale, especially in modular and serial projects	– Cost-effectiveness – Productivity	Benros and Duarte (2009), Noguchi and Hernàndez-Velasco (2005)
	Faster construction improves financial performances	– Profitability – Enhancement	Elnaas (2014)
	Greater control over the quality of the work	– Quality – Reliability – Continuity of performance	Jaillon and Poon (2008), Jiang et al. (2018), Goodier and Gibb (2005)
Social	Greater safety for workers	– Safety	Elnaas (2014), Jaillon and Poon (2008), Jiang et al. (2018)
	Increased speed in housing production	– Productivity	Elnaas (2014), Švajlenka et al. (2017)
	Less inconvenience on site	– Safety – Productivity	Elnaas (2014)
	Shorter duration of the site	– Lower environmental impact	

(continued)

Table 27.1 (continued)

Sustainability	Advantage	Performative advantage	Source
	Greater maintainability	– Facilitated management – Facilitated maintainability – Durability	Luther et al. (2007)

The **manufacturing**[5] world is reluctant to accept IBS, above all because they require considerable investments of time and money in design, prototyping, for machinery production and for the stock of components (Elnaas 2014). Furthermore, for small to medium size builders, it would be impossible to realize advanced systems on their own, but they should externalize them, thus reducing their profit margins (Jiang et al. 2018; Chao et al. 2015). In addition, IBS has a fixed place of production and the costs of carrying out the work are directly linked to the distance from the construction site (Elnaas 2014), thus restricting the possible catchment area of the company. Furthermore, IBS could have a disconnection between production times (which tend to be continuous and homogeneous) and the construction site (which is instead heterogeneous and cyclical). Lastly, IBS requires skilled and trained labour (Jiang et al. 2018; Chao et al. 2015) and can hardly be entrusted to general-purpose subcontracts, requiring significant investments in training for personnel.

The **designer** generally does not accept IBS as he fears they may limit his creativity (Boafo et al. 2016) and the early freeze design (Elnaas 2014) does not allow him to carry out design actions in parallel with the implementation phases of the project, thus requiring considerable investments of time in an accurate design, for which the professional studios are not equipped and for which the extra design costs would not be recognized.

Another aspect that hinders the spread of IBS can be found in the scarce **institutional sustainability**, i.e. the lack of characterization of the standard to allow and support innovation: off-site industrialization sometimes struggles to find adequate acceptance by local authorities (Elnaas 2014) and a precise location within standards designed for traditional on-site construction (Jiang et al. 2018). Furthermore, the aspects of "plus enhancement" are often not recognized, as in the case of CAM or Uni PdR13 (Ginelli et al. 2019).

27.3 Acceptance and Sharing of Innovation Processes

Some non-acceptability factors have a conjuncture and structural nature, linked to the Country-system and to the structure of the economy itself. Only targeted policies and long-term investments, especially public ones, can influence these factors.

[5] i.e. construction companies.

However, there are non-acceptability factors, related to an erroneous and fearful vision of innovation, which can be overcome through an adequate transmission of information, team planning and the dissemination of virtuous and paradigmatic examples of achievements. Furthermore, in the transformation of the market, innovative approaches to living are bringing out (Fig. 27.1), which are stimulating the interest also of sectors previously unrelated to innovation for which the trinomial project/process/production is emerging as a new correspondence and the anticipation of decisions is becoming a cornerstone of the design approach.

These new approaches outline a design "process" that is increasingly in a "platform" of processes and actors, understood as a virtual place of meetings and connections able to generate the project. Below, an example concerning a product innovation linked above all to process innovation that has been able to share objectives and examples of innovative management that, starting from market changes, is a potential supply to seize the opportunities offered also by technical innovations.

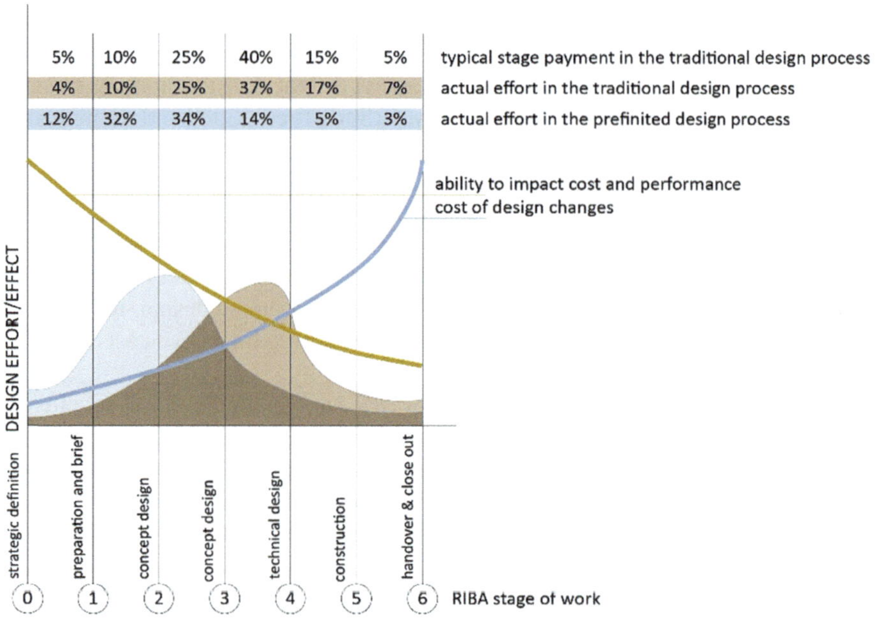

Fig. 27.1 Elaboration of Macleamy curve with data (UK market) Avalon Building, from www.avalonbuild.co.uk (visited on March 2022) that shows sharply the transition from a traditional process (brown) to an innovative one (light blue)

Table 27.2 Satisfied requirements of cHOMgenius project

Requirement (UE 305/2011)	Responding features
Mechanical resistance and stability	Project of specific seismic isolator, auto-centring and easy-replaceable, tested on seismic and safety simulator
Safety in case of fire	Structure, partitions and insulation are completely non-flammable
Hygiene, health and the environment	Internal surfaces are easily cleanable and waterproof
Safety and accessibility in use	Electrical plants are low voltage (24 V), and all plants are flexible and adaptable
Protection against noise	Reverb is excellent thanks to corrugate sheet and specific acoustic panels. Facade insulation is compliant with legislative prescriptions
Energy economy and heat retention	Off-grid management of energy guarantees zero energy consumption, also allowed by high specific heat of the envelop (more than 20 h of shift)
Sustainable use of natural resources	The use of reuse container as structural system brings incorporated energy of structure to 0 and the 30% of reduction of total CO_2. All the components are easy disassembling and reusable or, at least, recyclable

27.3.1 An Affective Example of Product Innovation and Sharing Project in Italy

The project "cHOMgenius. PrototypeSystem&SharedProject. Extraordinary solutions for smart living",[6] partially funded by Regione Lombardia Smart Living, is a two floors permanent residential building, entirely dry-joint with clamping techniques, built off-site using HC shipping container, off-grid and entirely dismantling and reversible. The very high satisfied requirements (Table 27.2) achieved are the result of the sharing of objectives, right from the first design phases, of 22 companies, in addition to the scientific support of the DABC of the Politecnico di Milano and UNI.

Product innovation lies above all in the transfer of techniques and products from the mechanical industry and in the development of nodes that, thanks to the continuous and effective interaction of all the actors involved, have allowed significant improvements in performance and reduced production times, as in the case of the interface between window and opaque casing described in Fig. 27.2.

Thanks to the effective application of the decision anticipation principle, based on an open, collaborative and cooperative design method, cHOMgenius has developed techno-typological solutions that demonstrate (Fig. 27.3) the great flexibility of the aggregative variants and the complete customization of spatial configuration, functional and finishing solutions (Pozzi 2021).

[6] https://www.dabc.polimi.it/en/ricerca/ricerca-competitiva/chomgenius-prototypesystemshared project/.

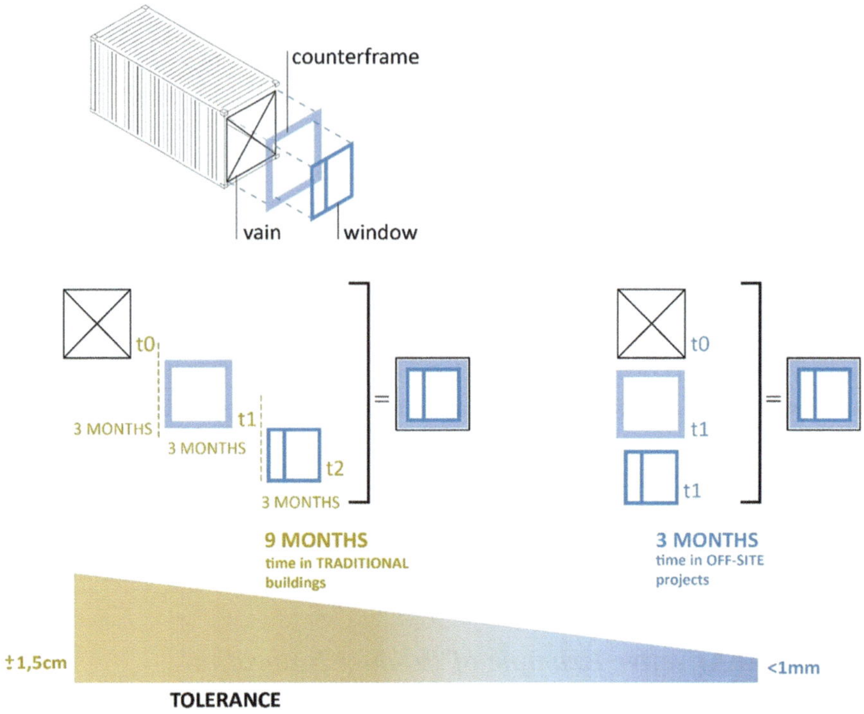

Fig. 27.2 IBS and tolerance: small off-site tolerance (< 1 mm) allows saving time in many operations. The example above is the interface between wall and window: in a traditional process (left with ± 1.5 cm tolerance) after the realization of the vain in the wall, you have to measure it and start making the counterframe. After the counterframe positioning, you have to measure it and start making the window (9 months in total). If you work with mechanical off-site precision all these operations can be made in parallel reducing by one-third the time

Fig. 27.3 Images of cHOMgenius project prototype in Busnago MB (from the top left, clockwise): transportation of the first module, factory assembly of the structure of the "other space", external south view, internal view from the mezzanine, internal view of the double-floor residential space and internal view from the second floor. All images by the authors

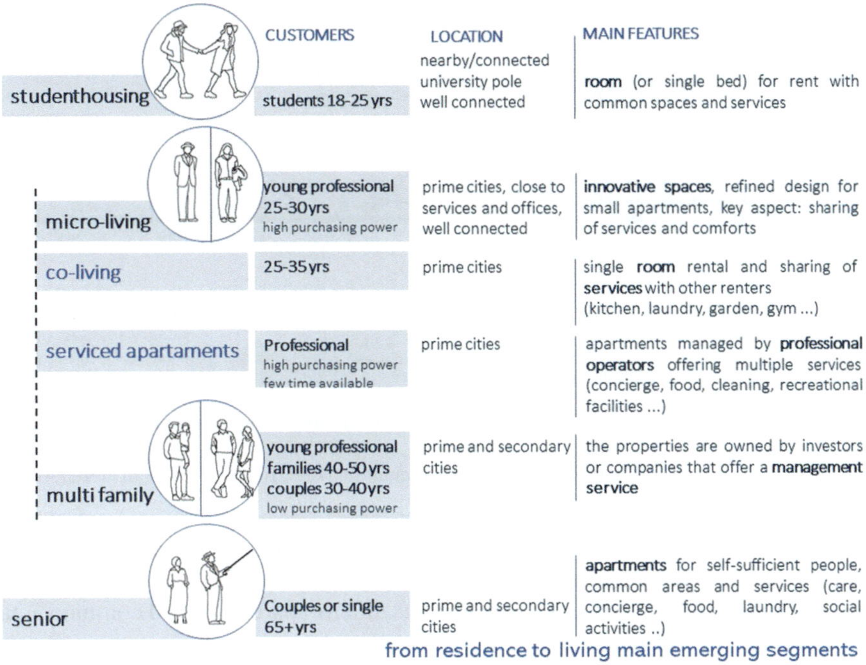

Fig. 27.4 Reorganization of the real estate market for new housing demands. Elaboration from ANCE (2019)

The residential market has partially diversified the demand for living spaces, especially in four emerging segments (ANCE 2019): student housing, micro-living, multifamily and senior housing (Fig. 27.4), where flexible, dynamic common spaces and services are crucial.[7] In Europe, living reached 83.4 billion in investments in 2020 (+10% in 2019) and diversification is a new investment strategy (Aberdeen Standard Investments): from a survey on 40 investors, 63% intend to expand into multifamily, 34% in student housing, 10% in co-living.[8]

From the management point of view, these models introduce hybrid housing typologies between domestic environment and services, to satisfy transitory needs for a short fixed time, where high performances are required at low management

---

[7] According to Savills' report "Global Living Report 2020" real estate investments in the residential sector worldwide accounted for 27% of global real estate investments, (+11% compared to ten years ago), in particular with the following investments:

- Multifamily: $223 billion transacted globally in 2019.
- Senior Living: $21.4 billion in investments in 2019.
- Student housing: three most active investors Blackstone iQ Student Accommodation for 6 billion dollars.

[8] Investimenti immobiliari, tutti pazzi per il "living" (mark-up.it).

Fig. 27.5 Main aspects underlying the "Industrial rental market". Elaboration from ANCE (2019)

costs, in a "package" that provides for furniture, energy supply contracts, maintenance, management (Glumac and Islam 2020). An example is the city-pop[9] model, developing in Milan and other European cities: micro-apartments and common areas (co-working, restaurant, minimarket) with optional services that can be managed via application (laundry, cleaning, booking, consumption, service management). A further example is the Ambient Assisted Living technologies in senior housing,[10] to manage the environment and monitor vital activities. The future of the housing market will also deal with the new structural ways of working remotely: for tertiary buildings, there will be a resizing of spaces, accompanied by an important divestment, and for workers, there will be the possibility of opting for locations far from companies and the need for adequate living spaces to support smart working.

27.4 The Missing Ring: The Innovative Management of Industrialized Buildings as an Open Conclusion

Some Contemporary "green" demands require high-performance and reversible buildings, low management costs, high maintainability and durability. In parallel, the housing market is orienting towards less stationary forms of housing that favour a temporary short-term lease (Fig. 27.5).

The conclusion of this contribution is, after having underlined and connected these two instances, that they must find a point of contact and converge for mutual benefit. It is in fact essential for a manager of estate asset, with dynamic management of the users, to rely on properties able to absorb new models of life and make the

[9] www.citypop.com.

[10] As examples www.seniorhousingitalia.it or www.amatilive.com.

management sustainable, especially from an economic point of view. Specifically in Italy (but not only), the obstacles to industrialization are often linked to the idea of a home as a durable (almost eternal) resource and to a stationary form of living that has made the house an experience of belonging. The lease, with a strong vocation of reduced temporariness, undermines these constraints and gives to builders an "industrial" way of building as "industrial" is the rent required today (ANCE 2019).

This convergence allows the client more awareness to find the most suitable systems that guarantee a flexibility and functional transformation and quality over time that allow an adequate management of the assets (Ginelli and Perriccioli 2019). It allows the world of production to interface with informed clients who need building stocks to be managed (and therefore to be realized) with adequate systems suitable for the scale of the intervention, thus initiating economies of scale that requires the use of IBS. Finally, the scale of the intervention allows designers to invest time in the design of reversible functional spaces, which can foresee the changing needs of users and which allows them to prototype and test, before the construction site, the appropriate technical solutions.

Thanks to cHOMgenius project, we have demonstrated that an effective IBS for housing is possible in Italy too and the highlighted new tendencies of the market confirm the changing of a paradigm for housing. The future research and experimentation we call for are related to the institutional dimension of sustainability: in order for these new processes and these new production systems to generate effective and incisive operating models and constructive syntax, and it is however essential that the legislator creates the appropriate conditions, providing clear indications on the choice of innovative construction systems, instead of, as also outlined by the PNRR, generically financing the sector, thus leaving room for those same lobbies that have hindered a real transformation of the construction market in past cycles.

References

ANCE (2019) Un mercato industriale dell'affitto. Strategic seminar June 26th, 2019

Benros D, Duarte JP (2009) An integrated system for providing mass customized housing. Autom Constr 18(3):310–320

Bergan TL, Gorman-Murray A, Power ER (2020) Coliving housing: home cultures of precarity for the new creative class. Soc Cult Geogr 22(9):1204–1222

Boafo FE, Kim JH, Kim JT (2016) Performance of modular prefabricated architecture: case study-based review and future pathways. Sustainability (Switzerland) 8

CDP Cassa Depositi e prestiti (2018) Smart housing. Le nuove dimensioni dell'abitare. Report. Oct 4th, 2018

Chao M, Qiping S, Wei P, Kunhui Y (2015) Major barriers to off-site construction: the developer's perspective in China. J Manag Eng 31(3)

CRESME (2020) XXIX rapporto congiunturale e previsionale CRESME. scenari e previsioni per il mercato 2020–2025. Cresme, Roma

Elnaas E (2014) The decision to use off-site manufacturing (OMS) system for house building projects in the UK, Unpublished thesis. The University of Brighton

Ginelli E, Perriccioli M (2019) Design for living: strategy and tactics to face changes. In: Lucarelli MT, Mussinelli E, Daglio L, Leone MF (eds) Designing resilience. Maggioli Editore, Santarcangelo di Romagna

Ginelli E, Lazzati G, Pirillo D, Pozzi G, Vignati G (2019) Il progetto cHOMgenius: relazioni virtuose tra progetto, prodotti e imprese. U & C. UNIFICAZIONE E CERTIFICAZIONE 5

Glumac B, Islam N (2020) Housing preferences for adaptive re-use of office and industrial buildings. Sustain Cities Soc 62, art. 102379

Goodier CI, Gibb AGF (2005) Barriers and opportunities for offsite in the UK. Paper presented at the systematic innovation in the management of project and processes, p 148

Goodier C, Gibb A (2007) Future opportunities for offsite in the UK. Constr Manag Econ 25(6):585–595

Haas CT, O'Connor JT, Tucker RL, Eickmann JA, Fagerlund WR (2000) Prefabrication and pre-assembly trends and effects on the construction workforce. Center for Construction Industry Studies, University of Texas, Austin, TX, USA

Housing Communities and Local Government Committee (2019) Modern methods of construction. 15th report of session 2017–19. House of Common, London, UK

Jaillon L, Poon CS (2008) Sustainable construction aspects of using prefabrication in dense urban environment: a Hong Kong case study. Constr Manag Econ 26(9):953–966

Jiang R, Mao C, Hou L, Wu C, Tan J (2018) A SWOT analysis for promoting off-site construction under the backdrop of China's new urbanisation. J Clean Prod 173:225–234

Lehmann S, Fitzgerald GB (2013) Wood in the city: social acceptance of prefabricated multi-storey timber buildings using low-carbon construction systems. In: Motivating change: sustainable design and behaviour in the built environment, vol 1, pp 385–418

Losasso M (2010) Percorsi dell'innovazione. industria edilizia, tecnologie, progetto. CLEAN, Napoli

Lu W, Chen K, Xue F, Pan W (2018) Searching for an optimal level of prefabrication in construction: an analytical framework. J Clean Prod 201:236–245

Luther M, Moreschini L, Pallot H (2007) Revisiting prefabricated building systems for the future. Paper presented at the ANZASCA 2007. Proceedings of 41st annual conference. Towards solutions for a liveable future: progress, practice, performance, people

Noguchi M, Hernàndez-Velasco CR (2005) A 'mass custom design' approach to upgrading conventional housing development in Mexico. Habitat Int 29:325–336

Pozzi G (2021) Industrialhousing. Or the way industrialisation can improve housing buildings. Mimesis, Sesto San Giovanni

Russo Ermolli S (2007) The environmental benefits of the off-site manufacturing. Paper presented at the Portugal SB 2007. Materials and practices: challenge of the industry for the New Millennium, Minho, pp 974–981

Švajlenka J, Kozlovská M, Spišáková M (2017) The benefits of modern method of construction based on wood in the context of sustainability. Int J Environ Sci Technol 14:1591–1602

Tam VWY, Tam CM, Zeng SX, Ng WCY (2007) Towards adoption of prefabrication in construction. Build Environ 10:3642–3654

Chapter 28
3D Printing for Housing. Recurring Architectural Themes

Giulio Paparella and Maura Percoco

Abstract Our present era asks architecture to confront new questions; visions and scenarios that project social, economic, and environmental issues toward that particular intersection between the green transition and the prefiguration of housing solutions for the city of tomorrow. In this drive toward a sociocultural renewal, digital architectural tools play a crucial role in the optimization of resources, customization of building components, and promotion of participative designing-building processes. As an innovative technique of digital fabrication, Additive Manufacturing makes this mass production economically accessible, also on-site, and using local materials. While the topic of a 'home for everyone' has started to be addressed, experiments and applications often focus primarily on technical aspects. To be understood, controlled, and aimed at truly improving quality of life, these innovations require a reflection on the paradigms that inspire digital design. Can the adoption of 3D Printing change design theory and the ways of conceiving the spaces of the habitat of tomorrow? More in detail, is it already possible to identify some particular architectural features? Using a selection of case studies, this paper critically interprets and analyzes these questions. The return of recurring architectural themes—the concept of instant architecture, the relation between natural–digital ecosystems, or the issue of self-determination—offers different ways of looking at 'printed' architecture.

Keywords Large-scale 3D printing · Architectural design for dwelling · High-tech processes and low-tech materials

28.1 Introduction

The complexities and criticalities of our current era force us to think in a new way.

G. Paparella · M. Percoco (✉)
Sapienza University of Rome, Rome, Italy
e-mail: maura.percoco@uniroma1.it

G. Paparella
e-mail: giulio.paparella@uniroma1.it

E. Arbizzani et al. (eds.), *Technological Imagination in the Green and Digital Transition*, The Urban Book Series, https://doi.org/10.1007/978-3-031-29515-7_28

There is a need for requires visions and scenarios that project social, economic, and environmental problems toward the intersection between the green transition and the prefiguration of near-future forms of housing and the city. Perhaps more than other disciplines, architecture is asked to do its part by reversing the indiscriminate exploitation of resources and production of waste.

The numbers tell the story. The built environment has 'consumed' some 50% of all raw materials extracted. In Europe, the construction sector is responsible for more than 35% of waste production, and greenhouse gas emissions from raw materials extraction, production, construction, and building renovation account for 5–12% of total emissions (DGIMIE 2021). Roughly 75% of all buildings in Europe are energy-inefficient (European Commission 2020). However, in the coming thirty years we will have to reach emissions neutrality to respect the objective of the European Green Deal!

Moreover, if we consider that the world population is forecast to grow by 2 billion people in the next thirty years and potentially reach almost 11 billion in 2100 (United Nations 2020), what scenario can we expect?

In addition to the environmental impact of the construction sector, we cannot ignore the socioeconomic effects induced by population growth. The spread of alternative living models—in response to economic restrictions, flexible working and, increasingly, the dramatic consequences of conflicts and natural disasters—is a clear signal.

Facing these challenges, as designers, we are asked to radically rethink the paradigms, processes, and logics that have contributed to defining this dramatic reality.

We must 'reverse the course' by rethinking architecture, setting aside models and habits, and adopting a transversal and systemic approach based on a holistic mediation between environment, society, and economic issues.

Faced with this challenge to manage multiple aspects, to reduce complexity to simple problems without losing sight of the whole, the visionary nature of architecture plays a central role, today more than ever thanks to the support of IT tools. In this effort to achieve an ecological, inclusive, and collaborative sociocultural renewal, the use of digital tools during different phases of the building offers significant opportunities in terms of resource optimization, mass customization of building components, and the activation and growth of participatory processes, during both design and construction.

Specifically, Digital Manufacturing tools, initially developed for other sectors, allow designers to faithfully materialize a series of computational design experiments, beginning with digital project modeling. Among large-scale Subtractive or Formative Digital Manufacturing tools, Additive Manufacturing[1] brings a particular advantage: It allows for the production of highly customized building components, also on-site and using local materials, employing an automated and economically accessible process.

[1] Additive manufacturing (sometimes referred to as rapid prototyping or 3D printing) is a method of manufacture where layers of material are built up to create a solid object (Redwood et al. 2017).

This intersection between architecture and industry is certainly nothing new. During the last century, the prefabrication of modular building components has contributed to reducing construction costs, although within the limits of standardization. In parallel, reinforced concrete construction enabled the development of modern housing models.

Given the impact of these technological-building innovations, it is essential that we attempt to predict the effects of the recent introduction of Digital Manufacturing technologies.

What advantages and objectives justify the application of 3D Printing tools in architecture? Is it possible to identify some of the architectural themes this innovative building method allows us to develop? Will the adoption of large-scale 3D Printing change the design theory and conception of the city and the shape of space? Theoretical-design research finds ample room in these considerations.

28.2 3D Printing for Architecture. What Questions Is It Answering?

Recent years have seen a significant rise in the number of studies of Additive Manufacturing in architecture; in 2016 only 3% of 3D Printing investigations involved this specific sector (Kianian 2017). Among the precursors, in 2006 Professor Behrokh Khoshnevis (University of Southern California) began to develop the Contour Crafting construction method for integrating the layering of fluid dense materials with standard robotic building processes (Bosscher et al. 2007). In Italy, the engineer Enrico Dini experimented with Binder Jet printing technology to produce load-bearing construction elements from cement powders (Gardiner and Burry 2009). Thanks to further national and international research, these initial insights would be consolidated in the following years. The aim is to investigate the characteristics of 3D Printing for the construction sector: flexible production brings significant advantages in terms of mass customization for on-site fabrication, accuracy of production, automation, and cost-effectiveness, all of which can be challenging to obtain from other Digital Manufacturing techniques (Lim et al. 2012).

Today numerous ongoing investigations are exploring large-scale 3D Printing. They focus primarily on technical aspects, such as material 'printability' and the structural and energy performance of pavilions and small infrastructures. In recent times, these preliminary studies have fostered a diffuse and growing interest in applying 3D Printing as a possible solution to the 'home for all.'

In 2018, Locatelli Partners applied Additive Concrete to construct the 3DHousing05 prototype in Milan (Locatelli Partners 2018); the minimum housing unit *UnaCasaTuttaDiUnPezzo* (AHouseMadeFromOnePiece) by D-Shape and the architect Marco Ferreri, built using Binder Jet printing technology for cement powders (D-Shape 2010); the first printed village consisting of customizable housing for low-income people by New Story + Icon (New Story 2019); the TECLA—Technology and Clay project by MCA—Mario Cucinella Architects and WASP for sustainable raw earth housing units constructed on-site from local materials (Andreoli 2019); the recent project in Africa by 14Trees demonstrates that affordable housing can be achieved quickly using Additive Concrete (Holcim 2020) (Fig. 28.1).

In this frenetic race of experiments, focused largely on quantitative aspects such as speed of construction and cost-effectiveness, there seems to be little space for a parallel meditation on architectural aspects. 3D Printing technology is certainly one possible answer. But to what question? What idea of the future is being proposed?

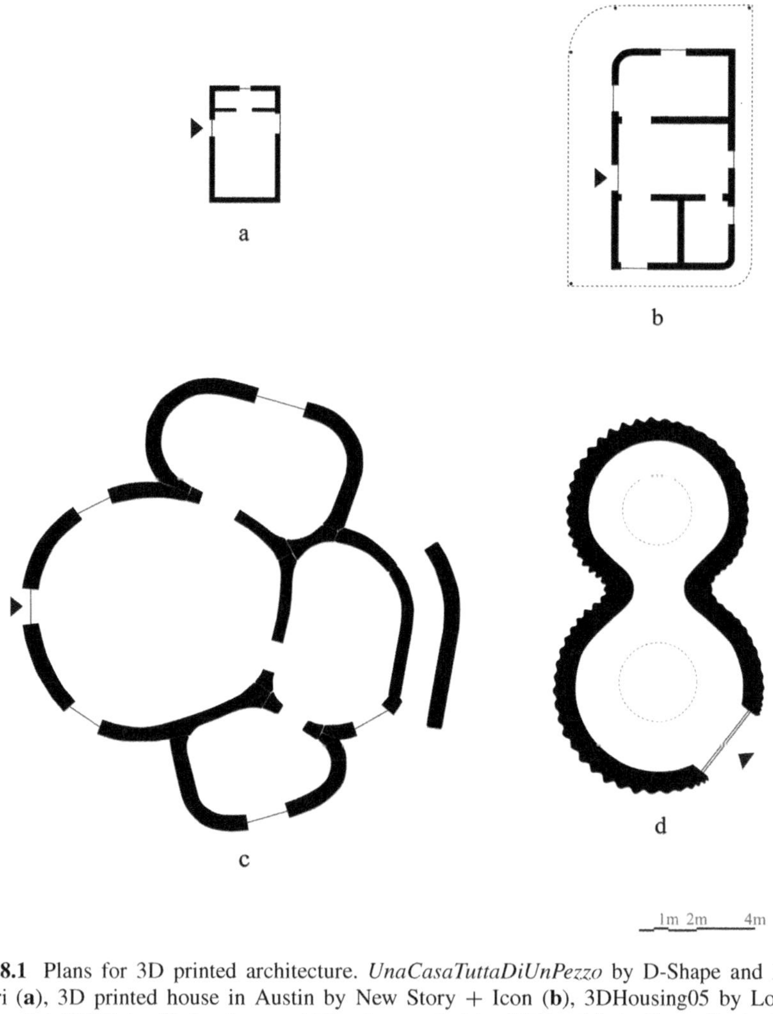

Fig. 28.1 Plans for 3D printed architecture. *UnaCasaTuttaDiUnPezzo* by D-Shape and Marco Ferreri (**a**), 3D printed house in Austin by New Story + Icon (**b**), 3DHousing05 by Locatelli Partners (**c**), TECLA—Technology and Clay is a project by MCA—Mario Cucinella Architects and WASP (**d**). Drawings by the authors

What image will the 'printed' house have? Will it belong to the habitat of a greener and digital tomorrow?

28.3 Recurring Architectural Themes of 'Printed' Living

Today's technical and constructive innovations require a reflection on the paradigms that inspire design practice, now 'digitalized.'

These innovative tools must be understood and controlled by designers to generate real social progress with the ability to improve everyone's quality of life. As it was at the end of the last century for other technical-constructive innovations, contemporary dwelling is probably also the true test bed. Since most current studies analyze performative aspects of 3D printed elements in the absence of a real functional context, this research defines projects explicitly developed for contemporary living as case studies. A critical analysis of these pioneering examples aims to deduce some recurring architectural themes.

28.3.1 'Here and Now.' Instant Architecture

'Here and now' is more than just a mantra that invites us to live in the present with more awareness. In a world often constrained by the past or projected into the future, 'here and now' represents an alternative way of thinking about architecture. It allows us to imagine a different idea of the city, other forms, and ways of living and building through which to seek alternative solutions for our fragile habitat. There are many signals that this renewal process is already underway, from themes of self-construction and participation to the preservation of nature, and the return from the metropolis to the countryside. In this evolutionary process, what role can 'printed' architecture play?

Instead of considering digital architecture as a mechanical result of an operation of 'copy and paste,' 'printed' architecture is rooted directly to the site of intervention. The sensitivity of designers is crucial as it brings subjective data into a digital project.

Hence, at a time when technique still prevails over content, and when the tool often ends up overshadowing reasons and forms, 'printed' architecture helps us rediscover the key role of the designer, and echoes how 'the whole is greater than the sum of its parts.'

Data usage may be a collaborative action strategy for identifying a shape that is 'suitable,' informed but non-formal, to optimize the structural, thermal, or functional improvements of a component. Hence, it is clear that we cannot define in advance whether printed architecture is 'good' or 'bad' architecture; there are only valid designers or other 'code executors.'

Nowadays, designing also means taking a step back, not looking for appearance and spectacularity, and putting ourselves at the service of the community. In a society exhausted by its devotion to the viral image, where architecture is successful only when it is 'Instagrammable,' we must move the act of building away from the expressive whims of the individual. This period of multiple changes invites us to approach design with caution and, at the same time, with immediacy and decisive action: 'here and now' (Fig. 28.2).

Architecture presently responds to social and environmental questions with a more resilient approach, and it increasingly captures a possible answer in the theme of temporariness. Moreover, this low-density design strategy is 'printed' architecture's contemporary answer to the question of urbanization, far from the usual high-density approach. Any 'disposable' logic is rejected, while privilege is given to buildings designed to be 'circular,' recyclable, or relocatable elsewhere, or even degradable within the host ecosystem.

Fig. 28.2 'Here and Now,' instant architecture. TECLA—Technology and Clay is a project by MCA—Mario Cucinella Architects and WASP (a ©WASP, b, c ©Iago Corazza). 3DHousing05 is a project by Locatelli Partners (d–f ©Luca Rotondo, courtesy Locatelli Partners)

28.3.2 Ecosystems in Dialogue. Between the Natural and the Digital

The vision of future technological development often leads us to think of an apocalyptic scenario of machine domination, in increasingly urbanized and alienating contexts, distant from the 'natural world.' In some experiments, however, there is a strong influence among traditional building techniques and Additive Manufacturing. Together with other digital tools, such as 3D Scanning technology, 3D Printing allows for technological-constructive hybridization thanks to design solutions 'modeled' directly on the existing. This intersection of worlds (past and present), materials (natural and artificial), and tools (analogue and digital) is evident in the Digital Twin design methodology, also based on the potential to digitalize low-tech materials and components (Boje et al. 2020). Architecture behaves like an 'organism.' This is evident in the use of organic building materials and the emulation of the natural lifecycle (conception, birth, growth, and development, to the transformation of the already built).

The advancement of knowledge, which has led science to discover new materials during the past century, has introduced an elevated specialization of building components. The result is that structural parts are disconnected from non-structural ones; walls are divided into a multiplicity of layers made of different materials. This 'specialization' of building elements may have answered the needs of a past era by supporting the search for spatial flexibility (indeterminacy). However, this strategy requires complex production and recycling processes. On the contrary, many 3D Printing approaches exploit the geometric properties of building elements to ensure adequate structural or thermal performance in association with natural materials (Fig. 28.3). In this sense, the potential of virtual modeling tools is fully exploited by identifying and optimizing building shape: geometry once again plays a pivotal role in architectural design, far from any decorative purpose.

28.3.3 Digital Self-determination

An online community, only apparently distant, can conceive of ideas, generate projects, and constructive actions, even at a neighborhood level. This apparent incongruity led to an investigation of current advancements to design and production strategies. In our global world, characterized by strong political and economic interdependencies, where the well-being of entire populations is in the throes of an international crisis, it is strategic to think of integrated design-construction processes both in dialogue and as potentially independent.

The potential to use local materials and resources is crucial also from this perspective, and this logic is also explored in other sectors and at other scales of application. Indeed, the Makers community has been sharing digital knowledge for years and placing it at the service of the community using Digital Manufacturing

Fig. 28.3 Ecosystems in dialogue. Between the natural and the digital. Sombra Verde Pavilion by AIRLAB and SUTD (a–c ©Carlos Bañón AIRLAB @SUTD). Digital Adobe by IAAC Open Thesis Fabrication 2017–2018 (d, e ©Giulio Paparella)

machines (Fig. 28.4). Scaling this logic from the neighborhood to the city, the Fab City model is based on sending and receiving data instead of exchanging goods: processes whose sustainability comes from reducing transported volumes (Fab City 2021).

In architecture, the preference for local materials opens new frontiers of expression for contemporary forms of digital regionalism: the global design solution is locally diversified according to specific needs and available resources. This logic is applicable also at the scale of furniture; for example, plastic materials can be recycled and reused to produce street furniture, favoring the appropriation and livability of public spaces (The New Raw 2021).

Finally, from the point of view of participatory processes, printed architecture is more inclusive and capable of supporting contemporary forms of shared living during the design and construction phases: several cost-effective variations of the proposed architectural solution allow for the implementation of the 'city for all.' During the construction phase, Mixed Reality technology can also be used to remotely guide non-highly qualified workers. These settlements can accommodate low-income families without neglecting spatial quality and the personalization of individual housing units.

Fig. 28.4 Digital self-determination. Maker Economy Starter Kit by WASP (©WASP)

28.4 Conclusions and an Open-Ended Question

Among digital tools for the green transition in architecture, 3D Printing occupies a crucial position because it permits the production of energy-efficient components, and the optimization of material quantities related to structural performance. Furthermore, since Additive Manufacturing can communicate with other 'IT ecosystem' tools, it is also decisive to the digital transition. In terms of social repercussions, this opens up potential scenarios in which Additive Manufacturing responds sustainably to housing issues arising out of the now widespread phenomenon of urbanization that increasingly characterizes the areas around large cities; these new sustainable building processes allow us to think of alternative horizontal models of urban expansion that can replace outdated high densification solutions.

In conclusion, it is essential to reiterate the fundamental role of the designer in coordinating and caring for the coherence of the final outcome as part of the green and digital transition of architecture, which includes 3D Printing. Design must be integrated with its quantitative–qualitative aspects.

For architecture, the innovative aspect is the high level of inclusiveness that increasingly characterizes the building process. Many ongoing experiments promote activities that involve people in the phases of design and construction, thus helping to 'build' a community and the spaces in which its members will live together; in short, the utopia of CoDesign-CoBuild-CoLiving is finally being realized.

As our past teaches us, the image of architecture is constantly changing, and the advent of 3D Printing technology is the latest demonstration. Albeit in an embryonic phase, this initial architectural shift already shows some recurring aspects: continuous and optimized structure, often hybridized with other construction techniques and materials; fluid and wrapping plans; elevations whose surfaces are 'marked' by 'informed' lines of layered printing; openings conceived as excavations made by inserting frames during the printing process; roofs that are either seamlessly joined to the walls or formally autonomous.

Only by monitoring the evolution of these experiments will we manage to understand if these still unripe traits will be consolidated into a true language of 'printed' architecture.

References

Andreoli M (2019) TECLA, un habitat eco-sostenibile stampato in 3D. WASP. https://www.3dwasp.com/casa-stampata-in-3d-tecla/. Accessed 10 Feb 2022

Boje C, Guerriero A, Kubicki S, Rezgui Y (2020) Towards a semantic construction digital twin: directions for future research. Autom Constr 114:103179

Bosscher P, Williams RL II, Bryson LS, Castro-Lacouture D (2007) Cable-suspended robotic contour crafting system. Autom Constr 17(1):45–55

Directorate-General for Internal Market, Industry, Entrepreneurship and SMEs (2021) Buildings and construction. https://ec.europa.eu/growth/industry/sustainability/buildings-andconstruction_en. Accessed 10 Feb 2022

D-Shape (2010) UnaCasaTuttaDiUnPezzo. https://d-shape.com/Prodotti/unacasatuttadiunpezzo/. Accessed 3 Feb 2022

European Commission (2020) Energy efficiency in buildings. https://ec.europa.eu/info/sites/default/files/energy_climate_change_environment/events/documents/in_focus_energy_efficiency_in_buildings_en.pdf. Accessed 3 Feb 2022

Fab City Organization (2021) Locally productive, globally connected. https://fab.city/. Accessed 3 Feb 2022

Gardiner JB, Burry MC (2009) D_Shape-construction scale additive manufacturing; reflection on current projects and the state of technology development. In: 6th international conference on innovation in AEC 2009

Holcim (2020) 14trees Pioneers 3d printing technology in Africa for affordable housing and schools. https://www.holcim.com/14trees-pioneers-3d-printing-technology-africa-affodable-housing-and-schools. Accessed 3 Feb 2022

Kianian B (2017) Wohlers report 2017: 3d printing and additive manufacturing state of the industry, annual worldwide progress report. Chapters titles: The Middle East, and other countries

Lim S, Buswell RA, Le TT, Austin SA, Gibb AG, Thorpe T (2012) Developments in construction-scale additive manufacturing processes. Autom Constr 21:262–268

Locatelli Partners (2018) 3D housing 05. https://www.3dhousing05.com. Accessed 12 Feb 2022

New Story (2019) Housing innovation for those who need it most. https://newstorycharity.org/innovation. Accessed 10 Feb 2022

Redwood B, Schöffer F, Garret B (2017) The 3D printing handbook: technologies, design and applications. 3D Hubs, Amsterdam

The New Raw (2021) Print Your City—Amsterdam. https://thenewraw.org/Print-Your-City-Amsterdam/. Accessed 3 Feb 2022

United Nations (2020) Population. https://www.un.org/en/global-issues/population. Accessed 20 Feb 2022

Chapter 29
Photovoltaic Breakthrough in Architecture: Integration and Innovation Best Practice

Guido Callegari, Eleonora Merolla, and Paolo Simeone

Abstract In the new context of the trialling and the development of the materials, buildings systems and innovative processes required to meet new challenges posed by environmental transition in Europe and across the globe, the construction sector urgently needs to define more sustainable development models to achieve decarbonisation, as is the case in other sectors. In this context, recent experiences of incorporating photovoltaics into architecture are a clear sign of a change in focus on how systems are integrated into architectural design: a new way of viewing the technological innovation of PV modules which is ever more closely linked to the architectural design right from the initial concept stages. The study we present is based on a critical analysis of the current international state of the art of architectural design incorporating photovoltaics, selecting case studies which illustrate best practice for technological innovation to demonstrate possible scenarios for future developments. Therefore, all the principle approaches identified by the international research will be described as well as the impact that these technological developments are having on architectural style and quality of life in cities. With regard to the aesthetic and formal properties that are the dominant feature of recent practice for the integration of photovoltaics, the study will highlight further areas of research with a view to defining a component of the building shell in which the generation of energy from renewable sources represents just one of the potential components of a system integrated into the architectural style. In addition, the intention is to demonstrate that the architectural designs analysed can be considered to be the result of a close relationship between designers, applied research and the industrial sector; therefore, technological innovation of photovoltaic products will inevitably be linked to a deeper and fundamental innovation of processes leading to these results.

G. Callegari · E. Merolla · P. Simeone (✉)
Polytechnic University of Turin, Turin, Italy
e-mail: paolo.simeone@polito.it

G. Callegari
e-mail: guido.callegari@polito.it

E. Merolla
e-mail: eleonora.merolla@polito.it

E. Arbizzani et al. (eds.), *Technological Imagination in the Green and Digital Transition*, The Urban Book Series, https://doi.org/10.1007/978-3-031-29515-7_29

Keywords Building-integrated photovoltaics BIPV · Energy-active façade · Collaborative design process · Technology transfer · Sustainable architecture

29.1 Introduction

The aim of this paper is to provide some observations that have emerged from research into the impact of technological innovation processes in defining new formal paradigms in architecture (Conato and Frighi 2018). The research in question is specifically focussed on interpreting the impacts of form on the perception of architecture, and in particular the building envelope, deriving from energy and environmental "transitions" paradigms introduced in Europe beginning with the Green Deal. Will the objectives of the European policies lead to new generations of architecture in which formal characteristics are an expression of new potential ways of interpreting the concept of environmental sustainability? We can already see that change is happening in terms of the perception of architecture and in particular in terms of building systems when analysing certain specific areas where there have been recent regulatory and market developments, such as Building-integrated photovoltaics (BIPV). This paper concisely sets out the findings of in-progress research focussed on how photovoltaics are incorporated into architecture by analysing the evolutionary process from the 1970s to the present day and outlines an initial framework of approaches to design through an atlas of "energy transition architecture" in Europe in order to produce a taxonomy of BIPV design.

The strategic role of the energy sector in European decarbonisation is fundamental for achieving climate neutrality by 2050 (IRENA 2021). The shift to energy communities has also been supported by new regulatory standards which in some cases also have implications for architecture. One example is the standard Photovoltaics in Buildings EN 50583:2016 which was the first to include the integrated photovoltaic module in a multifunctional construction component, in accordance with the Construction Products Regulation (EU) CPR 305/2011. This new interpretation has assisted with the move from PV to smart BIPV systems as innovative technological components contributing to tackling current decarbonisation challenges (IEA 2020, 2021). These processes have stimulated the market through R&D to produce new generation smart materials which can generate electricity, extending the surface area of the building envelope used for this purpose. This approach has led to a change of interpretation in how building systems are incorporated into the architectural design, as shown in the case studies analysed in this paper.

29.2 Investigation Outline

29.2.1 Research Field

The work of the research group at the Politecnico di Torino Department of Architecture and Design is centred on developing and designing smart BIPV system building envelope components characterised by prefabrication, recyclability and modularity.

The background to the work features two areas of research which are linked but separate: an industrial development project for a private client (Department of Architecture and Design 2020) and contributing as a partner on the Green Deal project H2020-LC-GD-2020 to develop a BIPV component for a demonstration project (ARV-Climate Positive Circular Communities 2020). The aim of the paper is to illustrate the results obtained from the state-of-the-art analysis in order to show new scenarios regarding the current AEC and BIPV sectors in the European context.

29.2.2 Criteria and Indicators

The research led to the definition of an analytical framework regarding the architectural, constructive and technological integration of BIPV components in the architectural design, based on the following analysis criteria:

- aesthetic and formal characteristics: the main technologies for customising the appearance of PV have been analysed;
- PV morphological integration: a classification of BIPV components that can be integrated on the vertical envelope has been identified; the main integration strategies will be explained;
- smart grids and smart buildings: selected case studies will be shown as virtuous examples of plus energy buildings.

29.2.3 Analysed Sources

Reports and scientific articles published by European research institutions have been analysed in detail, as well as major online databases for sharing BIPV best practices were consulted; interviews have also been conducted, as a means for comparison and critical analysis, with BIPV innovation technology researchers from a number

	scientific papers	selected content
1	D'Ambrosio V., Losasso M., Tersigni E (2021) Towards the Energy Transition of the Building Stock with BIPV: Innovations, Gaps and Potential Steps for aWidespread Use of Multifunctional PV Components in the Building Envelope	Product customisation
2	Attoye D. E., Tabet Aoul K. A., Hassan A (2017) A Review on Building Integrated Photovoltaic Façade. Customization Potentials	BIPV façade applications
3	Pelle M., Lucchi E., Maturi L., Astigarraga A., Causone F., (2020) Coloured BIPV Technologies: Methodological and Experimental Asses¬sment for Architecturally Sensitive Areas	Product and manufacturer overview
4	Sánchez E., Izard L., (2015) Performance of photovoltaics in non-optimal orientations: An experimental study	PV façade benefits
5	Munari M. C., Roecker C., (2019) Criteria and policies to master the visual impact of solar systems in urban environments: The LESO-QSV method	Visual impact of PV system
6	Norwood, Z., Theoboldt, I., Archer, D.E. (2016) Step-by-step deep retrofit and building integrated façade/roof on a 'million program' house	case study of BIPV façade for retrofit

	reports	selected content
1	Zanetti I., Bonomo P., Frontini F., Saretta E., van den Donker M.N., Verberne G., Sinapis K, Folkerts W., Vossen F., (2017) Building Integrated Photovoltaics: Product overview for solar buildings skins. Status Report 2017	Customized products overview
2	Corti P., Bonomo P., Frontini F., (2020) Building Integrated Photovoltaics: a practical handbook for solar buildings. Status Report 2020	BIPV case studies
3	IEA SHC (2013) Designing Photovoltaics System for Architectural Integration. Criteria and guidelines for product and system developers, Task 41.A.3/2	Product and project overview
4	Kraubitz, T., Scheibstock, P., Guillen, G. (2018) Plus Energy Buildings and Districts, keystone papers 1 for GIZ's Sino-German Urbanisation Programme	case study of plus energy building with BIPV
5	IEA PVPS (2019) Coloured BIPV: Market, Research and Development, Task 15, Subtask E, 2019	PV Aesthetic features and technology

	interview	topic
1	Pietro Florio / *JRC, Scientific Research Project Officer*	the role of applied research in BIPV technological innovation
2	Alessandro Virtuani / *Senior researcher at EPFL & co-founder Officina del Sole*	the importance of BIPV for European decarbonisation challenges
3	Pierluigi Bonomo / *Researcher- Head of BIPV Advanced Building Skin Team at SUPSI*	new formal configurations interpreted by architecture
4	Enrico Ferramondo Marchesi /*innovation manager of living lab NEST, Zurigo*	collaboration between research, industry and architects to speed up the transfer of BIPV within the market and the architecture

database

1 solarchitecture.ch **2** bipv.ch **3** solaragentur.ch **4** bipv.eurac.edu/en

Fig. 29.1 Main analysed sources for this research

of European organisations, such as the NEST[1] research lab in Zurich, SUPSI[2] and EPFL.[3]

A comparison of the leading European producers of BIPV modules, such as SwissINSO, AGC Glass and Ertex Solar, has been made to complete the analytical framework. A summary of the sources consulted is presented in Fig. 29.1.

29.2.4 Selected Case Studies

As a result of the analysis, 78 BIPV integrated façades in Europe have been identified and analysed, of which 67% relate to the residential construction sector. Both first generation (c-Si), second generation (a-Si, CIGS, CIS, CdTe) and third generation (OPV, DSSC) PVs have been considered. It is relevant to show how—as a result of the technological innovation of the PV appearance started in 2010—39.4% of the surveyed façades adopt completely camouflage solutions or coloured PV cells (Fig. 29.2).

[1] Modular research and innovation building Next Evolution in Sustainable Building Technologies (NEST) of Empa and Eawag, Zurich, Switzerland.

[2] University of Applied Sciences and Arts of Southern, Lugano, Switzerland.

[3] Swiss Federal Institute of Technology, Lausanne, Switzerland.

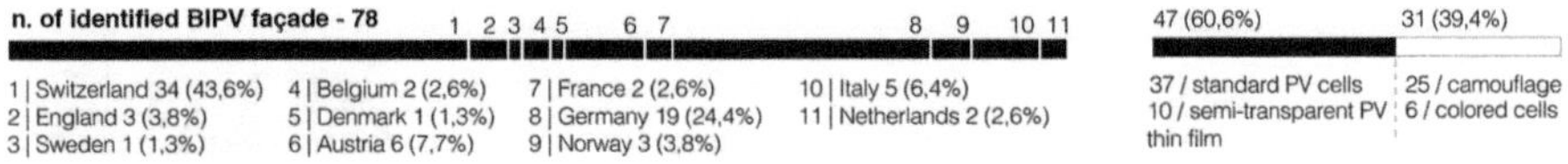

Fig. 29.2 Analysis of the current state of the art of BIPV integrated façades in Europe. Original graphics by authors

29.3 Output (or Results)

29.3.1 Aesthetic Evolution of BIPV

Recent experiences of PV integration into the building envelope represent the current culmination of a technology which has evolved over time, since the early trials in the 1980s and 90s. The research sets out the main stages of the evolution of BIPV, with a particular focus on façade integration. From the first instances of inserting PV cells into glass-glass modules to later colouring techniques, the evolution of PV has been driven by continuous scientific research and experimentation by architects, leading to examples of PV integration which are completely organic with the architectural design (Fig. 29.3).

The customisation of colourings, shapes and configurations of PVs has increased interest in technical innovation of photovoltaics in terms of integration into the architectural design. Whilst these designs must achieve ever more stringent decarbonisation targets, all stakeholders now have an enormous range of aesthetic and formal solutions available (Fig. 29.4).

29.3.2 PV Integration Forms and Strategies: Best Practices

Integrated photovoltaic systems offer new construction solutions which the architectural design can employ in order to interpret the increased energy efficiency requirements with an expressive architectural language that features a high degree of technological awareness. The conventional building elements such as cladding panels, sunscreen, parapets and accessories can today be enhanced with multifunctional components that are highly customised, generating electricity that can be fed to the energy community's network, fulfilling the building's energy requirements. The research has thus identified a series of PV façade integration categories: cladding system for cold façade, solar shading systems, balconies and solar glazing (Fig. 29.5).

Research has found that the most widespread integration approach involves the PV component being inserted into the architectural design in a bounded way, interacting with the other elements of the envelope and shaped by the system of solar shades and balustrades. In this way, the integrated photovoltaic system can cover part of the building's energy requirements (Fig. 29.6a). However, on the other hand, a fully

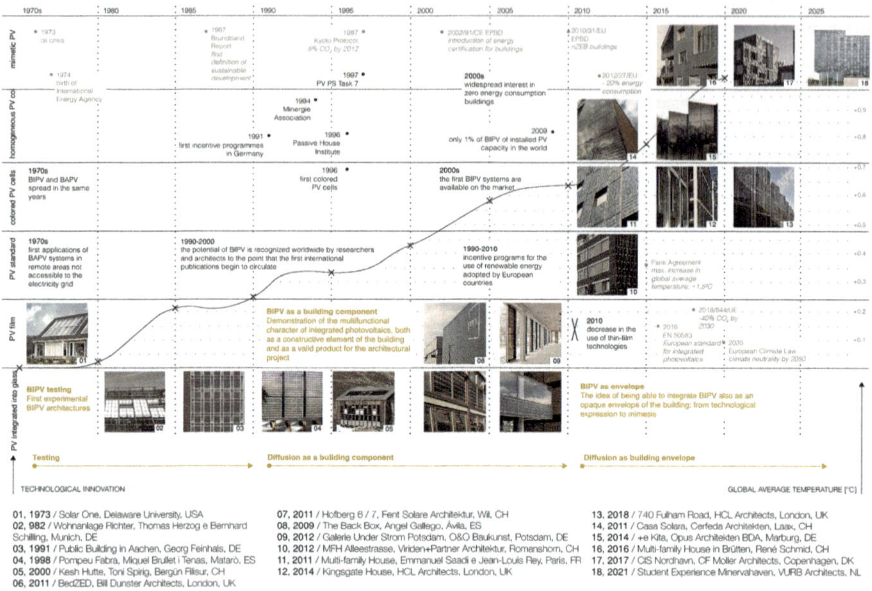

Fig. 29.3 Evolution in technological innovation of building-integrated photovoltaics. The graphic depicts the evolution of integrated PV in architecture, identifying the main stages and the respective dominant PV technologies. The relationship between the increasing global average temperature curve (NASA 2021) and the current BIPV experiences is shown. These reference buildings refer to the 78 surveyed façades cited in Fig. 29.1. Original graphics by authors

active envelope configuration can be highlighted where the photovoltaic component constitutes the main cladding material and, through camouflage components, it creates a more conventional architectural language or, through components which are identifiably as technological, a more innovative image with emphasised PV system.

This strategy ensures that the building generates power more uniformly across the whole day, thanks to the variety of PV exposure to the transit of the sun, sometimes even achieving a positive energy balance so that the surplus can be fed back to the grid (Fig. 29.6b).

The research has selected best practices which represent the different strategies for PV façade integration, extended and bounded, respectively, to demonstrate the quality of the architectural design. A few case studies selected by the authors are examined in more detail below (Figs. 29.7, 29.8, 29.9 and 29.10).

29.3.3 Buildings as Small Power Plants

The active envelope concepts of BIPV architecture can change the distribution model for the local power network, viewing buildings as energy community power stations.

a / MAIN PRODUCTS CONSULTED FOR THIS RESEARCH

product	manufacturer	PV technology	color*	dimension [mmxmm]	shape	cells visibility	efficiency ** [%]
Kromatix™	SwissINSO	c-Si	C1	custom	rectangular	NO	16,5-17,5
Solaxess	Solaxess / CSEM	c-Si / CIGS	B2	custom	custom	NO	11,4
VGS Design PV	Ertex Solar	c-Si	A1-C2	custom; max 2400x5000	custom	NO	n/a
ColorBlast	Kameleon Solar	c-Si	C2	custom	custom	NO	8-15
Kaleo	CSEM	c-Si	B2	custom	custom	NO	n/a
DSD-PV	Dutch Solar Design	c-Si	C4	900x1200	rectangular	NO	n/a
SunCol	Glassfer&Sunage	c-Si	C5	custom	custom	NO	n/a
Lacobel T Active	AGS Glass	c-Si	C2	n/a	n/a	NO	n/a
eForm unichrome	Sunovation	c-Si	C4	max 3500x1500	rectangular	NO	15
colored PV cells	LOF Solar	c-Si	A1	custom	n/a	YES	15
a-Si module	Onyx Solar	a-Si	A2	custom; max 4000x2000	n/a	/	n/a
PS-CT-series	PolySolar	CdTe	A2	standard 1200x600	rectangular	/	6-12
PS-ASG-series	PolySolar	a-Si	A2	standard 1400x1100	rectangular	/	5-10

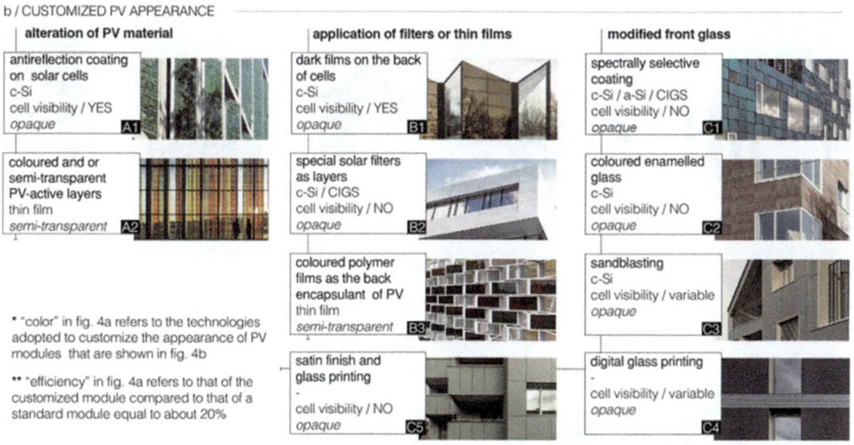

Fig. 29.4 PV products overview. The table shows the main available products and related aesthetic and technological features (**a**). The graphic illustrates the current colouring technologies for PV components and the range of different formal interpretations (**b**). Original graphics by authors

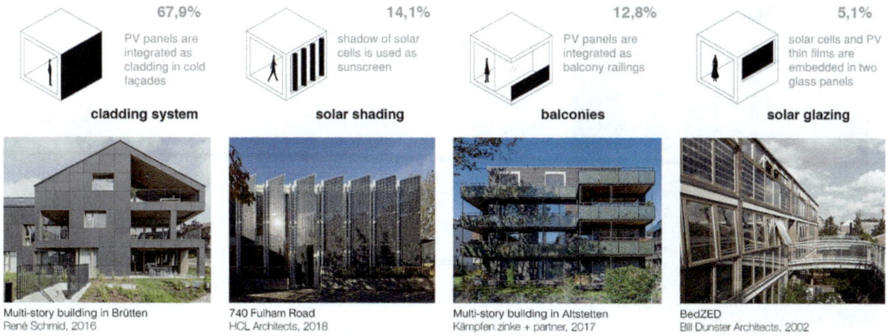

Fig. 29.5 PV façade integration forms. The percentages refer to the 78 studied façades. Original graphics by authors

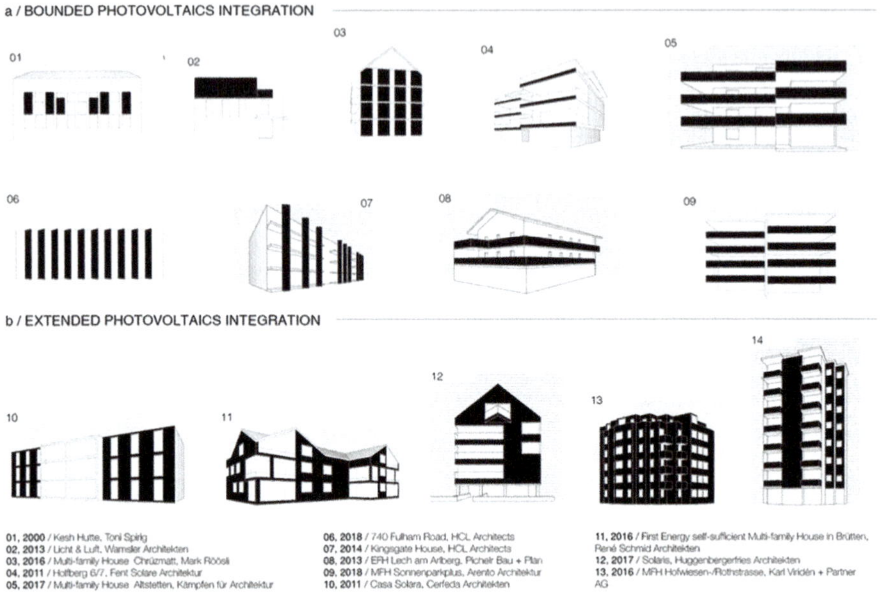

Fig. 29.6 PV façade integration categories. The areas shaded darker denote PV integration strategies in the architectural design, such as claddings, balconies and shades. Original graphics by authors

KINGSGATE HOUSE (UK)
HCL Architects

Best Affordable Housing Scheme at 24 Housing Awards 2014

BIPV INTEGRATION

INTEGRATION FORM	SOLAR SHADING
INTEGRATION STRATEGY	BOUNDED
PV SURFACE	188 m²
TILT ANGLE	90°
ORIENTATION	SOUTH
TECHNOLOGICAL SYSTEM	*SHADOWVOLTAIC SYSTEM BY COLT*

PV TYPOLOGY

MANUFACTURER	ROMAG
PV TECHNOLOGY	POLYCRYSTALLINE
PV MODULE	GLASS/GLASS MODULE
PV APPEARANCE	EMPHASIZED PV
TRANSPARENCY	SEMI-TRANSPARENT
COLOURING TECHNOLOGIES	ANTIREFLECTION COATING ON SOLAR CELLS
NOMINAL POWER	-
MODULE EFFICIENCY	17,60-15,20%

ENERGY DATA

ENERGY PRODUCTION	13145 kWh/yr
NOMINAL POWER	-
SELF-CONSUMPTION	20%

Fig. 29.7 Kingsgate House—case study. Original graphics by authors

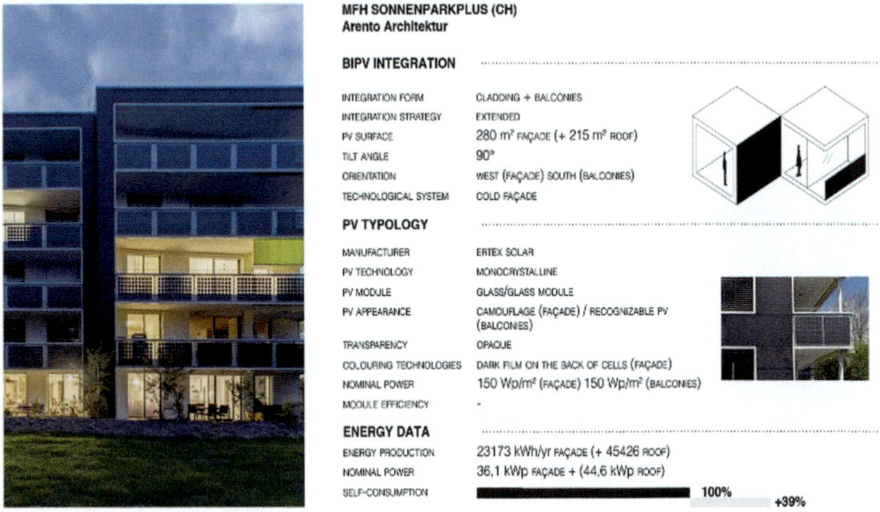

Fig. 29.8 MFH Sonnenpark Plus—case study. Original graphics by authors

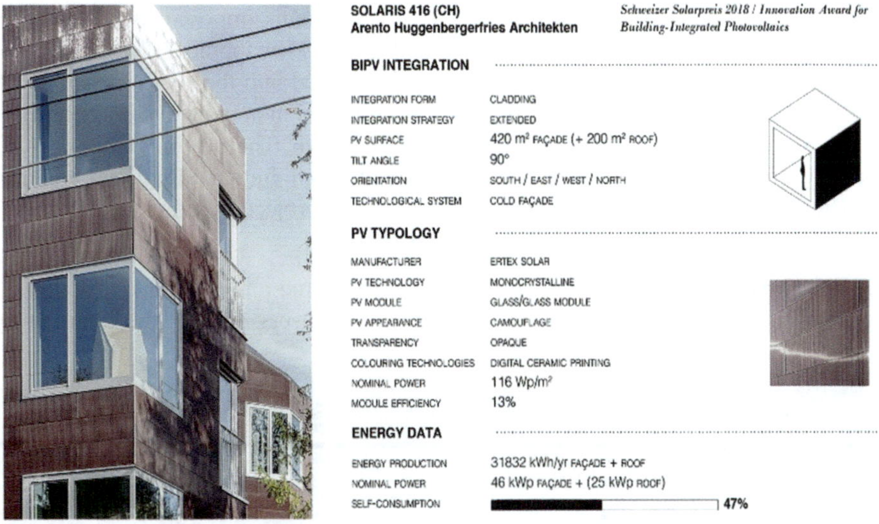

Fig. 29.9 Solaris 416—case study. Original graphics by authors

With this smart grid principle, the power surplus generated from buildings can be used to recharge electric vehicles or fed back to the network from which these buildings receive electricity when they cannot generate it independently.

Aktiv-Stadthaus (Fig. 29.11). The *Aktiv-Stadthaus*, a building designed by HHS Planer + Architekten in Frankfurt (2015), is based on the Effizienzhaus Plus energy

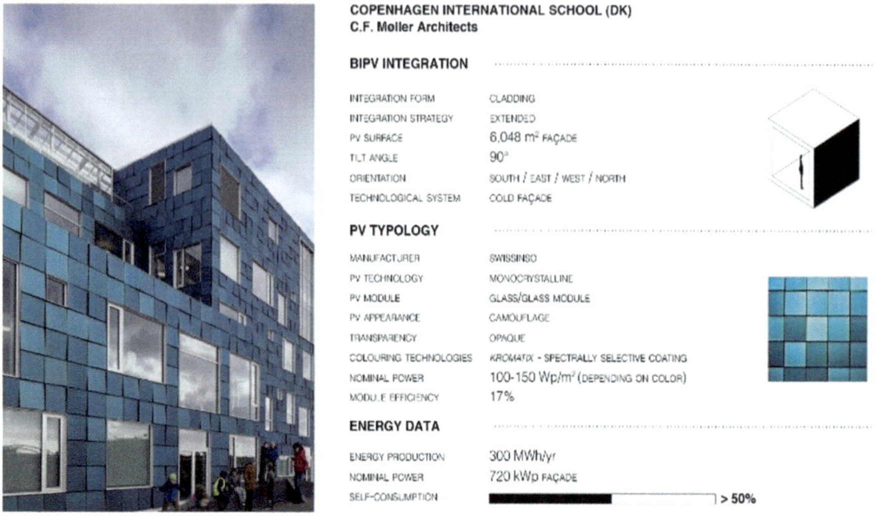

Fig. 29.10 International School—case study. Original graphics by authors

efficiency standard and complies with the requirements imposed by the German Federal Ministry of Transport, Building and Urban Development (BMVBS). The electricity generated by the PV components integrated into the south façade, along with the photovoltaic system on the roof, is collected in storage systems and used to charge electric vehicles or fed back to the network. The project falls under the subsidy scheme launched by the Federal Office for Building and Regional Planning (BBR), which finances research and development projects relating to the energy consumption of buildings (Kraubitz et al. 2018).

MFH Hofwiesen/Rothstrasse (Fig. 29.12). The same active envelope approach, generating electricity to meet the requirements of users and for the community,

Fig. 29.11 Newly built apartment building Aktiv-Stadthaus

has been adopted in a Swiss pilot project to demonstrate a possible renovation and energy efficiency improvement method for existing building stock; the *MFH Hofwiesen-/Rothstrasse* residential complex in Zurich (2016), designed by Viridén + Partner AG (SD, Pd 2017), financed by the Canton of Zurich as part of the Federal Energy Office's (UFE) programme, is notable for the use of camouflaged photovoltaic modules integrated on all sides.

Stacken (Fig. 29.13). The renovation work carried out on the *Stacken* residential complex (2017) in Gothenburg (Norwood et al. 2016) in Sweden, also supported by public and private financing, demonstrates the potential of integrated photovoltaic systems for renovating the existing building stock through the application of a BIPV façade with external insulation retrofit.

Fig. 29.12 Renovation of an apartment building

Fig. 29.13 Renovation of Stacken apartment building

29.4 Conclusions

The recent shift to energy communities for the generation of power represents a conversion for end users too, who become "generators" rather than "consumers" of energy via a decarbonisation, digitalisation and decentralisation model; for the authors, this model is also evidence of a social evolution which architecture is beginning to interpret and translate to new formal configurations; the potential relations between the building envelope and the building systems mean that architecture can be equated to energy infrastructures of high architectural quality.

In addition, this research paper seeks to demonstrate how the move over the last 20 years from the European target for a high energy efficiency building stock to the current aim for NZEB low carbon footprint buildings represents not so much a challenge as a change of paradigm in the AEC sector, with many consequences on potential innovation models that the market, the research sector and the profession are still seeking to interpret. It is based on this approach that researchers, manufacturers and designers are collaborating to pursue common objectives, representing the stakeholders in a technological innovation process which is having repercussions on the construction sector.

The construction sector—slower in innovation compared to other industrial sectors (Bellicini 2019)—now therefore has a greater opportunity of technological innovation that the market is gradually accepting. Within this transformation scenario, the contribution of the research has tried to highlight further directions of technological innovation aimed at contributing to an increasing industrialisation and prefabrication of the construction sector and in the specific case of the BIPV market. Among these, the authors support the thesis that integrated PV systems can move away from the dry construction systems to which they now belong in order to be conceived as three-dimensional prefabricated components belonging to Industry 4.0, integrated with additional systems and directly installed on the building envelope in an off-site production logic. Based on this reflection on the relationships between the building envelope, plant components and architectural languages—and within the broader framework of digitisation and technological innovation outlined above—the applied research activity is conducted by the Politecnico di Torino Department of Architecture and Design for the development and industrialisation of smart BIPV systems solution.

References

ARV-Climate Positive Circular Communities H2020-LC-GD-2020 Green Deal Call Building and renovating in an energy and resource efficient way (LC-GD-4-1-2020)

Bellicini L (2019) Le resistenze delle costruzioni all'aumento della produttività e all'innovazione. CRESME 2019. http://www.cresme.it/it/articoli/23/le-resistenze-delle-costruzioni-allaumento-della-produttivita-e-allinnovazione.aspx

Conato F, Frighi V (2018) Il ruolo dell'innovazione nella definizione di nuovi paradigmi formali in Architettura. Techne 16:105–112. https://doi.org/10.13128/Techne-22965
IEA (2020) Clean energy innovation. International Energy Agency. https://www.iea.org/reports/clean-energy-innovation/innovation-needs-in-the-sustainable-development-scenario. Accessed 21 Feb 2022
IEA (2021) ETP clean energy technology guide. International Energy Agency. https://www.iea.org/articles/etp-clean-energy-technology-guide
IRENA (2021) World energy transitions outlook: 1.5°C pathway. International Renewable Energy Agency. https://www.irena.org/publications/2021/Jun/World-Energy-Transitions-Outlook. Accessed 20 Feb 2022
Kraubitz T, Scheibstock P, Guillen G (2018) Plus energy buildings and districts. Keystone papers 1 for GIZ's Sino-German urbanisation programme. https://www.burohappold.com/projects/giz-china-keystone-papers/. Accessed 27 Feb 2022
NASA (2021) https://climate.nasa.gov/. Accessed 24 Feb 2022
Norwood Z, Theoboldt I, Archer DE (2016) Step-by-step deep retrofit and building integrated façade/roof on a 'million program' house. Passipedia. https://passipedia.org/planning/refurbishment_with_passive_house_components/step-by-step_deep_retrofit_and_building_integrated_facade_roof_on_a_million_program_house. Accessed 27 Feb 2022
Research project for the development and executive design of an innovative building component prototype BiPV (Building integrated PhotoVoltaic) (2020–on going). Department of Architecture and Design, Politecnico di Torino
SD, PD (2017) Urbaner Leuchtturm: Das sanierte und aufgestockte Mehrfamilienhaus gilt als schweizweiter Vorreiter. First. Bauen und leben mit holz, pp 32–37

Chapter 30
Reworking Studio Design Education Driven by 3D Printing Technologies

Jelena Milošević, Aleksandra Nenadović, Maša Žujović, Marko Gavrilović, and Milijana Živković

Abstract The advances and proliferation of digital technologies impact architectural practice asking for a revision of not only design production but also the education of future professionals. Using a case study from the University of Belgrade—Faculty of Architecture, this paper examines the efficient application of 3D printing as a design tool and opportunities for the implementation of this technology in architectural education. The research goal was to establish an educational framework for the studio course that was appropriate to local settings, starting with a review of educational approaches and usage of 3D printing in architectural design. Starting with the premise that there is a bidirectional relationship between design and its tool, educational framework for architectural design studio was proposed, tested in real educational settings, and evaluated. The results indicate that the use of 3D printing in studio course proved to be an effective tool for design exploration and presentation that supports (1) linking the logical way of thinking that requires parametric modeling with concept-based thinking; (2) change in mindset that occurs in the design process when students have a physical model in front of them to assess; and (3) improvement of deep understanding of spatial cognition among students as well as their competencies related to the use of the specific technology in the design process. The paper demonstrates how 3D printing technology improved educational methods, impacted students' experiences in the design process, and elevated design exploration to previously unattainable levels of materiality, detail, complexity, accuracy, and aesthetics.

J. Milošević (✉) · A. Nenadović · M. Žujović · M. Gavrilović · M. Živković
University of Belgrade, Belgrade, Serbia
e-mail: jelena@arh.bg.ac.rs

A. Nenadović
e-mail: aleksandra@arh.bg.ac.rs

M. Gavrilović
e-mail: gavrilovic@arh.bg.ac.rs

M. Živković
e-mail: milijana.zivkovic@arh.bg.ac.rs

E. Arbizzani et al. (eds.), *Technological Imagination in the Green and Digital Transition*, The Urban Book Series, https://doi.org/10.1007/978-3-031-29515-7_30

Keywords Design process · Design tools · Design studio · Architectural education · 3D printing

30.1 Introduction

As a core methodology in architectural education (Salama 2017), studio design must constantly evolve to facilitate students to build competencies relevant to future practice. Studio allows students to learn to design and be designers (Dutton 1987) by studying curriculum topics and theoretical concepts in a practical context (Schon 1987) and simulating professional scenarios in an academic setting (Laurillard 2012). Although the basic structure of architectural design studio appears to be quite resilient to diverse cultural, social, and production changes over time (Schon 1987; Nicol and Pilling 2000), the impact of digital technologies asks for a rethinking of both design process and education in terms of new operation tools. Various publications discuss impact of digital technologies on architecture (Kolarevic 2005; Leach et al. 2005; Gramazio and Kohler 2008; Menges and Ahlquist 2011; Carpo 2012, 2017; Willmann et al. 2019), as well as technology-assisted learning (Anderson 2016) and its implementation in architectural studio pedagogy (Guler 2015; Masdeu and Fuses 2017; Ioannou 2018; Milošević 2021; Jones et al. 2021).

This paper explores the application of 3D printing (3DP) technology in architectural design studio education. The following research question arises from the premise that design and its tools have a bidirectional relationship: How can we employ 3DP tools in studio design to create a learning environment that allows future architects to better prepare for technological and professional challenges? In response to the research question, the objectives of the study are to (1) analyze diverse approaches of the implementation of 3DP technologies presented in the literature; (2) describe a studio design framework that includes the use of 3DP technologies and its implementation; and (3) summarize the challenges and opportunities of the proposed approach.

To address the research questions, an integrated literature review method was used to analyze, critically assess, and synthesize representative literature on the topic and generate new perspectives and framework. Furthermore, the new framework developed based on the literature review was empirically tested in the real educational setting and evaluated qualitatively (Groat and Wang 2013).

30.2 Literature Review

The literature on applying 3DP technology in architectural design education was searched using the following keywords: design studio, 3D printing, rapid prototyping, architectural education, in two main databases, Web of Science and Google Scholar. A total of fifteen relevant references were included in the content analysis. The themes identified in papers were concise into three main categories of research explained in the following sub-paragraphs.

30.2.1 Effects of Implementing of 3DP Technology in Design

The effects of the introduction of 3DP technologies into the architectural design curriculum have been reviewed by several authors (Loy 2014; Kim et al. 2021; Chiu et al. 2015; Lugo Nevarez et al. 2016; Kwon et al. 2017; Greenhalgh 2016; Boumaraf and İnceoğlu 2020; Budig et al. 2014; Paio et al. 2012; Gu et al. 2010; Bøhn 1997; Kristiánová et al. 2018). For example, some studies indicated that rapid prototyping (RP) technology piqued the interest of students who were previously accustomed to the manual creation of physical models and 3D modeling for design through physical models (Loy 2014; Kim et al. 2021). Furthermore, students confirmed in several studies that the use of 3DP helped them develop innovative thinking, enhanced learning motivation (Chiu et al. 2015; Lugo Nevarez et al. 2016; Kwon et al. 2017; Greenhalgh 2016), and considerably improved their design capabilities (Boumaraf and İnceoğlu 2020; Budig et al. 2014).

Many students' designs were more complicated as they adopted 3DP technology for prototyping. RP enabled them to materialize physical models with far more conceptual and geometric complexity than traditional methods (Greenhalgh 2016; Budig et al. 2014). Findings show that the use of RP, in some cases, significantly improved students' spatial cognition since they were able to perceive their design proposals in the physical environment (Paio et al. 2012). Also, making complex models on smaller scales made it easier for students to focus on the overall design concept than the details (Budig et al. 2014).

However, several authors noted that students had not used the full potential of a given technology (Gu et al. 2010). Previous was, in many cases, due to the time constraint and tight schedules that studio design projects often imply. Some studies indicate that students still tend to use 3DP technology for the final presentation of projects instead of for research (Bøhn 1997; Kristiánová et al. 2018).

30.2.2 Implementing 3DP Technology in the Studio Course

Additive manufacturing is thought to be one of the rising technologies in education that will help students learn and foster creative thinking (Chiu et al. 2015). The students' perceptions of 3DP technology in the architectural studio could be linked to their previous experience with model-making in project creation. Integrating 3DP made students accustomed to digital modeling more interested in constructing physical models using 3DP rather than traditional building methods in a workshop (Loy 2014).

Students with less CAM experience had more difficulty learning about the 3DP process and RP technology (Sampaio et al. 2013), and they should be given lectures to improve their skills (Kwon et al. 2017). Depending on their academic level, students are likely to be exposed to different teaching methods. Students with less expertise should be guided through the concepts and objectives initially, but if no methods are offered, they will be challenged to solve problems and be more proactive. More open teaching methodologies and experiments can be employed with more advanced students. They could be primarily introduced to concepts and a brief description of the problem and have greater flexibility through the project development phase (Celani 2012).

Also, Fernandes (Fernandes and Simoes 2016) explained how students in higher education with various learning styles react to using 3DP as a collaborative learning resource in their classroom. The study found that most students prefer to test their theoretical knowledge using 3DP models. It gives them more freedom and technical experience than simply having a theoretical approach to the subject (Fernandes and Simoes 2016).

30.2.3 Methods of Implementing 3DP Technology in the Curriculum

Currently, the design process is highly dependent on using information and digital technologies (Paio et al. 2012). It is generally agreed that the implementation of RP in curricula enforced innovative thinking and improved the sense of materiality and space. Additionally, using 3DP continuously fosters practical aspects of design studio methodology while model-making represents a learning-by-doing mode (Kristiánová et al. 2018).

A seven-step pedagogical model was introduced at the City University of Hong Kong to all freshmen from various fields of study enrolled in the same class. It is based on classic instructional design theory and the Conditions of Learning by Sampaio et al. (2013). The aim was to bring in 3DP technology in the educational process and analyze its practical problems. It is considered that 3DP is one of the emerging technologies in education that would support student learning and encourage innovative thinking (Chiu et al. 2015).

Another example is from the Singapore ETH Centre for Global Environmental Sustainability, where the research project "Design of Robotic Fabricated High Rises" explores the possibilities of robotic high-rise construction. This design studio aims to shift the physical model as a crucial explorative tool combined with computational design, with robotic technology used to fabricate it. Rather than simply developing forms, the design research studio focuses on designing techniques that merge design computation with robotic manufacture (Budig et al. 2014).

30.3 Case Study

The case reported is from the University of Belgrade—Faculty of Architecture (UB–FA). It focuses course Studio Design Project: Spatial Structures, which is taught annually during the fall semester at the Master Studies of Architecture–Module Architectural Engineering (MASA–AE). The course is designed to introduce architectural students to the challenge of designing spatial structures. In this course, students acquire theoretical and methodological knowledge and skills required for project development following ARB Criteria at Part 2 (ARB 2010) through practically oriented design research.

30.3.1 Course Preparation

Findings of the literature review related to techniques, concepts, and learning perspectives of 3DP technology served as a starting point for establishing an educational framework for reworking the studio design curse. As a result, two aspects of the studio design curriculum were adopted: (1) project task and (2) teaching method. It was essential to specify engaging, a problem-based assignment that fosters the exploration of complex designs using digital technology (Greenhalgh 2016; Budig et al. 2014), facilitating the acquisition of competencies relevant to future professionals (Foque 2011). Furthermore, teaching methods standardly applied in design studio education were complemented with workshops and skill-up classes in which students developed and improved skills in using digital tools for design production (Fernandes and Simoes 2016). These were organized in collaboration with the external experts to introduce, to a certain degree, a collaborative manner of work in a studio environment essential for future practice (Gnaur et al. 2015).

30.3.2 Course Implementation

The classes, which took place twice a week, included instruction, open discussions, the presentation of students' works, and workshops to enhance students' skills. Students develop their expertise through an active process of information gathering analysis, exploration, synthesis, testing, discussions, reflections, refinement, presentation, and evaluation in the collaborative learning space of the design studio. The process was broken down into five phases to ensure the achievement of learning outcomes: (1) analysis, (2) model explorations, (3) conceptual urban and architectural design development, (4) conceptual structural design development, and (5) post-production. Each phase had its goals and outcomes and diverse tools for performing activities.

Digital tools (including fused deposition modeling (FDM) 3DP devices, selective laser sintering (SLS) 3DP device, 3DP pan, and 3D scanner) were chosen regarding the (1) design problem, (2) size (Leach 2017), and (3) stage of the design process, and the function of the physical model (Fig. 30.1). Accordingly, for form exploration (phase 2), tools that enable fast production of physical models and evaluation of ideas were favored. In this case, the less precision and quality of the models were acceptable. To produce small-scale prototypes and functional models (phase 3), more sophisticated tools that construct precise models of material suitable for testing are required. Finally, models for design presentation (phase 5) were made using precise devices and materials with desired aesthetic qualities. Also, reverse engineering proved to be a good way to support the iterative nature of the design process.

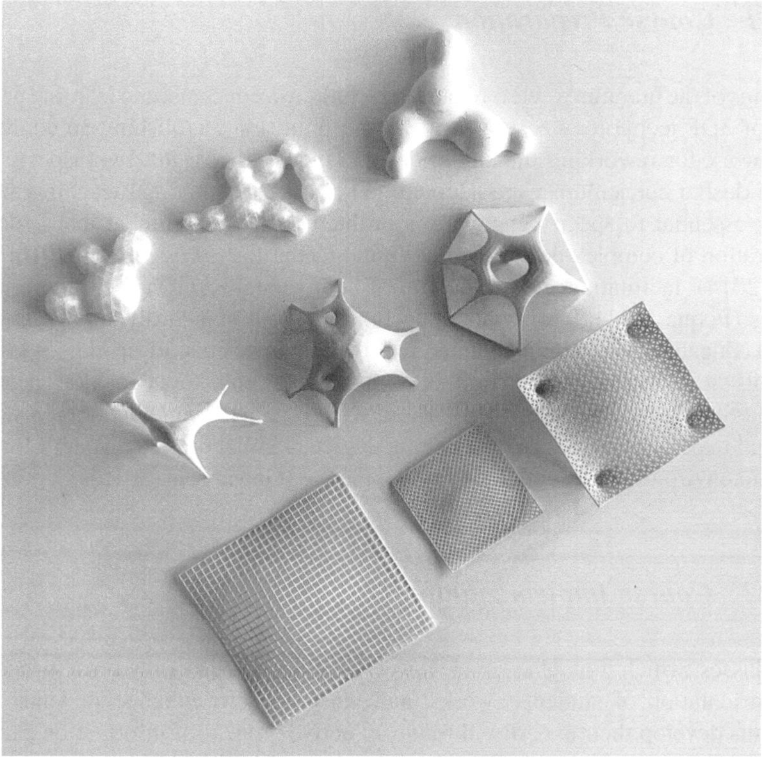

Fig. 30.1 Models produced with different 3DP devices used for exploration, assessment, and presentation of designs

30.3.3 Course Results and Assessment

The outcomes of the educational process are two types of experiences: (1) operational experience and (2) subject experience. Operational experience is related to practicing a design approach that can be reused in the continuation of the studies or professional practice. Accordingly, the framework enabled students to acquire knowledge and skills architects should possess to act competently in future working environments. On the other hand, subject experience concerns developing knowledge and skills by working on a particular topic. In this respect, the framework supported students in creating designs that display simultaneous consideration of diverse aspects—context, form, function, structure, materialization, and fabrication—using the holistic design approach.

The course was evaluated qualitatively using a questionnaire on the pedagogical work regularly filled out at the UB–AF at the end of each term. Students were very satisfied with the instructions and course materials; the consistency between classes and the scope of the course; their active participation; critical thinking and creativity; the volume and quality of recommended literature and learning resources; and their results, according to the results of the survey. Students were particularly motivated by the studio's research orientation and the opportunity to explore innovative concepts and technologies. However, students indicated that the course duration and hours of classroom activities were a bedside of the course. Furthermore, some students said that finishing tasks on time was difficult and time-consuming. Accordingly, better time management should be suggested, as learning new techniques and changing students' learning and design methods requires time. The course results were displayed at the UB–FA final exhibition and as a web exhibition, which students found exciting and as a way to show their work to a larger audience.

30.4 Discussion

The paper provides a structure for an architectural design studio that integrates 3DP technologies and tests a new framework in a real-life educational context. Our teaching process was outlined for other educators and researchers to observe our experience, compare it to theirs, and consider alternative paths. It is crucial to analyze the findings in light of the study's and course's research limitations in this regard:

- The research is restricted to a single teaching experience. For generalization, more work is needed, including a comparison of distinct findings across diverse educational contexts and study programs.
- The course has technical constraints due to a lack of more sophisticated equipment that allows students to enhance their learning through hands-on activities such as building and testing large-scale prototypes or more sophisticated models made of

diverse materials. Therefore, more resources are required to further improve the course in this respect.

The following advantages of implementing a 3DP studio design course could be identified:

- Technologically advanced creative learning environment motivated students to link the logical way of thinking that requires parametric modeling with concept-based thinking.
- When students have a physical model in front of them to analyze, they have a change in a mindset that occurs during the design process, in which they work on relevant challenges.
- Students improved their understanding of spatial cognition and their competencies related to using this technology in the design process for effective exploration, assessment, and communication of ideas.

30.5 Conclusions

The findings show that using 3DP tools in a studio design course can aid design exploration, assessment, and presentation. Shared educational experience demonstrates how 3DP technology can improve learning methods, impact students' design process, and elevate design exploration to previously unattainable levels of materiality, detail, complexity, accuracy, and aesthetics. The paper offers an example of how using technological resources could improve studio structure and facilitate achieving the desired learning outcomes, such as students developing competencies that will help them operate professionally in changing work contexts with the support of digital technologies. Finally, future studies that will include interdisciplinary research on 3D printing technology in studio design education to develop product design at various scales, typological frameworks, and timeframes could be advantageous.

References

Anderson T (2016) Theories for learning with emerging technologies. In: Veletsianos G (ed) Emergence and innovation in digital learning: foundations and applications. Athabasca University Press, Edmonton

ARB (2010) Prescription of qualification—ARB criteria at part 2. https://arb.org.uk/wp-content/uploads/2016/05/ARB_Criteria_pt2.pdf . Accessed on 15 Feb 2022

Bøhn JH (1997) Integrating rapid prototyping into the engineering curriculum—a case study. Rapid Prototyping J 32–37

Boumaraf H, İnceoğlu M (2020) Integrating 3D printing technologies into architectural education as design tools. Emerg Sci J 4(2):73–81. https://doi.org/10.28991/esj-2020-01211

Budig M, Lim J, Petrovic R (2014) Integrating robotic fabrication in the design process. Architectural design: made by robots: challenging architecture at a larger scale, vol 84, no 3, pp 22–43. http://doi.org/10.1002/ad.1752

Carpo M (2012) The digital turn in architecture 1992–2012. Wiley, New Jersey

Carpo M (2017) The second digital turn: design beyond intelligence. The MIT Press, Cambridge

Celani G (2012) Digital fabrication laboratories: pedagogy and impacts on architectural education. Nexus Netw J 469–482

Chiu PH, Chiu Lai KW, Fan TK, Cheng SH (2015) A pedagogical model for introducing 3D printing technology in a freshman level course based on a classic instructional design theory. In: 2015 IEEE frontiers in education conference (FIE), pp 1–6. http://doi.org/10.1109/FIE.2015.7344287

De Sampaio C, Spinosa R, Vicentin J, Tsukahara D, Silva J, Borghi S, Rostirolla F (2013) 3D printing in graphic design education: educational experiences using fused deposition modeling (FDM) in a Brazilian university. In: Todd MK (ed) High value manufacturing: advanced research in virtual and rapid prototyping. CRC Press. http://doi.org/10.1201/b15961-7

Dutton T (1987) Design study pedagogy. J Architectural Educ 16–25

Fernandes SF, Simoes R (2016) Collaborative use of different learning styles through 3D printing. In: 2nd international conference of the Portuguese Society for Engineering Education (CISPEE). IPCA—Instituto Politécnico do Cávado e do Ave, Barcelos, p 8

Foque R (2011) Introductory panel: new priorities, new competences. Learning for the future: new priorities of schools of architecture in the era of uncertainty. EAAE & ENHSA, pp 91–93

Gnaur D, Svidt K, Thygesen MK (2015). Developing students' collaborative skills in interdisciplinary learning environments. Int J Eng Educ 275–266

Gramazio F, Kohler M (2008) Digital materiality in architecture. Lars Müller Publishers, Baden

Greenhalgh S (2016) The effects of 3D printing in design thinking and design education. J Eng Des Technol 14(4):752–769. https://doi.org/10.1108/JEDT-02-2014-0005

Groat L, Wang D (2013) Architectural research methods, 2nd edn. Wiley, New Jersy

Gu N, Wyn JM, Wiliiams A (2010) Utilising digital design and rapid prototyping tools in design education. In: New frontiers: proceedings of the 15th international conference on computer-aided architectural design research in Asia CAADRIA 2010. Association for Research in Computer-Aided Architectural Research in Asia, Hong Kong, pp 249–258

Guler K (2015) Social media-based learning in the design studio: a comparative study. Comput Educ 87:192–203. https://doi.org/10.1016/j.compedu.2015.06.004

Ioannou O (2018) Opening up design studio education using blended and networked formats. Int J Educ Technol High Educ 15. https://doi.org/10.1186/s41239-018-0129-7

Jones D, Lotz N, Holden G (2021) A longitudinal study of virtual design studio (VDS) use in STEM distance design education. Int J Technol Des Educ 31:839–865. http://doi.org/10.1007/s10798-020-09576-z

Kim S, Shin Y, Park J, Lee S-W, An K (2021) Exploring the potential of 3D printing technology in landscape design process. Land 10(259). http://doi.org/10.3390/land10030259

Kolarevic B (2005) Architecture in the digital age: design and manufacturing. Spon Press, New York

Kristiánová K, Joklová V, Mečiar I (2018) Physical models in architectural education and the use of new technologies. In: 11th annual international conference of education, research and innovation, pp 2177–2183. http://doi.org/10.21125/iceri.2018.1482

Kwon YM, Lee Y-A, Kim SJ (2017) Case study on 3D printing education in fashion design coursework. Fashion Text 4(1). http://doi.org/10.1186/s40691-017-0111-3

Laurillard D (2012) Teaching as a design science: building pedagogical patterns for learning and technology. Routledge, New York

Leach N (2017) Size matters: why architecture is the future of 3D printing. Archit Des 87(6):76–83. https://doi.org/10.1002/ad.2241

Leach N, Turnbull D, Williams C (2005) Digital tectonics. Wiley, London

Loy J (2014) eLearning and eMaking: 3D printing blurring the digital and the physical. Educ Sci (4):108–121. http://doi.org/10.3390/educsci4010108

Lugo Nevarez H, Thomas Pitcher M, Perez O, Gomez H, Espinoza P, Hemmitt H, Anaya RH (2016) Work in progress: designing a university open lab 3D printer model. In: New Orleans jazzed about engineering education. ASEE, New Orleans, LA, pp 1–11

Masdeu M, Fuses J (2017) Reconceptualizing the design studio in architectural education: distance learning and blended learning as transformation factors. Archnet-IJAR 11(2):6–23
Menges A, Ahlquist S (2011) Computational design thinking: computation design thinking. Wiley Sons, London
Milošević J (2021) Blended design studio model in architectural engineering education: a learning experiment. Int J Eng Educ 35(5):1300–1312
Nicol D, Pilling S (2000) Changing architectural education. Taylor and Francis Publications, London
Paio A, Eloy S, Vasco MR, Resende R, João de Oliveira M (2012) Prototyping Vitruvius, new challenges: digital education, research and practice. Nexus Netw J 14(3):409–429. https://doi.org/10.1007/s00004-012-0124-6
Salama AM (2017) Spatial design education: new directions for pedagogy in architecture and beyond. Ashgate, Surrey
Schon DA (1987) Educating the reflective practitioner. Jossey-Bass, San Francisco
Willmann J, Block P, Hutter M, Byrne K, Schork T (2019) Robotic fabrication in architecture, art and design 2018: foreword by Sigrid Brell-Çokcan and Johannes Braumann. Springer, New York

Chapter 31
The New Technological Paradigm in the Post-digital Era. Three Convergent Paths Between Creative Action and Computational Tools

Roberto Bianchi

Abstract The current evolutionary situation of the post-digital era is influenced by a scientific research based on an almost unconditioned technological positivism and requires a more agile a flexible built environment, adapted to the unpredictable changes and transformations of our planet. The building sector is going through an important evolutionary phase, mainly characterized by the use of new digital technologies which have considerable impact on the organizational dimension of the design and implementation, in order to adopt innovative solutions with the least environmental impact. While the architect is replaced by machines in the worksite, the use of digital technological instruments needs a new approach toward the project and the building procedures. Within this change, the draftsman, besides engaging with multidisciplinary specialized skills ranging from computational engineering to biotechnologies and neuro-sciences, has to be proactively able to control the current space, time, and technological processes which are offered by evolved and complex computerized instruments. These go beyond formality to create new constructive realities, which are being experimented. The inorganic computational instruments of the post-digital era afford the "Total Designer" a new opportunity to rethink composition not only in terms of productivity and efficiency but to also gain a deeper understanding of empathy and experience. It is pertinent in this situation to consider the possible convergence between design thinking and technical thinking based on the development of digital algorithms. With the help of some research and case studies, this contribution analyzes the relationship between creative action and computational instrument through three main directions—representation/simulation, communication/perception, and manufacture/materialization—that encompass the essential elements characterizing the role of the designer, a draftsman that is not only aware but sensitive to modern technology.

Keywords Representation · Simulation · Communication · Perception · Manufacture · Materialization

---

R. Bianchi (✉)
Mercatorum University, Rome, Italy
e-mail: roberto.bianchi@unimercatorum.it

E. Arbizzani et al. (eds.), *Technological Imagination in the Green and Digital Transition*, The Urban Book Series, https://doi.org/10.1007/978-3-031-29515-7_31

31.1 Introduction

Architectural research has undergone several phases since the arrival of digital technology, which has often questioned the boundaries of such a discipline in the relationship between design and construction and between art and science. Following a period of fideism and inferiority toward digital instruments and techniques, we now witness a shift of the cultural and technological paradigm, which brings mankind back to being the main focus of architecture and design.

As part of this interaction between technology and human activities, the project is described as 'intelligent,' 'flexible,' 'interactive,' and 'empathetic.' According to Biraghi «[...] in the era of the digital panoptic—without shape, immaterial and ubiquitous—the Bentham panopticon is unexpectedly brought back» as research of «disciplinary control and demonstration of digital freedom» (Biraghi 2019, p.162).

Having experienced more than 30 years of digital technology, we are in the midst of a "second digital turn," a new paradigm engendering a new way to rethink the project on a comprehensive level, compared to the current computational instruments «[...] we are learning that computers can work better and faster when we let them follow a different, nonhuman, postscientific method; and we increasingly find it easier to let computers solve problems in their own way—even when we do not understand what they do or how they do it» (Carpo 2017, p 7).

As observed by the Khannas, in this scenario designers experiment with a hybrid approach to the project, where «technologies merge together and with humans», and where «technology transcends a purely instrumental level to reach the existential realm» (Khanna 2013, p 6,8) It is a type of inductive-experimental analytic and compositive process that gathers information around the digital instruments and re-elaborates it in a creative way, where prediction can be based on sheer information retrieval, and form-finding by simulation and optimization can replace deduction from mathematical formulas.

The inorganic computational instruments of the post-digital era afford the "Total Designer" (Ortega 2017) a new opportunity to rethink composition not only in terms of productivity and efficiency (Paris 2021, p 131) but to also gain a deeper understanding of empathy and experience.

It is pertinent in this situation to consider the possible convergence between design thinking and technical thinking based on the development of digital algorithms. With the help of some research and case studies, this contribution analyzes the relationship between creative action and computational instrument through three main directions—representation/simulation, communication/perception, and manufacture/materialization—that encompass the essential elements characterizing the role of the designer, a draftsman that is not only aware but sensitive to modern technology.

31.2 New Creative Design Approach

The codes of computational language—'dynamic relationships,' 'flow diagrams,' 'non-Euclidean geometries,' 'the integration of natural models,' etc.—have often been examined to experiment with new creative approaches to the project, following a new constituent logic that allows a simultaneous, systemic, replicable, variable, precise, and controlled action.

While architects such as Eisenmann, Ghery, Lynn, Cook, and Hadid explored the software's potential at the dawn of the digital revolution primarily in the abstract language of the neo-organic forms, we have recently been witnessing a new use of the calculation instrument in the process of conception, elaboration, and production of a project where the individual user is the main interest. With the use of modern multi-platform software with 'user-friendly' interfaces, it is possible to produce new computational principles of mass personalization of the project which can encourage creative processes enabling the visualization, comprehension, and production of an artifact in its highest 'forms.' In the phases of elaboration and production, the designer can simultaneously control the complexity of space in its figurative, dimensional, performant, and management aspects.

The "Total Designer" comes to light with a holistic vision able to: combine artistic creativity and computational science; use digital technology with critical thinking and awareness; manage the softwares' space, time, and technology processes; recognize the models of access, control, and transmission of data; relate to further 'specialist' knowledge; adopt a cooperative and circular process; engender digital models in line with the forms of living. The designer implements an analytic design process, inductive and experimental, sourcing information from the data accessible on the Internet and from digital instruments. They re-elaborate the information through creative action, a constituent device outlining the conceptual approach and the cultural background of the project.

The design process has changed dramatically from the traditional linear sequence of conception, representation, and production to a circular, repetitive, and dynamic process that emerges from continuous feedback between conceptual ideas, models, prototypes, redefinitions, and the perfection of the creative idea. A process 'loop' combines the intuitive ability and the experiential knowledge with the automatic digital tools. A system that reduces the gap between conception and production allows the creative invention to extend to the prototype phase where the computational tolerances require a higher level of definition, optimization, and control of morphology, dimension, and performance.

An inclusive constituent process is outlined. As an actor and director, the designer abandons the modernist authorial dimension here and joins a cooperative and collective network of convergent multidisciplinary skills (Fig. 31.1).

Fig. 31.1 Digital era and second digital turn: theories of some protagonists (author's elaboration)

31.3 Creative Act and Digital Instrument: Three Axes of Change

The post-digital era offers an opportunity to rethink design through a combination of converging paths between design thinking and technical thinking derived from the elaboration of digital algorithms.

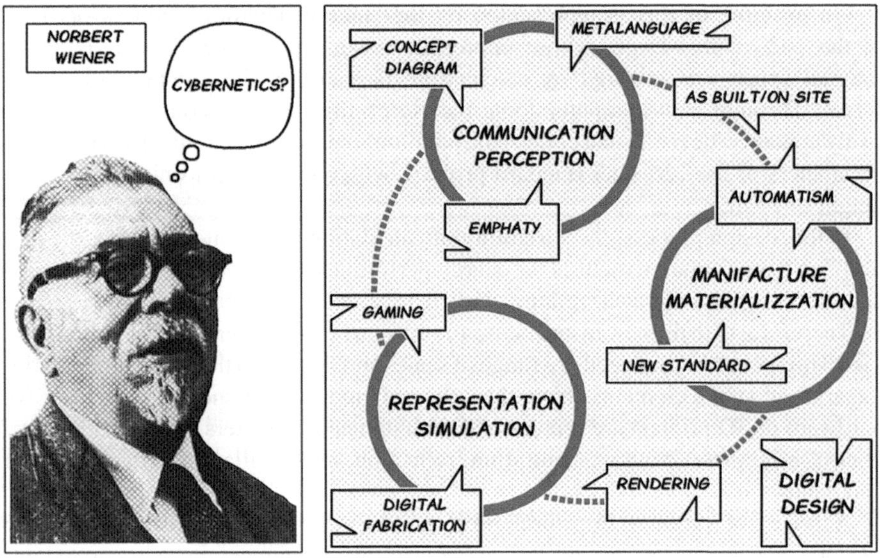

Fig. 31.2 Circular process in the second digital turn. Three main axes of change: representation/simulation, communication/perception, manufacture/materialization (author's elaboration)

Draftsmen experiment with new action modalities that redefine the relationship between the creative act and the computational tools on three levels: representation/simulation, communication/perception, and manufacture/materialization. As a sensitive draftsman aware of changes, the Total Designer operates along three main axes (Fig. 31.2).

31.3.1 Representation/Simulation

The new software allows us to achieve unprecedented levels of precision that, through hyper-realistic virtual images, enable the conception, visualization, measurement, and control of space as well as technical and performance parameters in terms of energy consumption and costs.

On the one hand, sophisticated digital models operating with precision and in real time are created in the realms of immaterial abstraction (rendering) and physical material (digital fabrication), while on the other hand, new creative modalities of elaboration and project management are explored.

As well as in the artistic domain, where softwares such as 'processing' define original visual languages outlining new writing codes—synthesizing graphic project and systemic materiality—, also in the architectural domain, the latest programs contribute to defining new spatial setups and creative languages, digital programs which combine figurative images and calculation diagrams with the processuality.

An example of this is represented by the Bahrain Pavilion, built in occasion of the 2020 Dubai Expo, whose unprecedented metal filigree characterizing the internal space was inspired by calculation diagrams of the computational softwares. Through new computer tools the Kerez firm simultaneously and interactively conceive, develop, and control the pavilion's space with technical-performance parameters of the 126 steel standard profiles, with a diameter of only 11 cm and a height of 24 m.

In this design process, the variability of calculation algorithms, whether they are deriving from formal, structural, or energy performance parameters, they define new degrees of figurative transformation and creative interpretation.

Therefore, with the new instruments, the tension between computation solidity and visual image can be analyzed at a range of scales and depths. In the process of creating the artifact, subtle elements (from the software) and solid elements (from the CNC hardware or 3D printer) are represented with optical parameters that reproduce and control brightness, contrast, saturation, reflection, and oscillation between surface and depth.

A series of proto-architectural projects, Projectors, by the firm MILLIØNS explores how image processing and decomposition techniques can be integrated into the numerical material methods of 3D printing, to decrease error tolerance as well as optimize workflow. The process begins by rendering platonic shapes in different planes and with different optical effects and then dissecting the structural basis of the image rendered. The content of the image is then reorganized through algorithmic operations along the three Cartesian axes. Different values and parameters are manipulated in such a way that clusters of pixels can be shuffled and rearranged to draw out a new image. With the help of 3D printing, ideas are transformed into full-scale prototypes from the digital model. Even color acquires a tectonic parameter, in which the colored pigments are obtained through the numeric coding of the composition (Koreitem 2021, p 11).

31.3.2 Communication/Perception

«A moiré is a figural effect produced by the superposition of two regular fields. [...] They shift abruptly in scale, and repeat according to complex mathematical rules» (Allen 1985, p 8). In his definition of moiré, Stan Allen explains how it is possible to perceive a new paradigm shift in the passage from analog to digital: "from mechanical to speculative production." (ivi, p 8). The diagrams he uses in his studies about "Field Conditions" reveal a new communicative-perceptive digital approach to "modeling program and space" that moves from the "Figure" of "Classical and Modernist Compositions" of the "Mechanical Era Modern" to the "Field/Ground" of the "Field Composition" of the "Digital Era."

As a result of the ideogram sequence, linked to the generative rules of the digital calculation, a connection is formed between the 'figure' and the 'abstraction,' establishing a scheme, a pattern, which explains the design concept. A repeating image with multiple combinations and variations of figurative molds, which are effectively sparks of design elaboration—a "productive indeterminacy" (Schumacher 2004). A « "field condition" shifts the focus of attention from the singular to the multiple, from the individual to the collective», where «[…] It recognizes constraints as opportunities» (Ortega 2017, p 26).

At the start of the digital age, architects such as Eisenmann, Tschumi, and Van Berkel used diagrams to represent the abstract language of the project, but in this new era depicting the "field condition," software-made models have evolved into dynamic images that carry out aspects of conception, communication, perception, and experience, in addition to simplifying and coding the complexity of the design process.

Ensemble uses three-dimensional scanning in its Ca'n Terra project to convey the idea of a new "material" and "experiential" topography of the real world: "Throwing millions of laser points on the wrinkles of the continuous stone surface, we register with millimetric precision the solid structure" (from Ensemble's project description).

In "Thousand Plateau," Gilles Deleuze states that diagrams «are not a reenactment of any form, on the other hand, construct a new form of reality» (Deleuze and Guattari 2001).

Beyond its role as a technological medium, the digital instrument articulates an alternative mode of communication and perception simulated through the definition of 'metalanguage' that, in contrast to hyper-realistic renderings—which reduce the conceptual dimensions of the project—, still leaves open and open-ended issues in the process of the design's theoretical synthesis. A paradigm shift occurs when we observe the figurative-communicative-perceptive-experiential process of the project: a 'new field condition' that is creative and inductive, founded on a renewed empathetic and multisensory interaction between the draftsman, the device, and the immaterial surfaces of the artifacts, adding to a "new materialism" (Paris 2019, p 12).

31.3.3 Manufacture/Materialization

We aim to publish all proceedings papers in full-text xml. Our xml templates for LaTeX are based on CMR, our xml templates for Word are based on Times. We ask you to use the font according to the template used for your papers. Papers using other fonts will be converted by our typesetters.

Using the plug-in platforms for dynamic planning and robotics and 3D printing, a creative idea becomes tangible as a result of the steps of 'manufacturing' and 'materializing' that are comparable to the processes of rapid prototyping and mechanized assembly.

As shown by some recent experiments such as DFAB House, the first house that was digitally designed and built with the help of robots and 3D printers, coding the data processed by the parametric software allows interaction in real time in the realm of creative interpretation directly 'on-site' in the steps of installation of the project where «[…] the numerization implicit in digitalization translates all of the information into a common binary code, an abstraction that promotes links and connections between previously unrelated elements» (Ortega 2017, p 25).

With the new digital techniques—where computational design issued by the variability of the algorithms corresponds to flexibility and simultaneity in the building processes carried out by robots—, the creative process is facilitated by a direct intervention during the realization phase, creating architectural works where the design and building phase are always more coherent, synchronic, and integrated with the reference environment.

The simultaneous, accidental, improvised, rapid, and combination aspects of this new 'creative dimension' contribute significantly to the definition of a "new standard" (Carpo 2017), which has relevance to the parametric manufacture and materialization processes through concepts such as 'series' and 'repetition' in terms of 'versions' and 'variations'. In the concept development process, the abstraction created by modeling the algorithms that control the genetic behavior of the project amplifies, simplifies, and optimizes the infinite interpretive variables. The steps of installation as well as the tectonic and syntactic aspects of the artifact are affected by these factors. Thus, they mark a transition from a culture of identity and repetition (mechanical production) to a new "ecology" of multiple and different performances (speculative production).

By using the 'Superlinear design process,' Ryuji Fujimura develops the project in a 'new aesthetics of play' which performs directly in the manufacture and assembly phase while simultaneously controlling parametric-performant data. Experiments are conducted in an experimental manner and are followed by a prototype that documents the steps. Each of them is the result of the progressive change of a single part of the project. It starts with a simple model that is modified proportionately with subsequent modifications: There is an input (a question or a request), an output (the formal solution), and a time progression describing the development of the artifact materialization traced by the prototype.

The design process, which produces a series of morphological variations that can be represented with the incubation steps of a fish before hatching the egg, can be summarized in the sequence: prototyping (giving an approximate form based on known information and requirements); feedback (removing inconsistencies between the form and the context by adding rules); and the iteration of these first two steps (inferring the original purpose from the derived set of rules, and increasing the certainty of the resulting form).

A 'Superlinear design process' is still in experimentation, and it would allow the creation of a new collective creativity, more open to variation and mass digital personalization directly 'on-site.'

31.4 Conclusions

The rapid development of computational techniques has profoundly transformed the design process, as well as the approach to design and the role of the draftsman, since the beginning of the digital era.

In the early days of the digital age, Ben Van Berkel wrote in his book "Move," "Architecture will be the fashion designer of the future" (Van Berkel 1999, p 27). In this phase, the draftsman—a creative artist able to "dress the future"—explores the software's potential through a neo-organicist abstract language, within a design process that maintains the traditional linear path of conception, representation, and realization.

During the "second digital turning point," about twenty years after Van Berkel's statement, Spanish architect Louis Ortega outlines the draftsman with a holistic and inclusive vision, in his book "The Total Designer," someone able to control creativity and technical performance «[...] who has the ability to work with space while drawing on a profound awareness of its relationship with time in its many dimensions [...]» (Ortega 2017, p 7) A draftsman, being part of an inclusive system of convergent multidisciplinary skills, is both director and actor who utilizes the digital potential to satisfy individuals' real needs. An ongoing 'as built' system is being developed via a circular sequential process where creative action and computational instrument exchange dynamic feedback in the stages of representation/simulation, communication/perception, and manufacture/materialization.

As draftsmen, once we observe the ongoing and rapid evolution of software and services around the virtual economy, we can imagine the role of a 'Meta Designer' changing as well. An exceptional draftsman who sees beyond the surface appearance of an object to conceive, simulate, perceive, and create in a simultaneous, circular, interactive, immersive, and experimental way a manifold human-sized reality.

Over a short period of time, we went from a "digital" to a "virtual" to an "augmented reality" design, and the distance between the creative act and the finished product has practically vanished in favor of a high level of control and definition.

A dimension of virtual reality, connoted by innovative solutions aiming for the least environmental impact, offers new possibilities and knowledge for design and architecture today as a new kind of sense of space and time between things and objects.

In the post-digital era, we can conclude that the new technological paradigm is represented by actions that sought to redefine, through a holistic and aware approach of the designer, a balance and a parallel dialogue between technical knowledge and humanistic culture. An opportunity to rethink the project's identity and activity within a new interaction and cooperation between science, art, space, and time, resulting in a reconciliation between creative expression, performance, and individual needs (Fig. 31.3).

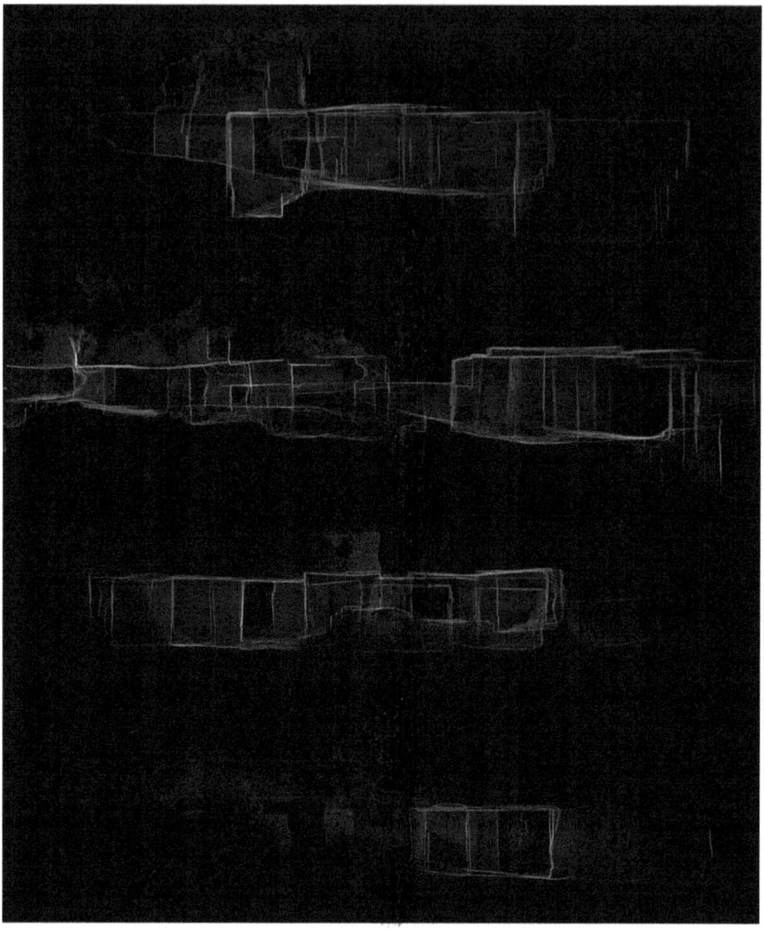

Fig. 31.3 Ensamble Studio, Ca'n Terra, Menorca, Spain, 2018. The layout was redefined after a high-precision 3D survey using infrared light

References

Allen S (1985) Field Conditions, in Points + Lines
Biraghi M (2019) L'architetto come intellettuale. Einaudi editore, Torino
Carpo M (2017) The second digital turn: design beyond intelligence. The MIT Press, Cambridge
Deleuze G, Guattari G (2001) A thousand plateaus: capitalism and schizophrenia. The Athlone Press, London
Khanna AP (2013) L'età ibrida. Il potere della tecnologia nella competizione globale. Codice, Torino
Koireitem Z (2021) Reality is averrated. Domus 1062, November, pp 8–11
Ortega L (2017) The total designer. Actar, Barcelona
Paris S. (2019) Design, technology, empathy. a contemporary issue in the conception and production of artifacts. Common Ground Research Networks, Champaign
Paris S (2021) Progetto e tecnologie, tra scienza e nuovo umanesimo. Innovazioni nella formazione e ruolo dei progettisti. TECHNE 21:124–132. Firenze University Press, Firenze

Schumacher P (2004) Hadid digitale. Paesaggi in movimento. Testo & Immagine, Torino
Van Berkel B (1999) Move. UN Studio & Goose Press, Amsterdam

Chapter 32
Technological Innovation for Circularity and Sustainability Throughout Building Life Cycle: Policy, Initiatives, and Stakeholders' Perspective

Serena Giorgi

Abstract The introduction of innovative technologies across the design decision-making leads to a change of entire management of operational and organizational models, lengthening the design time, as many more predictive and cognitive phases are introduced. Nevertheless, the traditional character of construction sector obstacles the introduction of new technologies which need an acceptance process that must be triggered. The paper identifies how the non-tangible technological innovation, towards sustainability and circularity, is promoting by policies and how it is perceived by stakeholders of supply chain, providing inspiration for further actions to increase diffusion in practice. The results, shown in this paper, come up by a dialogue at national and international level to stakeholders in the occasion of research works and participation to national and international working groups and co-creation groups, fulfilled by the author. To this end, at first, some emblematic policy measures, from national and international level, addressing the introduction of technology to enable circularity and sustainability in the building sector are shown. Secondly, the point of view of stakeholders regarding the difficulties linked by technological innovation is highlighted. Finally, necessary initiatives to introduce and diffuse acceptance of technologies within construction sector are discussed.

Keywords Circular economy · Enabling technology · Green transition · Stakeholder networks · Co-creation process

32.1 Introduction

Research and innovation in the field of architectural technology currently have, on the one hand, the commitment to respond to green policies which require greater attention to circularity and environmental sustainability and, on the other hand, the

S. Giorgi (✉)
Polytechnic University of Milan, Milan, Italy
e-mail: serena.giorgi@polimi.it

E. Arbizzani et al. (eds.), *Technological Imagination in the Green and Digital Transition*, The Urban Book Series, https://doi.org/10.1007/978-3-031-29515-7_32

responsibility to support practices at building project and process, understanding the change attitudes and the necessary drivers.

In fact, it is necessary to guide the various actors involved in construction processes towards a change of mentality, supported by technologies that allow a circular and sustainable approach to design, construction, maintenance, and demolition of buildings.

The European Commission is promoting technological innovation in all economic sectors, setting organizations responsible for identifying, co-financing, and coordinating specific activities (an example is the commitment of the European Institute of Innovation and Technology—EIT), and supporting priority areas research and innovation (as a case of Key Enabling Technologies—KETs).

A generative and responsive technology within construction sector can establish new and more correct relationships between humanity and nature (Perriccioli 2020). Technological innovation in the construction sector is necessary to predicting the transformation and impact that the design action causes on the environment, in order to respect the aims of circularity and sustainability.

Technological innovation has been already achieved by some industries, in order to increase energy and resources efficiency, circularity of resources, productivity, highly products' customization, the commercial and social value of products, and for the acquisition and data management to monitor input and output flows.

Nevertheless, the construction sector is resistant to change toward and high technological innovation, and the reason is intrinsic to the construction sector itself, which has a strong traditional character.

Technological innovation, in construction sector, can concern tangible technologies, such as constructive solutions aimed at reversibility, but mainly, by non-tangible ones, constituted by digital technologies. In fact, modeling, monitoring, and simulating processes, the Internet of things (IoT) and digital platforms are priority areas of research, which help and increase the ability to prefigure design scenarios, measure their impacts, increase the adaptability of spaces, and activate circular material flows to extend the useful life of the products and buildings.

The objective of this contribution is to investigate technological innovation for a green transition (in term of circularity and sustainability) in the construction sector, showing some political measures and initiatives that some European countries are promoting, and highlight how stakeholders of value chain are experiencing this growing demand for innovation.

32.2 Technological Innovation for Circular and Sustainable Transition

Technological innovation, towards resource efficiency and circular economy, is not limited to the production processes or new materials, but also regarding the reconfiguration of the entire building life cycle (from design phase, construction phase, use phase and end-of-life management).

In literature, it is possible to identifies research areas of non-tangible technological innovation to support design and construction phases, aimed at design optimization, reduction of environmental impacts, material flow management, circular/reverse logistics of resources.

In the context of circularity and sustainability, building information modeling (BIM) is identified as enabling technology for monitoring the use of resources during the whole building life cycle, sharing information between operators and simulating the potential reuse of building materials early in the project (Akanbi et al. 2018, 2019; Charef and Emmitt 2021). Together with BIM, material passports are identified as BIM-interoperable tools to keep knowledge of all building materials in the long term and preserving their (economic) value (Luscuere 2017; Munaro et al. 2019).

In order to predict the environment impacts of activities on the built environment, the use of life cycle assessment (LCA) during the design phase is considered fundamental, especially to activate material flows circular dynamics (Campioli et al. 2018; Lavagna et al. 2020, Eberhardt et al. 2019).

To improve material flow management and to allow knowledge of potential components to be reused in a new project, traceability systems based on digital technologies are considered able to trace geometric and mechanical characteristics of the components, the location of building, the residual value of the materials, and the expected building and material life cycle (Minunno et al. 2018; Rašković et al. 2020). Furthermore, when the product reaches the end-of-life stage, traceability technologies enable the collection of products, the potential of reusing, remanufacturing, or recycling and the economic feasibility (Alcayaga et al. 2019). Moreover, digital platforms for exchanging materials are considered useful digital technologies to support material revers logistics (Baiani and Altamura 2018) and to pursue collaborative process and networks among different actors (Konietzko et al. 2019; Talamo et al. 2020).

32.3 Investigation Method for Policy, Initiatives and Stakeholder's Perspective

Given the main technological innovation dealt with in the literature to support the green transition of construction sector, this article focuses on some emblematic policies, the stakeholders' perspectives, and the necessary initiatives to support the technological innovation in the building sector.

The results, shown in this paper, come up by a direct dialogue with building stakeholders at national and international scale, occurred in different occasion from 2019 to 2022, fulfilled by the author. In particular, the occasions of dialogue occurred in the field of:

- in part for PhD thesis research and in part for a short-term scientific mission funded by the COST-Action "CA15115—Mining the European Anthroposphere (MINEA)" which were opportunities for direct interviews with numerous building stakeholders, like policymakers, designers, investors, researchers, and operators of building process;
- the participation at national Working Groups of the Italian Circular Economy Stakeholder Platform (ICESP), which allows to focus the opinion of Public Administration (PA), manufacturer, research organisms, and environmental consultants;
- the participation of co-creation groups, organized by SocKETs project, funded under the H2020 framework program, with the aim of discussing barriers and opportunities related to the introduction of key enabling technologies for circular economy in construction sector, involving industry representatives, market operators, researchers, policymakers, civil society, and citizens;
- the participation and organization of roundtable with stakeholders, organized by Re-NetTA Project, funded by Fondazione Cariplo and coordinate by Dept. ABC of Politecnico di Milano, with the aim of defining new organizational business models related to circular economy strategies in the construction sector, involving mainly manufacturer, seller, general contractors of tertiary building sector and social cooperatives.

At first, some policy measures, from national and international level, which address the introduction of technology to enable circularity and sustainability in the building sector are shown. Secondly, the point of view of stakeholders regarding the technological innovation is underlined. Finally, some important initiatives to introduce and diffuse acceptance of technologies are discussed. To conclude, the foreseeable risks and misunderstandings to be avoided in a field of innovation technologies are highlighted.

32.3.1 Policy Measures Addressing Technological Innovation for Circularity and Sustainability in the Building Sector

To support the design and construction phases toward circularity and sustainability transition, emerging policy measures regards non-tangible technology, rather than tangible ones, which remain less applied at legislation level (Giorgi et al. 2022).

There are some policy measures which imposed the use of BIM during the design phase, particularly regarding public building. The introduction of BIM is aimed at enabling efficient information sharing throughout the operators, reducing the risk of building design errors and waste during the construction phase.

For example, Denmark, through the "ICT Regulation" n. 118 of 06-02-2013 and n. 119 of 07-02-2013, establishes the obligation to submit digital building models for public tenders that exceed a cost limit. The government also requires that digital information related to the project must be processed during the construction phase and organized as construction project documentations useful for future building management.

Also, Italy, following the Ministerial Decree 560/2017, in public construction, considers the use of digital BIM tools mandatory, in order to put interoperability and usability of building project information, by every operator during the design, construction and management process. The mandatory use of BIM tools concerns projects that exceed a cost limit. This cost limit decreases every year and by 2025 will affect the majority of public project. Another example regards the UK which, since 2016, through the "Government Construction Strategy 2016–2020," has established the obligation to incorporate and increase the use of digital technology in public construction contracts, asking for a 3D BIM Level 2, to facilitate the reduction of construction waste.

Even if BIM is a technology useful for mapping the information of buildings, legislative measurement does not yet introduce the mandatory digital systematization of information, using material passports to keep all information in a common digital platform, accessible by enabled and qualified users, as experimented in the Netherlands by Madaster Platform (Baiani and Altamura 2020; Giorgi 2020).

To achieve sustainable building processes, in the field of circular economy transition, the legislation of some countries promotes the introduction of technologies capable to quantify the environmental impact of building life cycle. In particular, the introduction of LCA tools during the design phase represents an important support for forecasting and optimizing materials and energy in-flows and out-flows. For example, the Netherlands with the legislative decree "Milieuprestatieberekening van gebouwen" (art. 5.8 and 5.9), called MPG, set the mandatory reporting of buildings' environmental performance through a LCA study, for new homes and office buildings (with a surface greater than 100 m^2). Moreover, the LCA must respect a maximum environmental impact limit value defined by legislation.

A similar initiative has been established by Belgian policy plan, which has promoted the development of a shared methodology for calculating buildings environmental impacts (MMG). Policy measure providers also a digital tool called Tool to Optimize the Total Environmental impact of Materials (TOTEM) based on the LCA methodology, to support designers, investors, and policymakers (OVAM, 2018).

In Italy, LCA has been introduced to building process by the Green Public Procurement, (Legislative Decree 50/2016) and the related Minimum Environmental Criteria—CAM. In the first version, CAM referred to LCA through the request for Environmental Product Declarations (EPDs) which are based on LCA assessment, to demonstrate compliance with some mandatory CAM criteria. The updating of CAM (DM 23 June 2022 n. 256), incentivize the use of LCA and LCC assessment with a rewarding logic, giving reward credits, for public tenders, to design firms and construction companies that use LCA as a decision support tool.

To improve a circularity management of materials/waste (e.g., reuse, recycling), policy measures have developed traceability system to follow along the building process.

In particular, Belgium legislation sets a system for materials traceability, called Tracimat (VLAREMA, article 433) which controls the material flow in output of demolition process, requiring the drawing up of pre-demolition audit and the waste monitoring. Tracimat system is currently mandatory only for non-residential buildings > 1000 m^3 and residential buildings > 5000 m^3, but in order to spread it on the whole building stock, the legislative framework establishes economic incentives (as well discount on gate-fees at recycling plants). Tracimat is based on a digital platform where all operators of the building process can have an access. This digital platform creates a link between operators and the entire traceability process, put available all necessary information and documentations.

32.3.2 *Point of View of Stakeholders Regarding the Technological Innovation*

The introduction of innovative technologies across the design decision-making leads to a change of entire management of operational and organizational models, lengthening the design time, as many more predictive and cognitive phases are introduced. Multiple interdisciplinary interactions are needed from the initial stage, through the involvement, in the decision-making process, of owners, designers, builders, and manufacturers, to co-create building solutions in a "horizontal" way.

If the technological innovation represents the fundamental support for achieving certain environmental, economic, and social objectives, inevitably the addition of evaluations and informative monitoring constitutes an extension of the design and decision-making phase, requesting sometimes new professional figures, new roles, and new operators along the construction process. For example, traceability systems lead to the need to complete and follow the entire procedure, such as completing the pre-demolition audit, performing data analysis, monitoring the demolition work, and checking the correct separation of materials. Other example, LCA leads to the need to make an inventory of materials used in the project, monitoring the input and output, collect environmental data, calculate, and interpretate the LCA results.

Following the legislative boost, the knowledge of technology is quite widespread, but sometimes stakeholders show that companies often do not fully exploit the innovative technologies equipment to full potential. Mainly in the case of BIM (when it is not mandatory), the real potential of the tool for keeping information along the building life cycle and for increasing the interoperability between operators along the building process is not applied.

For example, the use of the LCA is still complex, and the application is often performed as a post-design analysis. Consequently, the potentiality of impacts assessment for orienting choices toward sustainability is not achieved, and the design solution is not optimized.

When the introduction of technology (for green transition) is set by legislation, the success of technological diffusion depends also on the capacity to verify the right application by governmental authorities. Nevertheless, sometimes the Public Administrations (PA) are not trained to have the skills to control the right accomplishment of process innovation.

In order to face the green transition, PA highlight the limited information capacities, the need for simplification of the legislation, the scarcity of financial resources (ICESP 2022), highlighting the need for a choice of priority actions to be implemented.

The opened dialogue with stakeholders of the building value chain allowed to highlight that, beside a first spirit of interest for innovation of the construction sector, especially by young entrepreneurs, there is a general concern about the commitments that technological change entails in practice. In particular, there is a general lack of aware on the advantages offered by the new technologies, and a general difficulty in identifying the benefit.

The main concern of design firms, construction companies' association, and order of architects regards the different ability for introducing new technologies, based on company size. In fact, the introduction of BIM technology and LCA requirement implies the need to increase the digital technology capacity of all operators involved, sometimes through specific fee-paying trainings. It is clear that medium and small firms (SMEs), very widespread on the national territory (and which form part of the national economy) feel threatened and, without any national subsidy, unable to sustain the request for economic resources and time necessary to acquire the technologies and specialized personnel.

Consequently, SMEs fear that they will not be able to remain competitive on the construction market and be destinated to disappear.

Moreover, stakeholders highlight that the ability to learn the use of new technologies is generational, so the introduction of technologies often creates a gap between generations, creating a great loss of generational cognitive transmission.

Moreover, stakeholders underline the need of understand in advance the benefit of circular business and sustainability choices, with a clear network of operators, and a defined market. Consequently, it is necessary to introduce accessible and inclusive training models to allow all dimensions of firms to keep up with technological innovation and to prepare PA to play a role of assessment and support for circular and sustainable practices' activation.

32.4 Initiatives for Encouraging Knowledge Sharing and Acceptance of New Technologies

In order to spread a greater awareness of technological innovation in support of the project, in an inclusive way, involving big firms, SMEs and PA, important initiatives are represented by the green deal approach, which promotes the activation of Living Labs to create an exchange of knowledge and expertise, as an accelerator in the transition to a circular construction practice.

The research institutions and universities play a key role in encouraging the experimentation of innovative technologies and promoting "knowledge sharing" through the involvement of a large panel of stakeholders who activate new forms of interaction and collaborative innovation, with significant social repercussions.

Belgium and the Netherlands have been activated, with the support of their government the "National Green Deal," aimed at removing obstacles to circularity in legislation and practices through cooperation and experimentation between stakeholders in the areas of "Circular Building Living Lab". The goal of these Living Labs is to creates places for the exchange of research experiences and results, to develop policies and practical recommendations, based on an "experimental field" of practical experiences and research results that are shared, disseminated, and questioned, to accelerate the transition to a circular economy in construction.

The National Green Deal developed in Belgium in 2019 under the boost of government and 300 companies and institution involved in the building value chain, promotes the activities of experiences exchange and "learning by doing". Also in the Netherlands, the Green Deal Circular Buildings (GDCB) started from early 2015 and last around four years. The Dutch GDCB are working with the aim of providing suitable tools to assess the circularity of a building. To this end, the GDCB has developed a circular passport, which describes the circularity of buildings and a circular manual that supports and explains how buildings become circular.

The green transition therefore requires a great effort that not only implies technological innovation, but also cultural innovation, based on educational and training programs, which are expressed in a new way of operating and looking at needs.

32.5 Conclusion

Technological innovation in the construction sector allows to carry out complex assessments to achieve process optimizations, to increase knowledge of materials/resources, to facilitate the exchange of information and materials throughout the building life cycle.

By the technological support, humans can achieve the activation of new sustainable supply chain dynamics based on circularity. Nevertheless, the utilization of technologies for cognitive and predictive assessments must be conduct with humans' awareness, knowledge, and culture; otherwise, the use of technologies become a mean to govern complex system with a "black box approach", without knowing how technologies generate responses to imputed actions.

Moreover, the enabling of collection of big data for monitoring and keep information on built environment must be content in the limit of data management.

Consequently, in the field of scientific research, it is necessary to understand how to activate innovation training practices both for operators, who should change their practices, and for PAs, who should verify their correct performance. Furthermore, it is important to target accessible and inclusive technological innovation to all construction operators. It is therefore necessary to understand whether the diffusion of simplified means (e.g., simplified tools to trace materials, to keep buildings information, to calculate environmental impacts, etc.) accessible and usable by all is the key for the diffusion of innovation, or whether it can become a means of trivializing virtuous practices, for example, for the sole purpose of satisfying innovative legislative requests.

References

Akanbi LA, Oyedele LO, Akinade OO, Ajayi AO, Davila Delgado M, Bilal M, Bello SA (2018) Salvaging building materials in a circular economy: a BIM-based whole-life performance estimator. Resour Conserv Recycl 129:175–186

Akanbi LA, Oyedele LO, Omoteso K et al (2019) Disassembly and deconstruction analytics system (D-DAS) for construction in a circular economy. J Clean Prod 223:386–396

Alcayaga A, Wiener M, Hansen EG (2019) Towards a framework of smart-circular systems: an integrative literature review. J Clean Prod 221:622–634

Baiani S, Altamura P (2018) Superuse e upcycling dei materiali di scarto in architettura: progetto e sperimentazione. J. Technol. Archit. Environ. 16:142–151

Baiani S, Altamura P (2020) La seconda vita dell'esistente, Tool digitali per il life Cycle Approach e il resilient design. In: Perriccioli M, Rigillo M, Russo Ermolli S, Tucci F (eds) Il progetto nell'era digitale. Tecnologia, cultura, Natura, pp 133–137

Campioli A, Dalla Valle A, Ganassali S, Giorgi S (2018) Designing the life cycle of materials: new trends in environmental perspective. J. Techne 16:86–95

Charef R, Emmitt S (2021) Uses of building information modelling for overcoming barriers to a circular economy. J Clean Prod 285:124854

Eberhardt LCM, Birgisdottir H, Birkved M (2019) Life cycle assessment of a Danish office building designed for disassembly. Build Res Inf 47:666–680

Giorgi S, Lavagna M, Wang K, Osmani M, Liu G, Campioli A (2022) Drivers and barriers towards circular economy in the building sector: stakeholder interviews and analysis of five European countries policies and practices. J Cleaner Prod 336:130395

Giorgi S (2020) Strumenti di supporto al progetto per dinamiche di economia circolare in edilizia. In: Perriccioli M, Rigillo M, Russo Ermolli S, Tucci F (eds) Il progetto nell'era digitale. Tecnologia, cultura, Natura, pp 243–245

ICESP (2022) La transizione circolare di città e territori nel panorama italiano. Position Paper. Retrieved from https://www.icesp.it/GdL/5

Konietzko J, Bocken N, Hultink EJ (2019) Online platforms and the circular economy. Innovation for sustainability. Palgrave Macmillan, Cham, pp 435–450

Lavagna M, Dalla Valle A, Giorgi S, Caroli T, Campioli A (2020) Circular processes and life cycle design for sustainable buildings. In: Carmelina B, Francesco C, Spina Lucia D (eds)New metropolitan perspective. Springer, pp 1448–1457

Luscuere LM (2017) Materials passports: optimising value recovery from materials. Proc Inst Civ Eng: Waste Resour Manag 170:25–28

Minunno R, O'Grady T, Morrison GM, Gruner RL, Colling M (2018) Strategies for applying the circular economy to prefabricated buildings. Buildings 8:125

Munaro MR, Fischer AC, Azevedo NC, Tavares SF (2019) Proposal of a building material passport and its application feasibility to the wood frame constructive system in Brazil. IOP Conf Ser Earth Environ Sci 225(1)

OVAM (2018) Environmental profile of building elements. Retrieved from https://www.ovam.be/sites/default/files/atoms/files/Milieuverantwoord-milieugebruik-bouw.pdf

Perriccioli M (2020) La dimensione ecologica del progetto nell'era digitale. In: Perriccioli M, Rigillo M, Russo Ermolli S, Tucci F (eds) Il progetto nell'era digitale. Tecnologia, Cultura, Natura, pp 15–17

Raskovic M, Ragossnig AM, Kondracki K, Ragossnig-Angst M (2020) Clean construction and demolition waste material cycles through optimised pre-demolition waste audit documentation: a review on building material assessment tools. Waste Manag. Res. 38(9):923–941

Talamo C, Lavagna M, Monticelli C, Atta N, Giorgi S, Viscuso S (2020) Re-NetTA. Remanufacturing networks for tertiary architectures. In: Regeneration of the built environment from a circular economy perspective. Springer, Cham, pp 303–314

Chapter 33
Fair Play: Why Reliable Data for Low-Tech Construction and Non-conventional Materials Are Needed

Redina Mazelli, Martina Bocci, Arthur Bohn, Edwin Zea Escamilla, Guillaume Habert, and Andrea Bocco

Abstract The paper proposes considerations stemming from the analysis of twenty-two buildings that show different approaches to 'vegetarian architecture'—a theoretical stance based on principles learnt from agriculture and nutrition. The first phase consisted in a systematic investigation of the constructional characteristics of each building, and the cataloguing of their components. The 'cradle to gate' embodied energy (EE) and 'embodied carbon' (EC) were then calculated, based on two open access databases: ICE and Ökobaudat. The applicability of these databases was considered, as they do not cover low industrialised bio-based construction materials. For some materials, data are missing; while in others, EE values are overestimated since high energy-intensive manufacturing processes seem to be assumed. In a second phase, the uses and production process of some non-conventional materials was investigated, evidencing their variability. Building technologies that are not just aimed at low operational energy but at a more holistic understanding of low

R. Mazelli (✉) · M. Bocci · A. Bohn · A. Bocco
Polytechnic University of Turin, Turin, Italy
e-mail: redina.mazelli@polito.it

M. Bocci
e-mail: martina.bocci@polito.it

A. Bohn
e-mail: arthur.bohn@polito.it

A. Bocco
e-mail: andrea.bocco@polito.it

E. Zea Escamilla · G. Habert
ETH Zürich, Zürich, Switzerland
e-mail: zea@ibi.baug.ethz.ch

G. Habert
e-mail: habert@ibi.baug.ethz.ch

E. Arbizzani et al. (eds.), *Technological Imagination in the Green and Digital Transition*, The Urban Book Series, https://doi.org/10.1007/978-3-031-29515-7_33

environmental impact represent a paradigm shift in 'sustainable' construction practices. Despite ongoing actions and policies, as long as these materials and techniques are not suitably represented in reliable and accessible databases, it will be difficult to make such a shift happen. Manufacturers and contractors who produce and use such materials would benefit from the availability of easily applicable, scientific data demonstrating environmental advantages offered by non-conventional materials.

Keywords Environmental impact · Embodied energy · Embodied carbon · Non-conventional construction materials · Databases

33.1 Introduction

Embodied emissions will likely constitute the majority of emissions generated by new buildings built between now and 2050 (Simonen et al. 2017). It is increasingly recognised that embodied impacts can constitute more than half of the total life cycle impacts from new buildings, and they will grow both proportionally and in real terms with the reduction of operational impacts (Rasmussen et al. 2018). 'Net-zero' or 'near-zero' operational greenhouse gases (GHG) emissions mean that the GHG emissions budget becomes almost entirely allocated for embodied GHG emissions (Röck et al. 2020; Habert et al. 2020; Moncaster et al. 2019).

Current EU regulations mainly cover the operational energy performance of buildings, while the embodied impacts remain largely unregulated, despite the significant carbon reduction potential (Toth and Volt 2021). However, since 2017, the Netherlands has required all new residential and office buildings larger than 1000 m^2 to account for embodied impacts based on a simplified LCA. Switzerland has introduced LCA requirements for public buildings (Swiss Society of Engineers and Architects (SIA) 2017). Denmark's National Strategy for Sustainable Construction will phase a LCA requirement into the building code, enforcing maximum CO_2 emissions of new buildings larger than 1000 m^2 from 2023 and for all new buildings from 2025 (The Danish Housing and Planning Authority 2021).

France's new *Réglementation environnementale* RE2020 includes carbon thresholds for offices and educational buildings starting 1 July 2022; limits will be progressively lowered. France is the first country to apply a dynamic LCA approach to the construction sector (Ministère de la Transition écologique (MTE) 2022a; Ministère de la Transition écologique (MTE) 2022b).

These increasing efforts for setting legally binding limits to the embodied environmental impacts highlight the crucial need for developing accurate, equitable, and easily accessed data for construction materials and techniques.

33.2 Data and Methods

33.2.1 Case Studies

The reflections proposed in this paper stem from the analysis of twenty-two ecologically oriented buildings showing different approaches to 'vegetarian architecture'—a theoretical stance based on principles learnt from agriculture and nutrition (Bocco Guarneri 2020), which advocates:

- natural, renewable, locally available construction materials, free of toxic chemicals, and as little processed as possible (Wolley 2017, 2016; Berge 2009; Ghavami 2014; Walker et al. 2009; Harries and Sharma 2016);
- minimization of energy-intensive, high-tech components;
- labour-intensive, small-scale production processes, and simple constructional techniques;
- passive solar design.

The case studies are located in Europe and Japan and cover a variety of functions—residential, commercial, educational, workspace—and different patterns of use. Their gross internal areas (GIA) range from 23 to 3 232 m^2. Both refurbishment of traditional buildings and new constructions are included to exemplify techniques that make use of bio-based and other natural materials.

33.2.2 Data

The systematic investigation of each building's technical and constructional features allowed to draw inventories and 3D models representing as-built situations. The 'cradle to gate' embodied energy (EE) and 'embodied carbon' (EC) were calculated by adding up the components manually, using a process-based LCA methodology and a purpose-designed spreadsheet. Impact coefficients were retrieved from two open access databases—the Inventory of Carbon and Energy (ICE) (Jones and Hammond 2019; Hammond and Jones 2011) and Ökobaudat (ÖBD) (BMI 2021).

Later, the most relevant non-conventional materials (NOCMAT) used were identified. NOCMAT encapsulate sustainable use of novel technologies and innovative uses of more established materials; many of them have their roots in traditional vernacular construction, including bio-based materials, and other natural materials such as stone, earth, lime (Ghavami 2014). For these materials, EE and EC values are either unavailable or show inconsistencies in the databases used. The production processes and the uses of such materials show a wide variability, which leads to a variability of the associated environmental impacts.

33.2.3 *Reference Databases*

ICE and ÖBD were chosen because open access and user friendly. More detailed sources are available but are proprietary and/or require a high level of expertise for their use. Using ÖBD and ICE to calculate the two basic environmental indicators was a sensible compromise between data availability and results uncertainty. Using the same set of values for all calculations guaranteed consistency.

ÖBD is managed by the German Federal Ministry of the Interior, Building, and Community; datasets must comply with EN 15,804. Data entries are constantly added, and the entire database is updated once a year. Datasets are based on the background database GaBi. Additional datasets based on EcoInvent background data are provided. In 2020, the new DIN EN 15,804 + A2 was adopted, which includes separate reporting of fossil, biogenic, and luluc GWP.

ICE is managed by Circular Ecology and the University of Bath. Version 2.0 (2011) was based on ISO 1404 and 14,044; 53% of sources dated before 2005. Carbon sequestration was excluded. Version 3.0 (2019) no longer includes energy factors. Carbon storage data are available for timber only. The values are the average of several sources, usually EPDs complying with EN 15,804. Data are updated for some materials. In our study, ICE V2.0 was used for all EE values and some EC values.

33.3 Results

33.3.1 *On Case Studies' Embodied Energy and Embodied Carbon Values*

While a rigorous internal methodology allowed for a detailed comparison between case studies (Bocco and Bocci 2022), it was difficult to verify whether these buildings have a lower environmental impact than conventional ones (Bocco Guarneri 2020). Systematic reviews (Simonen et al. 2017; Rasmussen et al. 2018; Birgisdottir et al. 2017; Dixit 2017; Hoxha et al. 2017; Röck et al. 2019; Säynäjoki et al. 2017; Schwartz et al. 2018) have not yet reached the degree of harmonisation which would offer benchmarks. The completeness of the underlying inventories is doubtful; the variations are up to two orders of magnitude (Rasmussen et al. 2018).

The average results of the analysed 'vegetarian' buildings do not appear significantly lower than those found of more conventional buildings. Moncaster et al. (Moncaster et al. 2019) found an average of 125 $kgCO_{2eq}/m^2$ for retrofitted buildings and of 254 $kgCO_{2eq}/m^2$ for new ones; we obtained 127 and 328, respectively, with ICE, and −132 and −74 with ÖBD (Fig. 33.1). The question about the reliability of databases, methodologies, and benchmarks stays open.

bldg. no.	GIA	weight	EE (ÖBD)		EE (ICE)		GWP (ÖBD)		GWP (ICE)	
	m²	kg	MJ/kg	MJ/m²	MJ/kg	MJ/m²	kgCO2eq/kg	kgCO2eq/m²	kgCO2eq/kg	kgCO2eq/m²
1	114	106 983	2.38	2 235	0.41	381	-0.18	-170	0.02	20
1(r)		80 934	3.15		0.54		-0.24		0.03	
2	23	25 738	4.75	5 318	1.36	1 524	-0.51	-570	0.12	131
2(r)		21 744	5.63		1.61		-0.60		0.14	
3	572	911 500	2.50	3 985	1.05	1 666	-0.09	-136	0.07	115
3(r)		204 818	11.13		4.65		-0.38		0.32	
4	411	469 598	3.35	3 823	2.30	2 633	-0.08	-92	0.15	174
4(r)		305 981	5.14		3.54		-0.12		0.23	
5	103	106 869	1.78	1 843	0.36	369	-0.46	-474	0.09	97
6	65	166 427	5.01	12 821	4.95	12 664	-0.09	-241	0.11	286
7	125	343 872	8.73	24 019	4.09	11 252	-0.37	-1 015	0.23	627
8	76	129 560	4.44	7 550	3.52	5 978	0.12	210	0.26	446
9	176	29 476	8.60	1 441	8.62	1 443	0.16	28	0.42	70
10	153	330 945	4.67	10 105	3.37	7 286	0.03	55	0.32	687
11	61	67 683	4.57	5 067	3.56	3 949	0.04	46	0.22	242
12	156	340 417	2.51	5 471	4.72	10 290	0.02	39	0.18	389
13	2 212	1 966 360	7.90	7 019	4.98	4 429	-0.05	-41	0.35	307
14	183	267 401	4.69	6 858	3.84	5 614	-0.01	-12	0.27	393
average			5.43	6 449	3.59	4 261	-0.07	-89	0.23	277

Fig. 33.1 GIA, weight, EE, and GWP of each case study, calculated both with ÖBD and ICE

33.3.2 On Vegetal Materials

The analysis highlighted a divergence of the profiles for timber and timber-based products in the two databases. Data for highly industrialised products such as timber-based boards and window frames are widely available and cover a good range of variations; this comes less for untreated solid wood. ÖBD provides very high values for EE, reflecting German processing: kiln drying; industrial debarking and sawing machinery; and an average distance between forest and sawmill of 144 km (Fig. 33.2). While timber was widely used in most of our case studies, it was often low processed and underwent little treatment, if any. In most cases, timber was air-dried, while most sources for both databases consider kiln drying at high temperatures—an energy— and carbon-intensive process that alters timber's properties. Where elements were hand-sawn (5), debarked and cut on site (21), and untreated (5, 6, 21), the impact risked to be overestimated (Fig. 33.3). In case 21, the whole structure—which makes a relevant portion of the building's weight—was manually debarked, and most of the lumber came from the site or district forest. In 5, timber was obtained from the ecovillage's forest and transported with horses to the site, where it was assembled with hand tools. No data are available for *brettstapel* (7, 13, 17, 22) as opposed to various entries for laminated timber: the lack of glue and nails reduces the environmental impacts of the first. Wood chips (21) and loose wood fibre (10) are not covered, as opposed to wood fibre boards for which ÖBD includes five entries and ICE two.

Straw was employed in ten case studies in a range of ways: load-bearing bales (6, 7, 21), thatch (2), bale infill (12, 20), bale retrofit (3), loose insulation (1, 6, 15, 21); ropes, mats (1), and chaff (2) are also used. Only values for standard size bales

Fig. 33.2 Timber dried at low temperatures in a greenhouse at Kitokuras sawmill (Kagawa prefecture). *Photo* Andrea Bocco Guarneri

Fig. 33.3 Manual debarking of trees at Biotal site. Untreated logs are used as structural columns in the building. *Photo* Christoph Bosch

with a density of 100 kg/m^3 are given in the databases, forcing an approximation for products like jumbo and round bales and loose straw. Not only does the crop cultivation vary (in 5 and 21 it is harvested from local organic farms), but also the baling process is expected to influence the overall impact. In many of these buildings, straw represents a large portion of the weight: divergences can therefore influence the overall environmental impacts. Reported values for straw bales are disorienting: in ÖBD, the EE value for straw (17.13 MJ/kg) is seventy times higher than that found in ICE 2.0, and higher than that of structural steel (14.14 MJ/kg) and more than three times higher than that of fired, solid bricks (4.85 MJ/kg). The reliability of the EE value for straw in ICE may be low, since it is based on four references only, the latest dating from 2003; but that in ÖBD (based on an EPD provided by FASBA) is high compared to recent studies: e.g., nearly 5 times higher than Upstraw School of Natural Building's 2021 EPD (Up-Straw—School of Natural Building (SnaB) 2021).

Databases seem also ill-suited for representing other vegetal materials. No data are present in either database for bamboo, not only for whole culms but also for products such as mats, panels, and laminated bamboo. In 9, locally harvested bamboo culms make up the entire structure (Fig. 33.4). Loose hemp shiv is not covered, which makes it difficult to assess, for instance, hemp-lime building components (8, 13). Only hemp mats are found in ÖBD (Fig. 33.5): these include 15% polyester fibres and are impregnated with soda.

Fig. 33.4 Structure of the Bamboo Ark consists of radially arranged frames, composed of a base truss and arches. The culms were harvested from a nearby grove. *Photo* Toki Hirokazu

Fig. 33.5 Bamboo mats in Iya Valley, Japan. *Photo* Andrea Bocco Guarneri

Analogously, there are no data for loose flax fibres (21), just for mats. In ÖBD, the same assumptions as for hemp fibre mats are applied. ICE V2.0 reports that most of the impacts are due to the polyester binders and fire retardants (Schmidt et al. 2004). No data are present in either database for reed mats (4, 6) and jute (6), while building paper (10, 13) is covered by ÖBD only.

33.3.3 On Other Natural Materials

Earth construction techniques are poorly represented: ICE just reports data on earth (it is unclear whether this is rammed or bulk earth); ÖBD includes data on rammed earth, adobe, plaster, earth panels, and bulk earth. Data sheets do not provide enough information on basic features, such as the presence of additives or fibres, or the size of the adobes. The entry for earth plaster in ÖBD has an ambiguous description, which does not even seem to refer to an earth-based product. ÖBD also assumes processes such as the artificial drying of adobes (Fig. 33.6).

Fig. 33.6 Light-earth external skin applied onto battens in a new house in Darmstadt. *Photo* Franz Volhard

No data are provided for earth paints, which are usually ready-made products (4, 8).

In many case studies, earth products were made at the construction site (Fig. 33.7): adobes in 10 and 12; manually compressed earth blocks in 13; earth plasters in 2, 5, 6, 10, 12, and 21 made with various mixtures, in some cases including fibres, sand, or lime; clay mortar in 4; tamped earth floors in 1, 4, 5, 6, 12, and 14 (coated with earth finishing in 6 and 12). Earth was also used with straw for infilling wooden or bamboo frames (1,2,4,15,21).

Fig. 33.7 Construction of the lecture theatre of the WISE at the Centre for Alternative Technology. The attractive load-bearing rammed-earth walls, 500 mm thick, are pneumatically tamped and left unfinished on both faces. *Photo* Pat Borer

Emblematic is the case of stone: while being one of the most common materials in vernacular constructions, its use has drastically shifted from massive blocks that require minimal (if any) dressing to 1 ~ 3-cm-thick cladding slats. Both databases just provide values for thin elements that underwent cutting and finishing processes. In ÖBD, entries for 2 ~ 4-cm-thick granite and limestone elements assume processing—steel grit, grinding road, saw, and multi-blade saw. ICE V2.0 acknowledges that data sources were generally poor, except for stone slates. Quarried stone blocks are then associated with a risk of overestimating their embodied impacts if the custom values are employed. Even in cases when thin paving stone slabs were used, ecological considerations resulted in specifying little-processed elements: in 7, 50 ~ 70-mm-thick soapstone flooring slabs were obtained on site by cutting conglomerate rock boulders; in 21, 9-mm natural stone tiles were hand-cut with pliers and left unpolished.

Animal-origin products are also little represented, if at all: for felt (7), the only data available is the EE value in ICE V2.0; for sheep's wool (8, 18), no values are available in either database.

33.4 Conclusions

The quality of the environmental assessment of buildings depends heavily on the quality of the data used, which remains a major challenge. Most of the processes mentioned here could be modelled with tools such as EcoInvent: but doing so requires skills and time that are beyond the possibilities of an average designer or contractor. ÖBD and ICE, while open source and easy to use, are focused on conventional materials and do not satisfactorily cover bio-based and little-processed materials. Even when they do, the values provided are regional or global averages, not representative of specific production patterns in terms of processes or efficiency, electricity mix, and transportation distance (Zea Escamilla and Habert 2014).

A proper assessment requires significant expertise, time, and financial resources, which are less likely available for alternative construction materials. Furthermore, meeting the impact thresholds set by current and future regulations pushes towards using certified conventional construction materials, a tendency that is not consistent with decarbonizing the building trade. Data availability is then a key issue in the implementation of environmental reduction policies.

The development of appropriate data for non-conventional construction materials should be supported by a bidirectional technological transfer between the research and industry sides. Current research efforts should be aligned to make the results reliable and widely available. This will rebound on an easier introduction of these materials in the construction market and help achieve the ambitious emissions reduction targets.

References

Berge B (2009) The ecology of building materials, 2nd edn. Architectural Press, London

Birgisdottir H, Moncaster A, Wiberg AH, Chae C, Yokoyama K, Balouktsi M, Seo S, Oka T, Lützkendorf T, Malmqvist T (2017) IEA EBC annex 57 'evaluation of embodied energy and CO2eq for building construction.' Energy Build 154:72–80. https://doi.org/10.1016/j.enbuild.2017.08.030

BMI (2021) Deutsche Bundesministerium des Innern, für Bau und Heimat. Ökobaudata 2021

Bocco A, Bocci M (2022) Reflections on the environmental impact of "vegetarian" buildings, and on the reliability of databases. Constr Technol Archit 1:395–404. https://doi.org/10.4028/www.scientific.net/CTA.1.395

Bocco Guarneri A (2020) Vegetarian architecture: case studies on building and nature. Jovis Verlag GmbH

Dixit MK (2017) Life cycle embodied energy analysis of residential buildings: a review of literature to investigate embodied energy parameters. Renew Sustain Energy Rev 79:390–413. https://doi.org/10.1016/j.rser.2017.05.051

Ghavami K (2014) Non-conventional materials and technologies for sustainable engineering. In: Proceedings of 14th NOCMAT, 24–27 Mar 2013, João Pessoa. Trans Tech Publications, Dürnten

Habert G, Röck M, Steininger K, Lupisek A, Birgisdottir H, Desing H, Chandrakumar C, Pittau F, Passer A, Rovers R, Slavkovic K, Hollberg A, Hoxha E, Jusselme T, Nault E, Allacker K, Lützkendorf T (2020) Carbon budgets for buildings: harmonising temporal, spatial and sectoral dimensions. Build Cities 1:429–452. https://doi.org/10.5334/bc.47

Hammond G, Jones C (2011) Inventory of carbon and energy. Circular Ecology and University of Bath

Harries K, Sharma B (2016) Nonconventional and vernacular construction materials. Woodhead Publishing, Sawston

Hoxha E, Habert G, Lasvaux S, Chevalier J, Le Roy R (2017) Influence of construction material uncertainties on residential building LCA reliability. J Clean Prod 144:33–47. https://doi.org/10.1016/j.jclepro.2016.12.068

Jones C, Hammond G (2019) Embodied carbon—the ICE database. Circular Ecology and University of Bath

Ministère de la Transition écologique (MTE) (2022a) Décret n° 2022a-305 du 1er mars 2022a relatif aux exigences de performance énergétique et environnementale des constructions de bâtiments de bureaux et d'enseignement primaire ou secondaire en France métropolitaine, Journal officiel, n°0052 du 3 Mars 2022, Texte n°48

Ministère de la Transition écologique (MTE) (2022b) Centre d'études et d'expertise sur les risques, l'environnement, la mobilité et l'aménagement (Cerema). Réglementation Environnementale des Bâtiments Neuf (RE2020)

Moncaster AM, Rasmussen FN, Malmqvist T, Houlihan Wiberg A, Birgisdottir H (2019) Widening understanding of low embodied impact buildings: results and recommendations from 80 multinational quantitative and qualitative case studies. J Clean Prod 235:378–393. https://doi.org/10.1016/j.jclepro.2019.06.233

Rasmussen FN, Malmqvist T, Moncaster A, Wiberg AH, Birgisdóttir H (2018) Analysing methodological choices in calculations of embodied energy and GHG emissions from buildings. Energy Build 158:1487–1498. https://doi.org/10.1016/j.enbuild.2017.11.013

Röck M, Saade MRM, Balouktsi M, Rasmussen FN, Birgisdottir H, Frischknecht R, Habert G, Lützkendorf T, Passer A (2019) Embodied GHG emissions of buildings—the hidden challenge for effective climate change mitigation. Appl Energy 258:114107. https://doi.org/10.1016/j.apenergy.2019.114107

Röck M, Balouktsi M, Saade M, Nygaard Rasmussen F, Hoxha E, Birgisdottir H, Frischknecht R, Habert G, Passer A, Lützkendorf T (2020) Embodied GHG emissions of buildings—critical reflection of benchmark comparison and in-depth analysis of drivers

Säynäjoki A, Heinonen J, Junnila S, Horvath A (2017) Can life-cycle assessment produce reliable policy guidelines in the building sector? Environ Res Lett 12:013001. https://doi.org/10.1088/1748-9326/aa54ee

Schmidt AC, Jensen AA, Clausen AU, Kamstrup O, Postlethwaite D (2004) A comparative life cycle assessment of building insulation products made of stone wool, paper wool and flax. Int J LCA 9:122. https://doi.org/10.1007/BF02978571

Schwartz Y, Raslan R, Mumovic D (2018) The life cycle carbon footprint of refurbished and new buildings—a systematic review of case studies. Renew Sustain Energy Rev 81:231–241. https://doi.org/10.1016/j.rser.2017.07.061

Simonen K, Rodriguez BX, De Wolf C (2017) Benchmarking the Embodied Carbon of Buildings. Technol Archit Des 1:208–218. https://doi.org/10.1080/24751448.2017.1354623

Swiss Society of Engineers and Architects (SIA) (2017) SIA 2040: Effizienzpfad Energie

The Danish Housing and Planning Authority (2021) National strategy for sustainable construction. Ministry of the Interior and Housing, Copenhagen

Toth Z, Volt J (2021) Whole-life carbon: challenges and solutions for highly efficient and climate-neutral buildings. BPIE

Up-Straw—School of Natural Building (SNaB) (2021) Environmental product declaration In accordance with ISO 14025 and EN 15804:2012+A2:2019 for: Straw as insulation material-UK

Walker P, Ghavami K, Paine K, Heath A, Lawrence M, Fodde E (2009) 11th International Conference on Non-conventional Materials and Technologies—NOCMAT 2009 Proceedings. University of Bath, Bath

Wolley T (2016) Natural building: a guide to materials and techniques. The Crowood Press, Ramsbury
Wolley T (2017) Building materials, health and indoor air quality. Routledge, Abingdon
Zea Escamilla E, Habert G (2014) Environmental impacts of bamboo-based construction materials representing global production diversity. J Clean Prod 69:117–127. https://doi.org/10.1016/j.jclepro.2014.01.067

Part III
Session | Environment

Technological innovation has long been the driving force of the material progress of civilizations, improving health, safety, and quality of life while creating more closely connected, well-integrated communities; advanced technologies have identified more efficient sources of energy, but also led to extreme growth in the consumption of the resources needed for their maintenance.

A new balance between development and the ecosystem calls for a revision of the drivers of innovation, in terms of their efficiency in transforming the anthropogenic environment and narrowing the great divides (of wealth, health and technology). The sustainability and decarbonization of all production sectors is based on soft skills applied to governing processes and optimizing skills and technologies, so as to leave behind an anthropic ecosystem of high technological intensity for a managed environment of low intensity and high efficiency. At the frontier of innovation, technology's impact is reduced and reformulated, with an emphasis on intangible resources and planning capabilities for transforming the built environment.

The session is designed as a forum for discussing R&D models and planning strategies for a low-tech environment, with advanced integration of carbon-neutral buildings/plants, so to revive the built environment through a low-intensity, high-efficiency approach to energy and environment concerns. Particular attention will be placed on contexts of energy and economic poverty in which the digital and technological divides are barriers to development and inclusion.

Chapter 34
Technological Innovation for the Next Ecosystem Transition: From a High-Tech to Low-Tech Intensity—High Efficiency Environment

Carola Clemente

Abstract Technological innovation is the driver of the progress of the material culture of human civilization. A new balance between development and ecosystem requires the revision of the innovation drivers, in terms of efficiency and transformations of the anthropic environment to reduce the great divides. Sustainability and decarbonization of all production sectors are based on process management skills and optimization of technical knowledge and technologies, to move from a highly technological anthropic ecosystem to a low intensity and high efficiency managed environment. The frontier of innovation is marked by the reduction of the impact of technology and its remodelling, enhancing intangible resources and the design abilities of transformation of the built environment. It is therefore urgent to focus on the R&D models and project strategies for a low-tech environment and highly advanced carbon neutral building/plant integration, the regeneration policies of the built environment with low intensity and high energy and environmental efficiency, with the aim the recovery and inclusion of marginal contexts of energy poverty and economic, where digital and technology divide represent barriers to development and inclusion. The traumatic awareness of the material "limit" of the availability of resources involves a paradigm shift in the global system of the supply chain and resource management on which we have based the development of the "technosphere" and perhaps represents the definitive culture shock necessary to redefine a new relationship between man and the environment.

Keywords Anthropocene · Environment · Great divides · Ecological transition · Technological ecosystems

---

C. Clemente (✉)
Sapienza University of Rome, Rome, Italy
e-mail: carola.clemente@uniroma1.it

E. Arbizzani et al. (eds.), *Technological Imagination in the Green and Digital Transition*, The Urban Book Series, https://doi.org/10.1007/978-3-031-29515-7_34

34.1 Complexity of Transition Scenarios

Technological innovation is the engine of the progress for the material culture of civilizations. It has improved health, security, and quality of life, and created more connected and integrated communities. However, while evolved technologies have identified more efficient energy sources, they have led to an extreme growth of the consumption of the resources needed to maintain them (Ziman 2003).

The intensity of human activity has so conditioned our planet's life over the centuries that the quantity of its manmade, artificial matter has equalled and exceeded that of the natural matter present on the globe.

Ecology has defined the portion of the physical environment created and organized as a result of human activity in the sphere of urban settlements and connected structures as the anthroposphere (Kuhn and Heckelei 2010); by extension, this term is used to indicate the set of humans and their artefacts which, beyond the transformations of the territory, also includes all environment phenomena, whether intentionally or unintentionally caused, such as pollution or developed technologies more generally, to which some studies apply the term "technosphere" (Zalasiewicz et al. 2017a).

Environment is therefore the landscape of all the systemic conflicts of this development, which is only erroneously limited in time to the first Industrial Revolution (Zalasiewicz et al. 2017b, 2017c).

What is clear is that the environment, intended as the complex physical, chemical, and biological system in which all living matter evolves, is conditioned on a local and global scale by the effects of human action, and that the intensity of this action has grown exponentially since the eighteenth century, the moment when humans' impact on ecosystems progressively increased along with world population growth, producing substantial alternations of the natural, eco-systemic equilibria (Haff 2016, 2017).

The power of the human phenomenon over the environment has been of such scope and intensity that Paul Crutzer (Crutzen and Stoermer 2000; Crutzen 2002) has defined this latest historical phase as tantamount to a geological epoch, using the term Anthropocene (Crutzen 2006, 2021).

The opening to discussion created by this session does not expect to find a solution to the evident imbalance that exists between human development and the conservation and restoration of the biosphere, but the aim has been to understand, through the research experiences we have gathered, whether there is hope for relating the demands of development, well-being, and security with the health of our host environment, or whether our disciplines are working on defensive positions, without in the least questioning the established dynamics of development of the sector of construction and of transformation of the territory.

Federico Butera recently expressed a highly critical thought on one of the pillars of current technological and economic life—the circular economy—where it is used to a distortive effect to continue to generate new material and new products only because one day they may be recycled and reused (Butera 2021). The real paradigm shift lies in ceasing to produce certain families of materials above all, except under the condition of continuing to produce them exclusively from secondary raw materials—that is, recycled material.

This principle also holds for the technological innovation connected to all the apparatus of systems necessary—or held to be necessary—for our well-being and security; it is in fact indispensable to determine how ready and willing we are to truly change established ways of designing and managing our buildings, and how willing we are to detach ourselves from devices, even those deemed reliable and efficient, but that are now clearly anti-historical. We must therefore determine when we will contemplate abandoning the traditional systems of the environmental management of our buildings so as finally to achieve an effectively integrated planning commensurate with our actual needs, leaving behind the principle of "redundancy" looked to by much of mainstream planning, and to definitively embrace a principle of "adequacy".

Integrated technological planning of the building/installations/urban surroundings system allows the quality of the indoor environment and the quality induced by our building on the outdoor environment to be conceived in a way suited to the needs of use. Only in this way can the material impact of producing the building be reduced: not by remodulating our demand, but by effectively optimizing the material response offered by the built environment.

An environment of low technological intensity does not imply a low-tech approach to realizing it, typical of a demand for more or less successful degrowth, but the use of high-precision planning technologies to carry out interventions with high added value in environmental terms.

The frontier of innovation is marked by technology's reduced impact and by its remodulation, capitalizing on intangible resources and the design capacities for transforming the built environment.

A new balance between development and ecosystem requires revising the drivers of innovation, in terms of efficiency of the transformations of the manmade environment and of closing the "great divides" (Wealth, Health and Technology) (Stiglitz 2016), which are more and more dramatically clear and dangerous for the geopolitical as well as socio-demographic balance of many areas of the planet (Information Resources Management Association 2020*)*.

The fast growth of human technological development has in fact never permitted the egalitarian redistribution of its benefits throughout the planet's social and geo-demographic stratification but has instead widened the divide: locally among the various social groups and globally among geographically or culturally distant communities (Barca and Lorefice 2021).

The key for human activities to coexist with the environment, then, lies not in perfecting technology as relates to its performance, but in reducing its impact or in diminishing its use—and, as a last resort, in reducing the consumption of resources.

This session would aim to set the boundaries of a new technological and constructive imagination, for a path of innovation oriented towards increasingly low-impact/high-value technologies. This is also to be achieved using precision components for increasingly ZEB—Zero Energy Buildings in order to maintain the management of a **low-tech environment**, possibly capable of autonomously healing the injuries suffered over time, in which technology is not required to go about correcting the distortions produced by human activities.

The new technological imagination needed to achieve the objectives of sustainability and decarbonization of all production sectors requires not only high-value/low-impact technologies but also the formation of soft skills governing the processes and the optimization of competences and technologies, in order to transition from a manmade ecosystem with high technological intensity to an environment managed with low intensity and high efficiency.

34.2 Research Outcomes and Contributions

This session compares R&D models and design strategies for a low-tech environment and advanced, carbon neutral building/plant integration, and policies to regenerate the built environment with low intensity and high environmental and energy efficiency.

This first meeting allowed a variety of research experiences to be compared, which dealt with the complexity of managing the built environment's transformation processes, while examining the issue's various scales, from the urban to that of the building component and of managing the materials.

On the urban scale, certainly of particular interest are the issues of Energy Communities (Paola Marrone and Ilaria Montella—Roma Tre) and of PED—Positive Energy Districts and Neighbourhoods for Sustainable Urban Development (Rosa Romano—University of Florence, Emanuela Giancola—2UiE3—CIEMAT, and Maria Beatrice Andreucci—Sapienza University of Rome). They have also highlighted the importance that the interconnection of every urban planning and construction choice has for the proper commensuration of needs to be met and services to be distributed on the territories.

The issue of the technical policies to guide the process of technical programming and of planning is dealt with in the papers by Claudio Piferi (University of Florence) and Anna Dalla Valle (Polytechnic University of Milan)—two very different contributions, both examining the initial phases by process. The former reconstructs the history and evolution of the leading, multi-year programme for financing university residential construction, a strong example of long-term technical planning that has proved able to evolve in a mature and aware manner, also by learning from its own critical areas, thus bringing about real progress in the technical, environmental, and functional quality of the buildings constructed in the context of that programme. The latter provides an interesting account of the issue of LCA at the initial levels of the construction programming process, as a preliminary form of assessment of the building's technical feasibility.

The paper by Antonella Violano and Monica Cannaviello (University of Campania Luigi Vanvitelli) sets out an example of an experience of integrated management of a public service and of its stock of carbon neutral instruments.

Resilience and emergency are two faces of the fragility of environmental equilibria; on these issues, the paper by Vincenzo Gattulli (Sapienza University of Rome) with Elisabetta Palumbo (Bergamo University) deals with the issues of the resilience of human settlements in one of the most fragile and at-risk environments in the western region of the Indian Ocean, while the paper by Maria Vittoria Arnetoli and Roberto Bologna (DIDA—Department of Architecture, University of Florence) deals innovatively with the issue of post-disaster temporary emergency settlements in terms of impact and of actual circularity of their management.

The session also included numerous presentations that may be framed within the two future projections of the environmental management of constructions, which is to say the sector's digital and green transition; it is highly interesting that a large portion of these papers originates from doctoral research, with young PhD candidates therefore becoming interpreters of the more innovative thrusts within the sector (cf. Irene Fiesoli, University of Florence; Francesco Sommese and Gigliola Ausiello—University of Naples Federico II; Tecla Caroli—Polytechnic University of Milan; Nazly Atta (Polytechnic University of Milan).

The researchers who chose to share their work effectively covered all this session's topics; the only regret is that no deeper analysis has been offered as relates to contexts of energy and economic insecurity, in which the digital and technology divides present fundamental barriers to development, especially at a time like the one we are living in, which might further worsen many people's inability to meet their own energy needs.

34.3 Conclusions

Representing one of the largest economic systems and markets, the European Union has always aimed to play a driving role in supranational environmental strategies. The 8th EU Environmental Action Programme of 2020 (European Parliament and of the Council 2022), which guides European environmental policy until 2030 within the framework of the long-term strategy to 2050, intends to speed the transition towards a climate-neutral economy, an economy efficient from the standpoint of managing resources, that aims to be "regenerative"—which is to say able to restore to the planet more than it has taken from it.

All the documents shared on an international level (UE and UN) recognize that the well-being and quality of life of human beings depend on the health of the ecosystems in which we operate.

The European Green Deal (European Commission 2019) sets six priority goals:

- to achieve the goal of reducing greenhouse gas emissions by 2030 and climate neutrality by 2050;
- to strengthen the capacity for adaptation, strengthen resilience, and reduce vulnerability to climate change;
- to progress towards a regenerative growth model, by dissociating economic growth from the use of resources and from environmental degradation, and speeding the transition towards a circular economy;

- to pursue the goal of "zero pollution", comprising air, water, and soil pollution, and to protect the health and well-being of Europe's citizens;
- to protect, preserve, and restore biodiversity and strengthen natural capital—in particularly the air, water, soil, and forests, freshwater, wetlands, and marine ecosystems;
- to reduce the environmental and climate pressures connected to production and consumption (particularly in the sectors of energy, industrial development, construction and infrastructure, mobility, and the food system).

Unlike the UN's 2030 Agenda which has a global vision more conditioned by the weight of the global south, this document has been written with a view to the European setting, marked certainly by a situation of economic crisis, but just as certainly not comparable with the extraordinary historic phase we have been experiencing in recent years. First the pandemic, followed by the geopolitical clash that is destabilizing the entire continental economy and a large portion of the systems economically interconnected with our own, raise important questions to be answered, and have at last brought to the general attention the backwardness of the implementation of environmental and energy transition policies that scientifically, for the entire community, no longer hold any secrets and are instead still quite far from being actually operative and metabolized by the "market". Precisely when our scientific community met for this conference, the European Council adopted general guidelines on reductions of emissions and on their social impact through the implementation of the package of measures termed "Fit for 55" (European Commission 2021) which is to say to achieve, as an intermediate result, the EU's target of reducing net greenhouse emissions by at least 55% by 2030. The proposed package is instrumental for aligning the regulations of the community and of the Member States with a view to the 2030 deadlines, which is to say that a consistent, balanced framework and a new Social Climate Fund for achieving the EU's climate targets is to be provided (Council of the European Union 2022), able to:

- guarantee a socially fair and proper transition;
- maintain and strengthen the innovation and competitiveness of the EU's industry while at the same time insuring conditions of parity for economic operators in third countries;
- support the EU's leading position in the global fight against climate change.

The urgency of achieving the Green Deal's goals, during a setting of crisis linked to a global economic situation rapidly evolving after the events of 2022, has clearly linked environmental demands with the social equity of the increasingly indispensable ecological and energy transition, in order to keep it from generating new poverties and phenomena of social de-cohesion depriving of the benefits of the ecological transition precisely those layers of the population that will suffer most from the greater costs expected for energy procurement in the near future.

Will it be the war economy to allow us to discover the unsustainable costs of traditional energy sources?

Will it be the scarcity of valuable resources to remind us that the first kilowatt earned is the one not consumed? Will it drive home that the whole quantity of devices we will be forced to cast off will have to be the new mines from which to extract those metals, rare earths, and special alloys that we have discovered originate from once-productive scenarios that are now scenarios of war that can no longer be drawn from?

Anyone who has done research over these past twenty years on the environmental and energy efficiency of the construction sector has seen many developed designs break apart against the ruthless comparison between the high "cost of efficiency" and the low-cost availability of energy and of many raw materials. This has always led to the belief that investing in reducing energy consumption and in diminishing the material footprint of construction was unaffordable.

The sudden unavailability of certain routes for the supply of raw materials, including those produced for decades at an incalculable environmental cost, and the need to differentiate the sources of energy procurement, at last risks creating that culture shock, on top of the economic one, that might allow the construction industry as well to find justification once again in developing efficient and low-impact—and perhaps actually regenerative—solutions as required of us by the **Green Deal**. This will be done by seeking to give back to the planet more than what is taken from it but also by interrupting the taking of many of the natural resources currently deemed still preferable to regenerated ones.

In this phase, so particular as it is, discussion sessions like this one are important for consolidating, beyond the scientific assumption, an increasingly solid awareness of the urgency of implementing, on the territory and in the market of technical operators and above all of public stakeholders, the result of these research efforts through their engineering and industrial development, in order to contribute towards building real energy and environmental communities that are collaborative and regenerative towards the host environment.

References

Barca F, Lorefice F (2021) Disuguaglianze conflitto sviluppo. Donzelli Editore

Butera F (2021) Affrontare la complessità. Per governare la transizione ecologica. Edizioni Ambiente, Italia

Council of the European Union (2022) Interinstitutional File: 2021/0206 (COD)—ST 10775/22—'Fit for 55' package—proposal for a regulation of the European Parliament and of the Council establishing a Social Climate Fund—General approach, Brussels. https://data.consilium.europa.eu/doc/document/ST-10775-2022-INIT/en/pdf

Crutzen PJ (2002) The "Anthropocene". J Phys IV Fr 12:10–11. https://doi.org/10.1051/jp4:20020447

Crutzen PJ (2006) The "Anthropocene". In: Ehlers E, Krafft T (eds) Earth system science in the Anthropocene. Springer, Berlin, Heidelberg. https://doi.org/10.1007/3-540-26590-2_3

Crutzen PJ (2021) The "Anthropocene" (2002). In: Benner S, Lax G, Crutzen PJ, Pöschl U, Lelieveld J, Brauch HG (eds) Paul J. Crutzen and the Anthropocene: a new epoch in earth's history. The Anthropocene: politik—economics—society—science, vol 1. Springer, Cham. https://doi.org/10.1007/978-3-030-82202-6_4

Crutzen PJ, Stoermer EF (2000) The Anthropocene. In: IGBP newsletter, vol 41, pp 17–18. http://www.igbp.net/download/18.316f18321323470177580001401/1376383088452/NL41.pdf
European Commission (2019) Communication from the commission—the European green deal. COM(2019) 640 final Brussels. https://eur-lex.europa.eu/legal-content/EN/TXT/?uri=CELEX%3A52019DC0640&qid=1666537322699
European Commission (2021) Communication from the Commission to the European Parliament, the Council, the European Economic and Social Committee and the Committee of the Regions 'Fit for 55': delivering the EU's 2030 climate target on the way to climate neutrality, COM(2021) 550 final, Brussels. https://eur-lex.europa.eu/legal-content/EN/TXT/?uri=CELEX%3A52021DC0550&qid=1666535452461
European Parliament and of the Council (2022) Decision (EI) 2022/591 of the European Parliament and of the Council of 6 April 2022 on a General Union Environment Action Programme to 2030. Official J Eur Union L 114 65:22–36. https://eur-lex.europa.eu/legal-content/EN/TXT/?uri=CELEX%3A32022D0591
Haff PK (2016) Purpose in the Anthropocene: dynamical role and physical basis. Anthropocene 16:54-60. ISSN: 2213-3054. https://doi.org/10.1016/j.ancene.2016.07.002
Haff PK (2017) Being human in the Anthropocene. Anthropocene Rev 4(2):103–109. https://doi.org/10.1177/2053019617700875
Information Resources Management Association (2020) Wealth creation and poverty reduction: breakthroughs in research and practice. IGI Global
Kuhn A, Heckelei T (2010) Anthroposphere. In: Speth P, Christoph M, Diekkrüger B (eds) Impacts of global change on the hydrological cycle in West and Northwest Africa. Springer, Berlin, Heidelberg. https://doi.org/10.1007/978-3-642-12957-5_8
Stiglitz JE (2016) The great divide: unequal societies and what we can do about them. Norton & Company
Zalasiewicz J, Williams M, Waters CN, Barnosky AD, Palmesino J, Rönnskog AS, Edgeworth M et al. (2017a) Scale and diversity of the physical technosphere: a geological perspective. Anthropocene Rev 4(1):9–22. https://doi.org/10.1177/2053019616677743
Zalasiewicz J, Waters CN, Wolfe AP, Barnosky AD, Cearreta A, Edgeworth M, Ellis EC et al. (2017b) Making the case for a formal Anthropocene epoch: an analysis of ongoing critiques. Newsl Stratigr 50(2):205–226. https://doi.org/10.1127/nos/2017/0385
Zalasiewicz J, Waters CN, Summerhayes CP, Wolfe AP, Barnosky AD, Cearreta A, Crutzen P et al. (2017c) The working group on the Anthropocene: summary of evidence and interim recommendations. Anthropocene 19:55–60. https://doi.org/10.1016/j.ancene.2017.09.001
Ziman J (ed) (2003) Technological innovation as an evolutionary process. Cambridge University Press

Chapter 35
Technological Imagination to Stay Within Planetary Boundaries

Massimo Palme

Abstract Technological imagination has been, until now, a stronger driver of development and has permitted to scale economy and even to obtain increasing returns of investments. However, times are a changing. Humanity faces now societal and environmental changes that are pushing the planet Earth toward a danger zone, overpassing recommended limits for several critical processes, such as bio-geochemical fluxes of nitrogen and phosphorus, greenhouse gases concentration in the atmosphere, biodiversity loss and land use change. The role of technology applied to built environment design should be redefined to stay within the so-called safe operation space for humanity, considering the limited resources we have and the need of low-energy solutions for buildings and cities. This chapter introduces the key concepts for the understanding the new role that we must assign to technological imagination to face the challenge of the Anthropocene epoch and discusses how to achieve the seven transitions objectives for transforming our world in a sustainable way.

Keywords Anthropocene · Safe operation space · Planetary boundaries · Strong sustainability · Kondratiev waves · New economy · Smart city · Environment

35.1 Introduction: The Planetary Boundaries

We live in a transition world. Since the decade of the '70 in the past century, science and society have been focused on global problems, such as global warming, resources depletion, environmental pollution and sustainability of development. Key moments to be remembered are at least: the publication of the book "Limits to growth" in 1972 (Meadows et al. 1962); the publication of the report "Our common future" in 1987 (Brundtland et al. 1987); the creation of the Intergovernmental Panel for Climate Change in 1988 (World Meteorological Organization 1988); the United Nation Convention on climate change in 1992 (United Nations 1992); the adoption of the Kyoto protocol in 1997 (United Nations 1998); the official proposal of the

M. Palme (✉)
Universidad Técnica Federico Santa María, Valparaíso, Chile
e-mail: massimo.palme@usm.cl

E. Arbizzani et al. (eds.), *Technological Imagination in the Green and Digital Transition*, The Urban Book Series, https://doi.org/10.1007/978-3-031-29515-7_35

term "Anthropocene" to refer to our geological epoch (Crutzen and Stoermer 2000; Crutzen 2002) in early XXI century; the proposal of the 17 Sustainable Development Goals (United Nations 2016); and the adoption of the Paris Agreement in 2015 (United Nations 2015). The transition to a sustainable world, a world in which future generations can satisfy their own needs, required of urgent actions to be taken, both under environmental and societal point of view, to face the challenges of the present. Respect to environmental challenges, climate change and other critical processes (Steffen et al. 2015) must be limited to stay in the so-called safe operating space (Rockström et al. 2009a) or within the planetary boundaries (Rockström et al. 2009b) (Fig. 35.1). Respect to societal challenges, the end of poverty and hunger, the provision of wealth to everyone, the right to a safe urban environment and the reduction of inequalities are the most important goals to be reached soon (United Nations 2022).

35.2 A social Dilemma: Growth, Wealth and the Role of Technology

Unfortunately, our world is at the same time a globalized world and a very uneven world. Economic growth is still highly depending on fossil fuel consumption, and the vision of a dominant model in finance and industry has led to an accelerated increase in inequalities, both between countries and inside each country, especially in growing economies. In 1954, the economist Simon Kuznet formulated its hypothesis that development implies an inequality increase during the early stages and a reduction of inequalities during the mature stages (Wang-Erlandsson 2022). However, the Kuznet hypothesis and the proposed curve of development are now under revision (Kuznets 1953), incorporating environmental issues and considering the interrelationships between countries in a very connected world (Jha 1996). Moreover, new economic theories, such as the proposition of Thomas Piketty (Stern 2018), pointed out that inequalities are increasing even in developed countries, due to the new disbalance among capital and labor rentabilities.

Technological imagination can be one of the driving forces to a more sustainable world. Early in the XX century, Russian economist Nicolai Kondratiev (Piketty 2013) formulated a theory of technological "waves" that push forward the economic cycles. He identified and predicted five waves, which already occurred and now can be analyzed.

As pointed out by several academics from different disciplines (as for example the economist Leo Nefiodov and the urban planner Micheal Batty), the next wave is approaching and could relate to aspect such as: environment protection, health, digital transition, smart cities or a mix of them. Nefiodov (Kondratiev 1935) suggest that health care should be the most important economic driver in the sixth wave. Batty (Nefiodov and Nefiodov 2017) think that the smart city and the digital transformation are the main vectors. However, things are much more connected. The pandemic episode of 2020–2021 underlined the existence of a deep link among nature, cities

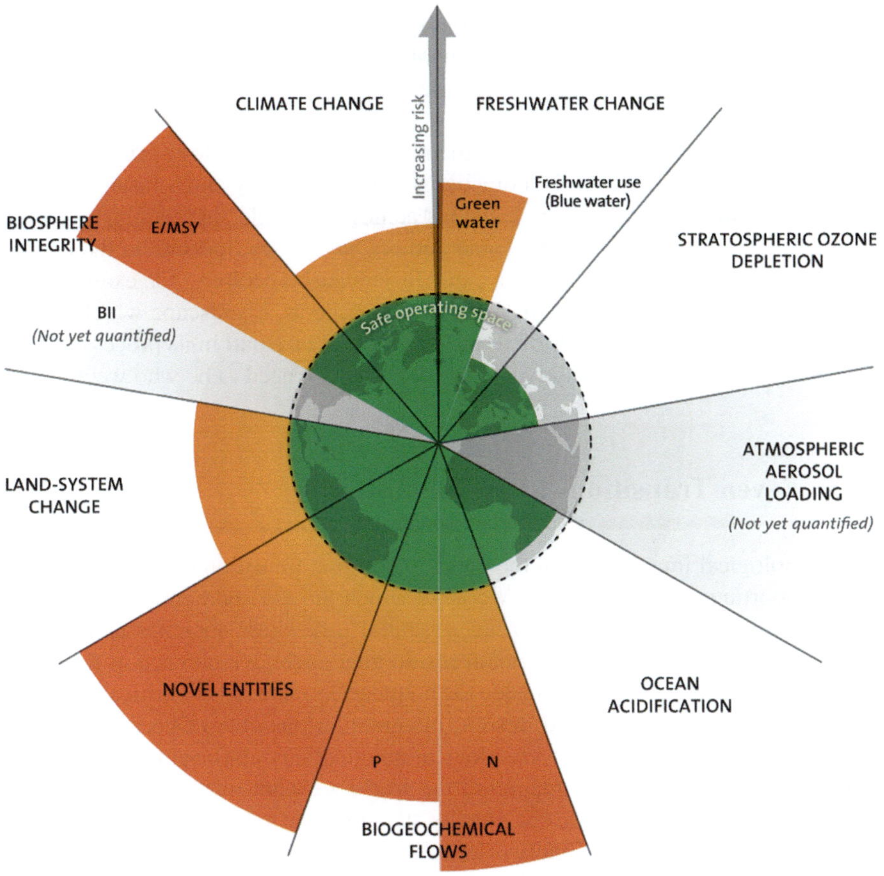

Fig. 35.1 Safe operation space for humanity and critical processes of the Earth system. Image from: Azote for Stockholm Resilience Centre, based on analysis in Wang-Erlandsson et al. 2022 (Wang-Erlandsson 2022), Persson et al. 2022 (Kuznets 1953) and Steffen et al. 2015 (Steffen et al. 2015). Reproduced with permission

and health (including mental health). Environmental issues are entangled with any possible technological innovation, being the biotechnology only one aspect of that complex relation.

However, looking at Batty's proposition and reinterpretation of Kondratiev theory, it can be observed that the new waves have the characteristic to be more and more intense and frequent, pointing out the risk of collapse of the structural stability (Batty 2018) in which we live in. A new economy is also necessary. Sustainability concept can be regarded as weak or strong (Naumer et al. 2010) depending on the interpretation of its components. If an economical benefit can justify a reduction in environmental or social "shares" of sustainability, we are using the weak concept. If, on the contrary, an economical benefit can be achievable only inside a social bearable world, and a social benefit can be achievable only in a healthy environment, we are using the strong concept to define what "sustainable" should mean. Under this perspective, to squeeze productive efficiency to increase rentability of financial investments, it is clearly not a solution.

Jevons paradox (Kauffman 1993) remember us that if technology is used to maximize output power, the result is a drastic increase in entropy and a faster resources depletion. Moreover, no change of scale is achievable while approaching planetary boundaries. For all these reasons, some authors have been working on the ecological meaning of "sustainability" concept. In landscape ecology, for example, Wu (Beckerman 1994) defined the sustainable landscape as a landscape which has the capacity to carry with the needs of all the species that take part in its processes. Such a landscape, complex and adaptive, is the substrate we need to be who we are.

35.3 Seven Transitions to Sustainability

So, technological imagination in the green and digital transition should be regarded as the opportunity to drive us to a world in which the environment is protected by specific policies; economy is a strong, essential tree where societies and cultures can flourish; and diversity is the global result, guaranteeing resilience of the global ecosystem and helping us to stay inside the boundaries of the Earth's critical processes (Fig. 35.2). In architecture and design, imagination has always been the base of development. Today, however, technological imagination must be guided. The vision should be retrospective: first imagining the world we want for the future, and then go back to the present to figure out, step by step, the way to build that desired world. We need for a low-tech, low-energy-dependent, smart built environment to face the challenges of the future. Our buildings and cities are more than the deposition of technomass (Sorrell 2009) on a biological substrate. They are the drivers of our behaviors, showing us the way to follow. Architects and planners have the immense responsibility to make the way to a sustainable scenario, not a business as usual one.

In the '80s, Nobel laureate Murray Gell-Mann (Wu 2013) stated that sustainability will only be achieved by humanity under passing through seven transitions: demographic transition, technological transition, economic transition, social transition, institutional transition, informational transition and ideological transition. All described transitions are necessary to build a world that could be called "sustainable." Population should converge to a fixed limit (someone estimates around 11,000 millions of individuals) and then move around this attractor; technology should help improving quality of life and focus on human health; economy should transform itself (or return to be) into the science of "taking care" of our home; human society must accept its very peculiar character of being at the same time diverse and globalized; institutions have to be changed to adapt us to our new complex world, possibly eliminating older structures as national states and substituting them with both global and regional new institutions; informational transformation has already pushed human possibility to communicate over limits that even Murray Gell-Mann cannot imagine

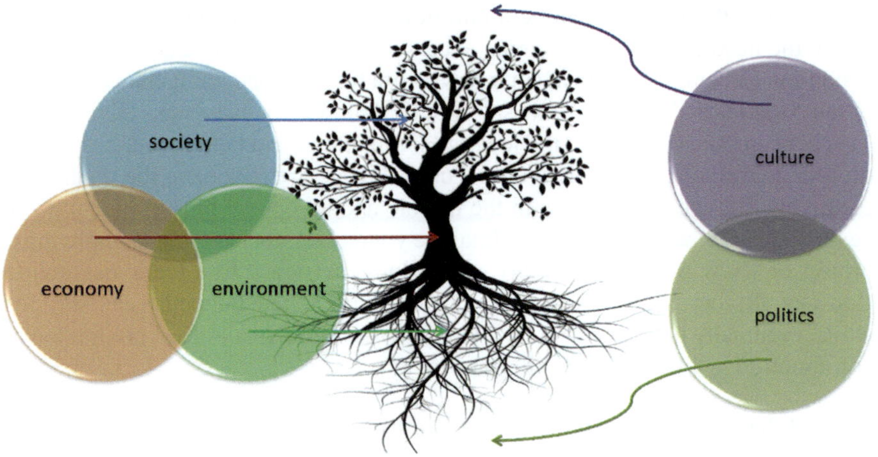

Fig. 35.2 Vison of sustainability as something bigger than a range of opportunities. Drawing by Manfred Max Bergman 2017 (Gell-Mann 1994). Reproduced with permission

in the '80s; ideological transformation is probably the most important transition and the most difficult to be achieved. We still live in a very conservative world, as demonstrated for example by the existence of academic static disciplines; mentors and masters to be served; dictators dominating and sometimes owning entire nations; labor division often following strange class and race criteria; political and economic power concentration in very few hands.

35.4 Conclusion

In this chapter, I have briefly introduced the idea of technological imagination as a fundamental tool for humanity to stay within planetary boundaries. In the past, creativity and innovation have been regarded as important driving factors of economic growth, permitting to shift scales and to even obtain increasing returns (Inostroza 2014). However, that times were privileged times, and now, we face a very different situation. Saturation of markets and resources depletion are envisioning times of transitions and great challenges. As described by David Harvey (Bergman 2017), the global capitalistic economy, based on finance more than on industrial production, is quite close to its end. A new economic thinking and a new ideology are the final points to be reached in the transition to a completely new world.

The role that technology and especially built environment-related technology will have during the transition, an after its completion, is still to be determined in detail. Nevertheless, I can figure out for such technology a very important role in balancing the reduction in excessive consumption of goods, energy, water, food and the increase in social health, happiness and cultural development of humanity. It is not a non-sense: the need to reduce the human excesses justified only by extreme capitalistic vision can be combined with a better way of life for everyone. Indigenous thinkers of South America call it "sumak kawsay" (Gell-Mann 1994), that means "living good." This concept must substitute the idea of "living better and better" that has failed in driving us toward a desirable world. For architects and planners, the challenge is enormous. But we must stay with the trouble and accept the idea to imagine, once again, that a new world is possible. Imagine yourselves, for example, completing the project of "New Babylon," the city of freedom envisioned by situationist architect Constant (Romer 1986) in the '60s, a place where a new humanity can stay for a while, eventually unchained, so far away from the mindless circle of production and consumption in which we are still trapped.

Acknowledgements This work was supported by the Chilean National Agency for Science and Development (ANID), under FONDECYT Project number 1200275.

References

Batty M (2018) Inventing future cities. MIT Press, Cambridge
Beckerman W (1994) Sustainable development: is it a useful concept? Environ Values 3:191–209
Bergman M (2017) A revolutionary welcome message for a world in transition. In: World sustainability forum 2017, Cape Town, South Africa
Brundtland G et al (1987) Our common future. Oxford University Press, Oxford
Crutzen P (2002) Geology of mankind. Nature 415:23. Retrieved from www.nature.com/articles/415023a. Accessed on Oct 2022
Crutzen P, Stoermer E (2000) The Anthropocene. Glob Change Newsl 41:17. www.igbp.net/download/18.316f18321323470177580001401/1376383088452/NL41.pdf. Accessed on Oct 2022
Facultad Latinoamericana de Ciencias Sociales (2014) Dossier: en busca del sumak kawsay. Íconos. Rev Cien Sociales 42. Quito 2014
Gell-Mann M (1994) The quark and the jaguar: adventures in the simple and the complex. Henry Holt and Company, New York
Harvey D (2014) Seventeen contradictions and the end of capitalism. Profile Books, London
Inostroza L (2014) Measuring urban ecosystem functions through "Tecnomass"—a novel indicator to assess urban metabolism. Ecol Ind 42:10–19
Jha S (1996) The Kuznets curve: a reassessment. World Dev 24(4):773–780
Kauffman S (1993) The origin of order: self-organization and selection in evolution. Oxford University Press
Kondratiev N (1935) The long waves in economic life. Rev Econ Stat 17:105–115
Koolhas R (2016) Constant: new Babylon. Hatje Cantz Publisher
Kuznets S (1953) Economic change: selected essays in business cycles, national income, and economic growth. W.W. Norton, New York
Meadows D et al. (1962) Limits to growth. Universe
Naumer H, Nacken D, Scheurer S (2010) The sixth Kondratieff—long waves of prosperity. www.allianzglobalinvestors.de/capitalmarketanalysis
Nefiodov L, Nefiodov S (2017) The sixth Kondratieff: a new long wave in the global economy. CreateSpace Independent Publishing Platform
Piketty T (2013) Capital in the twenty-first century. Harvard University Press

Rockström J et al. (2009a) A safe operating space for humanity. Nature 461:472–475. Retrieved from www.nature.com/articles/461472a. Accessed on Oct 2022
Rockström J, Steffen W et al. (2009b) Planetary boundaries: exploring the safe operating space for humanity. Ecol Soc 14(2):32. Retrieved from www.ecologyandsociety.org/vol14/iss2/art32. Accessed on Oct 2022
Romer P (1986) Increasing returns and long-run growth. J Polit Econ 94(5):1002–1037
Sorrell S (2009) Exploring Jevons' paradox. In: Energy efficiency and sustainable consumption. The rebound effect. Springer
Steffen W et al. (2015) Planetary boundaries: guiding human development on a changing planet. Science 347:1259855. https://doi.org/10.1126/science.1259855. Accessed on Oct 2022
Stern D (2018) The environmental Kuznets curve. In: Reference module in earth systems and environmental science. Elsevier
United Nations (1992) Framework convention on climate change. Retrieved from https://unfccc.int/resource/docs/convkp/conveng.pdf. Accessed on Oct 2022
United Nations (1998) Kyoto protocol to the United Nations framework convention on climate change. Retrieved from https://unfccc.int/resource/docs/convkp/kpeng.pdf. Accessed on Oct 2022
United Nations (2015) Paris agreement. Retrieved from https://unfccc.int/files/essential_background/convention/application/pdf/english_paris_agreement.pdf. Accessed on Oct 2022
United Nations (2016) Sustainable development goals report 2016. Retrieved from https://unstats.un.org/sdgs/report/2016/. Accessed on Oct 2022
United Nations (2022) Sustainable development goals report 2022. Retrieved from https://unstats.un.org/sdgs/report/2022/. Accessed on Oct 2022
Wang-Erlandsson L et al. (2022) A planetary boundary for green water. Nat Rev Earth Environ 380(3)
World Meteorological Organization (1988) Report of the first session of the WMO/UNEP Intergovernmental panel on climate change. Geneva, 9–11 Nov 1988. Retrieved from www.ipcc.ch/site/assets/uploads/2018/05/first-final-report.pdf. Accessed on Oct 2022
Wu J (2013) Landscape sustainability science: ecosystem services and human well-being in changing landscapes. Landscape Ecol 28:999–1023

Chapter 36
Quality-Based Design for Environmentally Conscious Architecture

Helena Coch Roura and Pablo Garrido Torres

Abstract The discussion is focused on the view that technological innovation, in order to move toward a green transition, must take advantage of the opportunities and focus on quality-based design rather than quantity-based design. It is argued that true technological innovation requires a major mental shift. It is not about making processes better and more efficient, but about rethinking them outright. Three strategies are proposed at the urban and building level that support this higher quality as a way to go toward a sustainable approach as stated in the Brundtland report. The strategies proposed to reduce the amount of resources required to satisfy citizens' demands are based on a more in-depth study of their real needs. Innovative design and technological solutions could lead us to a healthier and more environmentally friendly life.

Keywords Environmental impact · Energy · Architectural urban design

36.1 Introduction

The Brundtland report (United Nations 1987), in 1987, defines sustainable development as:

"Humanity has the ability to make development sustainable to ensure that it meets the needs of the present without compromising the ability of future generations to meet their own needs. The concept of sustainable development does imply limits—not absolute limits but limitations imposed by the present state of technology…" with a broad vision that at no time denies growth. In the definition of sustainable growth as

"Growth has no set limits in terms of population or resource use beyond which lies ecological disaster. Different limits hold for the use of energy, materials, water, and land. (…) The accumulation of knowledge and the development of technology can enhance the carrying capacity of the resource base," it points to technology as the vehicle capable of achieving this development.

H. C. Roura (✉) · P. Garrido Torres
Universitat Politécnica de Catalunya, Barcelona, Spain
e-mail: helena.coch@upc.edu

E. Arbizzani et al. (eds.), *Technological Imagination in the Green and Digital Transition*, The Urban Book Series, https://doi.org/10.1007/978-3-031-29515-7_36

The current meaning of technological innovation and the possibilities it has for generating growth that is more about quality than quantity are discussed.

Architecture deals with, works with, and organizes these resources, matter, energy, water and land, and the use of new technologies must be promoted from the point of view of making better use of these resources. The challenge is to satisfy the needs of citizens by offering a design based on the search for quality.

When considering the impact of architecture on the environment, the first question is what impact means.

Greenhouse gas emissions directly affect the earth's atmosphere and are directly related to energy consumption. However, the energy consumption of users comes from the demand needed to live in comfort. One could open a discussion on what the term "comfort" means. Huizinga (Huizinga 1996) pointed out that in the Middle Ages, comfort was merely a matter of survival, something that today differs greatly from our models, and above all that it is not the same in all countries.

Cities, buildings, etc. designed by architects are responsible for the demand, generated in the early stages of the design process.

The energy demand generated by the architecture directly influences consumption, although user management can vary the final amount greatly. However, if the initial design does not take into account that the users can make good use of the resources and forces them to spend large amounts of energy, it can be considered an initial design mistake.

It is evident that if demand is high and, consequently, consumption rises, the impact on the environment will be greater regardless of the type of energy source used. A discussion could also be opened on the energy sources used and their economic and environmental impact on the environment. A debate could also be opened on the energy sources used and their economic and environmental impact on the environment. Whatever the energy source, from the most polluting to the least polluting, the reduction strategy will have less impact on the environment.

We, as citizens, must put pressure on governments, energy companies, and all involved agents to adopt cleaner types of energy. We, as users, must find ways to efficiently manage the energy needed for our daily lives. Architects, engineers, and designers must take advantage of the opportunities and focus technological innovations on a green transition.

The same approach can be used when considering the other resources mentioned: water, materials, and land.

36.2 Discussion

Strategies applicable to architectural and urban design, acting on quantity, intensity, and sharing, are proposed.

The first strategy is based on viewing growth as an increase in quality rather than quantity.

When industry has made the choice to reduce the physical dimension, be it volume, weight, material, etc., and has made investments in improving design, results have been achieved that are on everyone's mind.

Excellent examples of this decision can also be found in urban and architectural decisions such as the dwellings designed by Josep Llinas in Barcelona, where he opted to manage the buildability in such a way as to allow interstices and spaces between the buildings. The final quality of the dwellings is unquestionable, as they all have exterior windows. The dwellings not only allow natural light to enter and the possibility of cross ventilation, but the decision also affects the urban space, managing to sponge a very narrow street in a dense urban area of Barcelona's old town.

The second strategy to grow in quality is based on adapting the intensity of the resources used in relation to the needs of the space. Would it make sense to build a holiday home in a paradisiacal location and lock yourself in an airtight box? Jon Utzon's house in Majorca is located on the seafront, on a cliff. Enjoy the smell of the sea, the breezes, and the sound of the waves. The project creates a sequence of intermediate spaces, porches, balconies, etc., enclosing only certain rooms. The dining room and some lounges are outdoor spaces, protected from the sun but in constant contact with nature.

In less benign climates, similar actions can also be implemented, as in this school in the UK, by Alsop Architects, where only the classrooms are air-conditioned and a space in between, protected from the rain, is left for children's play.

The presence of intermediate spaces, with less energy demand to acclimatize them and less material resources for their construction, is a strategy that is increasingly used by other architects such as Lacaton and Vassal or Glenn Murcutt.

The third strategy is based on the idea of sharing, uses, materials, and functions. Does it make sense to enjoy an individual swimming pool, or one by family?

The trend toward an increasingly individual life leads us to want to have a swimming pool, television, or garden for each one, leaving aside the possibilities of sharing experiences and forgetting its social function. The Arab baths and the Roman Thermae had a function, beyond the bath, surely very similar to the role currently played by sports clubs or modern spas, social meeting points today.

This discussion has focused on considering that technological innovation, in order to orient itself toward a green transition, should take advantage of the opportunities it offers us to commit to focusing on the design of spaces based on quality instead of the quantity of resources necessary for a healthy life and environmental friendly.

References

Huizinga J (1996) The autumn of the middle ages. University of Chicago Press (First published in 1919)

United Nations (1987) Report of the Brundtland commission "our common future". Presented to the general assembly on 4th Aug 1987, New York

Chapter 37
Digital Transformation Projects for the Future Digicircular Society

Irene Fiesoli

Abstract The rapid technological development leads us to identify innovation with technology itself. This becomes the core piece of the innovation process in all sectors. In reality, digital transformation has the power to change the meaning of things (Epifani in Digital sustainability: why sustainability cannot disregard digital transformation. Digital Transformation Institute, Rome, 2020) and therefore needs to cultivate a strategic vision of systems and scenarios that can be implemented only through creative design. Designers, thanks to their ability to see, show, predict (Zurlo in Le strategie del design. Disegnare il valore oltre il prodotto. Libraccio editore, Milan, 2021), and design the future, have the role of meeting the challenges posed by digital evolution. This dichotomy between digital and sustainability is analyzed in the article thanks to the workshop "Space Transformation/Industrial Living Environment," a pilot project for the valorization of productivity in the Valdelsa Senese area that involves, in interdisciplinary groups, students from the various design fields of the School of Architecture of the University of Florence. Another example of planning is the project SMAG—SMArt Garden (Tuscany Region Call RSI—POR FESR 2014–2020), which develops a product-service system equipped with an advanced technological set-up able to control vital parameters of public or private green spaces, using the Internet of things. These examples underline how the physical and digital worlds are interfacing more and more and getting closer. In this scenario, the role of the project is even more important because it allows to manage and direct the innovation and change processes in the direction of a "digicircular" transformation (Epifani in Digital sustainability: why sustainability cannot disregard digital transformation. Digital Transformation Institute, Rome, 2020).

Keywords Digital transformation · Circular environment · Service design · Design for sustainability · Interdisciplinary approach

---

I. Fiesoli (✉)
University of Florence, Florence, Italy
e-mail: irene.fiesoli@unifi.it

E. Arbizzani et al. (eds.), *Technological Imagination in the Green and Digital Transition*, The Urban Book Series, https://doi.org/10.1007/978-3-031-29515-7_37

37.1 Introduction

The rapid technological development leads us to identify innovation with technology itself, wherein technology becomes the core piece of the innovation process in all sectors. But in reality, digital transformation is a process of radical redefinition of operational conditions both in technological systems and the social and economic ones.

It has the power to change the meaning of things (Epifani 2020), and for this, precise reason must also be analyzed from the societal point of view, in order to understand both the positive and negative implications for the society, the ecosystem, and people.

It appears necessary to form a strategic vision of the system and potential scenarios, which can only be achieved through a creative process. The designers, thanks to their ability to see, show, foresee (Zurlo 2012), and plan the future, will thus have the role of gathering all the challenges posed by the digital age, foreshadowing alternative scenarios, and creating new approaches to innovation capable of responding to the increasing demand for competitiveness and sustainable development.

This dichotomy between the digital world and sustainability, and the capacity to develop strategic projects precisely at the intersection between these two worlds was tested both in the "Space Transformation/Industrial Living Environment" workshop, supported by the School of Architecture at the University of Florence, as well as in the Research and Development project "SMAG—SMArt Garden," funded by the Region of Tuscany. These examples emphasize the interest in drafting a usable and replicable path—even in other territorial contexts—for the development of a future design that would be more thoughtful and sustainable.

37.2 Transformation of Spaces in the Industrial Context: An Interdisciplinary Workshop Between Design, Urban Development, Landscape Design, Urban Planning, and Architecture

Parting from the world of education, the "Space Transformation/Industrial Living Environment" workshop is a pilot project on productive area enhancement. In particular, the identified area is the Valdelsa in the Region of Siena.

Territorial competitiveness is increasingly tied to their ability to project a clear sense of identity to the outside world—easily communicable—and to reinforce the quality of its overall image, starting from the local identity in all its different forms. In this way, the Valdelsa represents a particular mixture between industrial areas and the agricultural-touristic, between construction and nature, with a high quality of spaces and images. It has thus become apparent that a strategic reinforcing of the area's image may be in order. An image tied to the aspect of urban development, the quality of constructions and green areas, to the furnishings' elements and communication.

Furthermore, the close relationship between the urban and industrial contexts in this area has created a particular need for a reimagining of living conditions with more sustainability, inclusivity, and accessibility.

All these elements have led us to form an interdisciplinary workshop that would include students from various disciplines within the University of Florence: design, urban development, landscape design, urban planning, and architecture. The methodology used was one comparable to design thinking, which saw its first phase of research-action, characterized by a high level of the scientific method. The following phase was one of immersion within the territorial context, in order to analyze its various stakeholders and directly interact with the potential users, understanding their individual relationships, problems, and needs. Finally, we began the phase of converging design, which led to the development of interdisciplinary strategic projects.

The groups of designers, each composed of at least one student from all of the participating disciplines (design, urban development, landscape design, urban planning, and architecture), had as their principal objective to support and assist in the evolution of the material and non-material assets of the Valdelsa in a direction that would reinforce this area's capacity to attract and maintain the essential components of the demands of the territory—said demands coming from both physical persons and economic organizations—for a sustainable development of the area itself. Five possible design trends have been developed, with various groupings therein, as detailed below:

The project would broaden the traditional concept of tourism, forming unexplored ties with users of unconventional interests (film commission, urban exploration, wedding tourism). This would be achieved through an identification of alternative locations (thematic routes), where narrative and instructional devices (totems) would be installed, equipped with QR codes to access an application specially created for this project.

(1) Sharing territorial accessibility—"Valdelsa Unconventional," by Fabiana Sannino, Barsanti Mirko, Ann Kereselidze (Fig. 37.1).

The goal of the project is to create a network of routes throughout the Valdelsa territory, realized via resting spots, shaded areas, and panoramic points overlooking landscapes that would allow the public access to extraordinary views of a vast and varied landscape, while directing the visitors' means of access (no longer constraining them to the use of motor vehicles), simultaneously including and protecting the rural areas, farms, and any fauna present.

(B) Finding meeting points—"LOTTO," by Valeria Labruna, Francesca Matteoni, Vittorio Scarnati (Fig. 37.2).

This project aims to create meeting points for the employees of the companies and the local population, thus forming a meeting between people and nature. The name of the project is due to the shape of the trail that connects all the points of interest, which

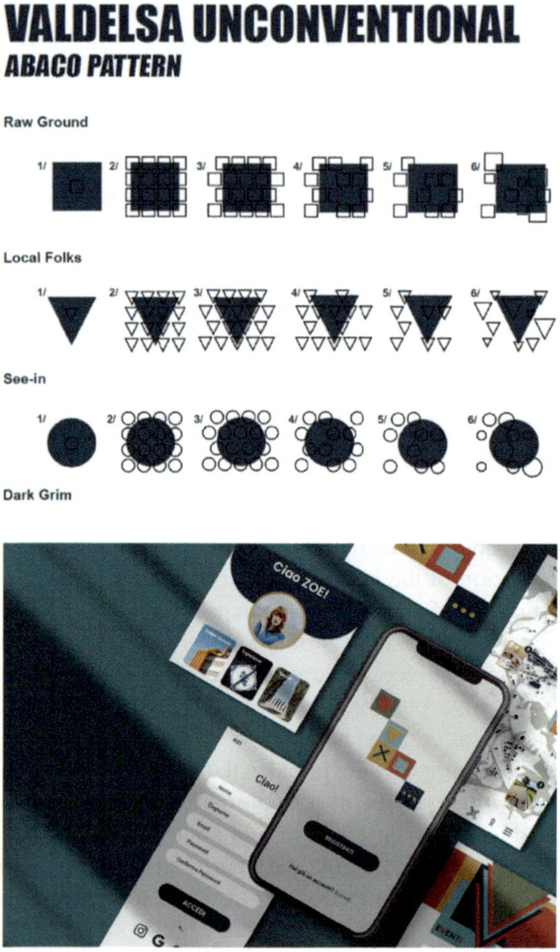

Fig. 37.1 Patterns used for the wayfinding of the project and its application (Sannino et al. 2021)

is shaped like the number 8 (OTTO in Italian). Nature, the sense of community, the union between business and territorial identities are some of the elements present in each meeting point, and together they aim to propose a route that would promote wellbeing and properly value the area.

Soundscape—translated into Italian to mean something closer to sound landscapes.

The term was introduced in order to represent the sound environment which we are immersed in. Sound influences us psychologically, even when we do not realize it, which is why introducing a sensory experience in a working context such as the one in the Valdelsa could improve the wellbeing of the worker, as well as representing a territorial value.

Fig. 37.2 Map with areas of design intervention in order to enhance green spaces (Labruna et al. 2021)

(C) Nurturing healing processes—"Square Root," by Elena Dionori, Qianwen Diao, Margherita Poli, Silvia Roseto (Fig. 37.3).

There is an osmotic relationship of mutual contamination between all users of this space, an equal interaction between what is constructed and the land, where all that is taken from nature is restored to it. A living nature, which changes and grows without restrictions, colonizing space autonomously. Square Root is a project, which, like the elementary squared form, aims to make the area into a unified organism, where all composing parts collaborate and co-exist. Thus, it plans activities based on the design of the landscape, forming its fabric via small islands, which, placed within the system, form a park that adapts to the conforming of the territory and self-maintains over time.

The phases of naturalization would be: planting of various arboreal essences, the phase of growth and flowering of the various trees and plants, creation of islands,

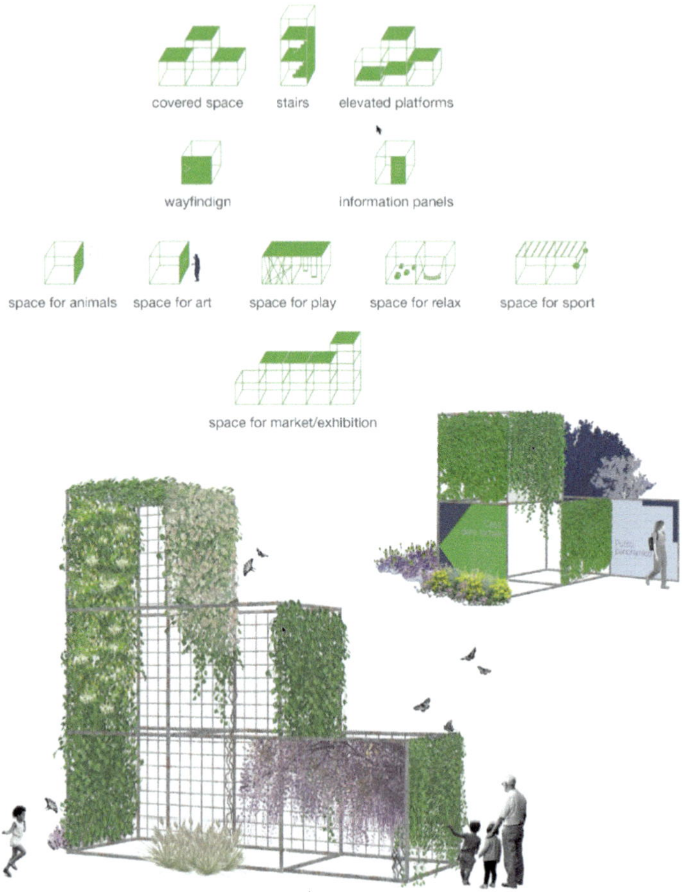

Fig. 37.3 Modular system of urban furniture for the industrial area of Casole d'Elsa and settings of some modules (Dionori et al. 2021)

defined natural points within the wooded areas, with the aim of maintaining equilibrium between the empty and populated areas within the fabric. The choice to insert nature within crevices of a densely populated and strongly compromised area would go beyond the limits of urban center/urban countryside and would instead would put the two in a close union.

(D) Narrating stories and emotions—"ConTact," by Camilla Canessa, Martina Mastropietro, Alessia Pasqualetti (Fig. 37.4).

The project was born with the aims to introduce the companies to the community by recreating a physical, multisensory, and tactile dimension of reality. This is why they have designed a festival inclusive to all, that would cover work, society, play, and respecting ourselves and the planet. The festival would consist of a collection of separate interventions that would serve as urban acupuncture, with the aim of

Fig. 37.4 Coordinated images of the Festival "contatto" visible in the poster and brochure on the right. At the top, a map with the points of intervention to redevelop the environments just in view of carrying out the activities of the festival (Labruna et al. 2021)

promoting the wellbeing of the individual and the community, creating areas that make an industrial zone accessible to all, not just during the working hours. From tactile playgrounds where the youngest could find multisensory and creative stimulation, to areas both within and outside companies where workers (and not just them) could find areas to unplug and reconnect with themselves, or even socialize. Storytelling and emotions would be the key to a strategy of a meeting between territory and areas of productivity.

(E) Weaving the web of the territory—"Textere," by Asia Ferri, Ilaria Fiorentino, Giorgia Giovi, Antoine Tallarico (Fig. 37.5).

The objective of this activity would be to mend and reinforce the system of connections and exchange between isolated areas of various natures that make up the land, aiming to contrast fragmentation and its negative effects on the landscape. The project is highly aware of the industrial nature of the place and does not aim to disturb it, but to enhance it, offering the possibility to broaden the feeling of belonging among all the stakeholders within the Valdelsa. The project for landscapes of limits does not aim to recreate the conditions of a closed city and its borders, but to instead revolutionize the concept of limits, wherein they would no longer be the separators of space, but generators of relations and opportunities. In this view, the landscape of limits becomes a point of interface and suturing between two opposing realities, a mediating space that takes the characteristics of both, and then constructs a third dimension in which the different identities overlap.

Fig. 37.5 Scheme of plant structure intervention: expanding public space (on the right), creating urban gardens and rain gardens (on the left) (Ferri et al. 2021)

Thus, the land becomes a design space, to represent the social relations of the various players, and as the engine for a network of interconnections between the places in which those players are found (De Matteis and Guarrasi 1995). In the last few years, schools of thought have formed that have posed factors such as creativity, the arts—even artisanal ones—and the culture at the center of a possible new entrepreneurial development in the areas. In fact, culture is the engine of creativity, and creativity on the other hand is the basis of all social and economic innovation (Santagata 2009) that could stimulate research and investment in the technological field. It is precisely this productive and creative regional network that led to that competitive national advantage termed Made in Italy, which is understood as a productive and creative capacity not exclusively identified in the production of material goods but also in the socio-cultural systems tied to the historical-architectural beauty, cultural and environmental factors, such as the eno-gastronomic culture or the production of foodstuffs. This is an intangible heritage that does not find its roots in a single company, but in the totality of the territorial system (Maffei and Simonelli 2002).

This is the reflection upon which the concept of territorial capital is based, defined as the system of elements—tangible and intangible ones—that a land has at its disposition, understood as elements of strength or true, actual bonds (Farrel et al. 1999). The territorial capital thus represents a departure point for territorial innovation processes, where starting from an analysis of territorial capital, one might begin to promote developmental activities which would aim to design sharing spaces for the large group of political, economic, and social players which all act at the local level (Fig. 37.6).

Today, we talk about a territorial management style that would put together decisions aimed at empowering infrastructure and general territorial services, so as to favor industrial and agricultural development and new seats of productivity and tourism, without harming the environment. Consequently, we are developing a new economic model that aims to internalize the demands coming from the ground up, moved by strong ethical motivations and the awareness that certain practices need to change. The green economy seems to be a good response to the demands for change. This economy would need to both reduce the environmental impact of production

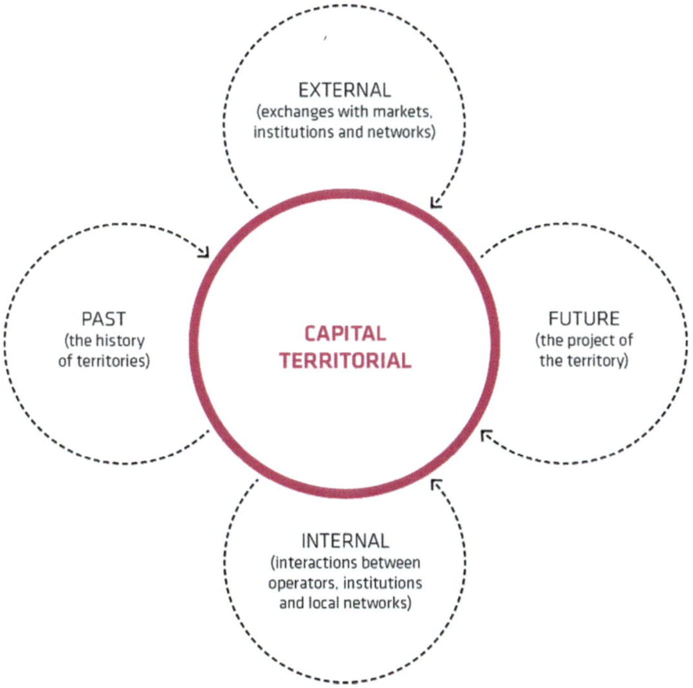

Fig. 37.6 Representation of territorial capital (Farrel et al. 1999)

and increase the level of employment. For this, the economy must become an intelligent one that produces goods and services which could improve the quality of life and have a lower impact on the environment because they were produced mindfully, taking into account their full life cycle.

"The new politics would need to have the characteristics of systems that fulfill operativity and productive as well as infrastructural investments on a local, urban, regional, national, and international level; complementary contributions from the systems of SMEs and large company campuses connected by networks of innovators; an increasing role of universities and public research on large projects based on the social, cultural, and economic peculiarities of the territories." (Cappellin et al. 2017, p 41). We presuppose an environment of cultural integration and economic collaboration in which we might assert innovative practices for the process and transformation of said environment, which would be founded on a kind of multisectorial, integrated, and interdisciplinary approach (Morelli and Sbordone 2018).

37.3 A Smart System for a Dialogue with the Vegetation of the Cities: The SMAG Research Project

The research project used as a case study is SMAG—SMArt Garden (Regione Toscana Bandi RSI—POR FESR 2014–2020). The partnership is made up of Nuvap (project leader), UpGroup and Travertino Sant'Andrea, Arredo di Pietra, as scientific partners from the University of Florence—DIDA Department of Architecture of the University of Florence and GESAAF Department of Forest Sciences, Consortium Ubiquitous Technologies—Cubit, Co-Robotics.

The project (Fig. 37.7) develops a system based on products and services equipped with an advanced technological set-up that is capable of controlling vital parameters of public or private green spaces—temperature, humidity, irrigation, activity of pathogen agents and harmful animals, pollution, or environmental benefits—using the Internet of things. In particular, thanks to the development of a multisensor system and a cloud platform for the management of the detected data, it is possible to intervene on factors such as: health of the garden and of the people who frequent it, irrigation, lighting, video control, anti-intrusion, anti-diffusion, loads and alarms, atmospheric and sound pollution, and more generally on the overall sustainability of the system. By means of particular sensors positioned on plants or in their vicinity, it is possible to concretely monitor different aspects of the life of a garden, such as: external environmental conditions (temperature, humidity, pressure, CO2, particles); soil conditions (humidity, density, PH, organoleptic composition); conditions of the plants (state of growth, presence of problematic elements such as insects, poor structure of the plant); remotely accessible information, images and noises; and implementation of actions through electronic drives (irrigation, soil, and plant nourishment).

Fig. 37.7 SMAG interconnections (Alessio Tanzini, Marika Costa, University of Florence, Department of Architecture, Laboratory of Design for Sustainability 2020)

These particular sensors have been inserted into intelligent products such as furniture systems and outdoor accessories in stone material made by two Tuscan companies: Travertino Sant'Andrea and UpGroup. The sensors and drives send data to a specific platform through a control unit, using different access technologies, wireless and wired.

The management platform collects the data coming from the control unit, records them, and analyzes them through advanced algorithms that enhance the performance of the green spaces monitored, allowing the problems of the maintenance processes of these spaces, whether public or private, to be managed in a predictive and systematic way. Through this control app that connects with the control unit, the various products become effectively intelligent, able to relate to the various actors who interact with the system: on the one hand with the maintenance technician regarding aspects concerning health and maintenance of greenery; on the other hand with people, from an emotional/experiential point of view—by sensitizing them to the green space they are living in and to how much that particular place contributes to the improvement of our ecosystem (Marseglia 2020, pp 216–229).

The SMAG project brings out the enabling role of technologies, and platforms in particular, in accompanying and supporting innovative processes not only through a technological push but also the social and economic one. With a view to the general sustainability of the ecosystem, the SMAG project falls within what is called the platform economy, which stimulates the formation and continuous growth of a number of nodes—or subjects—suppliers of relevant information populating the overall system, giving life to a "fragmented society" or a fragmented ecosystem, in which fragmentation is understood positively as polarization and unequal distribution of social and economic conditions and opportunities (Guarascio 2018).

37.4 Conclusions

Given the discoveries and the practical examples reported in support of the thesis, it is possible to define today's reality in an almost "biological" nature that developed a neuronal system of connections between things and people that has never existed before (Zannoni 2018). A hybrid ecosystem, despite the fact that its evolution and transformation is still ongoing, manages to expand the concept of the network to define it as a system of synapses in which everything is potentially connected and interactive.

However, the environment must not succumb here, but rather it must become the pin that makes this progression work, because as we can see from the examples provided: without land, there is no identity, and without identity, there is no context; therefore there are also no stakeholders to analyze, let alone connect. The land functions as an anchor that allows all innovations to be brought to reality and made applicable, avoiding parts of the negative effects inherent in new technologies.

Any innovation, even technological and digital, without a land for application and an ultimate goal aimed at sustainability, today has no point, if not in a dimension of mutual co-existence. In the future, therefore, to develop a society and a territorial system aimed at sustainability without neglecting the component of innovation, including the technological and digital ones—especially in relation to production processes—it will be important to try to work to create concrete bridges between these two spheres. Technology should not be demonized but used as a strategic driver aimed precisely at improving our ecosystem with a view to sustainability, both environmental, and social and economic. Technological tools as concrete mediums, yet to be tested in their infinite potential, can bridge the gap, make processes more sustainable and users more aware—as well as active—even in complicated and technologically complex systems such as the smart system that is currently being formed and that will lead us, through a global redesign, toward the future Society 5.0 (Ruffinoni 2020), also termed the Super Smart Society (Takahashi 2018).

This consideration together with the examples provided underlines how the physical and the digital world are interacting in an ever-closer way and the boundaries between the two are becoming more porous. In this view, the role of the project is even more important since it allows us to manage and direct the processes of innovation and change in sustainable directions, in which the digital can become the ally and engine of a "digicircular" transformation (Epifani 2020). Thus, the two fundamental pivots of contemporary innovation and scientific debate are to be united: digital transformation and circular economy, in a single word that aims to make one become the conceptual as well as design support of the other, in a new, desirable, and certainly more sustainable harmonic whole.

References

Arquilla V, Simonelli G, Vignati A (2005) Design, imprese, distretti. Un approccio all'innovazione. Edizione POLI.design, Milan

Cappellin R, Baravelli M, Bellandi M, Camagni R, Capasso S, Ciciotti E, Marelli E (2017) Investimenti, innovazione e nuove strategie di impresa. Quale ruolo per la nuova politica industriale e regionale?, Egea, Milan

De Matteis G, Guarrasi V (1995) Urban networks, Pàtron, Bologna.

Epifani S (2020) Digital sustainability: why sustainability cannot disregard digital transformation. Digital Transformation Institute, Rome

ESPAS—European Strategy and Policy Analysis System (2017) Tendenze globali fino al 2030: l'UE sarà in grado di affrontare le sfide future? Lussemburgo

European Commission (2017) White paper on the future of Europe. Reflections and scenarios for the EU27 by 2025. COM(2017) 2025, Bruxelles

Farrel G, Thirion S, Soto P (1999) La competitività territoriale. Costruire una strategia di sviluppo territoriale alla luce dell'esperienza LEADER", Innovazione in ambito rurale, Quaderno n. 6, Fascicolo 1, dicembre, 1999. Retrieved from https://ec.europa.eu/agriculture/rur/leader2/ruralit/biblio/compet/contents.htm

Grandinetti R, Rullani E (1996) Impresa trasnazionale ed economia globale. Carocci, Rome

Guarascio D (2018) Report sull'economia delle piattaforme digitali in Europa e in Italia, Inapp

Maffei S, Simonelli G (2002) I territori del design. Made in Italy e sistemi produttivi locali. Il sole 24 ore, Milan

Magnaghi A (2000) Il progetto locale. Bollati Boringhieri, Turin

Mantellini M (2018) Bassa risoluzione. Einaudi, Turin

Manzini E (2018) Politiche del quotidiano: progetti di vita che cambiano il mondo. Edizioni di comunità, Rome

Marseglia M (2020) SMAG—Smart Garden. In: Lotti G (eds) Impresa 4.0/sostenibilità/design. Ricerche e progetti per il settore Interni. Franco Angeli, Milan, pp 216–229

Martini ER, Sequi R (1995) La comunità locale. Approcci teorici e criteri di intervento. Carrocci Editore, Milan

Montanari F (2011) Territori creativi. L'organizzazione delle politiche a supporto della creatività. Egea, Milan

Morace F (2013) Che cos' è il future. Mind, Milan

Morelli N, Sbordone MA (2018) Il territorio delle relazioni Il Design infrastructuring per i contesti locali, MD Journal, n. 5 – Design e Territori, Ferrara, pp 176–185

Ruffinoni W (2020) Italia 5.0. Un nuovo umanesimo per rilanciare il Paese. Mondadori Electa, Milan

Rullani E (2004) La fabbrica dell'immateriale. Produrre valore con la conoscenza. Carocci, Rome

Santagata W (2009) Libro bianco sulla creatività. Per un modello italiano di sviluppo, Università Bocconi Editore, Milan

Takahashi T (2018) Behavioral economics of addiction in the age of a super smart society: society 5.0. Oukan J Transdisc Fed Sci Technol 12(2):119–122

Thomson M, Koskinen T (2012) Design for growth and prosperity. DG Enterprise and Industry, EU
Zannoni M (2018) Progetto e interazione. Il design degli ecosistemi interattivi, Quodlibet, Macerata
Zurlo F (2021) Le strategie del design. Disegnare il valore oltre il prodotto. Libraccio editore, Milan

Chapter 38
The Regulatory Apparatus at the Service of Sustainable Planning of the Built Environment: The Case of Law 338/2000

Claudio Piferi

Abstract It is wrongly assumed that the environmental sustainability of the building organism is only achievable thanks to the contribution of the systems and their ability to reduce harmful emissions and generate energy from alternative and natural sources. So, system projects have assumed an increasingly considerable importance both in terms of the quantity of documents and the cost of the building. The quantity and complexity of the most recent plant engineering solutions amplify the difficulty of dialogue between the different design levels (architectural, structural, and system design) forcing the professionals involved to compromise that end up disregarding the expected quality. Although it is now clear that the design levels must progress hand in hand from the first hypotheses, and that all must contribute equally to the overall sustainability of the intervention, this does not always happen by preferring to derogate from the system designers the choice of environmentally sustainable solutions. In summary, more and more, often we rely on the technical solutions of the machines used, rather than on the technological qualities of the project. This inevitably involves problems in the construction, operation, and eventual decommissioning phase of the building, especially in the public sphere where the low economic resources of the contracting stations are increasingly used in the purchase, maintenance, management, and disposal of plant engineering tools. A well-structured regulatory system can help to minimize these criticalities: this is the case of law 338, enacted in 2000 with the aim of increasing the availability of residences for university students, which is distinguished for the attention to the environmental issue, orienting the realization of accommodation places towards solutions able to contain waste, soil consumption, etc. The paper aims to describe and analyse the attention paid by the specific legislation to environmental sustainability.

Keywords Environmental sustainability · Law 338/200 · Built environment · Buildings and urban regeneration · Energy efficiency

C. Piferi (✉)
University of Florence, Florence, Italy
e-mail: claudio.piferi@unifi.it

E. Arbizzani et al. (eds.), *Technological Imagination in the Green and Digital Transition*,
The Urban Book Series, https://doi.org/10.1007/978-3-031-29515-7_38

38.1 Introduction

If the regulations on energy saving are now established in our country, just think of the innovative law no. 373/76, replaced by law no. 10/91, and with Legislative Decree 192/05 and with Legislative Decree n. 311/07, less present are the rules for a 360-degree sustainable technological design. The Minimum Environmental Criteria[1] represent an undoubted innovation as they not only aim at reducing environmental impacts, but also at promoting sustainable production and consumption models aimed at identifying the best design solution, product, or service under the environmental profile over the life cycle assessment given the market availability. Law no. 338/00 is the first national example of an organic program aimed at increasing the number of residences and accommodation places for university students and is strongly characterized by the attention paid to actions and policies aimed at the sustainability of the built environment (Piferi 2021). In Germany, the construction and administration of university residences are entrusted to a state-run non-profit organization, the Deutsches Studentenwerk (DSW), partly funded by the Lander and partly by private companies, which is legally responsible for the construction and management of student accommodation (Germany Visa 2021), while in France, the organization responsible for allocating and managing student accommodation is the Centre National des Œuvres Universitaires et Scolaires (CNOUS), which is financially autonomous and reports to the Ministère de l'Enseignement Supérieur et de la Recherche (MESR 2011). In Spain, in addition to public universities, the construction of university residences is mainly delegated to private investors, including foreign ones (JLL 2018), and in Ireland, the public Housing Finance Agency (HFA) has been allowed to finance Higher Education Institutions (HEI) to build Purpose-Built Student Accommodation (PBSA), bespoke student accommodation built by private investors (CEB 2021). In Italy, Law 338/00 provides state co-financing for measures aimed at upgrading regulations, extraordinary maintenance and renovation, new construction and purchase of buildings already intended or to be intended as university residences.

38.2 Framing

The legislator immediately understands how the relationship between generations in training between the ages of 19 and 26 and sustainability is now indissoluble: young people are generally more aware and interested in ecological issues and climate change, as well as naturally predisposed to the use of new technologies. The university residence can be configured as an exemplary place where to experiment and implement solutions in support of environmental sustainability. In the twenty years

[1] In Italy, the effectiveness of CAM is guaranteed by art. 18, L. No. 221/15 and by art. 34, Legislative Decree No. 50/2016, containing "Criteria for energy and environmental sustainability", amended by art. 23, Legislative Decree No. 56/17 and subsequent amendments.

and in the five calls (2002, 2007, 2011, 2016, 2021), the law has further evolved, becoming a real best practice also in this area. Perhaps, the most important aspect lies precisely in the fact that the law goes into detail on the expected energy performance and addresses the issue of sustainability with a circular approach, taking care of the environmental, economic, social, and cultural aspects of the operation, because only the satisfaction of all these components can guarantee the real sustainability of the intervention. The strong orientation of the legislation towards the recovery of the existing building heritage, rather than towards the construction of new buildings, underlines this conviction. Informed that new buildings involve a considerable use of energy, that Italian cities are characterized by the presence of abandoned buildings, and that the system of "disseminated" universities over the territory, typical of the Italian reality, is mainly structured in the historic centres of city (Del Nord 2014), the legislator, in art. 1, specifies that the law was created to allow the state to contribute to the implementation of interventions necessary for the removal of architectural barriers, for the adaptation to the current provisions on safety and for extraordinary maintenance, recovery, and renovation of buildings already existing, as accommodation or residences for university students, as well as for new construction and purchase of areas and buildings to be used for the same purpose: the priority, therefore, appears to be to recover the existing building stock, compared to new construction interventions This has allowed the recovery of abandoned buildings, and the urban and social regeneration of depressed and degraded areas (Baratta and Piferi 2015): if we add the inevitable efficiency improvements to the theme of urban and social redevelopment energy and the low (or zero) economic investment that the proponents have had to support, we can understand the circular sustainability of the program. The recovery of the existing building stock is encouraged by the possibility, offered by the law, to bring the value of the property as part of the co-financing amount and consequently obtain state funding equal to the total amount of the work. Also, the interventions on the existing building patrimony of value and constrained have considerable advantages in terms of compliance with dimensional standards, providing for decreasing percentages on legal obligations. More than 76% of the 323 projects in progress concern existing buildings: almost 13,000 of the over 38,000 accommodation places involved are in existing buildings and around 11,500 are newly built (Figs. 38.1 and 38.2).

Very significant are the projects of the residences of "Crociferi" in Venice and of "Benedettini" in Palermo, which have redevelop abandoned historic buildings and, at the same time, regenerate neighbourhoods characterised by significant social complexity or that were slowly depopulating (Figs. 38.3 and 38.4).

The focus on the recovery of existing assets was further strengthened in 2021 when the law was specifically amended. Art. 15, Legislative Decree no. 152/21, amends art. 1, adding paragraph 4-bis which states: "In order to pursue the objectives identified in the communication of the European Commission of 11 December 2019 on the European Green Deal, implemented in the National Recovery and Resilience Plan, restructuring, transformation, also through demolition and reconstruction interventions, and the purchase of existing structures and buildings with the aim of pursuing

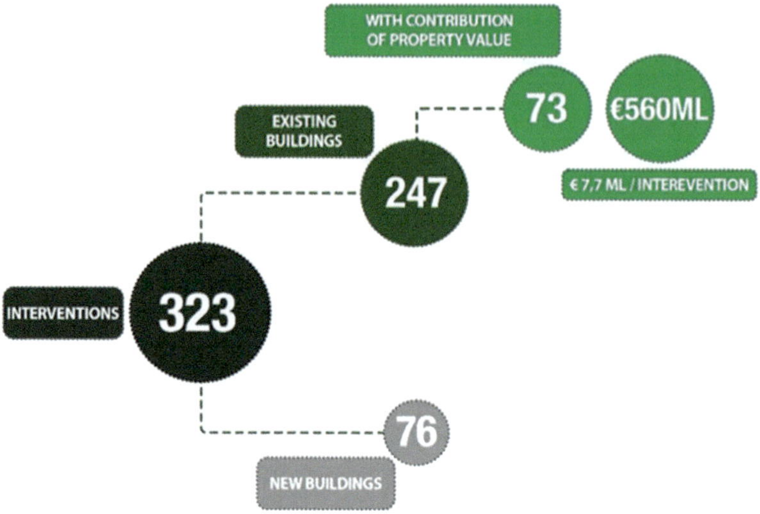

Fig. 38.1 Interventions in progress financed by law no. 338/00: new construction and recovery of the existing building heritage (*Author* arch. Andera Sichi)

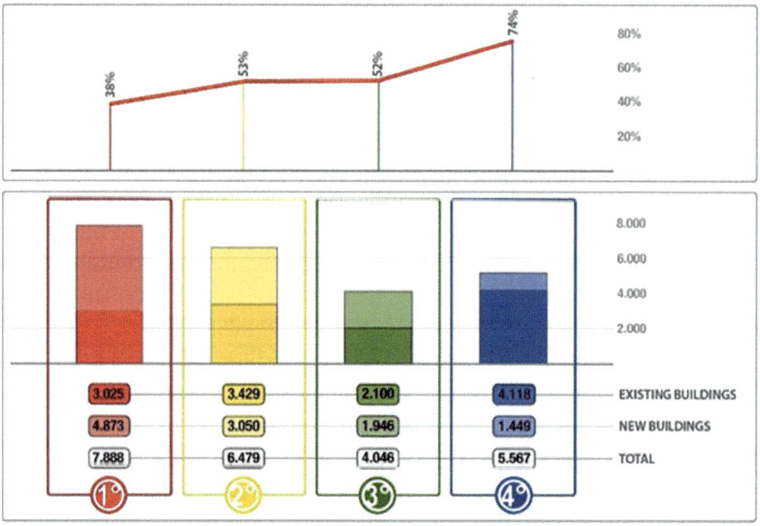

Fig. 38.2 Accommodation places financed by law no. 338/00 divided by calls for tenders and between former nine and existing buildings (*Author* arch. Andera Sichi)

Fig. 38.3 "Benedettini" University Residence, Palermo (*Source* TESIS archive)

Fig. 38.4 "Crociferi" University Residence, Venice (*Source* TESIS archive)

high environmental standards in the construction and management of the interventions". The D.M. n. 1256/21, the fifth call of the law, sharpens this concept, specifying in paragraph 1, art. 4, that in implementation of the provisions of the annex relating to the approval of the PNRR assessment of 8 July 2021, new construction of green-field buildings is not admissible, except for interventions included in existing campuses, or in neighbouring areas to university settlements. In the first issue,[2] the law, among the technical performance requirements to be respected, introduces, energy saving, maintenance, and well-being (thermo-hygrometric, olfactory, visual, etc.), etc. The annex specifies that the building and its systems must be designed and built so that energy consumption during use is optimized in relation to the conditions of well-being for the occupants. So, it is necessary to control solar irradiation and ventilation

[2] Annex B "Guidelines relating to technical and economic parameters", Ministerial Decree No. 118/2001.

and to comply with the indications of the legislation regarding heat loss. Furthermore, it specifies that the energy needs must be met preferably using renewable and alternative sources. The annex defines some factors, subsequently implemented by specific legislative apparatus aimed precisely at energy saving, to be taken into consideration for a correct energy conception of the building, i.e.:

- local climatic conditions;
- geomorphological characteristics of the area and exposure;
- typological and local settlement characteristics;
- control of solar radiation in summer: reduction to 30% of the heat due to summer heat radiation in the absence of external protections and possible adoption of selective glass;
- use of free heat supplies in winter but their limitation to 20% of the energy requirement calculated for each room;
- use of solar systems such as capturing walls, greenhouses, solar panels, etc.;
- exploitation of natural ventilation;
- innovative systems for the exploitation of natural lighting;
- control of the thermal behaviour of opaque and transparent components, with the elimination of interstitial and surface condensation;
- plant efficiency, with recovery of residual energy and reductions in exhaust emissions;
- introduction of integrated regulation and control systems between the electrical system, heating, and air conditioning.

To underline the importance reserved to the topic, the law specifies that newly constructed buildings must consider the principles of environmental protection, even in the absence of indications in urban planning tools and building regulations: these principles must be respected, whenever possible, also in the interventions of extraordinary maintenance, recovery, or renovation of existing buildings. Finally, particular importance is given to the feasibility study which must include an exhaustive characterization of the site (depending on the climate, availability of renewable energy sources, availability of natural light, etc.) and the environmental factors that may be influenced by the intervention, to direct it towards their respect (air, water balance and water cycle, soil and subsoil, ecosystems and landscape, historical typological aspects).

This attention to the issue of environmental sustainability was further expanded in the fourth call (2016) and is certified with the introduction of a specific category of projects, those aimed at energy efficiency of existing buildings, for which the Ministry identifies a specific line of co-financing equal to 15 million euros.

38.3 Analysis and Structuring of the Data Collected with the Research Activity

The research activity carried out with specific agreements with the MUR and with CDP S.p.A., through the investigation and monitoring of the interventions financed with law no. 338/00, has also made it possible to identify and catalogue the energy solutions promoted by the subjects, investigating the most significant ones and verifying the expected results with those achieved. About 490 projects were examined: for each request for co-financing a specific verification form was prepared to understand and validate the environmental and technological solutions, both traditional and innovative, adopted by the subjects (Fig. 38.5).

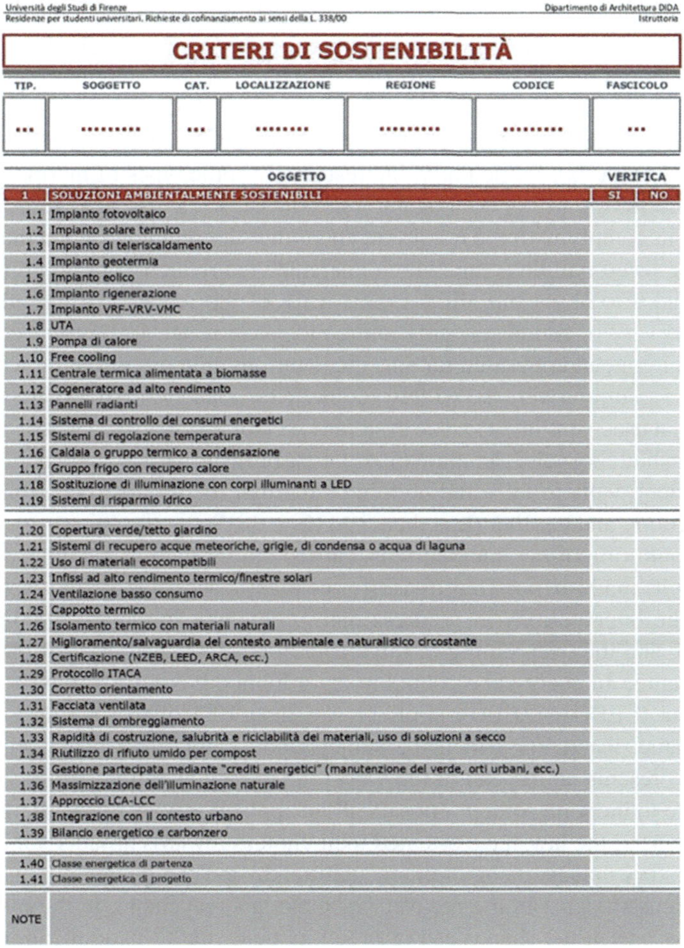

Università degli Studi di Firenze — Dipartimento di Architettura DIDA

Residenze per studenti universitari. Richieste di cofinanziamento ai sensi della L. 338/00 — Istruttoria

CRITERI DI SOSTENIBILITÀ

TIP.	SOGGETTO	CAT.	LOCALIZZAZIONE	REGIONE	CODICE	FASCICOLO
...	....	...	....	....	....	...

	OGGETTO	VERIFICA SI	VERIFICA NO
1	SOLUZIONI AMBIENTALMENTE SOSTENIBILI	SI	NO
1.1	Impianto fotovoltaico		
1.2	Impianto solare termico		
1.3	Impianto di teleriscaldamento		
1.4	Impianto geotermia		
1.5	Impianto eolico		
1.6	Impianto rigenerazione		
1.7	Impianto VRF-VRV-VMC		
1.8	UTA		
1.9	Pompa di calore		
1.10	Free cooling		
1.11	Centrale termica alimentata a biomasse		
1.12	Cogeneratore ad alto rendimento		
1.13	Pannelli radianti		
1.14	Sistema di controllo dei consumi energetici		
1.15	Sistemi di regolazione temperatura		
1.16	Caldaia o gruppo termico a condensazione		
1.17	Gruppo frigo con recupero calore		
1.18	Sostituzione di illuminazione con corpi illuminanti a LED		
1.19	Sistemi di risparmio idrico		
1.20	Copertura verde/tetto giardino		
1.21	Sistemi di recupero acque meteoriche, grigie, di condensa o acqua di laguna		
1.22	Uso di materiali ecocompatibili		
1.23	Infissi ad alto rendimento termico/finestre solari		
1.24	Ventilazione basso consumo		
1.25	Cappotto termico		
1.26	Isolamento termico con materiali naturali		
1.27	Miglioramento/salvaguardia del contesto ambientale e naturalistico circostante		
1.28	Certificazione (NZEB, LEED, ARCA, ecc.)		
1.29	Protocollo ITACA		
1.30	Corretto orientamento		
1.31	Facciata ventilata		
1.32	Sistema di ombreggiamento		
1.33	Rapidità di costruzione, salubrità e riciclabilità dei materiali, uso di soluzioni a secco		
1.34	Riutilizzo di rifiuto umido per compost		
1.35	Gestione partecipata mediante "crediti energetici" (manutenzione del verde, orti urbani, ecc.)		
1.36	Massimizzazione dell'illuminazione naturale		
1.37	Approccio LCA-LCC		
1.38	Integrazione con il contesto urbano		
1.39	Bilancio energetico e carbonzero		
1.40	Classe energetica di partenza		
1.41	Classe energetica di progetto		
NOTE			

Fig. 38.5 Monitoring board for sustainable solutions (*Source* TESIS archive)

The research first identified and catalogued the solutions and technologies put in place to increase the energy efficiency of buildings and reduce their consumption and verified their actual presence and potential. In some cases, the inspections, carried out on over 170 residences, also made it possible to document and verify the effectiveness of the solutions adopted. If we take into consideration the 93 requests for co-financing accepted in the fourth call, we see that the subjects, to reduce the energy needs of existing buildings, many of which characterized by a starting energy class equal to F or G, have focused mainly on the potential of the systems, even if there is no lack of broader architectural solutions. Among the most adopted system solutions, it is necessary to mention the use of photovoltaic panels, solar thermal systems and heat pumps, VRF or VMC ventilation systems, and the replacement of lighting devices with LED bodies, (between 40 and 50% requests for co-financing): almost all the solutions also adopt software and systems for controlling energy consumption, which regulate the temperature, the electricity consumed, the opening of the windows, etc.: there are fewer interventions that provide for the use of condensing boilers (15%). For the "architectural solutions", the most adopted ones are the use of thermal insulation on open spaces (not many with the use of natural materials) and the use of high thermal performance windows with triple glazing and systems solar collection (between 40 and 60% of requests). Particularly, interesting are some types of thermal insulation that provide for the use of nanotechnologies to allow the use of reduced thickness insulators to comply with the requirements of the superintendence for interventions concerning listed buildings and which guarantee the reduction of the transmittance of the facades to about one third. About 15% of the requests proposed solutions for the recovery of rainwater, grey and condensate water, to be reintroduced into the water circuit, and provided for specific solutions aimed at reducing water waste: garden roof solutions were not widely adopted (less than 5%). Also, there were many requests that used ventilated facade systems (7%), often integrated with a brise-soleil system for shielding solar radiation (15%). Few projects have exploited the thickness of the ancient walls to raise the building's thermal inertia or orientation to favour natural lighting and cooling (less than 5%) (Fig. 38.6).

The monitoring of the interventions, carried out through inspections on the residences built, made it possible to verify, in many cases, the results achieved: on average, the interventions in progress had a reduction in energy requirements of about 30% and the overcoming of three energy classes. Over 40% of the interventions guarantee an energy class equal to A, or higher, but only 10% have an energy certification (LEED, NZEB or similar): less than 10% of the buildings (of architectural value, listed, for which the subjects could not adopt some improvement solutions) presents, at the end of the works, an energy class lower than E. It seems appropriate to mention some proposals of the applicants who, correctly interpreting the intentions of the law, have oriented their choices towards broader solutions such as participatory management, or the recognition by the energy credit manager of the most virtuous students in the recovery of materials recyclable, in the maintenance of greenery and in the shared use of urban gardens, the re-naturalization of areas, the elimination of asphalts, and the use of dry construction solutions to reduce waste

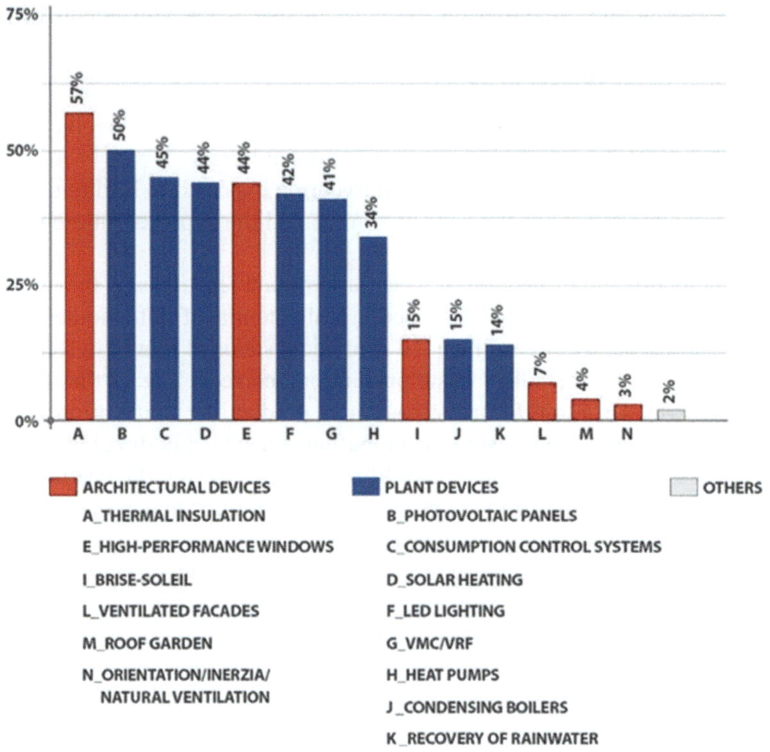

Fig. 38.6 Solutions adopted for the energy efficiency of university residences (*Author* arch. Andrea Sichi)

during construction and facilitate the possible disassembly and recycling of building components.

38.4 Results of Research and Regulatory Developments

In addition to confirming the availability of a specific financing quota for energy efficiency interventions, the evolution of law no. 338/00 in the field of environmental sustainability is inseparably linked to the PNRR which dedicates a specific paragraph precisely to university residences and to reform of the law and obliges the regulatory apparatus to comply with the environmental and digital parameters established in Europe.[3] To access the resources, attention to the climate issue and energy saving are further emphasized from the perspective of the Do No Significant Harm principle.

[3] Mission 4: education and research. Reform 1.7: Student housing and reform of student housing legislation.

The DNSH principle[4] is based on the concept that when you intend to plan an intervention, you need to ask yourself questions whose answers define the actual sustainability of the project in terms of:

- mitigation of climate change: is the intervention expected to involve significant greenhouse gas emissions during construction and use?
- adaptation to climate change: is the intervention expected to lead to a worsening of the negative effects of the current climate and the expected future climate on itself or on people?
- sustainable use and protection of water and marine resources: is the intervention expected to harm the good status or ecological potential of water bodies, including surface and groundwater or the good ecological status of marine waters?
- circular economy, including waste prevention and recycling: is the intervention expected to lead to a significant increase in the production, incineration, or disposal of waste, [...] or to lead to significant inefficiencies [...] in direct use of indirect natural resources at any stage of their life cycle or cause significant and long-term environmental damage from the point of view of the circular economy?
- prevention and reduction of air, water, and soil pollution: is the measure expected to lead to a significant increase in emissions of pollutants into the air, water, or soil?
- protection and restoration of biodiversity and ecosystems: is the measure expected to significantly harm the good condition and resilience of ecosystems or harm the conservation status of habitats?

In addition to the evaluation of the efficacy and usefulness of the project, understood as the ability to achieve the objectives assigned in the indicated times and as the convenience for the reference "community", the efficiency of the project is included, understood as the degree of achievement of the objectives with the minimum possible consumption of resources and the sustainability/durability of the project, understood as the ability of the project to sustain itself over time and in the subsequent management and implementation phases. The previously described research activity also allowed the identification of some parameters that will be used to assess eligibility for co-financing and to define the ranking of interventions. Regarding the admissibility of the interventions, the call specifies that at least four of the following parameters must be respected:

- the reduction of the consumption of material resources through:
  - the use of material with a content of recycled or recovered material for at least 20% by weight estimated on the total of all materials used, excluding plants;
  - the use of prefabricated mechanical assembly solutions for at least 60% of the volume of the works;
- the reduction of energy consumption:
  - for interventions on existing assets, increase of three energy classes;

---

[4] EU Regulation no. 2020/852.

- the achievement of any energy class A, for new buildings, energy positive building or energy positive district;

- the reduction of water consumption, so the adoption of recovery and reuse systems for at least 50% of rainwater;
- the reduction of soil consumption, so no increase in the surface ratios between artificial and non-artificial coverings.

In relation to the assignment of the score for the definition of the ranking, however, environmental impact indicators are introduced, such as:

- the energy sustainability indicator: use of renewable energy (solar, wind, geothermal water, biomass);
- the energy efficiency indicator (energy class of new buildings and, in the case of intervention on existing assets, increase in energy class);
- the consumption reduction indicator of material resources (cubic metres of material to be reused or recycled of materials and products/cubic metres of total material used);
- the sustainability indicator (adoption of environmentally sustainable products and technical solutions).

38.5 Conclusions

The law calls for the adoption of solutions aimed at limiting energy consumption, regulating the functioning of the energy systems used, resorting when possible to renewable energy sources, intervening on the regulation and improvement of the local microclimate, adopting solutions aimed at reducing the consumption of drinking water, with the adoption of integrated plant solutions and systems that also promote energy saving, together with the incentive for the reuse of water resources, suitably purified to reduce the polluting load in the environment. The regulatory system, therefore, defines design criteria aimed at a circular sustainability that contemplates the plant solutions but integrates them with a broader cultural approach. Although plant evolution has played, and still plays, a very important role in energy efficiency and the achievement of environmental sustainability of building interventions, law no. 338 has shown, in the twenty years of its application, how the issue of sustainability can also be tackled with a circular approach, allowing particularly "enlightened subjects" to experiment with innovative solutions. The research activity highlighted how the solutions can be chosen, implemented, and specifically calibrated according to the specific needs of end-users and facility managers.

References

Baratta A, Piferi C (2015) Le residenze universitarie come strumento di rigenerazione urbana. L'esperienza del Programma 338/00, in 3° edizione di abitare il futuro. Giornate internazionali di studio, Napoli, 1–2 ottobre, Clean edizioni, pp 485–493
CEB (2021) CEB to support construction of student housing in Ireland with a € 75 million loan. https://coebank.org/en/news-and-publications/news/ceb-to-support-construction-of-student-housing-in-ireland-with-a-75-million-loan/
Del Nord R (2014) L'innovazione di processo come strumento per promuovere la qualità delle opera. In: Del Nord R (ed) Il processo attuativo del piano nazionale di interventi per la realizzazione di residenze universitarie, Edifir, Firenze, pp 19–27
Germany Visa (2021) Student accommodation and housing in Germany. www.germany-visa.org/student-accommodation-housing/
JLL (2018) Investors wake up to Spain's student bed opportunity. www.jll.co.uk/en/trends-and-insights/investor/investors-wake-spains-student-bed-opportunity
Ministère de l'Enseignement Supérieur et de la Recherche (2011) La concrétisation du changement, Paris
Piferi C (2021) La qualità dell'abitare a servizio del diritto allo studio. Edizioni Opera Universitaria, Trento, p 54

Chapter 39
From Nature to Architecture for Low Tech Solutions: Biomimetic Principles for Climate-Adaptive Building Envelope

Francesco Sommese and Gigliola Ausiello

Abstract Building envelopes represent the interface between indoor and outdoor environmental factors. In recent years, attention to climate adaptive building envelopes has increased. However, some types of adaptive envelopes don't always offer low-tech solutions, but require energy for their activation and high operating and maintenance costs. Nature has always proposed a large database of adaptation strategies that are often complex, multi-functional, and responsive. Transferring the functional principles of natural organisms and their associated adaptive modalities to technologies is the challenge of the biomimetic discipline (from Greek bios, life, and mimesis, imitation) applied to the field of architecture. In this article, various examples of biomimetic architecture that illustrate the relationships between biology, architecture, and technology, were considered. Various analyses of the operating principles of natural organisms are carried out, particularly with regard to self-adapting materials, in order to transfer them to the building envelope, and to propose technological solutions capable of passively adapting to external climatic conditions. Among all natural organisms, plants are prefereble to animals because, like buildings, they remain stationary in a specific location. Despite this, plants have developed different adaptation mechanisms to survive in certain environments. Buildings with biomimetic adaptive envelopes, characterized by passive and low-tech solutions inspired by plants, help limit energy consumption, and improve not only the indoor microclimate but also the outdoor environment. In line with the ecological transition, this work highlights the importance of biomimetic as a strategy to orient the new paradigms of built space design towards innovative and sustainable models of low-tech solutions.

Keywords Biomimetics · Adaptive building envelope · Responsiveness · Environment · Smart materials

F. Sommese (✉) · G. Ausiello
University of Naples Federico II, Naples, Italy
e-mail: francesco.sommese@unina.it

G. Ausiello
e-mail: gigliola.ausiello@unina.it

E. Arbizzani et al. (eds.), *Technological Imagination in the Green and Digital Transition*, The Urban Book Series, https://doi.org/10.1007/978-3-031-29515-7_39

39.1 Introduction

The natural state of the climate is altered by accentuated anthropic factors (IPCC 2021).

Adaptation actions in all sectors are essential, especially in the building sector, which is held responsible for about 40% of energy consumption (IEA 2019). To ensure good indoor comfort while limiting the impact on the environment, traditional building envelope solutions are not particularly efficient due to their staticity. So, responsive solutions to changing external conditions are needed. The building envelope represents the interface between the indoor and outdoor space and plays an important role in the energy balance. In recent years, the building envelope has been the subject of studies and research to define solutions to improve energy efficiency, ensure optimal indoor comfort, and limit environmental impact. The latest technological solutions propose the adaptive envelope as an interface capable of responding and adapting to changing external environmental factors. Nature provides adaptive solutions because natural organisms have always used adaptive mechanisms to survive environmental stress. Transferring the adaptation strategies of natural organisms to building envelope technologies is the challenge of biomimetics, which is considered an emerging discipline in the field of engineering and architecture (Badarnah and Kadri 2015).

The combination of natural strategies with smart and self-activating materials make it possible to obtain low-tech solutions, that do not consume energy but are based on self-activating mechanisms (Sommese et al. 2022). In this way, the autonomous response to weather changes provides indoor comfort and, at the same time, the absence of energy consuption limits the impact on the environment, expecially the heat island in the city.

The focus of this work is on the use of self-activating smart materials for biomimetic adaptive envelopes. Some projects are analyzed and classified according to the type of material used and the response to a specific external environmental stimulus.

39.2 Environmental Challenges and Smart Materials

The most important external environmental factors to consider are solar radiation and high temperatures. Due to man-made pollution, the CO_2 content in the atmosphere is also particularly high. All these elements, and others that we do not consider in this article (such as chemical concentration or electrical voltage), are the environmental stimuli that allow materials, to activate and generate a response to such phenomena. In fact, smart materials represent a class of innovative materials that allow to develop architectural organisms capable of responding to environmental stimuli (Fiorito et al. 2016).

Table 39.1 Classification and characterization of smart and self-activating materials

Smart materials	External stimulus	Response	Adaptation mechanism	Visibility	Control mechanism
Photocatalytic	UV radiations	Chemical reaction	Static	No	Intrinsic
Wood	Humidity level	Shape change	Dynamic	Yes	Intrinsic
Shape memory	Temperature variation	Crystal structure change	Dynamic	Yes	Intrinsic

They are able to respond to stimuli in real time but some of them are also self-activating, meaning they have a reaction capacity inside the material and do not require external activation systems, such as sensors and actuators (Otsuka and Wayman 1999). In particular, they perceive the stimulus, respond to it, and return to their original state when the stimulus is removed (Tabadkani et al. 2020). In recent years, these materials have influenced the construction industry, in fact, various studies and experiments are present in the literature. Among the most widely used smart materials are shape memory alloys (SMA), shape memory polymers (SMP), electrochromic and piezoelectric materials, and others (Addington and Schodek 2005). The most common smart materials are based on an electrical stimulus (López et al. 2015). In this study, only materials that can act through an intrinsic change in properties and therefore without external implementation systems, were considered. The choice of self-activating smart materials orients the design towards low-tech and low-energy solutions, to limit consumption in line with the European goals of climate neutrality by 2050.

Table 39.1 shows the classification of some smart and self-activating materials used in the examples of biomimetic envelopes that are described in the following sections.

For each smart material, the environmental stimulus that triggers the reaction to the external environment was identified (Casini 2014). In addition, the type of response and the control and adaptation mechanisms are defined. The activation mechanism is classified as static when there is no visible movement and dynamic when there is no visible mechanism (Kuru et al. 2018). Intrinsic control is when the material response is due to the change in its properties; extrinsic control is when the change is due to external devices such as actuators or sensors that cause the material to move (Tabadkani et al.2020; Loonen et al. 2013).

Fig. 39.1 Lotus left and Italian Pavilion Expo 2015. *Credits* the author G. Ausiello

39.2.1 Biomimetic Solutions with UV Radiation Reactive Materials

Photocatalytic materials are widely used in the construction industry. Metal compounds such as titanium dioxide (TiO_2) react to UV radiation by oxidizing the material they come into contact with. The Italian pavilion at Expo 2015 is one of the latest examples of this solution. The interaction between concrete and titanium dioxide nanoparticles gives the surface the ability to self-clean and antipollution (Ausiello 2018). By absorbing light energy, nanometric crystals of titanium dioxide trigger photochemical reactions, and act as photoactive particles on the order of micrometres. The photochemical reactions, which take place in the nanometre range, and are therefore not visible to the naked eye, mimic natural processes that are reproduced and amplified by titanium dioxide becoming a conductor from semiconductor.

The ability to "self-clean" is called the "lotus effect", in relation to the natural behavior of the leaves of this plant which, immersed in muddy water, are always shiny and clean (Ausiello 2018). For several years these leaves have served as a model for experimenting with this property, which is reproduced by the photochemical effect induced by titanium dioxide, which leads to a nanometre scale change in the "roughness" of the surface. When exposed to sunlight, the surface becomes nanometrically hydrophobic and the water droplet remains in relief, resting only on the minute bristles of the lotus leaves, or on the nanoprotrusions of the surface of the concrete panels, then flows on the leaf or the surface of the panels, dragging dust and insects with it, which are thus naturally removed (Fig. 39.1).

39.2.2 Biomimetic Solutions with Humidity Reactive Materials

Wood is the natural material par excellence. It is not always classified among the intelligent materials, but it has the properties of hygroscopicity and anisotropy that allow it to be considered a self-activating material, that responds to the moisture content of the environment. An important example of this ability is the cones of conifers, which open and close the bracts depending on the humidity to protect the fruit (pine nuts) they contain. Inspired by the functioning of the pinecone is the project of the exhibition pavilion HygroSkin designed by the ITKE and ITC institutes of the University of Stuttgart. It is a meteorosensitive pavilion that uses the hygroscopic and anisotropic properties of wood to create an architectural skin that can open and close autonomously in response to climatic fluctuations (Krieg et al. 2014). In particular, the openings respond to changes in relative ambient humidity in the range of sunny to rainy climates in a mild environment (Correa et al. 2013). In this way, light penetration and the visual permeability of the interior spaces can be controlled.

39.2.3 Biomimetic Solutions with Temperature Variation Reactive Materials

In recent years, shape memory alloys (SMA) have gained notoriety in the architectural field. SMAs can return to their original shape or size when exposed to certain factors, including temperature. The material deforms due to an external force and contracts or resumes its original shape when heated above a certain temperature (Mohd Jani et al. 2014). SMA are considered intelligent materials in response to temperature variation due to its shape memory effect and pseudoelastic effect (Tamai and Kitagawa 2002). Air Flow(er) is a ventilation device designed to regulate air flow and internal temperature, without the use of electricity or mechanical equipment (LiftArchitects). It represents an interesting example of low-tech solutions.

The design principle of the Air Flower was based on the imitation of the thermonastic movement of plant organisms, in particular the yellow Crocus Chrysanthus, which responds to the increase and decrease in temperature that occurs in day-night passages (Rascio et al. 2017). Crocus flowers petals open when the temperature increases and closes when it decreases (Andrade et al. 2021). In the case of Air Flow(er), the kinetic reaction of the plant organism, due to the thermonastic movement, is provided by the shape memory alloy wire (Fig. 39.2).

When the alloy is exposed to low temperatures (martensite), it deforms more easily and then the four doors can open to allow the passage of air. However, when the alloy is exposed to higher temperatures, it returns to its original shape and the doors close, blocking the air flow.

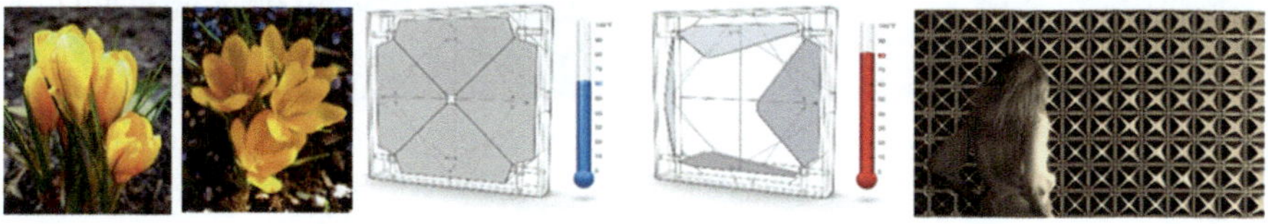

Fig. 39.2 Crocus chrysanthus and Air Flow(er) prototype

Fig. 39.3 Pho' Liage

A recent study of shape memory materials led to the definition of the prototype Pho' Liage, by ArtBuild studio (Fig. 39.3). It is a prototype of a kinetic device for shielding against natural light, which does not require human or mechanical action to activate. This system guarantees solar shielding through Thermo-reactive materials and the storage of thermal energy through photosensitive materials (ArtBuild). This solution is inspired by the opening and closing mechanism of the flowers, in response to variations in the intensity of solar radiation. In particular, the reference is to the troponastic (or thermonastic) mechanism, understood as the behavioral adaptation of some plant organisms, which has allowed to define of a solution capable of reacting autonomously by exploiting the chemical-physical and environmental interaction (Trelcat et al. 2022). This example, which does not require energy for activation, can also be classified as a low-tech solution.

Thermo-bimetals are among the materials that react to temperature difference.

They refer to a double layer conformation of the metal with different coefficients of thermal expansion, able to make it to fold in response to the increase in temperature (Al-obaidi et al. 2017). The use of laser-cut bimetal lining is illustrated by the Bloom system (Fig. 39.4), a self-supporting paraboloid-shaped shell structure. When the shell is heated at the surface, the structure curls autonomously. In this way the curling of the shape shades and allows air to pass through, thus ventilating certain areas. This solution emulates the mechanism of opening and closing of stomata in natural organisms. Moreover, this solution is configured as a passive system, as it opens and closes its slats according to the heat of the sun, without using artificial energy (Barozzi et al. 2016).

Fig. 39.4 Bloom system

39.3 Results

Table 39.2 summarises the examples analyzed in the previous section. They are classified according to various factors, including the required performance, controlled environmental factors, environmental stimuli (input), material, response, and organism inspiration.

The examples described above are all inspired by the plant world. Indeed, these have always inspired architects and designers (Ausiello et al. 2020). Nature provides a database of adaptation solutions to respond to changing environmental stimuli, and smart materials make it possible to transform these solutions into low technologies to be applied to the building scale.

The methodological approach of biomimetic design (Fig. 39.5) thus provides, as a starting point, the definition of the environmental challenge, then the identification of the biological organism to emulate (in this case plants), and finally the identification of the smart and self-activating material capable of reacting passively to environmental stimuli, as happens in nature. In this way, it's possible to move from biology to technology and define solutions, techniques, and materials that can emulate the functioning of the identified natural organism. Figure 39.5 shows the relationships between the different phases of the methodological approach, deriving from the analysis of the cases studied in this work.

39.4 Future Scenarios

In the panorama of urgent actions to limit environmental impacts, low-technology solutions are becoming increasingly necessary. This paper highlighted the combination of biomimicry and smart materials as an expression of a solution to be implemented in architecture to propose climate adaptive envelope solutions. Nature becomes a model and guide for the realization of new technological solutions and biomimetics is an essential tool in design, that facilitates the transition from the industrial to the ecological era. Biomimetic does not necessarily use natural resources, but reproduces the functioning of natural organisms in response to specific needs (Ausiello et al. 2020). The use of smart materials allows you to design buildings with advanced properties that can adapt to changing weather conditions while also saving energy and improving indoor comfort. Smart materials can completely revolutionize the construction industry to steer new building paradigms towards ecological transition and ensure sustainable solutions.

Table 39.2 Results and characteristics of case studio analysis

	Requirement	Environmental factor	Input	Inspiration organism	Material	Output
Italian Pavilion Expo 2015	CO_2 capture	CO_2	UV radiation	Lotus effect	Photo-catalytic	Inaudible
HygroSkin Pavilion	Modular light radiation and visual permeability	Water	Relative humidity	Pine cone	Wood	Elastic-flexural
Air Flow(er)	Regulate air flow and internal temperature	Air	Temperature variation	Flowers	Shape Memory Alloy Wire (SMA)	Kinetics
Pho' Liage	Solar shading	Light	Temperature variation	Flowers	Thermo-reactive materials (TBM) Photosensitive materials (PV)	Dilation
Bloom	Airflow shading	Air Light	Temperature variation	Pores/stomata	Thermo bimetal	Folding

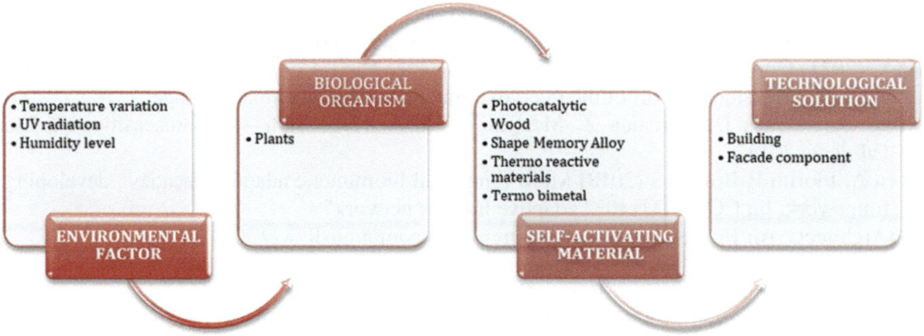

Fig. 39.5 Biomimetic approach from nature to technologies

Acknowledgements This work is a part of F. Sommese PhD research in Civil Systems Engineering, at DICEA (Department of Civil, Architectural and Environmental Engineering) of Federico II Univeristy of Naples (Italy), supervised by professor Gigliola.

Author contributions Francesco Sommese: conceptualization, methodology, writing, editing and visualization. Gigliola Ausiello: Supervision.

References

Addington M, Schodek DL (2005) Smart materials and new technologies : for the architecture and design professions. Architectural

Al-obaidi KM, Ismail MA, Hussein H, Abdul AM (2017) Biomimetic building skins: an adaptive approach. Renew Sustain Energy Rev 79(May):1472–1491

Andrade T, Beirão JN, de Arruda AJV, Eysen C (2021) Toward adaptable and responsive facades: using strategies for transforming of the material and bio-based materials in favor of sustainability. In: Cuaderno 149—Cuadernos del Centro de Estudios en Diseño y Comunicación

ArtBuild, Pho'liage ®—a biomimetic façade which increases building energy efficiency

Ausiello G (2018) Calcestruzzo fluido per architetture fluide. Luciano, editor, Napoli, p 144

Ausiello G, Compagnone M, Sommese F (2020) Imitare per costruire: dalla natura alla biomimetica. In: New horizons for sustainable architecture, pp 1–10. ISBN: 978-88-96386-94-1

Badarnah L, Kadri U (2015) A methodology for the generation of biomimetic design concepts. Archit Sci Rev. https://doi.org/10.1080/00038628.2014.922458

Barozzi M, Lienhard J, Zanelli A, Monticelli C (2016) The sustainability of adaptive envelopes: developments of kinetic architecture. Procedia Eng 155:275–84. Retrieved from https://doi.org/10.1016/j.proeng.2016.08.029

Casini M (204) Smart Building Involucro 2.0. DEI Tipogr. DEI Tipografia del genio civile, editor. DEI Tipografia del genio civile, Roma

Correa D, Krieg OD, Menges A, Reichert S, Rinderspacher K (2013) Hygroskin: a climate-responsive prototype project based on the elastic and hygroscopic properties of wood. In: ACADIA 2013: Adaptive architecture—proceedings of the 33rd annual conference of the association for computer aided design in architecture, pp 33–42

Fiorito F, Sauchelli M, Arroyo D, Pesenti M, Imperadori M, Masera G et al (2016) Shape morphing solar shadings: a review. Renew Sustain Energy Rev 1(55):863–884

IEA (2019) Energy efficiency 2019. France. https://iea.blob.core.windows.net/assets/8441ab46-9d86-47eb-b1fc-cb36fc3e7143/Energy_Efficiency_2019.pdf

IPCC (2021) Climate change 2021: the physical science basis. Contribution of working group I to the sixth assessment report of the intergovernmental panel on climate change

Krieg O, Zuluaga D, Christian Z, Menges A (2014) HygroSkin—meteorosensitive pavilion. Fabricate 2014

Kuru A, Fiorito F, Bonser S (2018) Multi-functional biomimetic adaptive façades : developing a framework. In: COST TU1403 "adaptive facades network"

LiftArchitects. Air Flower. http://www.liftarchitects.com/#/air-flower

Loonen RCGM, Tr M, Cóstola D, Hensen JLM (2013) Climate adaptive building shells: state-of-the-art and future challenges. Renew Sustain Energy Rev 25:483–493

López M, Rubio R, Martín S, Croxford B, Jackson R (2015) Adaptive architectural envelopes for temperature, humidityÃ, carbon dioxide and light control. In: 10th Conference on advanced building skins, pp 1206–1215

Mohd Jani J, Leary M, Subic A, Gibson MA (2014) A review of shape memory alloy research, applications and opportunities. Retrieved from https://doi.org/10.1016/j.matdes.2013.11.084

Otsuka K, Wayman C (1999) Shape memory materials. In: Otsuka K, Wayman CM (eds) Shape memory materials. Cambridge University Press. https://books.google.com/books/about/Shape_Memory_Materials.html?hl=it&id=DvItE9XUlN8C

Rascio N, Carfagna S, Esposito S, La Rocca N, Lo Gullo MA, Trost P et al. (2017) Elementi di Fisiologia Vegetale. EdiSES, editor. EdiSES, Naples

Sommese F, Badarnah L, Ausiello G (2022) A critical review of biomimetic building envelopes: towards a bio-adaptive model from nature to architecture. Renew Sustain Energ Rev 169:112850. https://doi.org/10.1016/j.rser.2022.112850

Tabadkani A, Roetzel A, Li HX, Tsangrassoulis A (2021) Design approaches and typologies of adaptive facades : a review. Autom Constr 121:103450. Retrieved from https://doi.org/10.1016/j.autcon.2020.103450

Tamai H, Kitagawa Y (2002) Pseudoelastic behavior of shape memory alloy wire and its application to seismic resistance member for building. Comput Mater Sci 25(1–2):218–227

Trelcat S, Chariot C, Troublé R, Roulin D (2022) Architecture e design. Art E Build Architectes, archiSTORM Hors-Série 29. Retrieved from www.archistorm.com

Chapter 40
Soft Technologies for the Circular Transition: Practical Experimentation of the Product "Material Passport"

Tecla Caroli

Abstract The change of mind regarding waste conceived as resource is possible if the transition toward circularity is actualized. Design and build in a reversible perspective allow to enable circular strategies on a constructive system at the end of its service life. These strategies and solutions toward reversibility are defined Hard and Soft Technologies. If the first ones are the reversible constructive technics used on different elements of the building; the second ones consist in the operative aspects, regarding to tools and methods that allow the realization of reversible processes, projects and products. Specifically, the MP is a Soft Technology that facilitates the second use of a constructive systems that has residual performances once disassembled. As tool, the Material Passport collects the technical and operative information about the product, tracing the resources employed, maintaining their value over time and reducing the grade of uncertain about the circular potentiality of disassembled elements. The paper demonstrates the importance of Soft Technologies for the development of Hard Technologies and reports the outcomes of a practical experimentation developed. The application consists in the development of the Material Passport of a product of the manufacturing company Cel Components and in the definition of reversible practices for the generation of the circular transition of the company. The MP is processed within the BIM model of a company-produced panel, used for façade, false ceiling, internal partition or flooring systems, to elaborate a framework for the data entry. The application provides a user-friendly framework for designers and manufacturers.

Keywords Circularity · Sustainability · Reversible technologies · Material passport · BIM

T. Caroli (✉)
Polytechnic University of Milan, Milan, Italy
e-mail: tecla.caroli@polimi.it

E. Arbizzani et al. (eds.), *Technological Imagination in the Green and Digital Transition*, The Urban Book Series, https://doi.org/10.1007/978-3-031-29515-7_40

40.1 Introduction

In recent years, life cycle design has been promoted by implementing the design for disassembly and deconstruction (Design for Disassembly—DfD) of prefabricated constructive systems to facilitate the reuse, remanufacturing and recycling of building elements and systems at end of their service life (Jaillon and Poon 2014). The challenge is to solve the problem of the huge quantity of building components that once disassembled has still residual performances, to demonstrate that reversible technologies could optimize the construction process emissions.

The Reversible Technologies are defined as Hard and Soft. The Hard Technologies are the off-site constructive technics (materials, components, assembly methods), that are designed and built to be disassembly; the Soft Technologies consist in operative—management and assessment—aspects (phases, operators and their role, tools, evaluation methods, technical documentations and standards) that support the construction process. The paper focuses on Soft Technologies, considering that they are the driver and support for the actualization of reversible, circular and sustainable processes, projects and products. A lack of preventive measures and limited knowledge of construction products characteristic increases the waste generated, hinders its control and affects the cost and time of waste management (Assefa and Ambler 2017; Jin et al. 2018). Planning and monitoring reversible processes provide one of the best opportunities to reduce waste generation and strengthen reuse and recycling practices from the early stage of construction planning and throughout the entire value chain (Gorgolewski 2017).

The aim of the paper is to highlight the role of the Material Passport, as tool to facilitate the application of circular solutions to constructive systems that are disassembled, and to report a practical experience of the elaboration of MP for a sandwich panel.

40.2 Soft Technologies: Operative Strategies Toward Circularity

To realize reversible project and/or product it is necessary to develop a reversible process (Fig. 40.1). It is possible thanks to the control and improvements of the operative aspects, regarding to the role of the operators involved, their network and collaboration, the methods and tools used to support the actions.

How the products are conceived and manufactured influence the effective reversibility of a constructive system. As said before, to obtain the reversibility for the activation of potential circularity, it is necessary develop simultaneously Hard and Soft conditions referred to technical and operative requirements. To this end, the product conception should change and follow the aim to maintain the value of the resources employed working on the product performances toward a "Product as Service" (Vezzoli et al. 2014) conception. The aim is to exploit the technical and

Fig. 40.1 Business models and respective product conception and waste production

operative reversibility of the product and manage it as a service and not as a good (consuming product). It means that the focus is on the performances offered that achieve the extension and increase of the product value. This vision can be achieved thanks the direct involvement of the manufacturer and the other operators for the development of the take-back and second-use activities. Exploiting the extended-life of reversible products and conceiving the used products as resources and never as waste, they can be reused and counted in different life cycles and resources remain in the closing-loop process. In this way, a circular economy model is created in which nothing is lost that occurs thanks to the design and construction of reversible technologies.

The focus on the performance of the product represents the opportunity to use the reversibility of constructive systems to fulfill temporary users' needs and maintain the value of resources, involved in the constructive systems, during time. The challenge of manufacture services rather than goods is still difficult to develop due to a consolidated linear market. The manufacturer aims to develop the product performances, considering its extended use, the combination of reversible elements and materials that can be recovered entirely or in parts, as they are (reuse) or with operations that improve their performances (remanufacturing) or again, re-processing them to produce new products (recycling). In this way, the products at the end of their service life are not waste, but organized deposits of materials (material banks) exploiting the duration of their technical life.

Despite this, the activation of reversible practices determines an implementation of the operational aspects leads by manufacturers, developing the Material Passport of products and creating the "second-use" program within the company. Such practices will encourage the transition to circular business models in which products will be conceived only as services.

40.3 Material Passport: Product Value Retention

Waste is sometimes described as material without information. Furthermore, the experimentation of circular solutions to reversible and long-lasting construction system is possible thanks to the identification of resources available, their trackability and recognition of their potentiality.

The importance of tagging or attaching information about a product is recognized in industrial design; by providing adequate information about a product or material it may be possible to pass this to future generations. Technical specifications can be encoded in what are sometimes referred to as "materials passports" (or resource passports). The Material Passport consists in a document containing a detailed inventory of all the materials, resources and components of a product or building, as well as detailed information about their location—providing materials with identities that are independent of their current use (Gorgolewski 2017). It can provide details of all a product's qualities, nutrients and properties as well as production information such as location, date and name of manufacturer. This enables future reuse or recycling to occur when a product reaches the end of its initial life. For some components it can also aid in maintenance throughout its lifetime.

40.3.1 Methodology and Tool Information

Thanks to the reversibility of products, it is possible to activate circular solutions. A huge quantity of product knowledge is concentrated during the design phase. It is essential that this knowledge is passed on to those who are ultimately responsible for management once construction is complete. This information is now retained in the Material Passport. Indeed, to ensure that resources are retained, the technics used to manufacture a product are not enough. It is necessary the use of tools that allow to reversibility products to have a future, exploiting their potentiality.

The collection of information makes a product permanently available, in that it is identified and recognize its value during time, regardless of the construction system to which it is connected. The recognition of the product identity is a manufacturer responsibility. On the contrary, an undocumented product would be deleted and would be worthless.

In order not to lose the opportunity to activate circular economy strategies that the reversibility of construction systems offers, it is necessary to process the Material Passport of the products that are used in a reversible process. It defines and requires a certain transparency of production process and other manufacturing activities.

The information collected regarding to:

- material composition;
- manufacturing company and suppliers information;
- process information (phases and tools);
- logistic information (operators and resources involved)
- element/system performances on technological level;
- environmental impacts;
- economic impacts;
- technical and operative reversibility;
- circular economy strategies applied/applicable;
- future uses.

The collection of product information is the first step, how they are collected, or better the framework to follow, have to follow the aim to be a user-friendly tool for all operators involved in the process. Currently, there is not a reference framework to follow and so the practices developed are differently and not comparable because they do not follow a same goal.

To be repeatable and effective the Material Passport should be supported by BIM software (Revit, ArchiCAD, Rhino) to import the information on the product model and collect the information about all constructive systems of a building in a parametric file. This is a digital twin of the constructive systems (products and building parts), based on the BIM model, which is already created during the design process. This virtual, three-dimensional representation includes everything that is needed to construct or renovate the building.

The access to other operator of the constructive process allows the implementation of the Material Passport information, defining a framework that can become recognizable and simple in the constructive sector. The practice about the elaboration of the Material Passport of a product (element/system/building) is essential for the achievement of a reversible process that recognize and maintain the resources involved for the activation of circular strategies.

In order to understand the feasibility of the elaboration of a material passport of a real product, it is developed a collaboration with Cel Components, a sandwich panel manufacturing company.

40.4 Manufacture and Manage the Circular Transition

The practical experimentation consists in the definition of action to generate the circular transition of the manufacturing company, focusing on a specific product: a sandwich panel. The choice on Compocel AL panel is the result of a technical and marketing analysis thanks to the collaborations of the operators of the company. Analyzing the level of reversibility of the panel, it was possible to define the technical (Hard) improvement of the panel used as wall or cladding systems, and the operational (Soft) improvements to activate a circular approach on the panel after—temporary—uses considering the durability of the product.

After the decision to develop the reversible and circular transition, it has been verified which practices are possible to activate and implement. Working on the Hard Reversibility is difficult since, after years of experimentation and improvements in production techniques, the company assumed consolidated technics and the costs is too high to chance in the short period the production processes. For these reasons, the study focused on Soft Reversibility that regarding to the elaboration of the second-use program in the company, the codification of all information of the products of the company and the elaboration of the Material Passport with the integration of BIM methodology.

The paper reports the elaboration of the last one: the Material Passport.

Product		
Name		
Dimensions		
Source		
Image		
Indicators		
PROCESS		
Product Labels & Certification		
Standard & codes		
Business models		
Actors	Manufacturer	Name Location Company registration number (P.IVA) REA (Repertorio Economico Amministrativo) Number Ateco Code Telephone Site PEC
	Transportation & Logistics	Packaging Vehicles Traceability
	Ownership Design for disassembly and reversible structures Material supply Installation/assembly Use Disassembly Function	
Logistics	Design Data	Design for Durability Design for Disassembly Design for Adaptability Design for Circularity
	Production Data	Supplier Information Manufacturing process & techniques Installation and handling instructions Time of delivery Waste
	Digitalisation	BIM model Technical Drawings
Material flows	Input flows Output flows Waste and Recovery	Raw material Production scrap End-of-life

BIOLOGICAL		
Materials	Skin	Aluminum
	Core	Aluminum
	Glue	Polyurethane Resin
Renewable/non-renewable		
Untreated/treated		
Recycling & re-use potentials		
PHYSICAL		
Dimensions	Measurements	
	Thickness	Panel Skin Core Cel
	Mass Weight Area Volume	
	Density	Skin Core Adhesive
Structure data	Compressive strength Moment of inertia Modulus of elasticity (E) Resistance of delamination	
Building physics	U-value Thermal conductivity	
Fire reaction		
Optical	Colour Surface Transparency	
Lifespans & durability		
Recycling and re-use potentials		
Chemical		
LCA Assessment	Environmental Emission trading	Global Warming Potential (kgCO2eq/m2) Energy Demad (MJ/m2)
LCC Assessment	Economic Emission trading	Recommended retail price per unit Additional Processes Manufacturing costs (sensitive data) Costs for maintanance and operation Costs for recycling Transportation and handling costs Transport insurance
Warranty		
Lifespans & durability		
Recycling & re-use materials		

Fig. 40.2 Reference framework of material passport for the collection of data

40.4.1 Collection of Data and Reference Framework

During the knowledge phase, it is investigated which are the current Material Passport (Fig. 40.2) experiences, it is researched a reference framework that also consider the other building levels and contain information about technical (Hard) and operative (Soft) aspects.

To this end, the Framework followed is developed by BAMB (Heinrich and Lang 2019).

As it is possible to see in the table besides, the information regarding to the product and company data about the process, physical, chemical and biological characteristics.

The collection of data was achieved thanks to the help of technicians and marketing managers of the company that providing the data. In addition, to activate the potential recovery of the products, within the matrix were provided some information about the panel and possible methods of assembly/disassembly and recovery.

40.4.2 Material Passport Digitalization Toward the Replicability

To remember, the Reversible Technologies are the technical and operative tools that allow to a constructive system (elements, systems and buildings) to return to its own original states. To this reason the Material Passport is assumed as a Reversible

Technology tool that allows to collect and track the date related to a constructive system, during the lifespan of this latter.

As said before, there is not a standard to develop a Material Passport, so it is decided to assume the BAMB model as reference and improve it toward the creation of a BIM model of a constructive product. It was possible thanks to the help of a BIM Manager.

The challenge is to digitalize the information collected and to create a replicable framework, useful for professionals (designer, manufacturer, etc.) that wants to develop a tracking tool for reversible constructive system (product or project).

The software used is "Revit", since Autodesk is the most widely used architectural and engineering design companies worldwide. The aim was to use a parametric software that works in BIM, generating the Industry Foundation Classes (IFC) file collecting only the families or the entire model, to exchange files with other operators (exchange files between operators) as.pdf file, not editable.

The Level of Development used was LOD 500, in that the level of detail of the element's geometry allows to associate the information contained in the MP. In order to develop a tool able to be used both manufacturing operators and designers it is the right way to increment the usability and reduce the limitations.

Considering Revit interface, all parameters collected in the previous framework (Fig. 40.2), divided in group, are added in the model. For each group it was necessary to write all the parameters, after that it was possible to start to insert them into Revit: Family Types (see Table 40.1).

The difficulty found was that the MP entries of the BAMB model used did not correspond to the Revit model and therefore the equivalents were chosen for each parameter. For this reason, for each category of the matrix are selected the right voices for the Revit Parameters.

After entering all the data within the model, the data entry mode (parameters and contents) was exported to a .txt file: a file that collects the matrix entries inserted in the Revit model.

The file exported allows to replicate this process for other families and therefore other construction products, in such a way as to obtain the same reference framework

Table 40.1 Matrix parameter and Revit parameter for family types

Matrix parameter	Revit parameter
Product: Name, dimensions, source, image	Identity data
Process: Product labels and certification, Standard and codes, business models	Data
Process: Actors	General
Process: Logistics	Other
Process: Material flows (input flows)	Primary end
Process: Material flows (output flows, waste and recovery)	Secondary end

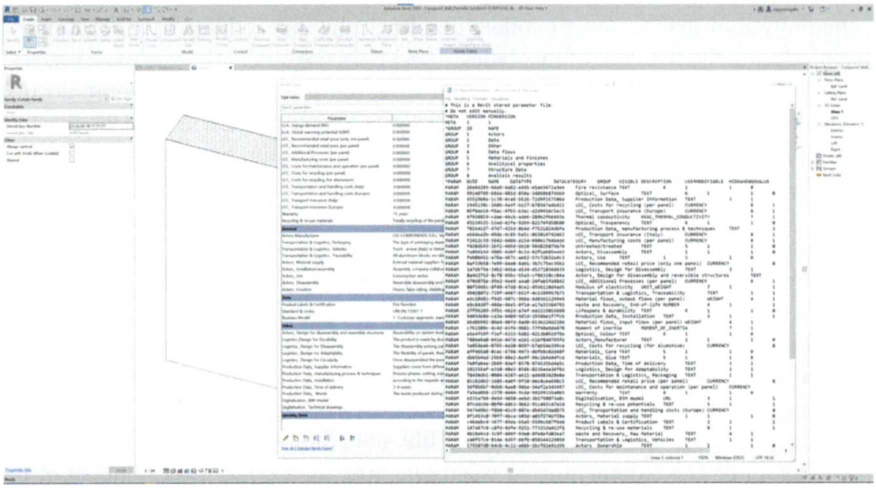

Fig. 40.3 Screenshot from Revit: elaboration of .txt file

for the Material Passport and then enter the specific data relating to the product you are working on.

The .txt then collects all "Shared Parameters" and can be used as a model to replicate the method to fill out the BIM model and add the Material Passport information. The aim is to create a Soft Tool that facilitate the Hard Reversibility of constructive systems.

The widespread use of the Material Passport as a data collection and sharing tool is useful if it is a homogeneous tool among operators and in the application sector. Thanks to the development of the Material Passport for the sandwich panel it was possible to provide a Framework as a compilation model within a BIM model (Fig. 40.3).

40.5 Conclusion

The built environment needs an important regeneration. The approach to design and build must change reference to the whole life cycle of the building: to conceive the end of service life products as resources; to develop reversible technologies, activating the potential circularity of disassembled constructive systems; to generate new recovery markets to create sustainable networks among stakeholders.

At the end of the practical experimentation and the planned activities, it was possible to verify the results obtained by evaluating the effectiveness and efficiency of the transformations activated to achieve the Soft Reversibility.

According to the vision of the company to be circular and reversible, the next proposed actions to actualize will focus on the production of reversible elements (layers of the panel dry assembled) and the creation of Second-use program to take-back the products disassembled. The practices cited will generate the development of a new business model in which the Product is conceived as a Service and the Manufacturer will become the responsible of the product (Extended Producer Responsibility—EPR).

The paper testified the importance to develop tools and methods that support the reversible actions and generate the real circular transition. Thanks to the census of the elements that compose the building through the Material Passport, it is possible to estimate the behavior of a constructive system over its whole life cycle and calculate its value during time. Integrating other methodologies to the BIM, for example the Life Cycle Costing, it is possible to estimate the increase of value achieved with the use of reversible technologies, and to support the design phase, guiding the choices toward a more conscious design and construction processes.

Furthermore, a future development of the Material Passport integrated in the BIM model of a product, with a LOD500, will consist in the addition of the parameters relating to the levels of reversibility, total and/or partial, considering the quantity of reversible construction systems (site, structure, skin, space plan, service, stuff). It would allow to provide indications about the effective reversibility that offer a low (partial reversibility) or high (total reversibility) increase in the value of the construction product. Finally, the BIM model integrated with the Material Passport could provide information related to the pre-demolition audit of the building system, preventing unnecessary waste, allowing the second use of the elements and maximizing the value and sustainable use of the building material.

Acknowledgements The study reported in the paper is included in the PhD research of the author supported by Department of Architecture, Built Environment and Construction Engineering of Polytechnic University of Milan. The acknowledgments may include the company Cel Components that supported the development of one of application of the PhD research cited.

References

Assefa G, Ambler C (2017) To demolish or not demolish: life cycle consideration of repurposing buildings. Sustain Cities Soc 28:146–153

Gorgolewski M (2017) Resource salvation: the architecture of reuse. Wiley Blackwell, Oxford, pp 66–74

Heinrich M, Lang W (2019) Materials passports—best practice. Innovative solutions for a transition to a circular economy in the built environment. Technische Universität München, BAMB, Munch

Jaillon L, Poon CS (2014) Life cycle design and prefabrication in buildings: a review and case studies in Hong Kong. Autom Constr 39:195–202

Jin R, Gao S, Cheshmehzangi A, Aboagye-Nimo E (2018) A holistic review of off-site construction literature published between 2008 and 2018. J Clean Prod 202:1202–1219

Vezzoli C, Kohtala C, Srinivasan A (2014) Product-service system design for sustainability. Sheffield

Chapter 41
Imagining a Carbon Neutral University

Antonella Violano and Monica Cannaviello

Abstract Universities are the main centers where the drivers of innovation for sustainability and decarbonization of the built heritage are investigated and developed. But are existing university buildings sustainable? If zero carbon buildings are to be our goal in 2050 (EU Green Deal), what is the current carbon footprint of these buildings? How can we enhance post-occupancy evaluation and drive technological and energy retrofits for participatory environmental design? This is the focus of the research carried out within the MedEcoSuRe (Mediterranean University as Catalyst for Eco-Sustainable Renovation) Project, funded by the European Union under the ENI CBC MED Program, which analyses and compares a number of sustainability assessment methods for existing university buildings (Green Metric, Stars, GRI, …) in order to develop the most effective indicators, not only to highlight the really virtuous buildings, but also to identify the strengths and weaknesses of the university building stock and to implement the most appropriate redevelopment strategies. According to the Renovation Wave Strategy, these approaches are aimed at improving not only the energy performance of buildings but will also improve the quality of life of people living in and using university buildings. The research considered multiple aspects concerning not only the environmental and functional performance of buildings, but also the direct satisfaction of users (providing a safe, healthy and comfortable environment for students, teachers and staff) and the strategies to manage energy, water, green and material resources during the operational phase (Xue et al. in Sustainability 12(1):294, 2020). The evaluation of environmental and functional performance of educational buildings should ensure that the effectiveness of buildings is maximized not just in terms of occupancy costs but also with respect to user satisfaction (Ekekezie et al. in Int J Progressive Res Sci Eng 2(8/202):388–397, 2021).

Keywords Sustainability · Assessment tool · University building · Indicators

A. Violano (✉) · M. Cannaviello
University of Campania "L. Vanvitelli", Aversa, Italy
e-mail: antonella.violano@unicampania.it

M. Cannaviello
e-mail: monica.cannaviello@unicampania.it

E. Arbizzani et al. (eds.), *Technological Imagination in the Green and Digital Transition*, The Urban Book Series, https://doi.org/10.1007/978-3-031-29515-7_41

41.1 Introduction

Universities have the primary task of investing in research and development of innovative technologies aimed to mitigate climate change. As centers where the drivers of innovation are studied and designed, they intrinsically have the role of demonstrators of the feasibility and effectiveness of policies for sustainability and decarbonization of the built environment (Fig. 41.1).

From a scenario analysis to think structurally about the challenges that universities might face in the long run, an interesting characterization of the typological identity found emerges: there are universities with an orientation mainly close to society (open to life acting as a financially strong cooperative partner), universities rather distant from society (conservative maintaining a niche existence) and universities with a mainly instrumental role (market-oriented generating profitable knowledge) (Barth et al. 2011).

The Committee for International Cooperation (CIC) highlighted that universities' commitment to sustainability is academic and involves its three missions: Education, Research, and Third Mission, and it is not implementable separately by the interested actors (del mar Alonso-Almeida et al. 2015).

The scientific work carried out by the DADI-Vanvitelli and ANEA research groups as part of the Project "Mediterranean University as Catalyst for Eco-Sustainable Renovation" (MedEcoSuRe), funded by the European Union under the ENI CBC

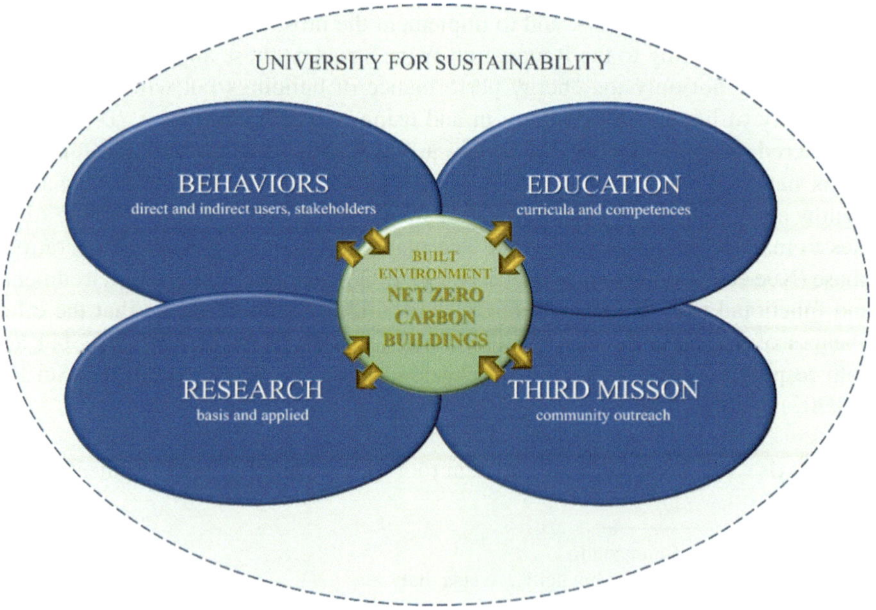

Fig. 41.1 University for sustainability: the central role of the net zero carbon built environment

MED Program, focuses on the environmental aspects of sustainability, in particular the management of energy and natural resources in university buildings.

In WP4—Policy and Project tools for Energy Efficiency retrofit in Higher Education Buildings, DADI-Vanvitelli research group different analyzed and compared some university sustainability assessment methodologies in order to extrapolate the most effective indicators to assess the environmental and energy performance of existing buildings, not only to highlight the truly virtuous buildings, but also to identify the strengths and weaknesses of the university building stock in order to implement the most appropriate renovation strategies that would be able to make them sustainable in the fullest sense of the term. According to the Renovation Wave Strategy, these strategies are intended to improve not only the energy performance of buildings but also the quality of life of people who live in and use university buildings. This is consistent with what was already stated by the Stockholm Declaration (1972) and reiterated by the Talloires Declaration (1990), which advocated the direct correlation between people and their living/studying/working environment, giving university buildings a key educating role in achieving environmental sustainability.

41.2 Tools for Assessment of Sustainability in Universities

The international strategies promoted by the European Green Deal and the New European Bauhaus lead us to question the environmental energy performance of the built heritage. If carbon neutral buildings are to be our goal in 2050, we need to understand: what is the current carbon footprint of university buildings? And how can we measure it? Almost all the investigated tools deal with the theme in a complex way, not separating the environmental energy assessment of the built environment from the ways of using it and from the awareness taught in these places of knowledge, giving strength to the concept that the habitat in which the human being lives, conditions in a biunivocal way his behaviors.

However, in many cases, there is a strong gap between the sustainability taught in the different courses of study, and the real performance (in terms of ecological footprint) of the buildings where they take place. For this reason, a series of operational tools, tested in different cultural areas of the world, have been studied in order to highlight not only the recurring non-negligible features, but also the strategies to enhance the best practices to implement a Cross Border Strategic Plan for University Building Retrofitting (WP 4.2). The research highlighted that several tools for assessing the sustainability of universities have been developed around the world over the past two decades.

Dalal-Clayton and Bass (2002) describe three main approaches to measure and analyze sustainability:

- Accounts (raw data that are then converted to a common unit: monetary, area or energy),
- Narrative assessments (that combine text, maps, graphics and tabular data and might use indicators),
- Indicator-based.

Indicator-based appraisal is certainly preferable for tackling the sustainability assessment challenge of university buildings. This kind of approach involves a comprehensive process of prioritization and ensures better strategy advancement, performance follows up and genuine decision-making and most importantly describes strengths and weaknesses (Adenle et al. 2020).

In most of the instruments analyzed, the indicators are generally divided into thematic categories, which attempt to assess, through multi-objective (qualitative–quantitative) criteria, all the aspects that make a University more or less sustainable. Usually, the indicators should cover the entire system to address: Education (referring to Courses and Curricula), Research, Campus operations, Community outreach and Assessment and reporting (Lozano 2006).

In the MedEcoSuRe research project, net zero carbon buildings assume a central role in this quadrilateral of convergence toward sustainability (Fig. 41.1), promoting cross-sector dialogue between institutions on sustainability and stimulating environmentally conscious behavior and learning.

In the United States, for the past two decades, academics and environmentalists have sought to evaluate places of knowledge based on their sustainable practices and policies, primarily through the tools proposed by three organizations: Association for the Advancement of Sustainability in Higher Education (AASHE), The Princeton Review and Sierra Club (Albis 2017). Among the tools developed, the one proposed by AASHE (Sustainability Tracking, Assessment and Rating System™–STARS®) is one of the most exhaustive, as well as being one of the first assessment systems specifically geared to assessing the sustainability of universities (Adenle 2020). This is a voluntary and transparent self-assessment framework, active since 2006, based on a well-structured set of indicators and used to assess a wide range of actions from energy use to transportation, procurement to academic offerings in the field of sustainability, against six main categories: Institutional Characteristics, Academics, Engagement, Operations, Planning and Administration and Innovation and Leadership. Instead, the Princeton Review, which publishes an annual green guide with rankings of America's sustainable universities (see https://www.princetonreview.com/press/green-guide/press-release-2022), assigns the score through a Green Rating. In the questionnaire administered to students, the questions regarding the energy-environmental performance of buildings are as follows:

1. Are school buildings that were constructed or underwent major renovations in the past three years LEED certified?
2. Does the school have a formal plan to mitigate its greenhouse gas emissions?
3. What percentage of the school's energy consumption is derived from renewable resources?

Therefore, Princeton Review includes more energy-related questions than any other topic (Albis 2017).

Assessing the efforts made toward sustainable development by universities is also covered by the Global Reporting Initiative (GRI): a voluntary tool, born in a predominantly corporate environment (Hahn and Kühnen 2013), which offers a comprehensive set of standards for reporting impacts related to the three dimensions, economic, environmental and social, aimed at 40 different sectors, divided into 4 main groups. Universities belong to Group 4: Other services and light manufacturing–Educational services Education services at all levels, including online education. This tool can also be used by universities (Lozano 2011), to communicate to the outside community how they address the dual mission of providing students with new skills to create a more sustainable society and reducing the environmental impact of their activities. In this second mission, the role of buildings and how they are designed, upgraded and managed takes on strategic importance. Although at the global university level the adoption of reporting standards through the GRI framework is not yet sufficiently widespread, European universities can still be considered pioneers in the adoption of such standards (del mar Alonso-Almeida et al. 2015).

41.3 Approach and Methodology

The critical and contextualized analysis of the main tools for assessing the sustainability of universities was carried out in order to:

- extrapolate more-representative indicators, to evaluate the energy-environmental performance of university buildings,
- attribute a weight to each indicator in relation to the role that sustainability of the built environment has in relation to the global rating,
- highlight, with a SWOT analysis, the critical elements, bringing out the significant aspects not yet evaluated,
- select other evaluation indicators for the aspects that have not been evaluated,
- analyze the criterion of connection between the tools to assess global sustainability,
- obtain a Rating system for sustainable buildings.

The SMART approach (Specific, Measurable, Achievable, Relevant and Limited in Time) (Alshuwaikhat et al. 2017) is the basis of the methodology for selecting indicators, made particularly laborious by the great complexity and interdependence of energy-environmental phenomena and socioeconomic impacts with which the tool must measure itself (Fig. 41.2).

The main macro-categories of indicators considered relevant refer to: energy use (for heating, cooling, ventilation, domestic hot water production and lighting), waste management, water resource management, air quality and integration of renewable energy sources. The indicators of effectiveness, referring to the above categories, are specifically selected in order to:

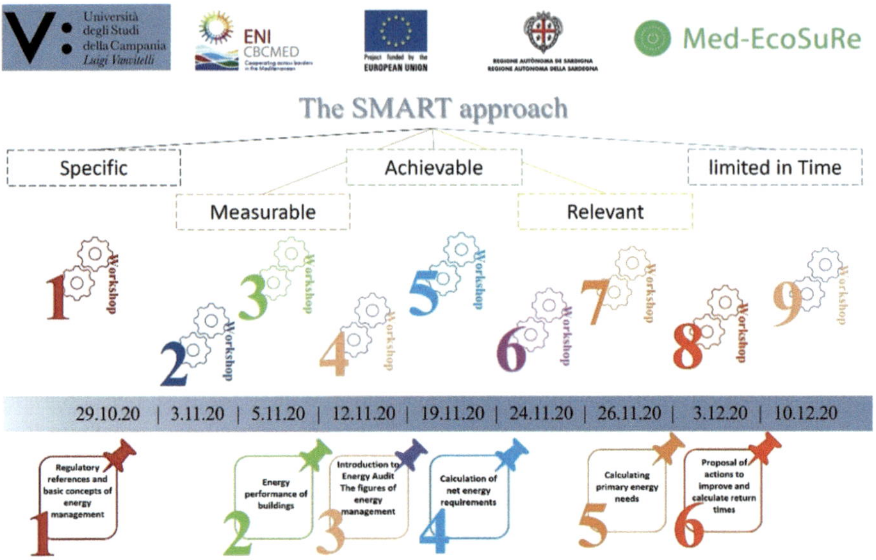

Fig. 41.2 The SMART approach in the structure of the MedEcoSuRe workshop

1. contribute to the pursuit of specific objectives (quantified if they lend themselves to quantification);
2. be consistent with proposed actions/strategies;
3. monitor the status of implementation of actions/strategies in terms of physical results, outcomes and, if possible, impact at the appropriate level;
4. contemplate economic and social implications.

For this reason, a prescriptive taxonomy of objectives is not possible, nor it is possible to provide a univocal indication of the indicators that must be used to evaluate the effectiveness of the proposed actions. Instead, the recommendation is to univocally identify the limits of validity of the selected indicators. This will allow the start of a comparison of certain interest and legitimacy in the perspective of a consolidation and improvement of the programming action of eco-oriented actions of the universities.

From the analysis of the assessment tools, it emerges that a point of weakness is certainly the applicability of the same tool in different contexts, even if under similar conditions and purposes. Climatic and geomorphologic conditions, but also the binding legislative context and local government policies dictate the rules of use of economic and environmental resources.

During the International Workshop "Energy Efficiency Action Plan in the Higher Education Building Sector", organized by DADI-Vanvitelli within the MedEco-Sure Project, the challenge was not so much to analyze as to compare the energy-environmental performance of three pilot buildings, located in Italy, Palestine and

Tunisia, using the same assessment tool. Similarly, in the choice of retrofit interventions to improve the existing performance, the knowledge of the local building systems, of the relevant technical and historical elements and of the local technological culture has been fundamental, guiding, with technical sensitivity, the proposed design solutions in relation to the perception of the cultural/architectural value intrinsically present in the studied buildings.

41.4 Results: Analytical, Propositional and Debate Aspects

The literature review of the major tools indicated that the most comprehensive tool for assessing building performance is Sustainability Tracking, Assessment and Rating Systems (STARS®). Only Sustainability index Model–DPSEEA (Waheed et al. 2011), Sustainable Campus Assessment System (SCAS) (Hokkaido University 2013) and STARS® have extensively included spatial indicators at both indicator and sub-indicator levels (Adenle et al. 2020).

The indicators used by STARS® to assess the energy-environmental performance of university buildings, and the use of renewable energy, are contained in the Operation category and are listed in Table 41.1, where for each credit are indicated Points available, Applicable to, Minimum requirement (Table 41.1).

With respect to indicators OP3 and OP4, the score is attributed, according to STARS® 2.2 Technical Manual, to the buildings that were constructed or underwent major renovations (in the previous five years) were designed and built in accordance with a published green building code, policy/guideline and/or rating system.

Green building codes, policies/guidelines and rating systems may be:

- Multi-attribute
- Single-attribute: focusing predominantly on one aspect of sustainability such as energy/water efficiency, human health and well-being, or sustainable sites.

Third-party certification under a multi-attribute green building rating system developed/administered by a WorldGBC member Green Building Council (GBC) is weighted more heavily for scoring purposes (Table 41.2).

"Each rating system also has criteria related to LEED/sustainable certified buildings. STARS and Sierra Club go as far to measure percentage of certified sustainable building space" (Albis 2017).

Global Reporting Initiative (GRI) uses the following indicators to assess energy sustainability:

- GRI 302-1 Energy consumed within the organization
- GRI 302-2 Energy consumed outside the organization
- GRI 302-3 Energy intensity
- GRI 302-4 Reduction in energy consumption
- GRI 302-5 Reduction in the energy requirements of products and services.

Table 41.1 Credit, applicability, criteria and scoring in STARS® 2.2 Technical Manual

Credit, number and title	Points available	Applicable to	Minimum requirement
OP3 Building Design and Construction	3	Institutions that have new construction and/or major renovation projects completed within the previous five years	Own new or renovated buildings that were designed and built in accordance with a published green building code, policy/guideline, or rating system
OP4 Building Operations and Maintenance	5	All institutions	Own buildings that are operated and maintained in accordance with a sustainable management policy/program or a green building rating system focused on the operations and maintenance of existing buildings
OP5 Building Energy Efficiency	6	All institutions	Have data on grid-purchased electricity, electricity from on-site renewables, utility-provided steam and hot water, and stationary fuels and other energy products
OP6 Clean and Renewable Energy	4	All institutions	Support the development and use of clean and renewable energy sources

The Green Metric, promoted in 2010 by the University of Indonesia and whose reference for Italy is the University of Bologna, was also studied as part of the research. This tool groups indicators into six macro-categories to which a specific weight is attributed:

1. Setting and Infrastructure (15%)
2. Energy and Climate Change (21%)
3. Waste management (18%)
4. Water use (10%)
5. Means of transport (18%)
6. Education and research (18%).

Table 41.2 Relation between credit ad rating system in STARS® 2.2 Technical Manual

Credit	Type of rating system	Rating system
OP3 Building Design and Construction	Multi-attribute GBC rating systems	BREEAM, CASBEE, DGNB, Green Star, LEED BD+C, LEED ID+C, Living Building Certification, Parksmart
	Multi-attribute non-GBC rating systems	Green Globes NC
	Single-attribute rating systems	EDGE, Fitwell, Living Building Petal Certification, Net Zero Energy, Passive House/Passivhaus, WELL, ZCB-Design
OP4 Building Operations and Maintenance	Multi-attribute GBC rating systems	BREEAM-In Use, CASBEE for Existing Buildings, DGNB, Green Star Performance, LEED O+M, Parksmart Pioneer
	Multi-attribute non-GBC rating systems	BOMA BEST, Green Globes EB
	Single-attribute rating systems	EDGE, ENERGY STAR, Fitwell, TRUE, WELL, ZCB-Performance

Within the Italian Network of Sustainable Universities (RUS), a simplified methodology based on the verification of some minimum requirements related to automation, energy, water, indoor comfort, lighting and security, developed by the Energy Working Group of RUS, coordinated by the Polytechnic of Turin, and has been adopted.

Assessing the sustainability of university buildings has to take into account multiple aspects that relate not only to the environmental and functional performance of buildings, but also to direct user satisfaction (providing a safe, healthy, comfortable environment for students, teachers and staff).

The evaluation of environmental and functional performance of educational buildings should ensure that the effectiveness of buildings is maximized not just in terms of occupancy costs but also with respect to user satisfaction (Ekekezie et al. 2021).

However, the analysis of the analyzed tools showed that the centrality of the direct user and his perception of sustainability and comfort is not among the evaluation indicators. Moreover, the evaluation of the green potential of the building, that can be defined as the "capacity to refurbish a conventional building into a green building (green refurbishment) through architectural interventions" (Ben Avraham and Capeluto 2011) is delegated to other assessment tools.

In light of these considerations, in the MedEcoSuRe research we see the need to

- investigation of direct users: identification of critical issues in relation to specific modes of use
- verification phase of the green potential of buildings.

In fact, the research considered multiple aspects concerning not only the environmental and functional performance of buildings, but also the direct satisfaction of users (providing a safe, healthy and comfortable environment for students, teachers and staff) and the strategies to manage energy, water, green and material resources during the operational phase (Xue et al. 2020).

A questionnaire was administered to the students based on indicators both related to indoor environmental quality (air quality, temperature, ventilation, room acoustics, natural and artificial lighting) and related to obvious performance/functional deficiencies of the building as directly encountered by the end user. These aspects played an important part of the Participatory Energy Audit (Violano et al. 2021) in which students were involved during the workshop. However, it was not possible to collect all the data necessary for a thorough evaluation because the workshop took place during the COVID-19 pandemic and the students were not allowed to attend the university continuously and under normal conditions to do an appropriate direct evaluation. For this reason, the research work has been delayed and is currently in progress.

41.5 Conclusions

Among the many aspects that affect the environmental and functional performance of buildings, the research also brought out the need to consider the satisfaction of direct users (providing a safe, healthy and comfortable environment for students, teachers and staff). In fact, evaluating the energy and environmental performance of places of knowledge should ensure that the effectiveness of buildings is maximized not only in terms of occupancy costs, but also with respect to user satisfaction (Ekekezie et al. 2021). Providing a fundamental tool to support the decision-maker (energy manager), this user satisfaction analysis should help identify functional and environmental inadequacies of building performance in universities, and most importantly, it should be aimed at improving the quality of life of people living in and using university buildings.

This research yielded three significant findings. It analyzed and discretized several possible approaches to assessing the sustainability of universities, comparing some of the most widely used tools. Second, it has unequivocally shown that the way buildings are designed (energy and environmental performance) is relevant to overall sustainability. Finally, it provides methodological indications for the decision-maker (primarily the Energy Manager) in the energy and environmental upgrading of university buildings, proposing criteria for defining priorities both in the choice of interventions and in the buildings on which to intervene.

In conclusion, the green transition requires the improvement of the quality of the built environment by incorporating the principles of sustainability (Humblet et al. 2010), the creation of healthy living and learning environments by establishing policies and regulations that encourage sustainable practices in daily activities and decision-making processes (Alsharif et al. 2020) and a reduction of environmental impacts that depend directly on the policies and actions of universities (Creighton 1998). Guidance can be derived from this analysis to support administrators and energy managers in both assessing priorities for action and the best strategies that can be implemented.

Acknowledgements The contribution concerns the international research project MedEcoSuRe, funded by ENI CBC MED Program 2014–2020. It is the result of a common reflection of the involved authors; in particular: M. Cannaviello on "1. Introduction", "2. Tools for assessment of sustainability in universities"; A. Violano on "3. Approach and methodology", "4. Results: analytical, propositional and debate aspects" and "5. Conclusion".

References

Adenle YA, Chan EHW, Sun Y, Chau CK (2020) Exploring the coverage of environmental-dimension indicators in existing campus sustainability appraisal tools. Environ Sustain Ind 8:100057. http://doi.org/10.1016/j.indic.2020.100057

Albis J (2017) University sustainability: assessing college sustainability rating systems. J Environ Sustain 5(1), Article 3. Available at: http://scholarworks.rit.edu/jes/vol5/iss1/3. Accessed on 14/03/2022

Alsharif MA, Peters MD, Dixon TJ (2020) Designing and implementing effective campus sustainability in Saudi Arabian universities: an assessment of drivers and barriers in a rational choice theoretical context. Sustainability 12:5096. https://doi.org/10.3390/su12125096

Alshuwaikhat HM, Abubakar IR, Aina YA, Adenle YA, Umair M (2017) The development of a GIS-based model for campus environmental sustainability assessment. Sustainability 9(3):1–23. https://doi.org/10.3390/su9030439

Barth M, Adomßent M, Albrecht P, Burandt S, Balsen AF, Godemann J, Rieckmann M (2011) Towards a "sustainable university": scenarios for sustainable university development. Int J Innov Sustain Dev 5(4):313–332. Available at: http://doi.org/10.1504/IJISD.2011.043322. Accessed on 14/03/2022

Ben Avraham O, Capeluto IG (2011) A tool for determining the green potential of existing buildings. Paper presented at the PLEA 2011—27th conference on passive and low energy, architecture. Louvain-la-Neuve, Belgium

Creighton SH (1998) Greening the Ivory tower: improving the environmental track record of universities, colleges, and other institutions. MIT Press, Cambridge, MA, USA

Dalal-Clayton B, Bass S (2002) Sustainable development strategies, 1st edn. Earthscan Publications Ltd., London, p 2002

del Mar Alonso-Almeida M, Marimon F, Casani F, Rodriguez-Pomeda J (2015) Diffusion of sustainability reporting in universities: current situation and future perspectives. J Cleaner Prod 106144–154. https://doi.org/10.1016/j.jclepro.2014.02.008

Ekekezie CU, Okolie KC, Okongwu MI (2021) Evaluation of the building performance and environmental sustainability principles of university buildings in South-East Nigeria. Int J Progressive Res Sci Eng 2(8/202):388–397. Available at: https://www.journals.grdpublications.com/index.php/ijprse/article/view/390. Accessed on 14/03/2022

Foster I, Kesselman C (1999) The grid: blueprint for a new computing infrastructure. Morgan Kaufmann, San Francisco

Hahn R, Kühnen M (2013) Determinants of sustainability reporting: a review of results, trends, theory, and opportunities in an expanding field of research. J Clean Prod 59:5–21. https://doi.org/10.1016/j.jclepro.2013.07.005

Hokkaido University (2013) Sustainable campus assessment system. Retrieved from http://www.eauc.org.uk/theplatform/sustainable_campus_assessment_system. Accessed on 14/03/2022
Humblet EM, Owens R, Roy LP (2010) Roadmap to a green campus. Green Building Council (USGBC), Washington, DC, USA
Lozano R (2006) A tool for a graphical assessment of sustainability in universities (GASU). J Clean Prod 14(9–11):963–972. https://doi.org/10.1016/j.jclepro.2005.11.041
Lozano R (2011) The state of sustainability reporting in universities. Int J Sustain High Educ 12(1):67–78
Syed Yahya SNN, Ariffin ARM, Ismail MA (2014) Green buildings in campus: an assessment of green potential for existing conventional buildings. J Build Perform 5(1):2014
Talloires Declaration (1990) University leaders for a sustainable future. Available at: https://ulsf.org/talloires-declaration/. Accessed 14/03/2022
Violano A, Ibrik I, Cannaviello M (2021) Human-centered design: participated energy retrofit for educational buildings. Sustain Mediterr Constr (SMC) J 13:106–116. Available at: http://www.sustainablemediterraneanconstruction.eu/SMC/The_Magazine_n_13_files/1313.pdf. Accessed on 14/03/2022
Waheed B, Khan FI, Veitch B (2011) Developing a quantitative tool for sustainability assessment of HEIs. Int J Sustain High Educ 12(4):355–368. https://doi.org/10.1108/14676371111168278
Xue Z, Liu H, Zhang Q, Wang J, Fan J, Zhou X (2020) The impact assessment of campus buildings based on a life cycle assessment–life cycle cost integrated model. Sustainability 12(1):294. http://doi.org/10.3390/su12010294

Chapter 42
Life Cycle Assessment at the Early Stage of Building Design

Anna Dalla Valle

Abstract In view of the urgent need to construct informed and advanced vision of the built environment in terms of environmental impacts, Life Cycle Assessment (LCA) is even more emerging as the most recognized supporting tool for Architectural, Engineering and Construction (AEC) practices. This is proved by Level(s), a voluntary framework established in Europe that is fully life cycle-based, looking buildings beyond energy performance to the whole life cycle, while fostering the implementation of circular economy strategies. To face buildings complexity, it recommends applying life cycle approach with an increasing level of detail and accuracy, shifting from the assessment of carbon emissions to complete cradle to grave LCA. In this context, many calls for competitions at the reach of environmentally sustainability include Level(s) measures as reference frame to deal with. The paper provides insights of building LCA application performed during the preliminary design phases, since crucial for the decision-making process especially if operating into competition aimed at minimizing environmental impacts. In particular, a sample of building projects developed to address an international architecture competition specifically committed to decarbonization issues in compliance with Level(s) is discussed. Starting from a concrete in situ scenario, the attention is on integrating dry assembled solutions composed of environmental-friendly materials. Results show range of carbon footprint of low-carbon buildings in relation to building shape and volume, outlining building parts that generally contribute to highest release of CO_2 and providing effective technological solutions. The aim is to support AEC practitioners in the design and implementation of buildings embracing a life cycle approach starting from the early design process.

Keywords Life cycle assessment · Preliminary design · Low-carbon building · Decision-making

A. Dalla Valle (✉)
Polytechnic University of Milan, Milan, Italy
e-mail: anna.dalla@polimi.it

E. Arbizzani et al. (eds.), *Technological Imagination in the Green and Digital Transition*, The Urban Book Series, https://doi.org/10.1007/978-3-031-29515-7_42

42.1 Introduction

Technological innovation is commonly defined as a purposeful activity aimed at developing new products/services as well as new methods to produce, distribute and use them. However, if it intends to have a positive impact on human life, the associated impacts on the environment has too long been neglected, now turning out to be at the forefront of the debate. An evidence is the stressing of advanced technologies to ensure energy efficiency but producing some kind of rebound effects on the environment in terms of resource consumption and waste generation (Huang et al. 2018). Hence the pressing need to develop and apply technological innovations looking at the whole life cycle to take a holistic and systemic approach.

In the field of construction, due to the high environmental, economic and social impacts, this vision is even more shared, promoting the adoption of life cycle criteria, methods and metrics (Röck et al. 2021). At European level such effort is sustained in practice by Level(s), a common EU framework of core indicators for assessing the sustainability of buildings over their full life cycle (European Commission 2021). It offers an extensively tested system for supporting stakeholders across the construction and real estate value chain in measuring and achieving improvements from design to end-of-life, by identifying sustainability hotspots to develop future-proofing buildings, contributing to key EU policy goals (e.g., European Green Deal, Circular Economy Action Plan, Renovation Wave and green transition toward carbon neutrality). In particular, the core sustainability indicators concern carbon, materials, water, health, comfort and climate change impacts throughout the entire building life cycle, acknowledging Life Cycle Assessment (LCA) to estimate effects and define effective ecological transition actions (Sala et al. 2021).

42.2 Call for Competition as Driver Toward Life Cycle Design

In this context, building design becomes a favorable testing ground for experimentation, because here technological innovations are still malleable unlike once completed.

Besides creativity that clearly play a crucial role, design is seen as a form of structured decisions driven by rationality and a set of criteria. Accordingly building project is meant as multi-criteria decision problem to be solved by choosing the design option that maximally meets all requirements. This is why it is pivotal the integration of life cycle criteria into the design decision-making process (Trigaux et al. 2021).

As a result, many design competitions start calling for the application of LCA to verify the implementation along the whole life cycle of solutions aimed at the environmental sustainability. In compliance with Level(s), the claim is for a progressive integration of LCA into design process, starting from simplified assessments based on a single environmental indicator up to complete assessments based on the whole set of indicators (Hollberg et al. 2020). Aim of the paper is thus to check up on the life cycle design pursued to meet competition requirements at the forefront of environmentally sustainability. In particular, the focus is on the early stage decision-making, where leading strategical ideas are arguably placed (Bueno et al. 2018; Najjar et al. 2019) to later affect the entire building process (Rezaei et al. 2019). To this end, a sample of competing projects is discussed in detail, providing insights on LCA application within Architectural, Engineering and Construction (AEC) practices to foster the development of low-carbon buildings proposal.

The design call at issue is the Architecture Student Contest (Saint-Gobain 2022), a two-stage international competition, intended for collecting visions about the revitalization of the district around the East railway station of Warsaw (Poland). Participants are expected to design new housing and to transform the existing old factory into a community hub for meetings, cultural events and leisure activities. Beyond the standard technical parameters of previous editions (thermal/acoustic comfort, indoor air quality, fire safety, natural daylight, energy efficiency), the novelty is the request of building carbon emissions. Such calculations have to assess the whole building life cycle by means of OneClick LCA tool. Moreover, special attention is paid on the achievement of circular buildings and resource efficient solutions. Buildings have to be designed for longevity, resulting flexible in use and easily adaptable over time, with durable and reversible technological solutions, to become material banks for future generations. The embedded resources must rely on efficient products made with minimum use of non-renewable materials and maximum share of recycled content, valuable at their end-of-life for reuse (preferred option) or recycle scenarios. Note that the selection of a single environmental indicator (Global Warming Potential) is in line with the decarbonization goals of construction sector as well as with Level(s), drawing upon the suggested LCA methods and metrics toward circular economy.

The life cycle design promoted by the call is following deepened through the analysis of a sample of 12 architectural projects submitted to the first stage at Italian level. Specifically, with respect to the whole proposed district, one building is in the spotlight for each project for understanding how LCA affects the decision-making starting from early design phases. The goal is to outline the way of practice to reduce/optimize the embodied carbon of buildings for supporting AEC practitioners in the development of low-carbon buildings.

42.3 Building LCA Application for Low-Carbon-Oriented Decision-Making

The major challenge of the competition is to reach the greatest improvement to building performance (meeting all design requirements) while reducing the project carbon footprint. To succeed the call, the effort is to strike out against the conventional notion in which best performance imply the use of more resources: buildings need to be high-performing and resource efficient, carefully assessing design solutions for limiting the environmental impacts along the whole life cycle. For this purpose, all design teams carried out LCA analysis starting from the concept phase to orient the decision-making process and select the most suitable technological solutions. Building LCA have been developed in compliance with EN standards (EN15978) by experiencing the recommended LCA tool and focusing on carbon footprint (CO_2eq). As required, the evaluation considers the entire life cycle from cradle to grave (production, transport, replacement, end-of-life), including all building material flows (operational energy is out of scope) and assuming 60 years of service life.

For defining emission reduction, the designed buildings are compared with "Business As Usual" (BAU), modeled by accounting the same shape, function and location but with standard performance and technological solutions. In this way, the baseline models vary in terms of mass consistent with the investigated projects but keep equal constructive solutions, appropriately sized with the minimum performance demanded. BAU models are based on concrete in situ scenario and consists of: concrete foundation; concrete frame structure; concrete ground slab assembly including insulation (ground floors U-value at 0.30 W/m^2 K); hollow clay bricks wall assembly with insulation and render finishing (external wall U-value at 0.20 W/m^2 K); triple glazed aluminum frame window (windows U-value at 0.90 W/m^2 K); concrete roof assembly with concrete tiles (roof U-value at 0.15 W/m^2 K); concrete slab assembly for intermediate floors and balcony; ceramic tiles and plasterboards for slab finishes; concrete wall assembly and hollow clay bricks assembly for internal wall (10% and 90% respectively); concrete assembly for stairs and elevator shafts. All designed and BAU buildings are intended for student housing.

To meet competition requirements, the preliminary phase is distinguished, besides the design of the architectural space and functional arrangement, by the evaluation of different technological solutions in order to minimize impacts compared to baseline buildings. As opposed to BAU wet solutions, all design teams opted first of all for dry construction technologies, off-site prefabricated elements, modular construction and lightweight systems in order to ensure circularity criteria. In particular, 75% of the investigated projects switch in favor of steel scenario, while the remaining 25% for wood scenario, later optimizing step-by-step the different construction technologies. Before getting into detail, it is important to emphasize that the life cycle-oriented decision-making process allowed an average reduction of carbon emission of −26% for the designed buildings compared to BAU. Figure 42.1 shows the carbon footprint reduction for each building project, ranging from −7% in less virtuous cases to −44% in the most virtuous cases. Normalizing the total building carbon emission per gross surface area, it is possible to identify the average BAU impacts at 534 kg CO_2e/m^2 (range 380–659 kg CO_2e/m^2), while the average design impacts at 391 kg CO_2e/m^2 (range 271–522 kg CO_2e/m^2).

Despite the correlation between building mass and carbon emissions is not evident (Fig. 42.2), shape and volume affect to some extent the impacts. This is proved by the fact that buildings with about the same surface-volume ratio (i.e., envelope surface and heated volume) present different values of kg CO_2e/m^2. For instance, with reference to 0.27 ratio, biggest is the volume, lower are the impacts. However, the

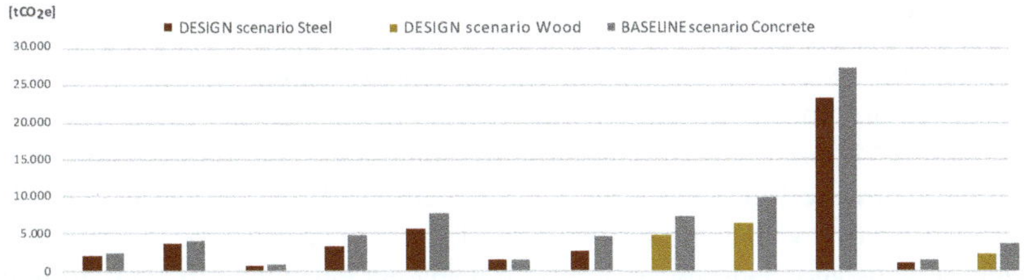

	B01	B02	B03	B04	B05	B06	B07	B08	B09	B10	B11	B12
Gross Floor Area [m^2]	3.988	7.300	1.402	8.593	19.952	3.500	8.525	13.596	16.662	52.540	2.304	6.837
Height [m]	25	22	15	24	26	25	21	23	25	17	25	25
Surface-Volume ratio	0,25	0,21	0,37	0,27	0,27	0,30	0,22	0,34	0,18	0,24	0,40	0,27
DESIGN [tCO_2e]	**2.082**	**3.438**	**697**	**3.176**	**5.417**	**1.346**	**2.548**	**4.607**	**6.284**	**22.628**	**951**	**2.186**
BASELINE [tCO_2e]	2.289	3.927	864	4.752	7.582	1.442	4.578	7.151	9.697	26.795	1.518	3.528
% emission reductions	*-9%*	*-12%*	*-19%*	*-33%*	*-29%*	*-7%*	*-44%*	*-36%*	*-35%*	*-16%*	*-37%*	*-38%*

Fig. 42.1 Reduction of carbon emissions (cradle to grave) between designed building and BAU model [tCO_2e]

display of scattered and oscillating values and thus the failure of trend evidence means that technological solutions grave deepest in the definition of impacts compared to surface-volume ratio. Hence the importance for the design teams of paying much attention to the decision-making selection process of the technological solutions to propose at building scale.

Concerning construction technologies, design efforts for carbon reduction were concentrated, depending on the projects, on specific parts or on the whole buildings, always assessing different alternatives by performing LCA from cradle to grave, to take a comprehensive view of impacts. The optimization process focused especially on building parts above ground, maintaining BAU solutions for foundations and ground slabs. Figure 42.3 (sx) displays the average contribute in terms of CO_2eq

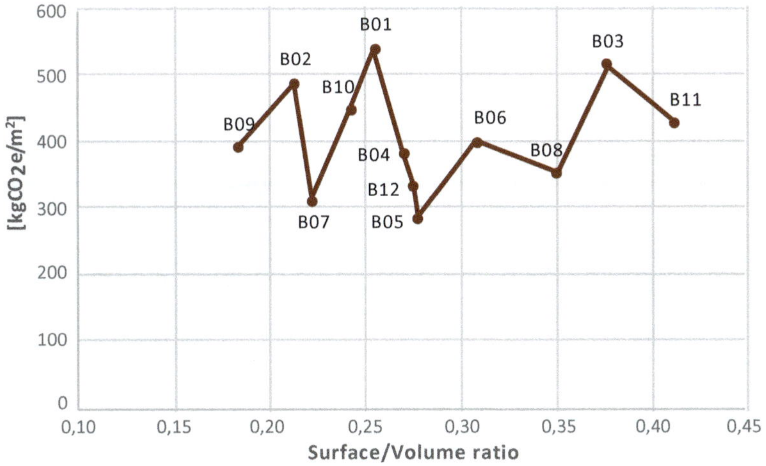

Fig. 42.2 Building carbon emissions (cradle to grave) by *S*/*V* ratio [kg CO_2e/m^2]

of the different building parts on the total carbon emissions, useful during preliminary design to steer decision-making process and optimize building parts with the greatest environmental impacts. Results reveal that the largest share is associated to intermediate floors (34.8%), including resistant slabs and both paving and ceiling finishes. They are followed in turns by: structures (15%), including frames and, if any, load-bearing walls; internal walls (12.1%); vertical envelope (11.1%), including external walls and cladding assembly; transparent envelope (10.7%); ground floors (8.1%); roof (4.5%), including resistant slabs and finishes; foundations (3.2%) and finally accessories as balconies (0.4%). Note that summing up vertical and transparent envelope the share rises at 21.8%, deserving thus close attention of designers, unlike balconies that appear negligible since not provided by many projects due to the climatic conditions of the application context. Anyway, it is worth mentioning that the percentage contribution to the total carbon footprint entails a variability of values among the different building projects because of the different technological solutions. Below an overview of the selected solutions, according to their importance within the decision-making process designed to reduce building carbon emissions along the entire life cycle.

Contrary to BAU, more than half of designed buildings chose metal-concrete intermediate floors, even if intending to provide a full dry constructive technology at present not available within the used LCA tool. The remaining opted for CLT slab assembly and, to a lesser extent, for wooden joist assembly or for traditional solutions in hollow-core or concrete slab assembly. Intermediate floors are finished on the top with parquet flooring and/or ceramic tiles, both including underlayment membrane, and on the bottom mainly with plasterboard or suspended ceiling and in only one case with wood cladding. As advanced, BAU concrete structures have been replaced, as appropriate, with steel or wood frame (in the latter switching also internal load-bearing wall in CLT panels) but maintaining concrete assembly for

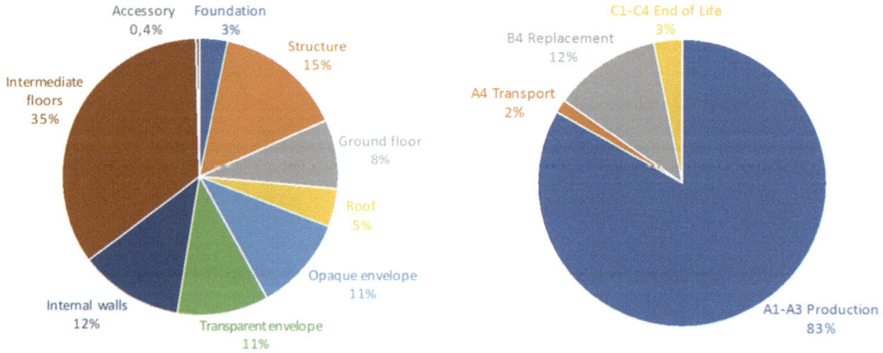

Fig. 42.3 Average carbon emissions (cradle to grave) per buildings parts (sx) and per life cycle phases (dx) [% of impacts]

stairs and elevator shafts due to missing dry solutions. In all building projects with underground spaces, basements are in concrete sandwich wall assembly including insulation.

Internal wall solutions rely for the majority of buildings on steel stud wall assembly and to a lesser degree on wooden stud wall assembly, both including insulation. Dry construction technologies are also selected for vertical envelope, in which timber frame external wall results the prevailing options, even if envisaged in most cases with metal frame (not present into the database). Alternatively, few projects choose insulated steel sandwich wall assembly. If technological solutions for external walls are mostly shared among design teams, they are miscellaneous for cladding system, ranging from the most common fiber cement sheet or natural stone cladding to wood cladding or curtain wall with aluminum frame. Concerning windows, the inclusion of triple glazed wood frame windows in lieu of aluminum frame allows to decrease carbon emissions. Nevertheless, more significant benefits are provided in buildings with double glazed windows both in aluminum and PVC, once ensured the availability on the market of products able to meet the required performance.

With regards to horizontal envelope, all projects modeled ground floors as equal to BAU, only changing the concrete roof solution into steel or wooden frame assembly and, to a lesser extent, into CLT compact roof. Regarding roof finishing, there are no predominant solutions among the investigated projects, composed of a wide range of alternative options: sheet roofing in steel, aluminum or fiber cement; OSB sheeting board and bitumen membrane; double layer of asphalt roofing membrane; slate roof tiles. Finally, for shifting from wet to dry technological solutions, balconies are mostly designed with wooden assembly, even when supported by steel structure (missing solution), to avoid excessive penalization of impacts by keeping the standard concrete balconies.

The designed buildings result from the different combination of the presented technological solutions, reducing carbon emissions in a more or less distributed way between the various building parts. However, looking at the whole life cycle, it is important to emphasize not only the incidence of the different construction technologies, but also how they affect the different phases of life cycle. To this end, Fig. 42.3 (dx) provides an overview of the average percentage share allocable to each phase assessed within the preliminary LCA in relation to the overall carbon emissions. Since operational energy is outside the system boundary and the assessment focuses on material flows over the entire life cycle, the production phase (A1–A3) is as expected responsible for major impacts, counting about 83% of total building CO_2eq. It is followed by the replacement phase (B4) that accounts for approximately 12% and afterwards by the end-of-life phase (C1–C4) and the transportation phase (A4) which represent 3% and 2%, respectively. Of course, these percentage shares vary slightly among the different building projects depending on the selected technological solutions, which notably afflict production and replacement phases. Indeed, while carbon impacts of the transportation and the end-of-life phases are mostly constant, the production phase ranges from 74 to 89% and the replacement phase from 8 to 19% according to technological choices.

42.4 Conclusions

The results discussed above derives from the LCA application at early stage of building design, intended for orienting the decision-making process toward the decarbonization of construction sector and thus the creation of low-carbon buildings. In particular, they deal with a sample of building projects submitted to a design contest which targets carbon emission reduction along the whole life cycle as well as the implementation of circular strategies, in line with Level(s) goals and metrics. Findings are helpful for supporting AEC practitioners in the development of sustainable building and practice, by identifying hotspots over the entire life cycle and outlining effective technological solutions. Notwithstanding, for ensuring informed choices, it is necessary to bear in mind some key issues, related to both the modeling of the foreground system (quantitative data) and background system (environmental data).

Firstly, the constraints imposed by the use of the recommended LCA tool (Carbon Designer) in the definition of impacts (Meex et al. 2018), starting from the modeling of the building geometrical parameters. The tool setting is structured in a user-friendly way that allows automatic calculation of all building reference surfaces from very few parameters in input, such as gross floor area. This served well especially during the concept phases, while offering the possibility of manually adjusting parameters to better reflect the designed building mass. However, it is worth stressing that it is a simplified model, particularly when buildings have complex shapes both on floor (e.g., envelope efficiency factor variable between different sides) and in height (e.g., overlapping floors with difference shapes), but also merely in the presence of loggias instead of balconies. In fact, in the latter case, the loggia slab become the roof of the lower floor, inducing user to deduct the related surface from inter-floor slabs and to add it to the roof, in order to obtain a more representative evaluation (zeroing the balcony surface and considering insulation).

The second key aspect is the limited range of technological solutions to choose from.

If, on one hand, it is useful to orient the decision-making process especially in early design phase, on the other, it risks of restricting design freedom and thus the implementation of innovative solutions. A first attempt to solve the gap has been made by giving users the option to change the layer thickness and rarely also specific material. Nevertheless, in view of current construction sustainability goals and challenges, it is recommended to enrich the set of dry solutions (to date very limited), to enable the edit of external walls, including metal stud in lieu of wood stud, and to expand at least the range of insulation materials to make environmentally-informed decisions about. Moreover, concerning envelope solutions, it would be extremely valuable to display impacts of the chosen solutions in conjunction with transmittance values (pre-calculations to be refined over design process).

In addition, it is important to note that the LCA application at early stages of building design is based on a very few input data, relating in particular to the geometrical modeling of buildings (from which surfaces of different parts are derived) and the selection of technological solutions. This leads to the automatic calculation of material bill of quantities, associating them not only the environmental impacts during the production phase, but also those of the downstream phases by using default values. However, if this approach is effective in the preliminary phase, it is always necessary to check the representativeness of the pre-set quantitative data and gradually detail them as the design process advances (Dalla Valle 2021). For instance, transport distance, replacement cycles and assumptions underlying the end-of-life scenarios turn out to be of significant concern.

Also of note, the calculation of the carbon footprint does not consider the contribution of biogenic carbon of wood-based products, which is evaluated separately inside the LCA tool.

Whilst conducted at preliminary stage and thus with higher degree of uncertainty, it is possible to identify among the investigated projects more or less accurate assessments of building impacts in term of carbon emissions. Indeed, for each building part, the tool allows to select multiple technological solutions simultaneously, enabling users to assign to each one specific percentage of the total. This functionality has been exploited only by few design teams who, for example, specified the share of green roof, rather than of the glass partition wall including the aluminum framing. In this way, depending on the effort of designers, even at the preliminary phases, it is possible to define different levels of detail, that inevitably affect the obtained results and therefore the reduction in the rate of building carbon emissions.

References

Bueno C, Pereira LM, Fabricio MM (2018) Life cycle assessment and environmental-based choices at the early design stages: an application using building information modelling. Arch Eng Des Manag 14(5):332–346. https://doi.org/10.1080/17452007.2018.1458593

Dalla Valle A (2021) Change management towards life cycle AE(C) practice. SpringerBriefs, Cham. https://doi.org/10.1007/978-3-030-69981-9

European Commission (2021) Level(s) European framework for sustainable buildings. https://ec.europa.eu/environment/levels_en

Hollberg A, Genova G, Habert G (2020) Evaluation of BIM-based LCA results for building design. Autom Constr 109:102972. https://doi.org/10.1016/j.autcon.2019.102972

Huang L, Krigsvoll G, Johansen F, Liu Y, Zhang X (2018) Carbon emission of global construction sector. Renew Sust Energ Rev 81:1906–1916. https://doi.org/10.1016/j.rser.2017.06.001

Meex E, Hollberg A, Knapen E, Hildebrand L, Verbeeck G (2018) Requirements for applying LCA-based environmental impact assessment tools in the early stages of building design. Build Environ 133:228–236. https://doi.org/10.1016/j.buildenv.2018.02.016

Najjar MK, Figueiredo K, Evangelista ACJ, Hammad AW, Tam VW, Haddad A (2019) Life cycle assessment methodology integrated with BIM as a decision-making tool at early-stages of building design. Int J Constr Manag 1–15. https://doi.org/10.1080/15623599.2019.1637098

Rezaei F, Bulle C, Lesage P (2019) Integrating building information modeling and life cycle assessment in the early and detailed building design stages. Build Environ 153:158–167. https://doi.org/10.1016/j.buildenv.2019.01.034

Röck M, Baldereschi E, Verellen E, Passer A, Sala S, Allacker K (2021) Environmental modelling of building stocks—an integrated review of life cycle-based assessment models to support EU policy making. Renew Sust Energ Rev 151:111550. http://doi.org/10.21203/rs.3.rs-638476/v1
Saint-Gobain (2022) Architecture student contest: edition 2022 Warsaw. https://architecture-student-contest.saint-gobain.com/edition-2022
Sala S, Amadei AM, Beylot A, Ardente F (2021) The evolution of life cycle assessment in European policies over three decades. Int J Life Cycle Assess 1–20. https://doi.org/10.1007/s11367-021-01893-2
Trigaux D, Allacker K, Debacker W (2021) Environmental benchmarks for buildings: a critical literature review. Int J Life Cycle Assess 26(1):1–21. https://doi.org/10.1007/s11367-020-01840-7

Chapter 43
Design Scenarios for a Circular Vision of Post-disaster Temporary Settlements

Maria Vittoria Arnetoli and Roberto Bologna

Abstract The construction sector has a considerable impact on the environment in terms of both exploited natural resources and greenhouse gas emissions. Therefore, converting the production process from linear to circular is essential. In increasingly vulnerable human settlements, post-emergency recovery can become an opportunity to develop innovative circular design strategies. The research focuses on how to strengthen the resilience of risk-prone territories through pre-disaster strategic planning based on a systemic approach. Post-emergency management of 2009 and 2016–2017 earthquakes in the inner areas of Central Italy is assumed as a case study. In particular, the tender specifications that guided the recovery revealed a deep lack of preventive programmes on the post-use phase of the settlements, which remain suspended between temporary and permanent. Starting from the analysis, the paper proposes a matrix of alternative scenarios for the end-of-life of temporary structures. The matrix allows connecting the recovery phase with the objectives of social cohesion and territorial regeneration policies, adapting the response to the needs of the specific context. Assuming that the artefact's technological requirements depend on the different perspectives of their life cycle, the scenarios are oriented towards different degrees of reversibility, addressing the complete disassembly, with the reuse and recycling of components, up to the reconversion of temporary assets as local facilities and as resources for green and digital transition. Integrating post-disaster into ordinary tools would trigger virtuous synergies to optimise public funding use. In this framework, post-disaster temporary housing can become a field of experimentation for disaster-resilient communities and circular economy.

Keywords Emergency · Post-disaster · Temporary housing · Strategic planning and scenario · Circular design

M. V. Arnetoli (✉) · R. Bologna
University of Florence, Florence, Italy
e-mail: mariavittoria.arnetoli@unifi.it

R. Bologna
e-mail: roberto.bologna@unifi.it

E. Arbizzani et al. (eds.), *Technological Imagination in the Green and Digital Transition*, The Urban Book Series, https://doi.org/10.1007/978-3-031-29515-7_43

43.1 Introduction

Complex and interconnected emergency phenomena, derived from natural and anthropogenic origin, determine an increasingly widespread and articulated need for transitional living. Moreover, cities have a growing vulnerability and poor coping capacity. This requires a new approach to the issue of emergency management, addressing the overcoming of the distinction between the *ordinary* and *extra-ordinary time*.

To respond to this framework, the contamination between Urban Resilience and Disaster Risk Reduction is an innovative field to be explored (UN 2015). The challenge is to combine responses focused on restoring the built environment (*hard solutions*) with strategies aimed at strengthen the adaptive capacity of physical and social systems at risk and reduce the environmental impact of emergency solutions (*soft solutions*), turning emergency into a "window of opportunity" for increasing sustainability (Brundiers and Eakin 2018). Post-disasters should be addressed not to "bounce back" but to "bounce forward" towards a more resilient society able to face up future extreme events by an improved combination of resistance and adaptability (Alexander 2015).

Within the Italian context, the identified field of investigation is the post-earthquake temporary settlements, consisting of a structural component, the housing modules, and an infrastructural one, the set of services that connect the area to the technological and mobility existing networks.

The current process is strongly top-down and the institutional actors (NDCP—National Department of Civil Protection, SOR—Special Offices of Reconstruction, Regions, Municipalities) operate in a sectorial way.

The housing solutions, even if called temporary, have a strong environmental impact: at the product level, the components are not designed to be reintroduced into further cycles of use, this leads to an unacceptable waste of resources; while at the settlement level, the foundations and the infrastructures are not realised as reversible, consuming natural soil. From a socio-economic point of view, the significant public investments are subtracted from the reconstruction effort.

Therefore, the study aims to integrate Circular Design paradigms in pre-disaster planning (Fig. 43.1). Temporary Architecture reaffirms its role as a privileged field of experimentation for the development of production models and technical solutions based on circular economy and digitalisation processes (Antonini et al. 2020).

The study identifies the inner areas of Central Italy affected by the 2009 and 2016–17 seismic events as a priority case study due to the pre-existing overlapping of environmental vulnerability and socio-economic marginality, compounded by the current management of reconstruction and the future reconversion or dismantling of temporary settlements.

Starting from this framework a design tool focused on the entire life cycle of the temporary solutions is proposed.

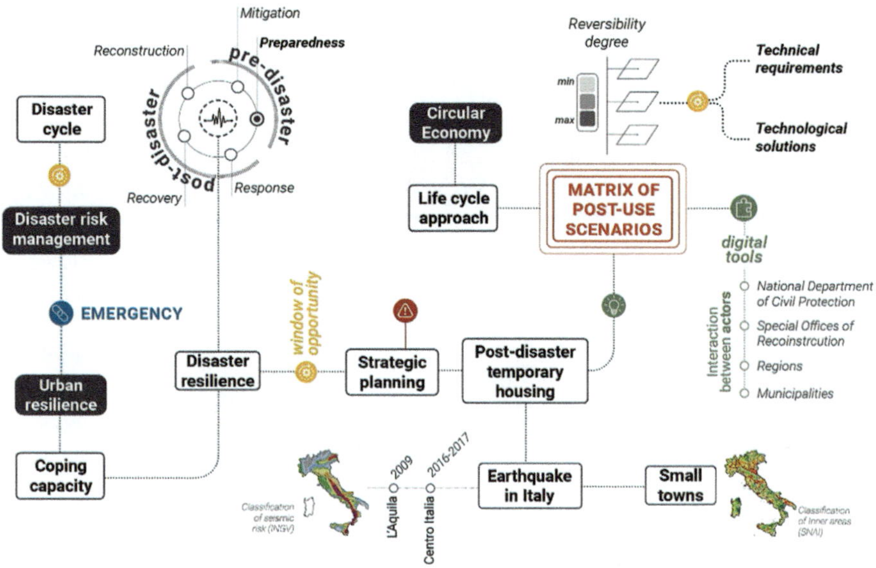

Fig. 43.1 Research map. *Source* Authors

43.2 From Temporary to Permanent: The Italian Context

In Italy, the need for post-disaster transitional housing derives mainly from the frequent earthquakes to which most of the territory is exposed, combined with other destructive events due to the high hydrogeological fragilities, such as floods and landslides.

The long reconstruction period, partly motivated by the pervasive presence of the historical heritage and the inefficiency and ineffectiveness of the public implementation processes, require housing solutions with adequate standards for a time of use of at least ten years.

Despite the numerous disasters faced in the recent past (Guidoboni and Valensise 2013), the response is still characterised by many issues, including poor humanisation of the spaces, lack of adaptability to the intervention context, and costs that are comparable or even higher than those of ordinary buildings.

A typical Italian problem, on which the research focuses, is the unresolved and contradictory relationship between the structures' temporary and permanent nature: solutions classified as temporary remain for decades. Striking examples are the "*baracche*" in Messina, built after the 1908 earthquake and still inhabited in seriously precarious conditions, or the *Bucaletto* district in Potenza, built in response to the 1980 earthquake, now hosting public facilities and under renovation. These are just two examples of the constellation of temporary settlements spread across Italy, which turn into areas of degradation exacerbating social inequalities and absorbing public investments (Fig. 43.2).

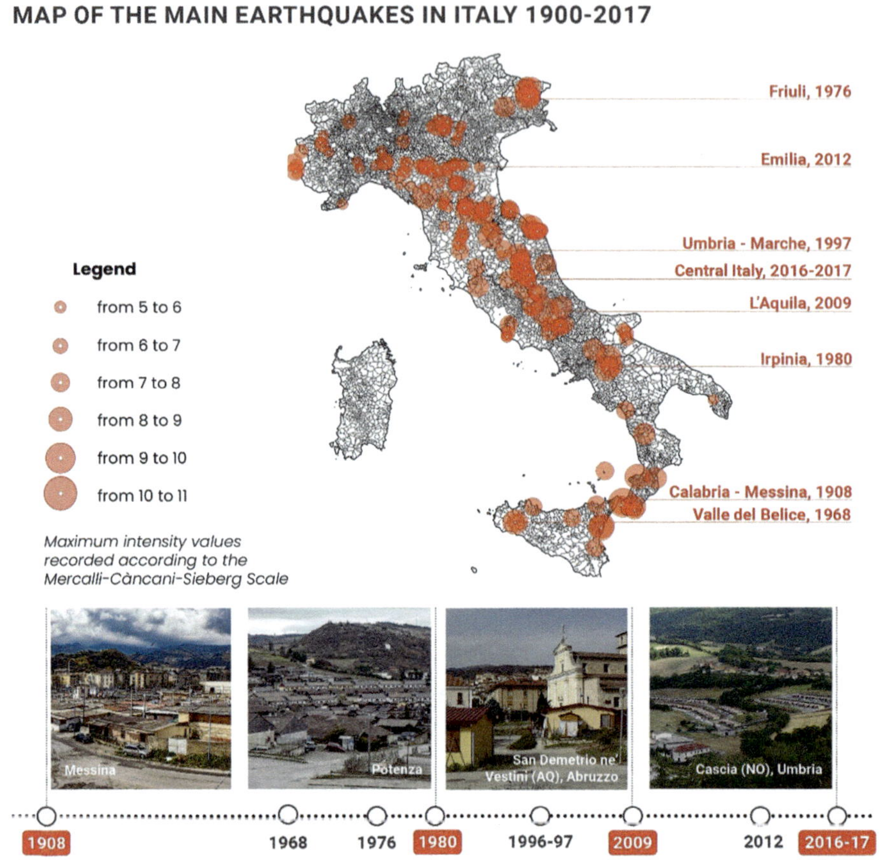

Fig. 43.2 Main earthquakes in Italy (1900–2017). *Source of data* National Institute of Geophysics and Volcanology; image elaborated by authors

At the root of this "permanent temporariness" phenomenon is a chronic lack of preventive and strategic planning. Concerning the disaster cycle, the weak role of the preparedness phase leads to responses designed to meet the immediate recovery needs, without a systemic vision that can orient the entire life cycle.

The procedural tools adopted by the NDCP for the provision and construction of temporary housing modules following the 2009 (*Capitolato Speciale 2008*) and 2016–17 (*Capitolato Tecnico 2014*) earthquakes were examined concerning their ability to guide the post-use phase (Bologna 2020). The two tender specifications define the product requirements without including them in the more complex and dynamic process of reconstruction and parallel socio-territorial regeneration of the affected areas. In both documents, the following are the reversibility requirements: disassembly of the system, recyclability of materials and reusability of elements and/or components.

However, these have not been sufficient to achieve the environmental sustainability of the interventions, remaining expressions of intent.

The current conception of the housing module frames temporary structures in a linear process that flows through the phases of pre-use (production, transport, construction) and use (management and maintenance), interrupting at the post-use, which remains suspended without a direction.

Post-emergency use should inform product design from the beginning (EEA 2017).

On the contrary, postponing the evaluation of end-of-life possibilities after a period of use of at least a decade brings to the less sustainable scenario: landfill disposal.

The broader question is «reorganise public action around the themes of multi-temporality and circularity of planning» (Balducci et al. 2021) and the production system.

To overcome this gap, it is necessary to be able to respond to the different needs of specific contexts by proposing different alternative scenarios.

43.3 Lessons Learnt from Central Italy Earthquakes

Thirteen years after the L'Aquila earthquake and five years after the one in Central Italy, the critical issues that emerge today regarding the future of temporary assets are outlined below. The relationship between module and area is fundamental in post-disaster planning, which has to deal with a changed territorial structure, disrupted firstly by the earthquake and then by temporary interventions, which in small towns often equal or even exceed the size of the original centre.

The aspects highlighted derives from the analysis of scientific literature, research experiences (NDCP-ReLUIS 2019–2021) and direct talks with institutional stakeholders (NDCP, SOR); which were accompanied by field analyses that allowed defining a bottom-up view from local administrations, inhabitants and associations.

According to the investigations, the future destination of the temporary housing structures already installed on the territory is oriented according to two possible directions: dismantling or renovation. Regarding the infrastructural component, restoring the areas to their pre-event condition, mostly agricultural or natural, is economically unsustainable. While the degree of obsolescence reached by the structures is essential to evaluate whether or not reconversion actions can be undertaken.

Currently, the prevailing choice is to preserve settlements, except in the case of areas under landscape constraints or belonging to natural parks. The high cost of demolition and waste management and the fact that two different actors carry out the provision and management (NDCP and Municipalities) also suggests that settlements should be maintained.

In these evaluations, the municipalities are operating individually without a large-scale direction. The privileged functional destination of the reconversion seems to be tourism. However, this does not consider that the areas are oversized compared to the tourist flows involving these territories.

Moreover, the lack of public services is not considered, nor are the green and digital transitions goals, for which small towns could be a living laboratory. Sectoriality leads to a parallelism between public fundings that act on these areas, separately, as earthquake-affected territories and as inner areas. This strict division ignores that a large portion of the *crater* coincides with the 2014–2020 National Strategy for Inner Areas (Agency for Territorial Cohesion 2014) pilot project areas.

Moreover, as the reconstruction process is carried on, the modules are progressively inhabited not only by earthquake victims but also by other categories of users (elderly, young couples, immigrants). In this way, the most vulnerable population groups are concentrated in the temporary settlements, leading to greater marginalisation.

It is no longer acceptable that post-emergency interventions exacerbate pre-existing inequalities by failing to act as a driver for development, especially in the light of territorial cohesion purposes.

The lack of cross-cutting policies has led to the construction of structures with no end-of-life plan. The *MAP—Moduli Abitativi Provvisori* (Housing Provisional Modules) provided for the L'Aquila earthquake have already reached a high degree of obsolescence, while there are more significant margins for intervention on the *SAE—Soluzioni Abitative in Emergenze* (Emergency Housing Solutions) used in Central Italy.

Given the vast volume of modules spread across Abruzzo, Lazio, Marche and Umbria regions, it is urgent to consider alternatives to landfill or abandon.

In addition, once the modules are dismantled, the areas' infrastructures will remain without a functional destination.

These findings call for learning from past experience by promoting a radical change in post disaster management: from linear to circular.

43.4 Matrix of Post-use Scenarios in a Circular Vision

For the criticalities highlighted, the approach to post-disaster temporary solutions must be rethought in the twin dimensions of process and product, according to the paradigms of circularity as a pathway to resilience.

This requires defining tools and methods for putting into practice the principles of *Design for Disassembly, Reuse and Recycle* (DfDRR), assuming that the technical requirements of the products defined in the design phase depend on their final destination.

Thus, a matrix of post-use scenarios is proposed as a preventive, strategic and inter-scalar tool that integrates the planning and the design level (Fig. 43.3). As part of the preparedness phase, the matrix allows to orient the response according to the needs of the specific contexts.

Moreover, the matrix is addressed to the multiple stakeholders involved in emergency management, primarily the NDCP and Municipalities, but also the private operators like manufacturers.

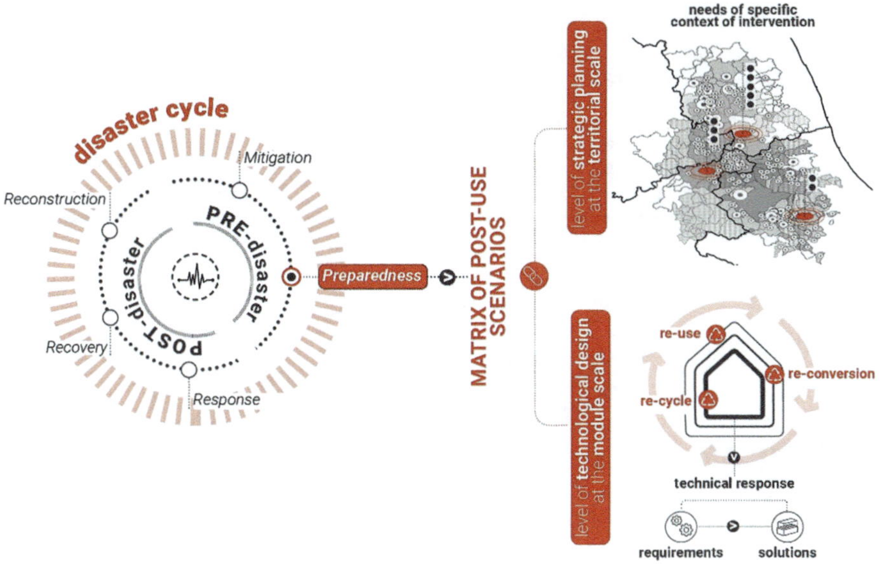

Fig. 43.3 Disaster cycle and inter-scalarity character of the matrix. *Source* Authors

The environmental issue changes the relationship between matter and time: the centre of the design activity becomes the product's entire life cycle (Campioli et al. 2018).

The holistic approach moves toward an adaptive and evolutionary vision of the settlement's life, reinforcing the integration between the phases of the disaster cycle.

Differently from the second life options explored by Johnson (2010), the matrix orients the end-of-life of the product from a technical perspective. The following four macro-scenarios (MS) move in a spectrum of design solutions from temporary to permanent (Fig. 43.4):

- *dismantle*: the structure is disassembled and its elements/components put into reuse and recycling processes, focusing on the features of the materials used;
- *disassembly*: the module is disassembled and reassembled elsewhere for further uses, with the same functional destination (residential) or others;
- *permanent core with temporary devices*: the device is composed of an infrastructural component designed as permanent, on which the temporary structures are integrated;
- *reuse*: the building is not designed as temporary but as an ordinary facility for the local community, designed to adapt in response to changing local needs.

To each MS corresponds different design parameters, which are translated into sets of technical requirements that correspond to different technical solutions. The main design features are flexibility and adaptability to allow subsequent transformations in the long-term period.

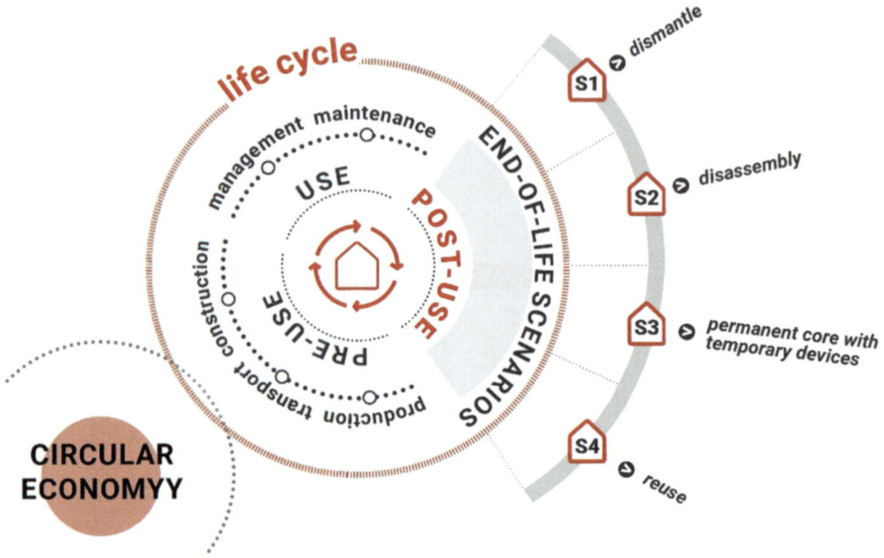

Fig. 43.4 Life cycle and post-use scenarios. *Source* Authors

The matrix embodies also the variety of solutions adopted in international contexts by analysing them with respect to the end-of-life trajectories of post-disaster temporary solutions (Arslan 2007; Askar et al. 2019; Johnson 2008; Seike et al. 2018), framing the Italian case in a wider horizon.

Emergency temporary housing needs to be rethought entirely at the different scales of the project, from settlement to single component.

For the vast scale, the challenge is to make the emergency response part of the policies operating in a specific territory. This can only be done during ordinary planning, as the emergency causes a time compression that does not allow the development of site-specific strategies.

The definition of the needs of territories at risk is the basis for planning emergency interventions combining national provisions and local realities.

The territorial planning dialogues directly with the technological system in the matrix (Fig. 43.5). The design approach is reshaped using time as an essential design resource.

The technical response is diversified in the four macro-scenarios, moving from temporariness to permanence.

For dismantle and disassembly scenarios, designing the supply chain and production process in a cradle-to-cradle logic, according to which the modules are derived from and destined for recycling circuits, would make it possible to test replicable solutions for other construction sectors.

While looking at the permanent core and the reuse, the requirements would be oriented towards standards closer to traditional constructions in terms of durability and adaptability of the structures.

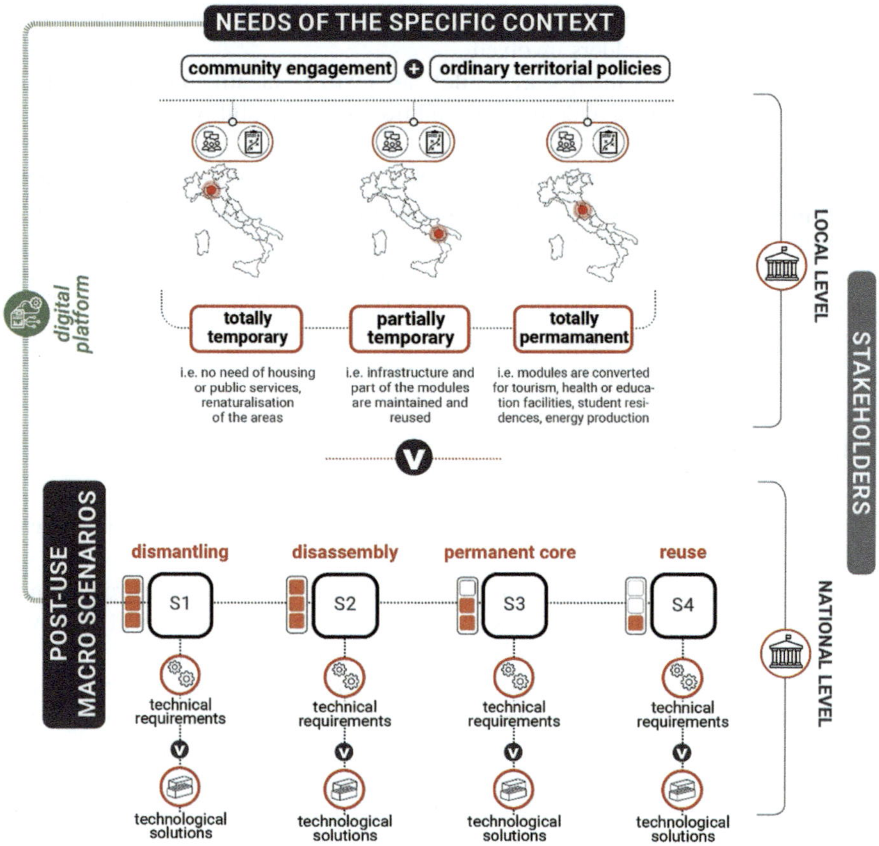

Fig. 43.5 Matrix of scenarios. *Source* Authors

In both directions, pre-disaster planning would make the emergency response a laboratory where different stakeholders such as universities, research centres, manufacturers, recycling sites could find common ground for collaboration and exchange.

A further dimension concerns the management of artefacts. This is closely linked to the logic with which the product is placed on the market, whether through traditional sale-purchase contracts, through which the public body takes possession of it or according to models that favour forms of sharing economy and product-service systems, acting on the concept of ownership.

In conclusion, the matrix is proposed as an open and dynamic tool, a link between the multiple actors, disciplines and sectors involved in the post-disaster temporary response, capable of operating directly on its multidimensional character.

The result achieved defines a methodological approach, the next step is to develop each scenario and implement the matrix as a digital open-source platform that can be continuously updated and immediately integrated into the tools currently in use by the different stakeholders involved.

In the next phase of the research, pilot cases will be identified both in the Italian and international context to define the conditions of applicability and replicability of the matrix.

The application of the matrix with respect to different post-emergency management processes will allow to identify the potentialities and limits of the tool. At the same time, the selection parameters for each scenario will be defined using an inductive method, starting from local contexts with different socio-economic, environmental and regulatory profiles. This will lead to the definition of a set of basic and specific technical requirements.

43.5 Conclusions

The work presented is aligned with the evolution of emergency management towards environmental sustainability.

The purpose of the matrix is to determine the product's technical requirements according to the different scenarios chosen. Hence, the methodological framework leads to a practical result that directly affects the circular economy.

Therefore, due to their diffusion, the challenge is to transform temporary structures from a public asset difficult to manage from a legislative-administrative point of view and to reconvert from an urban-architectural one, into a resource for technological experimentation.

The process strategies and technical solutions developed are transferable to non-emergency construction, constituting potential good practices to be replicated.

The transformation most stressed in this paper is the green one. Again, digital tools also play a crucial role in the renovation of the current model. They can be used in the disaster cycle to strengthen the interaction between actors in the different phases and in the new-generation industrialisation of building production.

Transitional Living is investigated as a complex phenomenon interacting with the further challenges of Anthropocene societies, for which it is increasingly urgent and vital to define a new balance with nature by reducing their impact.

Looking at the design of the emergency management process, the ultimate goal is, then, to identify the preconditions for putting into practice a vision of circularity of resources.

Acknowledgements The author Maria Vittoria Arnetoli is co-founder and member of the interuniversity and interdisciplinary Italian research network *TEMP*—with Ilaria Tonti | Polytechnic of Turin, Giovangiuseppe Vannelli | University of Naples 'Federico II', Francesco Chiacchiera | Marche Polytechnic University, Marco Pizzi | University of Perugia, sharing research experiences on the topic of post-disaster temporariness.

Account ig: @temp_researchnetwork

References

Agency for Territorial Cohesion (2014) SNAI Strategia Nazionale per le Aree Interne

Alexander DE (2015) Disaster and emergency planning for preparedness, response, and recovery. Oxford research encyclopedia of natural hazard science. Oxford University Press, Oxford, pp 1–20

Antonini E, Boeri A, Giglio F (2020) Emergency driven innovation. Low tech buildings and circular design. Springer, Berlin

Arslan H (2007) Re-design, re-use and recycle of temporary houses. Build Environ 42:400–406

Askar R, Rodrigues AL, Braganca L, Pinheiro D (2019) From temporary to permanent; A circular approach for post-disaster housing reconstruction. In: IOP conference series: earth and environmental science, vol 225. SBE19 Brussels—BAMB-CIRCPATH "Buildings as material banks—a pathway for a circular future", 5–7 Feb 2019, Brussels, Belgium

Balducci A, Caramaschi S, Coppola A, Curci F, Di Giovanni G, Di Venosa M, Fontana C, Franz G, Gritti A (2021) Nei Territori Sismici: Principi Fondamentali per Una Legge Quadro per Le Ricostruzioni. In: Coppola A, Del Fabbro M, Lanzani A, Pessina G, Zanfi F (eds) Ricomporre i divari. Politiche e progetti territoriali contro le disuguaglianze e per la transizione ecologica. il Mulino, pp 57–69

Bologna R (2020) Operational dimension of post-disaster housing temporality and technical control tools. TECHNE 20:213–221

Brundiers K, Eakin H (2018) Leveraging post-disaster windows of opportunities for change towards sustainability: a framework. Sustainability 10(5):1–18

Campioli A, Dalla Valle A, Ganassali S, Giorgi S (2018) Designing the life cycle of materials: new trends in environmental perspective. TECHNE 16:86–95

EEA—European Environmental Agency (2017) Circular by design. Products in the circular economy. EEA report no 06

Guidoboni E, Valensise G (2013) L'Italia dei disastri. Dati e riflessioni sull'impatto degli eventi naturali 1861–2013. Bononia University Press, Bologna

Johnson C (2008) Impacts of prefabricated temporary housing after disasters: 1999 earthquakes in Turkey. Habitat Int 31:36–52

Johnson C (2010) Planning for temporary housing. In: Lizarralde G, Johnson C, Colin D (eds) Rebuilding after disasters. From emergency to sustainability. Routledge, London, pp 70–88

Seike T, Kim Y, Hosaka Y, Ida S, Masuda T (2018) Intraregional reuse of emergency temporary housing in Japan. Sustain Cities Soc 42:650–662

UN—United Nations (2015) Sendai framework for disaster risk reduction 2015–2030

Chapter 44
Towards Climate Neutrality: Progressing Key Actions for Positive Energy Districts Implementation

Rosa Romano, Maria Beatrice Andreucci, and Emanuela Giancola

Abstract Positive Energy Districts (PEDs) represent an emerging urban transition paradigm, an advanced framework to effectively attain decarbonization targets, as well as a holistic approach to foster more resilient and livable cities. However, implementing PEDs is challenging, demanding substantial planning, design, and operations changes. Mainstreaming PEDs calls for innovative legal, institutional, business, and organizational frameworks, as well as an active involvement of the main actors (i.e., cities, municipalities, communities, investors, industry players, and service providers), to co-design and jointly progress ambitious agendas, multiscale plans, flexible instruments, and adaptive structures. Benefitting from the authors' cooperation within the Horizon 2020 project, Cooperation in Science and Technology COST Action 'Positive Energy Districts European Network' (PED-EU-NET in PED-EU-NET | COST ACTION CA19126, 2020), the proposed contribution addresses relevant issues and opportunities characterizing the development of PEDs in Europe, relating attention to effective implementation, context-specificity, replicability, and upscaling. Among the results achieved in the first year of the COST research activities, the authors present an understanding of the PEDs policy landscape in Europe, and a catalogue of the key lessons learned from PEDs in progress. In detail, some comprehensive and interrelated aspects (stakeholder-oriented strategies and technological and system innovation) that have emerged towards enabling conditions for upscaling PEDs structure are analyzed. Through the investigation of existing framework conditions, barriers, and enablers of piloting projects, as well as emerging impacts at international level, the authors provide original insights, and formulate key recommendations for take-up and advancement towards climate neutrality, making a timely and original input to enhanced scholarly understanding of PEDs.

R. Romano (✉)
University of Florence, Florence, Italy
e-mail: rosa.romano@unifi.it

M. B. Andreucci
Sapienza University of Rome, Rome, Italy
e-mail: mbeatrice.andreucci@uniroma1.it

E. Giancola
UiE3-CIEMAT, Madrid, Spain
e-mail: emanuela.giancola@ciemat.es

E. Arbizzani et al. (eds.), *Technological Imagination in the Green and Digital Transition*, The Urban Book Series, https://doi.org/10.1007/978-3-031-29515-7_44

Keywords PEDs · Climate neutral cities · Energy communities · Energy poverty · Renewable energy

44.1 Introduction

This document The European Union's SET-Plan Action 3.2 introduced, in 2018, the acronym PEDs, to describe 'Energy-efficient and energy-flexible urban areas, or groups of connected buildings, which emit net zero greenhouse gas emissions and actively manage an annual local or regional surplus production of renewable energy.

They require integrating different systems and infrastructures, as well as interaction between buildings, the users, and the regional energy, mobility, and ICT systems, while securing the energy supply and a good life for all, in line with social, economic, and environmental sustainability' (EU Commission 2018). Accordingly, in the last years, ongoing initiatives such as the EU's Horizon 2020 'Smart Cities and Communities (SCC) Lighthouse Projects', the Joint Programming Initiative Urban Europe (JPI UE), and the European Energy Research Alliance–Joint Program Smart Cities (EERA–JP SC) have been active to increase the knowledge about the integration of innovative solutions for the planning, deployment, and replication of PEDs. PEDs are already a common objective of ongoing EU research projects (e.g., SPARCS, POCITYF, ATELIER, +CityxChange, Making City, etc.) that have joined the discussion on key issues, such as a commonly shared definition, or the role and relevance of different stakeholders towards implementation, as recently highlighted by the Horizon 2020 project Cooperation in Science and Technology COST Action PED-EU-NET (2020–2024).

In detail, PED-EU-NET aims to drive the deployment of PEDs by harmonizing, sharing, and disseminating knowledge and breakthroughs on this innovative urban and sustainable model across different stakeholders, domains, and sectors, at national and European levels. The Action has been structured in four interlinked thematic Working Groups (WGs) around the core PED vision that are focused on: (WG1) mapping existing concepts, strategies, projects, socio-technical innovations related to PEDs in Europe; (WG2) developing guides and tools to support the implementation of PEDs; (WG3) exploring the success factors of PED Labs, working up common protocols for monitoring and evaluation of PEDs and PED Labs across Europe; (WG4) ensuring the transfer of knowledge and the translation of experience beyond the network. The goal is to establish a PED innovation ecosystem to facilitate open access of knowledge, exchange of ideas, pool resources, experimentation of new methods, and co-creation of novel solutions across Europe. Additionally, this COST Action wants to support the capacity building of new generations of PED professionals, early-career investigators, and experienced practitioners, mobilizing relevant actors from and across Europe to collectively contribute to the long-term climate-neutral target (PED-EU-NET 2020).

Starting from the results achieved in the first year of this scientific cooperation—and in particular during the 1st PED-EU-NET Urban Stakeholder Workshop held in Rome in October 2021—the paper presents some comprehensive and interrelated aspects of enabling conditions for mainstreaming and upscaling PEDs, focusing on stakeholder-oriented strategies, and technological innovations to ensure energy production, efficiency, and flexibility at the district scale. The objective is twofold, i.e., to explore existing framework conditions, barriers, and emerging impacts at international level, and provide key recommendations for uptake and advancement towards climate neutrality, making a timely and original contribution to enhanced scholarly understanding of PEDs.

44.2 Stakeholder-Oriented Strategies and Innovative Governance Model for PEDs

Among the key aspects suitable to describe the characteristics of a PED project, the implementation phase ('PED phase') is considered particularly relevant, as challenges and barriers for PED related projects vary from the planning stage to the implementation stage. For this reason, a 'PED phase' can be defined according to the ambition and the development stage it is experiencing (i.e., 'planning', 'implementation', 'implemented/in operation') (Zhang et al. 2021). This categorization refers to projects where the construction of the energy systems has been completed (e.g., in Drammen and Stor-Elvdal Municipality, in Norway, already in operation), or not yet commissioned (e.g., in Turku, in Finland, and Bergen in Norway, currently under planning), or integrated into existing energy networks (e.g., in Lund in Sweden, and Odense in Denmark, currently under implementation). Accordingly, previous studies have highlighted some conditions that seem to facilitate inter-organizational collaboration and presented them as critical factors to ensure the PED process a lasting commitment to collaboration and coordination between interested parties and other actors. The initiators must facilitate engagement throughout the process and be enabled by profitable and innovative business models, well-conceived governance structures, and information and communication technologies (Civiero et al. 2021).

The PED project is, in fact, a complex process that requires a high degree of coordination, considering that some actors tend to shift their positions, or rethink their involvement, when moving from one phase to another (Hamdan et al. 2021). From a governance perspective, PEDs offer an appropriate arena for collaboration between different sectors (residential, manufacturing, commercial, public, etc.) and users (owners and tenants, intervening as individuals or as members of communities) to enable a holistic and inter-sectoral approach to energy planning as a key integrative part of sustainable urban development (Bhowmik 2020). For this reason, one of the crucial challenges and barriers to implementing a PED project also depends on the engagement of the different stakeholders in its planning and implementation, and not only in the adaptation of the technical solutions (Bossi et al. 2020).

In that direction, the experiences, and cases studies on 'PED Gaps' presented during the recent 1st PED-EU-NET Urban Stakeholder Workshop confirmed how the implementation phase of PEDs can be envisioned as a direct function of active collaboration between cities, private actors, and citizens. Those findings suggest a potential for strengthening the levels of cooperation and mutual trust between all stakeholders, and especially the energy and buildings sectors. Shifting the focus of private stakeholders from pure profit to mutual benefits appear crucial to strengthen this inter-sectoral cooperation, even if this requires significant changes in current practice and business models, as well as the introduction of new technologies, especially the ones that would allow energy storage for a flexible energy supply and use. In addition, the PED-EU-NET Urban Stakeholder Workshop results also suggest that the technologies that target flexible energy demand in buildings are considered very important features by the stakeholders. The high costs of technologies and the weak financial incentives in today's context were identified as the strongest barriers to the adoption of PEDs. Consequently, the introduction of new business models that reduce costs and mitigate investment risks is considered vital to support PEDs deployment.

The removal of the existing legal and regulatory barriers, which were identified to particularly impact the development of PEDs in Italy, was, also considered crucial to create stronger incentives for mainstreaming districts and neighborhoods able to produce and use energy in a more flexible and 'smarter' way, thus paving the way to a wide-scale adoption of PEDs in cities (Fig. 44.1).

Considering the large variety of stakeholders who are meant to participate in this process (end-users; public authorities; financial institutions, and property developers, etc.), it is also imperative to identify both the desired configuration of the energy systems, and the different roles the actors are willing to take (Ahlers et al. 2019).

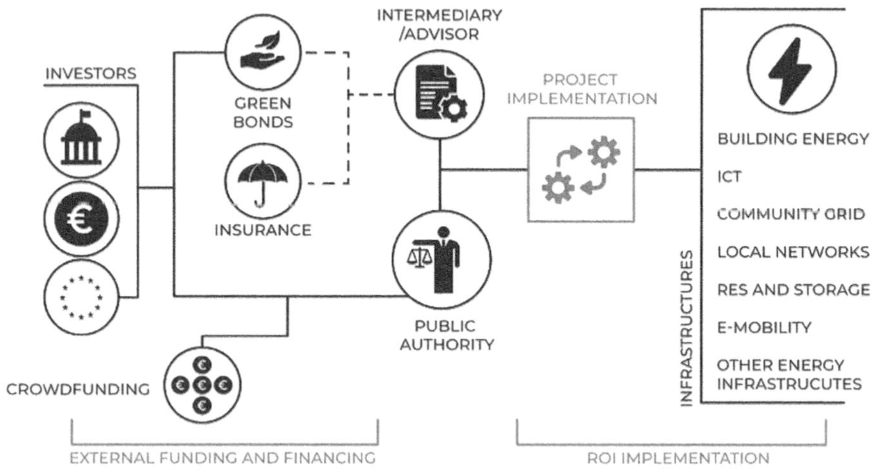

Fig. 44.1 Power distribution of channel at 1555 nm along the link of 383 km (Ahlers et al. 2020)

The different roles of the various stakeholders in all phases of PEDs implementation may be attributed to factors such as the challenges of building refurbishment, the integration of new energy systems, and the surrounding regionals and urban energy infrastructures. However, with the development of PED projects, the stakeholder constellation has been changing according to the different interests, with a consequent transfer of the responsibilities. This point is critical since users need to become more involved in the management of neighborhood systems and may also be exposed to technical challenges. All this requires a major global planning exercise from a technical and spatial planning point of view, a well-conceived governance structure, as well as solid stakeholder engagement strategies (Cheng et al. 2022).

In that direction, the project Sustainable energy Positive and zero cARbon CommunitieS (SPARCS) demonstrates and validates technically and socioeconomically viable and replicable, innovative solutions for rolling out smart, integrated positive energy systems for the transition to a citizen-centered zero carbon and resource efficient economy. Among the examples currently in progress, SPARCS is developing and piloting new models for co-creation, energy communities and stakeholder engagement to bring residents in the new Kera district in Espoo (Finland), to the center of the energy ecosystem, maximizing local production and encouraging prosumer models to enhance the utilization of distributed power generation (Hukkalainen et al. 2020).

Another interesting example is the project +CityxChange (Fig. 44.2) that is finalized towards the co-created Europe-wide deployment of PEDs through integrated planning and design, creation of a common energy market, and community exchange with all urban stakeholders. In detail, this experimental approach establishes mechanisms that need to be followed up to better understand local needs and develop tailored actions to engage citizens and stakeholders in an open innovation process finalized to optimize the energy consumption of the urban district (Ahlers et al. 2019).

44.3 Technological Innovation to Ensure PEDs Energy Self-sufficiency

PEDs can generally be defined as parts of a city that generate more energy than they consume annually and that promote higher self-consumption and self-sufficiency (Hedman et al. 2021).

This increased dependence on intermittent RES intensifies the need for flexible options to ensure reliable power system operations via integrated solutions consisting of energy storage, smart urban energy networks, ICT, and e-mobility (European Commission 2018). Consequently, it is necessary to optimize the building integration within the district, local, and distant renewable and low carbon energy sources into a resilient energy system, moving beyond nZEBs. Therefore, in the framework of PEDs, the building systems should be specifically designed and should present tailored load profiles (e.g., the long-term load forecast cannot be based on its

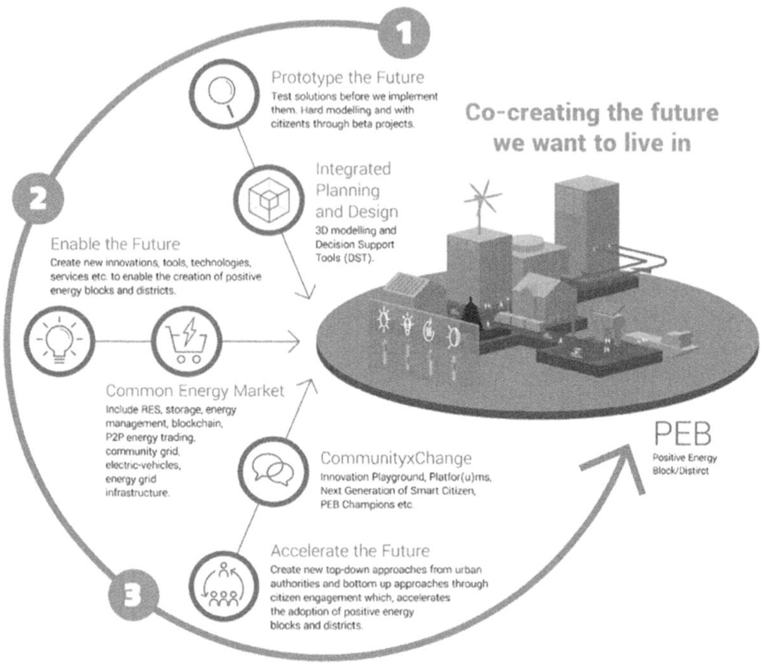

Fig. 44.2 +CityxChange integrative approach and framework for the development of positive energy blocks (Ahlers et al. 2020)

historic pattern anymore) that will enable a high penetration of carbon-free renewable sources such as wind and photovoltaics (PV) for the electricity demand as well as heat pumps, thermal waste, geothermal, solar thermal, etc. for heating and cooling. Additionally, district-scale buildings, should use 'energy cooperation' or 'energy pooling' to empower communities to own the locally produced renewable energy. Consequently, a building designed to be a functional unit of a PED must be built, or refurbished, according to the possibility of integrating a series of technical solutions, such as batteries, electric vehicles (EV), and grid-responsive control systems (Zhou et al. 2021). Equally, smart grids and digital tools for demand-side management such as smart meters, smart chargers, and Building Automation and Control Systems (BACSs) can facilitate flexibility in the PED energy system, which can help align energy demand with supply (Lyons 2019).

According to this innovative and sustainable urbanization concept (Fig. 44.3), the construction and renovation process should also be increasingly accompanied by digital solutions, such as building information modeling (BIM), 3D printing, digital twins, Internet of Things, or augmented reality. These innovative digital technologies help save costs and time and require suitable novel skills and the latest available software to be most effective.

Moreover, digitalization enables the uptake of smart energy services for building users of a PED, allowing the development of demand-side management strategies

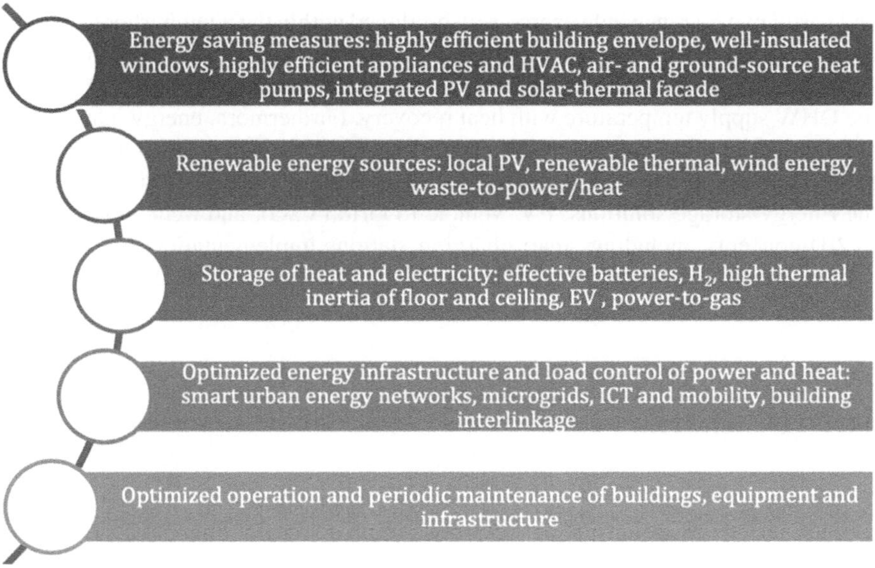

Fig. 44.3 Key research fields of technological innovations needed for deploying PED (EU Commission 2018)

to integrate flexible and decentralized renewable energy systems into its energy structure (Fig. 44.4).

Some interesting case studies of innovative technologies and system integration have been developed in numerous European Projects such as ZenN, Syn.ikia, Sharing Cities, MAtchUP. In all these examples the energy demand in the district cluster is low

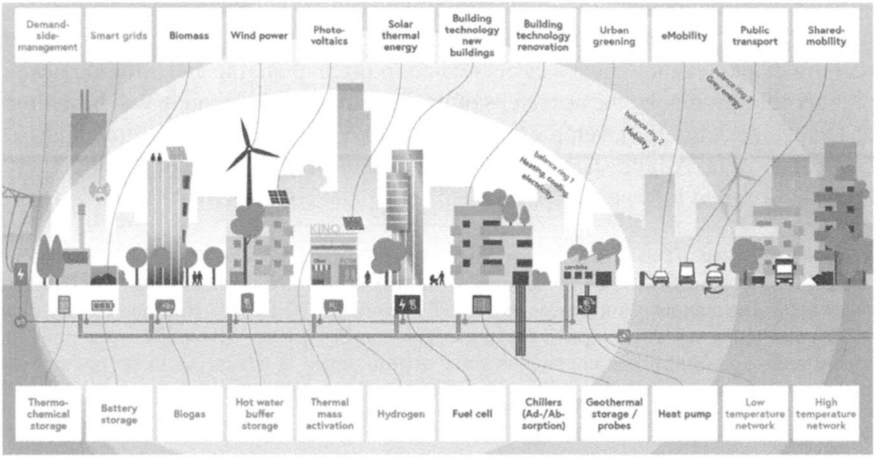

Fig. 44.4 Positive energy district model (BMVIT 2022)

and partly met by renewable energy self-produced within the neighborhood. Another measure applied to achieve the target PED is an enhanced envelope/insulation, followed by real-time measuring of consumption and heating controls reduction of the DHW supply temperature with heat recovery. Furthermore, energy management systems (Neighborhood/Building) and boiler upgrades are widely utilized. Regarding the RES, the technologies most utilized are Heat Pumps and PV. Finally, concerning the Energy storage solutions, EV, Vehicle to Grid (V2G), and Vehicle to Building (V2B) concepts, including smart charging stations implementation, are the major solution employed, followed by batteries and thermal storage (Castellanos and Oregi 2021).

44.4 Conclusion

Increasing evidence from science, governance, and practice show that climate change adaptation and mitigation and the decarbonization of our environment are of greatest urgency and should be targeted at the macro, urban, and district levels. Cities are consequently required to take an active role in adopting sustainable energy production and use for people and the environment (Saheb et al. 2018). It is now stepping up efforts towards city-wide transformation with the pioneering concept of PEDs, which builds on the paradigm of smart cities, supporting municipalities that make energy savings and provide it with a good and sustainable metabolism (Ruhlandt 2018). For achieving this relevant goal PEDs design should equally be considered powerful opportunities for users and stakeholders to promote equity and inclusion, favoring vulnerable communities where far too many people still struggle facing energy poverty (Jessel et al. 2019). At the same time, it will be necessary to develop innovative envelope and plant technologies capable of transforming existing urban districts into efficient and low-impact energy ecosystems.

Finally, the PED model requires a synergy between the management of social, economic, and technological aspects related to urban planning and building design.

Accordingly, among the next steps of the PED-EU-NET research, will be defining PED specific KPIs to develop effective integrated frameworks and guidelines for European Public Administrations, with the aim of implementing and realizing 100 PEDs in different European geographical and social contexts by 2025.

Acknowledgements This article is based upon the individual work of the authors as members of the COST Action 19126 Positive Energy Districts European Network PED-EU-NET (https://pedeu.net/). The authors acknowledge the COST Action 19126 PED-EU-NET and other international cooperating initiatives—particularly the IEA Annex 83 (IEA EBC 2020)—for providing opportunities for fruitful and continuous collaboration aimed at advancing PEDs research and deployment.

References

Ahlers D, Driscoll P, Wibe H, Wyckmans A (2019) Co-creation of positive energy blocks. IOP Conf Ser Earth Environ Sci 352:012060. http://doi.org/10.1088/1755-1315/352/1/012060

Ahlers D et al (2020) In: Vandvyvere H (ed) Positive energy districts solution booklet. https://smart-cities-marketplace.ec.europa.eu/insights/solutions/solution-booklet-positive-energy-districts. Accessed 18 Mar 2022

Bhowmik AK et al (2020) Powers of 10: seeking 'sweet spots' for rapid climate and sustainability actions between individual and global scales. Environ Res Lett 15

BMVIT (2022) Towards plus-energy neighborhoods (folder). https://nachhaltigwirtschaften.at/en/sdz/publikations/folder-plus-energy-neighbourhoods.php#contentDescription. Accessed 18 Mar 2022

Bossi S, Gollner C, Theierling S (2020) Towards 100 positive energy districts in Europe: preliminary data analysis of 61 European cases. Energies 13(22)

Castellanos SR, Oregi X (2021) Positive energy district (PED) selected projects assessment, study towards the development of further PEDs. Environ Clim Technol 25(1):281–294. https://doi.org/10.2478/rtuect-2021-0020

Cheng C et al (2022) A systematic approach towards mapping stakeholders in different phases of PED development. Extending the PED toolbox. In: Littlewood JR, Howlett RJ, Jain LC (eds) Sustainability in energy and buildings 2021. Smart innovation, systems and technologies, vol 263. Springer, Singapore. http://doi.org/10.1007/978-981-16-6269-0_38

Civiero P, Pascual J, Arcas Abella J, Bilbao Figuero A, Salom J (2021) PEDRERA. Positive energy district renovation model for large scale actions. Energies 14:2833. http://doi.org/10.3390/en14102833

EU Commission (2018) SET-plan action no.32 implementation plan—Europe to become a global role model in integrated, innovative solutions for the planning, deployment, and replication of positive energy districts. https://setis.ec.europa.eu/system/files/2021-04/setplan_smartcities_implementationplan.pdf. Accessed 18 Mar 2022

Hamdan HAM, Andersen PH, de Boer L (2021) Stakeholder collaboration in sustainable neighborhood projects. A review and research agenda. Sustain Cities Soc 68(1):102776

Hedman Å et al (2021) IEA EBC Annex 83 positive energy districts. Buildings 11(3):130

Hukkalainen M et al (2020) Detailed plan of the Espoo smart city lighthouse demonstrations. https://www.sparcs.info/sites/default/files/2020-09/SPARCS_D3.1_Detailed_plan_Espoo.pdf. Accessed 18 Mar 2022

IEA EBC (2020). IEA EBC—Annex 83—positive energy districts. https://annex83.iea-ebc.org/. Accessed 18 Mar 2022

Jessel S, Sawyer S, Hernández D (2019) Energy, poverty, and health in climate change: a comprehensive review of an emerging literature. Front Public Health. http://doi.org/10.3389/fpubh.2019.00357

Lyons L (2019) Digitalisation: opportunities for heating and cooling, EUR 29702 EN. Publications Office of the European Union, Luxembourg. JRC116074. http://doi.org/10.2760/00116

PED-EU-NET (2020) PED-EU-NET | COST ACTION CA19126. https://pedeu.net/action/. Accessed 18 Mar 2022

Ruhlandt RWS (2018) The governance of smart cities: a systematic literature review. Cities 81:1–23

Saheb Y, Shnapp S, Johnson C (2018) The zero energy concept: making the whole greater than the sum of the parts to meet the Paris climate agreement's objectives. Curr Opin Environ Sustain 30:138–150

Zhang X, Penaka SR, Giriraj S, Sánchez MN, Civiero P, Vandevyvere H (2021) Characterizing positive energy district (PED) through a preliminary review of 60 existing projects in Europe. Buildings 11:318. https://doi.org/10.3390/buildings11080318

Zhou Y, Cao S, Hensen JLM (2021) An energy paradigm transition framework from negative towards positive district energy sharing networks—battery cycling aging, advanced battery management strategies, flexible vehicles-to-buildings interactions, uncertainty and sensitivity analysis. Appl Energy 288:116606

Chapter 45
Remanufacturing Towards Circularity in the Construction Sector: The Role of Digital Technologies

Nazly Atta

Abstract Among the different circular strategies, remanufacturing proves to be particularly interesting since it aims to maintain the value of building components overtime extending their lifespan by guaranteeing multiple consequent cycles of use, overcoming in this way the most common down-cycling logics. However, unlike other industrial fields which already benefit from remanufacturing, the construction sector delays to adopt this practice due to barriers of different nature, namely organizational, information, technical, regulatory and economic. Among these barriers, the first two can now be addressed more effectively thanks to the support of Information and Communication Technologies. The latter offer the possibility of real-time monitoring, remote communication and scenario modeling, opening up to innovative solutions for remanufacturing. Hence, the paper aims to investigate how the application of ICTs can support the cognitive and organizational processes related to remanufacturing of building components. In particular, the paper explores the application of sensing technologies, digital twins and information platforms and assess their potential to support the implementation of circular service-based remanufacturing models in the construction sector.

Keywords Circular economy · Construction sector · Remanufacturing · Building processes · Information and communication technologies

45.1 The Role of ICTs Towards Circular Remanufacturing Models in the Construction Sector

The last decade has witnessed the rapid spread of Circular Economy (CE) principles in several industrial fields at the global scale, strengthened by the introduction of the UN Sustainable Development Goals (SDGs) advocating Sustainable Production

N. Atta (✉)
Department of Architecture, Built Environment and Construction Engineering (DABC), Polytechnic University of Milan, Milan, Italy
e-mail: nazly.atta@polimi.it

E. Arbizzani et al. (eds.), *Technological Imagination in the Green and Digital Transition*, The Urban Book Series, https://doi.org/10.1007/978-3-031-29515-7_45

and Consumption (Goal 12) (UNEP 2022). The CE concept promotes the development of sustainable technological and business solutions able to guarantee value creation while reducing the resource throughput (Ness et al. 2019). Unlike traditional linear models, in which components are fabricated, built, used and demolished, CE approaches have emerged as sustainable alternatives following the closed-loop strategy, i.e. material components are fabricated, built, used, disassembled, reworked and reused. Hence, circular models aim at promoting closed cycles based on multiple re-uses of materials, components and products, minimizing in this way the creation of waste. In this context, the construction sector—representing one of the most crucial economic industry—is responsible of the 30% of all the waste generated in EU (ECSO 2019) with 923 million tons of waste produced in 2016 (Eurostat 2017). Indeed, based on volume, Construction and Demolition (C&D) waste represents the largest waste stream in the EU, equal to one third of all waste produced (European Commission 2016). Being one of the most resource- and waste-intensive economic activities, the construction industry counts a share of 38% of the total global energy-related CO_2 emissions (UNEP 2020) and 40% of all material consumption (Ness et al. 2019). To address such environmental issues, in the last decades the construction sector mainly focused on down-cycling and recycling practices as alternatives to material transfer to landfill.

However, recycling processes for converting waste into new products often imply the consumption of large amounts of water and energy, as well as the related generation of greenhouse gas emissions (Rios and Grau 2020). Hence, if on one hand the recycling practices contribute to the conservation of raw materials, on the other hand they result in high resource consumptions, generating significant environmental impacts (Rios and Grau 2020). Going beyond down-cycling and recycling practices, in recent years remanufacturing has emerged in several industrial fields as a strategy for closing the loop, extending the lifespan of products by restoring them at the end of their use-cycle for starting a subsequent new one, thus preserving embedded resources and limiting environmental impacts. Through remanufacturing processes, used products are returned to "as new" with a limited waste of materials, water and energy. Several industrial fields, such as for instance automotive, aerospace, machinery, electronics, rail, marine (Parker et al. 2015), already benefit from circular business models based on remanufacturing, in which components are disassembled, remanufactured and reused multiple times in closed-loop cycles.

Unlike these industrial fields, the construction sector delays to adopt this virtuous practice due to barriers of different nature, i.e., organizational, information, technical, regulatory and economic. Among these barriers, the first two can now be addressed more effectively thanks to the support of Information and Communication Technologies (ICTs), especially Internet of Things (IoT). As evidenced by ICT-based remanufacturing practices consolidated in the above-mentioned industrial fields (Butzer and Schötz 2016), the application of ICTs has the potential to support circular business models by improving cognitive and organizational processes. With reference to the construction sector, ICTs and IoT offer new capabilities of real-time monitoring, remote communication and scenario modeling, opening up to innovative solutions for remanufacturing. IoT is meant as a digital network of connected physical objects that are equipped with a variety of identifying, sensing, communication and processing technologies, becoming the so-called "smart products". These novel abilities open up to the development of innovative smart circular systems within building practices. Based on these premises, the present paper proposes an overview of digital technologies that can contribute to solve the bottlenecks of current practices, supporting the spreading of reuse and remanufacturing within the construction sector. In particular, addressing remanufacturing as a key strategy for achieving long-term manufacturing sustainability in the construction sector, the next paragraphs explore the potential of sensing technologies, digital twins and information platforms.

45.2 Sensing Technologies for Supporting the Provision of Service-Based Business Models in the Construction Sector

Supporting circularity within the construction sector, Internet of Things (IoT) and its sensing technologies (e.g. wireless sensors, actuators, RFID, smart tags, mobile devices, etc.) and digital wireless networks (e.g. Wi-Fi, NFC, Bluetooth, etc.) currently offer virtual identities and real-time communication capabilities to physical products, that acquire the so-called "smartness" becoming "smart products" (Wang et al. 2020). By equipping products with a unique identifier, it is possible to collect data during the product use phase, allowing a real-time IoT-enabled traceability. Moreover, through the empowerment with sensing and communication capabilities, smart products can monitor and report their own condition and they can communicate over the Internet with other devices or/and with people by means of data visualization tools and digital smart interfaces. These new capabilities of products are unlocking new ways of value creation by allowing information gathering and analysis after the construction products leave the production site or the distribution facility (Alcayaga et al. 2019). Especially, these capabilities enable the transfer within the construction sector of Service-Based Models (SBMs), already consolidated in different business sectors with proven benefits (Bressanelli et al. 2021), allowing an increase in resource efficiency and a reduction of the overall life cycle costs, contributing to the transition towards a circular economy. SBMs imply a paradigm shift towards practices no longer oriented to the sale but rather to the offer of products "as a service". According to this new vision, the ownership of products is not transferred to the customer but it is retained by the supplier. Hence, the customer is no more the buyer of a physical good but it becomes the purchaser of a service, shifting from being the "owner" to become the "user" of the product. Hence, SBMs allow to decouple the value creation from the resource throughput, since the value is linked always to the same product that is sold as a service for several use-cycles (to the same user or to different users), preserving the resource-embedded value over time. In this perspective, the adoption of the IoT facilitates the implementation of SBMs within construction practices, supporting the collaborative consumption of products by offering advanced information management functionalities (e.g. remote monitoring, dynamic data storage, data processing and analytics). The IoT-based SBMs experimented in recent years are varied and characterized by different contractual terms and payment systems (e.g. rental, leasing, sharing, outsourcing, performance-based, functional result, etc.), however they all require the implementation of product-lifecycle-extension smart strategies, namely: smart use, smart maintenance, smart reuse, smart remanufacturing (Table 45.1) to guarantee the value creation overtime (Alcayaga et al. 2019). Table 45.1 introduces these smart strategies for circular IoT-based SBMs in the construction sector, highlighting the needed enabling technologies and their role towards an effective product and information management.

In particular, it is possible to classify the SBMs in three main categories, according to different delivery strategies and value retention approaches, i.e.: product-oriented, use-oriented, result-oriented SBMs (Alcayaga et al. 2019; Bressanelli et al. 2021). In particular:

- *Product-oriented* SBM refers to the selling of the product as a service plus a set of "quality services" (Alcayaga et al. 2019) (e.g., preventive maintenance, periodic quality testing, condition monitoring, sub-component upgrading, etc.) during the product use-cycle. The set of additional services enables to guarantee the steady availability and reliability of high quality products.
- *Use-oriented* SBM includes leasing and renting contracts (the same product is used sequentially by different users) and sharing approaches (the same product is used simultaneously by different users) (Alcayaga et al. 2019). Instead of the traditional single payment method, the use-oriented SBM involves new pay-per-use or pay-per-period formulas.
- *Result-oriented* SBM involves contracts based on the results, thus on the product performance. The user relies on global service outsourcing strategies. Hence, the supplier delivers a performance result (agreed with the user) and there is no particular technical or aesthetic specification expressed by the user about the product that supports the service delivery (Alcayaga et al. 2019). Result-oriented SBM involves payment systems based on the results, i.e., pay-per-performance methods.

According to this classification, Table 45.2 shows the suitability of the IoT-based strategies (introduced in Table 45.1) with respect to the three different typologies of SBMs.

Albeit still in their ancillary investigational phases, some experimentations of SBMs in the construction sector have been recently carried out at the European level, such as the virtuous case of smart façade leasing developed by TU Delft (Azcarate-Aguerre et al. 2022). The project focuses on the development of two full-scale Façade-as-a-Service (FaaS) prototypes and the related circular business model (Azcarate-Aguerre et al. 2018). The service-based model proposed by TU Delft is based on the use of the façade as performance-delivering tool: instead of purchasing the façade panels as products, the customer pays for their performance,

Table 45.1 Enabling technologies for circular SBMs in the construction sector. Adapted from Alcayaga et al. (2019)

Smart strategy	Execution frequency	Enabling technologies	Improvements towards circular models
Smart use	Constant during product use-cycles	• Smart products • ID tags, sensors, actuators • BMS and IoT • Visualization tools	• Remote monitoring to collect data on product use-profiles for improving product efficiency and safety • Data accessibility though a digital information system with dynamic database to store and access overtime product data • Analysis of product usage data for estimating product residual lifespan
Smart maintenance	Regularly during product use-cycles	• Real-time monitoring systems • BMS • Dynamic databases • Visualization tools • Data analytics	• Diversification of the service offer and availability of new ICT-based services • IoT-based preventive maintenance strategies: maintenance is performed adaptively according to product behaviors (increased product availability, reduced expenditures and downtimes) • Interventions for product lifespan extension
Smart reuse	After a use-cycle	• BMS • Data visualization tools • Dynamic databases • Data analytics tools	• Estimation of product residual performance and evaluation of the reusability potential of smart products • Data availability to support the decision-making related to the possible reuse and remanufacturing actions • Increased efficiency of reuse processes, e.g. reduction of materials losses and logistic costs by accessing reliable product data (location, technical features, use level, etc.)

(continued)

Table 45.1 (continued)

Smart strategy	Execution frequency	Enabling technologies	Improvements towards circular models
Smart remanufacturing	After multiple use-cycles	• BMS • Data visualization tools • Dynamic databases • Data analytics tools	• Advanced estimation of product residual performance • Improved ability to estimate and simulate the performance of remanufactured products • Reduction of the uncertainty related to the remanufacturing processes in terms of timeframes, costs, waste generation, etc. • Availability of data for improving the design process of new products (design for remanufac-turing/disassembly strategies)

Table 45.2 IoT-based strategies for product-, use- and result-oriented SBMs

Smart strategy	Addressed SBM type/s	Possible payment systems
Smart use	All the three types of SBMs (Product-/use-/result-oriented)	• Pay-per-use/performance/period formulas • All inclusive (product plus services) payment • Fixed fee with incentives/penalties
Smart maintenance	Marketed in all-inclusive formulas or as extra service in use-/result-oriented SBMs	• All inclusive (product plus services) payment • Fixed fee with incentives/penalties • Pay-per-performance
Smart reuse	Suitable for use-oriented SBMs but it can also address the other two SBM types	• Pay-per-use or pay-per-period formulas • Fixed fee for the product preparation-to-reuse activities, such as cleaning, logistics
Smart remanufacturing	Suitable for product- and use-oriented SBMs but it can also address the result-oriented SBMs	• Pay-per-use or pay-per-period formulas • Deposit/credit or buy-back formulas • Fixed fee for product preparation-to-reuse activities (cleaning, logistics, etc.)

outsourcing their management and upgrade to the supplier/manufacturer. According to the project team (Azcarate-Aguerre et al. 2022, 2018), the adoption of this business scheme can guide to a wider market adoption of circular products as well as reuse/remanufacturing practices, limiting environmental and economic impacts.

45.3 Digital Twins for Promoting Design-for-Remanufacturing and Life Cycle Cost Estimations

The term Digital Twin (DT) refers to a 3D digital copy of the actual physical asset. The DT represents a virtual model of the product matching the real geometry, structure, physical characteristics and functional attributes (Wang et al. 2020). By implementing this digital-physical one-to-one correspondence, the DT opens up to new opportunities to simulate remanufacturing interventions and estimate the related needed time and costs, reducing in this way the uncertainty on remanufacturing feasibility. The DT can also be digitally integrated with smart sensors, IoT technologies, artificial intelligence, machine learning and big data analytics in order to (i) replicate the specific behavior of the real product and automatically updating the 3D model when changes in the physical world occurs, (ii) detect the real-time use conditions of assets and analyze their maintenance status, (iii) plan data-driven predictive maintenance interventions to be carried out to extend the lifespan of products. By means of these capabilities and exploiting the physical-digital bi-directional data flows, the DT can support the delivery of the SBMs introduced in the previous paragraph, with particular regard to product-oriented (preventive maintenance services) and result-oriented (real-time monitoring of product performance) models.

Moreover, the adoption of DTs has the potential to facilitate the integration of Design for Disassembly and Design for Remanufacturing criteria (Table 45.3) within current building product design practices. Design-for-D/R practices aim to ensure the disassemble-ability of the built assemblies with the final goal of facilitating the maintenance, repair, remanufacturing and reuse of their components.

By being the initial stage of the product lifecycle, the product design directly affects the subsequent phases (i.e. manufacturing, delivery, use and end-of-life). Hence, the development of a proper and accurate DT in the design stage has the potential not only to optimize the design processes but also to support the operational management of (i) maintenance intervention during the product use-cycles and (ii) remanufacturing interventions at the end of each product use-cycle by identifying the most sustainable and cost-effective solution, performing scenario simulations and what-if analyses. In the long run, Design-for-D/R approaches can contribute to a more efficient use of material resources and energy, reducing the life cycle cost of buildings and contributing to environmental savings.

Table 45.3 Examples of design-for-D/R criteria for supporting circular SBMs. Adapted from Denis et al. (2018)

DfR/D criteria	Design verifiers
Accessibility	The element is free to move according to the disassembly direction
Transportability and move-ability	Maximum dimensions and weight of the element for transportation and movement
Modularization	Standardization of dimensions
Reversibility of connections	Connections can be unfastened without damaging the elements
Disassemble-ability	Ease of disassembly and limited time to disassemble
Sequential dependency of disassembly actions	The order of the disassembly actions (sequential dependency) is easy to be recognized

45.4 Information Platforms for Facilitating the Creation of Remanufacturing Supply Networks and Digital Marketplaces

Information Platforms (IPs) are increasingly recognized in literature as promising tools towards circularity in the built environment. IPs exploit the novel capabilities of data management offered today on the market by ICTs and IoT for the creation and management of stakeholder networks, so that the various actors involved in Service-Based Models (SBMs) can share and exchange data and information.

By facilitating a long-term multi-stakeholder engagement, IPs support the overcoming of main communication barriers to the uptake of SBMs within the construction sector. In particular, the development of IPs can contribute to face some key challenges for SBMs implementation related, firstly, to the lack of stakeholder collaboration and communication tools and, secondly, to the lack of tracking and displaying tools able to visualize the demand-offer of available remanufactured/to-be-remanufactured products. In this regard, on the one hand, IPs support the relationship between manufacturer, remanufacturer, service provider and customer, facilitating their interaction (e.g. through smart apps, queries on shared databases, communication interfaces, mobile devices, real-time notifications systems, etc.). On the other hand, digital platforms can be used for the development of online brokerage websites of e-commerce for the purchase and sale of goods and/or services (digital matchmaker marketplaces), facilitating the demand-offer matching and reshaping the traditional way of selling products. Specifically, Marketplace Platforms (MPs) or Transaction Platforms are aimed at facilitating the online buying and selling by creating an e-commerce for B2B or B2C transactions of products and/or services, according to different procurement features. Although sharing the intent, MPs features may vary in relation to the following aspects:

- *Product categories.* MP can be mono-product or multiple products;
- *State of use of products.* MP can sell only remanufactured ready-to-use products or they can sell only/also products to be remanufactured;
- *Broker role of the platform.* MP can play a significant role in the relationships between buyer and sellers, by defining contract terms and payment modalities, or they can merely act as a virtual place for demand-offer matching;
- *Geographic market scale.* MP can act worldwide or they can focus on few neighboring countries or they can act only in a single country;
- *Type of contract.* MP can offer sales contracts and/or renting contracts;
- *Accepted stakeholders.* MP can accept only companies for B2B contracts or they can only/also accept single buyers/sellers for B2C or C2C relationships.

With respect to SBMs, MPs can be considered as a "new virtual stakeholder" (Moro Visconti 2021) that connects conventional partners (manufacturers, construction companies, service providers, customers, spare-parts suppliers, etc.) increasing their interaction skills and expanding their ways of communicating.

45.5 Conclusions

Contributing to overcome inefficiencies in the management of both product-related information and supply chain relationships, ICTs and IoT can reduce the high uncertainty that still characterizes remanufacturing processes in the construction field.

The paper outlined the contribution of digital technologies in supporting the implementation of circular SBMs based on remanufacturing in current construction practices. Firstly, sensing technologies and ICT-based monitoring systems can efficiently track building components and products and collect information on their levels of use and degradation during their use-cycles, useful to outline their residual performance and to understand possible remanufacturing actions to carry out. Secondly, digital twins and data analytics tools enable to assess the usage conditions of the products to be remanufactured allowing the estimation of the number and entity (time and cost) of the required rework operations to be performed. Lastly, information platforms and cooperation tools facilitate the implementation of collaborative remanufacturing process by connecting stakeholders, supporting new collaborative businesses and facilitating the demand-offer matching. The adoption of service-based business scheme (SBMs), empowered by the use of these digital tools, represents a booster for a faster market uptake of reused and remanufactured construction products, allowing to limiting resource consumption and environmental impacts, while reducing the initial investment for building owners. Indeed, business schemes focused on the performance delivery rather than the product sale, if properly supported by ICTs, can prove to be successful approaches for initiating circular practices, reducing the economic risks that currently hold back construction industry stakeholders.

References

Alcayaga A, Wiener M, Hansen EG (2019) Towards a framework of smart-circular systems: an integrative literature review. J Clean Prod 221:622–634
Azcarate-Aguerre JF, Klein T, den Heijer A, Vrijhoef R, Ploeger H, Prins M (2018) Façade leasing: drivers and barriers to the delivery of integrated façades-as-a-service. Real Estate Res Q 17(3):11–22
Azcarate-Aguerre JF, Klein T, Konstantinou T, Veerman M (2022) Facades-as-a-service: the role of technology in the circular servitisation of the building envelope. Appl Sci 12(3):1267
Bressanelli G, Pigosso DC, Saccani N, Perona M (2021) Enablers, levers and benefits of circular economy in the electrical and electronic equipment supply chain: a literature review. J Clean Prod 298:126819
Butzer S, Schötz S (2016) Map of re-manufacturing processes landscape. European Remanufacturing Network. Available at: https://www.remanufacturing.eu/assets/pdfs/ERN_DeliverableReport_WP3_Processes_final_for_upload-1.pdf
Denis F, Vandervaeren C, De Temmerman N (2018) Using network analysis and BIM to quantify the impact of design for disassembly. Buildings 8(8):113
ECSO 2019 European Construction Sector Observatory (2019) EU construction sector: in transition towards a circular economy. Trend paper series, April 2019
European Commission (2016) EU construction and demolition waste management protocol, Sept 2016
Eurostat (2017) http://appsso.eurostat.ec.europa.eu/nui/show.do?lang=en&dataset=env_wasgen
Moro Visconti R (2021) Networking digital platforms and healthcare project finance bankability. Sustainability 13(9):5061
Ness D, Xing K, Kim K, Jenkins A (2019) An ICT-enabled product service system for reuse of building components. IFAC-PapersOnLine 52(13):761–766
Parker D, Riley K, Robinson S, Symington H, Tewson J, Jansson K et al (2015) Remanufacturing market study. Available at: https://www.remanufacturing.eu/assets/pdfs/remanufacturing-market-study.pdf
Rios FC, Grau D (2020) Circular economy in the built environment: designing, deconstructing, and leasing reusable products. Encycl Renew Sustain Mater 5:338–343
UNEP (2022 accessed) Issue brief SDG 12. Ensuring sustainable consumption and production patterns. Available at: https://wedocs.unep.org/bitstream/handle/20.500.11822/25764/SDG12_Brief.pdf?sequence=1&isAllowed=y
UNEP—GlobalABC (United Nations Environment Programme—Global Alliance for Buildings and Construction) (2020) 2020 global status report for buildings and construction. Available at: https://globalabc.org/sites/default/files/inline-files/2020%20Buildings%20GSR_FULL%20REPORT.pdf
Wang Y, Wang S, Yang B, Zhu L, Liu F (2020) Big data driven hierarchical digital twin predictive remanufacturing paradigm: architecture, control mechanism, application scenario and benefits. J Clean Prod 248:119299

Chapter 46
Territorial Energy Potential for Energy Community and Climate Mitigation Actions: Experimentation on Pilot Cases in Rome

Paola Marrone and Ilaria Montella

Abstract One of the conditions toward mitigation and a zero-emission economy is to plan the transition to a sustainable urban energy system. The dimensional and typological variety of urban pattern, and the functional contribution of inhabitants, represent an important potential to reduce energy consumption and climate-changing gases. Despite this evidence, many studies focused on the energy transition have given limited attention to issues of scale, space, and context in urban settings and how they can shape different energy systems. This article deals with renewable energy communities in the urban context and, by presenting some results of research that, through pilot cases in Rome, aims to test mitigation and adaptation solutions in proximity spaces. In particular, it investigates how the different forms of already built urban fabrics, together with social and environmental resources, can influence the form and implementation of the decentralized energy system and vice versa.

Keywords Energy transition · Renewable energy communities · Urban pattern · Energy potential · Indicators

46.1 Introduction and Reference Context

The acceleration toward an energy system based on fossil fuels, which technological progress has encouraged to pursue over time, has been the usual condition for the economic and technological advancement of the European economy. The trend

P. Marrone · I. Montella (✉)
University of Roma Tre, Rome, Italy
e-mail: ilaria.montella@uniroma3.it

P. Marrone
e-mail: paola.marrone@uniroma3.it

E. Arbizzani et al. (eds.), *Technological Imagination in the Green and Digital Transition*, The Urban Book Series, https://doi.org/10.1007/978-3-031-29515-7_46

reversal that the European guidelines,[1] the sustainable growth strategy,[2] and the NRRP[3] are currently calling for, leads to new and shared decarbonization and energy transition objectives toward 2050 and with the reducing emissions goal of 55% by 2030 compared to 1990 levels.

The recently approved Plan for the Ecological Transition (PTE),[4] awaiting the European Legislative Package "Fit for 55", is part of the European Green Deal and links up with the Italian National Recovery and Resilience Plans, proposes new targets for the reduction of emissions and primary energy, and electricity production from renewable sources (RES).

The recommendations contained in Energy Efficiency First outline, from principles to practice, guidelines, and examples for its implementation in decision-making in the energy sector through mitigation measures, increasing the share of RES energy and promoting energy efficiency toward more cost-effective solutions from production to grid transport, to the supply and demand management and end-use energy savings.

However, despite the fact that cities represent a place of integration of multi-sectoral actions, being structured at the same time by urban patterns that are already densely built up, they do not allow easy changes required by these new decarbonization scenarios, identifying, among the most difficult points, that of electrification of primary energy with integration of RES. Potential transition scenarios, in which polycentric decentralization of energy production is one of the founding strategies, attribute to the built-up area renewed active functions as an energy infrastructure (Sandroni et al. 2021), despite persisting criticalities, which are also highlighted by the PTE, such as the difficulties in adapting the network infrastructure, the spatial and relocation requirements relating to storage, and the issue of delicate evaluation procedures for the inclusion of plants located in areas adjacent to protected areas.[5]

However, studies on energy transitions have often dealt with socio-technical, technological, or governance dynamics while devoting little attention to space and the interdependence between urban processes and energy transitions (Huang and

---

[1] The reference is to the European Climate Law—Regulation (EU) 2021/1119 establishing the framework for achieving climate neutrality, which transposes the adaptation goal of the Paris Agreement, the European Green Deal intent (COM(2019)640), the Clean energy package and the Integrated National Energy and Climate Plans (NIPECs) that provide for greenhouse gas reductions.

[2] COM(2020)575 final.

[3] NRRP—National Recovery and Resilience Plan.

[4] Compared to the emission reductions envisaged in the NIPECs (328 million ton of CO_2 eq in 2030), the PTE reduces the target to 256 million ton of CO_2 eq. Compared to the European PRIMES 2007 baseline energy scenario, it indicates a reduction of primary energy from 43 to 45%. Furthermore, it sets a target for RES of 72% by 2030 and 95–100% in 2050, with a gross efficient power from RES of 70–75 GW in 2030, for photovoltaics 200–300 GW. Source: https://temi.camera.it/leg18/post/la-proposta-italiana-di-piano-nazionale-per-l-energia-e-il-clima.html.

[5] Decree Law No. 77/2021 addresses the issue of assessment and permitting with acceleration provisions for RES plants in areas adjacent to protected areas, with reference to projects covered by NIPECs and NRRPs.

Castán 2018) even though energy systems are spatially arranged with technological components that, although integrated, change their nature by defining "geographical patterns" of grid, connections, storage, and control (Bridge et al. 2013).

In this framework, REC's,[6] being a decentralized and widespread energy system model on the territory potentially acceptable by the inhabitants, represent a production scenario compatible with local, environmental, and social resources (Sandroni et al. 2021; Bolognesi and Magnaghi 2020).

Recently, several trials of RECs[7] are emphasizing the need to experiment with appropriate technologies and optimal energy mixes in relation to the energy potential of each territory (EEA 2021; RSE SpA 2021), highlighting the need for the choice of an energy system to be based on compatibilities with a wide range of ecosystem services since the ecological impacts of energy systems tend to occur locally, while energy policies are designed at the national and global level (Holland et al. 2016).

Through the study of some pilot cases in Rome, the article addresses the limits and potential of RECs in the urban context, identifying a framework of indicators to support the assessment of the elements that may define their appropriateness.

46.2 Approach and Methodology

In the energy transition process, both cities, with their already built-up urban patterns, and inhabitants, with their energy needs, are crucial. This happens because of the synergy between energy demand, which is a function of the built environment and users' consumption, and energy production from RES, which requires space for technological components and its sizing according to consumption. This dense relationships and actions at the local level becomes decisive in the energy transition process and enables the implementation of RES on a large scale (Carreón and Worrell 2018), emphasizing that the energy potential of a territory can come from the synthesis of a proper balance between environmental, social, historical, and cultural resources.

Moreover, since among the objectives of the RECs are the principle that rewards the balance between local production and consumption, energy saving actions to be implemented both through interventions on the existing building stock and on the thrifty governance of energy needs by users also become central.

The methodology adopted in the research was structured in three phases and considered the following questions:

- Phase 1: on what elements does the feasibility of REC's in urban contexts depend?
- Phase 2: what potential and limitations does an urban fabric express, in terms of energy production, final consumption, mitigation actions that can be implemented taking into account environmental and landscape values?

[6] RECs are defined and regulated by Legislative Decree No. 199 of 8 November 2021, which transposed the REDII Directive (2018/2001UE).

[7] Legambiente interactive map: https://experience.arcgis.com/experience/1d992eb312f942959b55c611dd0ce968.

- Phase 3: how do the physical elements of the context, such as buildings and open spaces, impact on the feasibility of RECs, also in view of the space required for their role as providers of collective services?

In order to study the elements that condition the feasibility of RECs in urban contexts (Figs. 46.1 and 46.2), a framework of components and indicators (Fig. 46.3) concerning "production-consumption-savings-integration" (Phase 1) has been assumed, excluding in this first phase the economic feasibility indicators and the incentives related to the mechanism that rewards the energy produced and self-consumed instantaneously in the REC.

In order to evaluate the components of the energy system and their possible application in the built environment, the analysis was carried out on three pilot cases in

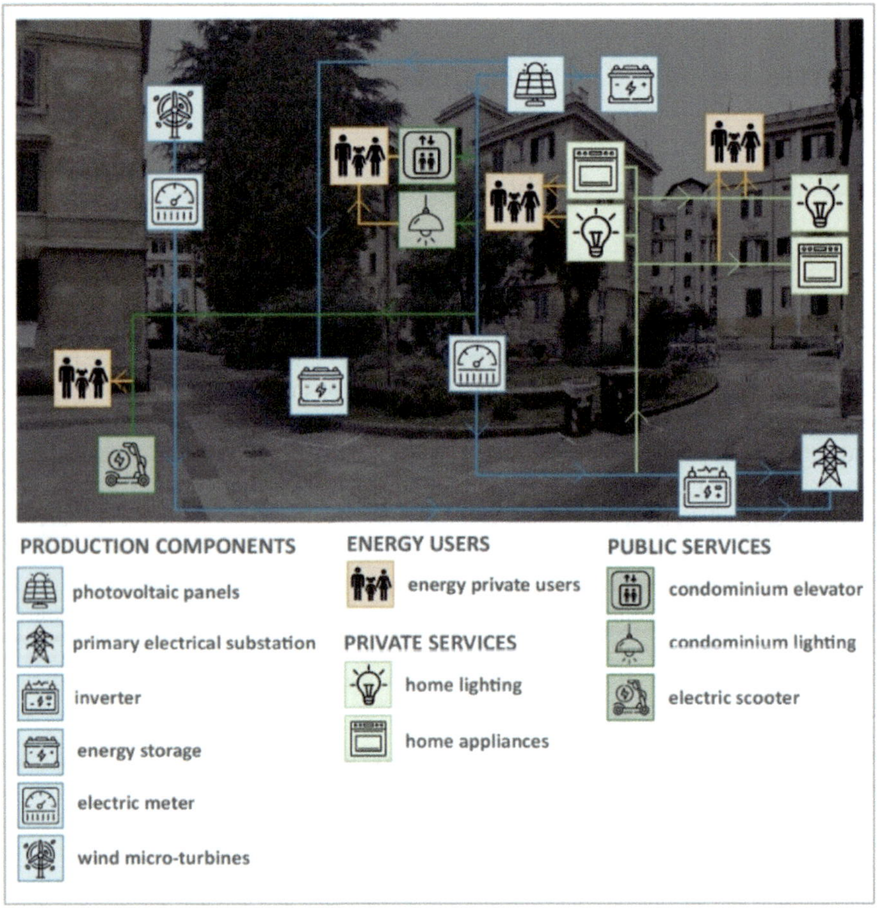

Fig. 46.1 Renewable energy communities: elements—connections—prosumers. Private use of energy

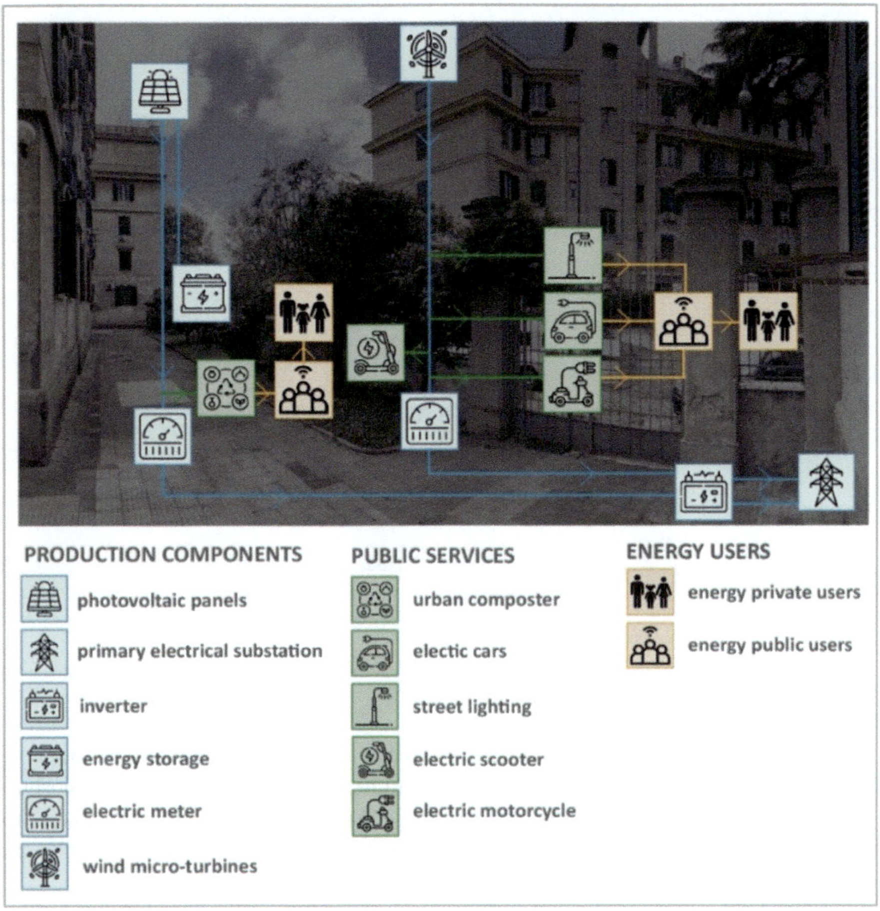

Fig. 46.2 Renewable energy communities: elements—connections—prosumers. Collective use of energy

Rome (Testaccio, Balduina, and Prima Porta[8]) (Fig. 46.4), chosen because representative of different types of building fabric, characterized by a different relationship between built-up areas and open spaces (Phase 2).

[8] Testaccio: social housing. Period of construction, 1983–1912; high territorial and building density; high coverage ratio; good endowment of public services; street spaces with parking functions and scarcity of green; condominium courtyards with high functional flexibility; presence of public components of high environmental quality (floodplain areas) inaccessible or not equipped.

Balduina: small buildings with public spaces on the lot. Period of construction, 1955–1980; high territorial and building density; very high coverage ratio; sufficient endowment of public services; street spaces with parking functions and scarcity of greenery; presence of some tree-lined avenues; minimal areas pertaining to the buildings; only decorative greenery; presence of public components of high environmental quality (floodplain areas) usable or inaccessible and not equipped.

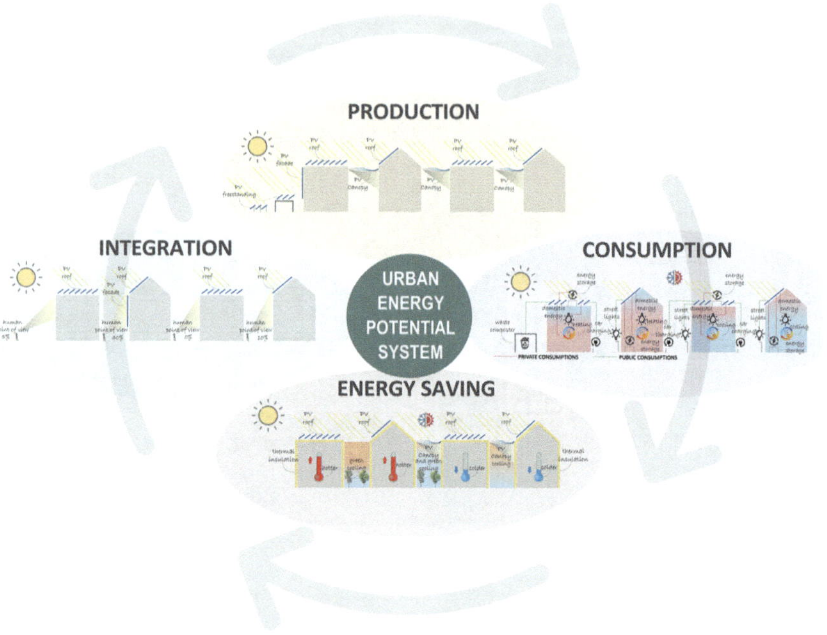

Fig. 46.3 Urban energy potential system. Production, consumption, energy saving, and integration

The aim of the analysis was to identify and catalog the surfaces available to accommodate the elements of the energy system and to understand how the neighborhood spaces could relate to the buildings, becoming the place for mitigation and adaptation actions oriented not only to the energy sector, but also to sustainable mobility and the enhancement of ecosystem services (Fig. 46.5).

The pilot cases were analyzed according to the indicators of production, consumption, saving, and integration.

For production, the surfaces of buildings and open spaces available for the location of the production system have been identified while also considering the interoperability with existing grids and all spatial and technological requirements related to the inclusion of storage systems, location of energy storage and MV/LV transformers, and considering the impact that, in a built-up urban center, brought shadows can have on radiation; for consumption, the kWh of domestic consumption and collective use of the inhabitants have been assumed; for saving, the kWh that can be saved by mitigation measures and by improving the energy efficiency of the buildings have

Prima Porta: informal ex-abusive settlements with semi-public, residual or interstitial spaces. Construction period 1959–2003; moderate territorial and building density; moderate coverage ratio; sufficient provision of public services; restricted road spaces with parking functions; scarcity of equipped green spaces; private appurtenant spaces; presence of public components of high environmental quality (floodplain areas) that are inaccessible or not equipped.

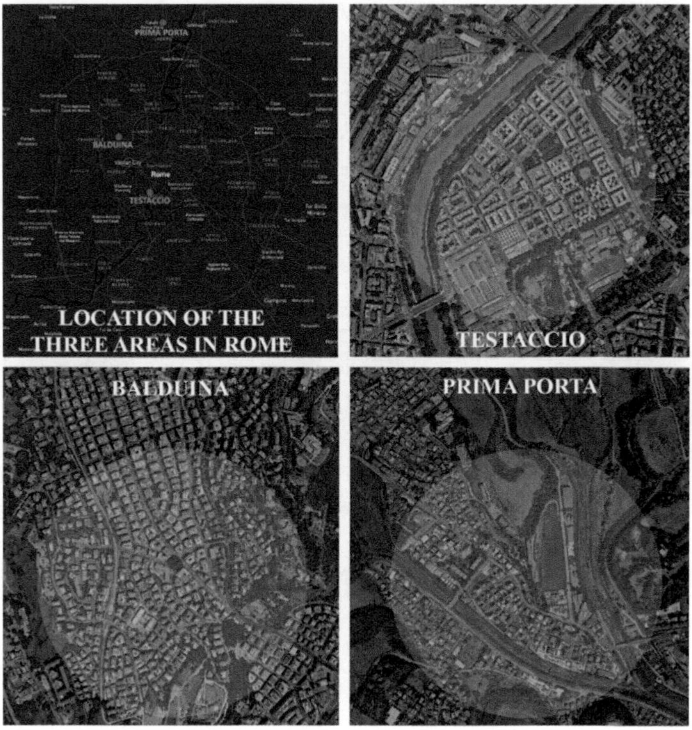

Fig. 46.4 Location of the pilot cases and pilot cases plan

been assumed; for integration, the elements characterizing the landscape and the compatibility constraints have been considered.

In order to answer the third question, in Phase 3, a REC was simulated in the three pilot cases to assess how much RES energy could be produced by the different urban patterns, to evaluate the percentage of final consumption could be satisfied by RES, to quantify the possible savings through mitigation actions, to hypothesize additional collective services useful also in the perspective of adaptation and mitigation.

The simulation was carried out with the RECON tool[9] assuming: to use PV technology, positioned on the roof, occupying no more than 60% of the available area, and to produce a maximum of 200 kWp for both the RECs configuration and the collective self-consumption.[10]

[9] The Renewable Energy Community ecONomic simulator (RECON) tool developed by ENEA, and the electricity and production consumptions were calculated with the collaboration of ENEA's Smart Cities and Communities Laboratory. I Thermal consumptions are calculated considering the thermal consumption for multi-family residential building reported by NIPECs.

[10] Energy communities and collective self-consumption are two configurations introduced by RED II.

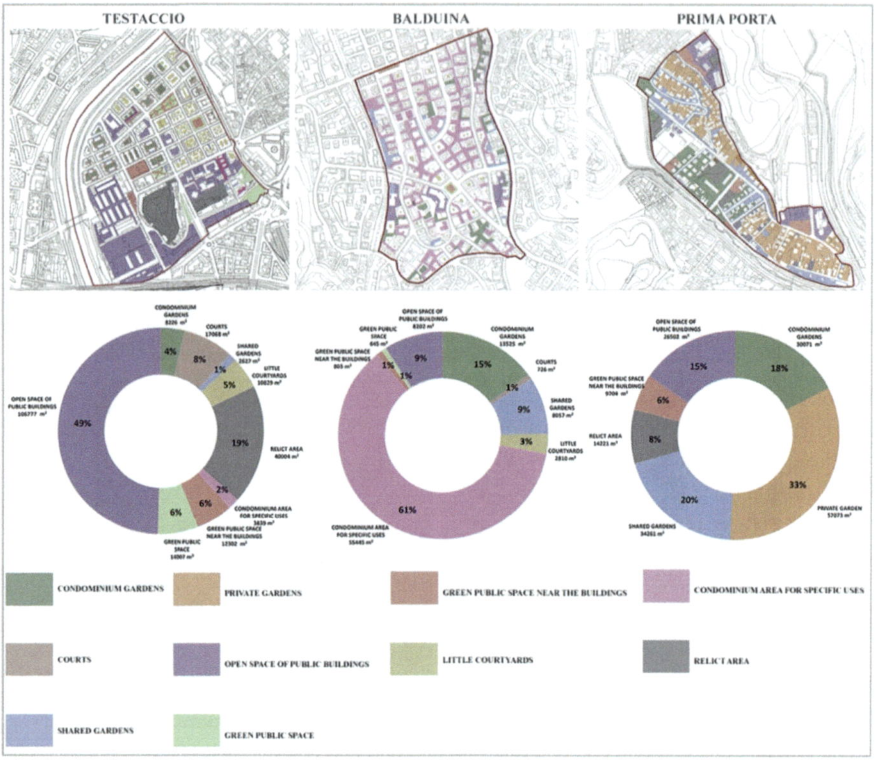

Fig. 46.5 Taxonomy and counting area of spaces of proximity

Comparing the results of the three pilot cases, some considerations are made to support the development of indicators, in a later stage of research,[11] using GIS systems and a multi-criteria evaluation model applied to the thematic areas of energy, mobility, and ecosystem services, to assess their interrelationships in terms of mitigation and adaptation.

46.3 Results and Discussion

To support the feasibility assessment of RECs in urban contexts (Phase 1), a first result concerns the definition of the indicator system (Fig. 46.6). The classes identified, in addition to the role of the urban space, also considering the role of the inhabitants, as prosumers who govern production and energy needs related to them, and that of

[11] The study presented here is part of a broader PRIN research project focusing on the role of neighbourhood spaces in climate mitigation.

encouraging the collective use of energy through synergistic actions for mitigation and adaptation.

Starting with the classification of indicators, an initial characterization of the pilot cases was carried out in Phase 2, and the results are reported here for the first three components of the system.

Production: a taxonomy and quantification of proximity open spaces has been defined (Fig. 46.5), identifying the prevailing types and their suitability to accommodate the components of the energy system or the collective services it supplies. For Testaccio, characterized by a higher percentage of public spaces and relict areas, it is possible to hypothesize the use of these areas for the production systems location, rather than on the roofs of buildings. For the Balduina and Prima Porta, the scarcity of available public areas and the prevalence of private areas pose the difficulty of locate technological elements of the production system and the collective services they supply (Fig. 46.7).

Consumption: an analysis of the population and composition by households and age was made, suggesting the need for an important energy production system, particularly in Testaccio, which has a high population density despite the high number of one-person households (Fig. 46.8).

The simulations, carried out considering a production capacity of 200 kWp for the three pilot cases, showed that in Testaccio only one block is sufficient to build a REC, while in Balduina 5 buildings are needed and in Prima Porta 17 houses (Fig. 46.9). With this scenario, the simulations showed that in Testaccio and Balduina, it is possible to cover with RES about 35/30% of the electricity consumption, while in Prima Porta about 60% (considering however that in Prima Porta there are no condominium lifts) (Figs. 46.10, 46.11 and 46.12).

Evaluating the first results of the simulations, therefore, it can be assumed that in Testaccio, with the presence of public or relict areas and the high presence of public proximity spaces, an energy production system that makes use of them for the location of the RECs' technological systems would be desirable.

In Balduina, considering the scarce availability of public proximity spaces and the typological conformation characterized by predominantly condominium areas, it can be assumed smaller RECs composed of a few buildings. In Prima Porta, characterized by a less dense urban fabric and the prevalence of private proximity open spaces, it can be assumed a REC configuration that envisages the synergy between private users also with the intention of increasing the endowment of collective services powered by shared energy.

Energy Saving: the analysis of the age of construction has highlighted in Testaccio buildings built between the early 1900s and the 1960s, in Balduina between the 1940s and 1970s, and in Prima Porta between the 1960s and 1980s, and therefore, it can be imagined that given the construction characteristics, the implementation of RECs must also be associated with deep-renovation to improve energy efficiency. In view of the public ownership of many buildings and the more favorable *S*/*V* ratio, it is assumed that such interventions can be more easily achieved in Testaccio while, in Prima Porta and Balduina, they will depend largely on the capacity for synergy and intervention between private users (Fig. 46.13).

PRODUCTION
1_ ENERGY PRODUCTION POTENTIAL: how much energy can the territory produce?
1a. RENEWABLE ENERGY SOURCES
1b. ENERGY PRODUCED BY THE BUILDING
1b.1 Energy produced on roofs / on vertical surfaces • Factors related to the production system • Environmental factors • Local factors related to the context • Positioning factors • Factors related to storage • Factors related to the inverter
1c. ENERGY PRODUCED BY TERRITORIAL RESOURCES
1c.1 Ground free surfaces / "FV Canopy" road surfaces • Factors related to the production system • Environmental factors • Local factors related to the context • Positioning factors • Factors related to storage • Factors related to the inverter
CONSUMPTION
2_QUOTE OF SHARED ENERGY: how much of the consumption can be satisfied by the energy potential (point 1)?
2a. SHARED ENERGY FOR PRIVATE CONSUMPTION
2a.1 Domestic energy consumption (including electricity consumption related to heating, cooling and DHW)
3_PHYSICAL SELF-CONSUMPTION RATE: which and how many services can be satisfied by the energy potential?
3a. SELF-CONSUMED ENERGY FOR COMMON SERVICES
3a.1 Condominium and neighborhood energy consumption
ENERGY SAVING
4_ENERGY SAVING POTENTIAL: how much energy can be saved in the area?
4a. ENERGY SAVING OF BUILDINGS
4a.1 Energy performance index of buildings
4b. COOLING CAPACITY (THROUGH DESEALING, GREEN PLANTING, CANOPY)
4b.1 Cooling capacity UGI - Urban Green Infrastructure 4b.2 "FV Canopy" shading capacity 4b.3 Albedo of surfaces and variation for cooling
INTEGRATION
5_COMPATIBILITY BETWEEN RES TECHNOLOGIES and HERITAGE: what elements are to be considered to assess the suitability of the heritage / territory to accommodate systems for RES production?
5a. COMPATIBILITY AND INTEGRATION
5a.1 Type of technology and its visibility from the outside 5a.2 Level of integration with the buildings / context 5a.3 Impairment of recognisability of a place 5a.4 Impairment of the ecological network 5a.5 Land consumption
SOCIAL VALUE AND PARTICIPATION
6_SOCIAL CAPITAL: what is the degree of involvement of the local community?
6a. SHARE CAPITAL AND PARTICIPATION
6a.1 Associations that can plan actions bottom-up
6a.2 Initiatives and forms of self-organization aimed at actions of common interest

Fig. 46.6 Indicator system according to production, energy saving, consumptions, and integration categories

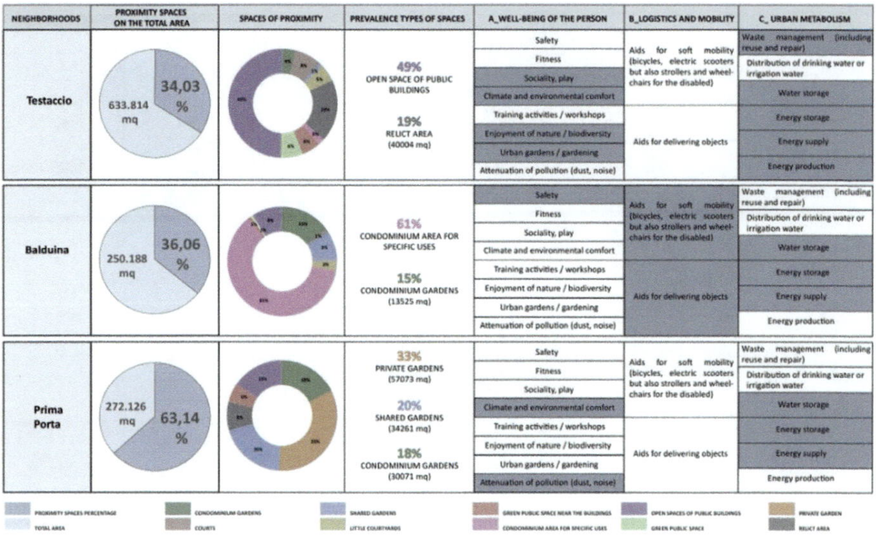

Fig. 46.7 Production (energy produced by territorial resources): study of spaces and functional tendency

Furthermore, in view of the values of the urban heat island intensity indicator in the summer periods (Fig. 46.14), further cooling measures to reduce energy demand may be very effective in the three districts. In fact, the differences between average temperatures in central Rome compared to those in the rural area (Asdrubali 2022) show particularly high daytime and nighttime UHII especially in Testaccio and, surprisingly, given the UHII values, also in Prima Porta.

46.4 Conclusions

To support the feasibility assessment of RECs in urban contexts (Phase 1), a first result concerns the definition of the indicator system (Fig. 46.5). The classes identified, in addition to the role of the urban space, also considering the role of the inhabitants, as prosumers who govern production and energy needs related to them, and that of encouraging the collective use of energy through synergistic actions for mitigation and adaptation.

Although the technological development of renewable energy sources and the progressive reduction of their costs has encouraged their diffusion, in Italy, it is still far from being possible to achieve the production targets to support the energy transition.

RECs represent an interesting topic both because they are supported by European and national policies and because, thanks to their social implication, they allow a bottom-up contribution to the instances of economic transition. Citizens, having

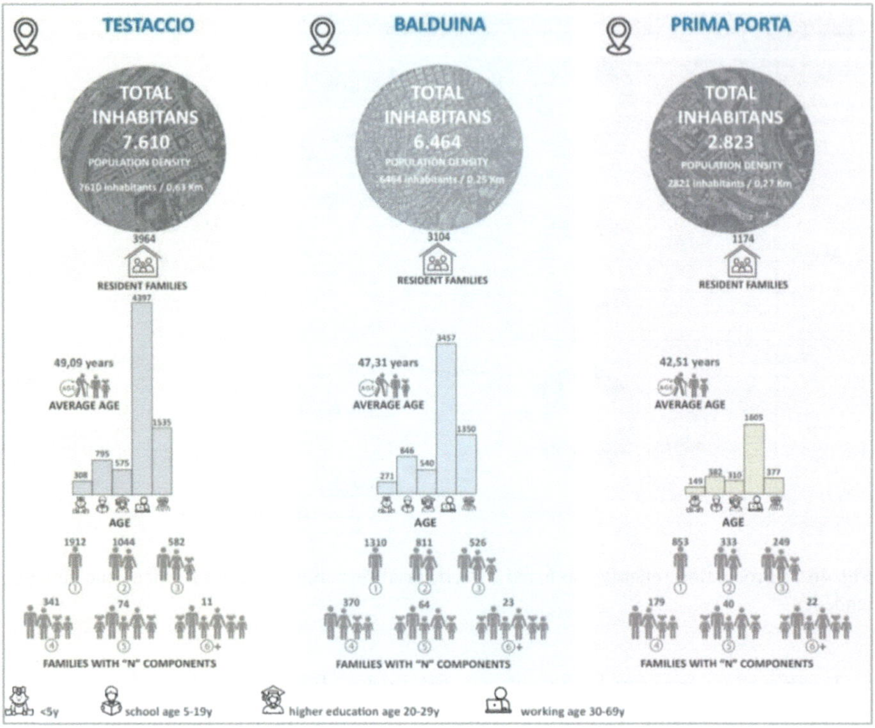

Fig. 46.8 Consumption (shared energy for private consumption): population study—inhabitants—households

become prosumers, become aware of the importance of their role by implementing reparative, generative, potentially evolutionary and socially innovative activities (Montebugnoli and Padella 2021).

For this reason, implementing a REC requires assessments that are not only technical or economic in nature, but also concern the protection and enhancement of the territory's resources, health and well-being, and the provision of services (Maestosi and Meloni 2021; Civiero et al. 2021).

Starting from three neighborhoods in Rome, the research has developed an analysis of how cities, with buildings and proximity spaces, can contribute to decentralized energy systems. The ongoing study has set the objective of identifying indicators of the energy potential of the built urban fabric, with reference to the scope of energy production, energy consumption by the community, possible actions to reduce consumption, and the possibilities of integrating systems.

From the first application of the indicators on the three case studies, it was shown how, by addressing the analysis with different levels of thematic depth, it is possible to aspire to integrated projects respecting the territorial and heritage potential. The development currently underway involves the combination of analytical and quantitative GIS-based methods (Fig. 46.15), with AI.

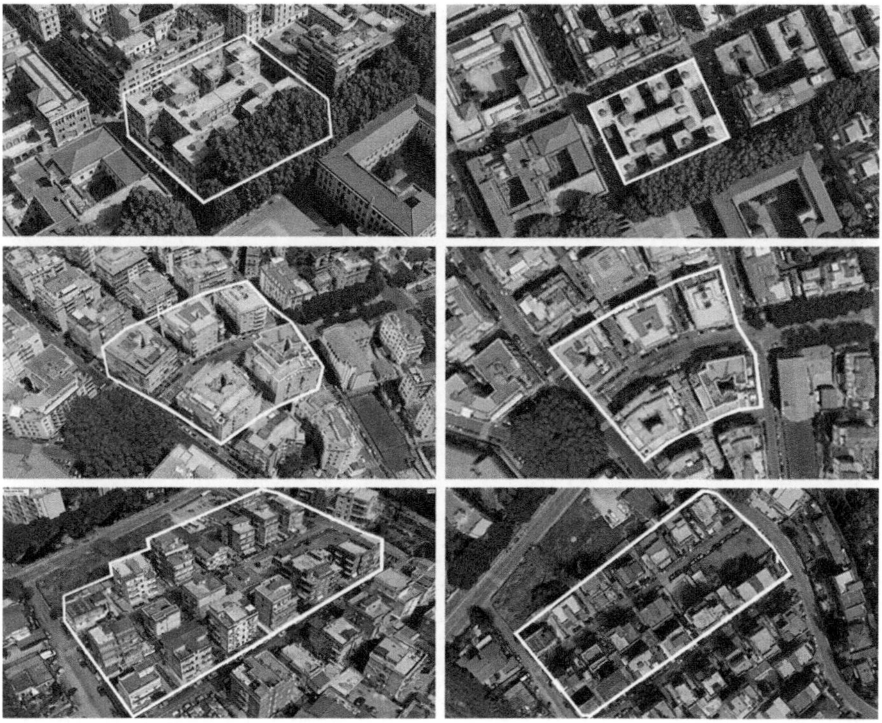

Fig. 46.9 Different dimension of RECs

In particular, information on individual areas (size, inhabitants, greenery, soils, areas available for photovoltaic production in relation to exposure, inclusion of technological elements related to energy production and new services, etc.) was included in a GIS system, used as a quantitative basis with the aim of developing a digital tool for risk prediction, and an assessment to support institutions and planners.

In fact, the digitized data are included in the risk model and algorithmic assessment model to prefigure, from a proximity and climate mitigation perspective, optimal interactions for sequential transition scenarios.

The indicators, aspiring to define the energy potential of a territory, offer support in the evaluation from different points of view and, since energy demand is determined by the built heritage and energy production requires the availability of public and private spaces, a decisive role is assumed by the inhabitants in governing the balance between energy demand and supply, as advocated by the RECs, and in resolving possible conflicts between different interest groups at local level also through social innovation initiatives.

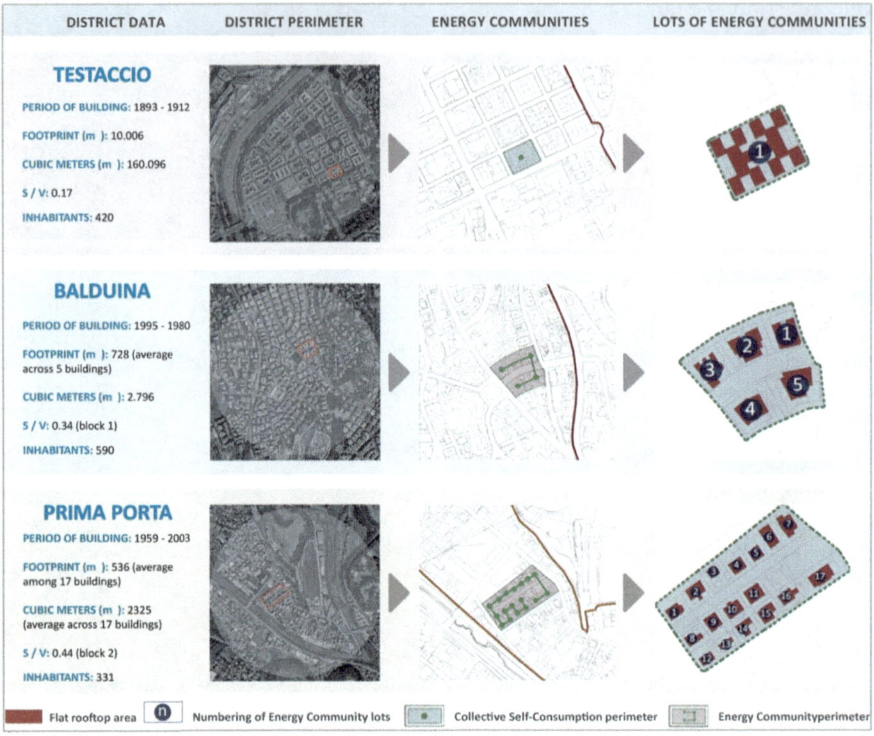

Fig. 46.10 Production (energy produced by the buildings): study of different size of renewable energy communities and simulation results

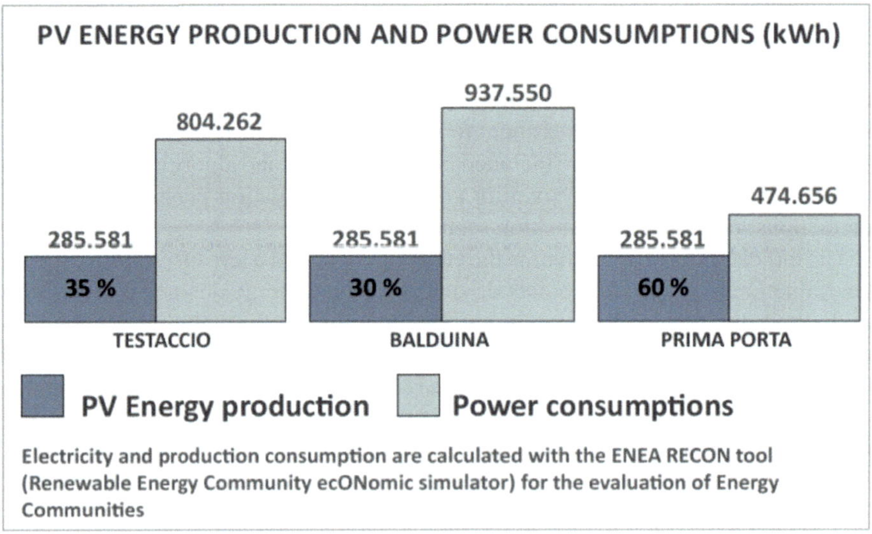

Fig. 46.11 PV energy production and power consumptions (kWh)

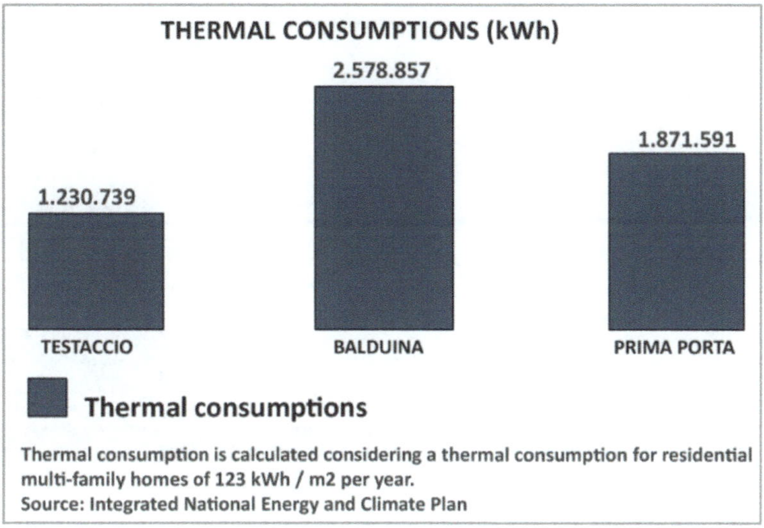

Fig. 46.12 Thermal consumptions (kWh)

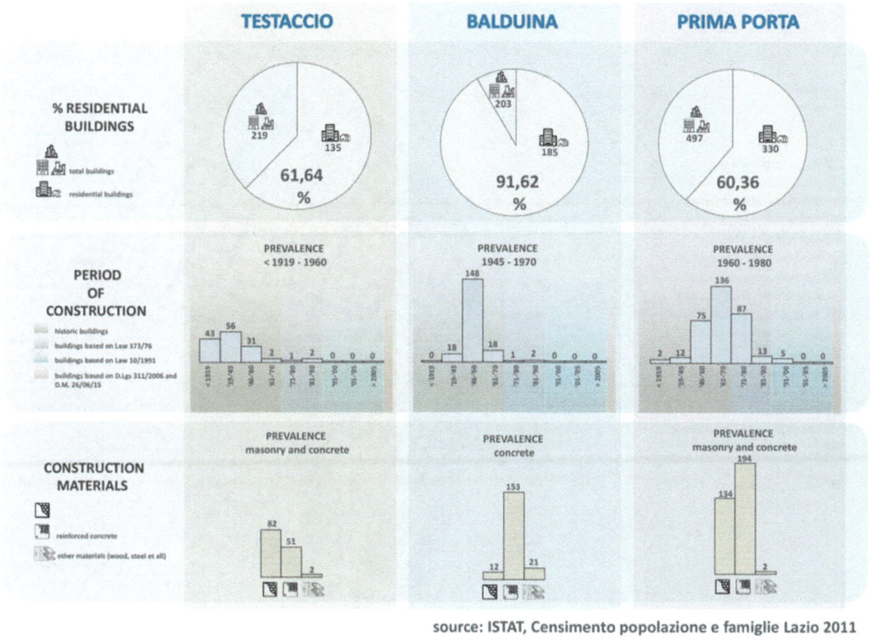

Fig. 46.13 Energy saving (energy saving of buildings): study of buildings—period of construction—materials

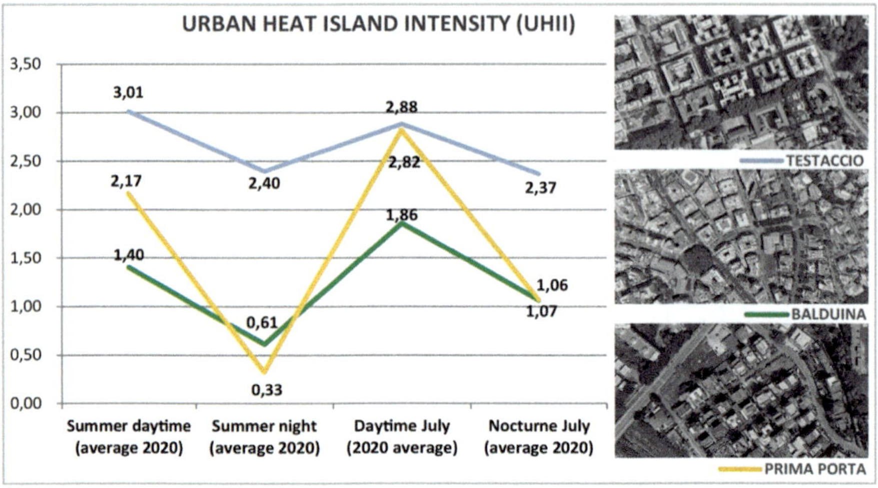

Fig. 46.14 Energy saving (cooling capacity): study of urban heat island intensity (UHII). Graphic reworking starting from the data of the map included in the references

Fig. 46.15 Image of the GIS input of proximity spaces, surveyed according to the identified taxonomy, and street types in Testaccio

Acknowledgements The paper concerns partial results related to the ongoing research project PRIN (Project of Significant National Interest) 2017: "TECHSTART-key enabling TECHnologies and Smart environmenT in the Age of gReen economy. Convergent innovations in the open space/building system for climaTe mitigation".

Authors' Contributions Introduction and reference context, P.M.; Approach and methodology, P.M. and I.M.; Results and discussion I.M.; Conclusion, P.M. and I.M.

References

Asdrubali F (2022) Quanto fa caldo nei quartieri di Roma? www.mapparoma.info. Available at: https://www.mapparoma.info/mappe/mapparoma36-quanto-fa-caldo-nei-quartieri-di-roma/?fbclid=IwAR0koJ2lML-cKkYIQntltaUqMUrlSQnc3N6Rf862Z0n0yHJ4c1IuV93db8E. Accessed 20 May 2022

Bolognesi M, Magnaghi A (2020) Verso le comunità energetiche (in Italian: Towards energy communities). Scienze del Territorio, Special issue Abitare il territorio al tempo del covid numero, pp 142–150. http://doi.org/10.13128/sdt-12330

Bridge G, Bouzarovski S, Bradshaw M, Eyre N (2013) Geographies of energy transition: space, place and the low-carbon economy. Energy Policy 53:331–340. https://doi.org/10.1016/j.enpol.2012.10.066

Carreón JR, Worrell E (2018) Resources, conservation and recycling urban energy systems within the transition to sustainable development. A research agenda for urban metabolism. Resour Conserv Recycl 132:258–266. https://doi.org/10.1016/j.resconrec.2017.08.004

Civiero P, Pascual J, Abella JA et al (2021) Pedrera. Positive energy district renovation model for large scale actions. Energies 14. http://doi.org/10.3390/en14102833

EEA—European Environment Agency (2021) Urban sustainability in Europe opportunities for challenging times, no 07, p 80

Holland RA, Scott K, Hinton ED et al (2016) Bridging the gap between energy and the environment. Energy Policy 92:181–189. https://doi.org/10.1016/j.enpol.2016.01.037

Huang P, Castán V (2018) Interdependence between urban processes and energy transitions: the dimensions of urban energy transitions (DUET) framework. Environ Innov Soc Transitions 28:35–45. https://doi.org/10.1016/j.eist.2018.03.004

Maestosi PC, Meloni C (2021) Transizioni urbane sostenibili-Sustainable urban transitionz. Maggioli Editore

Montebugnoli A, Padella F (2021) Le transizioni gemelle, vol 49, pp 1–61 Available at: https://centroriformastato.it/le-transizioni-gemelle/. Accessed 20 May 2022

RSE SpA - Ricerca sul Sistema Energetico (2021) Le comunità energetiche in Italia. Note per il coinvolgimento dei cittadini nella transizione energetica. Alkes Editrice

Sandroni C, Angelucci V, Armanasco F et al (2021) Comunità dell' Energia: selezione casi di studio ed analisi delle barriere tecniche, economiche, autorizzative e regolatorie. RSE - Ricerca Sistema Energetico

Chapter 47
Integrated Design Approach to Build a Safe and Sustainable Dual Intended Use Center in Praslin Island, Seychelles

Vincenzo Gattulli, Elisabetta Palumbo, and Carlo Vannini

Abstract A flexible multi-purpose center for a dual intended use—hospitality and observation and research related to climate change—has been designed in the fragile environment of Praslin Island, Seychelles. The technical solutions adopted for a low environmental impact LCA based in the designed center during the life cycle will be illustrated: starting from the local supply raw materials, the self-disassembling construction system, the described process is compatible with the site use that the owners have foreseen. Specific logistic systems have been chosen both to the transportation of the material on the site, and to the integrated structural and architectural solutions. In addition, a reconstruction of the natural characteristics of the building site has been developed both by google-earth observation and with a survey directly on the site through processing acquired images. The multi-disciplinary perspective through which the project has been conceived shows beneficial effects in terms of reduced impact on the original and resilient natural environment. Future developments of the work will be devoted to the optimization of this multi-disciplinary approach.

Keywords Sustainable new construction · Life cycle assessment · Resilient built environment · Seychelles

V. Gattulli · C. Vannini
Sapienza University of Rome, Rome, Italy
e-mail: vincenzo.gattulli@uniroma1.it

C. Vannini
e-mail: carlo.vannini@uniroma1.it

E. Palumbo (✉)
University of Bergamo, Bergamo, Italy
e-mail: elisabetta.palumbo@unibg.it

E. Arbizzani et al. (eds.), *Technological Imagination in the Green and Digital Transition*, The Urban Book Series, https://doi.org/10.1007/978-3-031-29515-7_47

47.1 Introduction

The Seychelles is extremely vulnerable to growing threats due to the greenhouse effect.

Additionally, considered an ecological paradise for its natural beauty, the Seychelles Island attracts many tourists and hence needs to meet their increasing request for tourist accommodation service.

Tourism is an important lever for poverty alleviation of naturalistic and desirable tourist destinations (e.g., offering job opportunities for local residents). Several authors recognize the effectiveness of tourism as an adequate method to boost the economic status of local communities, such as small islands (Tsung and Fen-Hauh 2019).

The Seychelles has played a leading role in implementing all 17 Sustainable Development Goals defined by the 2030 Agenda (Seychelles National Climate Change Strategy—NCCS) (Escamilla et al. 2018).

More specifically, within the Seychelles' commitment to tackle climate change and minimize its impacts, the promotion of sustainable practices in the tourism industry plays a key role.

Among the many African certification programs, Seychelles Sustainable Tourism Label (SSTL) is one of the three developed by the Government (Sebestyén et al. 2021).

SSTL is a voluntary sustainable certification program for tourism accommodation in Seychelles, locally developed and introduced by the Ministry of Tourism in 2012. It is a points-based certification scheme, and is third-party assessed.

In order to be certified by the SSTL, a hotel needs to meet three conditions:

- satisfy all the 22 mandatory criteria;
- reach a minimum score in each category, according to business size (5 points for enterprises with rooms from 1 to 24; 6 from 25 to 50 rooms and 7 for more than 51 rooms);
- obtain an additional six points in any area.

There are 8 evaluation areas defined by the scheme: Management, Waste, Water, Energy, Staff, Conservation, Community and Guests.

Despite the scheme being recognized as good and comprehensive for tourism accommodation, it proves to be limited for different uses such as multi-purpose centers.

Sustainable construction is one of the keywords in the debate on environmental development of the most sought-after holiday destinations, which are mainly characterized by increasingly fragile contexts such as natural hazards and changing climate. In the last few years, the building life cycle impacts and energy efficiency have been examined sufficiently. A small number of them consider the environmental impacts from the view of structure design, which indeed could play an important role in reducing emissions.

Therefore, with the aim to design a sustainable and safe flexible center situated in a small location on Praslin Island, this study has adopted a multi-disciplinary approach to identify low environmental impact technical solutions taking into consideration several life cycle (LC) stages. Specifically, the LC stages analyses are: supply of raw materials, self-disassembling construction systems and a minimum-impact building site.

The method used to quantify the environmental impacts is the Life Cycle Assessment (ISO 14040-14044, EN 15804).

The final purpose is to understand how the Life Cycle Thinking approach, and in particular a Life Cycle Assessment (LCA) application can support a multi-disciplinary methodological design approach (structural and architectural), for the calculation of ecological footprints related to a flexible small center project located on Praslin Island.

47.2 Methodology Adopted at the Design Phase

The design proposal stands on a sloped site of 31.500 m^2 left in a state of total neglect.

After a first exploratory phase, focused on the analysis of the climatic context, the need to preserve the fragile island's environment and mitigation of the impact on natural resources to the maximum possible extent, was outlined in the design requirements shown in Table 47.1.

The general morphology of the plan was evaluated taking into particular consideration the natural ground feature and hence the position of the buildings which follow the shape of the ground (Figs. 47.1 and 47.2).

The key design concept was to promote a "modular" style of architecture based on a rapid assembly of structural components and consequently easier replicability (Fig. 47.3).

Table 47.1 Brief of factors and solutions satisfying design requirements

Needs	Design solutions
To reduce the amount of solar radiation	Orienting the building on the North–South axis
To promote natural ventilation	Correct choice of the position of the openings and elevating from the ground
To favor shading	Using of a single pitch sloping roof
To avoid flood problems	Elevating from the ground
To use of alternative resources	Providing tanks for water recovery
To reduce environmental impact	Using of eco-sustainable materials
To ease of assembly	Adopting of dry systems
To reuse at the end of life	Using of dry assembled materials

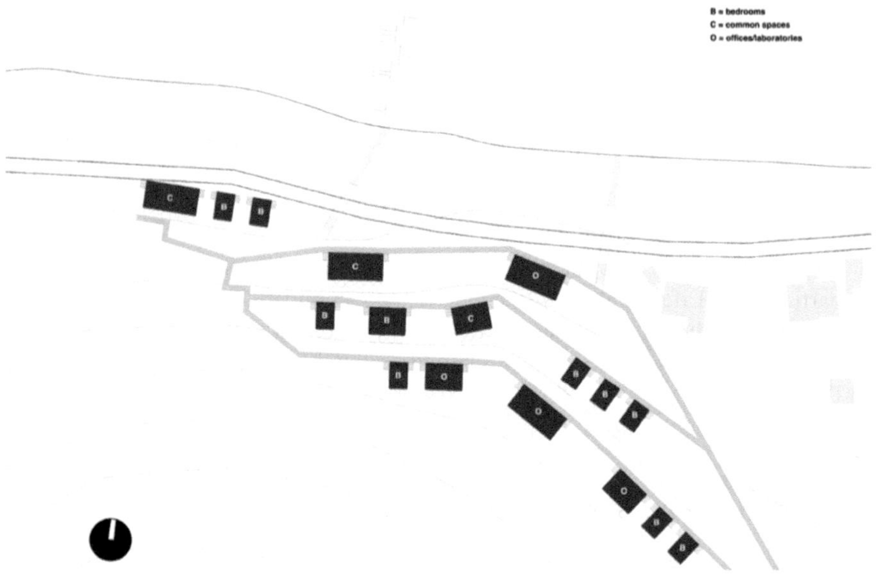

Fig. 47.1 Site plan (1:1000 scale)

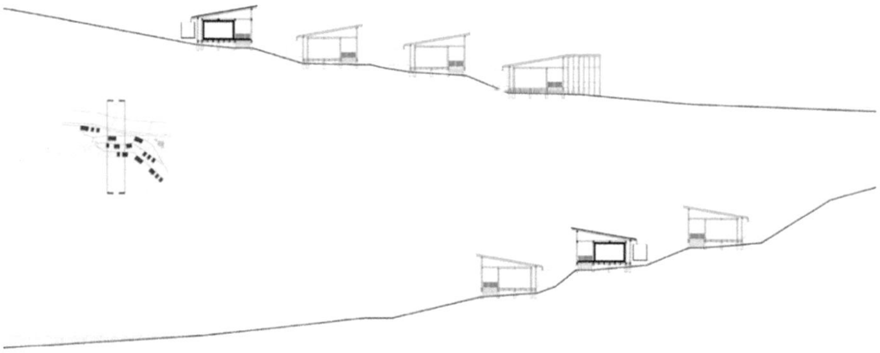

Fig. 47.2 Elevation section profile (1:500 scale)

47.2.1 Architectural and Structural Design Concepts

The environment, the climate, the morphology of the land and the scarce availability of materials on site were all considered in the choice of a modular structure that could be suitable for different uses and with a reduced environmental impact.

The pile dwelling shelter has been identified as the right building typology with the aim to improve the uncultivated and still scarcely used land. A raised structure that

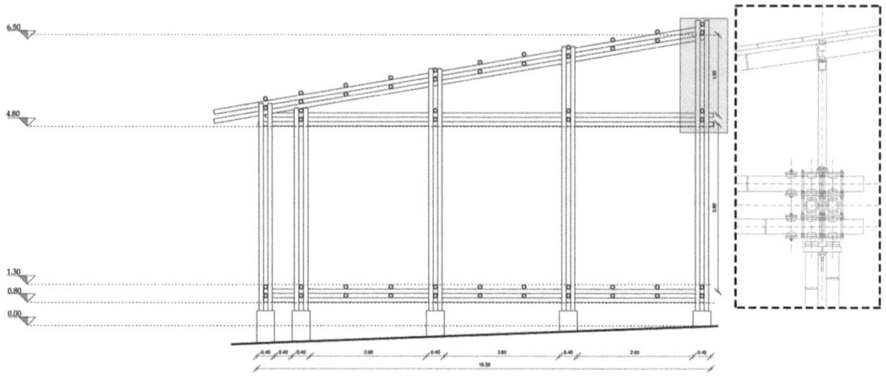

Fig. 47.3 Design schemes to illustrate the structural system realized by modular bamboo elements with a special metal joint

develops the surrounding environment without damaging it determines that earth movements or excavations on the ground are not necessary, while light materials such as wood and steel allow easy installation and quick disassembly, granting the involvement of local labor.

How can we build architecture that is aware of climate, light and air, through technical choices that are also structurally linked to the project? Lightness and slim elements shall be the main architectural features: the roofs rest on light uprights, wrapping them in minimal structures and covering transparent spaces. Permeability becomes the language of these spaces, while continuity is what they bring to the internal–external relationship: the architectures are drawn in the landscape and strengthened by their diversity. In this way a cordial and friendly relationship with nature is established, one capable of combining the use of poor and inexpensive materials (wood and steel industrialized elements, hemp fiber). Raising the house above the ground without the need for deep excavation protects the dry ground and the surrounding trees.

The structure of the module is made up of a symmetrical rectangular mesh with a constant trend and designed on a grid (6.40 × 9.70 m). All the components should be as much modular as possible with serial production in mind. The load-bearing elements (vertical and horizontal) are made up of bamboo rods and connected with a special joint—a metal cage composed of multilayer metal plates and bolts. The distances between the plates can be adjusted and adapted to the different dimensions of the raw bamboo, thus avoiding drilling holes. An elastic cushion is provided to protect the bamboo rod from the pressure exerted by the metal plate—this element increases friction, reducing the slippage and rotation of the wooden element. The vertical elements are raised and anchored to the foundation with metal brackets.

The module is designed to be used as a bedroom for guests. It has an external area, a covered and shaded patio that functions as an entrance and an internal area, with a space dedicated to the night and one to services. This module can be assembled creating different scenarios—for instance with a double or triple version—used as a tourist resort or as spaces for research and analysis labs.

This conscious design is based on improving energy efficiency so that the project provides several interventions on the building's envelope and systems, choosing a lightweight construction site and therefore technological solutions for customized industrial production (Table 47.1).

Despite the fragmentation and distribution on several levels of the settlement's lot, the modular units are spaced and rotated on the contour lines of the steep slope, creating a simple design in harmony with nature. It is a matter of designing economical buildings, which will save energy and which can easily adapt to this particular morphological connotation of the land and merge with the surroundings. It is also important to remember that the area is characterized by a tropical climate, with a high percentage of humidity and high temperatures, especially in the winter months. The construction is adapted to the needs of this particular climate—especially the needs for heat and ventilation—without resorting to artificial systems such air conditioners but instead paying close attention to the movement of the sun, the moon and seasons as well as designing buildings that can be in harmony with the movement of light and wind.

The orientation of the building and the exposure of the individual facades have a significant influence on the energy performances. By positioning the building along the North–South axis, the openings on these facades prevent direct sunlight from penetrating. In this way it is possible to keep the rooms cooler, to reduce exposure to the sun at noon and to take advantage of the winds and breezes for greater natural ventilation. To optimize ventilation and maximize breezes, in addition to lifting the building off the ground, openings such as windows and doors have been aligned to allow air flow and to minimize internal dividers for an unobstructed ventilation. The roof, with a single pitch, protrudes to shade the side walls and has a ventilated double skin.

The external skin shades the internal layer and absorbs solar heat based on its reflectivity while the cavity guarantees ventilation of the space between the roof and ceiling and acts as an insulator in the absence of winds.

In the context of the new policies for sustainable growth of Seychelles the opportunity to design an energy efficient experimental house in Praslin Island is very current and challenging. The proposal explores the potential of minimizing carbon emissions while maximizing environmental protection and natural ecological development through the use of the natural elements available in the area, such as sun, wind and natural materials (Fig. 47.4).

The end goal is to achieve a high-standard innovative house for the Seychelles context.

The topics that we have considered for the design of the module are the following:

Bamboo—a very special natural element which we use as structural for the house;
Modularity and Industrialization—modular elements to create an innovative industrialized bamboo construction system.

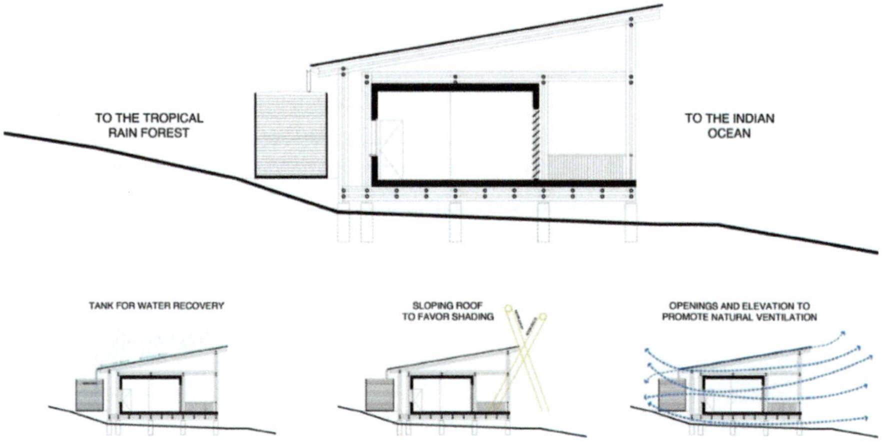

Fig. 47.4 Design schemes to illustrate the relevance of the project under sustainability point of view

Bamboo is the plant that absorbs the most carbon dioxide during its life cycle. It is strong enough after cultivation for 3 years and grows much faster than any other tree species. For the structure of the house we explored new ways of building using bamboo as a construction material. Sustainability is ensured not only by the use of natural materials such as bamboo but also by designing appropriate construction solutions—for instance dry-mounted connections that do not weaken bamboo through perforation nor fill it in with concrete and by allowing the replacement of bamboo poles if needed.

Also, an industrialized construction system can be achieved by designing light and easy to assemble aluminum connections, using same length bamboo poles and combining bamboo (known as vegetable steel) and steel together.

47.3 Choosing Materials

The aim of selecting materials including the following considerations:

- materials from renewable or replaceable sources,
- recycled materials,
- materials that are in plentiful supply,
- materials with a lower environmental impact across their whole life cycle.

The choice was thereby directed toward plant-based or local materials (Escamilla et al. 2018; Zea Escamilla et al. 2016), which consist of:

- bamboo for structural components,
- light-gauge steel produced in Africa for metal connections,

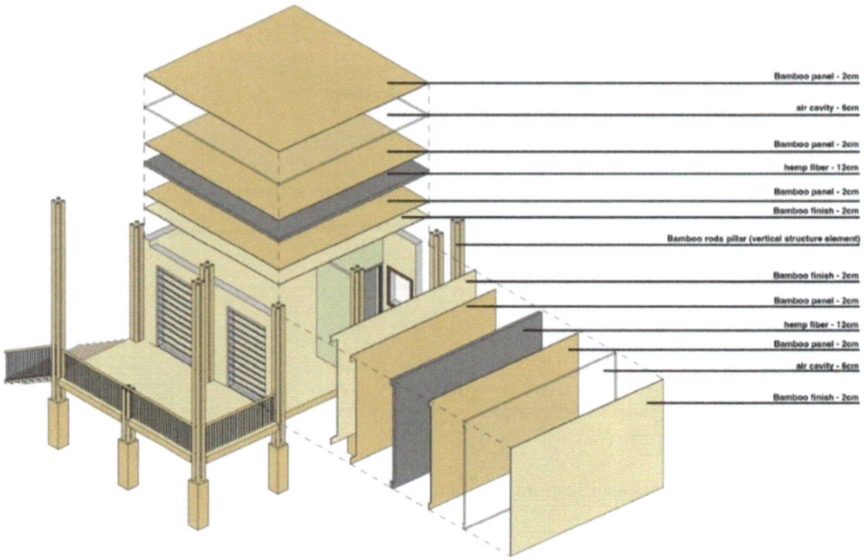

Fig. 47.5 Isometric view about technological solutions

- laminated bamboo for vertical and horizontal elements;
- hemp fiber for insulation (Fig. 47.5).

47.3.1 *Life Cycle Assessment Considerations of the Case Study*

To examine the environmental benefits of the constructive solutions adopted in the project, some relevant impact indicators were calculated by employing the Life Cycle Assessment method (Vogtländer et al. 2014). In particular, the aim of this additional analysis was to evaluate the environmental impact connected with the materials adopted, focusing attention on raw material supply and manufacturing process (Escamilla et al. 2018).

In this study, the use of specific impact values like Environmental Product Declaration (EPD) indicators developed by producers was preferred. The materials selected in the project were assumed to be sourced and processed locally. The data acquisition to model the Life Cycle Inventory (LCI) are shown in Table 47.2.

As there is little to no specific LCI data available for Africa, and specifically for the Seychelles, the data related mostly to EPDs of products with the same performances. Where EPD's product data was not matched with the product's performance literature data was used to determine the LCA impacts.

Table 47.2 Bill of materials for building with the mass and volume per material category for each building component

	Material	*Component*	*No. elements*	*Density* (kg/m^3)	*Volume* (m^3)	*Tot.* (m^3)
Structure	Bamboo	Beams	13	1080	0.611	7.94
		Columns	12	600	0.078	0.94
		Piles	9	600	0.026	0.23
			Thickness (m)	*Density* (kg/m^3)	*Quantity* (m^2)	*Total* (kg)
Vertical elements	Laminated bamboo	Internal wall finishing	0.02	700	61	823.5
		External wall finishing	0.02	700	61	823.5
	Hemp	Insulation	0.12	35	61	256.2
Horizontal elements	Laminated bamboo	Upper part	0.02	700	43	
		Lower part	0.02	700	43	5160
	Hemp	Insulation	0.12	35	43	180.6
Roof	Laminated bamboo	Lower part				
	Pine		0.2	450	43	5160
	Pine Hemp	Lower part	0.2	450	43	5160
		Insulation	0.12	35	43	180.6

47.3.2 LCA Data Assumption and Impacts

The GWP indicator of the designed technical solutions was assessed. Table 47.3 summarizes the LCA indicators considering the Global warming potential indicator for production and end of life phases, expressed in value impact (kg) and percentage contribution (%) to environmental impact analysis, respectively. GWP is expressed in kilograms carbon dioxide equivalents (kg CO_2 eq.) shows the problem of other gases and shows the problem of other gases (e.g., carbon monoxide, carbon dioxide, methane, HFC) standardized with reference to their lifespan in the atmosphere as compared to a unit of CO to a unit of carbon dioxide. The LCA indicators adopted in the assessment are based on EPDs' producers. According to these references, in relation to end of life phase, it was assumed the as scenario incineration for energy production. More specifically, 95% as incineration and 5% as dump (Table 47.2).

Figure 47.6 shows results of the whole building. The bars in this figure present the results for the studied technical solutions, and they represent the CO_2 emissions for each of them in the two life cycle scenarios: production and end of life (Table 47.3).

The results for the cradle-to-gate and end of life are presented in Fig. 47.6.

Table 47.3 Cradle-to-gate plus end of life impact assessment for the building component designs

Building component	Global warming potential (GWP)			
	Production		End of life	
	kg CO_2 eq	%	kg CO_2 eq	%
Structure	7949.9	65.2	−5726.3	70.4
Vertical elements	1871.7	15.4	−915.1	11.2
Horizontal elements	1181.3	9.70	−744.8	9.2
Roof	1181.3	9.70	−743.0	9.1
Total	12,184.3	100	−8131.1	100

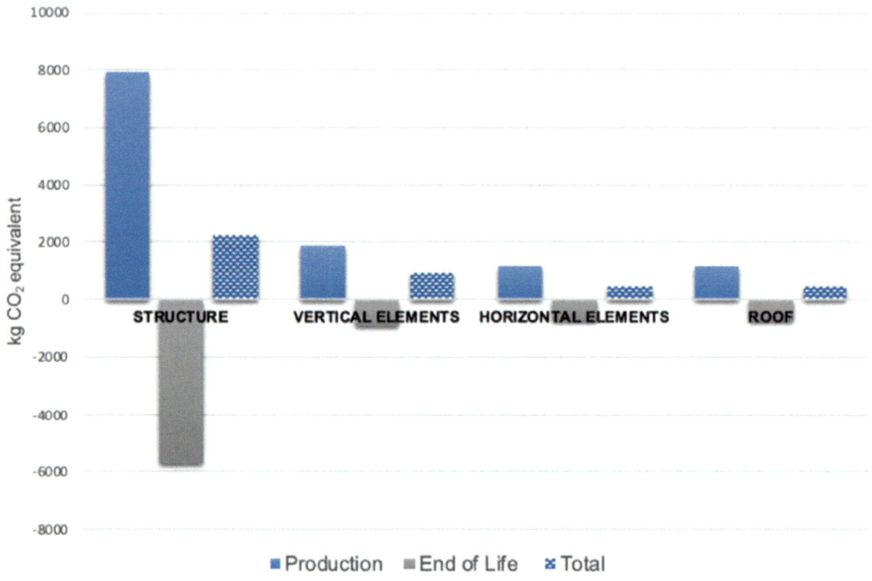

Fig. 47.6 Environmental impact in kg CO_2 equivalent

As can be observed in Table 47.3 and Fig. 47.6, for the bamboo-based structure (assumed for structural components like columns, beams and piles), the contribution from the life cycle phases ranges from 65% (production) to 70% (end of life) of their total environmental impact, whilst the laminated bamboo-based construction materials (adopted for walls) contribution ranges from 9.7% (production) to 15% (end of life) of the total environmental impact. These results support the idea that in order to obtain low carbon, it is necessary to optimize the amount of reuse materials post demolition, whereas the material production would require the optimization of manufacturing process.

Fig. 47.7 Sustainable dual intended use center development in Praslin Island, Seychelles

47.4 Conclusions

LCA-based design may have a great impact in approaching new solutions respectful European green deal. Climate change and environmental degradation are an existential threat to Europe and the world. To overcome these challenges, the European Green Deal will transform the EU into a modern, resource-efficient and competitive economy, ensuring: no net emissions of greenhouse gases by 2050; economic growth decoupled from resource use; no person and no place left behind.

The technical solutions adopted for a low environmental impact LCA based in the proposed designed center during the life cycle is a possible example of a new approach of conceptual design in fragile environment (Fig. 47.7).

Acknowledgements Part of the work has been developed during the master thesis in Project and Construction Management of Building Systems of Eleonora Cruciani. Her work is acknowledged.

References

Escamilla EZ, Habert G, Daza JFC, Archilla HF, Echeverry Fernández JS, Trujillo D (2018) Industrial or traditional bamboo construction? Comparative life cycle assessment (LCA) of bamboo-based buildings. Sustainability (Switzerland) 10(9). http://doi.org/10.3390/su1009 3096

Sebestyén V, Abonyi J (2021) Data-driven comparative analysis of national adaptation pathways for sustainable development goals. J Clean Prod 319. https://doi.org/10.1016/j.jclepro.2021.128657

Tsung HL, Fen-Hauh J (2019) Can community-based tourism contribute to sustainable development? Evidence from residents' perceptions of the sustainability. Tourism Manag 70:368–380. ISSN 0261-5177. https://doi.org/10.1016/j.tourman.2018.09.003

Vogtländer JG, van der Velden NM, van der Lugt P (2014) Carbon sequestration in LCA, a proposal for a new approach based on the global carbon cycle; Cases on wood and on bamboo. Int J Life Cycle Assess 19(1):13–23. https://doi.org/10.1007/s11367-013-0629-6

Zea Escamilla E, Habert G, Wohlmuth E (2016) When CO_2 counts: sustainability assessment of industrialized bamboo as an alternative for social housing programs in the Philippines. Build Environ 103:44–53. https://doi.org/10.1016/j.buildenv.2016.04.003

Part IV
Session I Climate Changes

Today's cities constitute both a problem and a solution to contrasting climate change they continue to grow, consuming more than two-thirds of all energy while producing equal levels of emissions. It follows that rendering the constructed environment adaptive and resilient under the effects of climate change, and climate neutral as well, is a key challenge to be faced, and won, over the next ten years, so as to achieve the goal for greenhouse-gas reduction by 2030. The use of new technologies, the limitation of land consumption and more attentive consideration of the how the ground is covered, and of urban density, along with the reuse and technological retooling, with respect to function, energy, and ecosystem factors, of decaying areas and buildings, represents the path taken by planning to arrive at the creation of healthy, resilient urban habitats capable of adjusting to ongoing changes, so as to promote prosperity, inclusiveness, and social equity.

The session is meant to encourage discussion of procedural models, strategies and planning solutions, technological and digital, useful to defining new images of resilient cities capable of contributing to reducing the effects of climate change while lowering their own ecological footprint. Particular importance will be placed on understanding and evaluating the impact of the pandemic in the formulation of initiatives that can be of aid in drawing up scenarios for the future evolution of the urban environment.

Chapter 48
Climate Change: New Ways to Inhabit the Earth

Eliana Cangelli

Abstract The text begins by proposing a critical reading of the climate crisis, aimed at pointing out the need for a collective consciousness allowing solutions to be found that are unbound by having to maintain the current economic model. After a brief history of the environmental issue identifying the commitment of the technology of architecture in this topic, a research framework linked to the environmental and technological design of the built environment is proposed. The paper closes with considerations on research prospects for the transformation of cities, of use for pursuing the reduction of climate change, as emerged at the conference.

Keywords Climate change · Built environment · Environmental design · Technology of architecture · Climate-neutral cities · Urban renewal · Public spaces

48.1 Inequalities: The Democracy of Climate Change and of Global Crises

Over the course of the Earth's history, climate has changed naturally,[1] bringing about various geological eras.

Today, however, we have been witnessing unnatural climate change proceeding at unprecedented speed, due to human activities on the Earth.[2] Human action is increasingly degrading the biosphere, atmosphere, oceans, the continents, and, locally, cities, rivers, and crops. It is human activities that produce the atmospheric gases that trap

[1] The climate changes bringing about the various ecological eras were due to slight variations in the Earth's orbit, which modified the amount of solar energy our planet received.

[2] "Since systematic scientific assessments began in the 1970s, the influence of human activity on the warming of the climate system has evolved from theory to established fact." IPCC Sixth Assessment Report, WGI, Technical Summary.

E. Cangelli (✉)
Sapienza University of Rome, Rome, Italy
e-mail: eliana.cangelli@uniroma1.it

E. Arbizzani et al. (eds.), *Technological Imagination in the Green and Digital Transition*, The Urban Book Series, https://doi.org/10.1007/978-3-031-29515-7_48

more of the sun's energy in the Earth system[3] and that trigger higher ocean temperatures, the melting of glaciers, higher sea levels, and a greater frequency of such tragic climate events as flooding, soil erosion, and the hurricanes lashing our cities.

Confirmation that the acceleration of climate change is due to our lifestyle came with the COVID-19 pandemic which, for a very short period of time, changed how we lived, worked, made things, and moved about.

For the period of the quarantine, COVID-19 changed our way of inhabiting the Earth.

The measures implemented to contain the virus showed how the sudden and obligatory change in our lifestyles quickly resulted in lowering CO_2 levels by 17% below 2019 averages as well as in a crisis of the global economy (Le Quéré et al. 2020).

The shock of the pandemic, then, served to lay bare, finally and for all, how urgent it is to change our development model, casting light on social inequalities and clearly demonstrating the impact that our way of living and inhabiting has on the environment and consequently on climate change.

Therefore, after the pandemic experience, the objectives of environmental sustainability and of reducing social inequalities have gained renewed meaning, highlighting the need to intensify the struggle against climate change and to reduce social inequalities for the protection of all. No one is excluded from the risks of climate change. As the recent past has shown, no one is excluded from the risks connected to new epidemics. Global crises affect everyone democratically, striking any place on Earth, any nation whatever its level of development and all social classes.

As Edgar Morin rightly states in his most recent book, "the planet-wide crisis born from the Coronavirus casts light on the community of destiny all humans share in an indissoluble bond with Planet Earth's biological destiny." (Morin et al. 2020).

There is always a silver lining, then: the pandemic has given life to a new and more solid collective consciousness, highlighting the need to come together to find solutions for global emergencies, for human beings' impact on the Earth, and for climate change which, in fact, which, if not combated, will lead to new equilibria on Earth in which the very survival of the human species might be cast into doubt.

But are the unspeakable prospects that climate change raises truly causing a global emergency of consciousness and the environment? Or is it the technical and economic crisis that spurs us to find solutions?

The objection may be raised that the question is an irrelevant one; it matters little whether the actions that may lead to mitigating the results of climate change are due to reasons of conscience or to economic motives. The important thing is that they are identified and implemented, and that effective solutions are found.

This is not the case, however. History teaches us that if the objective of environmental protection is not made foundational, even at the risk of having to make hard, unpopular choices, the inertia of the economic model guides the solutions and does not lead to real changes.

[3] Earth Science Communications Team at NASA's Jet Propulsion Laboratory | California Institute of Technology.

The current energy crisis is proof of this. After 50 years of energy policies promised and financed by the European Union, we find ourselves today, with the war in Ukraine, in the same situation of austerity as in 1973, then due to the cost of oil. Clearly, then, the policies put in place over the past 50 years to reduce atmospheric emissions and promote energy self-sufficiency connected to natural resources have been a failure, because they are subjugated to economic and financial relationships among the various nations.

History, then, can serve as an example so as not to repeat the same mistakes.

48.2 A Brief History of the Environmental Issue (in the Technology of Architecture)

The first environmental alarms raised around the world date to the 1970s, in fact, in the wake of the austerity and oil crisis that paralysed nearly all the Western countries.

At the time, the main concern was linked not to the use of fossil fuels, but rather to energy costs and the pollution produced by industries. In public opinion, the environmental emergency was linked to the image of sites producing smoke and pollutants poisoning the air and rivers.

Research to identify solutions to the environmental problem was carried out in the field of chemistry and industrial engineering and subsequently in the search for alternatives to generating energy from fossil fuels.

In 1976, the Seveso environmental disaster caused by a dioxin leak and the contamination of the resident population led to directives and protocols that contained environmental impacts in accordance with a targeted "end-of-pipe" approach—that is, not solving the problem in the production phase, but mitigating the environmental impacts after having generated them. Ten years later, the Chernobyl nuclear disaster (1986), followed in Italy by the nuclear energy referendum in 1987, shifted the focus to energy generation.

From the 1970s to the 1980s, then, a social awareness of the environmental emergency began to be come into being, and the environmentalist movements born in the 1960s thanks to the considerable contribution of Carson (1979) were given substance, no longer referring to industry alone but to the entire manmade environment, in a publicity effort carried forward by scholars and philosophers noteworthy among whom is Edgar Morin, who was among the first to propose a transdisciplinary approach to solving social and environmental problems.

It was in the following decade that the Italian academic world connected with architecture began to construct an environmental consciousness of its own: in 1985, Thomas Maldonado held the first course in Environmental Design at Polytechnic University of Milan. The year 1990 saw the conference *Chiudere il Cerchio* ("closing the circle") which included Zanuso's well-known paper *Progetto e Futuro* ("design and the future") (Zanuso 1987), joined and followed by publications developed in an international setting such as *Design with Climate* (Olgyay 2013) and *Environmental Design* (Fitch 1991) that to this day represent a milestone in Environmental Design.

These were years of great ferment for the technology of architecture and in particular for the Environmental Design springing from it, raising questions about issues of research and modes of advanced education on design production.

With the Maastricht Treaty (1993), the environment became an official sector of EU policy, and in 1999, the Amsterdam Treaty established the obligation of integrating environmental protection into all the Union's sectoral policies in order to promote sustainable development. These were the years when research shifted to the regulatory aspects and to the systems for assessing and certifying the environmental sustainability of architectural constructions and of industrial products, seeking to perfect methods that quantified the impact of design approaches geared to achieving energy efficiency and to reducing emissions into the atmosphere.

Relatively recently (2009), the close interrelationship between environmental policies and climate change was made crystal clear. Fighting climate change through strategies relating to city planning, industrial production, and infrastructure became a specific goal of the European Union with the Lisbon Treaty.

Over the past decade, technological research on the built environment has aligned with the objectives dictated by Europe, but has seen its incisiveness decline due to following policies provided without any more meaningful attempts to anticipate future scenarios.

48.3 Climate Change and Technological and Environmental Design[4]

If we wish to fully comprehend the contribution that technological and Environmental Design can provide to sustainability design and to fighting climate changes, we have to place society and economy more forcefully at the centre of our reasoning, albeit while taking into account numbers and statistics that derive from assessing the environmental impact of human activities on the planet.

The co-relationship between developing and implementing environmental policies and actions is a very close one, and thinking about new models for a different development also means thinking about how to re-establish the bond between economic, social, and environmental development, by verifying how the combination of these elements might be rearticulated starting from the characteristics of the local settings (Langer 1994).

In this sense, until the eco-efficiency of the territory and of the settlement arrangements is recognized with even greater conviction as a primary driver of economic and social development, the outcomes of environmental research will remain only theoretical, used by the intellectual elites and the academic community, or employed—in

[4] Cfr. Cangelli (2015).

the best of cases by certain political representations—as a temporary pretext for increasing their support, thus raising serious difficulties for their actual, applied experimentation.

Today, however, we are at the very beginning of a long-awaited transformation of settlement models that bases its evolution upon the concept of network (infrastructure, energy, social, and information), and research must therefore proceed with clarity towards developing these models which must be configured not as hypotheses subject to review, but as a guide for action, so as not to lose the problematic dimension of the concept of model that would only lead to defining "instruction booklets" founded upon unlikely certitudes (Augé 2012).

The recent research trends promoted in the European setting are inspired by the awareness that the integration of sources of widespread and renewable energy generation, for example through the construction of numerous Net Zero Energy Buildings, along with the rapid evolution of ITCs, can bring about new scenarios for the development of infrastructures, of the urban fabric, and of the construction product, linked in fact to a diversified mode of energy procurement and to the possibility of quickly transferring knowledge and information.

The future will see an evolution in the study of models for the development of technologies aimed at energy savings and at the sustainability of the processes of transforming the built environment—an evolution of thought that will by necessity have to dislodge all the current verticality in the disciplines in order to identify a new relationship between technological choices and cultural evolution in behavioural models and life styles, able to counter the element characterizing the current, involutional fossil-based system.

The role of technological and environmental research in this redefinition is strategic and must be aimed at innovating the baggage of specific skills and knowledge, while linking them to the current cultural and social orientations and intuiting their possible articulations.[5]

Over the past thirty years, the technological and Environmental Design has seen significant scientific production. Starting from the awareness that the pairing of technological innovation and the environment is the foundation for the development of the built environment, it has made a significant contribution towards proposing an integrated and multidisciplinary approach to design, substantiating it with material and immaterial content, in opposition to self-referential and formalist approaches oriented mainly towards the morphological and figurative aspects of architecture (Schiaffonati et al. 2011).

The area has developed its research activity with the objective of providing tools for the technological and environmental control of the design, construction, and management process, underscoring the dialectics between the project's environmental structure, technical choices, and the expressive purposes of architecture, and guiding educational goals towards understanding the close interdependence between

[5] Cfr. Cangelli (2012).

structure and form, material and figuration, so as to govern the relationship between conceptual and development activities in accordance with criteria of environmental sustainability.

During these years, technological and Environmental Design has prepared a considerable repertoire of theories and tools for design, organizing the technical information on materials and components; contributing towards extending codified procedures and methods for assessing the environmental quality of buildings and settlement arrangements; and defining codes and good practices that are slowly training and informing architects, and becoming part of their cultural baggage. The objective of this work, still in progress, is to proceed towards the spontaneous inclusion of the environmental issues informing the project in the creative process, in the awareness that architecture is a "moment of synthesis fertilized by everything that stands behind architecture: history, society, the actual world of people (…), geography and anthropology, climate, culture (…), as well as science and art" (Piano et al. 2010). The environmental approach, then, is just one of the elements that must be brought into synthesis in architectural design.

It is not, then, a matter of research aimed at defining instruments proposing a partial, deterministic approach to design, but of preparing an apparatus of theories and tools to give it substance.

The cultural heritage produced during these years can be consolidated, systematized, detailed, applied, and verified through design and its construction; its communication can be facilitated and its outcomes monitored. But a possible advance resides in understanding and foreseeing its possible developments. That is, it resides in setting out a "concrete utopia"[6] that aims to minimize the utopian, or speculative, component, and to maximize the concrete one, which is to say technology (Maldonado 2001), through a critical analysis of the present. What is needed today is that creative imagination identified by Morin as an engine of the progress of objective knowledge (Morin 1989).

The time is ripe.

48.4 Climate Changes | Effective Solutions for Resilient Urban Habitats

The Climate Change session of the Technological Imagination conference allowed the framework of reference to be updated, and the state of progress of the technology of architecture on these issues to be defined. The numerous papers that were received[7]

[6] For a critical discussion of the theory of "Concrete Utopia," see the chapter "Verso una prassiologia della progettazione" del testo "La speranza progettuale" by Tomas Maldonado, Einaudi Torino, 1970.

[7] At the session *Climate Changes | Effective solutions for resilient urban habitats*, 28 papers were admitted for publication after an intense peer review process in which 69 authors presented their contributions on the topic.

reinforce the conviction of the key importance of the environmental issue in the development of technologies and of technological approaches to the design of living and of cities.

Anna Pirani has been invited to open the session; heading the overall management of the Technical Support Unit of the Intergovernmental Panel on Climate Change Working Group I (WGI) that examines the physical science underpinning past, present, and future climate change, she has proposed a scientific reading of the climate change data, showing us climate's evolution in the Mediterranean area. According to Bruno Latour, climatologists like Anna Pirani, as nature's designated representatives, have always been political players and are now fighting a war in which the outcome will have global ramifications (Latour 2018); they play a necessary and important role in showing us, in concrete terms, the reality of the climate changes in store for us.

Other essential contributions to the opening of the session's proceedings were those by Peter Droege[8] and Paolo Rotelli,[9] which outlined strategies in progress for the resilience of cities and underscored that the ecological transition is not politically neutral: the setting of priorities, identification of the players, and allocation of funds are political choices that can intensify or repair inequalities.

The session opened discussion and dialogue concerning the procedural models, strategies, and design, technological, and digital solutions of use for defining new, resilient images of the city, capable of helping reduce the effects of climate change while lowering its ecological footprint.

In fact, cities today are both the problem and the solution for fighting climate change. Cities continue to grow, consuming more than two thirds of all energy, while at the same time, producing equal levels of emissions. Making the built environment adaptive and resilient to the effects of climate change, and even climate neutral, is the key challenge to be faced and won over the next ten years.

The session showed that scientific research has many tools available to it in order to be more incisive.

The availability of data and information about cities and of digital tools allow possible scenarios for the future development of the urban environment to be established.

Knowledge of climate risks and of the main environmental challenges that cities are called upon to face may be set out and resolved with the contribution of software packages to simulate and predict the state of urban arrangements and of buildings.

The technological evolution of the building sector permits brief construction and recovery times, and close control of the energy and environmental performance of buildings.

The selected texts show how researchers are aware and govern the strategies to be pursued in designing cities: reduced land consumption, to be implemented through the development of construction replacement and reuse practices, seeking to strike a balance between urban density and the distribution of services and functions; and

---

[8] Peter Droege is the Director of the Liechtenstein Institute for Strategic Development.

[9] Paolo Rotelli is a member of the ICC e and of the UN Green Economy Task Force.

increased green and blue infrastructure and the widespread development of sources of renewable energy production to give oxygen to the cities, to control the water resource and use it against climate change, and to reduce emissions into the atmosphere through the use of renewable resources, and the integration of practices to mitigate and adapt to climate change so as to pursue achievable designs for transforming cities.

Another element that may be gleaned from the distribution of the issues dealt with in the selected papers is the awareness that, in order to achieve the goal of creating climate-neutral habitats, we must act in an interdisciplinary fashion on various scales: territorial and urban scale; the scale of the building; and the scale of the production of construction components.

The collected texts, then, show how research on these issues is potentially able to explore and set out possible futures of the built environment, seeing in interdisciplinarity and in the systemic approach two essential elements for investigating the complexity of urban phenomena and looking to the transformation of our cities.

> The idea of transformation, richer than the idea of revolution, conserves its radical character, but binds it to conservation (of life, of the inheritance of cultures). Today, everything must be rethought. Everything has to start over. And in effect, everything has, without anyone knowing it. We are at the stage of beginnings—modest, invisible, marginal, and dispersed beginnings. Because there already is, on all continents, a creative ferment, a multitude of local initiatives going in the direction of economic, or social, or political, or cognitive, or educational, or ethical, or lifestyle regeneration. (Morin 2010)

References

Augé M (2012) Futuro. Bollati Boringhieri, Torino, IT

Cangelli E (2012) Progettazione ambientale ed evoluzione dei modelli insediativi. In: Tartaglia A, Gambaro M, Bolici R (eds) La ricerca tra innovazione, creatività e progetto/Research among innovation, creativity and design. University Press, Firenze, IT, pp 393–407

Cangelli E (2015) NZEB 2050 | Visioni possibili. J Plann Des Technol 3:144–153

Carson R (1979) Primavera silenziosa. Feltrinelli, Milano, IT

Fitch JM (1991) La progettazione ambientale: i caratteri ambientali dell'architettura. Franco Muzzio, Padova, IT

Langer A (1994) La conversione ecologica potrà affermarsi solo se apparirà socialmente desiderabile. https://www.alexanderlanger.org/140. Accessed 22 Dec 2023

Latour B (2018) Down to earth: politics in the new climatic regime. Polity Press, Cambridge, UK

Le Quéré C, Jackson RB, Jones MW et al (2020) Temporary reduction in daily global CO_2 emissions during the COVID-19 forced confinement. Nature News. https://www.nature.com/articles/s41558-020-0797-x. Accessed 17 Oct 2022

Maldonado T (2001) La speranza progettuale: ambiente e società. Einaudi, Torino, IT

Morin E (1989) La conoscenza della conoscenza. Feltrinelli, Milano IT

Morin E (2010). Elogio della metamorfosi. LASTAMPA.it. http://www.lastampa.it/redazione/cmsSezioni/cultura/201001articoli/51238girata.asp. Accessed 17 Oct 2022

Morin E, Prezzo R, Abouessalam S (2020) Cambiamo strada: le 15 lezioni del Coronavirus. Cortina, Milan, IT

Olgyay V (2013) Progettare con il clima: un approccio bioclimatico al regionalismo architettonico. Muzzio, Roma, IT
Piano R, Cassigoli R, Siciliano R (2010) La responsabilità dell'architetto. Passigli Editori, Bagno a Ripoli, IT
Schiaffonati F, Mussinelli E, Gambaro M (2011) Tecnologia dell'architettura per la progettazione ambientale. Techne J Technol Archit Environ 48–53
Zanuso M (1987) La cultura del progetto: dal meccanicismo all'organicismo. In: Gangemi V, Ranzo P (eds) Il governo del progetto. La Tecnologia per la formazione dell'architetto. Luigi Parma, Bologna, IT

Chapter 49
The Climate Report Informing the Response to Climate Change in Urban Development

Anna Pirani

Abstract The IPCC Climate Report was published in three volumes in August 2021, and February and April 2022. An overview of the key findings, sourced from these reports that are relevant for the resilient development of urban areas, is summarized in this paper. It is the authoritative, comprehensive assessment of the climate change, including the physical aspects, impacts and adaptation, and mitigation. Human-caused climate change has affected global and regional climate, including extremes and drivers of impacts with consequences for human and natural systems. Cities, urban areas, and settlements, for example in Europe, are particularly exposed to future risks. At the same time, there are multiple opportunities to address climate change both through adaptation and mitigation action with urban development. There are multiple synergies and co-benefits the responses to climate change and meeting the Sustainable Development Goals.

Keyword Climate change · Risks · Adaptation · Mitigation · European cities · Policy options · Sustainable urban development · SDGs

49.1 Introduction

The IPCC Climate Report, the Sixth Assessment Report (AR6) was published in three volumes in August 2021, and February and April 2022[1] (IPCC 2021a, 2022a, 2022b). These volumes are the outcome of an expert assessment of the published scientific and technical literature on the physical basis of climate change, the impacts,

[1] The three volumes of the AR6 are Climate Change 2021: Working Group I (WGI)—The Physical Science Basis; Climate Change 2022: Working Group II (WGII)—Impacts, Adaptation, and Vulnerability; and Climate Change 2022: Working Group III (WGIII) Mitigation of Climate Change, respectively. Their assessments cover scientific literature accepted for publication respectively by 31 January 2021, 1 September 2021, and 11 October 2021.

A. Pirani (✉)
Centre for Theoretical Physics, Trieste, Italy
e-mail: anna.pirani@ictp.it

E. Arbizzani et al. (eds.), *Technological Imagination in the Green and Digital Transition*, The Urban Book Series, https://doi.org/10.1007/978-3-031-29515-7_49

vulnerability and adaption to climate change, and the mitigation of climate change. The IPCC consists of an intergovernmental Panel of 153 member countries and an elected Bureau, which is the scientific body that provides guidance to the Panel on the scientific and technical aspects of IPCC assessments, working at the interface of the underlying scientific and technical community worldwide and the Panel.

The IPCC is a unique science-policy interface on climate change and is the authoritative source of information on climate change worldwide. IPCC reports are neutral with respect to policy choices, so not providing guidance on best policy choices. Instead, the reports provide a comprehensive assessment of the range of information and policy options that are available and relevant for policymaking, to inform science-based policy decisions and choices. The IPCC does not carry out its own research nor produce source datasets. Each volume of the Climate Report has been prepared by around 230–280 authors from about 65 countries that were selected by the Bureau, following a call for nominations from member governments and observer organizations of the IPCC. Across the three volumes, a total of 60,000 papers were assessed and close to 200,000 review comments were received.

The reports undergo a rigorous process of formal review. The first draft is reviewed by experts of the topics addressed in the assessment. The second draft is reviewed by experts as well as government representatives. In a third step, the Summary for Policymakers is reviewed by government experts. Finally, the Summary for Policymakers is approved line-by-line in an approval session of the IPCC that brings together authors, Bureau members, and government representatives in a final co-design step of the document that is intended for a global policymaker audience. All the review comments received on the first and second draft of the report are responded to by authors. The review comments and responses are available to the public after the publication of the reports.

49.2 Human-Caused Climate Change

The human influence has warmed the climate. The assessment of the current state of climate concludes that recent changes in the climate are widespread, rapid, and intensifying and are unprecedented in thousands of years. The surface temperature of the globe has warmed by 1.1 °C in the last decade (2011–2020) compared to the period before the advent of fossil fuel based and land use industrialization (1850–1900) (See (IPCC 2021b), Fig. SPM.1). This observed warming is all driven by emissions from human activities. Greenhouse gas emissions, including carbon dioxide, methane, nitrous oxide, lead to warming, while aerosols, such as sulphur dioxide and organic carbon, lead to a cooling effect that partly masks the warming (See (IPCC 2021b), Fig. SPM.2). Carbon dioxide concentrations in the atmosphere are the highest in at least 2 million years. The rate of global sea level rise is the faster in at least 3000 years. The extend of Arctic sea ice is at the lowest level since at least 1000 years. Glaciers are retreating at a rate that is unprecedented in at least 2000 years.

The human influence on climate is unequivocal, and climate change is affecting every region of the world (See (IPCC 2021b), Fig. SPM.3). The increased occurrence of hot extremes have been observed in almost all regions of the world since the 1950s and can be attributed to human-caused climate change. The changes we experience today will increase with further warming.

Unless there are immediate, rapid, and large-scale reductions in greenhouse gas emissions, limiting warming to 1.5 °C will be beyond reach. Future global warming depends on future emissions of carbon dioxide and other greenhouse gases (See (IPCC 2021b) Fig. SPM.8). Future scenarios of possible global warming outcomes depending on how future emissions could increase and evolve of the course of this century show that we will reach a global warming of 1.5 °C above pre-industrial levels, when averaged over 20 years, in the period 2021–2040. By the 2030s, every single year has a 50% chance of having an annual global surface temperature above 1.5 °C. In scenarios where low or very low greenhouse gas emissions by the end of the century, global surface temperatures may avoid the exceedance of 2 °C global warming, while scenarios with intermediate, high, or very high levels of greenhouse gas emissions are expected to exceed 2 °C global warming.

With every increment of global warming, changes get larger in regional mean temperature, precipitation, and soil moisture, as well as changes in climate and weather extremes and climatic drivers of impacts. Continued global warming is projected to further intensify the global water cycle across the ocean, land, atmosphere, and frozen regions, including global monsoon rainfall and the severity of wet and dry extreme events. The frequency and intensity of hot extremes and marine heatwaves, heavy precipitation, and drought in some regions are increasing. There is an increase in the proportion of intense tropical cyclones and a decrease in snow cover, permafrost and Arctic sea ice extent. For example, hot extremes that had a probability of occurring once in pre-industrial times are now observed to occur nearly five times more frequently and are expected to occur over 8.5 times as much with 1.5 °C global warming, nearly 14 times as much with 2 °C global warming, and nearly 40 times as much with 4 °C global warming (See (IPCC 2021b), Fig. SPM.5).

There is no going back from some changes in the climate system. Some changes could be slowed and others could be stopped by limiting warming. With 1.1 °C of human-induced global warming already happened now, we have set in motion the slow components of the climate system. They will continue to change over generations to come: glaciers react on timescales of decades or more, the ocean on timescales of hundreds of year, and the Greenland and Antarctic ice sheets on timescales of thousands of years. Historical emissions will lead to a committed sea level rise of 0.7 to 1.1 m by 2300. By lowering greenhouse emissions and limiting the level of future global warming, we increase the chance of avoiding triggering irreversible instabilities in the ice sheets that could substantially increase sea level rise more than that and give more time for coastal responses to enhance resilience.

Every ton of carbon dioxide emissions that is released into the atmosphere adds to global warming. There is a nearly linear relationship between cumulative carbon dioxide emissions and maximum global surface temperature increase caused by carbon dioxide in the atmosphere (See (IPCC 2021a), Fig. SPM.10). Stabilizing human-induced global temperature increase at any level requires net human-caused carbon dioxide emissions to become zero. Net zero carbon dioxide emissions are achieved when human-caused carbon dioxide emissions are balanced by human-caused carbon dioxide removals. Deep and sustained reductions of other gases would also be required. A sharp drop in carbon dioxide emissions would lead to rapid increase in air quality. The effects on global surface temperature would be discernible after about 20 years (See (Lee et al. 2021), FAQ4.2 Fig. 1). 2020 was the year of pandemic lockdown, starting in January and peaking in April when more than half the world's population was under some form of lockdown. While we saw a decrease in the emissions of carbon dioxide in the atmosphere, we have not seen any impact on climate. This is because the duration of the emission reductions was too short to see an impact on climate, though it led to fewer air pollutants being emitted, particularly from cars, leading to temporarily improved air quality over many cities and regions.

Every increment in global warming in future will result in increased risks to human and natural systems. With further global warming, every region is projected to increasingly experience concorrunt, multiple changes in climatic impact-drivers. These are indices that represent tolerance thresholds that can lead to severe consequences for people, agriculture, or wildlife when they exceeded. Climatic impact-drivers include changes in heat or cold, rain or drought, snow and ice, wind, coastal and oceanic, including open ocean, and other climatic and weather conditions. An example is a heat warning index of the number of days where maximum temperatures are above 35 °C that is relevant to agriculture and health that is projected to be exceeded more frequently at higher global warming levels. Higher latitude ecosystems will be affected by lower extreme heat thresholds as well.

49.3 Adaptation Actions in European Cities

In the European region, by the mid-century and at global warming of at least 2 °C and above, changes in climatic impact-drivers include increased warming and temperature extremes, increased droughts and aridity, decreased precipitation and snow cover, increased fire weather, increased mean and regional sea levels, and decreased wind speed. Users can explore and visualize indices and datasets assessed in the report, depending on their interests, by using the IPCC Interactive Atlas[2] tool. For example, the combination of dry climatic impact-drivers can be selected: aridity, hydrological drought, agricultural and ecological drought, and fire weather.

Impacts are magnified in cities, where more than half the world's population lives. Heatwaves amplify urban heat islands and air pollution to affect people's health.

Critical infrastructures such as transport, water, sanitation, and energy systems have been compromised by extreme events. European cities are hotspots for multiple risks of increasing temperatures and extreme heat, floods, and droughts.

[2] https://interactive-atlas.ipcc.ch/.

Key risks related to heat stress, morbidity, and mortality are assessed for future scenarios of medium or high levels of adaptation in the European region to show the level of additional risk to society with increasing global surface temperature (See (Bednar-Friedl et al. 2022), Fig. 13.29). The synthesis of assessed future risk is based on the assessment of climatic impact-drivers, vulnerability, and exposure. For scenarios of low to medium adaptation and high adaption, the risks of human heat stress, morbidity, and mortality increase with increasing average temperatures and heat extremes. Risk consequences will become severe more rapidly in Southern and Western Central Europe and urban areas. Thermal comfort hours during the summer will decrease significantly, by as much as 74% in Southern Europe at 3 °C global warming. Above 3 °C, there are limits to the adaptation potential of people and existing health systems, particularly in Southern Europe and Eastern Europe and areas where health systems are under pressure. While the consequences of heat are well studied, further studies of risks for cities and key infrastructure from hailstorms and lightning are needed.

By the 2050s, urban areas could be home to two-thirds of the world's population. The effectiveness and feasibility of transformative adaptation policies have been assessed for cities, settlements and key infrastructure in Europe. These include response options to impacts and future risks with a portfolio of options approach, combining nature- and engineering-based solutions. To manage flood risk, for example, it might be appropriate to install flood proofing on buildings, improve drainage along roads and create space for water within the city, at the same time as constructing flood defences. Establishing or restoring green and blue spaces—parks, green corridors, ponds, and wetlands—as well as urban agriculture can all be woven in to the built environment. Social safety nets for disaster management can help people overcome the impacts of climate change and can provide financial security. Additional benefits include public health improvements, especially from reducing heat stress and ecosystem conservation.

Current and planned adaptation options in European cities have been assessed (See (Bednar-Friedl et al. 2022), Fig. 13.6). 167 European cities provided data on their adaptation actions. Milan is taking a climate resilient development approach to new developments where by, from 2020, new buildings must be carbon neutral and reconstructions must reduce the existing land footprint by at least 10%. The Climate and Air Plan (CAP) and the city's 2019 Master Plan focus on low-carbon, inclusive and equitable development. The CAP is directed at municipal and private assets and individual to city-scale actions. In 2020, Milan released a revised Adaptation Plan and the Open Streets project to ensure synergies between the COVID-19 response and longer-term climate resilient development. However, implementation of Milan's plans remains a challenge, despite dedicated resources and commitment.

The constraints to adaptation in Europe for cities, settlements and key infrastructure have been comprehensively assessed. These are economic, social-cultural, human capacity, governance-institutions-policy, financial, information awareness and technology, physical and biological constraints (See (IPCC 2022c), Fig. A.I.55). The range of options available to deal with climate change impacts has increased in most of Europe since the last decade. However, observed adaptation actions are largely incremental, with only a few examples of local transformative action. There is an urban adaptation gap between current adaptation, planned adaptation actions and all currently possible adaptation options, for example for river-rain flooding, coastal flooding, and heatwaves, together with inequality in exposure relative to population income (highest 20% compared to lowest 20%) (See (IPCC 2022c), Fig. A.I.53).

49.4 Mitigation Actions to Reduce Greenhouse Gas Emissions

Mitigation actions to reduce greenhouse gas emissions are as much needed as adaptation actions and are critical to limit global warming and keep global warming of 1.5 °C within reach. We are currently not on track to limit warming to 1.5 °C. Global net anthropogenic greenhouse gas emissions continue to rise. Emissions in 2019 were about 12% higher than they were in 2010 and 54% higher than in 1990. However, emissions growth slowed from 2.1%/yr for 2000–2009 to 1.3%/yr for 2010–2019. The 2010–2019 average annual greenhouse gas emissions are at the highest levels in human history—64% from carbon dioxide released by the fossil fuel industry, 11% net carbon dioxide emissions from land use change and forestry, and 18% from methane emissions, the rest from nitrous oxide and fluorinated gases (See (IPCC 2022d), Fig. SPM.1).

The emissions gap to limit climate change remain large for 2030 between current pledges (NDCs) and optimal pathways for limiting warming to 1.5 °C and 2 °C. If net zero carbon dioxide emissions were reached by the early 2050s, there would be the chance to limit global warming to 1.5 °C. However, if net zero carbon dioxide emissions were reached by the early 2070s that would imply reaching a global warming of 2.0 °C by the end of the century. Actual policies implemented by the end of 2021 imply reaching a global warming of 3.2 °C by the end of the century. Greenhouse gas emissions pledges for 2030 (Nationally Determined Contributions) declared prior to the 26th Conference of the Parties held in 2021 would likely lead to a high temperature overshoot above 1.5 °C and require even faster global emissions reductions after 2030 as well as large net-negative emissions to bring warming back to below 1.5 °C by 2100. Such pathways are subject to climate-related risks and increased feasibility concerns.

Nonetheless, there is increased evidence of climate action. Progress on mitigation to-date is insufficient but not negligible. Some countries have achieved a steady decrease in emissions over several years at rates consistent with limiting warming to 2 °C. And a growing number of cities are setting net zero targets. Zero emissions targets have been adopted by at least 826 cities and 103 regions. An increasing range of policies and laws have enhanced energy efficiency, reduced rates of deforestation, and accelerated the deployment of renewable energy. Climate laws that result in reduced or avoided emissions are present in 56 countries, covering more than half of global emissions.

Large-scale transitions take time and an evolving mix of policies. The relative importance of different stages of policy development and implementation is regards strategic investment (research and development, demand-pull, and infrastructure and industrial development) relative to strategies of transition (emergence). Market effects (prices, taxes, market structures, planning, and regulation) are related to the chance of breakthrough and diffusion of policies. Norms and behaviour (standards, engagement, and dissemination programmes) enter into play at the culmination stage of policies (See (Pathak et al. 2022), Fig. TS.28). To-date, most progress has occurred in the electricity sector and associated clean and efficient technologies.

A notable success is around renewables. Since 2010, there have been sustained decreases in unit costs with reductions of 85% for solar energy, 55% for wind power, and 85% for batteries. In some cases, these costs have fallen below those of fossil fuels, with large increases in capacity already installed (See (IPCC 2022d), Fig. SPM.3). There have been increased investments in and diffusion of low-carbon technologies, especially for wind and solar energy, electric vehicles, energy-efficient appliances, and low-carbon heating. Reductions in energy intensity have occurred globally and in all but one world region, and reductions in carbon intensity have occurred in Europe, Eurasia, the Middle East, North America; and at a global scale.

There are options available now in every sector (demand and services, energy, land use, industry, urban, buildings, and transport) that can at least halve emissions by 2030 and keep open the possibility of limiting warming to 1.5 °C. With regards to demand and services, there is the potential to reduce global emissions by 40–70% by 2050. Cities and other urban areas account for more than two-thirds of global emissions, taking into account what is produced in the city and brought in from elsewhere. There is significant potential for emissions reductions in existing and rapidly growing cities, as well as new cities and urban areas. Resilient urban planning and targeted infrastructure are key for all cities. Other broad strategies have been assessed to be effective include sustainable production and consumption of goods and services, electrification, and improving carbon uptake and storage in cities, for example, with permeable surfaces, green roofs, trees, and lakes.

There is the opportunity to lower global emissions by between 40 and 70% by 2050 through demand-side and services actions. Of the 60 actions assessed in this report, on an individual level, the biggest contribution comes from switching to walking and cycling and using electrified transport. Other effective options include reducing air travel and adapting our houses. Shifting towards balanced, plant-based diets is an option. Changes to our lifestyles and behaviour can reduce our carbon footprint, as well as improve our health and wellbeing. To be effective, lifestyle changes will need to be supported by systemic changes across every aspect of society—including transport, buildings, industry, and land use.

In buildings, it is possible to reach net zero emissions in 2050. Action in this decade is critical to fully capture this potential through retrofitting existing buildings and using effective mitigation techniques in those that are yet-to-be-built. There is already an increasing number of zero energy or zero-carbon buildings in almost all climates. Ambitious policy packages are needed that may incorporate the use of renewables and the efficient design and use of space, energy, materials, and appliances.

49.5 Climate Actions and Wider Benefits for Sustainable Development

Accelerated climate action is critical for sustainable development. In all countries, actions to limit global warming that result in wider benefits to society can increase the pace and extent of emissions reductions. Accelerated and equitable climate action in mitigating and adapting to climate change impacts is critical to sustainable development. The co-benefits of adaptation options in urban areas for sustainable development have been comprehensively assessed in terms of their potential feasibility, synergies with mitigation, and the relation to implementing the Sustainable Development Goals (IPCC 2022e), Fig. SPM4.b). Green infrastructures (green roofs, parks, and street trees) help adapt by providing shade and water management as well as absorbing and storing carbon. For green infrastructure and ecosystem services and for sustainable land use and urban planning, there is medium confidence in terms of potential feasibility and high confidence in terms of synergies with mitigation. For sustainable urban water management, there is medium confidence for potential feasibility, but low confidence for synergies with mitigation options. These options have multiple co-benefits with the achievement of the Sustainable Development Goals.

The co-benefits of mitigation in urban areas for sustainable development have been assessed comprehensively, providing insights into the wider benefits of mitigation that can go hand-in-hand with achieving many of the sustainable development goals (See (IPCC 2022d), Fig. SPM.8). Mitigation options include urban land use and spatial planning, electrification of the urban energy system, district heating and cooling networks, urban green and blue infrastructure, waste prevention, minimization and management, and integrating sectors, strategies, and innovations. Equity and just transitions can lead to deeper ambition for accelerated climate action. Many countries and regions are already applying just transition principles. Green and blue infrastructure are also mitigation options. Green roofs and facades, networks of parks and open spaces, wetlands, and urban agriculture not only absorb and store carbon but, at the same time, achieve many other sustainable development goals. For example, they can reduce pressure on urban sewer systems, reduce flood risk and heat-island effects. They can also deliver health benefits from reduced air pollution.

49.6 Concluding Remarks

Mitigation to limit global warming, strong, rapid, and sustained reductions in carbon dioxide, methane, and other greenhouse gases are necessary and the climate we experience in future depends on our decisions now. Climate change consequences and the responses to climate change, both adaptation and mitigation, are critically important in cities, urban areas, and settlements. Climate change is the result of more than a century of unsustainable energy and land use, lifestyles, and patterns of consumption and production. The Climate Report shows how taking action now can move us towards a fairer, more liveable world.

There is an important opportunity for academics, practitioners, and professionals in the urban planning and architecture sector to meaningfully contribute to policy development and the solution space to address climate change together with multiple co-benefits for sustainable development. The seventh cycle of the IPCC will begin in mid-2023 and will include the preparation of a Special Report on climate change and cities that will be interdisciplinary, cutting across the physical science basis, adaptation and mitigation aspects. Experts from these sectors can directly participate in the assessment process, both through relevant publications from these communities and also as expert authors and reviewers of the report. Information is available on the IPCC website and social media.[3]

References

Bednar-Friedl B, Biesbroek R, Schmidt DN, Alexander P, Børsheim KY, Carnicer J, Georgopoulou E, Haasnoot M, Le Cozannet G, Lionello P, Lipka O, Möllmann C, Muccione V, Mustonen T, Piepenburg D, Whitmarsh L (2022) Europe. In: Pörtner H-O, Roberts DC, Tignor M, Poloczanska ES, Mintenbeck K, Alegría A, Craig M, Langsdorf S, Löschke S, Möller V, Okem A, Rama B (eds) Climate change 2022: impacts, adaptation, and vulnerability. Contribution of working group II to the sixth assessment report of the intergovernmental panel on climate change. Cambridge University Press, Cambridge, UK and New York, NY, USA, pp 1817–1927. https://doi.org/10.1017/9781009325844.015

IPCC (2021a) In: Masson-Delmotte V, Zhai P, Pirani A, Connors SL, Péan C, Berger S, Caud N, Chen Y, Goldfarb L, Gomis MI, Huang M, Leitzell K, Lonnoy E, Matthews JBR, Maycock TK, Waterfield T, Yelekçi O, Yu R, Zhou B (eds) Climate change 2021a: the physical science basis. Contribution of working group I to the sixth assessment report of the intergovernmental panel on climate change. Cambridge University Press, Cambridge, United Kingdom and New York, NY, USA, in press. https://doi.org/10.1017/9781009157896.

IPCC (2021b) Summary for policymakers. In: Masson-Delmotte V, Zhai P, Pirani A, Connors SL, Péan C, Berger S, Caud N, Chen Y, Goldfarb L, Gomis MI, Huang M, Leitzell K, Lonnoy E, Matthews JBR, Maycock TK, Waterfield T, Yelekçi O, Yu R, Zhou B (eds) Climate change 2021b: the physical science basis. Contribution of working group I to the sixth assessment report of the intergovernmental panel on climate change. Cambridge University Press, Cambridge, United Kingdom and New York, NY, USA, pp 3−32. https://doi.org/10.1017/9781009157896.001

IPCC (2022a) In: Pörtner H-O, Roberts DC, Tignor M, Poloczanska ES, Mintenbeck K, Alegría A, Craig M, Langsdorf S, Löschke S, Möller V, Okem A, Rama B (eds) Climate change 2022a: impacts, adaptation, and vulnerability. Contribution of working group II to the sixth assessment report of the intergovernmental panel on climate change. Cambridge University Press, Cambridge, UK and New York, NY, USA, 3056 pp. https://doi.org/10.1017/9781009325844

IPCC (2022b) In: Shukla PR, Skea J, Slade R, Al Khourdajie A, van Diemen R, McCollum D, Pathak M, Some S, Vyas P, Fradera R, Belkacemi M, Hasija A, Lisboa G, Luz S, Malley J (eds) Climate change 2022b: mitigation of climate change. Contribution of working group III

[3] Ipcc.ch.
Twitter—@IPCC_CH.
Facebook—@IPCC.

to the sixth assessment report of the intergovernmental panel on climate change. Cambridge University Press, Cambridge, UK and New York, NY, USA. https://doi.org/10.1017/9781009157926

IPCC (2022c) Annex I: global to regional Atlas. In: Pörtner H-O, Roberts DC, Tignor M, Poloczanska ES, Mintenbeck K, Alegría A, Craig M, Langsdorf S, Löschke S, Möller V, Okem A, Rama B, Götze S (eds) Climate change 2022c: impacts, adaptation, and vulnerability. Contribution of working group II to the sixth assessment report of the intergovernmental panel on climate change. Cambridge University Press, Cambridge, UK and New York, NY, USA, pp 2811–2896. https://doi.org/10.1017/9781009325844.028.

IPCC (2022d) Summary for policymakers. In: Shukla PR, Skea J, Slade R, Al Khourdajie A, van Diemen R, McCollum D, Pathak M, Some S, Vyas P, Fradera R, Belkacemi M, Hasija A, Lisboa G, Luz S, Malley J (eds) Climate change 2022d: mitigation of climate change. Contribution of working group III to the sixth assessment report of the intergovernmental panel on climate change. Cambridge University Press, Cambridge, UK and New York, NY, USA. https://doi.org/10.1017/9781009157926.001

IPCC (2022e) Summary for policymakers. In: Pörtner H-O, Roberts DC, Tignor M, Poloczanska ES, Mintenbeck K, Alegría A, Craig M, Langsdorf S, Löschke S, Möller V, Okem A, Rama B (eds). Climate change 2022e: impacts, adaptation, and vulnerability. Contribution of working group II to the sixth assessment report of the intergovernmental panel on climate change. Cambridge University Press, Cambridge, UK and New York, NY, USA, pp 3–33. https://doi.org/10.1017/9781009325844.001

Lee JY, Marotzke J, Bala G, Cao L, Corti S, Dunne JP, Engelbrecht F, Fischer E, Fyfe JC, Jones C, Maycock A, Mutemi J, Ndiaye O, Panickal S, Zhou T (2021) Future global climate: scenario-based projections and near-term information. In: Masson-Delmotte V, Zhai P, Pirani A, Connors SL, Péan C, Berger S, Caud N, Chen Y, Goldfarb L, Gomis MI, Huang M, Leitzell K, Lonnoy E, Matthews JBR, Maycock TK, Waterfield T, Yelekçi O, Yu R, Zhou B (eds) Climate change 2021: the physical science basis. Contribution of working group I to the sixth assessment report of the intergovernmental panel on climate change. Cambridge University Press, Cambridge, United Kingdom and New York, NY, USA, pp 553–672. https://doi.org/10.1017/9781009157896.006

Pathak M, Slade R, Shukla PR, Skea J, Pichs-Madruga R, Ürge-Vorsatz D (2022) Technical summary. In: Shukla PR, Skea J, Slade R, Al Khourdajie A, van Diemen R, McCollum D, Pathak M, Some S, Vyas P, Fradera R, Belkacemi M, Hasija A, Lisboa G, Luz S, Malley J (eds) Climate change 2022: mitigation of climate change. Contribution of working group III to the sixth assessment report of the intergovernmental panel on climate change. Cambridge University Press, Cambridge, UK and New York, NY, USA. https://doi.org/10.1017/9781009157926.002

Chapter 50
The Urban Riverfront Greenway: A Linear Attractor for Sustainable Urban Development

Luciana Mastrolonardo

Abstract The strategy for sustainable mobility of December 2020 by the European Commission defines the alignment of the transport sector with the European Green Deal, for a 90% reduction in greenhouse gas emissions related to transport by 2050. This involves linear infrastructure of sustainable mobility in our cities. The research has a focus on strategies to increase sustainable travel, with a view to improving the quality of public space and reducing the weight of heavy transport. A coordinated planning directs actions toward mitigation tools and sharing of public spaces through necessarily systemic interventions, which identify a common scenario, involving increased use of city greenways connected with urban node. Working on the level of cycling in Pescara case study (IT) means to act to a systemic approach involving different kind of action on infrastructure and on active participation of inhabitants. Among all a focus was developed on the urban greenway on the riverfront, crossing stretches of great environmental and landscape quality with its seven kilometers, which potentially could connect peripherals part of the cities, currently in a state of semi-abandonment. The Biciplan guidelines, meeting the objective by a project involving youth activism, could help achieve sustainability objectives and improve environmental performance, starting from its integrated enhancement, developing the axis in an urban sense, reconnecting the city and improving the peripheral mobility of the city. The consequence of coordinated planning and directing actions toward mitigation tools are followed in the reduction of emissions at the local level, contributing to proximity of travel.

Keyword Sustainable mobility · Public space · River ecomuseum · Greenway · Social innovation

L. Mastrolonardo (✉)
University G. d'Annunzio, Pescara, Italy
e-mail: l.mastrolonardo@unich.it

E. Arbizzani et al. (eds.), *Technological Imagination in the Green and Digital Transition*, The Urban Book Series, https://doi.org/10.1007/978-3-031-29515-7_50

50.1 Introduction

In December 2020, the European Commission defines the policy framework for aligning the transport sector with the ambition of the European Green Deal. It implies in particular a 90% reduction in transport-related greenhouse gas emissions by 2050 (European Comission 2019): achieving these reductions over the next decade underlies Europe's strategic choice to become the first climate-neutral continent. Furthermore, the New European Bauhaus introduces the relationship between the quality of public space and the generating matrix of culture, also with a view to re-appropriating the street and the "empty space of the habitat" (Vittoria 1965), that is the dynamic space of flows of mobility and sharing of the contemporary city. Sustainable mobility and decarbonization of transport are necessary to achieve the goals in the mobility sector, which produces 26% of CO_2 emissions. Italy is still one of the European countries in which air pollution causes more victims, and the traffic component is among those that produce the most pollutants. It is useful to remember that road transport is responsible for 80% of CO_2 emissions from transport, of which about 70% is produced for journeys of less than 50 km, and it is especially in urban contest that we must act decisively through acceleration of sustainable mobility.

The impact of climate change is associated with that of the COVID-19 pandemic in progress and its repercussions on urban space, places of proximity and mobility in general. In fact in the future, pandemics will be increasingly numerous and more widespread if there is not a transformative change in the global approach of activities that have impacts on ecosystems, biodiversity and habitats: prevention would be 100 times cheaper than the cost of responding to pandemics, which we have had in recent years (IPBES 2020). Planning, designing and creating urban spaces at the service of active mobility flows are necessary to face the challenges that await us for the next few years: mobility is certainly a field in which sustainability is called upon to confront each other in order to create solutions able to combine efficiency, respect for the environment and green economy, involving also the social innovation with the co-creation of solutions.

50.2 Strategies for Integrated Active Approach for Urban Mobility

Climate change and pollution require us to change our way of life, especially in cities, the places where the majority of humans reside and where the main social, cultural and economic relationships take place. Agglomerations that have become chaotic over time due to population growth and the need to provide answers to the needs of those who live there. The result is often the emergence of situations of social and ecological degradation, sources of stress and conflict. This is particularly evident in the sphere of mobility, which generates traffic, soil consumption, noise pollution and emissions that are harmful to people and the planet. Decay due mostly to the current car-based model, now questioned: a holistic vision for urban design emerges as necessary (Salici 2013). After the social distancing and the emptying of public spaces of the first lockdown, COVID-19, however, had the merit of giving us back a city more open to cycling, as photographed by the ISFORT (2020) research which shows a structural increase in the demand for individual mobility and active, despite the initial decrease in public transport.

The planning of infrastructure for cycling must be confronted with a necessarily holistic approach to sharing urban space between pedestrians/cyclists and motorized vehicles thanks to the adoption of traffic moderation systems. The reduction of the speed differential between motorized and non-motorized modes of transport induces a certain capacity for self-limitation of conflicts, with a consequent increase in urban safety and livability, not to mention the decrease in emissions linked to decreasing speed. The integrated approach provides essential actions for the modal split that must focus on:

- "hardware" component of the network of cycle paths hinged on the central axes and the main attractions, equipped with interchange parking lots with the main public transport terminals;
- "software" component of promotion, communication, participation, training, through policies for the diffusion of bicycles, also through active involvement and co-planning actions.

The integrated approach refers especially to the individual according to a holistic and eco-systemic perspective, taking into consideration his complexity and his active role in the construction of meanings (Mussinelli 2018). With a view to relaunching local services which, if recovered and relaunched make the city safer, it is necessary to allow the weakest users (the elderly and under 18) to be able to move around the city on a continuous, safe and priority path. In order not to let this crisis go to waste, it is necessary to relaunch accessibility to peripheral areas of collective and neighborhood importance, to reduce the chances of contagion but increase the chances of socializing, albeit at the right safety distance, to improve the quality of life of the inhabitants according to an eco-systemic perspective, taking into consideration his complexity and his active role in the construction of meanings.

50.3 A Case Study: Integrated Approach in Pescara

The Biciplan of Pescara is a functional tool for rethinking the public space and livability of places, crossed by the cycle paths making up the municipal cycle network (Mastrolonardo 2021). The constant expansion of the Pescara cycling network has allowed the city to confirm itself for the fourth consecutive year as a "cycling municipality" and to win the Urban Award 2019. The geomorphology of the Pescara area is particularly suitable for slow mobility, with great possibilities for strengthening the network even in the hilly areas that trace the boundaries with the neighboring municipalities. Examining the cycle flows and the consistency

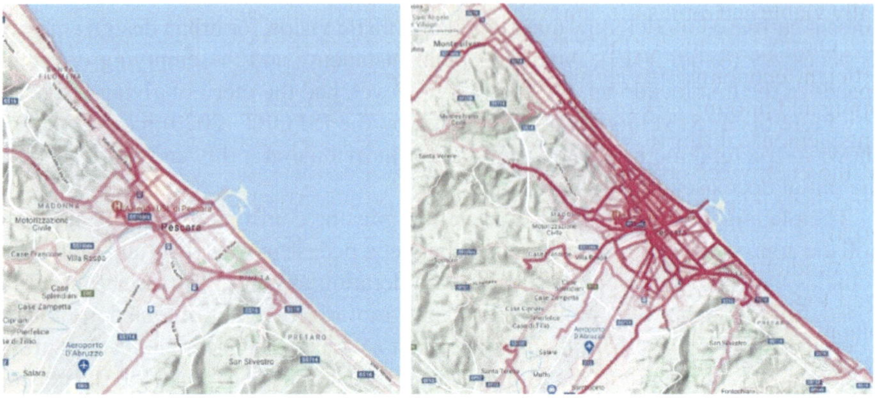

Fig. 50.1 Map of bicycle trips from July 15 to August 15, 2021, in the time slot 8: 00–9: 00 on the right, on the whole day on the right

of demand, the extreme diffusion of bicycle use and the capillarity of the routes traveled emerged, especially as regards the city center (Fig. 50.1), characterized by pedestrian areas and limited traffic areas, in which the circulation of cars provides for reduced speeds and a greater perception of safety by the weakest users. However, as can be seen from the study of the flows carried out, the use of bicycles is relegated to Sundays and free time, leaving the car undisputed queen of journeys, even short ones.

The definition of an overall strategy that guarantees, where possible, the development of a pleasant and safe primary network that leads above all to the most difficult suburbs, collides with urban difficulty due to reduced road sections, and a policy that is not inclined to the renunciation of parking spaces, which leads to a decrease in consensus in the short term (but far-reaching benefits in the long term).

Parallel to a system of priority travel axes for cyclists, the provision of a complementary and widespread network of shared routes is essential, which allows an entirely cycle-friendly city, deeply linked to the contexts it crosses and connects.

Therefore, urban cycling mobility plan defines:

- the network of priority cycle routes in the municipal area;
- the secondary network of cycle paths, within neighborhoods and inhabited centers: characterized by sharing and traffic moderation tools and overlapping cycle lanes, two-way cycle paths, urban roads with cycle priority and school areas are preferred;
- the network of cycle greenways.

Pescara lends itself particularly to being crossed by two of the elements that most distinguish it and which outline its position: the seafront and the riverside. The seafront has already been strengthened and is included as urban part of the regional "bike to coast". The riverfront instead is still of marginal importance and little explored and experienced by the community, it lends itself to a revitalization and redevelopment, as a function of an expansion and a connection with urban attractions and with the surrounding municipal territories, end it is recognized by Biciplan as Greenway. It potentially could coincide with cycle routes of national interest belonging to the Bicitalia network that cross Abruzzo to Rome: the Tibur Valeria (B07).

50.3.1 Urban Riverfront Greenway

A greenway has a double value both as a protection of environmental values and as a route reserved for non-motorized vehicles. In Italy, the Italian Greenways Association defines greenways as paths capable of connecting populations with the natural, agricultural, landscape and historical-cultural resources of the area and with the life centers of urban settlements (Salici 2013). With reference to the definitions provided by the sector literature, the G1 route responds perfectly to the characteristics reported, despite its almost complete inactivity (Fig. 50.2). The actions necessary for the revitalization and redevelopment of the route cannot be limited to maintenance interventions but must be configured as a system of activities that make it attractive in its entirety, signaling the access doors and improving overall safety.

This route is of fundamental importance, while enjoying a privileged position due to its proximity to the Pescara watercourse, uncontaminated also "thanks" to the presence of the motorway axis above, which prevented the city from overflowing the river bank, as happens near the coast. The riverfront is in a condition of no permeability toward the city context due to the presence of the Airport, at the border

Fig. 50.2 G1 greenway as described in Biciplan, with connection with the city and the reconnecting proposal to the seafront on the east and the other municipalities on the west

with San Giovanni Teatino. To give the axis greater relevance at the inter-municipal level, it is necessary to refer to the territories that fall within the broad area. In this perspective, the possibility of creating a super-cycle path that can allow the crossing of the Municipalities of Spoltore, San Giovanni Teatino and Chieti, retracing a section coinciding with the B07 Tibur Valeria which reaches up to Rome. It would be an opportunity to outline a real green thorn characterized by the succession of changing and morphologically distinct landscapes.

To this end, the strategy relating to the main green road in Pescara must necessarily include specific actions. These are outlined in relation to the four cardinal points, with respect to which the territory designates different scenarios and leads to different possibilities of intervention:

- East–West: extension of the route toward the sea along the river quay until it rejoins the C1 Bike to Coast Cycle Route—and toward the mountains (West)—crossing the municipal administrative boundaries so as to define a solid transport structure;
- North: connection with the left orographic bank of the river through the construction of at least two bridges capable of connecting the urban agglomerations with which the river axis relates—with specific reference to Villa Raspa and Santa Teresa di Spoltore—such as to make the route more accessible to cycle traffic which, through alternative routes, would hardly be able to reach the Pescara city center;
- South: overall redesign of an axis perpendicular to the riverfront, which allows users from Sambuceto and the neighboring city context to reach the green road at the first permeable axis, close to the airport area.

With regard to the G1 greenway, it is necessary to increase the relationships that, starting from the neighboring contexts, allow to reach a path from not insignificant landscape-environmental character (Fig. 50.3). To turn the spotlight on this axis and on its urban connections with the suburbs, an event to get to know the citizens of the greenway was organized, with all the associations linked to sustainable mobility, to make the inhabitants aware of an important and landscape-relevant piece of their city: the relationship between the city and the river, in fact, highlights a total alienation between the city and the river, unlike all the most important European cities. The navigability of the same river should also be implemented, making the river infrastructure stand out also as connecting arteries. Best practice is ongoing: a multinational company has placed a landing for boats in two points, close to its establishments, to respond to the need for sustainable mobility, after introducing this strategy in the home-to-work travel plan (Fig. 50.4).

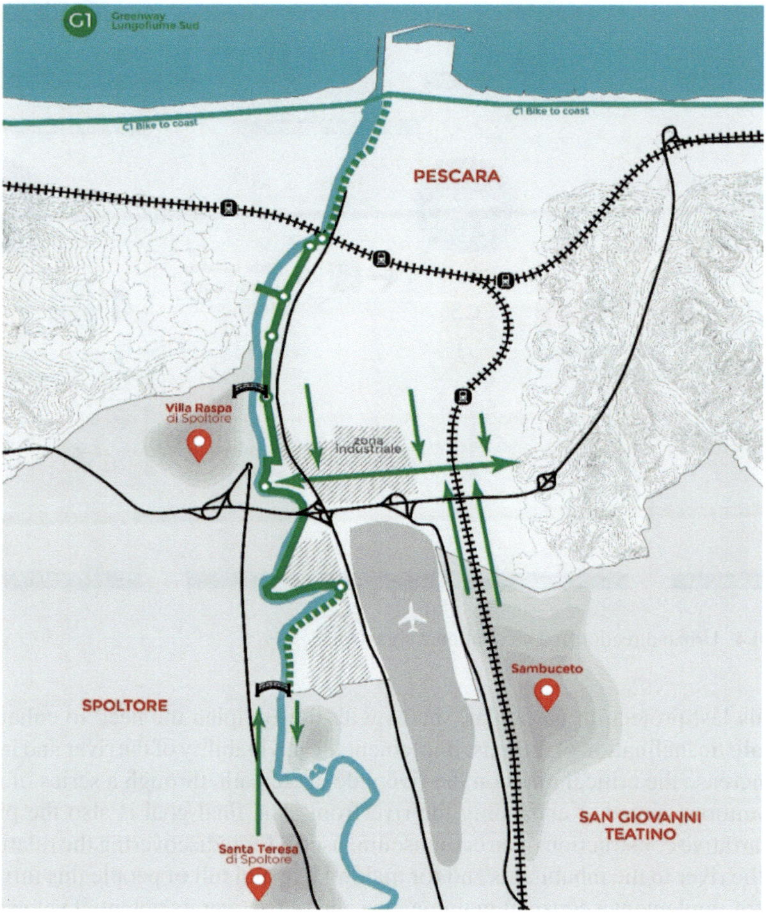

Fig. 50.3 Relations with the development of the greenway with proximity contexts

50.3.2 Joint Action for Re-functionalization to the Riverfront

The guidelines of the Biciplan with respect to the greenway have found fertile intersections in other joint actions:

- the Pescara river contract that has moved some resources from the National Resource Plan (PNRR);
- the founded project for coloring the wall along the river, financed from National Cultural Found: "The color that moves the world";
- the project for a Flvial Ecomuseum in Pescara, financed from National Found by ANCI.

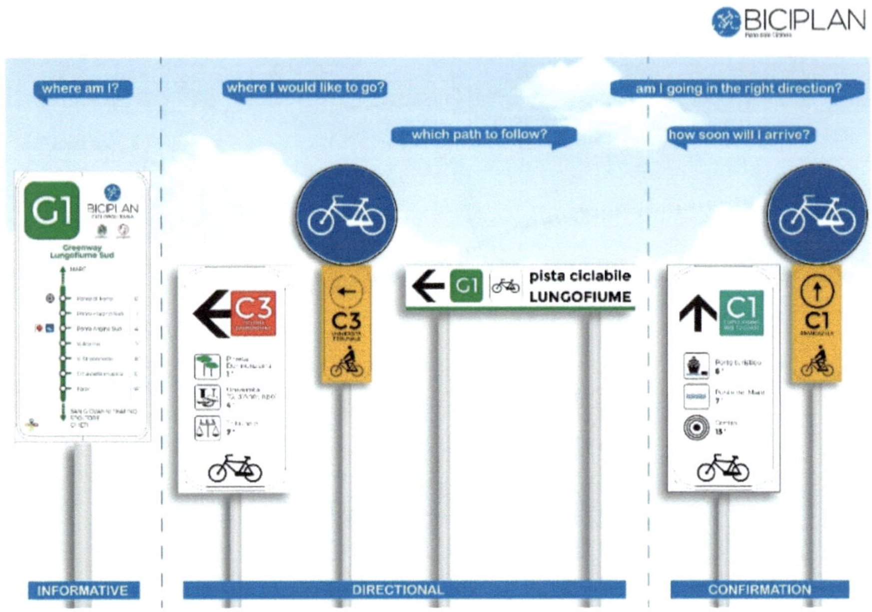

Fig. 50.4 Urban direction and motivational signs

This last project, in particular, shares with the Biciplan the need to enhance the naturalistic inclination of the axis, implement the navigability of the river and improve and increase the critical mass on the riverside cycle path, through a series of minute interventions aimed at enhancing the riverfront. The final goal is also the physical and narrative construction of an ecomuseum, useful for rediscovering the relationship with the river to the inhabitants and for making live and full of people this infrastructure, for implement a constant maintenance and rediscover its potential value respect to the city and to its naturality. The basis of the proposal winning ANCI tender is in the emblematic title of "Gol.e.n.a." (life, empowerment, nature, and activism).

The general objective is to carry out co-designed interventions for the redevelopment of the river area through the direct involvement of young people with a view to sustainability, rediscovery and enhancement of the river. The specific objectives of the project that intercept the need for a holistic approach to the transport artery are:

a. promote the eco-sustainable development of the territory and the dissemination of good practices for social, cultural and environmental promotion;
b. make citizens aware of the charm and potential of the river area;
c. strengthen the environmental, cultural and social dimension as an engine and opportunity for ethical and sustainable economic development.

The project is ongoing now, but all the physical and narrative activity along the river are carrying on aiming to accompany citizens, directly activating a specific target of young people, in a multidisciplinary and experiential "path" of community growth. The intention is to enhance the symbolic and identity relationship that binds the river to the city and to the citizens, highlighting the potential of this natural element as the object of a plan of youth leadership and redevelopment of the territory. Just as the river is an element in constant movement and evolution, in the same way, young people are the main element of ferment and innovation in a community and it is essential to focus on them for territorial development, making the greenway part of the urban mobility across the city, intercept the home-work and home-school journeys of the people and accompanying the development of slow mobility.

The project with the help of 20 young people already selected, between 18 and 35 years of age, develops a plan for the area which involves the creation of a River Ecomuseum as an opportunity for the city to:

- increase collaboration between public bodies and the third sector through a long-lasting multi-stakeholder co-design, also with tactical urban planning;
- encourage employment growth, also by stimulating forms of self-employment;
- return to the city and citizens a physical and highly symbolic space that is a source and opportunity for sharing and community.

50.4 Conclusions

The convergence of the objectives of different projects on the greenway as an urban cycle path, full of value and which can reconnect the city, represents the best way for the approach that the Biciplan of Pescara is the bearer of, giving back to the city: slow connections, the river, a public linear space. The research conducted for the drafting of the Biciplan focuses on space sharing tools, to increase sustainable travel even directed to the suburbs and reduce the impact on CO_2, improving the quality of public space.

The programming tool moves in the difficulty of a complex issue which is based on the obstacles linked to a highly sensitive issue and very high political impact, with personalized interests and visions and with many and very diversified stakeholders. Coordinated planning directs actions toward defining of alternative path for primary net and sharing of public spaces through traffic moderation tools, contributing to decarbonization and proximity to travel. The interventions want to redesign the language of the street between what is public and what is private for the emergence of a universal public sphere where active mobility takes place, rediscovering the city from other point of view.

Acknowledgements The author worked on Pescara Biciplan with Maurizio Oranges and Angelica Nanni, and she was involved as trainer in the co-creation project Gol.e.n.a. of Pescara fluvial Ecomuseum.

References

European Commision (2019) The European green deal. Available at: https://eur-lex.europa.eu/legal-content/EN/TXT/?qid=1576150542719&uri=COM%3A2019%3A640%3AFIN. Accessed 30 Dec 2021

IPBES (2020) Workshop report on biodiversity and pandemics of the intergovernmental platform on biodiversity and ecosystem services. In: Daszak P, das Neves C, Amuasi J, Hayman D, Kuiken T, Roche B, Zambrana-Torrelio C, Buss P, Dundarova H, Feferholtz Y, Foldvari G, Igbinosa E, Junglen S, Liu Q, Suzan G, Uhart M, Wannous C, Woolaston K, Mosig Reidl P, O'Brien K, Pascual U, Stoet, P, Li H, Ngo HT, IPBES secretariat, Bonn, Germany. https://doi.org/10.5281/zenodo.4147317

ISFORT (2020) Rapporto sulla mobilità degli italiani. Avaiable at: https://www.isfort.it/wp-content/uploads/2020/12/RapportoMobilita2020.pdf

Mastrolonardo L (2021) Sustainable mobility and beauty of public space. VITRUVIO—Int J Archit Technol Sustain 6(2):42–55. https://doi.org/10.4995/vitruvio-ijats.2021.16560

Mussinelli E (2018) Il progetto ambientale dello spazio public, Eco-WebTown. J Sustain Des 2(18)

Salici A (2013) Greenways as a sustainable urban planning strategy. Advances in landscape architecture, vol 10, cap 24

Vittoria E (1965) La città tradizionale di ieri e il paesaggio urbanizzato di domani" in Urbanistica n.4

Chapter 51
The Buildings Reuse for a Music District Aimed at a Sustainable Urban Development

Donatella Radogna

Abstract Reusing buildings must be 'convenient' for the environment and for people, therefore it must re-establish a balance between places and communities which, interacting, determine a continuous transformation of cities. The reuse of buildings is a sustainable development process that implies phases of adaptation and qualitative growth to create safe, healthy, useful, attractive and beautiful places. The objective of the guidelines for the music district of Pescara (research for the conservatory) is to establish activities linked to music, culture and socialization, for the expansion of the 'Luisa D'Annunzio' Conservatory (through the reuse of the former Muzii middle school owned by the municipality) and provides the city with inclusive and beautiful places for all. The needs of different users (students of the conservatory and citizens) and those expressed by the client (music teachers and musicians) are considered to ensure the sustainability of the initiative through the integration of activities fit for restoring economic and social 'gain', according to an ecological approach. In providing the addresses for the required spaces, it was important to hypothesize additional functions and spaces to reborn the city with inclusion and beauty. The reuse of the former middle school was deemed 'convenient' (just as music is effective in restoring social inclusion and cultural development) thanks to the resilience capacities found. The proximity between the former school and the conservatory does not require substantial connection works and the proximity to the urban parks, the sea and the most 'lively' area of the urban center demonstrates an aptitude of the place for social reception thanks also to pedestrian and cycle paths. The spaces of the former classrooms are suitable for music teaching and recording studios as other existing spaces are for a music hub and other functions for the conservatory and the city, with a view to sustainable urban development.

Keyword Building and environmental reclamation · Reuse · Regeneration · Beauty · Psycho-physical well-being

D. Radogna (✉)
University "G. D'Annunzio" of Chieti-Pescara, Pescara, Italy
e-mail: dradogna@unich.it

E. Arbizzani et al. (eds.), *Technological Imagination in the Green and Digital Transition*,
The Urban Book Series, https://doi.org/10.1007/978-3-031-29515-7_51

51.1 Introduction

Reuse acquired an increasingly broader meaning that from the rehabilitation of abandoned buildings extended to the regeneration of urban and extra-urban areas as well as the upcycling and downcycling of demolished materials, products and building elements. Reuse policies in fact have to face the environmental, economic and social dynamics of places (Ostanel 2017) and adjectives such as 'sufficient', 'necessary', 'useful', 'convenient' are very important and guide toward the design of places able to bring real benefit to people.

According to Dan Barash, the future of abandoned buildings can go in four directions (Barash 2019): 1. loss (disappearing through collapse or demolition), 2. oblivion (being forgotten and defining degraded scenarios), 3. re-imagination (being preserved and/or transformed in an appropriate way to outline innovative scenarios able to regenerate contexts) and 4. free transformation (dramatically and often unjustifiably modified).

The design of existing buildings has to protect above all people's lives (Settis 2017), accepting compromises with the reasons of ecology and safety. The aim of the reuse project is to establish or re-establish a balance between the buildings, the users and the contexts in order to understand how much must be conserved, transformed, demolished, as places and communities interact, determining a continuous transformation of cities. (Martinotti 2017).

The reuse of the built environment is therefore to be understood as a process of sustainable development, a path of evolution that does not deny either transformation or conservation but implies phases of adaptation and qualitative growth aimed at creating safe, healthy, useful, attractive and beautiful places. Based on these concepts, the drafting of guidelines for the expansion of the Luisa D'Annunzio Conservatory in Pescara was an important opportunity to promote a convenient reuse policy and provides the city with a musical district (Fig. 51.1).

The experience, done for a contract for third parties[1] (Barash 2019), required an important cooperation between the conservatory, the municipal administration and the University to create a place capable of both satisfying the needs expressed by the higher education school and of triggering processes of cultural inclusion and dissemination. Music culture has been recognized as having the ability to improve economic and social conditions in a place to be built according to the principles of the green transition (especially in the use of energy and materials).

The main objectives of the work carried out are to satisfy the needs expressed by the conservatory in the most sustainable way. The methodology adopted, at first, focuses on the needs to have new classrooms and an auditorium larger than the one existing in the main used building. Then, the abandoned school performance is analyzed to define those to be achieved with the interventions. The proposed solutions suggest the achievement of broader objectives because they consider all

[1] Third party agreement between the L. d'Annunzio Conservatory and the Department of Architecture of the G. D'Annunzio University of Chieti-Pescara for the drafting of the Guidelines for the L. D'Annunzio Conservatory expansion (2018).

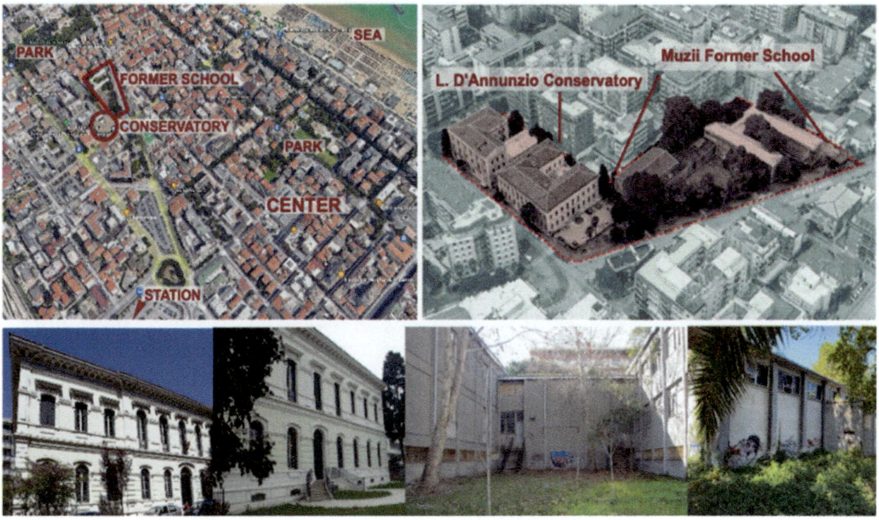

Fig. 51.1 Conservatory and the former school in the urban fabric

types of local resources to satisfy not only the needs of the conservatory but also those of the city. The closeness of the conservatory to the abandoned school, the city center, the sea, the urban parks, the cycle paths and the train and bus station was considered an important resilience capacity, to get started. The guidelines recall the objectives of digital transformation and ecological transition to guide the reuse of abandoned schools and thus equip the conservatory with new classrooms for music teaching, recording studios, an audio library and a hub (Fig. 51.2). A selective demolition and reconstruction project to build the new auditorium was suggested only for the large gymnasium of the former school, characterized by very serious structural problems and therefore not recoverable.

The guidelines in also providing indications for the design of in-between and urban spaces to promote regeneration through a strong connection between the conservatory and the city, is inspired by other effective European experiences such as the *Maison des Ensemble* (*Filippini Architecture and patrimoine*) in Paris and *Gillet Square* (*Hawckins Brown*) in London.

51.2 The Reuse of the Former Muzii Middle School

The hypothesis of reusing the school is mainly motivated by the recoverability of the technological system, the morphological-dimensional characteristics of the environmental system and the location of the building with respect to the urban fabric.

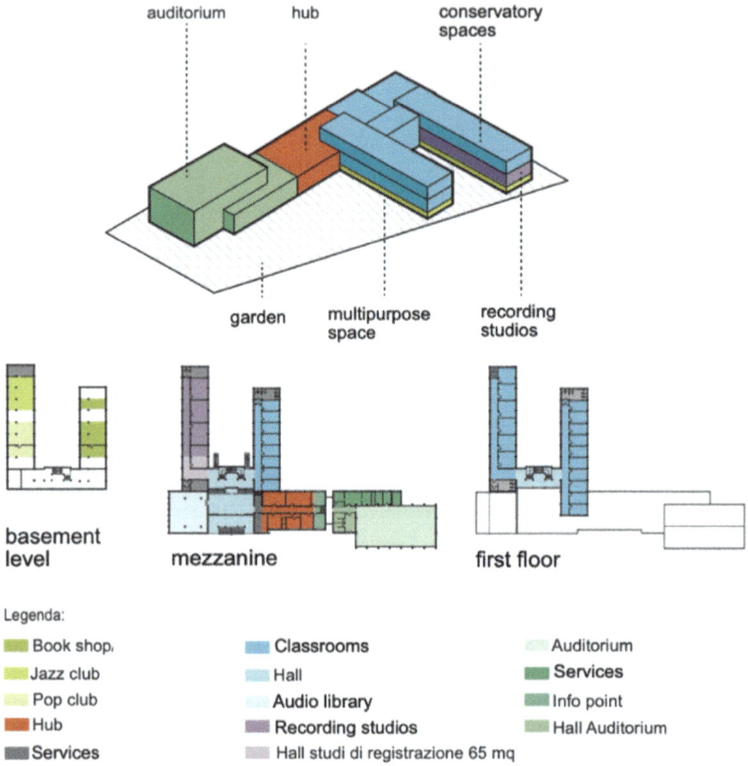

Fig. 51.2 New uses in the former school

The proximity of the school to the conservatory facilitates the organization of the new spaces because it does not require special connecting works and allows not to duplicate some functions. In addition, the connection (pedestrian, bicycle and vehicle paths) with the station, green areas, the sea and the most 'lively' area of the city center reveals an already existing integration with the city (Fig. 51.3).

In evaluating the reuse possibilities, the functions to be established were compared with the congenital characteristics and with the state of conservation of the environmental and technological systems to identify the constraints to modification and the latent potential. The main constraints are determined by the load-bearing structure in reinforced concrete, which does not allow the creation of large free spaces. Therefore, the functions that are not very compatible with this limit (auditorium) were located in the volumes to be demolished and rebuilt (large gymnasium). To contain the impacts of this hypothesis, the guidelines suggest to maximize the possibilities of reuse and recycling of the abandoned elements through selective demolition logic.

For the new classrooms, the recording studios and the hub, the evaluation of compatibility for reuse revealed a high suitability of the existing spaces, in terms of number, size and shape. The spaces of the former classrooms were found to be

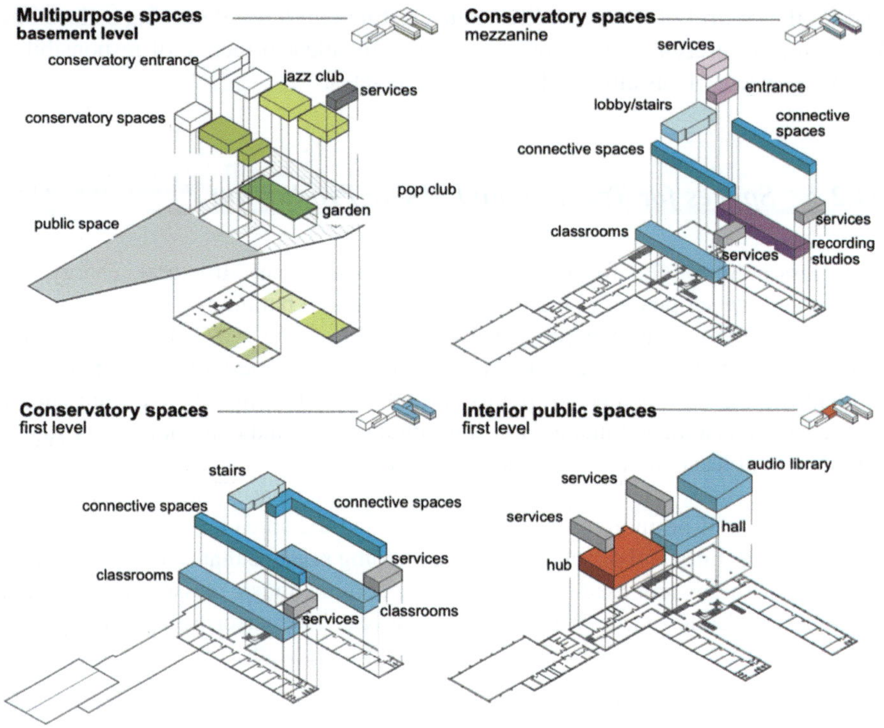

Fig. 51.3 Organization of the spaces at the different levels

appropriate to meet the needs of music teaching and recording studios as well as the other spaces, previously intended for other functions of the former school (corridors, secretariats, etc.), proved to be suitable to place a music hub and other service spaces.

For the load-bearing frame, retrofitting interventions are proposed, while for poor quality and very degraded walls, replacement with new high-performance systems is suggested, especially for safety, well-being, usability and environmental protection.

The results of the reuse options evaluations also suggest to transform the semi-underground environmental units (internal spaces near the 'street' and easy to be connected to public spaces) into a horizontal distribution and integration system as the main connection between the conservatory and the city. In this case, the transformation hypotheses affect not only the vertical closures but also the ground because they require a lowering of the external floor to improve the usability and comfort of the internal spaces and connect them directly with the open spaces and with the urban context.

The presence of the former school represents an important memory, even if it is not characterized by elements of particular value to be protected. The main values of the building are due to its position in the urban fabric and to the quantity, organization, size and shape of the internal and external spaces. The guidelines for the recovery and reuse project reveal how the constraints became the rules for the intervention choices and for the control of the future life of the existing buildings, decreeing a close link between past, present and future according to the clear principle of responsibility that the concept of sustainable development dictates.

51.2.1 Spaces for Training and Music Production

The classrooms of the former school require significant structural, energy and functional redevelopment, as the 90% of Italian school buildings.

The hypotheses outlined consider the contents of law 107/2015 on the 'Good school', in particular as regards the safety and innovativeness of school buildings. Among the eleven fundamental points around which the law revolves, the one on built heritage is aimed at making existing schools safer and equipping every Italian region with at least one innovative school building.

These measures offer the possibility of making the spaces intended for training places capable of promoting social acceptance and the spread of culture as well as 'soliciting' the birth and dissemination of a building practice that is finally 'greener'.

The hypotheses proposed to guide the project of the music classrooms aim at the creation of didactic spaces with high quality levels. The performance improvement is largely entrusted to the new closing and partitioning walls which will have to:

- Be structurally collaborative to deliver a good seismic response and not 'break' so as not to cause damage to users;
- Guarantee the conditions for the comfort and psycho-physical health of people;
- Allow flexibility of set-ups and spaces;
- Allow the containment of the impacts in the production and construction phases and energy consumption during the operation phase as well as maximizing the possibilities of reuse and recycling in a future disposal;
- Allow to facilitate maintenance actions, reducing intervention times and costs;
- Allow easy connection with existing building elements and specific equipment;
- Contribute to the containment of energy consumption;
- Determine a good architectural quality capable of making places attractive.

The proposals developed guide a project that is attentive to all phases of the building process to increase the sustainability of the actions starting from partial demolition operations, to the construction, operation and future disposal phases.

The new spaces for musical teaching are inserted in the two main bodies of the existing building, consisting of two levels above ground and a basement and connected to each other by the volume, transversal to them, containing other connecting spaces and the stairwell (Fig. 51.4).

The reorganization of the classrooms is conditioned by the relationships between the functions to be installed and by the opening to different types of users (inside and outside the conservatory, of different ages and with different times of attendance of the spaces), whose presence can ensure continuous use of new places.

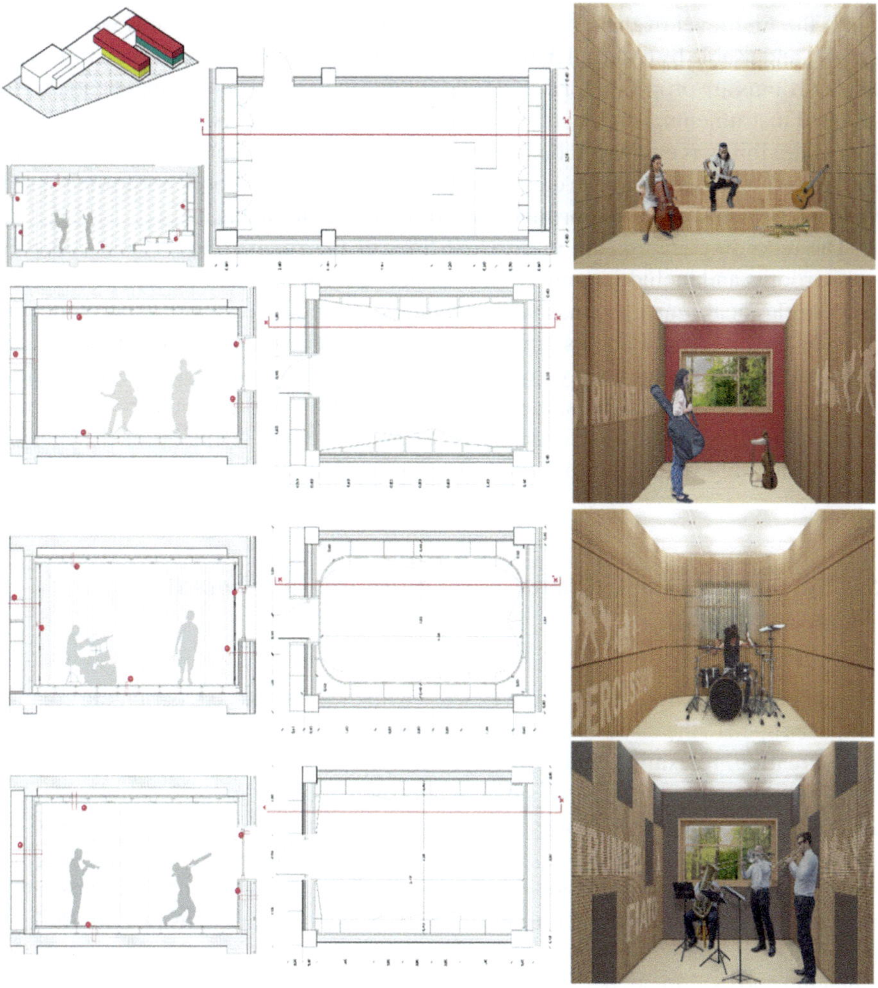

Fig. 51.4 Classrooms for playing the instruments

Foreseeing different types of users will allow to equip the conservatory with spaces that are economically and socially sustainable because they can be used both for internal activities and to satisfy external requests capable of giving an economic return. At the same time, the city will be provided with currently lacking services, intended both to meet the needs of musicians and to develop and spread the musical culture.

The guidelines propose classrooms for instrumental practice and study, classrooms for theoretical lessons as well as spaces for the production and digitization of music.

At the first level of the existing volumes, the location of the spaces with greater turnout and usable by external parties is foreseen while, at the second level, four types of classrooms are proposed for the exercise of musical instruments and rehearsals (for stringed instruments, wind, percussion, ensemble). For the classrooms, acoustic and spatial flexibility is envisaged both to meet the needs of different musical instruments and performances and to have classrooms capable of restoring both traditional spaces and larger spaces including primary and complementary areas.

The recording studios are on the first floor of the same volumes containing the music rooms and include a control room with two live rooms (one for pop music and one for classical music), a mastering room and a production room. The guidelines provide for diversified access and paths for the differentiated use of the spaces by users of the conservatory only, by external users only, by users of the conservatory and by external users at the same time. The recording studios could be rented out, to create micro-economies, forms of social welcome and diffusion of the musical 'world' not only in the playful-cultural dimension but also in the working one.

51.2.2 Spaces for Cultural Dissemination and Social Reception

To trigger regeneration processes in Pescara with the new Music District, the creation of spaces open to the city and including an audio library, a music hub and other spaces for recreation, relaxation and refreshment has been thought. For the audio library, the redevelopment of the former small gym is suggested through structural, energetic and functional adaptation works which also include the addition of a floor and the recovery of the materials of the demolished technical elements. The guidelines also suggest to give the possibility to listen and 'see' music at 360° through the installation of sound-video stations and equipment for immersive sound, consoles, multimedia totems and systems for augmented reality. The idea wants to allow people to live engaging and exciting experiences of watching videos or listening to recorded or live music (Favaro 2017). This new space should also include environments for listening to digital sources (workstations for listening to CDs) and analogue (workstations for listening to vinyls) (Fig. 51.5). According to the functional organization, the audio library allows access to a terrace that connects the new classrooms with the new auditorium.

The terrace is a garden roof with extensive evergreen vegetation, which in addition to acting as a thermal insulating and thermoregulatory element, contributing to the improvement of air quality, absorbing water and reducing the risk of flooding, allows to take advantage of the tax incentives provided by the 'Regulations for the development of urban spaces' (Budget Law 2018).

The recovery of the administrative offices and the connection and distribution areas of the former middle school (placed transversely to the volumes containing the classrooms and recording studios) provides for the organization of a hub and an

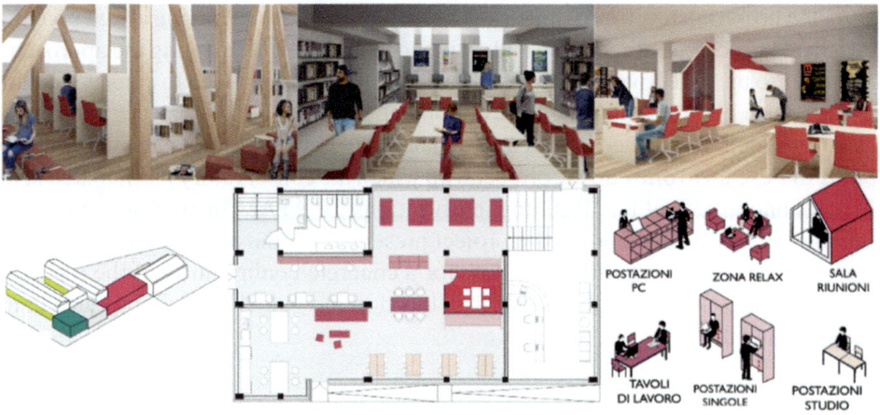

Fig. 51.5 Some spaces of the audio library and the hub

information and management point for all the activities of the new district. The hub is a place dedicated to music, in support of creativity, where students and professionals meet to develop projects to be placed on the market. A co-working space is proposed for those who want to make music or those who work or want to work in a music production can meet and discuss to realize their ideas. Recognized an important role for art and culture in regeneration strategies (Tarantino 2019), the integration between the new spaces for music and the city represents a key point for achieving effective results for a beautiful, inclusive and sustainable urban habitat (Fig. 51.5).

51.3 Conclusions

The hypothesis of reuse presented recalls some fundamental points of the European Green Deal. The building renovation works provide for high energy efficiency through the improvement of the behavior of the casings and the use of renewable sources. The proposed solutions include bio-dynamic facades able to make the most of the energy flows from the sun and wind. The control of natural light in the interior spaces of the classrooms is based on light deflectors (light-shelf).

Furthermore, the use of laminated wood and x-lam elements determines a consistent structural, comfort and circularity performance improvement.

The organization of the music district promotes the use of public transport and soft mobility as it provides a better connection between parking lots, e-bikes rent, cycle and pedestrian paths and the city bus station. The new activities that 'open' the conservatory to the city and integrate new functions linked above all to digitization, will be able to create new jobs also with reference to the future needs of the community.

In the proposal developed, musical culture is recognized as having a high potential to activate processes of recovery and resilience, thus also intercepting the approaches that refer to the objectives of the New European Bauhaus and lead to consider culture as the fourth pillar of sustainable development.

The guidelines have constituted a fundamental contribution to allow the Municipality of Pescara to draw up the project with which the conservatory participated in the MIUR announcement for 'Building interventions for AFAM institutions', obtaining funding of 10,000,000 euros. The project presented for the request for funding will therefore be carried out and will allow for a concrete confirmation of the repercussions that the scientific elaborations developed in the guidelines will produce in the actual context.

References

Barash D (2019) Ruin and redemption in architecture, Phaidon
Favaro R (2017) Suono e arte. La musica tra letteratura e arti visive, Marsilio
Martinotti G (2017) Sei lezioni sulla città. Feltrinelli
Ostanel E (2017) Spazi fuori dal comune. Rigenerare, includere, innovare, Franco Angeli
Tarantino A (2019) Lifestyle nella rigenerazione urbana: contesti, strumenti ed azioni, Cacucci, Bari

Chapter 52
Environmental Design for a Sustainable District and Civic Hub

Elena Mussinelli, Andrea Tartaglia, and Giovanni Castaldo

Abstract The paper presents the results of a design-based research focused on a specific context in the city of Milan (Municipality 4), where a holistic approach to a fruitive and environmental regeneration was experimented. The proposed design-based approach integrates the functional and fruitive reactivation of public space with analysis, simulations and assessments on the possible application of nature-based solutions (NBS) to increase urban resilience, comfort and public space usability. In addition to increasing the environmental benefits (ecosystem services) at the district scale, the project aims to strengthen the ecological connections at the wide area, stressing the necessity of a systemic approach in the GBI's development. The paper illustrates both the methodological and framing aspects of the experimentation and the project results, verified through a consolidated methodology for the assessment of the expected environmental benefits. The research project contributes in developing new approaches to the deep renovation of public space in urban and peri-urban contexts, that are a priority in the current Italian scenario.

Keyword Smart sustainable district · Civic hub · Environmental design · Nature-based solutions · Public space regeneration

E. Mussinelli · A. Tartaglia (✉) · G. Castaldo
Polytechnic University of Milan, Milan, Italy
e-mail: andrea.tartaglia@polimi.it

E. Mussinelli
e-mail: elena.mussinelli@polimi.it

G. Castaldo
e-mail: giovanni.castaldo@polimi.it

E. Arbizzani et al. (eds.), *Technological Imagination in the Green and Digital Transition*, The Urban Book Series, https://doi.org/10.1007/978-3-031-29515-7_52

52.1 Introduction

The design of the public space based on the principles of environmental sustainability, for the mitigation and adaptation to climate change and for the improvement of well-being, health and safety conditions has assumed a crucial role in the agendas of local administrations. The accessibility and usability of public space are in fact key-factors for the regeneration of the public city: ensuring qualified access to places, services and information is a primary challenge (DIAUD 2017) with multiple cultural, social and economic implications (AA.VV. 2013). The possibility of using the public space meeting individual and collective needs is in fact essential to guarantee an independent life for citizens and allow the development of adequate social relationships, promoting freedom, the well-being of communities (Hansen 1959) and social cohesion, according to an inclusive cultural model (Conti 2015). The full accessibility of the urban environment also derives from a complex set of factors that, in addition to allowing the overcoming of physical-spatial barriers, provide a healthy, safe and comfortable usability of urban space, rich in multisensory stimuli.

The research described in this contribution is part of this framework, experimenting a holistic approach to the fruitive and environmental regeneration of the urban public space. An approach that aims to improve the environmental characteristics for the use and accessibility of open spaces, according to a design-based methodology (Mussinelli and Tartaglia 2021). The adopted methodology is based on site-specific design simulations—characterized by the use of natural solutions (nature-based solutions—NBS and low impact development systems—LID)—and on an accurate assessment of the environmental, microclimatic and fruitive benefits deriving from the redesign of public space.[1]

The applied methodological approach is structured into the following phases:

- urban scale analysis in order to identify critical issues and opportunities for intervention for the improvement of the environmental ecosystem quality;
- definition of a study area at the district scale, identifying a sufficiently homogeneous context, albeit with variable geometry, in terms of settlement, morpho-typological, functional, environmental and infrastructural characteristics and having dimensions such as to present and/or allow the establishment of proximity relationships and territorial connections of significant social and eco-systemic value (Poliedra 2022; Mussinelli et al. 2021);
- study of the environmental, microclimatic and fruitive characteristics of the study area and identification of site-specific criticalities;
- definition of alternative intervention proposals;
- evaluation/quantification of the environmental, microclimatic and fruitive benefits generated by the proposals.

[1] The design experimentation was developed through the collaboration between the cultural associations Urban Curator TAT and Resilience Lab, with the patronage of the Municipality 4 of Milan, on the occasion of a scientific dissemination initiative held in November 2021.

52.2 Materials and Methods

52.2.1 Analysis of the Reference Urban Context

The research activity was focused on the South-East quadrant of Milan,[2] within the Municipality 4, a large-scale context with relatively homogeneous settlement, environmental and fruitive characteristics. It has been investigated through analyzes of the morphological structure of the built and open spaces, the land uses, the average surface temperature of the soils, the provision of public and private green areas and accessibility to public green areas.

The results of these analyzes are summarized through thematic maps highlighting the encountered criticalities and opportunities.

First of all, a limited presence of settlement sprawl phenomena emerges, with the persistence of large portions of agricultural land and wide-open spaces, unlike what happened in other quadrants of the city. This condition highlights a high potential for ecological reconnection both on a large scale and in the local area. In particular, it is analysed the relationship between the built systems of the urban fringe and the area of the *Parco Agricolo Sud Milano* (Peri-urban Agricultural Park), with significant service values with respect to the residential system and the large public housing districts of Corvetto and via Ripamonti (Fig. 52.1).

The analysis of the average temperature of the soils (Land Surface Temperature) shows lower values in correspondence of the *Parco Agricolo Sud Milano* which contributes to the mitigation of the urban heat island even in neighboring urban blocks and higher values, in some cases also critical, in the central areas of the urban sector characterized by high building densities and scarce green areas.

With reference to the analysis of urban greenery (Fig. 52.2), in addition to the mapping and analysis of green areas and the tree heritage[3]—which showed an overall endowment higher than the city average—attention was paid to the study of pedestrian accessibility to the so-called "neighborhood parks" (Zhang et al. 2011). Through the accessibility indicators to the green areas measured according to the buffer techniques and the time < 15 min. (Mussinelli et al. 2021), considering the population for each block of the quadrant, it was possible to define the road axes used mainly by the population to access these parks. Axes that therefore require a timely verification of their fruition and environmental quality (Fig. 52.3).

---

[2] The South-East sector of Milan has been studied by the Research Group ENVIREG—Environmental Regeneration of ABC Dept. of Politecnico di Milano since 2015 (e.g., PRIN 2015 "Adaptive Design and Technological Innovations for the Resilient Regeneration of Urban Districts in Climate Change Regime). ENVIREG Research Group: Elena Mussinelli, Andrea Tartaglia, Davide Cerati, Giovanni Castaldo, Daniele Fanzini, Roberto Bolici, Matteo Gambaro, Raffaella Riva.

[3] GIS analyzes of the green heritage were conducted using the database of Comune di Milano (years 2012 and 2014) referred to public green integrated with databases of private green heritage. For the tassonomy aspects, the main references of the classifications were ISTAT (2013) and ISPRA (2014).

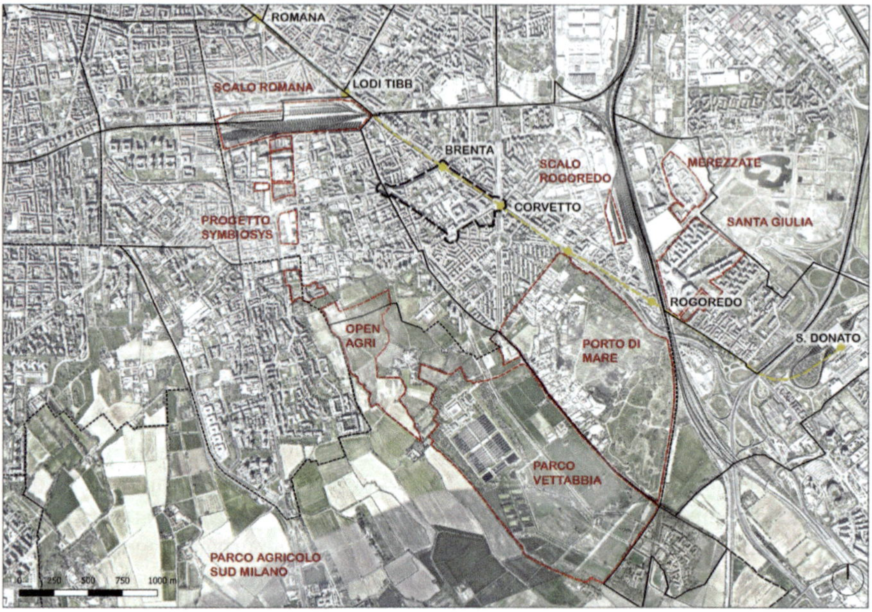

Fig. 52.1 Reference urban context—Authors' Elaboration

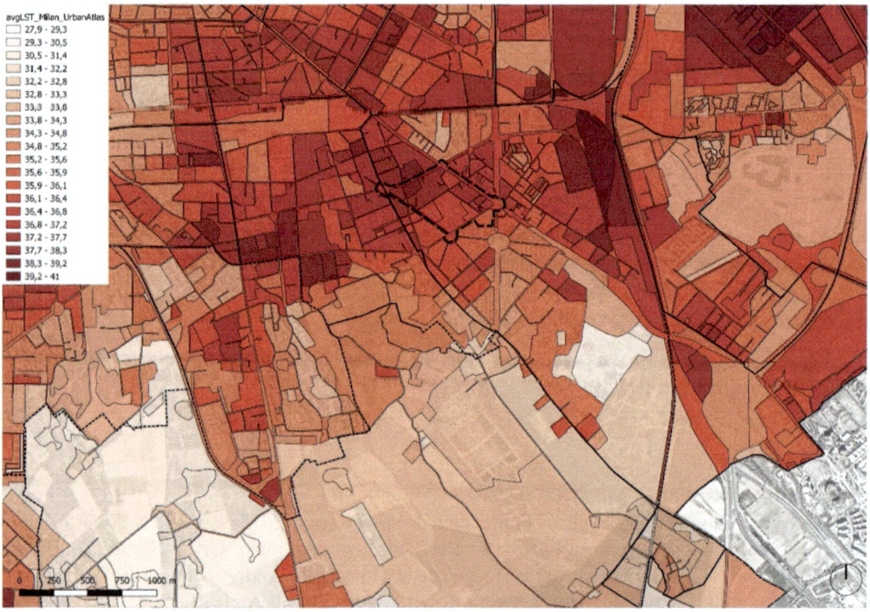

Fig. 52.2 Land surface temperature map—Authors' Elaboration. *Source* Sistema InformativoTerritoriale—SIT, Comune di Milano

Fig. 52.3 Accessibility indicator according to the buffer technique—Elaboration by Arch. PhD. Davide Cerati

In broad terms, it emerged that the green heritage, albeit extensive, presents multiple discontinuities, with entire large urban sectors without gardens and neighborhood parks and with numerous roads completely devoid of trees. The analysis of the tree-lined/non-tree-lined roads and that of the most used roads to access the neighborhood parks has made it possible to identify the roads that would need priority interventions to improve their environmental quality and to face the local phenomena of urban heat islands (Fig. 52.4).

52.2.2 Site-Specific Analysis of the Study Area (District) and Design Proposals

The design experimentation was conducted on a specific study area, identified on the basis of the evidence of the large-scale environmental analyzes above described[4] as well as on the recognition of the potential transformation of this area into a smart and sustainable district (Poliedra 2022).

The potential district identified is bounded by viale Brenta, corso Lodi, via Polesine and via Mincio, for a total land area of approximately 23 hectares. Inside there are five blocks defined by secondary crossing roads. The area is homogeneous from a

[4] In particular, due to the presence of high average surface temperatures, the absence of proximity parks, the consistent presence of road axes without trees.

Fig. 52.4 Measurement of accessibility to parks on the basis of ≤ 15 min. by walking—Elaboration by Arch. PhD. Davide Cerati

morpho-typological and functional point of view and is characterized by a particularly significant concentration of public and civic services.[5]

At this scale, environmental, microclimatic and fruitive site-specific analyzes were carried out aimed at pointing out the main critical issues to which the project proposals refer. About 40% of the land is occupied by buildings and 60% by open spaces, only 16.5% of which is devoted to green areas, while 21% is occupied by roads and parking lots and 23% is made up of equipped areas pertaining to the buildings. The analysis of the permeability of soils showed a clear prevalence of impermeable soils (83.5%). The arboreal heritage consists of 372 trees, of which 353 deciduous and 19 evergreen, mainly located along the road axes of Corso Lodi and Via Polesine and within the schools' gardens (therefore limitedly accessible) (Fig. 52.5).

The analyzes developed with reference to the microclimatic aspects concerned the land surface temperature and the level of sunshine of open public spaces, evaluated with the Lindberg and Grimmond methodology (2011). In addition to the average temperature of the soils, the characteristics of materials and the albedo of the portions of soil exposed to sunshine for more than 70% of the hours of the daylight were analyzed, identifying the public spaces and paths open to high use with the greatest criticalities (high land surface temperature, high and prolonged sunshine, low albedo of materials): viale Brenta, via Polesine and via Mincio (Fig. 52.6).

---

[5] In the study, area is located the following public buildings/services: Headquarter of Municipality 4; Headquarter of the Municipality of Milan; Primary School "Vittorino Da Feltre/M. Carlo Lorenzini", Middle School "Lombardini", School "Marcello Candia", Regional School Department Office, Municipal Swimming Pool "Mincio", Social Center "Arci Corvetto", Polifunctional Center "Polo Ferrara".

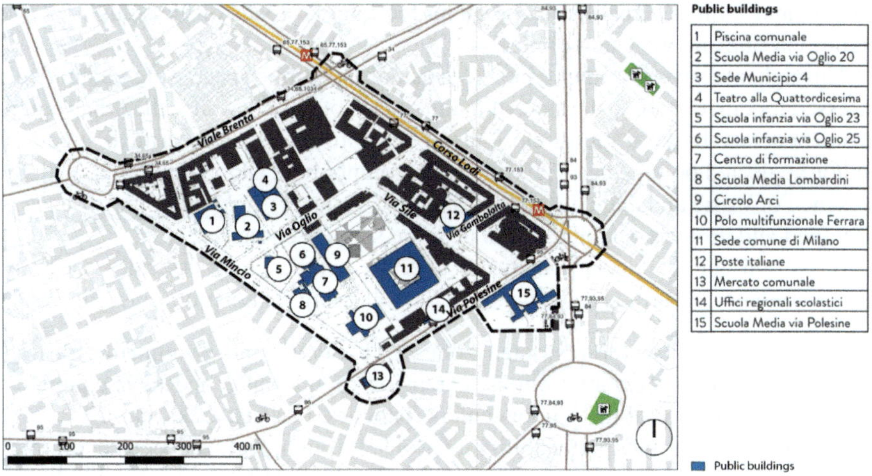

Fig. 52.5 Study area perimeter—Authors' Elaboration

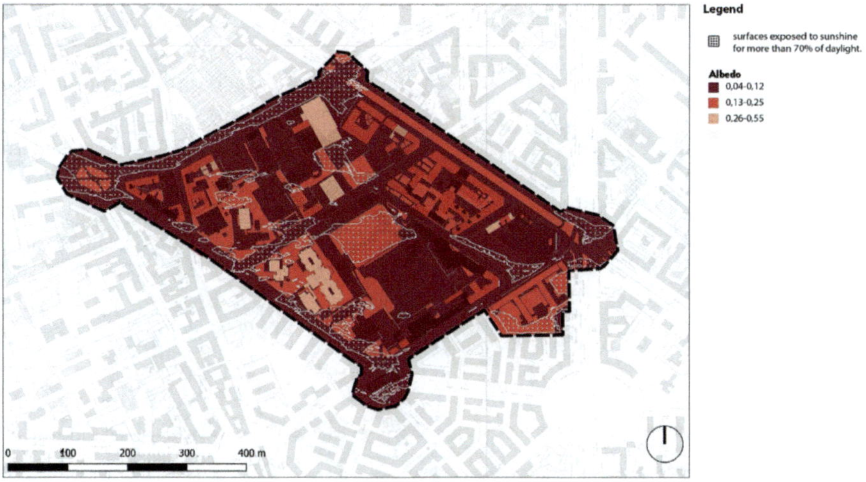

Fig. 52.6 Albedo analysis and portions of public space with high percentage of sunshine—Elaboration by Ing. Alessandro Grimaldi

With reference to the fruitive quality of the public space, the pedestrian and vehicular flows within the district were analyzed, with particular attention to the four schools present in the area; the extent and characteristics of user flows were then directly detected (by way of travel: pedestrian, cycle, public transport and vehicular traffic).[6] Analysis highlighted an overall prevalence of pedestrian transits to and from

[6] In the period 05/10/2021—08/10/2021, site surveys were conducted aimed at gathering information on the flows of users of the schools of the district. The data was collected in the time slots of

schools and limited vehicular traffic along the internal roads of the study area (via Oglio, via Sile and via Gamboloita).

Through site surveys, other characteristics of the public space were qualitatively assessed too, in relation to the state of maintenance of the sidewalks, the urban equipment (benches, baskets, racks, etc.), also by identifying specific critical situations and/or neglected points (e.g., via Mincio, via Gamboloita and viale Brenta, with the presence of waste and damages to the asphalted surfaces) and noting the absence of equipped and protected public spaces near the schools entrances.

Based on the conducted analyzes and the encountered critical issues, the experimentation identified and developed the following design proposals (Fig. 52.7):

- establishment of traffic-restricted zone (ZTL 30/10 km/h) within the district, with redefinition of road sections, extension of sidewalks and their equipment in correspondence of the school centers (benches, baskets, racks, etc.);
- improvement of the permeability of soils through high-reflectance draining asphalts along via Sile, Gamboloita and Oglio (about 24,500 sqm.) and bio-basins along via Oglio, Sile, Mincio and Brenta (about 1550 m^2);

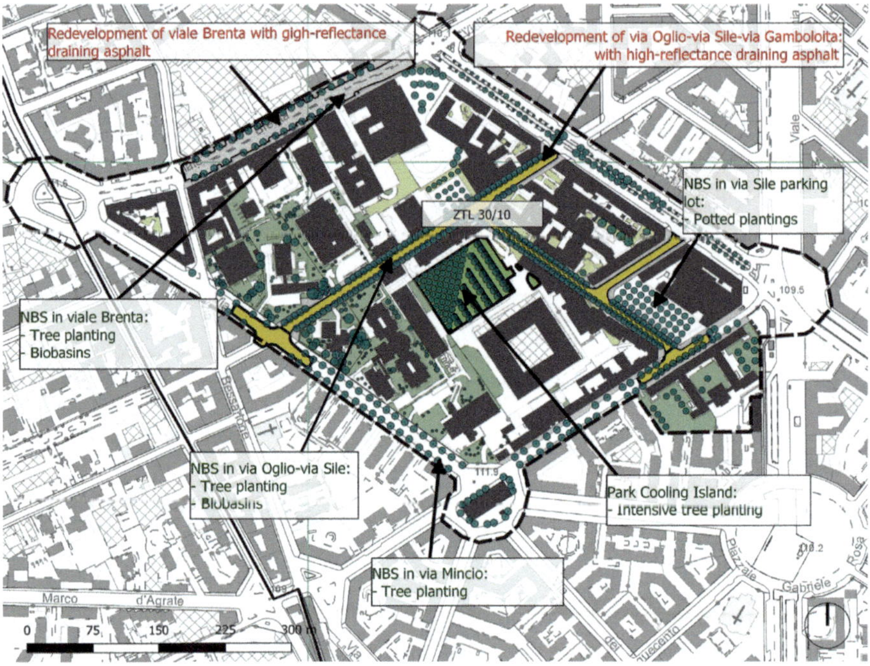

Fig. 52.7 Design proposal for the study area—Authors' Elaboration

entry/exit from schools or the beginning and end of the working day, sampling the number of users in transit at intervals of 30 min.

- increase in the arboreal heritage with the planting of 373 new trees, partly distributed along the road axes, partly concentrated in the intensive planting (Park Cooling Island) of a semi-enclosed free area (about 6,800 m^2) owned by the Municipality and currently neglected, located near the new Municipal Office Complex. There are also 39 potted plantings located in an area currently used as parking lot in via Sile.

52.3 Results

The environmental, microclimatic and usability benefits generated by these proposals were then evaluated.

First, the multiple benefits provided by tree planning have been assessed. The new trees along the road axes and the intensive tree-planting of the new Park Cooling Island allow the sequestration of 75.14 tons/year of CO_2 and the capture of 47.28 kg/year of O_3, of 38.25 kg/year of PM10, of 87.05 kg/ ear of SO_2 and 34.77 kg/year of NO_2. This reduction of air pollutants due to this vegetation (through deposition, absorption, etc.) directly improves the air quality at the local level (Silli et al. 2015) and contributes also to the reduction of rainwater run-off, with the interception of 432,450 L/year.[7] The interception of rainwater implies direct environmental benefits such as flooding risk mitigation and reduction; at the same time, the avoided management of rainwater in the sewer system entails savings in public expenditure (indirect benefit). In the meantime, plantations, through shading and evapotranspiration, significantly contribute to improve the thermos-hygrometric comfort conditions of public space, mitigating the effects of the urban heat island. The simulation carried out for the viale Brenta—developed through the application of ENVI-Met software—led to an estimation of a betterment of the perceived temperature (Universal Thermal Climate Index-UTCI) of approximately − 6 °C.

Secondly, the adoption of high-reflectance draining asphalts along the carriageways and sidewalks of the central streets of the district generates additional benefits. The approximately 24,500 m^2 of draining asphalt contribute to improve the run-off of the area, with the interception of approximately 4.3 mil. liters/year. In addition to this value, about 2.5 mil. of liters/year of rainwater intercepted by the bio-basins envisaged in the redesign of the road sections, as well as approximately 6.8 mil. of liters/year intercepted by the new green area must be considered.

[7] With respect to the reduction of air pollutants and the capture of CO2, models (UFORE) developed through specific software (i-Tree and i-Tree Design) were applied. The findings contributed to the production of table values defined according to a time scale of growth of trees. With regards to the water retention capacity of the bio-basins, an average runoff index of 0.6 was assumed, while for the air pollutants reduction, the values expressed by the Forest Service Tree Guides were used, considering the bio-basin enriched with low and medium stem shrubs. The algorithms employed for the quantification are those proposed by the Center for Neighborhood Technology (CNT 2010), to assess the value of ecosystem services associated with green infrastructures in urban areas (Mussinelli et al. 2021).

The use of asphalts with high albedo plays an important role in facing the urban heat island effect, specially identified in some portions characterized by high sunshine exposure and low albedo pavements. Different studies investigate the correlation between albedo of pavements in urban environments and microclimatic characteristics, showing a difference, in specific conditions, of approximately 5–6 °C between dark colored asphalts and light-colored materials (Chudnovsky et al. 2004).

Finally, the benefits provided by proposing a traffic-restricted zone with 30/10 km/h speed limit can be considered as well. This sort of traffic measure entails different advantages in terms of urban and environmental quality such as traffic reduction, improvement of the road safety, noise pollution reduction, air pollution reduction, improvement of the multifunctionality of urban streets (Staricco 2011).

52.4 Conclusions

In the presented case, the improvement of the environmental and microclimatic quality of the public space represents a key factor for the establishment of a real Smart and Sustainable District & Civic Hub,[8] based on full accessibility and usability of the articulated system of public spaces and services and cultural and civic venues already present in the area. The proposed set of site-specific solutions contributes to overcome the so-called "climatic-environmental barriers" (Tartaglia et al. 2019), that are the limitations in the usability of public space due to environmental risks and adverse microclimatic conditions. The use of NBS and the reorganization and equipping of the public space contribute to increase the quality, decorum and attractiveness of the places of collective use, while generating substantial ecosystem services.

Further research developments, currently being developed, aim at integrating the quantification of the environmental and microclimatic benefits generated by the proposal for the Hub of the Municipality 4 with the evaluation of the benefits obtainable through ecological reconnection interventions extended to the vast area of the peri-urban quadrant, thus recomposing some discontinuities currently present in the green and blue infrastructure system (environmental reconversion of the Corvetto overpass and regeneration of the Corso Lodi-Via Marocchetti axis, with the prevision of new green square in front of the Rogoredo station; strengthening of the green system along the axis of viale Omero toward the Vettabbia Park and the Porto di Mare Park).

The proposed methodology—based on the tight correlation between environmental local conditions and public spaces fruition—can represent a possible approach for a deep renovation of public spaces in urban and peri-urban contexts. This is a

[8] The Smart Sustainable District and Civic Hub is defined as "sustainable" with reference to all the social, environmental and economic aspects that contribute to the achievement and maintenance of conditions of quality of life, health and well-being of the communities and good environmental status with reference to mitigation/adaptation to climate change and the conservation of natural capital, while creating equity and equal opportunities for its inhabitants (Poliedra 2022).

priority in the current Italian scenario, with the presence of many urban areas characterized by poor environmental conditions worsen by increasing climate change effects.

References

AA.VV. (2013) Carta dello spazio pubblico. Istituto Nazionale di Urbanistica

Chudnovsky A, Ben-Dor E, Saaroni H (2004) Diurnal thermal behavior of selected urban objects using remote sensing measurements, Energy Build 36(11):1063–1074

Conti C (2015) Environmental accessibility as a tool for urban renewal. In: Mussinelli E (ed) Design, technologies and innovation in cultural heritage enhancement. Maggioli, Santarcangelo di Romagna, Italia, pp 45–52

CNT—Center for Neighborhood Technology (2010) Integrating valuation methods to recognize green infrastructure's multiple benefits. Chicago, IL

DIAUD—Disability Inclusive and Accessible Urban Development Global Network (2017) The inclusion imperative: towards disability-inclusive and accessible urban development. CBM

Hansen WG (1959) Accessibility and residential growth. Massachussets Institute of Technology

Lindberg F, Grimmond CSB (2011) The influence of vegetation and building morphology on shadow patterns and mean radiant temperatures in urban areas: model development and evaluation. Theoret Appl Climatol 105(3):311–323

Mussinelli E, Tartaglia A (2021) Organizing knowledge according to site-specific criteria. In: Bologna R, Losasso M, Mussinelli E, Tucci F (eds) From urban districts to eco-districts. knowledge, methodologies, programmes, pilot projects for climate adaptation. Santarcangelo di Romagna, Italia, Maggioli, pp 35–43

Mussinelli E, Tartaglia A, Raffaella R, Castaldo G Cerati D (2021) Knowledge, strategies, demonostrative projects for the environmental quality ad climate adaptation of publi outdoor spaces in the urban context of Milan. In: Bologna R, Losasso M, Mussinelli E, Tucci F (eds) From urban districts to eco-districts. Knowledge, methodologies, programmes, pilot projects for climate adaptation. Maggioli, Santarcangelo di Romagna, Italia, pp 123–161

Poliedra (2022) Libro bianco. SSD: un modello di Smart Sustainable District per progetti di rigenerazione urbana. Politecnico di Milano-Consorzio Poliedra

Silli V, Salvatori E, Manes F (2015) Removal of airborne particulate matter by vegetation in an urban park in the city of Rome (Italy): an ecosystem services perspective. Annali Di Boranica 5:53–52

Staricco L (2011) Multifunzionalità e conflittualità nelle Zone 30. TeMa,04/2011, pp 59–68

Tartaglia A, Mussinelli E, Cerati D, Castaldo G (2019) Questioni inerenti l'accessibilità dello spazio pubblico e il cambiamento climatico. Public space accessibility and climate change issues. In: Baratta AFL, Conti C, Tatano V (ed) Abitare Inclusivo. Il progetto per una vita autonoma e indipendente/inclusive living. Design for an autonomous and independent living. Arteferma, Conegliano, Italia, pp 48–57

Zhang X, Lu H, Holt J (2011) Modelling spatial accessibility to parks: a national study. Int J Health Geogr 10:1–14

Chapter 53
Earth Observation Technologies for Mitigating Urban Climate Changes

Federico Cinquepalmi and Giuseppe Piras

Abstract Since 2010 United Nations declared that for the first time in history up to 50% of mankind is living in urban areas, implying that challenges connected with global changes need to be evaluated primarily within urban systems, using the most advanced available technologies. Earth observation is nowadays the most promising field of research assisting urban planners, city managers, and building designers in their work of improving urban resilience to cope with climate change effects, and the long-term changes connected with extreme climatic events. Even though the deep understanding of the functioning of urban systems is a key factor for improving the quality of life at all levels, urban development is still poorly monitored globally, and reliable and comparable satellite urban data across countries is still limited, slowing down international comparative research. The Copernicus UE programme, replacing the Global Monitoring for Environment and Security (GMES) programme, recognizes the strategic importance of Earth observation for emergency management. Copernicus programme provides global, continuous, autonomous, high quality, wide-area Earth and Atmosphere observation. Copernicus links Space observations to ground-based and atmospheric data collection and processing, providing operational services in the fields of environment, ground infrastructures, civil protection, and security, supporting the implementation of a large number of sectorial and transversal public policies. Of the six thematic macro-areas of the present programme observation and ground monitoring of the European urban systems lies in the first thematic area of the land monitoring service. The enormous and continuous data generation from the Copernicus programme is allowing the construction of an accurate and up-to-date database to the state of health of our cities and surrounding environments, providing research materials simply inconceivable only a few decades ago.

F. Cinquepalmi (✉) · G. Piras
Sapienza University of Rome, Rome, Italy
e-mail: federico.cinquepalmi@uniroma1.it

G. Piras
e-mail: giuseppe.piras@uniroma1.it

E. Arbizzani et al. (eds.), *Technological Imagination in the Green and Digital Transition*, The Urban Book Series, https://doi.org/10.1007/978-3-031-29515-7_53

Keywords Sustainable urban management · Earth observation and satellite technologies · Climate change · Urban digital twin models

53.1 Earth Observation: A European Overview

Among innovative technologies applied to the management of the built environment as well as urban and territorial infrastructures, one of the most promising is unquestionably Earth observation. Since the so-called "First Space Race"[1] in the 1960s, satellites have progressively played a fundamental role in better understanding of our planet, its atmosphere, and its terrestrial and marine ecosystems together with human infrastructures. With a clear understanding of the ever-increasing importance of autonomous access to Space, the European Union has launched since 1998[2] its own Earth observation, satellite, and aerospace research programme, essentially related to global observation, involving the most relevant European based entities in order to understand the effects of climate change and thus better respond not only to extreme climatic events, likewise to other events potentially creating risk to the safety of EU citizens, such as forest fires or extensive pollution on land, at sea, and in the sky (soil, water, and air pollution) (European Commission 2010).

The first programme was launched by European Union in order to coordinate Space and Earth observation activities as Global Monitoring for Environment and Security (GMES) focused on ensuring the implementation of timely, reliable, and continuous information services, for the support of institutional and private decision-makers in global and regional matters related to environment and security (European Commission 2010). The programme based its activities on the analysis of Earth observation data, provided both by satellites and in situ monitoring networks, to be analysed, coordinated and made available to end-users, namely national, regional, and local authorities and agencies, environmental and civil protection organizations and the scientific community (Cinquepalmi 2021).

Certain European member States had already relevant experience in Space related activities and research, with dedicated Agencies and large scientific and Industrial Communities, but for those countries without real Space capabilities, the Commission

[1] The *Space Race* was during the second half of the twentieth century, a substantial part of the scientific and technological but also communication and ideological rivalry between the United States and the Soviet Union through the Cold War, aimed to achieve superiority on spaceflight capability. The advantage deriving from the development of this technology was considered crucial for military uses but soon demonstrated its fundamental importance also for the civil sector.

[2] The relevance of Earth observation within the European Union was recognized with the "Baveno Manifesto", a scientist and officials declaration laying the foundations of the EU Programme "Global Monitoring for Environment and Security (GMES), and committing Governments, Agencies, and the scientific community to the deployment of a shared Space programme, including environmental and security services towards satellites component (Joint Research Centre (JRC) 1998) (Cinquepalmi 2019b).

had to appeal to the principle of subsidiarity,[3] in order to fulfil the mandate of granting to European citizens the benefits of the newly established Earth observation system.

The initial GMES programme comprised an environmental and safety information services, combining the Member States' capacity in data collection on land, sea, and atmosphere (Cinquepalmi 2019a). Since 2014 European Commission, launched the Copernicus programme, established with the new EU Regulation, repealing Regulation (EU) No 911/2010 as an evolution of the previous GMES programme (European Commission 2010), and dedicated to the Polish mathematician and astronomer Mikołaj Kopernik.[4] Among the objectives of the Copernicus programme, the most relevant one is building up knowledge about the planet's state of health, needed for the implementation and support of several European and National key public policies (European Commission 2014).

53.2 Monitoring Urban Areas from the Sky: The Demographic Challenge

Among the monitoring and data collection activities of the EU Copernicus programme, perhaps one of the most relevant is the initiative dedicated to urban areas. Urban policies are presently considered among the most challenging sectors of public policies, having in mind that: *...according to the United Nations (World urbanization prospects—2014), approximately two-thirds of the world's population will be living in an urban area by 2050* (Kotzeva 2016).

This trend of global change in the analysis of urban systems is resulting in more evident in Africa and Asia, also considering the overall demographic decline corresponding to the so-called mature economies, namely Europe, North America, and Japan towards developing and emerging economies, mostly facing the challenge of the Ageing society (European Commission 2009). According to an evaluation made by the United Nations, around the end of the twentieth century, the cities up to one million inhabitants were 371, with a perspective growth up to 548 by 2018, the same evaluation states that this number will increase reaching 706 cities of one million inhabitants or more before 2030. Urban areas with more than 10 million inhabitants, the so-called Megacities (Fig. 53.1), were 33 in 2018, potentially reaching the number of 43 by 2030 (United Nations Development Programme (UNDP) 2009).

The largest world megalopolis is presently Tokyo with 37 million inhabitants, followed by New Delhi with 29 million, Shanghai with 26 million, and São Paulo

[3] Subsidiarity is a general principle of European Union law. According to this principle, the EU may intervene where the action of individual countries is evaluated as insufficient. The principle was established by the Treaty of Maastricht in 1992.

[4] Mikołaj Kopernik (1473–1543 in Latin transliteration Nicolaus Copernicus) was an astronomer and mathematician who graduated in canon law from the University of Ferrara in 1503, he is famous for having advocated, defended, and finally definitively promoted the heliocentric evidence against the hitherto supported geocentrism in the western world. His scientific demonstration is also commonly known as the "Copernican Revolution".

THE WORLD'S TEN LARGEST CITIES IN 2018 AND 2030				
City size		Population in 2018		Population in 2030
rank	city	(thousands)	city	(thousands)
1	Tokyo, Japan	37 469	Delhi, India	38 939
2	Delhi, India	28 514	Tokyo, Japan	36 574
3	Shanghai, China	25 582	Shanghai, China	32 869
4	São Paulo, Brazil	21 650	Dhaka, Bangladesh	28 076
5	Ciudad de México (México City) México	21 581	Al Qahirah (Cairo), Egype	25 517
6	Al Qahirah (Cairo), Egype	20 076	Mumbai (Bombay), India	24 572
7	Mumbai (Bombay), India	19 980	Beijing, China	24 282
8	Beijing, China	19 618	Ciudad de México (México City) México	24 111
9	Dhaka, Bangladesh	19 578	São Paulo, Brazil	23 824
10	Kimki M.M.A. (Osaka), Japan	19 281	Kinshasa, Democratic Republic of the	21 914

Fig. 53.1 Demographic evolution of megacities between 2018/2030. *Source* Cinquepalmi (2021)

and Mexico City with around 22 million each. Very close to 20 million inhabitants are Cairo, Mumbai, Beijing, and Dhaka. Kinshasa, the capital of the Democratic Republic of Congo, and Dhaka, the capital of Bangladesh is assumed to reach, respectively, 22 million inhabitants and 28 million inhabitants by 2030 (Cinquepalmi 2021).

53.3 Monitoring Urban Areas from the Sky: Copernicus and the Urban Atlas

Six priority thematic areas are identified within the Copernicus EU programme for Earth observation, namely soil (land monitoring service), sea (marine environment monitoring service), atmosphere (atmosphere monitoring service), climate change (climate change service), emergency management (emergency management service), and security (security service): urban systems are analysed as part of the land monitoring service thematic area.[5] The other Copernicus products in the land monitoring service area are very high-resolution mosaics; Riparian Zones; Urban Atlas EU-Hydro (Cinquepalmi 2019a).

One of the most interesting outcomes of the programme is presently the Urban Atlas,[6] a mapping of the Cities of Europe related to 513 million EU inhabitants, aimed to help decision-makers in their daily activities on public duty. The idea behind the Urban Atlas was to create a land monitoring service focused on defined "Hot Spots", defined by a harmonizing approach between land cover and land use maps, covering hundreds of cities of the pan-European area corresponding to the 46 countries presently part of the Council of Europe[7] (Fig. 53.2).

The Urban Atlas is monitoring urban areas essentially using land monitoring services data, applying to a number of uses such as air quality monitoring as well as the prior and postconditions of buildings and infrastructure assessment of after extreme climatic events, fires, or earthquakes. The entire system built on the combination between visual interpretation of very high-resolution (VHR) satellite images and (statistical) image classification and based on unit of analysis defined as Functional Urban Areas (FUA), a methodology developed by the OECD and primarily based on the economic functionality of the cities. The methodology proposed together by the OECD and the European Commission is based on a harmonized definition related to a reference grid of 1 square kilometre, which identifies urban areas as functional economic units (Functional Urban Areas—FUA). Using population density[8] and labour mobility flows as key information, urban areas can be described as densely populated "urban centres" and "hinterlands" whose labour markets are highly integrated with the cores (OECD 2012).

[5] The Copernicus products in the land monitoring service area are very high-resolution mosaics (general soil mapping) Corine Land Cover (CLC, which provides biophysical characteristics on land), Riparian Zones (which monitors long transition zones rivers and lakes), Natura 2000 (n2k that monitors the habitats protected from the naturalistic point of view), Urban Atlas (first service to create harmonized maps for land cover and land use on several hundred cities both in the EU that in the countries of the European Free Trade Association, namely Norway, Iceland, Liechtenstein, and Switzerland, EU Digital Elevation Model (EU-Dem, which combines data from different sources into a single system of data management.

[6] The Urban Atlas is a joint initiative of the EC Directorate-General for Regional and Urban Policy and the EC Directorate-General for Enterprise and Industry, in the frame of the EU Copernicus programme with the support of the European Space Agency and the European Environment Agency https://land.copernicus.eu/local/urban-atlas.

[7] The Council of Europe (CoE) is a multilateral organization founded on 5 May 1949, with the Treaty of London aimed to promote democracy, human rights, European cultural identity, and the possible way forward to social problems in Europe, presently composed by 46-Member States and based in Strasbourg, France. On 17 October 1989, the CoE was granted observer status at the United Nations General Assembly.

[8] The source of the population grid data for European countries is population density data disaggregated with the Corine Land Cover dataset, produced by the Joint Research Centre for the European Environmental Agency (EEA) (Joint Research Centre (JRC) 1998).

	Urban Atlas	2006	2012
Legend Code	Nomenclature		
11100	Continuous Urban Fabric (S.L.>80%)		
11210	Dicontinuous Dense Urban Fabric (S.L.: 50% - 80%)		
11220	Dicontinuous Medium Density Urban Fabric (S.L.: 30% - 50%)		
11230	Dicontinuous Low Density Urban Fabric (S.L.: 10% - 30%)		
11240	Dicontinuous Very Low Density Urban Fabric (S.L.: <10%)		
11300	Isolated Structures		
12100	Industrial, commercial, public, military and private units		
12210	Fast transit roads and associated land		
12220	Other roads and associated land		
12230	Railways and associated land		
12300	Port areas		
12400	Airports		
13100	Mineral extraction and dump sites		
13300	Construction sites		
13400	Land without current use		
14100	Green urban areas		
14200	Sports and leisure facilities		
20000	Agricultural + Semi-natural areas + Wetlands		
21000	Arable land (annual crops)		
22000	Permanent crops		
23000	Pastures		
24000	Complex and mixed cultivation patterns		
25000	Orchards		
31000	Forests		
32000	Herbaceous vegetarian associations		
33000	Open spaces with little of no vegetations		
40000	Wetlands		
50000	Water bodies		

Fig. 53.2 Urban Atlas—classes of surface analysis as identified within the Urban Atlas project and descriptive table of the evolution of the project. *Source* EU Copernicus programme, EC Commission 2013 (European Commission 2015)

The 17 classes originated by the application of the FUA methodology were defined since 2006 and summed to other 11 classes since 2012 and dedicated to agricultural land use, forestry and other vegetation land use, as well as wetlands and other water bodies, identified with unique codes and appropriate nomenclature as shown in Table 53.1 with a total number of 305 FUAs analysed in 2006, increased to 695 in 2012, passing from 640.500 km^2 in 2006 to 1.015.600 km^2 in 2012 as in Table 53.1.

Table 53.1 Urban Atlas—comparative table of the territorial coverage and description between 2006 and 2012 (EU Copernicus programme, EC Commission 2014 (Commission 2014))

Characteristic	Urban Atlas 2006	Urban Atlas 2012
No. of FUAs	305	695
Total area	640,500 km^2	1,015.600 km^2
Minimum mapping unit (MMU)	0.25 ha in urban areas, 1 ha in rural areas	0.25 ha in urban areas, 1 ha in rural areas
Min. mapping width	10 m	10 m
MMU change layer		Class 1 to class 1 = 0.1 ha Class 2–5 to class 1 = 0.1 ha Class 2–5 to class 2–5 = 0.25 ha Class 1 to class 2–5 = 0.25 ha
Positional accuracy	± 5 m	± 5 m
No. of classes	21	27
Min. overall accuracy for "artificial surfaces" classes	85%	85%
Min. overall accuracy (all classes)	80%	80%

53.4 Monitoring Urban Areas from the Sky: The Climatic Challenge

Without taking a side in the debate about global changes, we can certainly agree that the planet's climate is changing and that urban areas are seriously endangered by this phenomenon. The broader and shared position of the global scientific community foresees in the course of the twenty-first century a rise in average global temperatures of approximately 2/2 degrees Celsius, which will result in unprecedented climatic and ecosystem changes, at least compared to the last 10 centuries (The Royal society 2008).

Climate worsening will become a dominant feature across the planet, with extreme climatic events, such as heatwaves, storms, and floods, increasing in frequency and severity (The Royal society 2008). Numerous initiatives have been launched to address the issue such as International Human Dimensions Programme on Global Environmental Change (IHDP) Urbanisation Science Project, *Diversitas* Science Plan on Urbanisation, International Union for the Scientific Study of Population (IUSSP) Urbanisations and Health Working Group, U.S. National Academies' Panel on Urban Population Dynamics, U.S. National Academies' Roundtable on Science and Technology for Sustainability's Task Force on Rapid Urbanisation, UNESCO's initiative on Urban Biospheres. What most scientists and researchers predict is that higher temperatures will cause melting of glaciers and polar ice caps, with sea-level rise and serious consequences over low-lying territories across the planet, affecting especially the most fragile populations and territories along the coastlines

(Cinquepalmi 2021). The most relevant impacts to be considered towards cities in relation with extreme climatic events and connected with climate change are resumed in Fig. 53.3.

The satellite data of most immediate use for the management of urban systems are those resulting from thermic sensors of the atmosphere. The variations of the local temperature inside densely built urban centres, both caused by the human activities (air conditioning and heating systems at full blast, high concentrations of car traffic) or and by climatic large and local scale effects, are the cause of substantial disturbance to the citizens quality of life, resulting in the phenomenon of the Urban Heat Island (UHI) (Solecki et al. 2005).

Climatic phenomena	Main impacts expected on urban areas
INCREASE IN DAYTIME AND NIGHT-TIME TEMPERATURES	- Reduction of energy demand for heating - Increased energy demand for cooling - Decrease in air quality - Impact on winter tourism
INCREASE IN HEAT WAVES	- Reduction in the quality of life for people in warm areas
INCREASE IN HEAVY RAINFALL EVENTS	- Disruption of settlements, trade and transport due to flooding - Impact on urban infrastructure and real estate
INCREASE IN DROUGHT-AFFECTED AREAS	- Water shortages for inhabitants, industries and services - Reduction in energy production potential, especially hydroelectricity, but also nuclear and fossil fuels - Potential increase in migration phenomena
INCREASE IN TROPICAL STORMS	- Damage due to flooding and high winds - Disruption of public water supply - Potential increase in migration phenomena
SEA LEVEL RISE	- Flooding in coastal areas - Decrease in freshwater availability and saltwater intrusion - Potential increase of migration phenomena

Fig. 53.3 Expected impacts on urban areas from extreme weather and climate conditions. *Source* Revi et al. (2014)

The difference in temperature in such cases is substantially different between urban city centres and their surroundings, even by a few degrees Celsius, but above all remains significantly higher during the night than during the day. Such negative effect is amplified and worsened in case of absence of wind and periods of higher or lower temperature, i.e. in the middle of summer or winter. The association of the urban heat island with summer heatwaves results in very serious effects on human health, especially on the most fragile population groups and within economically deprived contexts, which are usually more frequent at those latitudes where such combination of factors is worsened by poor air quality due to the pollution, severe drought or extreme humidity (The Royal society 2008).

Air pollution at local and regional scales is another element detected by satellite sensors, helping scientists and policy makers to assess and possibly mitigate the phenomenon: ground temperatures together with winds and all other weather conditions help to understand the health condition of a city, combined with satellite sensors detecting nitrogen dioxide (NO_2) as shown in Fig. 53.4. Sensors dedicated to telemetry (remote sensing[9]) allows the acquisition of information in the wavelengths of visible light but also in the electromagnetic spectrum not visible to the human eye (microwaves, ultraviolet, X-rays, and gamma rays), capturing images even in cloudy weather conditions, and detecting even millimetric variation on land and water surfaces as well as on buildings, generating maps to be used by city administrators and civil protection as shown in Fig. 53.5 (Cinquepalmi 2019a).

53.5 Conclusions: Moving Forward Urban Digital Twins

The increasing complexity of the built environment and more generally of urban systems requires the adoption of an interdisciplinary approach in order to relate the quality of structures with the environmental quality and with the quality of services that all together must serve to ensure the citizen's quality of life. The exponential growth of digital computing capabilities since 1972, together with the advent of the most advanced technologies for digital data management and the diffusion of a digital culture of information has also led to the digital evolution of Life Cycle Assessment (LCA) methodologies, completing its functionality with two other enabling tools, one at the scale of the individual building, namely the Building Information Modelling (BIM) and the other at territorial scale, namely the Geographic Information System (GIS); the fourth enabling technology completing this technological framework is the Earth observation technology as well implemented by the Copernicus programme (Agostinelli et al. 2019).

[9] Remote sensing is a technique allowing the observation and study of objects located even at a great distance from the observer by using electromagnetic waves (emitted, reflected, or transmitted) thanks to optic-electronic instruments (sensors) installed on planes or satellites. Through the capture of digital tele-detected images, it is possible to observe, analyse, and study with incredible precision very large areas of the Earth's surface (Copernicus programme® 2016).

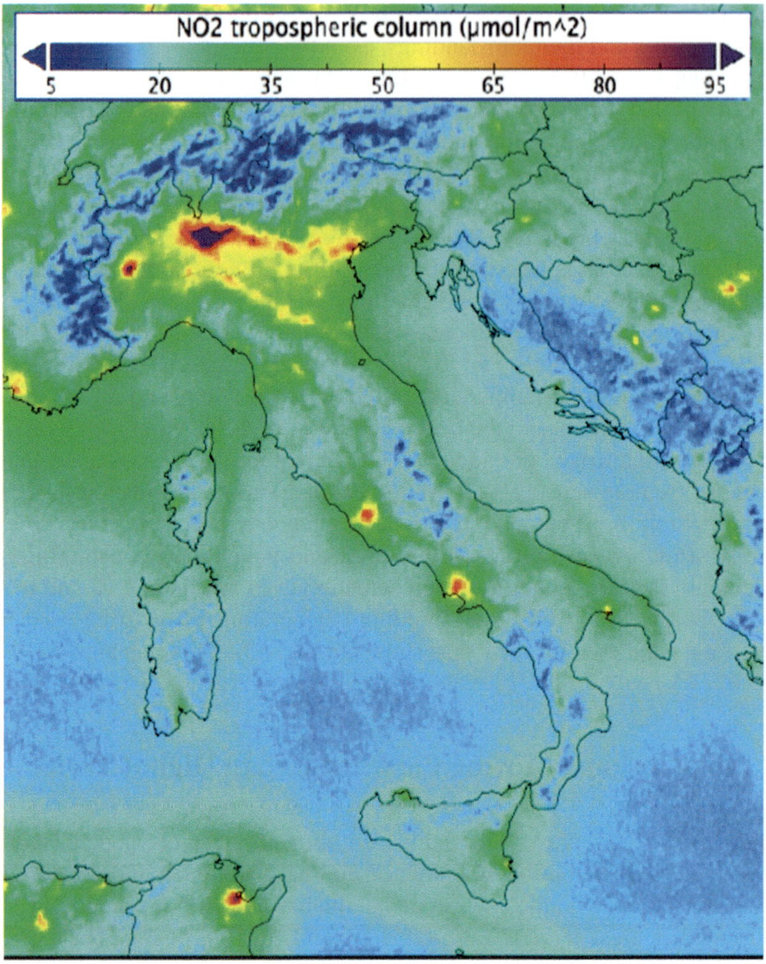

Fig. 53.4 Nitrogen dioxide over Europe evaluated in March 2019, with detail on Italy. The image shows the critical situation on the Pianura Padana as well as over Rome and Naples. Nitrogen dioxide pollutes the air mainly as a result of traffic and the combustion of fossil fuel in industrial processes. It has a significant impact on human health, contributing particularly to respiratory problems. [Contains modified Copernicus Sentinel data (©EU 2019), processed by ESA, CC BY-SA 3.0 IGO. *Source* Cinquepalmi (2019c)

The composite interaction of all these digital technologies makes available a comprehensive picture of built environments and infrastructures, but the evolutionary leap we are currently witnessing moves this complex interactive system of information towards a more dynamic scenario, leading to the creation of the Digital Twins[10]

[10] The so-called Urban Digital Twin Models are definable as the transposition of a real object into a virtual/digital representation, for the evaluation of its functionality and performance. (Ageing Report: Economic and budgetary projections for the EU-27 Member States 2009).

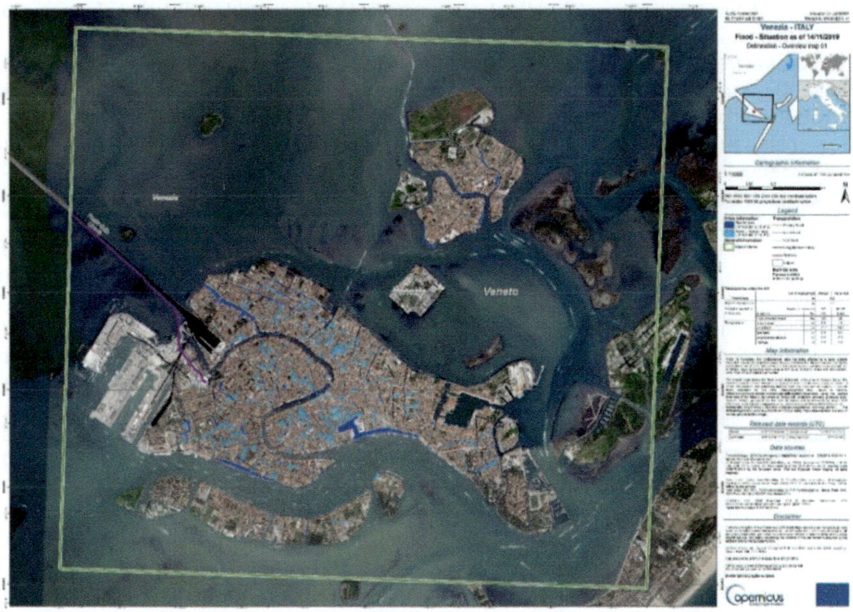

Fig. 53.5 High-resolution image of the flooding that occurred in Venice on 19th November 2019. The different levels of high tide flooding were scanned in real time by altimetry sensors. The restitution of the image represents the flooding with different

(DT) for built environment. Urban DT is designed as three-dimensional databases that are continuously fed with real time data representing both the building components as well as their qualities and other useful information, with the aim of giving a complete picture of the state-of-the-art and in the meantime simulating activities and management of processes of operation (Weekes 2019). The concept of DT implemented at the scale of urban environments is leading to the concept of City Digital Twin (CDT), a virtual model that takes the analysis from the single building (or even single housing unit) to a higher and more complex level, requiring the highest level of interdisciplinary integration to be jointly defined in the near future.

References

Agostinelli S, Cinquepalmi F, Ruperto F (2019) 5D BIM: tools and methods for digital project construction management. In: Building information modelling (BIM) in design, construction and operations III

Cinquepalmi F (2019a) The Copernicus program: Europe's eye on urban areas. Dwelling on Earth 2(51), Gangemi, International

Cinquepalmi F (2019b) Regulation (EU) no 377/2014 Establishing the Copernicus programme: from the "Baveno manifesto" to the present EU legislative framework for land and urban observation.

Rivista della Cooperazione Giuridica Internazionale, Quadrimestrale dell'Istituto Internazionale di Studi Giuridici, ANNO XXII—N. 63—Settembre—Dicembre, Aracne editore, Roma
Cinquepalmi F (2019c) Regulation (EU) no 377/2014 establishing the Copernicus programme: from the "Baveno manifesto" to the present EU legislative framework for land and urban observation. Rivista della cooperazione giuridica internazionale, pp 121–126
Cinquepalmi F (2021) Towards (r)evolving cities: urban fragilities and prospects in the 21st century. Didapress, Firenze
Commission E (2015) Copernicus Europe's eyes on Earth. Publications Office of the European Union, Luxembourg
European Commission (2014) Regulation (EU) No 377/2014 of the European Parliament and of the Council of 3 April 2014 establishing the Copernicus Programme and repealing Regulation (EU) No 911/2010. Text with EEA relevance. Official J Eur Union, Brussels
European Commission (2009) Ageing Report: Economic and budgetary projections for the EU-27 Member States (2008—2060). In: European economy, and demography report 2008: meeting social needs in an ageing society (SEC(2008)2911) https://ec.europa.eu/economy_finance/publications/pages/publication14992_en.pdf
European Commission (2010) Regulation (EU) no 911/2010 of the European Parliament and of the Council of 22 September 2010 on the European earth monitoring program (GMES) and its initial operations (2011 to 2013). Official J Eur Union, Brussels
Joint Research Centre (JRC) (1998) The Baveno initiative, draft discussion paper, an outline of structure and points for discussion. Prepared by the JRC for the third Baveno group meeting, ISPRA (VA)
Kotzeva M (ed) et al (2016) Europe statistics on cities, towns and suburbs. Eurostat, European Union Publications, Luxembourg. https://ec.europa.eu/eurostat/documents/3217494/7596823/KS-01-16-691-EN-N.pdf
OECD (2012) Redefining 'urban': a new way to measure metropolitan areas. OECD Publishing
Revi A et al (2014) Urban areas. In: Climate change 2014: impacts, adaptation, and vulnerability. Part A: global and sectoral aspects. Contribution of working group II to the fifth assessment report of the intergovernmental panel on climate change, Cambridge University Press, Cambridge
Solecki WD et al (2005) Mitigation of the heat island effect in urban New Jersey. Glob Environ Change Part B: Environ Hazards 6(1)
The Royal society (2008) Climate change controversies. A simple guide. London
United Nations Development Programme (UNDP) (2009) Human Development Report 2009. Overcoming barriers: human mobility and development
Weekes S (2019) The rise of digital twins in smart cities. SmartCitiesWorld, www.smartcitiesworld.net1

Chapter 54
A Systematic Catalogue of Design Solutions for the Regeneration of Urban Environment Contrasting the Climate Change Impact

Roberto Bologna and Giulio Hasanaj

Abstract The article illustrates a research for the definition of a catalogue of design solutions for climate change adaptation in the process of urban regeneration, reducing the vulnerability to climate change impacts and increasing the city resilience. Based on the analysis of relevant case studies of architectural and urban projects in the main biogeographical regions of Europe, the paper describes the research methodology applied for the construction of a catalogue of spatial and technological adaptive design models mainly focusing on the category of "nature-based solutions" but also considering "artificial solutions". In order to assess their effectiveness, different design alternatives are tested in a specific urban contest (a school courtyard in the City of Scandicci–Metropolitan City of Florence) prone to climate hazards of urban heat islands and pluvial flooding, simulating the impact on the more vulnerable user (children between 11 and 14 years old). For an adequate performance evaluation of multi-hazard effectiveness of the different adaptive design solutions, appropriate IT software and procedural models have been applied: ENVI-met microclimatic simulation software for thermal analysis and predictive method for hydraulic assessment. By comparing the results before and after the application, the climate-adaptive performance of alternative design solutions is measured through specific indicators. This approach is coherent to the design process management aiming to a predictive definition of performance evaluation through procedural models and digital instruments in order to properly address the complexity of architectural and urban project. The systematic catalogue of adaptive design solution offers useful tools and methods to designers and decision makers for the construction of climate change adaptation and mitigation plans in order to build a healthy and safe urban environment for citizens and drive an ecological and sustainable transition to green cities.

Keywords Systematic catalogue · Adaptive design solutions · Climate change · Urban regeneration · Resilience

---

R. Bologna · G. Hasanaj (✉)
University of Florence, Florence, Italy
e-mail: giulio.hasanaj@unifi.it

R. Bologna
e-mail: roberto.bologna@unifi.it

E. Arbizzani et al. (eds.), *Technological Imagination in the Green and Digital Transition*, The Urban Book Series, https://doi.org/10.1007/978-3-031-29515-7_54

54.1 Introduction

In the next years, the climate change will continue to affect urban settlements with increasing violence and worsening impacts in terms of heat waves, heavy rainfall, drought and windstorms (IPCC 2021). In this scenario, cities represent the main human habitat, where most of the world's population lives, and the place of concentration of those impacts due to the presence of sensitive elements (people, buildings and infrastructures) that exacerbate the climate risks (UN-Climate Change 2020).

The scientific literature–especially those on climate science, environmental design and urban ecology–highlights the opportunity to develop the urban design in terms of resilience and adaptation to the effects of climate change. Important international experiences demonstrate the added value of the adaptive design approach, which combines urban regeneration with the aim of reducing climate risks. The adaptation plans developed in the cities of Barcelona, Copenhagen, New York and Rotterdam have demonstrate the importance of environmental design as a driver for reducing vulnerability and enhancing the built environment (Losasso et al. 2020). In addition, the new European Climate Change Adaptation Strategy (EC 2020) has given to adaptive design a central role in reducing the impacts of climate change and emphasized the need to: (i) develop site-specific studies in various urban settlements to define adequate regeneration strategies to different climatic, environmental and socio-economic characteristics; (ii) identify appropriate tools and methods to monitor the performance of adaptive solutions. The challenge focus on preparing cities and public buildings such as hospitals, schools, universities, housing and urban spaces to cope with current and future extreme events (EEA 2020; Tucci 2019).

The paper describes the research concerning the climate change adaptation and in particular focuses on project interventions (see Sect. 54.2) and design solutions (see Sect. 54.3) for increasing resilience in the urban environment, with the aim of cities adaptation.

The importance of the topic is widely recognized by international agreements, frameworks and policies that include in their priorities the development of actions to tackle climate change impacts affecting the urban environment (UNFCCC 2015; UN General Assembly 2015).

The research was developed within the PhD programme of Architectural Technology at the Department of Architecture (DIDA) of the University of Florence and find its roots in the national research PRIN2015 "Adaptive design and technological innovations for the resilient regeneration of urban districts in climate change regime", which the authors have participated in. While the PRIN2015 research has defined strategic orientations and design principles to cope with the climate change adaptation on vulnerable urban districts in order to demonstrate the reduction of exposure to extreme climate events, the PhD research has developed an analytic catalogue of adaptive design solutions and evaluated their effectiveness to tackle the extreme climate events in public spaces.

54.2 Case Study Analysis of Climate Change Adaptation Projects

The research starts from the analysis of case studies of adaptation projects based on recognized international best practises in the field of climate change adaptation. The research has focused on European continent in order to: (1) provide a comparable framework of interventions; (2) analyse the most frequent climatic hazards in the different biogeographical regions; (3) study the interventions realized in urban contexts similar to those of Italian cities.

From the analysis, it has been possible to identify 14 case studies concerning: three urban parks, three climate neighbourhoods, three projects in public spaces and five projects on buildings. The interventions are located in 11 European cities, within nine different countries and three different biogeographical regions (Atlantic, Continental and Mediterranean) (Fig. 54.1). The belonging of a specific geographical region has allowed to identify various climatic hazards and determine different strategies and design solutions not always practicable in other climatic areas. The project interventions were analysed at different scales, from the urban district to the building, and through a critical reading that has highlighted technical and procedural design solutions characterized by a holistic and ecosystemic approaches.

The research has identified six projects located in central urban areas and eight in peripheries, underlining a wide diffusion of adaptive interventions in the suburbs. This outcome represents a good practise in the interventions of urban renovation of marginal and/or degraded areas because it permits to combine climatic targets with

Fig. 54.1 Location of case studies of adaptation project in the European continent. *Source* Author's elaboration

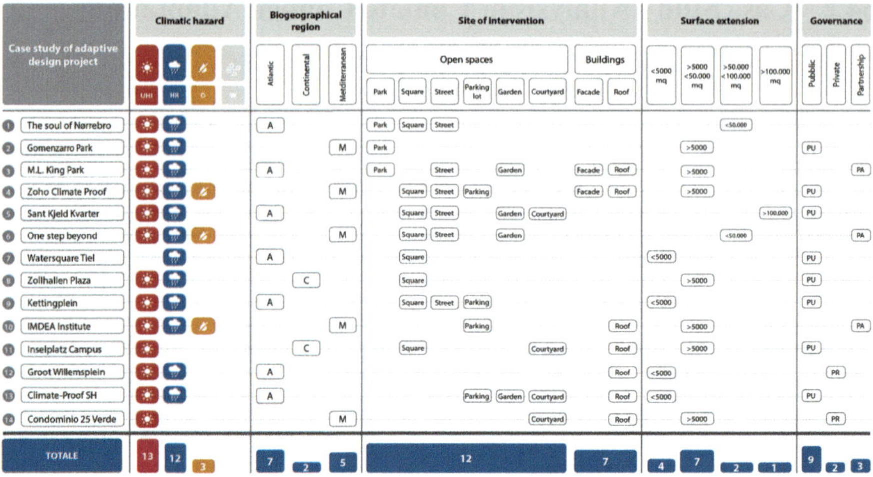

Fig. 54.2 Results of the case study analysis of climate adaptation interventions in European continent. *Source* Author's elaboration

economic, social and environmental objectives, providing benefits for the context where the project is located.

The results of the first part of research have shown that the most faced climatic hazards are urban heat islands (UHI; 13 cases out of 14) and pluvial flooding (PF; 12/14). Furthermore, it has observed that the analysed projects in most of cases (11/14) work to counteract both the hazards. Within this framework, the interventions on the climate-adaptive district of Zoho in Rotterdam, the redevelopment project "One step beyond" on the urban axes of Athens, and the Imdea Institute building in Madrid are exemplary cases because they attempt to respond also to drought issues (Fig. 54.2).

The adaptive design projects have found a wider application in the open spaces rather than in the buildings, and it has been interesting to note that most of the projects involve more than one element of urban space (Fig. 54.2). This result emphasizes two fundamental key parameters for the study of climate adaptation in urban contexts: on the one hand, the surface extension, and on the other, the systematic approach of the adaptive design interventions, that has confirmed the importance of the infrastructural actions, especially those with the use of nature-based solution such as green infrastructures.

Among the 14 case studies, nine projects were developed by public bodies, two by private companies and three by partnerships (Fig. 54.2). This result highlights the relevance of public governance in urban adaptation interventions. However, to achieve an effective level of climate resilience, the interventions in public space should be accompanied by incentive actions on private spaces.

54.3 Systematic Catalogue of Design Solutions for the Regeneration of Urban Environment Contrasting Climate Change Impacts

The analysis of the case studies led to identified 21 technological and spatial design solutions divided into two main categories, the "nature-based solutions" (NBS) and the "artificial solutions" (Fig. 54.3). The NBS include solutions "inspired and supported by nature, which are cost-effective, provide simultaneous environmental, social, and economic benefits, and help build resilience" (EC 2016). These solutions are excellent alternatives for restoring natural ecosystems and promoting sustainable development (Faivre et al. 2017; Mussinelli et al. 2018). Artificial solutions, on the opposite, are technical elements, coming from industrial processes (e.g., cool materials or permeable pavements) or architectural construction, characterized by a high level of engineering (e.g., urban floodable basins, urban canopies, water mirrors and underground water-collection tanks).

The 11 NBS identified include: trees, natural floodable basins, wetlands (ponds), bioswales, green roofs, vegetated ditches, urban farms, green walls, rain gardens, permeable green surfaces and drainage trenches. The artificial solutions collected are 10, among which we find: underground water-drainage boxes, underground water-collection tanks, cool materials, urban canopies, fountains, permeable pavements, water flow regulators, overground water-collection tanks, water mirrors and urban floodable basins.

The 21 design alternatives identified were further described in detail, forming a catalogue of technological and spatial solutions to address the climate adaptation

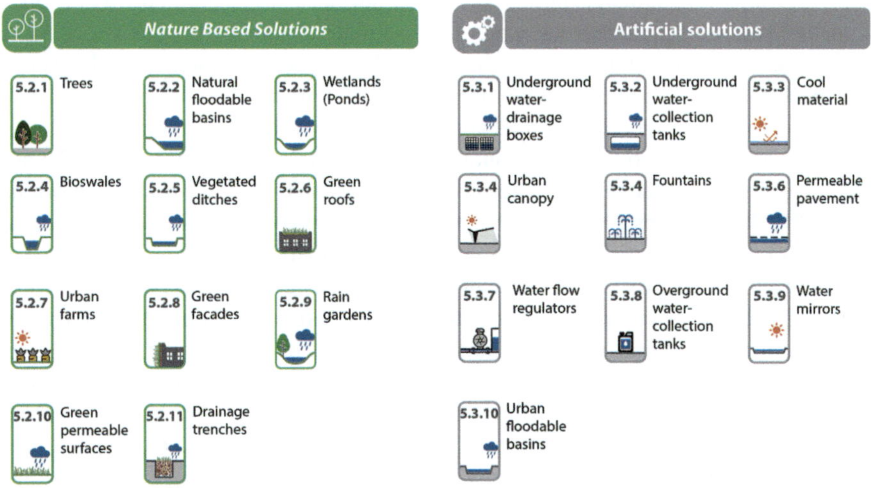

Fig. 54.3 Panel of adaptive design solutions identified from the case study analysis. *Source* Author's elaboration

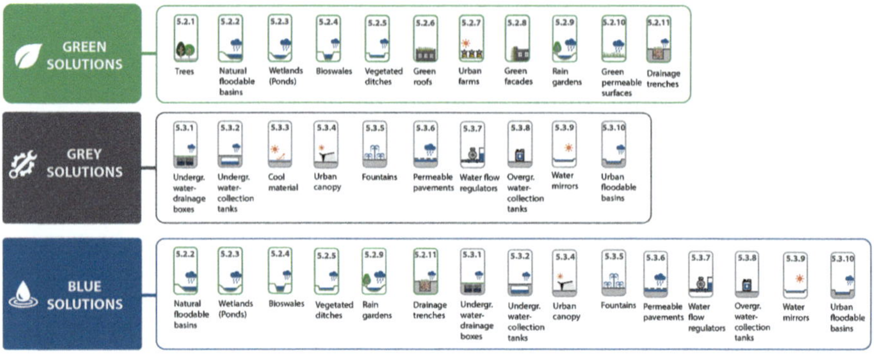

Fig. 54.4 Classification of the adaptive design solutions according to typology. *Source* Author's elaboration

project. From the creation of the catalogue, it has been possible to organize the different design solutions according to typology, category of intervention, site of intervention, contrasted hazards, working mechanisms and performance indicators.

The typology arranges the solutions into three groups: green, grey and blue solutions (Fig. 54.4). This further classification stems from the need to identify an appropriate group for design solutions, both natural and artificial, that have demonstrated a marked aptitude to tackle pluvial flooding, the blue solutions. Instead, green solutions are characterized by natural elements while grey solutions mainly include artificial components.

The category of intervention defines the correlation between the design solution and the spatial model of on site implementation, allowing to organize the solutions into 4 groups of elements: punctual, linear, surfaced and volumetric (Fig. 54.5). Punctual solutions include trees, water flow regulators, and overground water collection tanks. Linear solutions contain bioswales, vegetated ditches and drainage trenches. Surface solutions include urban canopies, green roofs, urban farms, green walls, green surfaces, cool materials and permeable pavements. Volumetric solutions contain natural floodable basins, ponds, wetlands, rain gardens, underground water-drainage boxes, underground water-collection tanks, fountains, water mirrors and urban floodable basins.

The site of intervention identifies elements of urban open spaces and parts of the building where adaptive solutions can be integrated (Fig. 54.6). The first includes parks, gardens, squares, parking lots, streets and courtyards; the latter refers mainly to the building envelope, including facades and roofs. It is interesting to observe that the most integrable solutions are trees, permeable green surfaces, cool materials and water flow regulators. On the contrary, the least integrable solution are essentially those with a higher demand of free space such as natural floodable basins, green roofs, and green walls.

The classification according to the climate hazards has shown that 16 solutions are able to respond to UHI, 17 solutions work against PF, four alternatives cope to tackled droughts and two work to counteract windstorms (Fig. 54.7). The result highlight

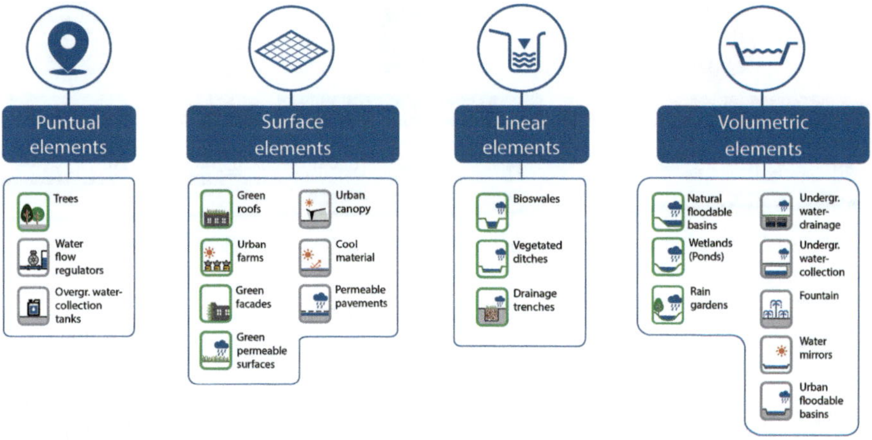

Fig. 54.5 Classification of the adaptive design solutions according to categories of intervention *Source* Author's elaboration

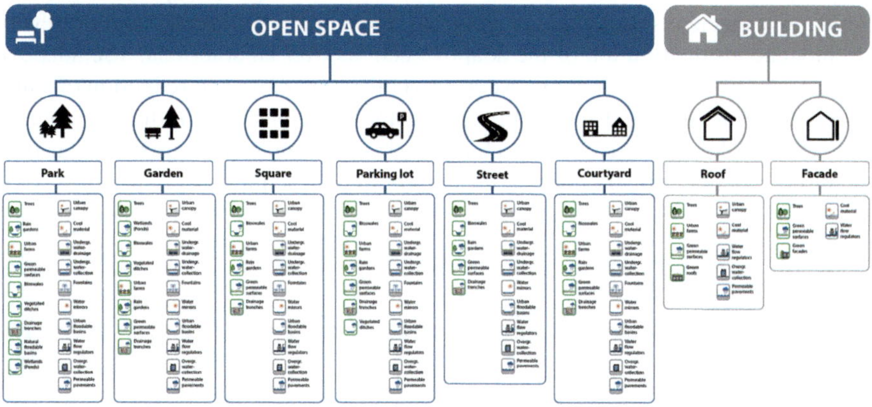

Fig. 54.6 Classification of the adaptive design solutions according to site of intervention. *Source* Author's elaboration

once again that the most addressed climatic hazards refer to thermal and hydraulic stress. Moreover trees, underground water-collection tanks, urban canopies and urban floodable basins are able to respond to three different climatic hazards. Instead, drainage trenches, underground water-drainage boxes, cool materials, fountains and water flow regulators are the unique solutions that address a single climatic hazard.

The analysis on the working principles of the design solutions has allowed to identify the adaptive mechanisms used to counteract extreme events (Fig. 54.8). The results have shown that the adaptive actions employed to counteract extreme temperatures include shading, reflection, absorption and thermal emission. Among these, absorption represents the most frequent principle, while shading, which occurs

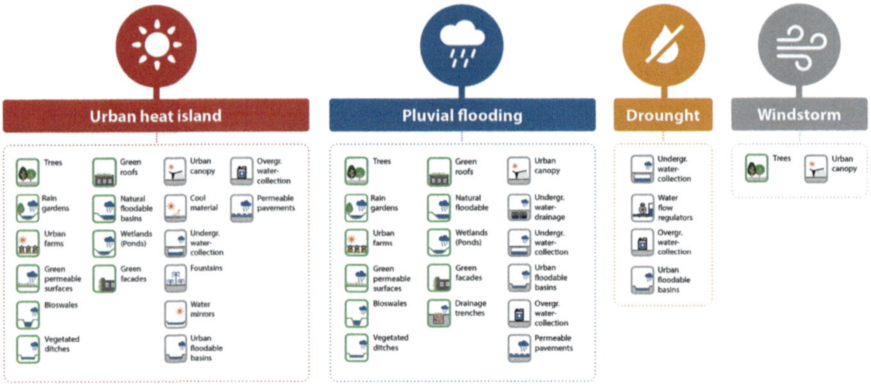

Fig. 54.7 Classification of the adaptive design solutions according to climatic hazards. *Source* Author's elaboration

more rarely and is implemented only by trees and urban canopies, represents a key factor for thermal reduction and increased well-being. Regarding pluvial flooding, the adaptive mechanisms include interception, collection, infiltration and runoff, with a homogeneous distribution of the adaptive actions over all solutions. The adaptive mechanisms of drought phenomena concern collection, reuse and reduction of water resources. While for windstorms, the only adaptive actions concern the windbreaks barrier effect. Adaptive mechanisms for droughts and windstorms are limited and do not allow to define primary and secondary working principles.

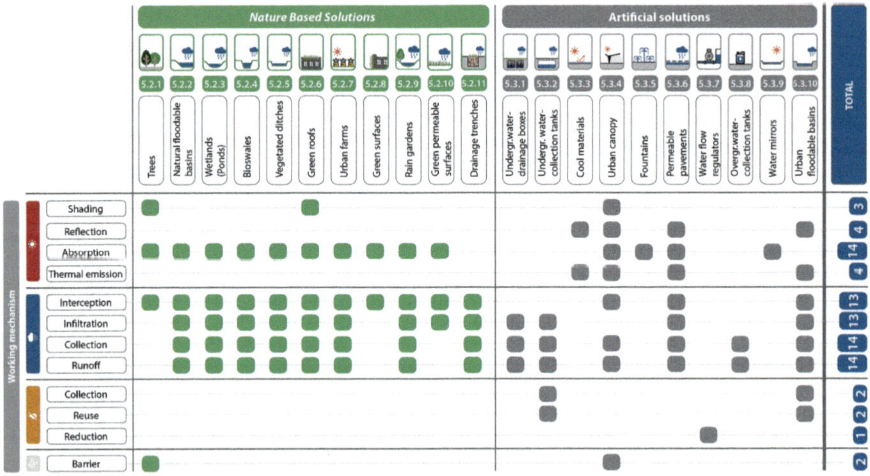

Fig. 54.8 Working principles of the adaptive design solutions according to the different climatic hazards. *Source* Author's elaboration

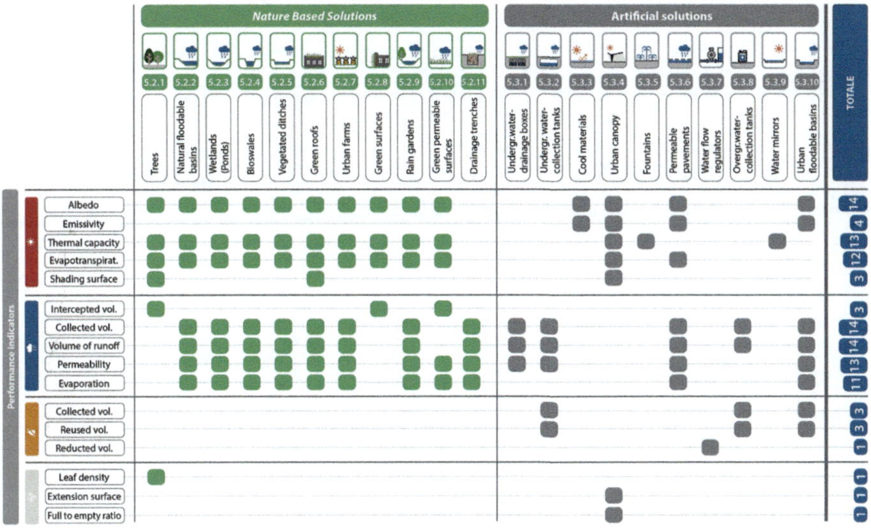

Fig. 54.9 Performance indicators of the adaptive design solutions according to the different climatic hazards. *Source* Author's elaboration

The performance indicators allow, on the one hand, to measure the effectiveness of the solutions, and on the other hand, to identify the key parameters to increase the climate-adaptive performance (Fig. 54.9). Regarding the UHI, the main performance indicators refer to: albedo, evapotranspiration and thermal capacity. In the PF, indicators include intercepted and collected volume of water, runoff, permeability (measurable through infiltration rate) and evaporation. For drought, the indicators refer to the volume of water collected, reused and reduced. While for windstorms include leaf density, extension surface and the full to empty ratio.

Matching all the information provided in the catalogue of design solutions for climate adaptation, it is possible to define many different multidimensional matrices. For example, it has been possible to identify the most appropriate solution for a given urban space and climate hazard (Fig. 54.10).

54.4 Performance Evaluation of Different Adaptive Design Solutions

In the last part of the research, has been measured the effectiveness of four different adaptive design solutions (trees, green surface, permeable pavement cool material and rain garden) from those collected on the catalogue. The aims of this phase have been the following: verify the application of the different technical alternatives identified, measure their effectiveness in terms of reduction of thermal and hydraulic hazards, and define a replicable methodology for evaluating performance effectiveness.

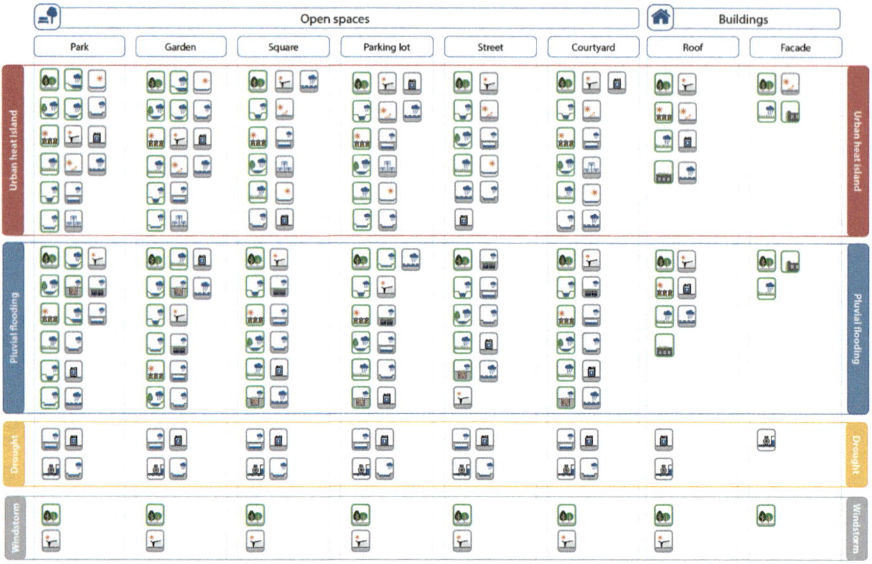

Fig. 54.10 Example of multidimensional matrix for a given urban space and climatic hazard *Source* Author's elaboration

The experimentation has identified as testing area the courtyard of the middle school "E. Fermi" in the Municipality of Scandicci (Metropolitan City of Florence) (Fig. 54.11) prone to climate hazards of UHI and PF, simulating the impact on the more vulnerable user (children between 11 and 14 years old).

54.4.1 Methods and Tools for the Performance Evaluation

The effectiveness of the adaptive solutions has been evaluated using the microclimate simulation software ENVI-met, for thermal analyses, and the predictive method, for hydraulic behaviour. The different solutions have been tested through simulations and measurements "ex ante-ex post" that has allowed to evaluate the effectiveness through comparison of performance indicators. The indicators used for the thermal analyses were potential air temperature, mean radiant temperature (MRT), Predicted Mean Vote (PMV) and Physiological Equivalent Temperature (PET). Among these, only the first one represents a physical indicator, while the others are indicators related to outdoor comfort. For the hydraulic analysis, the indicator used has been the runoff coefficient, which measures the percentage of incident rainwater on a given surface that is neither infiltrated nor evaporated and is transformed into surface water runoff (EEA 2021; Moccia and Sgobbo 2016).

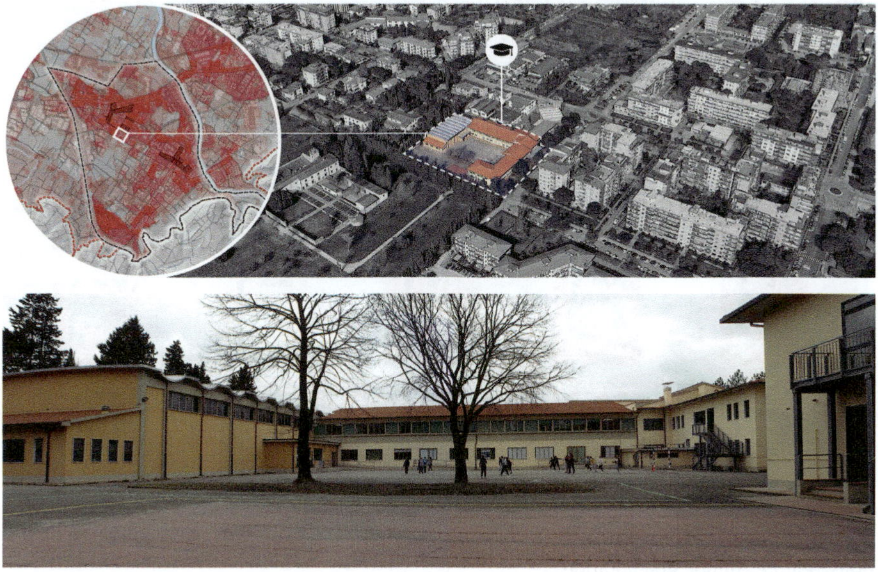

Fig. 54.11 Testing area of the schoolyard located in the Municipality of Scandicci. *Source* Author's elaboration

54.4.2 Results of Climate-Adaptive Performance on Four Design Solutions

The results of the thermal simulations have been extrapolated as graphical elaborations, in form of general plans (Fig. 54.12), and with tables (Fig. 54.13), referred to the centre of the testing area (coordinates $i = 30, j = 33, z = 4$). The outputs have been processed at 2 p.m. to ensure full solar exposure of the experimental area for the day August 1, 2021, corresponding with the presence of a heat wave.

Microclimatic simulations have shown that the major reduction values of the potential air temperature were obtained with the permeable cool material pavement in which is measured a reduction of − 0.46 °C, while the green surface has been the less effective solution, measuring a reduction of − 0.24 °C (Fig. 54.13). These results are justified due to the different amount of incident energy that is reflected from the solutions surfaces (albedo): in fact, for the cool material, the albedo value is 0.80, while for the green surface is 0.20. The results referred to the outdoor comfort indicators (MRT, PMV and PET) have shown similar results in terms of perception of well-being on the vulnerable user. The solution that provides the highest level of comfort is the rain garden, while the less effective is the permeable cool material pavement, which records in the MRT index a worsening value compared to the original state and increase the discomfort perception (Fig. 54.13).

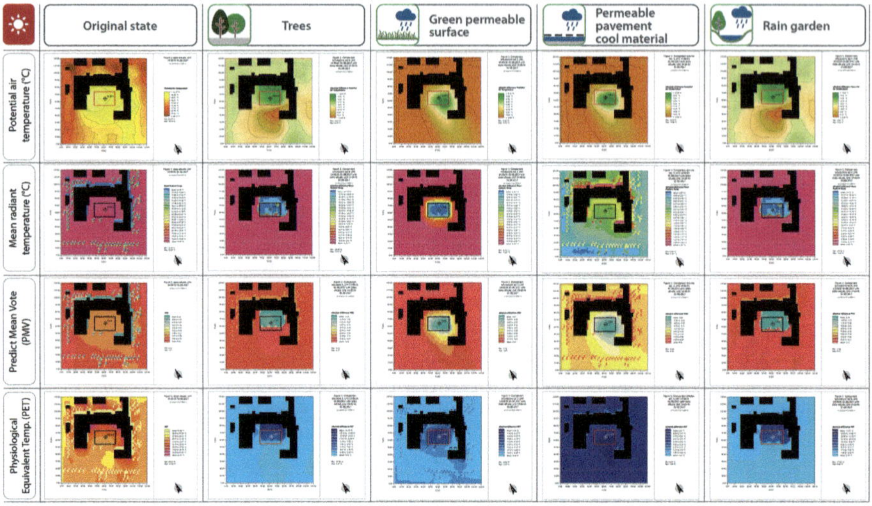

Fig. 54.12 Comparison of the thermal simulations: extrapolations in form of graphical elaborations. *Source* Author's elaboration

	Original state	Trees	Green permeable surface	Permeable pavement cool material	Rain garden
Potential air temperature (°C)	36,27	35,99	36,03	35,81	35,83
		- 0, 28	- 0, 24	- 0, 46	- 0, 44
		good	good	very good	very good
Mean radiant temperature (°C)	74,75	54,32	73,84	75,70	51,18
		- 20,43	- 0,91	+ 0,95	- 23,57
		very good	insufficient	insufficient	very good
Predict Mean Vote (PMV)	5,33	4,21	5,25	5,28	3,98
		- 1,12	- 0,08	- 0,05	-1,35
		very good	insufficient	insufficient	excellent
Physiological Equivalent Temp. (PET)	52,52	44,92	52,23	52,76	43,25
		7,60	0,29	0,24	9,27
		good	insufficient	insufficient	very good

Fig. 54.13 Comparison of the thermal simulations: extrapolations in form table. *Source* Author's elaboration

According to Nardino et al. (Nardino and Laruccia 2019) and Bruse et al. (Bruse and Fleer 1998), the measured reductions achieved by trees and rain gardens guarantee the change on thermal class of comfort and allow to demonstrate the effectiveness of the solutions. Conversely, the only application of cool material pavement and green surface doesn't provide an adequate performance to face the UHI effect.

Therefore, the widespread application of materials with high solar reflectance (high albedo), such as the cool material pavement, is surely able to reduce the air temperature but leads to a worsening on the user perceived comfort. In the gap between the reduction of physical and comfort indicators, the architectural project assumes a central role, which requires (case by case) an adequate site-specific assessment, considering the needs of vulnerable users, the context where it locates, and without being a simple substitution of parts with high-performance adaptive solutions.

In hydraulic evaluations, the analysis has been focused on the global value of runoff in the schoolyard, where surface testing area and runoff coefficient of the solutions deeply affect the performance. The testing area is 600 m^2 for all solutions except trees (where is considered the on-ground projection of their crown of 423 m^2), while the runoff coefficient is identified since scientific literature.

The results of hydraulic evaluations have shown that in the original condition, where the courtyard is covered with impermeable concrete, an overall runoff value of 88% is measured (Fig. 54.14). The most effective solution has been the rain garden with a runoff coefficient of 0.10, achieving a global value of runoff of 71% and a reduction of – 17%.

The less effective solution has been the permeable cool material pavement that, with a runoff coefficient of 0.60, achieves a global value of runoff of 82%, with a reduction of – 8%. Although the performances achieved by the four tested solutions are different, contrary to what observed for the thermal responses, all the design alternatives are effective to contrast the risks of PF.

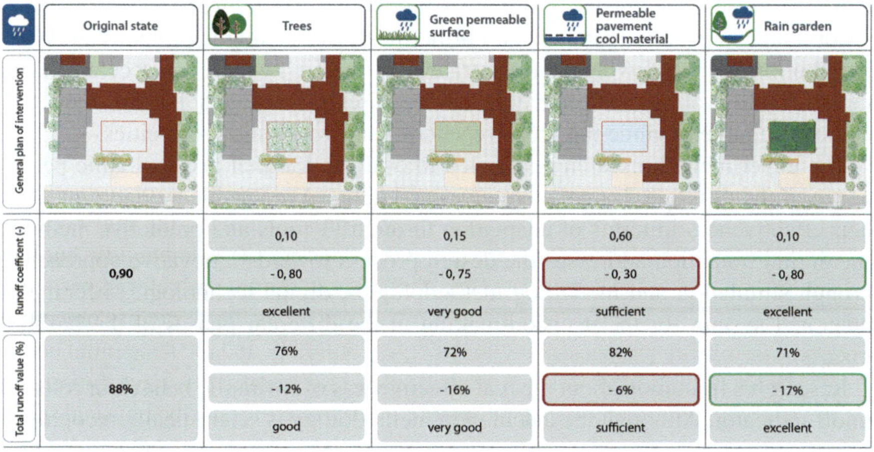

Fig. 54.14 Comparison of the hydraulic evaluations. *Source* Author's elaboration

54.5 Conclusion

The research moves from the analysis of case studies for climate adaptation in order to define a homogeneous framework of interventions in the main biogeographical regions of Europe. The identified projects represent good practises on climate adaptation, are recognized by the main international bodies, refer to the different urban areas of the contemporary city, and work to counteract one or more climatic hazards. These results have been fundamental to define the adaptive strategies and solutions for the regeneration of urban areas able to counteract the extreme events of climate change.

Further investigation should be carried out on the management, maintenance and costs of implementation of the different case studies on which, especially in the more complex interventions, significant investments are often required. Another observed critical aspect is the shortage of adaptation projects facing windstorm effects. This issue is also confirmed by the last report on the adaptation of cities in Europe and represent on the one hand, a limitation on knowledge, considering the increase in the number and intensity of such extreme phenomena, and on the other, an interesting field of experimentation where future research should be oriented.

The catalogue of design solutions–the main research output–identified 21 different technical and spatial alternatives to address projects based on climate resilience and adaptation. This result represents a useful tool for designers and decision-makers to identify the most appropriate and effective solution according to typological (NBS/artificial or green/grey/blue solutions), spatial (category and site of intervention) and technical-performative (climatic hazards, working mechanisms, performance indicators) parameters. Furthermore, the catalogue provides a relevant contribution for the new adaptation strategy of the EU, which has accelerated the implementation of adaptation actions and requires innovative solutions, tools, and methods suitable to guide the urban regeneration towards the green cities.

The experimental research results led to evaluate thermal and hydraulic performance of the different design alternatives and define a replicable model to measure their effectiveness in terms of adaptation through IT tools and predictive methods. The digital transition addresses the design process towards innovative approaches: through simulation and modelling methodologies, digital technologies become an integrated system for the decision-making in the different phases of the building process.

Research's limitation about the real effectiveness of hydraulic behaviour refers to runoff indicator. Although the calculation methodology is scientifically recognized, it represents an expeditious procedure, in which each surface area is associated with a coefficient obtained from the scientific literature that allows to reach a global value of the surface runoff. In this practical method the contribution of the sewerage system is not considered, and this leads to a less accurate assessment compared to hydraulic studies performed with the more complex software (Hec-ras, Epanet, etc.).

Finally, the catalogue of adaptive design solutions includes appropriate alternatives to tackle windstorm and droughts, able to reduce the vulnerability to the specific climate hazards, but their effectiveness was not evaluated.

References

Bruse M, Fleer H (1998) Simulating surface–plant–air interactions inside urban environments with a three-dimensional numerical model. Environ Modell Softw 13(3–4):373–384. https://doi.org/10.1016/S1364-8152(98)00042-5

EC (2016) Next steps for a sustainable European future—European action for sustainability. European Commission Communication. https://eur-lex.europa.eu/legal-content/EN/TXT/PDF/?uri=CELEX:52016DC0739&from=EN. Accessed 14 June 2020

EC (2020) Forging a climate-resilient Europe—the new EU strategy on adaptation to climate change. European Commission Communication. https://eur-lex.europa.eu/legal-content/EN/TXT/PDF/?uri=CELEX:52021DC0082&from=EN. Accessed 12 Dec 2021

EEA (2020) Urban adaptation in Europe: how cities and towns respond to climate change. EEA Report No 12/2020 European Environmental Agency. https://www.eea.europa.eu/publications/urban-adaptation-in-europe. Accessed 16 June 2021

EEA (2021) Nature-based solutions in Europe: Policy, knowledge and practice for climate change adaptation and disaster risk reduction. EEA Report No 01/2021 European Environmental Agency. https://www.eea.europa.eu/publications/nature-based-solutions-in-europe/. Accessed 22 Dec 2021

Faivre N, Fritz M, Freitas T, Boissezon B, Vandewoestijne S (2017) Nature-based solutions in the EU: innovating with nature to address social, economic and environmental challenges. Environ Res 159:509–518. https://doi.org/10.1016/j.envres.2017.08.032

IPCC (2021) Climate change 2021: the physical science basis. Contribution of working group i to the sixth assessment report of the intergovernmental panel on climate change. Cambridge University Press, Cambridge, United Kingdom and New York, NY, USA (in press)

Losasso M, Lucarelli MT, Rigillo M, Valente R (2020) Adapting to the changing climate. Knowledge innovation for environmental design. Maggioli Editore, Milano

Moccia FD, Sgobbo A (2016) Urban Resilience and pluvial flooding: the predictive study of the urban hydraulic behavior. In: D'Ambrosio V, Leone MF (eds) Environmental design for climate change adaptation. Innovative models for the production of knowledge. CLEAN, Napoli, pp 136–145

Mussinelli E, Tartaglia A, Bisogni L, Malcevschi S (2018) The role of nature-based solutions in architectural and urban design. TECHNE—J Technol Architect Environ (15):116–123. https://doi.org/10.13128/Techne-22112

Nardino M, Laruccia N (2019) Land use changes in a peri-urban area and consequences on the urban heat island. In Climate 7(11):133. https://doi.org/10.3390/cli7110133

Tucci F (2019) Adattamento ai cambiamenti climatici di Architetture e Città Green per migliorare la resilienza dell'Ambiente Costruito. Green City e adattamento climatico. In: Proceedings of the 2nd national conference of green City, Milano, 16–18 July 2019. https://www.greencitynetwork.it/wp-content/uploads/Adattamento-ai-cambiamenti-climatici.pdf. Accessed 26 Apr 2020

UN General Assembly (2015) Transforming our world: the 2030 agenda for sustainable development. https://sdgs.un.org/2030agenda. Accessed 30 Apr 2020

UN-Climate Change (2020) Climate change annual report 2019. United Nations Climate Change. https://unfccc.int/sites/default/files/resource/unfccc_annual_report_2019.pdf. Accessed 25 Dec 2021

UNFCCC (2015) The Paris agreement. In: 21st conference of the parties, Paris: United Nations. https://unfccc.int/documents/184656. Accessed 1 June 2018

Chapter 55
Digital Twins for Climate-Neutral and Resilient Cities. State of the Art and Future Development as Tools to Support Urban Decision-Making

Guglielmo Ricciardi and Guido Callegari

Abstract The increased effects of climate change in the built environment require a rapid and effective response to adapt urban settlements to the main impacts related to heatwave, extreme precipitation, sea-level rise, and so on. At the same time, there is not much time to reduce Greenhouse Gas (GHG) emissions that contribute to climate change and limit the mean temperature of the planet within the 1.5 °C imposed by the Paris Agreement. In this perspective, cities around the world have a key role toward carbon neutral and resilient targets. In parallel in the last years, we are witnessing the impacts of a big amount of data and information available at the city scale. There are many data coming from different databases that can be processed and managed to support the urban climate action planned and designed by decision-makers and urban practitioners, for example, to assess the carbon emission of the building sector or to simulate the effects of extreme precipitation or urban heat island and consequence behavior of the built environment. In this scenario, in the last years, among many different digital enable technologies available in the Industry 4.0 ambit, it has gained more attention in the field of urban planning and urban design the digital twin concept that could synthesize in a digital representation of the real-world data and information flow that could exchange from the physical side to digital representation and vice versa. The aim of the paper is to analyze the urban digital twin developed in last years in Europe to evaluate if and how they consider the climate change issue, in order to understand the state of the art, the applications developed for climate change and which is the level of experimentation in order to study and develop guidelines to build urban digital twin as a support tool for a climate-neutral and resilient city.

Keywords Climate change · Mitigation · Adaptation · Cities · Urban Digital Twin

---

G. Ricciardi (✉) · G. Callegari
Polytechnic University of Turin, Turin, Italy
e-mail: guglielmo.ricciardi@polito.it

G. Callegari
e-mail: guido.callegari@polito.it

E. Arbizzani et al. (eds.), *Technological Imagination in the Green and Digital Transition*, The Urban Book Series, https://doi.org/10.1007/978-3-031-29515-7_55

55.1 Introduction

During the last years, many cities are affected by the effects of climate change. Cities around the world are one of the main hot spots where the influence of the effects of climate change on extreme weather events such as heatwave, extreme precipitation events, drought, and sea-level rise are higher than in other environments. At the same time, the rapid and exponential increase of technological innovation has defined many effects for the human society. In this scenario, the digital transformation could play a key role to support planning and design practices in the urban context to prevent the effects of climate change and to reduce carbon emissions simultaneously.

The aim of the work is to examine one of the most recent innovations in the smart city paradigm, which is the digital twin tools, in order to understand the role to support urban scenario planning and design to manage the local climate action in European cities that have developed an urban digital twin. More in detail, the article explores the contribution of these tools to simulate the urban transformation project that takes into account the climate action. To deal with this aim, it has been explored the most recent scientific works published on the topic of integration of mitigation and adaptation and on the topic of urban digital twin. Due to the nature of the tools, to add more information it has been analyzed also gray literature and online platform where is visualize the development of digital twin in urban context. The paper is organized as follow.

In the second and in the third paragraphs are reported the theoretical framework related to climate change integration strategies in the urban context, digital transformation and Urban Digital Twin (UDT) as a process and tool to manage the high level of information required by mitigation and adaptation integration to support the scenario planning. The fourth paragraph entitled "the analysis of urban digital twin and the contribution to climate change scenario planning and design" shows the methodology and the results obtained by the work.

55.2 Digital Transformation and Integrated Climate Change Strategies in Urban Scenario Planning and Design

The exploration of the scientific literature on the topic of integration of climate change mitigation and adaptation shows a lack of integration that is linked to a historical dichotomy between these two strategies due to problems of spatial scale, time scale, institutional and administrative level, research traditions, and different disciplines (Grafakos et al. 2018).

The most recent publication of the IPCC (AR6) (Dodman et al. 2022) has underlined the importance to integrate mitigation and adaptation action both in plan and policies toward a Climate Resilient Development (CRD). This defines a new challenge for the urban design transformation in the next decades that may be faced with the complexity to integrate systems, sectors, information and data to define new urban scenarios in order to achieve the goal of CRD.

In urban planning and design practices, complexity could be managed and absorbed through the potential of digital transformation. This last process has defined a new space for our contemporary society, the "Infosphere" as defined by Floridi (2017) where people and systems have relied on Information and Communication Technologies (ICTs) to record, transmit, and use data of all kinds.

Digitalization has been promoted by the technological progress that has invested primarily in the manufacturing sectors. The digital transformation process combined with the increase of urbanization in cities around the world expected by 2050 (United Nations, Department of Economic and Social Affairs, Population Division 2019) and increase of the effects of climate change are three global megatrends as reported by European Environment Agency (2015). The increase of urbanization phenomenon connected to economic and environmental dynamics defines cities as high-complexity systems (Dembski et al. 2020), and the technological and digital progress during the last years has contributed to the highest development of the Smart City concept.

As reported by Losasso (2018), Smart City could be seen as: "a mirror of the instauration of the innovative relationship between people, urban spaces and new digital technologies". From the environmental point of view some authors have developed the concept of Sustainable Smart Cities or Smart Sustainable City without underlining a specific definition, but reflecting on some characteristics that are common, for example, Toli and Murtagh (2020), argue that: "Sustainable Smart Cities based on environment and social point of view, link technologies with governance to reduce the environmental impact of urbanism".

In this scenario, Balogun et al. (2020), underline the importance of digital transition that characterized the Industry 4.0 revolution as a paradigm and approach that offers novel opportunities to help cities to aid sustainable adaptation planning in urban areas.

As reported by Hurtado (2021) at the city level: "Planners need a more integrated way of simulating the complex potential futures of their cities in a more agile way". To prototype and test the possible future urban transformations there is the possibility to build a sort of "virtual city laboratory", a digital environment where the reality is reproduced in a control system in which planners, architects and decision-makers can do that with better outcomes, especially in the age of climate change where the time to act is fundamental to reduce the GHG emissions and to avoid more consequences. One of the main Industry 4.0 enables technology that is able to reproduce data about the built environment and test future possible transformation through scenario development is the digital twin that in the last years has gained more attention both in the most advanced smart city and academia context.

55.3 Urban Digital Twin (UDT) as a Tool to Support Urban Planning and Design Practices

The nature of the smart city concept is characterized by technological, human and institutional components that are useful to manage the different domains as well as transportation, environment, energy, waste, safety and education. For this reason, advancement in the technological components as urban or city digital twin could reinforce the potential of the smart city concept in the governance of the built environment.

As defined by Deng et al. (2021): "The digital twin is the inevitable goal of digital transformation". Many definitions of digital twin was reported in scientific literature by researchers in academia and in the gray literature (Arup 2019), by private IT company. Among these, the first conceptualization of digital twin has been made by Grieves in 2003, in the product life cycle management field, where a set of virtual information can describe a potential or real physical manufactured product from micro-scale to the macro level. More recently Batty (2018), introduce the concept of "Mirror representation" to describe the capacity of digital twin to reproduce the physical counterpart into the virtual world as well as Centre for Digital Built Britain in 2019 that synthesize digital twin in a digital representation of the real world. Key enabling technologies are part of the Industry 4.0 paradigm, as the Internet of Things, artificial intelligence, and machine.

Learning and big data analytics are introduced in the digital twin process to the real-time representation of the physical entities in the digital scheme and to understand the behavior of products, processes, and services to anticipate failure through real-time simulation and visualization. These key technologies permit to have a dynamic digital twin.

As reported by Deng et al. (2021), UDT aim to improve the efficiency and sustainability of logistics, energy consumption, communications, urban planning, disaster, building construction, and transportation. Many potential applications of UDT are reported by Shahat et al. (2021) based on five themes: Data management, visualization, situational awareness, planning and prediction, and integration and collaboration. In this categorization assume relevance the application related to policy evaluation, simulation and scenario evaluation could be carried out with citizen's engagement and multiple domains integration.

The next paragraph will describe the analysis of UDT developed in the European context (Fig. 55.1) in order to understand the contribution to support the scenario planning and design for climate change action.

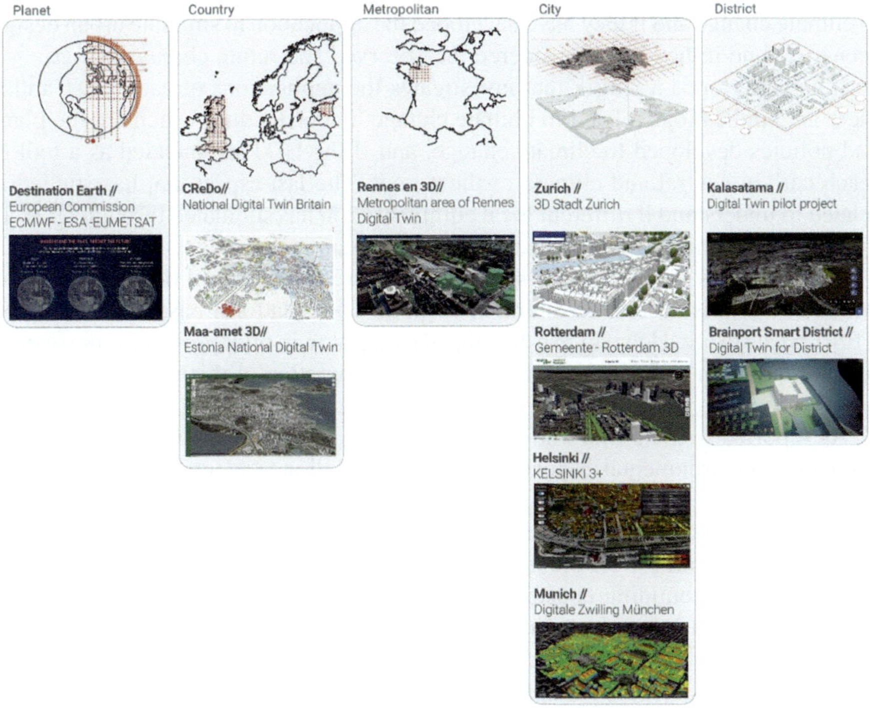

Fig. 55.1 From global to local scale: digital twin projects linked with climate change issue. *Sources* ECMWF, "Destination Earth" 2022. Accessed via https://stories.ecmwf.int/destination-earth/index.html#group-section-Digital-Twins-sAJHunyh2a; National Digital Twin Program, "CReDo" 2022. Accessed via https://digitaltwinhub.co.uk/credo/visualisation/; Republic of Estonia, Geoportal "Maa-amet 3D" 2022. Accessed via https://3d.maaamet.ee/kaart/; Rebecca Lyn Cooper "Urban Planning in 3D: How Creating a Digital Twin Leads to Smarter Cities" 2018. Accessed via https://meetingoftheminds.org/urban-planning-3d-creating-digital-twin-leads-smarter-cities-25212; Stadt Zürich Geomatik, "3D-Stadtmodell" 2022. Accessed via https://web.stzh.ch/appl/3d/zuerichvirtuell/; Gemeente Rotterdam, "Rotterdam 3D" 2022. Accessed via https://www.3drotterdam.nl/#/; Helsinki 3D, "Energy and Climate Atlas" 2022. Accessed via https://kartta.hel.fi/3d/atlas/#/; Geoportal Munchen, "Potenzialflachen Photovoltaik" 2022. Accessed via https://geoportal.muenchen.de/portal/energie/#/; Helsinki 3D, "City Information Model" 2022. Accessed via https://kartta.hel.fi/3d/#/; Brainport Smart District, "Digital Twin by Geodan" 2022. Accessed via https://brainportsmartdistrict.nl/en/project/digital-twin-by-geodan/

55.4 Analysis of Urban Digital Twin and the Contribute for Climate Change Scenario Planning and Design

The analysis focus on four different case study in Europe that has developed or under development an UDT in the last years and that they have accessible online (platform), scientific materials, reports, videos or institutional presentations that reveal the characteristics. The goal is to understand how they are considered the application related to climate change and if they have developed the application to simulate urban design scenarios, and if they have considered the observed and future climate change.

It has developed a matrix that investigates the applications of each UDT and if there are applications related to climate change, understanding which are the plans and policies developed to climate change, and if the UDT as intended as a tool to reach carbon neutral and climate resilient goal. The last aspect that has studied is related to understand if different local administration has simulated through UDT an urban scenario transformation that consider also climate change parameters.

All four cities analyzed are characterized by a 3D city model developed for the whole city that is accessible online through an open platform based on mainly on CESIUM software. Helsinki has developed a digital twin pilot project for the district of Kalasatama in order to support the urban regeneration of this part of the capital city.

As reported by Schrotter and Hürzeler (2020), the UDT of Zurich can be used to analyze environmental aspects (noise, air pollution, and mobile phone radiation modeling), energy performance of buildings and open-air spaces (solar potential analysis), urban planning transformation (visualization of construction projects, shadow, and visibility), the municipal development plan (current building development, maximum building capacity, possible target images in the form of various compaction scenarios), urban climate simulation (in order to understand the impacts in terms of heat island) and to support the architectural competition (e.g., visualize 3D new buildings developed in the competition). As reported by the gray literature explored for the case study of Rotterdam, the main possible application of UDT are related to making simulations, encouraging water safety, as a platform to support the gaming industry, to monitor the condition of public spaces, to support the municipal asset management, to support the city marketing applications, to make environmental analysis, to guarantee the interoperability among the 3D city model with BIM software and application in order to give the possibility to the users to develop and customize different analysis and applications.

Based on the 3D City Model of the city, the Helsinki 3D department has developed the "Energy and Climate Atlas" regard mainly to heating systems required from the built environment, conducted renovations and energy certification, wind simulation, consumption of electricity, district heating and water, and the solar energy potential. In addition, the application of the UDT of Helsinki is related also to the 3D historical reconstruction of the city center of Helsinki and the implementation and visualization of the 3D city development plan for 2050.

The UDT of the city of Munich is characterized by the 3D city model of terrain, buildings, and streets as other case studies analyzed, but is under development the implementation part related to the Connection Urban Twins project funded by the German Federal Ministry of the Interior, where they expected to develop an advanced UDT joined with Leipzig and Hamburg and based on the 3D city model developed yet and the data collected through different projects carried out in the last years.

The other aspect that is examined in the matrix is related to understanding if there are applications related to climate change. As reported above, for the case study of Zurich, the UDT offer the possibility to understand the impact of the Urban Heat

Island and the influence of new development considering some urban morphological aspects such as length, width, height, the position could have an influence on climate-ecologically relevant factors such as temperature, wind and cold air. Rotterdam UDT doesn't have an application that can directly consider indicators of climate change but give the possibility to obtain some data like energy saving potential, the green roof potential, the solar potential for installing renewable energy source systems, and visualize the impact on the surroundings directly.

The Energy and Climate Atlas based on the 3D City Model of the city of Helsinki has some indicators related to energy and climate factors calculated for each building, that take into account also the future climate change projections influence on space heating (kWh/(m^2 a)), heating saving potential (%), CO_2 emissions (kgCO_2/(m^2 a)), heating demand in a changing climate (MWh), and the refurbishment impact in terms of space heating demand comparing the original scenario and a possible advanced scenario.

The Munich UDT contains until now only the roof and façades solar potential in terms of application related to climate change. All cities, as reported in Fig. 55.2, have developed plans in order to achieve the carbon-neutral and climate-resilient goal with different time horizons.

The final aspect analyzed in the present work is related to understanding if climate change and local climate change parameters are used to support the simulation of urban transformation scenarios in order to be assessed by local authorities or decision-makers involved in the development process. As argued by Schrotter and Hürzeler (2020), for Zurich UDT: "climate issues can now be better integrated into the planning decision-making process". The same possibility is developed also from Helsinki UDT with different indicators that can support the regeneration of existing buildings, and thanks to a simulation conducted with parametric software and added to UDT, they have studied also the future scenario for Kalasatama district. The other two case studies, Rotterdam and Munich, are under development, and until now they don't have published the tangible results of their works, but both have the mission to simulate the future development scenario in order to measure the influence in terms of wind, temperature, solar radiation, and CO_2 emission of the new buildings and blocks.

55.5 Conclusions

European cities analyzed used a UDT to collect and visualize some local climate data to understand, among all, the solar surface potential of façades and roofs, the impacts of the built environment on Urban Heat Island phenomenon, the wind direction and so on. The Pilot Digital Twin for the Kalasatama district in Helsinki has a right attempt in order to simulate urban design transformation scenarios using some climate conditions even if they didn't consider all the indicators related to climate change.

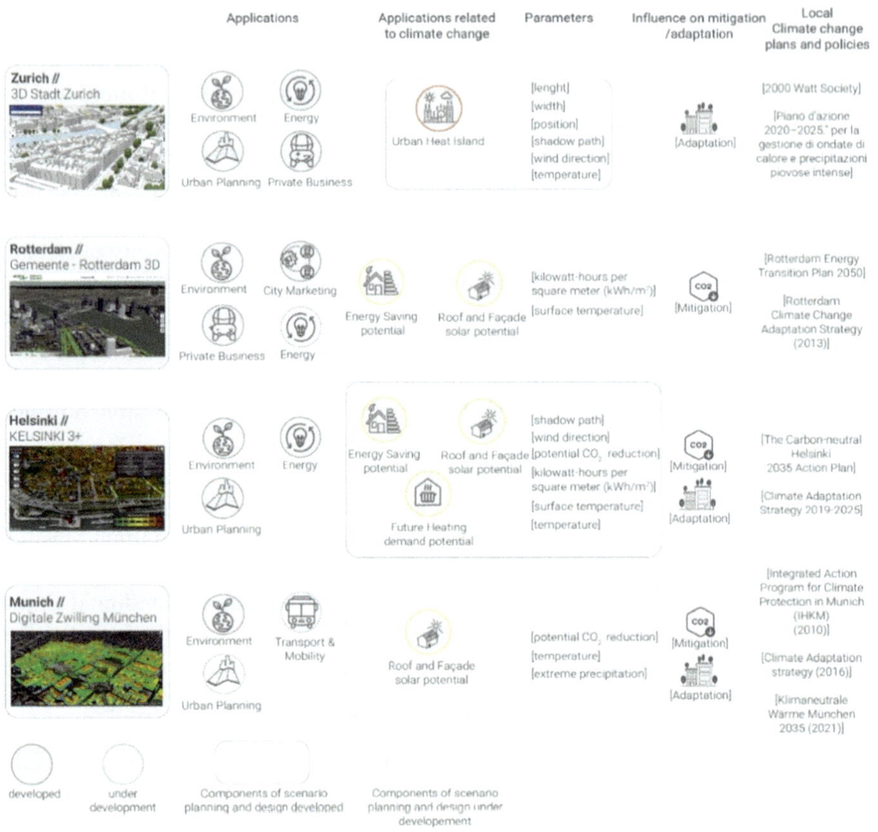

Fig. 55.2 Analysis of urban digital twin to support climate change scenario planning and design. *Sources* Stadt Zürich Geomatik, "3D-Stadtmodell" 2022. Accessed via https://web.stzh.ch/appl/3d/zuerichvirtuell/; Gemeente Rotterdam, "Rotterdam 3D" 2022. Accessed via https://www.3drotterdam.nl/#/; Helsinki 3D, "Energy and Climate Atlas" 2022. Accessed via https://kartta.hel.fi/3d/atlas/#/; Geoportal Munchen, "Potenzialflachen Photovoltaik" 2022. Accessed via https://geoportal.muenchen.de/portal/energie/#/

In the future, UDT could be the right platform to collect, elaborate, and visualize the results of the combination of different indicators to assess the possible mitigation and adaptation design measures could be implemented for buildings and open air spaces regeneration. This perspective required more studies in order to choose design measures and relative indicators, adequate parametric digital tools to simulate each indicator, right aggregation of different indicators, and rules to make results interoperable with UDT.

More efforts are required in the future to explore more in detail knowledge and skills used to carry out the development of the UDT and the simulation of urban planning and design scenarios for local climate action.

References

Arup (2019) Digital twin: towards a meaningful framework. https://www.arup.com/perspectives/publications/research/section/digital-twin-towards-a-meaningful-framework

Balogun A-L, Marks D, Sharma R, Shekhar H, Balmes C, Maheng D, Arshad A, Salehi P (2020) Assessing the potentials of digitalization as a tool for climate change adaptation and sustainable development in urban centres. Sustain Cities Soc 53:101888. https://doi.org/10.1016/j.scs.2019.101888

Batty M (2018) Digital twins. Environ Plan B Urban Anal City Sci 45(5):817–820. https://doi.org/10.1177/2399808318796416

Dembski F, Wössner U, Letzgus M, Ruddat M, Yamu C (2020) Urban digital twins for smart cities and citizens: the case study of Herrenberg, Germany. Sustainability 12(6):2307. https://doi.org/10.3390/su12062307

Deng T, Zhang K, Shen Z-J (Max) (2021) A systematic review of a digital twin city: a new pattern of urban governance toward smart cities. J Manag Sci Eng 6(2):125–134. https://doi.org/10.1016/j.jmse.2021.03.003

Dodman D, Hayward B, Pelling M, Castan Broto V, Chow W, Chu E, Dawson R, Khirfan L, McPhearson T, Prakash A, Zheng Y, Ziervogel G (2022) Cities, settlements and key infrastructure. In: Pörtner H-O, Roberts DC, Tignor M, Poloczanska ES, Mintenbeck K, Alegría A, Craig M, Langsdorf S, Löschke S, Möller V, Okem A, Rama B (eds) Climate change 2022: impacts, adaptation, and vulnerability. Contribution of working group II to the sixth assessment report of the intergovernmental panel on climate change. Cambridge University Press. In Press

European Environment Agency (2015) European environment—state and outlook 2015: assessment of global megatrends. https://www.eea.europa.eu/soer/2015/global/action-download-pdf

Floridi L (2017) La quarta rivoluzione. Come l'infosfera sta trasformando il mondo. Cortina Raffaello

Grafakos S, Pacteau C, Delgado M, Landauer M, Lucon O, Driscoll P (2018) Integrating mitigation and adaptation: opportunities and challenges. In: Rosenzweig C, Solecki W, Romero-Lankao P, Mehrotra S, Dhakal S, Ibrahim SA (eds) Climate change and cities: second assessment report of the urban climate change research network. Cambridge University Press, pp 101–138. http://www.cambridge.org/th/academic/subjects/earth-and-environmental-science/climatology-and-climate-change/climate-change-and-cities-second-assessment-report-urban-climate-change-research-network#yXvTlMYqbI9AbtfX.97

Hurtado P (2021) Smart city digital twins are a new tool for scenario planning. In: Digitalization and virtual simulation of entire cities can help planners experiment with solutions to meet complex problems. https://www.planning.org/planning/2021/spring/smart-city-digital-twins-are-a-new-tool-for-scenario-planning/

Losasso M (2018) Educational contents and innovative competencies for training and research in technology industry. Agathón 3(online). https://doi.org/10.19229/2464-9309/332018

Schrotter G, Hürzeler C (2020) The digital twin of the City of Zurich for urban planning. PFG J Photogramm Remote Sens Geoinf Sci 88(1):99–112. https://doi.org/10.1007/s41064-020-00092-2

Shahat E, Hyun CT, Yeom C (2021) City digital twin potentials: a review and research agenda. Sustainability 13:3386. https://doi.org/10.3390/su13063386

Toli AM, Murtagh N (2020) The concept of sustainability in smart city definitions. Front Built Environ 6:77. https://doi.org/10.3389/fbuil.2020.00077

United Nations, Department of Economic and Social Affairs, Population Division (2019) World population prospects 2019: highlights. ST/ESA/SER.A/423

Chapter 56
The Urban Potential of Multifamily Housing Renovation

Laura Daglio

Abstract Multifamily post-war middle-class housing in Italy represents a significant heritage which strongly characterizes urban landscapes. Although this huge stock has long been addressed by national policies as a major potential to pursue European climate targets, only the recent massive incentive measures (Superbonus 110%) have started to produce results for the energy upgrading of the buildings, offering alternatives and motivations (through the size of the public funding and the institution of the credit transfer) to the issues of the typical ownership fragmentation. However, these first partial outcomes are controversial from a life cycle, a social and an economic point of view. In addition, policies focus only on the energy performance of the single building, conceiving the interventions through a narrow-minded and generic attitude. The typological obsolescence and the multifaceted relationships between the building and the neighborhood are neglected, although important social, economic and energy efficiency benefits might emerge when addressing the renovation through a multi-scalar, multifunctional, and place-based approach. Stemming from the collection and analysis of ongoing initiatives and projects, possible models are outlined, enlarging the scenario of the transformations to include the urban scale. For example, underused private spaces can host new public or semi-public functions to contribute on the one hand to the management costs of the condominium and on the other hand to trigger local neighborhood regenerations. Moreover, widening the transformation perspective can envisage a group of buildings and the adjacent public spaces as a system to create energy districts where energy infrastructures introduce new amenities and added value.

Keywords Condominium · Energy community public private partnership

---

L. Daglio (✉)
Polytechnic University of Milan, Milan, Italy
e-mail: laura.daglio@polimi.com

E. Arbizzani et al. (eds.), *Technological Imagination in the Green and Digital Transition*, The Urban Book Series, https://doi.org/10.1007/978-3-031-29515-7_56

56.1 Introduction

Multifamily post-war middle-class housing in Italy represents a huge heritage where almost 40% of the population[1] live and which strongly characterizes urban landscapes.

Mainly constructed before the first energy saving (L. 373/76) and seismic (L. 64/1974) regulations were enacted and conceived for a now outdated society based on traditional family models even after the Fifties, it features a well-recognized technological as well as typological obsolescence; therefore, it has long since been the target of studies, proposals and initiatives also on a European level, aimed at its upgrading, though hindered by the peculiar ownership fragmentation of the buildings and the resulting difficulties affecting the property management.

Since 1998 (L. 449/97) consecutive national policies were launched and re-ratified year by year, thus characterized by a continuous instability and uncertainty, addressed at the renovation of the private building stock through tax incentives. These measures—beyond literally fostering the maintenance of the large percentage of buildings that had reached the 50 years of age as the recognized threshold of decay—had the aim of combating illegality as a longstanding plague in the building sector and of boosting national economy through the promotion of the construction industry; a typical governmental policy periodically adopted over the years, given the sector's significant contribution to Italy's GDP. Later, the compliance to European energy-saving and anti-pollution norms presented the occasion to introduce further incentive schemes for the energy upgrading of the asset (L. 296/2006, etc.), and in the wake of the occurrence of new tragic earthquakes, the policy was reinterpreted (L. 232/2016) to safeguard the building stock against the threat of seismic activity and in the light of the heightened awareness of risks deriving from their structural inadequacies.

Over the last 25 years, these still ongoing albeit renovated measures have achieved significant results, when considering the general/generic economic and environmental point of view, the fight against tax evasion and the job creation, as monitored by CRESME and by ANCE annual reports. However, conceived to be spread evenly on a territorial and building basis, these laws have produced inadequate and patchy results, especially on an environmental and societal basis. In fact, minor or isolated operations—such as the replacement of windows or boilers—are reported by ENEA yearly (ENEA 2020) as the prevailing interventions compared with a still too low number of deep renovations involving the envelope and building systems which would positively improve the energy performance and environmental impact on the architectural and urban scale. In addition, these measures struggle to support low-income families and affect the house rich and cash poor issues, whereas mainly wealthier strata of society are reported to employ and take advantage of the fiscal incentives.

[1] The current share of Italy's housing stock dating back to the period 1946–1971 amounts to some 10 million homes (in 2011 the national total amounted to some 31 million homes), in which 39.3% of Italian families reside (CENSIS 2011).

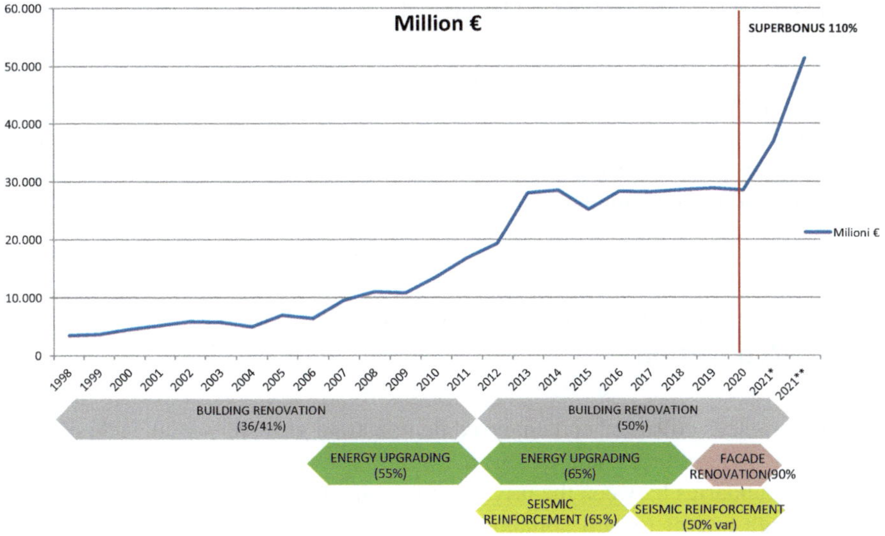

Fig. 56.1 Investments in buildings renovation in Italy since the enforcement of the first tax incentives regulations. The arrows highlight the changes in renovation measure and tax deduction rate (author's elaboration from: Camera dei Deputati 2021)

Although also developed and revised in order to overcome these limits, the recent Superbonus 110% policy has yet to produce the expected effects (Camera dei Deputati 2021; ENEA 2021) (see Fig. 56.1).

Beyond the controversial results in terms of life cycle (cf. the environmental impact of the insulation employed), a social (cf. the wealth of the incentive recipients) and an economic (cf. the payback periods borne by the state budget) point of view, the set of the energy efficiency measures focus only on the energy performance of the single building, conceiving the interventions through a narrow-minded and generic attitude. Moreover, on the one hand, the urban landscape consequences of the race toward the Superbonus 110% are completely neglected (triggering wide concerns (Sacerdoti 2021) about the fate of common modern architecture and related campaigns for its conservation) despite the expected transformation of the façades due to the added external layers; on the other hand, the Bonus Facciate 90% (now reduced to 60%) has been launched to improve the appearance of public streets and spaces.

Additionally, the typological obsolescence is neglected which is also related to the presence of often underused spaces which comprehend both semi-public condominium areas initially designed as representative of the bourgeoning middle class of the time (halls, corridors and entries) or in compliance to the then dominating architectural language (pilotis, etc.) and private big size flats conceived for large families now empty and to let with difficulties or underoccupied by empty nest retired elders.

Finally, the multifaceted relationships between the building and the neighborhood are overlooked, although important social, economic and energy efficiency benefits might emerge when addressing the renovation through a multi-scalar, multifunctional, place-based approach.

This paper has the aim of highlighting the need for a new attitude in rethinking the post-war residential built environment enlarging the scenario of the transformations to include the urban scale. Accordingly, ongoing initiatives and projects are disclosed outlining possible models and paths for a more integrated design also including an active role of the inhabitants.

56.2 The Urban Potential of Multifamily Housing Renovation

The experiences collected and analyzed demonstrate growing social, technological, regulatory and cultural changes toward an aware involvement of the dwellers, and blurred boundaries between the private and public domain.

Although a shift toward collaborative neighborhood communities was already taking place, from sporadic isolated to more coordinated initiatives,[2] as a result of a slowly growing shared economy (Manzini 2011) and to face the challenges of a lacking urban welfare, the pandemic lockdown with its forced smart working practices, the new multitasking dimension of the home together with the quest for a new balance between privateness and sociality, has fostered an unprecedented social and collaborative cohesion, also rediscovering the use of communal residential spaces.

Furthermore, the recent recognition from the European Union (Directive 2018/2001 and Directive 2019/944) of the importance of distributed generation for the energy transition, following technological innovation, several successful "bottom-up" projects developed over the decades, and the liberalization of electricity markets, has introduced another important step towards the creation of communities of energy prosumers and to enlarge the scale of the upgrading.

Finally, the convenience of the latest Superbonus and Bonus incentives (since 2020), has canvassed considerable discussions among the owners, highlighting the need for collaborative actions as well as sometimes unassailable cultural barriers about the opportunity of combining the individual gain with the common good.

Widening the perspective from the single building functional, technological and energy upgrading to respond to the social and energy demands on a larger scale, the following possible research and design paths can be explored for the reuse and reinterpretation of individual and shared spaces.

[2] E.g. http://www.viviconstile.org.

56.2.1 Colonizing Public Space

As the awareness raises of the pivotal role of Energy Communities to reach European decarbonization for the higher efficiency generated, and also for the enhanced cohesion provided by the economical income, simultaneously different barriers become clear on the one hand because of the national regulatory and market background, which is still transitional, on the other hand because of the existing urban texture constraints in terms of density, orientation, projected shadows, free space available for the new energy infrastructures. Experimental projects and case studies are ongoing and monitored concerning private condominiums (Legambiente 2020) and multifamily public housing addressing the challenge of merging multi-energy approaches with a multi-player perspective (Zatti et al. 2021) and underscoring the need for a channeled communication and for the active engagement of the inhabitants to pursue the general decarbonization targets within a renovation strategy (Minuto et al. 2021). However, even beyond the boundaries of the Energy communities definitory and regulatory Italian framework, the opportunity to interpret and enact the systemic nature of sustainable design, applying and further developing the digested principle that "No building is an island" (Addington 2007) emerges in the light of the lower power density and thus spatial impact (van Zalk and Behrens 2018) of renewables in comparison with the fossil fuel burning systems.

Accordingly, the conversion of the existing built environment to a low energy consuming stock, requiring both active and passive solutions should take into consideration the new installation of renewable energy infrastructures which might not be always located within the boundaries of the property they are conceived for (Fig. 56.2). This is also consistent with the smart grid concept involving a more efficient integration of the different energy production, storage, and distribution systems, maximizing the possible synergies among the available energy vectors and networks which extend beyond the limits of the private property plots. In terms of design applications, for example, small energy districts can be realized, grouping buildings and open spaces despite the ownership, so as to optimize the location of the necessary infrastructures (PV and thermal panels, heat pumps, geothermal power plants) not only following energy efficiency but also the urban morphology constraints and potentials. Since in dense urban environments there is often not enough space available in the private plot area for a possible renewable conversion (roof surface for solar panels, courtyards, or ancillary spaces for pumps or co-generators, etc.) to satisfy the demand of a single building, a system can be designed selectively grouping different properties with different layouts (including open spaces), even comprehending public areas. The new infrastructures through an integrated urban and energy design approach can become new amenities to set up and qualify underused public spaces with canopies, sport facilities[3] or even art installations.[4] The augmented functions of the energy devices in the urban landscape follows a multifunctional approach

[3] E.g. Modus, Mozart waste to energy plant, Brixen 2005 or NL Architects, WOS 8 Heat Transfer Station, Utrecht, The Netherlands.

[4] Cf. Hood Design, Solar Strand, Buffalo, USA, 2012.

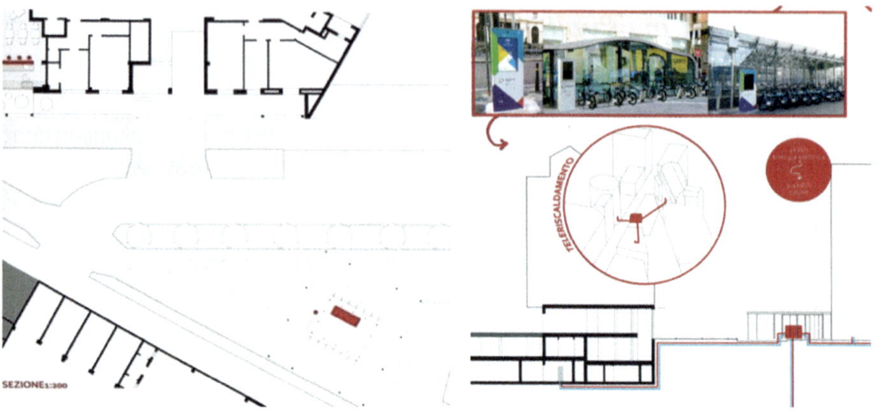

Fig. 56.2 Design exploration of a small district heating plant to be integrated with a bicycle rental facility in an existing open space in Milan (design exploration by Pamela Bosio, Elena Maria Rossi with Laura Daglio and Roberto Podda)

allowing to combine function form and use and overcoming common biases toward the impact of renewables in the landscape.

However, a similar paradigm shift as for the energy market and management is required for the property rights culture as well as for legislation.

56.2.2 Reinterpreting Semi-public Spaces as Common Facilities

Many of the postwar condominium complexes are characterized by the presence of communal indoor or outdoor spaces, whether in XIX century or in rationalist open block morphology: roof lofts, ground floor, abandoned janitor's quarters, large size halls, pilotis spaces, courtyards, small gardens, parking and flat roofs of single-floor garages, etc. These underused semiprivate areas can have a significant social and environmental potential for the building and for the local communities, especially after the issues harshened by the pandemic lockdown. The Municipality of Milan,[5] for example, has launched tenders and grants for the funding of new neighborhood welfare and commercial facilities to boost the economy, respond to services demand after the lack of public facilities harshened by the COVID-19 emergency and overcome the problems created by planned or market led monofunctional zoning.

[5] https://www.comune.milano.it/-/milano-punta-su-attivita-di-vicinato-imprese-sociali-e-start-up-per-lo-sviluppo-post-covid-19. Moreover, in the "Strade aperte" plan, launched in 2020 to rethink mobility and public space Milano Municipality also hints the possibility of considering the condominium private courtyard and gardens to increase the available surfaces as playgrounds for children (https://www.comune.milano.it/documents/20126/7117896/Open+streets.pdf/d9be0547-1eb0-5abf-410b-a8ca97945136?t=1589195741171).

Moreover, a changing housing demand moves towards the sharing of functions and services once encompassed in the traditional family home, a network-based concept of co-housing (Coricelli et al. 2018) to satisfy the demands both of the elderly as well as of the young.

As some pilot projects demonstrate, the condominium indoor spaces can host new shared facilities for the inhabitants as well as for the neighborhood, dedicated to the care of the elderly or of children (shared caregivers or babysitters) or for smart-working space. The upcycling of the underused area can offer a new income for the inhabitants in order to fund renovation works or the ordinary maintenance of the building or to reduce the management costs (tax and operational), simultaneously providing new welfare facilities, new jobs and thus a positive impact on the urban context. In addition to social sustainability benefits, also environmental advantages can emerge: the new energy-efficient enclosure of ground floor space or upgrading of the loft space, can contribute to the heat exchanges balance of the building envelope; moreover, the greening of impervious surfaces and flat roofs to create accessible shared gardens and orchards, as witnessed by some unfortunately too isolated experiences,[6] can contribute to the ecosystem quality of the urban area, when applied on a large scale. However, the implementation of public private partnerships agreements models and of a strategic shift in the design approach are required (Fig. 56.3).

56.2.3 Rethinking Private Spaces as Community Welfare Services

Finally, with a special focus on unoccupied or underused big size private flats, new upcycling chances can stem from the adoption of a service system approach. In fact, these residential spaces whether reorganized and divided in more than one unit one for the owner/s resident, the other/s to be let with controlled prices, can play a part on the one hand to respond to the rising demand of affordable, low rental housing, on the other hand, to host as well new neighborhood welfare services or small young enterprises offering an opportunity for a diversification of monofunctional residential districts, new jobs and providing a small income to the inhabitant. A financial plan distributing the owners' investment with a possible tax reduction together with the rental yields over the years can provide the economic feasibility of the operation.[7] Accordingly, a win–win strategy can be generated rethinking the traditional welfare system to combine private assets, new services initiatives with a supervision by the public, following an ongoing trend towards the growing role of private association and enterprises in the welfare sector (Fig. 56.4).

Although apparently with a limited punctual impact, the multiplication on the urban or local scale can show a significant contribution to respond to the unanswered

[6] Cf. Piuarch, Orto fra i cortili, Milan, 2015 (https://www.piuarch.it/it/progetti/orto-fra-i-cortili) or the OrtiAlti experience in Turin (https://ortialti.com/2021/02/orti-urbani-sui-tetti/).

[7] A more detailed description of the proposal can be found in Daglio et al. (2021).

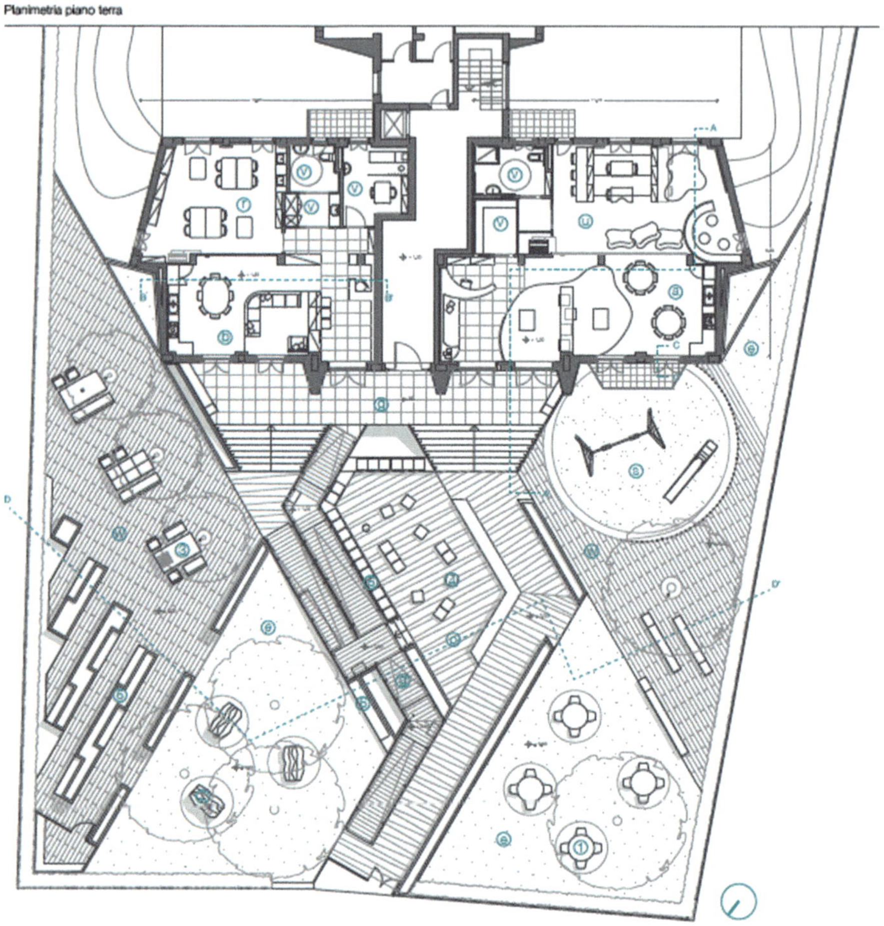

Fig. 56.3 Design exploration of a ground floor private flats and garden spaces conversion in kindergarten (design exploration by Elisa Beretta, Carlo Rossi, Valentina Zecchi with Laura Daglio and Roberto Podda)

social demand especially in metropolitan areas. Moreover, the specific renovation works required in the process, to be carried out possibly applying lightweight technologies for flexible and convertible layout reorganizations, can become a training field for new building enterprises to develop a specialization, new economies of scale and a transformation of the traditional supply chain to trigger an innovation path in the construction industry in general.

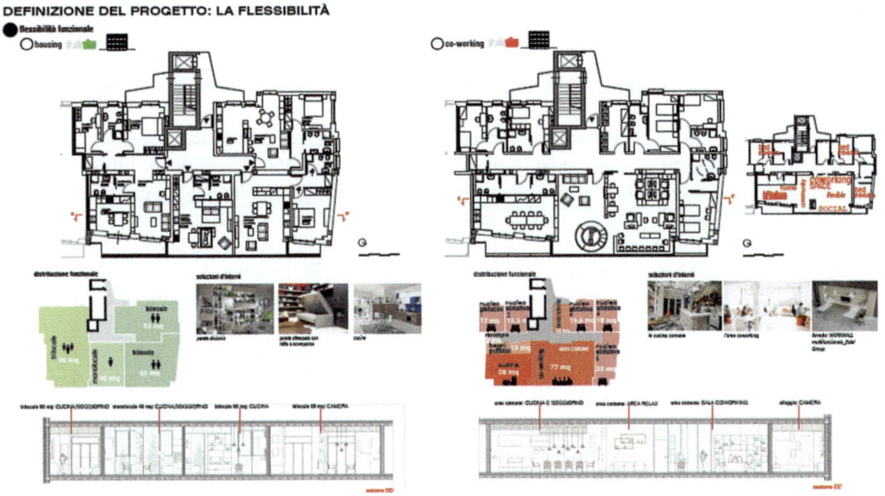

Fig. 56.4 Reuse of a large flat for health facilities in a private residential building in Milan (design exploration by Martina Lusi, Caterina Mancuso, Olga Massaro Valeria Trainini with Laura Daglio and Roberto Podda)

56.3 Conclusion

The dimension, complexity, and multifarious features and geographies of the multi-family post war private stock require innovative renovation design models to be integrated and interpreted beyond the traditional upgrading carried out so far in compliance with and following the tax incentive policies. This heritage should be conceived as a resource and opportunity on a multi-scalar level. On the one hand, a strategic dimension of design should be adopted, requiring the organization and definition of the overall process and the possible stakeholders involved, offering the power of technological imagination to anticipate new scenarios and to respond to contemporary demands. On the other hand, a rethinking of the boundaries between private and public ownership should take place as the basis for innovative forms of public–private partnerships, a move from the concept of asset to that of service, a paradigm shift which is recalled by distributed generation and prosumers community as well as by sharing economy. Finally, the regulatory framework should be enhanced and merged with local policies (Zanfi et al. 2021), taking into consideration the reality of local contexts to engender virtuous systemic links and win–win strategies.

References

Addington M (2007) No building is an island: a look at the different scales of energy. Harv Des Mag 26:38–45

Camera dei Deputati (2021) Il recupero e la riqualificazione energetica del patrimonio edilizio: una stima dell'impatto delle misure di incentivazione, n. 32/3, 2° edizione. https://temi.camera.it/leg18/dossier/OCD18-16007/il-recupero-e-riqualificazione-energetica-del-patrimonio-edilizio-stima-impatto-misure-incentivazione-dati-aggiornati-al-2021.html

CENSIS (2011) 45° rapporto annuale sulla situazione sociale del paese. CENSIS, Roma

Coricelli F, Quaglio C, Robiglio M, Rolfo D, Russi NP (2018) Re-housing. La casa come dispositivo di integrazione. Quaderni del Future Urban Legacy Lab—Politecnico di Torino, Torino

Daglio L, Marchigiani E, Zanfi F (2021) Case in affido: il riuso del patrimonio private in una prospettiva di locazione sociale. In: Coppola A, Del Fabbro M, Lanzani A, Pessina G, Zanfi F (eds) Ricomporre i divari. Politiche e progetti territoriali contro le disuguaglianze e per la transizione ecologica. il Mulino, Bologna, pp 163–174

ENEA (2020) Le detrazioni fiscali per l'efficienza energetica e l'utilizzo delle fonti rinnovabili di energia negli edifici esistenti. Rapporto Annuale 2019. https://www.efficienzaenergetica.enea.it/component/jdownloads/?task=download.send&id=452&catid=40%20&Itemid=101

ENEA (2021) Le detrazioni fiscali per l'efficienza energetica e l'utilizzo delle fonti rinnovabili di energia negli edifici esistenti. Rapporto Annuale 2021. https://www.efficienzaenergetica.enea.it/pubblicazioni/rapporto-annuale-detrazioni-fiscali/rapporto-annuale-detrazioni-fiscali-2021.html

Legambiente (2020) Comunità rinnovabili 2020. https://www.legambiente.it/wp-content/uploads/2020/06/rapporto-comunita-rinnovabili-2020.pdf

Manzini E (2011) Idee di benessere (ed idee sul benessere). Immagini del quotidiano nella transizione verso la sostenibilità. http://www.societadeiterritorialisti.it/wp-content/uploads/2011/03/03.04.16%20intro.%20idee%20di%20benessere%20%20dr%200%2016.4.2003.pdf

Minuto FD, Lazzeroni P, Borchiellini R, Olivero S, Bottaccioli L, Lanzini A (2021) Modeling technology retrofit scenarios for the conversion of condominium into an energy community: an Italian case study. J Clean Prod 282:1–17. https://doi.org/10.1016/j.jclepro.2020.124536

Sacerdoti M (2021) Milano, emergenza 'cappotto'. Abitare News. https://www.abitare.it/it/news/2021/12/12/ecobonus-a-rischio-novecento-milanese/

van Zalk J, Behrens P (2018) The spatial extent of renewable and non-renewable power generation: a review and meta-analysis of power densities and their application in the U.S. Energy Policy 123:83–91. https://doi.org/10.1016/j.enpol.2018.08.023

Zanfi F, Daglio L, Perrone A, Rusci S (2021) Bonus edilizi: diversificazione e integrazione con politiche urbane e territoriali. In: Coppola A, Del Fabbro M, Lanzani A, Pessina G, Zanfi F (eds) Ricomporre i divari. Politiche e progetti territoriali contro le disuguaglianze e per la transizione ecologica. il Mulino, Bologna, pp 149–162

Zatti M, Moncecchi M, Gabba M, Chiesa A, Bovera F, Merlo M (2021) Energy communities design optimization in the Italian framework. Appl Sci 11:1–16. https://doi.org/10.3390/app11115218

Chapter 57
A "Stepping Stone" Approach to Exploiting Urban Density

Raffaela De Martino, Rossella Franchino, and Caterina Frettoloso

Abstract Currently, there is an increasing need to use ecological-environmental strategies that can contribute to human well-being and biodiversity in urban contexts that are complex systems with an increasingly diverse demand for ecosystem services. This becomes important especially for areas with high urban density which are those that determine the greatest impacts. Recent studies have related the increase of the COVID-19 pandemic to urban density, and the results seem to converge on the hypothesis that the lack of availability of urban space and high population concentration are factors contributing to the spread of disease. However, density constitutes both a problem and an opportunity to creating open spaces on a human scale, favoring the development and articulation of spaces that can more easily creep in and create the necessary conditions both for functional and environmental improvement and for mending the built environment. A dense environment suggests to work according to new logics that, starting from the optimization of existing spaces able to provide ecosystem services, experiment especially the "micro" and "interconnected" formula. The issue of interconnection in cities is complex as the continuity of connections, necessary to ensure the reticularity of spaces, is often inhibited by urban density. The idea is to replace the "structural continuity" with a "functional continuity" according to the "stepping stones" approach, borrowed from the ecology of the landscape. In analogy to the ecological networks approach, therefore, it is proposed to use this logic to enhance existing open spaces and create new ones with the aim of implementing overall urban quality.

Keywords Ecosystem services · "Micro" and "interconnected" open spaces · Functional continuity

R. De Martino · R. Franchino · C. Frettoloso (✉)
University of Campania L. Vanvitelli, Aversa, Italy
e-mail: caterina.frettoloso@unicampania.it

R. Franchino
e-mail: rossella.franchino@unicampania.it

E. Arbizzani et al. (eds.), *Technological Imagination in the Green and Digital Transition*, The Urban Book Series, https://doi.org/10.1007/978-3-031-29515-7_57

57.1 The Stepping Stones Strategy

The uncontrolled growth of cities, with the consequent settlement, infrastructural and production phenomena, has determined profound alterations to the urban ecological structure, intensely transforming semi-artificial and/or semi-natural spaces. Nature is therefore often reduced to isolated fragments that are poorly connected and strongly compromised in their ecological functionality (Paolinelli 2003).

The concept of ecological network, even if usually applied to vast territorial areas, has been proposed for the reorganization of open spaces in urban areas; moreover, currently, this concept has further evolved as the ecological network has been incorporated into the "green infrastructure" model in which the supply of ecosystem services, understood as the set of direct and indirect benefits that man derives from ecosystems (Millennium Ecosystem Assessment 2005), is the main aim to be pursued.

In this perspective, traditional ecological networks evolve into polyvalent ecological networks. The polyvalency of the networks is expressed in their ability to provide ecosystem services both to biodiversity, but also to the various human activities in the area: agriculture, settlements, water use and transport (Malcevschi 2010).

However, the reticular approach requires that particular attention be paid both to the ecological/environmental quality of the network nodes and to maintaining the connections between these nodes. The issue of connectivity in cities is however extremely complex, as the continuity of connections, necessary to guarantee the reticularity of spaces, is often inhibited by urban density. The idea proposed in this work is therefore to replace "structural continuity" with "functional continuity" according to the "stepping stones" approach, borrowed from landscape ecology.

Structural connectivity concerns landscape structures regardless of any biological or behavioral attribute of the organisms that interact with them; therefore, it focuses on spatial and physical characteristics, such as shape, size and continuity of the connections. Conversely, functional connectivity includes species specific aspects and their interaction with landscape structures (Pe'er et al. 2014). Therefore, functional connectivity refers to the set of processes that take place and that binds the structural components of the connections to each other through energy flows and material transfers.

The stepping stones are structures formalized by the Pan-European Ecological Network (COE (Council of Europe), UNEP, ECNC 1996) which proposes a scheme of ecological network units formed by core areas, buffer zones, restoration areas which are areas of value or potential naturalistic value, ecological corridors, stepping stones that perform, albeit in different ways, connection functions.

The ecological corridors are in fact continuous physical connections that perform various vital ecological functions and ensure the self-regulating capacity of ecosystems, allowing key species to move between the mosaics of the ecosystem. The stepping stones, on the other hand, are fragments of habitats widespread in a territory and not directly connected, which represent important elements of the landscape for the stopping of species in transit or to accommodate specific micro-environments in critical habitat situations.

It is clear that in urban areas, physical connections are difficult to achieve due to urban density and territorial fragmentation. However, it is possible to appeal to a functional continuity that can be guaranteed by the stepping stones, i.e., by minor points of support between them sequentially (similarly to what the stones do along a ford line of a watercourse) able to perform a function of connection.

But what are the elements to connect in a city? The definition of an environmental system of connected green spaces, obviously includes not only large parks and equipped green areas, but also small gardens, squares, entire non-built areas or other open spaces (not green) such as pedestrian areas, parking lots and roads, the latter are particularly important in the urban matrix as connecting elements par excellence (Pagano 2006).

In addition to the types of open spaces, following recent studies that have highlighted their ecological importance, green building systems represent real competing infrastructures to strengthen the resilience of the urban environment. These "green" elements on buildings can in fact assume the function of "stepping stones" for fauna, in particular for some species at risk, integrating the natural elements existing around them and enriching the network of green corridors that may exist (Andri and Sauli 2012).

It should be noted that in relation to green building systems, the term biodiversity is often associated. In general, we mean the environmental value that a certain intervention can acquire from a naturalistic point of view. To achieve these objectives of increasing the environmental value, knowledge of ecological aspects and respect for the floristic coherence of the plant species to be used are fundamental requirements. By following these criteria, these works can determine an increase not only in the perceptual value, but also in the ecological and environmental value of the places. In fact, if correctly designed and implemented, they are able to activate processes "in favor of biodiversity", starting the formation of contexts capable of favoring many spontaneous animal and plant species (Andri and Sauli 2012).

57.2 Intercepting and Systemizing "Support" Spaces

The phenomena of urban densification accompanied by the expansion of the built environment have led to a sharp decrease in open spaces and, in general, in spaces dedicated to collective activities and pedestrians. The parallel loss of social awareness and the consequent loss of values based on the concept of sharing go some way to explaining the low level of not only social but also functional and environmental quality of our cities. However, as urban planner Monique Keller (Mühlberger de Preux 2017) explains, recalling the characteristics of medieval cities, "a densified city is not necessarily less green, noisier and more polluted than a traditional one" but requires a change of vision in which an attempt is made to recreate a human density that favors collective life.

When David Owen, in his book Green Metropolis (2009), states that "the key to New York's relative environmental efficiency is its extreme compactness", he meant to underline this peculiarity, which is then closely linked to the concept of proximity.

In fact, using the words of the economist and environmentalist Charles Komanoff, Owen observes how New Yorkers have given up the "supposed convenience" of the car in favor of the "real convenience of proximity" (Owen 2010).

As shown also by recent experiences in Europe, strongly focused on densifying the full by enhancing the void, there is a need to increase the technological-environmental dimension of these spaces and, in general, of the connective tissue, so that density constitutes an opportunity to create urban environments on a human scale and proximity ones. From a methodological-planning point of view, it means working on approaches and actions aimed at "reducing CO_2 emissions, combating the impacts of climate change, reducing the phenomena of environmental degradation and pollution, and improving conditions of well-being and quality of life" (Losasso 2017). The network of urban open spaces constitutes, with respect to energy-environmental issues in urban design, a strategic system of regulation and control that especially in a dense environment can more easily creep in and create the necessary conditions both for functional and environmental improvement and for mending the built environment.

In order to ensure that urban open spaces function as a real infrastructure at the service of the city, it is necessary to work on an approach that aims to reconnect the spatial and environmental fragmentation that usually characterizes dense urban environments, proposing alternatives that can be traced back to the concept of "functional continuity" described above. In fact, the ecological-environmental continuity, the comfort and safety of paths for all users qualify the urban open space network as a real ecosystem in which green and gray spaces integrate to respond to current environmental challenges: from the reduction of heat island phenomena thanks to the appropriate use of materials and the correct balance of absorbing and reflecting surfaces, to the improvement of stormwater management through sustainable drainage systems (Pregill 2020).

A dense environment is, however, usually characterized by the lack of spaces, in particular green ones, that suggests working according to new logics that, starting from the optimization of existing spaces able to provide ecosystem services, experiment especially the "micro" and "interconnected" formula. The system of open spaces will have to be organized according to a logic that integrates broader actions aimed at creating an infrastructure on an urban scale whose key elements are the relationships between the various nodes, with more punctual and site-specific actions, implemented according to the logic of urban acupuncture (Lerner 2014) and strongly dependent on the context. An infrastructure conceived in this way will be able to infiltrate the urban fabric and, where necessary, fill in gaps, helping to create the fundamental conditions for initiating both processes of conservation of existing natural resources and the interception of a series of intermediate spaces, also integrated with the built environment, which are functional connections between existing and potential spaces.

In this scenario, interstitial spaces play a strategic role, often lacking their own identity, which could instead acquire an important role in achieving what we have called the "stepping stone" approach: reserve spaces to traditional spaces that can constitute intermediate passages (even temporary and dynamic), a sort of "support" to increase the useful surface but above all to enhance or accommodate new ecosystem functions.

Moreover, national and international experience in various urban regeneration projects aimed at reconnecting buildings, people and the city in general with nature has shown how it is possible to increase biodiversity and continuity of use and the environment by working on the street system, roofs and building façades, and all those interstitial micro-spaces (including semi-private ones). It is a question of tracing trajectories that are not always coplanar, often starting from the re-naturalization of buildings and then arriving at the open space, crossing courtyards and narrow alleys that are not always completely transformable but on which it is possible to intervene in a specific way: de-paving or increasing the green component, sometimes introducing functional elements for collective activities, always taking into account the overall technological-environmental balance on the one hand and the innovation and naturalness of the proposed intervention on the other.

57.3 The "Mesosystem" Concept for Urban Regeneration

Following the considerations developed in the previous paragraphs, it is evident that it is of fundamental importance that the rebalancing interventions of highly anthropized areas must be structured in such a way that the territory is conceived as an organism endowed with dynamic equilibrium even if achieved through the technological control of complex functions.

The achievement of this goal is very ambitious because human activity, with the complex interrelated structures and relationships determines a significant trace in the environment that is a sign of degradation left as a burden to future generations. In order to limit this footprint, it is necessary to make as much as possible sustainable transformations from the environment in which they are located and direct the activity of rebalancing so that the footprint is contained as much as possible, and this is achieved by increasing the load capacity of the territory defined as the ability to absorb and control the phenomena of anthropization with a sustainable impact on the ecosystem.

Entering into the evaluation of the interrelationships between anthropogenic and natural phenomena with the aim of providing tools and methods for rebalancing can be helpful to look at the territory as a real ecosystem (Adler and Tanner 2013; Aitkenhead-Peterson and Volder 2010).

Starting from the premise that the natural ecosystem has a perfect functioning, the same does not happen for the urban ecosystem which is artificial, very complex and also in continuous transformation depending on many factors.

The urban ecosystem consists of artificial, semi-artificial and semi-natural biotopes and between the physical and biological components develop very complex relationships. It is, therefore, a transitory ecosystem because the anthropogenic activity does not allow it to be completely autonomous and to achieve a condition of stability.

In order to structure interventions of rebalancing, it can be effective to think about strategies based on the interconnection between microsystems. Structuring the urban ecosystem as a set of microsystems, both artificial and natural, that join together to form a mesosystem, we can aspire to establish an intermediate situation of stability. This can allow the urban ecosystem to achieve a certain "balance" by exchanging interactions with other ecosystems.

The context of Sarno is particularly significant for the purposes. This is achieved by reasoning about the transposition to the urban environment of the theory of ecological systems, also known as "development in a context" or theory of "human ecology", developed by Bronfenbrenner to understand the dynamic interrelationships between various personal and environmental factors that affect human development (Bronfenbrenner 1981). The transposition is to consider the mesosystem as an interconnection between microsystems even at the level of development of the urban environment.

In order to provide a practical application of the previously stated concepts, we examine the case study of the regeneration of a highly anthropized area in the Municipality of Sarno in the Campania Region of Italy. In particular, the area under study extends for about 3 km in a north-west direction, including in part the densely urbanized urban center and partially the rural area almost completely cultivated.

The context of Sarno is particularly significant for the purposes of this study due to the mass urbanization, which is characterized by natural spaces that have been damaged by anthropization (e.g., the Sarno river which has a poor water quality and whose banks are cemented in many places and often without vegetation) and intersected by a complex network of infrastructures (e.g.: roads, motorways, railways, bridges, water treatment plants, distribution systems for electricity) that have contributed to altering the natural landscape, fragmenting the territory as well as changing the eco-systemic quality of the context as a whole. Subsequently, the study has focused on identifying eco-oriented redevelopment strategies for the territory under study characterized by a dense urban matrix, with particular attention being given to the eco-systemic analysis and environmental improvement interventions for the restoration of the connection elements that can allow for physical and functional continuity between the urban and surrounding rural systems.

The proposed regeneration intervention concerns the creation of natural microsystems through the use of green resources to be integrated with pre-existing artificial ones articulated on different scales of intervention. In general, the proposed interventions are replicable in order to be implemented and to activate over time other natural microsystems that can allow as much as possible the achievement of a condition of ecosystemic "equilibrium".

Regarding the replicability of interventions on existing buildings for vertical ones it is necessary that the wall is blind or partially blind, for horizontal ones that the roofs are flat and accessible with a stairwell that allows direct access to the roof in order to make them usable not only to the inhabitants of the building but also by other citizens (Fig. 57.1).

At the territorial scale, the proposed interventions concern the redevelopment of some degraded urban spaces and their connection through a green way (Fig. 57.2).

To achieve these objectives, the proposed interventions use the green component also integrating systems of energy self-production and control of rainwater management both at the scale of the building and territorial. All this in order to operate in a perspective of zero land consumption and to prepare the territory to an ever-better adaptation to climate change.

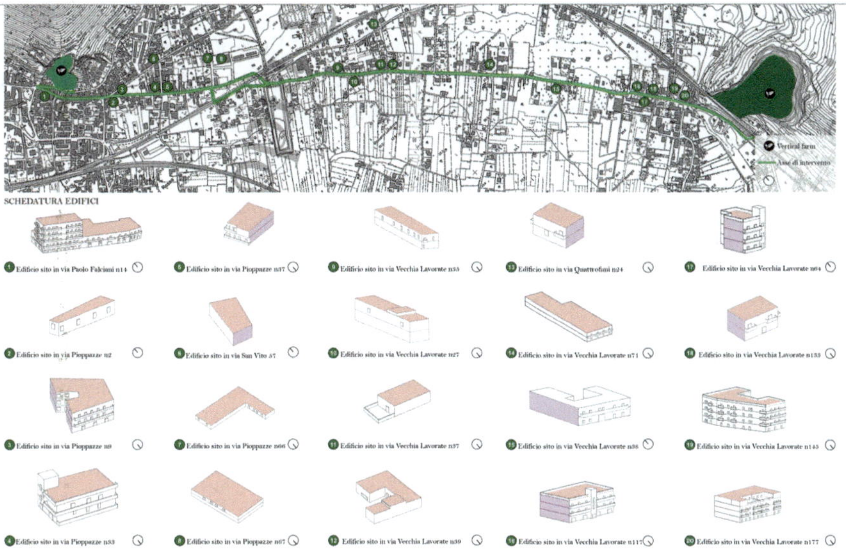

Fig. 57.1 Regeneration proposal for Sarno urban area: building cataloging (credits: M. L. Genito, S. Gravina)

Fig. 57.2 Regeneration proposal for Sarno urban area: green way and open spaces (credits: M. L. Genito, S. Gravina)

57.4 Conclusions

This paper, sharing the idea that density constitutes both a problem and an opportunity to creating open spaces on a human scale, suggests a strategy based on the replace of the "structural continuity" with a "functional continuity" according to an approach borrowed from the ecology of the landscape.

From an operational point of view, an important contribution to the presented approach may derive from the integrated use of digital data visualization and management tools, as well as forecasting models and monitoring systems, which are strategic both in the initial environmental status identification phase and in the subsequent evolutionary phases of the project.

All this in synergy with recent policies in the European sphere to respond to the current challenge centered on the relationship between digital technological innovation and environmental protection, consider environmental quality as a competitiveness factor.

The presented case study highlights how the objective to be achieved is to configure the territory as a fabric in which there is no boundary between the artificial and natural environment and in which each process is controlled so that its impact and consequently the irreversible degradation induced is as minimal as possible in relation to the constraints of the process itself.

Acknowledgements The paper is edited by all the authors. In particular: paragraph *The stepping stones strategy* is edited by Raffaela De Martino; paragraph *Intercepting and systemizing "support" spaces* is edited by Caterina Frettoloso; paragraph *The "mesosystem" concept for urban regeneration* is edited by Rossella Franchino.

References

Adler FR, Tanner CJ (2013) Urban ecosystems ecological principles for the built environments. University Press, Cambridge

Aitkenhead-Peterson J, Volder A (eds) (2010) Urban ecosystem ecology. In: Agronomy monograph 55

Andri S, Sauli G (2012) Verde pensile: prestazioni di sistema e valore ecologico. ISPRA, Manuali e Linee Guida, p 78

Bronfenbrenner U (1981) The ecology of human development: experiments by nature and design. Harvard University Press, Cambridge

COE (Council of Europe), UNEP, ECNC (1996) The Pan-European biological and landscape diversity strategy, a vision for Europe's natural heritage. Council of Europe, Strasbourg, France

Lerner J (2014) Urban acupuncture. Island Press, Washington

Losasso M (2017) Progettazione ambientale e progetto urbano. Eco Web Town II(16):7–16

Malcevschi S (2010) Reti ecologiche polivalenti—Infrastrutture e servizi ecosistemici per il governo del territorio. Il Verde Editoriale, Bresso (MI)

Millennium Ecosystem Assessment (2005) Ecosystem and human well being: a framework for assessment. Island Press, Washington

Mühlberger de Preux C (2017) Urbanizzazione centripeta: la città di domani—densa, verde e conviviale. Ambiente 4:32–38
Owen D (2010) Green metropolis. Why living smaller, living closer, and driving less are the keys to sustainability. Riverhead Books
Pagano C (2006) Reti di aree verdi, reti ecologiche e riqualificazione urbana. In: Fonti L (ed) Parchi, reti ecologiche e riqualificazione urbana. Alinea editrice, Firenze
Paolinelli G (2003) La frammentazione del paesaggio periurbano. Firenze University Press, Firenze
Pe'er G, Schmitz A, Matsinos YG, Schober L, Klenke RA, Henle K (2014) Connectivity: beyond corridors. In: Henle K, Potts SG, Kunin WE, Matsinos YG, Similä J, Pantis JD, Grobelnik V, Penev L, Settele J (eds) Scaling in ecology and biodiversity conservation. Pensoft Publishers, Sofia
Pregill P (2020) Urban connections in the contemporary pedestrian landscape. Routledge, Taylor and Francis Group

Chapter 58
Metropolitan Farms: Long Term Agri-Food Systems for Sustainable Urban Landscapes

Giancarlo Paganin, Filippo Orsini, Marco Migliore, Konstantinos Venis, and Matteo Poli

Abstract In the past decade, urban agriculture (UA) has attracted significant attention from urban planners and city managers as a sustainable, nature-based, and smart solution that may generate positive impacts for resilience, self-reliance, and social, economic, and environmental sustainability of cities. UA appears as an effective means to address climate change while also fostering urban transitions to sustainability in many ways, such as creating new commons, amenities, ecosystem services, reinventing urbanity, and encouraging community building by growing local food. Since UA is a strategy to support the re-configuration of more sustainable and resilient cities, it can be considered a seedbed for innovation. Based on these premises, the STRutture Agricole MEtropolitane (STRAME) research project aims at defining an innovative interpretation of the urban farming. The research proposes a vision of the UA based on an intermediate scale compared to the more investigated and developed mega-scale of large vertical farming and the microscale of urban gardens. Conceived as an adaptive infrastructure, STRAME—a system based on modular Vertical Farming units—is organized to be translated and applied in different urban and metropolitan scenarios. The "terrain vague" of metropolises (intended as residual urban spaces) and climate change are two challenges—the first of a physical-spatial type and the second environmental-social—in which STRAME wants to build a capillary system of highly efficient agricultural production. STRAME, starting from deep analysis of the background of UA, aims at defining a physical infrastructure integrated with a digital infrastructure (IoT), able of responding to the challenges posed by the agro-industrial chain in densely populated urban contexts. Its core is a system of modular elements to be used for the construction and commissioning of

G. Paganin (✉) · F. Orsini · M. Migliore · K. Venis · M. Poli
Polytechnic University of Milan, Milan, Italy
e-mail: giancarlo.paganin@polimi.it

F. Orsini
e-mail: filippo.orsini@polimi.it

M. Migliore
e-mail: marco.migliore@polimi.it

M. Poli
e-mail: matteo.poli@polimi.it

E. Arbizzani et al. (eds.), *Technological Imagination in the Green and Digital Transition*,
The Urban Book Series, https://doi.org/10.1007/978-3-031-29515-7_58

a medium-sized network of inter-connected vertical farming applicable in residual voids and in the open spaces in large residential districts.

Keywords Urban agriculture · Urban resilience · Food security · Green infrastructure

58.1 Introduction

Scientific literature highlights how over the past decade, urban agriculture (UA) has received ever increasing attention in relation to its potential of generating positive impacts for resilience, self-reliance, and social, economic, and environmental sustainability of cities.

Urban agriculture can improve the resilience of urban systems through food diversification to the household level and it can further develop the local and global food market in support of economic stability. At the same time, UA can promote social resilience through the enhancement of socio-cultural ecosystem services, for instance, empowering young adults to understand and be aware of the role and contribution of urban agriculture to the urban communities and environment. These could be conceptualized as a significant driver for sustainability and resiliency for urban communities, which can exist at multiple scales with the diverse function of urban agriculture (Salleh et al. 2020).

UA appears as an effective means to address global warming while also fostering urban transitions to sustainability in many ways, such as creating new commons, amenities, ecosystem services, reinventing urbanity, and encouraging community building by growing local food. The development of urban agriculture in interstitial areas and other residual areas of the cities is also related to another one among the many challenges of sustainability that is a better use of already existing resources (Mancebo 2018).

Urban agriculture refers to the process of growing and distributing edible plants in both public and private places of a city. It may take the form of micro-farms, kitchen gardens, and community gardens, but it may also include medium to large-sized urban vertical farms (UVF).

According to these premises, STRAME research project has identified as a field of investigation the topic of small/medium UVF that in literature appears less developed than the issues of micro UA and mega vertical farms. This scale is assumed to be the focus of research due to the fact that small/medium UVF can be easily adapted to the interstitial spaces of cities and, if properly connected to each other, they can constitute a real urban infrastructure.

58.2 The Urban Landscape of Food

Long-term decreasing stock of agricultural land per capita is an emerging global problem and a challenge for the future. Statistics on the future growth of the world population from the United Nations Food and Agriculture Organization (FAO) reveal that arable land per person is projected to decrease by 2050 to one-third of the amount available in 1970 (FAO 2016). The urban land covered area is expected to increase by four times over fifty years, from 300,000 km^2 in 2000 to 1,200,000 km^2 in 2050 (Angel et al. 2011) reducing in the same time the usable area for agriculture. On the other hand, the increase of the world population requires a corresponding increase in agricultural and food production, and the capacity of the agri-food industry shall increase up to 70% by 2050 to meet the food demand (FAO, World Food Programme 2015). The per capita arable land worldwide is decreasing impressively: it was estimated at 0.42 ha in 1960, and the forecast is 0.19 ha in 2050. In many countries farming more land is no more an option (Silva 2018). A possible option is therefore to include urban areas as part of production systems and many cities already reuse abandoned properties as urban farms and cultivate their own fresh vegetables to feed their citizens (Avgoustaki and Xydis 2020).

The anticipated population growth by 2050, with an expected 70% of it living in the cities, forces a profound revision of recent efforts on food security and the means to accomplish it (Nelson 2010). What is the future of food production and food security within highly dense urbanized areas? What are the practices that could render an urban environment resilient and sustainable today? How productive sites could speak not only about contemporary food production but also of the wider socio-political relations that sustain it?

UVF could envision a more 'entrepreneurial' model of urban transformation, where former productive urban spaces, interstitial, parasitic, residual and undesirable places; i.e., brownfield and contaminated areas, tunnels, terraces of commercial complexes, viaducts, underpasses, odd juxtapositions and remnant spaces between train lines, highways, subways, abandoned industrial zones, transport hubs could emerge as the new productive and ecologically resilient areas and culinary corridors, that are adapted to contemporary alimentary needs and destabilizing climatic realities.

Furthermore, from an environmental point of view, UA has been associated with the creation of habitats for pollinators (Goodman and Minner 2019; Goddard et al. 2010); modulating microclimates and hydrology (Oberndorfer et al. 2007); productive redirection of wastewater, organic matter, and biosolids (Armstrong 2000); blocking atmospheric nitrogen (Herridge et al. 2008) and carbon (Beniston and Lal 2012) which would otherwise contribute to climate change; and stem the loss of agricultural land linked to peri-urban and suburban development (Haight et al. 2016). In addition, UVF practiced on a large scale in urban centers has the potential to allow year-round food production without loss of yields due to climate change or weather-related events and eliminate the need for large-scale use of pesticides and fertilizers (Despommier 2005).

To understand the productive dynamics of contemporary cities within rapidly changing and highly dense urban environments, it is not enough to look at the structural and economic questions alone—we must also investigate the visual languages, conceptual approaches, and morphological terms determined by diverse cultural and social-economical contexts. The need to investigate the potential application of UVF leads us to a study and analysis that calls forth the creation of new areas of urban food production but also the development of new typological hybridizations both on an urban and architectural level that signify a re-entrance for new patterns of production-consumption-distribution, a re-invention of new metabolic perspectives and symbiotic relationships.

58.3 The Research for an Intermediate Scale for Urban Agriculture

Urban agriculture has become a new form of cultural and social expression relating to land use (Langemeyer et al. 2021). Its application can lead to different forms of results, including social cohesion (Coles and Costa 2018), environmental education, and the enhancement of undesirable urban areas (Contesse et al. 2018) (Fig. 58.1).

The theme of urban agriculture is present in several research projects whose objectives are environment and resource efficiency, sustainable use of land and nature-based solutions, climate change adaptation, innovative nature-based solutions in cities, smart agriculture, and food resilience. Regarding these objectives, included in almost all research and funding programs of the European Union, it is possible to identify projects that, both for methodological approach and results obtained, can become crucial for the definition of urban planning strategies aimed at the integration of urban agriculture at the intermediate scale (Fig. 58.2).

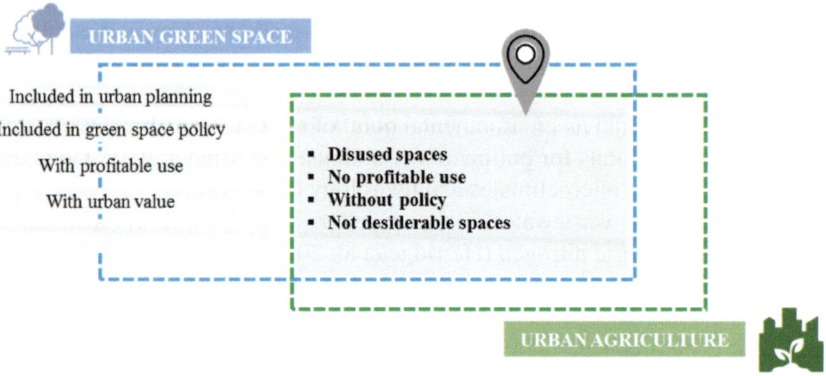

Fig. 58.1 Identification of urban spaces for the application of urban farming scenarios

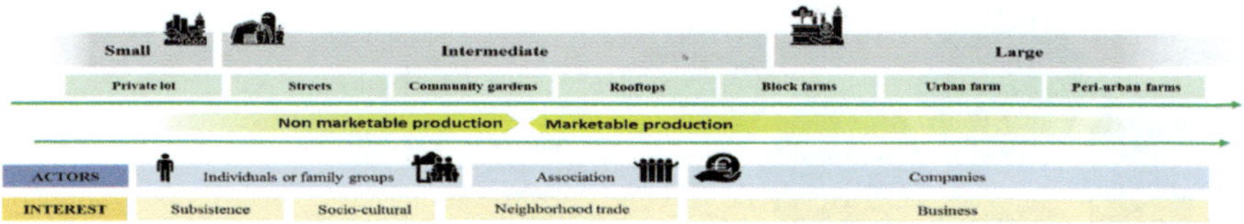

Fig. 58.2 Level, actors, and interests in urban farming projects

The results of a review of projects funding in Europe (Fig. 58.3) show that tendencies most oriented to the urban scale can be found among projects funded within the Urban Innovative Actions program (UIA). This program is an initiative of the European Union that provides urban areas throughout Europe with resources to test new and unproven solutions to address urban challenges. Among the proponents of these projects, there is always a municipality, and their purpose is often to test new practices that can transform into policies in territorial planning. Over the years, the Life program has financed many projects related to agriculture. Some of these projects can be referred to as urban farming and vertical farming, but in general, most of them aim at developing or improving specific cultivation techniques. Within the Horizon program, many more projects focused on the theme of this research can be found. However, most of such initiatives are aimed at improving cultivation techniques and/or accelerating industrial production transition towards more sustainable models.

One of the trends most observed in the projects examined is the focus on smart agriculture, also in the case of small-scale projects. The smart component is applied through diversified tools, which come from the new lifestyles and the tendency to control and govern agricultural systems to optimize their results (Fig. 58.4).

The intermediate scale, identifiable as the urban scale, shows considerable potential with social and environmental implications (Sullivan et al. 2019). On the social side, a process of raising awareness among citizens toward urban farming can be started. This is an opportunity that can generate income and new skills, or green jobs. Furthermore, resilient communities with more sustainable lifestyles can be activated.

	Number of projects examined	Overall budget [€]	EU Contribution [€]	EU Contribution [%]
Life programme	5	11.736.390	6.662.083	56%
HORIZON 2020 Horizon programme	17	71.094.190	60.898.349	84%
UIA initiative	4	16.766.242	16.766.242	100%
SusFood project	4	-	-	-

Fig. 58.3 Summary of the reviewed projects

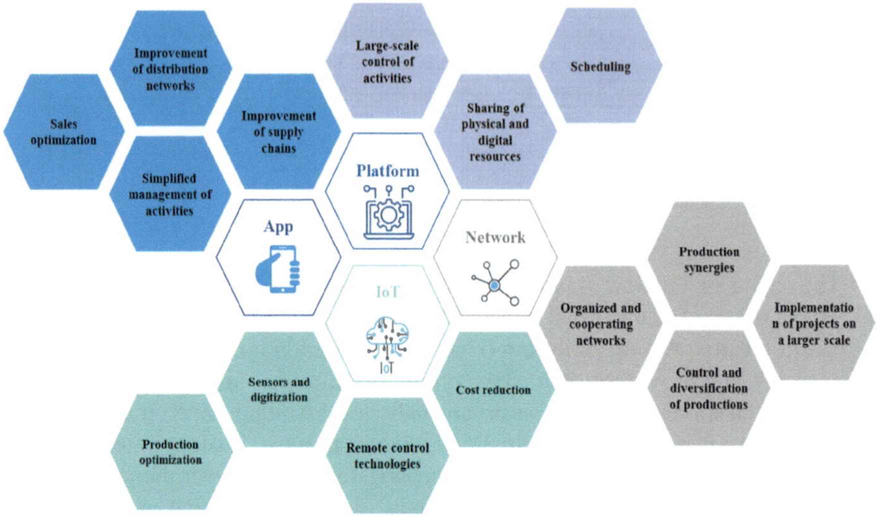

Fig. 58.4 Urban agriculture and smart applications

On the environmental side, the implications of urban farming processes at the intermediate scale are several and depend on the type of intervention implemented. Some of these implications are listed below:

- Increased agricultural production without additional environmental impact.
- Contribution to climate change mitigation by launching strategies that can limit the damage caused by atmospheric events.
- Human health improvement, through high controlled agriculture productions, without the use of pesticides.
- And finally, the possibility to proceed towards cities' re-naturalization, considering that urban farming applications must be imagined as settlements that relate and interact with the context.

58.4 STRAME: An Urban Adaptive Infrastructure

The STRutture Agricole MEtropolitane (STRAME) concept aims at the development of a modular system to build interconnected vertical farms in urban areas with the objective to set up an urban infrastructure able to operate in two configurations:

I. Routine operation: in this case, the urban infrastructure works as a set of distributed urban farms capable of providing food for consumption by local communities (at the scale of the neighborhood or block) or for the development of micro-businesses for the processing, sale, and delivery of vegetables.
II. Emergency operation: the urban infrastructure works as a distributed cultivation system in the city that can be supervised and operated as a single entity (thanks to

the presence of an IoT structure) by a city authority in the event of emergencies that could affect the food supply chain.

The STRAME concept will assume as meta-design constraints the following original features assumed in the research:

- Vision: a constellation of modular UVF landmarks as a physical and digital infrastructure able to increase the resilience of cities and urban areas with respect to disruptive risk factors;
- Scale: to develop a system at an intermediate scale between the mostly diffused micro individual farming and the mega vertical farms; the intermediate scale will allow the system to be more sustainable from a technical, economic, and social perspective;
- Location: the UVF system is placed in "void" areas such as residual spaces—a latent incubator of the ecological (bio)diversity, often located in fringe areas—or in open spaces system of social and public housing districts, for a new urban sustainable landscape as a shared fragment of a collective conscience.

The cultivation system will be based on hydroponic methods which are very well known since they have been studied from mid of the nineteenth century and became widely applied from the first quarter of the twentieth century (Jensen 1997). The purpose is to reduce the need of agricultural soil, by checking all the climatic, water, plant nutrients, and luminous parameters within the vertical farm, or by exploiting LED light integration systems to provide the right hours of light and the right photonic flow intensity specific for each crop scheduled within the UVF.

A remote-control system, based on the features of the Internet of Things, will be developed during the research in order to manage most of the routine checks without the presence of personnel also with the purpose to reduce internal pollution with respect to the external environment. A system of photovoltaic panels will supply electricity, and a dehumidification system will allow a recovery of water for watering plants, regulating internal temperature and humidity.

Concerning the interconnection of the farms, a specific part of the research will be devoted to defining standards and ontologies to enable, on the one hand, interoperability among platforms and, on the other hand, inter-domain interaction. The project will study, at the level of an advanced prototype, an IoT infrastructure that federates the heterogeneous IoT systems building up the complex environment represented by a (possibly distributed) vertical farm while leaving the data ownership and sovereignty in the hands of data owners, by offering means for security and privacy control.

The system will be characterized by some distinctive features:

- Operability: the vertical farm shall be easy to be operated by citizens and households, and therefore the technology adopted will be based on simplicity and easiness of use;
- Maintainability: in order to keep the farms in a state to perform the required functions, the system design will take into account the maintainability factors;

- Reversibility: the UVF modules will be designed in order to fulfill the requirements of reversibility intended as an uncertainty management factor that will allow to disassemble and move the modules if, on a middle time period, the requirements of the city will foresee to use the space of the UVF for other purposes (uncertainty management over time);
- Modularity: the system will be designed according to a modular scheme in order to manage the uncertainty related to the conditions of application in different contexts (management of context uncertainty);
- Interconnectivity: the modules will be equipped with an IoT infrastructure able to connect them with other similar modules in order to allow an overall vision of the productivity and availability of crops in the urban area, making easier the management of food supply in emergency scenarios;
- Agronomic performance: the possibility of replicating the UVF system in any place, even in the absence of agricultural land, electricity, using condensation water to irrigate plants.

The key aims of the STRAME project are:

- to increase the resilience of cities and urban areas in case of extreme events (pandemic, climate change-related issues or geopolitical events) providing—with the implementation of the urban vertical farms—a support for food supply continuity and microeconomic activities;
- to improve the adaptation strategy for preparing the cities to face the future challenges caused by climate changes or pandemic outbreaks;
- to encourage the creation of new jobs by proposing new business models for the construction, operation, and maintenance of the farms or for the set up of micro-businesses for the harvest, sale, and delivery of fruits and vegetables;
- to overpass the limit of many IoT systems that are often single-scoped, disjoint systems are generally defined as IoT silos.

Acknowledgements STRAME research proposal has been awarded with one of the prizes of the RIBA call, launched by the DAStU Department of the Politecnico di Milano to finance basic research, i.e., those interdisciplinary and innovative research, considered strategic for the scientific growth of the Department and that have no easy access to alternative funding channels.

References

Angel S, Parent J, Civco DL, Blei A, Potere D (2011) The dimensions of global urban expansion: estimates and projections for all countries, 2000–2050. Prog Plan 75:53–107

Armstrong D (2000) A survey of community gardens in upstate New York: implications for health promotion and community development. Health Place 6(4):319–327

Avgoustaki DD, Xydis G (2020) Indoor vertical farming in the urban nexus context: business growth and resource savings. Sustainability 12:1965. https://doi.org/10.3390/su12051965

Beniston J, Lal R (2012) Improving soil quality for urban agriculture in the northcentral U.S. In: Lal R, Augustin B (eds) Carbon sequestration in urban ecosystems. Springer, p 279

Coles R, Costa S (2018) Food growing in the city: exploring the productive urban landscape as a new paradigm for inclusive approaches to the design and planning of future urban open spaces. Landsc Urban Plan 170:1–5
Contesse M, van Vliet BJM, Lenhart J (2018) Is urban agriculture urban green space? A comparison of policy arrangements for urban green space and urban agriculture in Santiago de Chile. Land Use Policy 71:566–577
Despommier D (2005) The vertical farm: reducing the impact of agriculture on ecosystem function and services. Columbia University, New York, NY
FAO (2016) The state of food and agriculture 2016
FAO, World Food Programme (2015) The state of food insecurity in the world 2015
Goddard MA, Dougill AJ, Benton TG (2010) Scaling up from gardens: biodiversity conservation in urban environments. Trends Ecol Evol 25(2):90
Goodman W, Minner J (2019) Will the urban agricultural revolution be vertical and soilless? A case study of controlled environment agriculture in New York City. Land Use Policy 83:160–173
Haight D, Eyck Ten L, Arjomand S (2016) Cultivate New York: an agenda to protect farmland for growing food and the economy
Herridge DF, Peoples MB, Boddey RM (2008) Global inputs of biological nitrogen
Jensen MH (1997) Hydroponics. HortScience 32(6):1018–1021
Langemeyer J, Madrid-Lopez C, Mendoza Beltran A, Villalba Mendez G (2021) Urban agriculture—a necessary pathway towards urban resilience and global sustainability? Landsc Urban Plan 210:104055
Mancebo F (2018) Gardening the city: addressing sustainability and adapting to global warming through urban agriculture. Environments 5(3):38
Nelson GC (2010) Food security, farming, and climate change to 2050: scenarios, results, policy options. International Food Policy Research Institute, Washington, DC
Oberndorfer E, Lundholm J, Bass B, Coffman RR, Doshi H, Dunnett N, Rowe B (2007) Green roofs as urban ecosystems: ecological structures, functions, and services. BioScience 57(10):823–833
Salleh AM, Harun NZ, Halim SA (2020) Urban agriculture as a community resilience strategy against urban food insecurity. Environ Behav Proc J 5(13):369–376
Silva G (2018) Feeding the world in 2050 and beyond—part 1: productivity challenges. Michigan State University Extension
Sullivan CA, Bonnett GD, McIntyre CL, Hochman Z, Wasson AP (2019) Strategies to improve the productivity, product diversity and profitability of urban agriculture. Agric Syst 174:133–144

Chapter 59
Resilient Design for Outdoor Sports Infrastructure

Silvia Battaglia, Marta Cognigni, and Maria Pilar Vettori

Abstract Cities, and with them the criticalities and opportunities that characterize urban contexts, are one of the main challenges in the transition toward environmental and social sustainability today. Within the contemporary debate dominated by reflections on the effects of climate change, the culture of design is increasingly oriented to measure itself against the concept of resilience: the limitation of land consumption, together with the technological, functional and energetic reorganization of areas and buildings, is the path taken by design to make the built environment adaptable to the changes taking place, so as to promote development, equity and social inclusion. Public space, defined as a system of open urban spaces, is assuming an increasingly important role in urban and environmental regeneration processes. At the same time, the topic of sport and the public infrastructure of cities for practicing physical activity is an increasingly important factor for urban and social quality, requiring strategies capable of redefining places and the way they are used in line with objectives of environmental quality and collective well-being. The picture that emerges from studies and research on the European and Italian panorama of sport infrastructures highlights interesting and innovative trends that show, also in this sector, an increasing focus on the themes of urban, architectural and social resilience. On the basis of this premise, the contribution aims to analyze the recent evolution of the design of public space in relation to sports practices as an area where resilience policies are applied.

Keywords Sports infrastructure · Urban regeneration · Resilient design · Public space · Ecological footprint

---

S. Battaglia · M. Cognigni (✉) · M. P. Vettori
Polytechnic University of Milan, Milano, Italy
e-mail: marta.cognigni@polimi.it

S. Battaglia
e-mail: silvia.battaglia@polimi.it

M. P. Vettori
e-mail: mariapilar.vettori@polimi.it

E. Arbizzani et al. (eds.), *Technological Imagination in the Green and Digital Transition*, The Urban Book Series, https://doi.org/10.1007/978-3-031-29515-7_59

59.1 Introduction

In the era of environmental and economic crisis, characterized by climate change as a constant issue of planetary scope, and energy emergencies linked to resource scarcity, the capacity of societies to implement solutions to mitigate the causes of the primary problems and adapt to their main effects is assuming a central role (Martinelli and Mininni 2021).

Research and experimentation place at the center of the debate the need to address at different scales and in different contexts the issue of resilient design focused on the ability to give answers in terms of adaptability and flexibility framed in a vision geared toward sustainable development (Leone and Tersigni 2020).

Climate change, aggravated by the high rate of urban growth in the last century and the consequent increase in the production of greenhouse gases, requires capacities for "resilient development" (Adams and Watson 2010), i.e., development based, in addition to mitigation actions, on adaptation strategies capable of responding quickly and flexibly to actions arising from the changing environmental, social and economic needs (Ahern 2011; Ferrari 2021).

Transforming cities into resilient organisms, able to adapt to natural and anthropogenic pressures is a priority nowadays recognized by International Bodies, Public Administrations, Research Institutions (Rockfeller Foundation and Arup 2015) and the stakeholders involved in the project process. Adapting urban settlements and public space into resilient systems is a priority and transversal objective where design and innovation must be combined with reference to the conservative, adaptive and regenerative needs to minimize the impacts and vulnerabilities resulting from extreme climate events (Fabbricatti 2013; Newman et al. 2005)

Although the concept of resilience, which originated in the natural sciences in the 1970s, has been an integral part of the vocabulary of urban planning and regeneration for some years now (Landi 2012; Roberts 2000), understood as "the ability of the individuals, communities, institutions and economic systems that make up a city to survive, adapt and grow independently from any kind of shock or stress they may have had to suffer" (Arup 2015), in its widely accepted meaning, the concept of resilience, especially when applied to the public and collective spaces of the city, represents the capacity of a system to regenerate and reorganize itself following an adverse event or change (Fig. 59.1).

59.2 Sport, Space and Society: Resilient Design

In the context of a scientific debate on the concept of resilience that is involving different fields of knowledge, the development of a "resilience thinking" on the built environment can find innovative ideas in the design of public space, a privileged field of experimentation for projects that place social and environmental wellbeing at the center of their work.

Fig. 59.1 Sport, space, society, Portovenere, 2021, Ph. Maria Pilar Vettori

Many studies and research projects are focusing on defining new intervention strategies (Jha et al. 2013; Leone and Tersigni 2020; Losasso 2015), not only for the built system, but also for the unbuilt system, i.e., that open connective tissue that structures the city and constitutes its public spaces (Gehl 1971). Numerous programs, such as the Climate Change Adaptation Plan, envisaged by some cities (Barcelona, Copenhagen, Zurich, Paris) have precisely elected the public space as a privileged sphere of intervention for the adoption of specific resilient actions and strategies.

In this logic, the design of public space can be an important element for contrasting the negative impacts of climate change—such as the management of meteoric water, flood prevention, containment of temperatures, the reduction of urban heat islands—but above all it can promote interventions designed to improve urban health and foster a resilient habitat. These integrations are an exceptional way of doing and conceiving interventions for the defense of territories, as they can be interpreted as a possible way of enriching the articulation of the "Project of the Soil" (Secchi 1986) by developing new materials and new urban functions.

Within the dynamics of public space transformation, sport is an important tool for fostering resilience and regeneration, which can activate new uses and meanings of the places in which it is practiced.

Talking about public spaces broadens the view to include a series of aspects that are usually marginal in the design and planning processes. In this sense, the main themes that emerge as characterizing factors of a resilient project or initiative can be identified in the concepts of *flexibility*, *community* and resource *management*.

The definition of factors is primarily concerned with *flexibility*, understood as the ability to provide adaptable responses to constantly evolving needs.

This theme has a threefold significance: flexibility of the objectives in terms of vision and project drivers; long-term flexibility through an ability of the design process to work in steps; flexibility of the solutions in terms of multifunctionality and adaptability of the space in its life cycle.

Crisis factors of European cities have generated a large quantity of unused urban voids: awareness of their potential has led to the performance of various experiments on their use in economically sustainable, socially useful, and environmentally aware ways, which find an important opportunity in the combination of urban sports and public space (Castelli 2020; Cognigni and Vettori 2020).

Several programs implemented in Spanish cities are significant in this sense, such as "Esto no es un solar" in Zaragoza or the "Dispositivo de la Cebada", created in Madrid in 2011 on the initiative of the Todo Por La Praxis association. These place the synergy between different sporting activities and temporary cultural and social events at the center of a principle of physical and social urban resilience.

If placed at the center of projects, the *community*, understood both as a common identity and a sense of belonging to a group, and as a set of relationships, can strengthen existing links and spark new ones. One of the possible keys to the success of such an intervention is the active involvement of the residents in the initiative, with a view to stimulating a sense of belonging to the place where they live and a desire to take care of their neighborhood.

Finally, placing the emphasis on *resource management* involves stimulating the development of good practices. Resource management is seen as sensitivity toward the importance of recycling and reducing waste. In this sense, the resilience strategies of public space design, can represent a global reference landscape in terms of tools for achieving high standards of quality of life and environmental protection.

Within this scenario, the immaterial dimension linked to Information Technologies (IT) and Information and Communication Technologies (ICT) supports the management and decision-making processes in the various phases of the design process but also of the activities of users, redesigning decision-making and behavioral methods.

The dimension of computer technologies aimed at simulation and modeling actions allows to verify the contribution of the open space system to the increase of resilience through the ability to correlate data, scenarios, criticalities, opportunities, strategies and projects that reduce climate vulnerabilities through adaptation processes within urban districts most exposed to impacts. For example, the computational design offers useful features to incorporate relevant sets of information in open spaces, allowing them to be modelled and parameterized to evaluate, through simulations, the performance response of technical and design alternatives, also in the involvement with stakeholders and according to participation processes, as happened for Todo Por La Praxis.

59.3 Resilience and Outdoor Sports Infrastructure

Is sport an opportunity for urban resilience? The answer to this question is to be found in the characteristics that infrastructures dedicated to sport are assuming in relation to the contexts that host them, and to the increasingly evident need for communities to have healthy lifestyles.

As reported in the CONI 2016 sustainability report, it is important to consider how "Sport is a vehicle for inclusion, participation and social aggregation as well as a tool for psychophysical well-being and prevention. Moreover, it plays a fundamental social role as a tool of education and training that allows the development of skills and abilities essential for the balanced growth of everyone". Sport teaches resilience, and resilience mobilizes the resources identified to find positive and flexible responses to adaptation.

Public space, today an open and flexible place, together with a culture of leisure time, is based on a concept of the city in which all its constituent elements are geared toward the development of integrated programs and planning strategies that promote the idea of a healthy city.

Sport is an important tool for regeneration, whose integrative value can activate new uses, meanings and changes to the livability of the city (Faroldi 2019).

The themes characterizing the current role of sports infrastructures in the "promotion" and "production" of urban resilience of anthropized contexts, aimed at guaranteeing an ever-greater adaptive capacity of these elements of the urban system, lead to a forward-thinking reflection on the acquisition of adequate skills with which to tackle and manage the processes of planning, design, construction and management of the places and artefacts to be used for practicing sporting activities (Allegri and Vettori 2018).

In recent years, sport and the places connected to it have witnessed a profound transformation, which has produced food for thought on the concept and value of resilience. The uses, design and purpose of sports facilities have changed. The trend is to create places that can accommodate not only athletes but the whole community, with services defined by the needs and demands of society.

There are more and more examples of public spaces conceived to host informal sports practices, in synergy with other functions, taking inspiration, also from a morphological point of view, from the spatiality necessary for sport: the movement of the ground and the spaces, the alternation of permeable and impermeable materials, the coexistence of different categories of users, lead to a stratified use of the spaces (sporting activities, playgrounds, relaxation areas).

Now, physical activity is seen by more and more people as a means to achieving physical and psychological wellness: these changes are also reflected in the physical configuration of the spaces dedicated to sport in terms of type, characteristics and functionality. Alongside sports facilities designed to host competitive or structured sporting activities, new configurations are emerging, in which sport is practiced more informally and with greater flexibility. This trend leads to the creation of systems that integrate with the context, capable of redeveloping degraded areas, rethinking

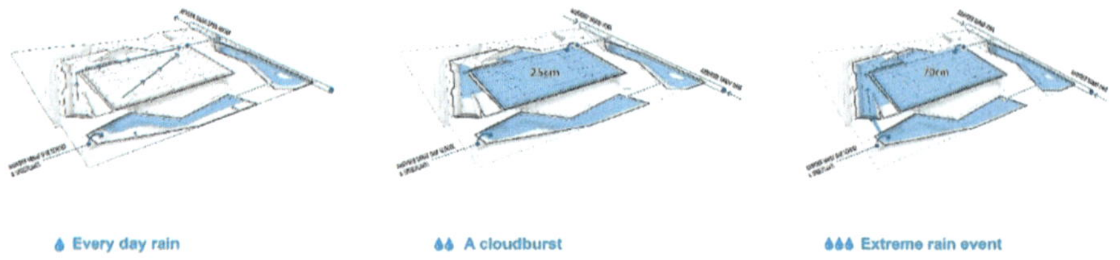

Fig. 59.2 Watersquare Benthemplein, Water system's concepts, Rotterdam, 2013, De Urbanisten, grafics by: De Urbanisten

existing spaces, and implementing and promoting a concept of *resilience* that is both the premise and the objective.

Squares, boulevards, urban parks, but also small residual spaces, roofs and wastelands take on physical configurations that can not only respond to a renewed concept of public space for sports but also configure possible solutions to combat the consequences of climate change (Santamouris 2013; White 2010).

One of the most investigated themes is the opportunity offered by these spaces, thanks to their large surface areas and high ceilings, to control the impact of rainwater in urban areas at risk of flooding, such as some of the projects carried out in Rotterdam as part of the *Stadvisie Rotterdam 2030* program: squares designed for games and recreational activities featuring a layout that can be changed depending on the climatic conditions. Equipped with volleyball, basketball and football pitches and terraces, the so-called "water squares" (*Bellamyplein* in Spangen, *Benthemplein* in Agniesebuurt, Tiel) are designed to "collect" rainwater in a disciplined way and to decant it to prevent flooding, thanks to the use of construction systems and appropriate materials (Figs. 59.2, 59.3 and 59.4).

With the same objective (water management), some interventions take advantage of the opportunities offered by the physical characteristics of different surfaces for sports spaces, the presence of green surfaces, or the necessary altimetric articulation of the spaces. They tend to work on the surface, the porosity of the materials, and the topography to allow their use even during extreme events. See, for example, the *Rabalder Parken* in Roskilde, Denmark, a system of open spaces for skateboarding, skating and BMX, sports which exploit the slopes of the land to create real basins in which to collect water.

59.4 Conclusions

In a perspective of uncertainty and unpredictability due to the qualification and quantification of the consequences of climate change on urban systems and the increase of the world population in cities, the difficulty of finding references for the direction of urban design is emerging.

Fig. 59.3 Benthemplein, Rotterdam, 2013, De Urbanisten, Ph. Ossip van Duivenbode

Multiple studies have been conducted to define new indicators of urban resilience (Carpenter et al. 2001; Normandin et al. 2009), investigate the role of materials and technologies in mitigation and adaptation strategies (Doulos et al. 2004), and to evaluate the effects of resilient strategies with models and instrumental analysis (D'Ambrosio and Leone 2015).

Understanding how to increase resilience, simulate its setbacks and measure its expected effects is a task that cannot be delegated to territorial governance. It requires reflection on the scale of the technological design, its methods and its tools to build a resilient city as a place that does not simply implement extemporaneous remedies to climate change but adapts to them by building new social, economic and environmental opportunities.

The relationship between resilient approach and the use of intangible technologies becomes ever closer, in the control and monitoring of interventions, in the need to simulate and model as well as to use data within the analytical and forecasting processes.

In this logic, the urban spaces of the community become places of change and not simply adaptation, with physical and social characteristics able to respond to the transformations. The community is the central element in the activation of the

Fig. 59.4 Watersquare, Tiel, 2014–2016, De Urbanisten, Ph. De Urbanisten

projects, promoting awareness of the need to address the social, urban and climate changes underway. This would be done through an approach that is inclusive but at the same time conscious of the need for a systemic and scientific vision, able to translate social innovations into behavioral innovations by redesigning public space to reflect the community and its aspirations.

References

Adams M, Watson, D (2010) Design for flooding: architecture, landscape, and urban design for resilience to climate change. John Wiley & Sons, Hoboken, New Jersey, USA

Ahern J (2011) From fail-safe to safe-to-fail: sustainability and resilience in the new urban world. Landsc Urban Plan 100(4):341–343

Allegri D, Vettori MP (2018) Infrastrutture sportive complesse e resilienza urbana. TECHNE J Technol Archit Environ 15:165–174

Arup & Partners (2015) City resilience framework, for "100 resilient cities" project of Rockefeller Foundation

Carpenter S, Walker B, Anderies JM, Abel N (2001) From metaphor to measurement: resilience of what to what? Ecosystems 8:765–781

Castelli C (2020) Resilienza e sport. Franco Angeli, Milano

Cognigni M, Vettori MP (2020) Spazio, Sport, Società. La pratica sportive nel Progetto dello spazio pubblico contemporaneo. TECHNE J Technol Archit Environ 19:142–152

D'Ambrosio V, Leone MF (2015) Controllo dei rischi del cambiamento climatico e progettazione ambientale per una rigenerazione urbana resiliente. Il caso applicativo di Napoli Est. TECHNE J Technol Archit Environ 10:130–140
Doulos L, Santamouris M, Livada I (2004) Passive cooling of outdoor urban spaces. The role of materials. Sol Energy 2:231–249
Fabbricatti K (2013) Le sfide della città interculturale. La teoria della resilienza per il governo dei cambiamenti. Franco Angeli, Milano
Faroldi E (2019) Architettura dello sport. Progettazione, costruzione gestione delle infrastrutture sportive. Maggioli, Santarcangelo di Romagna
Ferrari GF (2021) Le smart city al tempo della resilienza. Mimesis, Milano
Gehl J (1971) Life between buildings: using public space. Island Press, Washington, Covelo, London
Jha AK, Miner TW, Stanton-Geddes Z (2013) Building urban resilience: principles, tools and practice. The World Bank, Washington, DC
Landi A (2012) Il concetto di resilienza: origini interpretazioni e prospettive. Sociol Urbana Rurale 99:79–98. https://doi.org/10.3280/SUR2012-099008
Leone M, Tersigni E (2020) Progetto resiliente e adattamento climatico. Metodologie, soluzioni progettuali e tecnologie digitali. CLEAN, Naples
Losasso M (2015) Rigenerazione urbana: prospettive di innovazione. TECHNE J Technol Archit Environ 10:4–5
Martinelli N, Mininni M (2021) Città, Sostenibilità, Resilienza. Donzelli Editore, Rome
Newman P, Beatley P, Boyer H (2005) The resilient city. How modern cities recover from disaster. Oxford University Press, Oxford
Normandin J, Therrien M, Tanguay G (2009) City strength in times of turbulence: strategic resilience indicators. In: Proceedings of the joint conference on city futures
Roberts P (2000) The evolution, definition and purpose of urban regeneration. In: Roberts P, Sykes H (eds) Urban regeneration: a handbook. Sage, London, pp 9–36
Rockfeller Foundation and Arup (2015) City Resilience Framework New York USA
Santamouris M (2013) Using cool pavements as a mitigation strategy to fight urban heat island—a review of the actual developments. Renew Sustain Energy Rev 26:224–240
Secchi B (1986) Progetto di suolo. Casabella 520–521:19–23
White A (2010) Water and the city. Risk, resilience and planning for a sustainable future. Routledge, Abingdon

Chapter 60
Sustainable Reuse Indicators for Ecclesiastic Built Heritage Regeneration

Maria Rita Pinto, Martina Bosone, and Francesca Ciampa

Abstract In the context of anthropogenic impacts on pollution and global warming scenarios, reject from the construction sector accounts for 36% of European waste. This waste percentage includes disused and abandoned buildings that have lost the value of their function over time. In order to reduce the ecological footprint they generate, the paper rethinks Recovery in its circular meaning to put these buildings back into a normal circuit of usability, improving the creation of resilient urban habitats. In particular, decommissioned ecclesiastical buildings constitute a huge quantity and significant quality heritage, as by cultural, perceptive, morphological and material values. The sustainable reuse of this heritage must act on its double impacting value: the tangible one linked to the material culture of the buildings and the intangible one, linked to the identity values of sediment instances. Through a comparison desk research of more than 140 cases of reuse on a European scale, the contribution arrives at a system of indicators that allow evaluating the reuse sustainable compatibility of these buildings, able to promote prosperity, inclusiveness and social equity. These indicators make it possible to assess the appropriateness of design actions aimed at mediating between the conservation of the built heritage and the transformative needs of contemporary instances. The results provide scenarios tool of sustainable recovery, capable of transforming waste into a resource, extending the life cycle of the ecclesiastical heritage and thus mitigating its environmental impact, as well as the cost related to the loss of cultural values and identity for the community.

Keywords Sustainable reuse indicators · Recovery · Ecclesiastic built heritage

M. R. Pinto · F. Ciampa (✉)
Department of Architecture, University of Naples Federico II, Naples, Italy
e-mail: francesca.ciampa@unina.it

M. R. Pinto
e-mail: pinto@unina.it

M. Bosone
Research Institute on Innovation and Services for Development of the Italian National Research Council (CNR-IRISS), Naples, Italy
e-mail: m.bosone@iriss.cnr.it

E. Arbizzani et al. (eds.), *Technological Imagination in the Green and Digital Transition*, The Urban Book Series, https://doi.org/10.1007/978-3-031-29515-7_60

60.1 Introduction

In the European context, the role attributed to culture and its testimonies is consolidated into the Treaty on European Union (TEU). It referred to the desire to be inspired "by the cultural, religious and humanistic heritage of Europe, from which the universal values of the inviolable and inalienable rights of the person, freedom, democracy, equality and the rule of law have developed" (De Medici 2021). As indicated by the ex art. 2 of the Treaty (Article 3 in the current text), the European Union respects the richness of its cultural and linguistic diversity and supervises the preservation and development of European cultural heritage. In line with these principles, cultural heritage represents a strategic resource, able to determine positive externalities in the four sectors of sustainable development: economy, culture, society and environment (Pinto, 2004). To reduce the ecological footprint they generate, the contribution rethinks the rehabilitation, in its intrinsic circular meaning, to design scenarios capable of putting churches back in a normal circuit of usability, with the creation of sustainable urban habitats (Caterina 2013). The reuse of churches simultaneously affects the cultural heritage in cities and European cultural landscapes in the perspective of the circular economy as a model of sustainable development. The functional recovery of churches is able to generate positive impacts.

As highlighted in the Leeuwarden Declaration on Adaptive Reuse of the Built Heritage (ACE (Architects' Council of Europe) 2018), adaptive reuse interventions have positive impacts on the circularity of sustainable development processes, simultaneously managing to preserve and strengthen the original cultural values and the physical consistency of the heritage. The complex interrelationships between culture, economy, society and environment make adaptive reuse a necessarily multidimensional process in which new forms of economy, promotion of social cohesion, well-being and environmental protection can be tested and adopted (Bosone and Iodice 2021). Interpreting the adaptive reuse of cultural heritage as a "restorative, regenerative and sustainable form of conservation" (Gravagnuolo et al. 2017) requires the adoption of a circular development model capable of achieving economic growth and well-being, "by separating growth from resource consumption" (Foundation 2013). Moreover, such a model, by fully aware a new use value to the reused assets, allows extending their useful lifecycle, reflecting exactly one of the characteristics of the circular economic model. The reuse of churches can bring benefits in environmental and socio-economic terms, limiting the consumption of non-reproducible material resources, handing down the traditional construction culture, favoring the use of labor, limiting the processes of industrial production of materials and components compared to what is required by the construction of new buildings (Pinto et al. 2017). The original values system represents the so-called "intrinsic value", as accredited in the Burra Charter (Australia ICOMOS 1979). Intrinsic value is linked to the "spirit of place" and reflects the specific identity of a place, expressing the physical-spatial relationship that, over time, has bound communities to their environmental context. It is thus the result of a 'social construction' whose interpretation and evaluation require the participation of the local community in its various components.

In the processes of territorial enhancement and regeneration, the role of intrinsic value is fundamental in that it allows a direction to be set for the development and management of a site, a historic center, a city, consistent with the local history and with the value-bearing heritage embodied in its cultural heritage (Pinto et al. 2017). In these terms, reuse is an economic challenge for the construction sector, requiring a previous evaluation of the rehabilitated product through decision criteria for project (Robert 1991). The system of values attributed the potential for use identified in it, meanings for the church reuse process and contributing to the quality of location (construction, esthetic and dimensional) of existing buildings (Bianchi and De Medici 2023). The difficulties of reusing this type of building depend on the characteristics (for example, the presence of large rooms, the efficient thermal behavior, the reliability of lighting and natural ventilation systems, the use of durable materials), the skills of the workforce and the materials used not available on the market (Fitch 1990). This makes clear the need to identify suitable tools for the sustainable reuse of churches; therefore, the paper returns a system of indicators that allow evaluating the adequacy of the project actions, aimed at mediating between the conservation of the built heritage and the transformative needs of contemporary instances. These results provide scenarios of sustainable recovery, capable of transforming waste into a resource, lengthening the life cycle of the ecclesiastical heritage, and thus mitigating its environmental impact, as well as the cost linked to the loss of cultural and identity values for the community.

60.2 Methods and Materials

The care of building systems is the strategy of protection for the conservation and transmission of common goods to future generations. Based on the essential role of maintenance and reuse operations, they are guarantors of the ideal choice of a new functional purpose compatible with the character and meaning of the building on which to intervene (Bosone and Ciampa 2021).

In this research, the adoption of an analytical-making process aims at highlighting the necessity of adopting a systemic perspective to understand the complex interrelationships between culture, economy, society and the environment. In this perspective the adaptive reuse is considered as an interdisciplinary process, in which the creative and collaborative dialogue between different professions of expert and common knowledge leads to the elaboration of a tool able to re-interpret the needs expressed by the stakeholders as requirements for the reuse project (Pinto, 2004). This first co-design phase ensures that the tool can be used by the experts not only to preserve the building but also to valorize tangible and intangible values linked to the new use. The elaboration of an analytic tool becomes the opportunity to facilitate not only the involvement of religious and civil communities in knowledge and decision-making processes, but also the dialogue between them and expert knowledge, overcoming the traditional view of top-down projects imposed on passive and unaware communities. The opportunity to elaborate a preferred choice, through the analysis of a wide range of cases, derives from the ecclesiastical building witnessed as growing disposal of churches abandonment (Bullen and Love 2011). For this reason, the research focuses on a selection of 140 church reuse cases, considered good experimental practices of the literature, starting from the observation of experiences that stimulate a reflection on the contribution of the built environment rehabilitation project with respect to contemporary instances. The selected cases refer to virtuous practices of reuse of churches carried out in the last 30 years. These are case studies that, for the design made, have been awarded or indicated in literature and in international scientific journals as best practices. In particular, this selection of cases was also chosen to meet the current guidelines and guidelines described and addressed in the previous paragraph. These case studies have led to the triggering of transformation processes with experimental consequences and significant repercussions on the rehabilitation project innovation in the built environment management. The selection of these good practices has been structured with the aim of returning a scenario of the appropriate uses to safeguard and, when possible, increase the values of a disused church. This is aimed at deepening practices and behavioral correspondence to the adoption of different uses. Each card highlights the need to establish a previous description of the church to be reused, in order to study the opportunities for intervention offered and the criteria adopted for the new function inclusion (Fig. 60.1).

The card returns the system of requirement categories and features attributed to the intervention and the image of the potential use expressed by the function, this in order to provide useful indicators for the reuse design of deconsecrated churches. The terms of investigation that make up the structure of each individual intervention, return and discrete the card-type of four parts. The data relating to the first part are a particular report established for each own case. It concerns the location site, useful information to analyze the type of territorial geography distribution about the interventions conducted; the period of construction, useful information to indicate the time interval of existence of the building; and the typology, information aimed at understanding the original planning structure. The original use and any changes of it can be deduced from the type of cult in reference before the intervention, information necessary to understand any choices imposed by previous operations; the features of the building, information necessary for the description of the intervention following the insertion of the new function.

The second part contains data relating to the framework of the reuse project with reference to the period of intervention; the commission, given that it influences the interests acting on the choice of the new function inserted. It concerns also the new use, given that it affects the type of interventions to be carried out to adapt the structure to the new function; the presence of volumes added above ground or underground, given that it determines a transformation of the envelope of the structure and therefore of the image perceived by its users. The third part of the card consists of dimension the reuse project, which inevitably generates a new identity for the church deriving from the integration between the original characteristics of the building and the new use

S. GIUSEPPE CHURCH | *Spazio Kor*

Piemonte, Asti (AT)

THE BUILDING

Period of contruction
1660
Typology
Latin cross with rectangular single-nave church
Original use
Church
Previous use
Military building
Building features
The church has a rectangular floor plan with a single nave, punctuated by two deep communicating chapels on each side.

REUSE PROJECT

Commission
Asti Municipality

New use
Theatre and Museum

Additional volumes above or below ground level
yes ☐ no ☒

Period of reuse
2016

DIMENSION

Environmental Sustainability | **Social-Cultural Sustainability** | **Technological-Economic Sustainability**

REQUIREMENT

Morphological-dimensional features

- Preservation of the interior spaces original geometric configuration:
yes ☒ no ☐
- Preservation of the existing building dimension:
yes ☒ no ☐
- Preservation of aesthetic relationships with the context:
yes ☒ no ☐

Perceptive-cultural features

- Recognisability of the transformation:
yes ☒ no ☐
- Reversibility of transformation:
yes ☐ no ☒
- Acceptance of transformation:
yes ☒ no ☐
- Respect for collective memory:
yes ☐ no ☒

Material-constructive features

- Compatibility of transformation:
yes ☒ no ☐
- Conservation of technic elements:
yes ☒ no ☐
- Respect for the construction system:
yes ☒ no ☐
- Life-length of transformation:
yes ☒ no ☐

DESCRIPTION

The church has been reused as a theatre: the chancel welcomes the tribune and the nave the stage. The sacristy has been reused as foyer, ticket office and toilets for the audience; the rectory as dressing rooms, actor rooms and toilets. The elongated block behind the presbitery is a stagecraft museum.

ENVIRONMENTAL SYSTEM		TECHNOLOGICAL SYSTEM	
ENVIRONMENTAL UNIT	ORIGINAL SPATIAL ELEMENTS	TECHNOLOGICAL UNIT	NEW TECHNIC ELEMENTS
Theatre Hall	Single nave	Closure	Exterior vertical fixtures Ground floor Roofs
	Presbytery		
Foyer + Ticket Office + Toilets	Sacristy		
		Vertical int. partition	Vertical internal walls Vertical internal fixtures Protective elements
Dressing rooms + Artist rooms + Toilets	Rectory		
Exhibition Hall	Chapel	Inclined int. partition	Internal stairs

Fig. 60.1 Example of church card

that is made of it about Environmental, Social-Cultural and Technological-Economic Sustainability.

The acquisition of data, schematized within the previous parts of the card, allows analyzing each case through a decomposition into respective three features: Morphological-dimensional, Perceptive-cultural and Material-constructive (De Medici and Pinto, 2012).

The first one features, attitude of the building or its parts to represent stylistic and artistic canons recognized. The limitations concern the preservation of the interior spaces about the original geometric configuration and preservation of the existing building dimension. In addition, the preservation of the interior spaces original geometric configuration, in order not to change the balances present in the area, altering the vocations of development that determine its recognizability. In particular, the preservation of aesthetic relationships with the context, in order not to vary the morphological, dimensional and proportion relationships between the building and the environment surround.

The second one features, aptitude of the building or its parts to constitute evidence of historical events or periods; in many cases, the changes that over time have affected the built—due to variations in use or the changing needs of users—constitute overlapping stratifications, which highlight its role as a historical document. The limitations aimed at the permanence of this feature in the choice of new functions and in the interventions that follow concern the recognizability of the transformations in order to ensure the clear distinction between the new and the elements, avoiding camouflage. This principle aims to preserve the identity of the asset, allowing a clear dating of the interventions suffered by it. Also, the reversibility of the transformations, aimed at allowing the possibility of removing the added elements or materials or restoring those removed. In fact, it is necessary to provide both that the development of technological research makes available, over time, intervention solutions with a greater degree of compatibility, and that the changing needs of use of the good requires new changes. In these cases, it is necessary to be able to bring the building back to a condition as close as possible previous to the transformation, limiting the loss of matter and identity. The acceptance of the transformations, to ensure that variations in use, is shared not only by the client and the designers, but also by the direct, indirect and potential users of the building. This must be verified during the planning through the respect for collective memory, aimed at ensuring that transformations do not alter the recognizability of the good as a representative element of the identity of social groups, which attribute specific symbolic values to it.

The third one features, attitude of the building or its parts to represent the material and constructive culture of a place or an era. The limitations imposed by the compatibility of transformation concerns conserving technic elements in order to respect the technological and construction system. It aimed at ensuring the permanence of the materials present using building products compatible with them and technological solutions that allow minimizing the loss of material. In addition, it concerns the respect for the construction system, in order to protect the construction techniques present and the system of relationships that binds the elements that make up the building organism, testimonies of the material culture of a given historical moment or its evolution over time. The possibility to recognize the constructive conception of the asset, using materials and techniques that do not hide its traces. Moreover, the life-long transformation, with the aim of avoiding that the interventions carried out are subject to a rapid process of degradation, induces negative effects on up-cycling the elements (Figs. 60.2 and 60.3).

The fourth card part concerns the breakdown of the project description according to the systemic conception of Architectural Technology, divided into Environmental and Technological systems. The first one describes the link between the Environmental Unit and the original spatial elements, specifically what use are located into

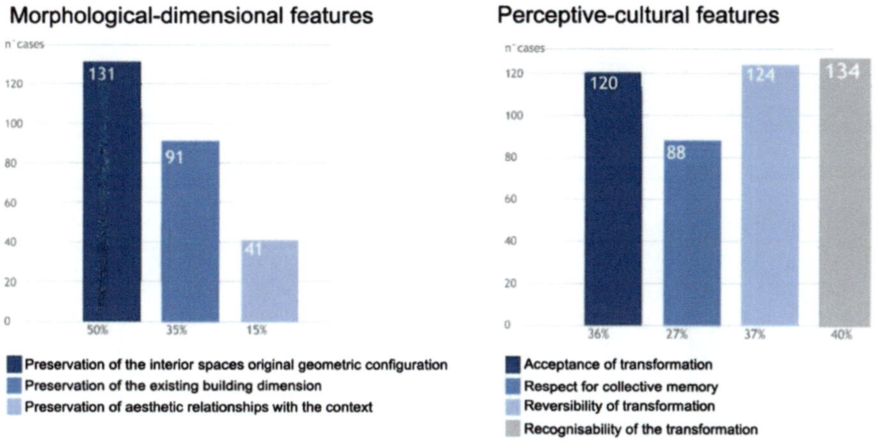

Fig. 60.2 Reuse features analysis (morphological-dimensional and perceptive-cultural)

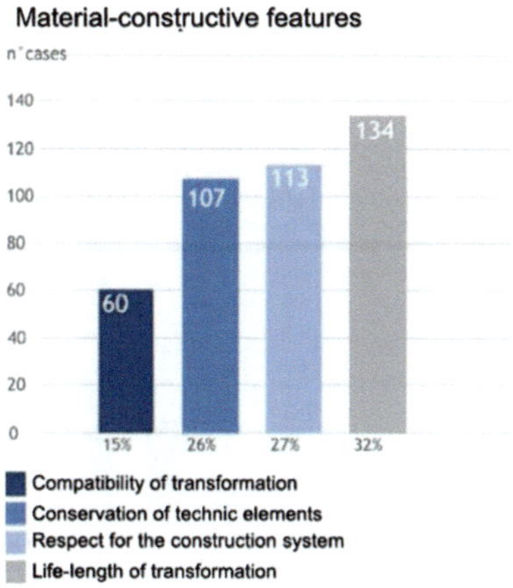

Fig. 60.3 Reuse features analysis (material-constructive)

original spatial elements, connoting new activities established and the physical transformations. The second one describes the link between technological unit and new technic elements, respectively returned by the physicality of the functional compartments and by the elements that allow their transformation. This comparison allows relating how spatial elements are more predisposed to accommodate certain functions rather than others.

They aimed at ensuring the compatibility of transformations to avoid that the use of inadequate materials and technologies or the variation of elements or spaces configuration can cause damage or accelerate its degradation process preserving the life-long transformation (Figs. 60.4 and 60.5).

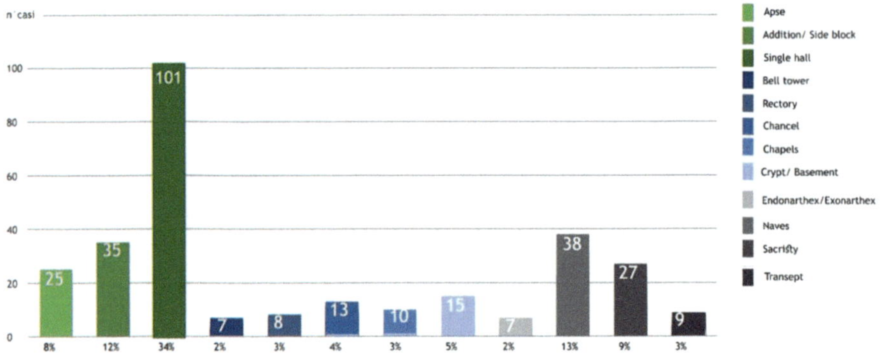

Fig. 60.4 Most frequent original spatial element analysis

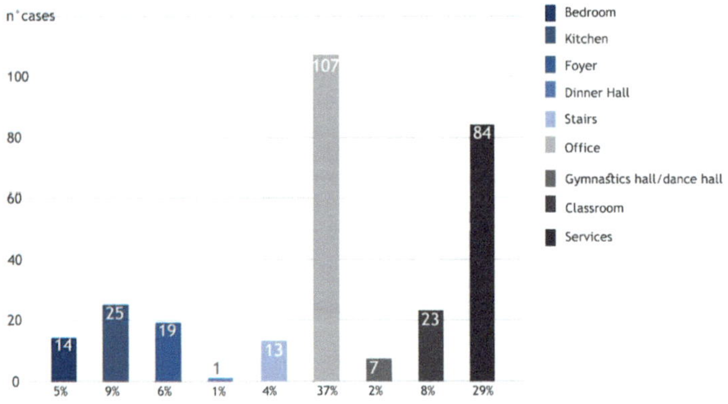

Fig. 60.5 Most frequent environmental unit analysis

This relationship in the functional field of space reuse is inextricably linked to the dominant technological units used in the reuse project to perform the function, detailed in the dominant technic elements to meet the performance requirements related to the new use (Figs. 60.6 and 60.7).

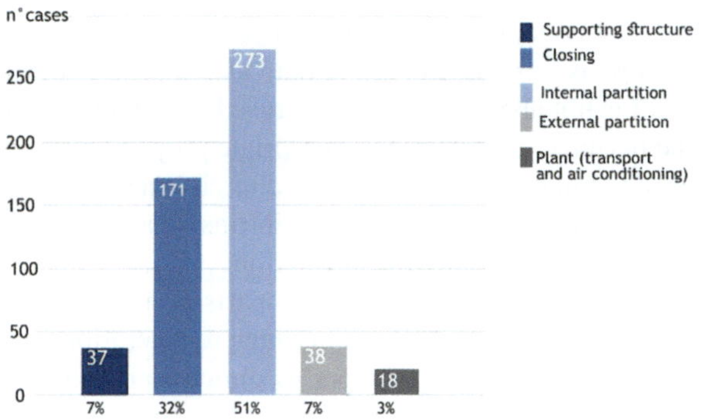

Fig. 60.6 Most frequent technological unit analysis

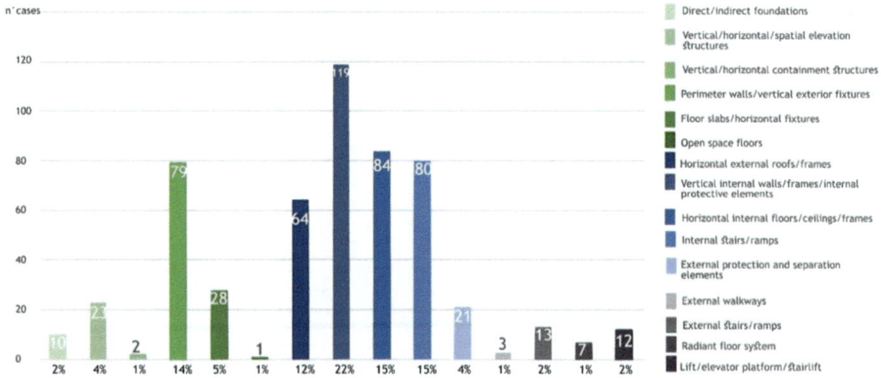

Fig. 60.7 Most frequent technic element analysis

60.3 Results and Discussion

These analyses reveal the existence of a link between the intended reuse and its transformative impact on the built cultural heritage. From the elaboration, the original spatial elements have been reused in accordance with the vocations of the environmental units, using technical elements not detrimental to the cultural image of the site, the techniques and materials existing in it. There is, therefore, the possibility of identifying and preferring certain design rather than others based on the new use and the environmental and technological units that derive from it. The insertion of a new function must inevitably consider the significance of the patrimony of ecclesiastical religious architecture as this value could remain also after the intervention of reuse. It is a meaning that is not abandoned with the ruse but influences the physical transformations of the heritage, determining appropriate project and compatible with the sacred character of the place. In particular, the church can also be deconsecrated and used with a new profane purpose, but it is this intrinsic value of the cultural heritage built to guide the most appropriate transformation.

The experimentation allows creating a comparison matrix from which sustainable reuse indicators for ecclesiastic built heritage regeneration can be extrapolated (Fig. 60.8). By using sustainability factors as dimensions, you can determine the requirement category associated with their features. The latter, compared with the actions determined by the design, referring to the more frequent technic and environmental elements determine the system of indicators. The latter present different forms of measurement demonstrating the degree of mixture and complexity of the processing obtained. The ability to aggregate them allows you to also dissociate them, using them if necessary only for the verification of a certain dimension of the project or at the end to compare its verifiability.

Dimension	Requirement Category	Features	Indicators	Unit of Measure	Actions: Intervention Project Choice — More frequent Technic Elements	More frequent Spatial Elements
Social-Cultural Sustainability	Perceptive-cultural	Recognisability of the transformation	Materials, finishing, technic elements or construction techniques clearly distinct from the original ones	% n./sqm	Vertical exterior fixtures/ Vertical and horizontal internal walls/frames Floor slabs/horizontal fixtures External and internal/protection and separation elements	Chancel Endonarthex/Exonarthex
		Reversibility of transformation	Materials, finishing or technic elements without permanently transform the existing building	% n./sqm	Internal stairs/ramps	Sacristy
					Radiant floor system	Single hall
					Vertical internal walls/frames/internal protective elements	Transept
					External protection and separation elements	Addition/Side block
		Acceptance of the transformation	Materials, finishing, technic elements or construction techniques which preserve the geometric configuration and the aesthetic value of the existing building	% n./sqm	Vertical exterior fixtures/ Vertical and horizontal internal walls/frames Floor slabs/horizontal fixtures	Chancel Endonarthex/Exonarthex
		Acceptance of the transformation	Materials, finishing, technic elements or construction techniques which preserve the geometric configuration and the aesthetic value of the existing building	% n./sqm	Vertical exterior fixtures/ Vertical and horizontal internal walls/frames Floor slabs/horizontal fixtures	Chancel Endonarthex/Exonarthex
		Respect for collective memory	Aesthetic elements characterizing the building identity value in the time	% n./sqm	Vertical internal walls/frames/internal protective elements/exterior and internal decorative	Naves
Environmental sustainability	Morfological-dimensional	Preservation of the existing building dimension	Volume preserved	% n./sqm	Vertical and horizontal exterior walls/frames Vertical and horizontal internal walls/frames	Addition/Side block
		Preservation of aesthetic relationship with the context	Spatial elements which integrate the building into its context	% n./sqm	Perimeter walls/vertical exterior fixtures	Bell tower
Technological-economic Sustainability	Material-constructive	Compatibility of transformation	Materials, finishing, technic elements or construction techniques without compromise the existing building degradation	% n./sqm	Open space floors	Naves
		Conservation of technic elements	Previous technical elements	% n./sqm	Vertical exterior and internal fixtures	Bell tower
		Respect for the construction system	Traditional techniques	Yes/No	Perimeter walls/vertical exterior and internal fixtures/ Vertical and horizontal internal walls/frames	Rectory
			Innovative construction system without compromise the existing building performances	Yes/No	Perimeter walls/vertical exterior and internal fixtures/ Vertical and horizontal internal walls/frames	Naves Chapels Apse
		Life-length of transformation	Sustainable and eco-friendly materials	Yes/No	Radiant floor system	Single hall
			Reused technic elements	% n./sqm	External stairs/ramps	Bell tower
			Recycled technic elements	% n./sqm	External protection and separation elements	Addition/Side block

Fig. 60.8 Sustainable reuse indicators for ecclesiastic built heritage regeneration

60.4 Conclusion

The indicators deriving from the elaboration of the cards have highlighted, in most cases, the need to mediate between the operations of conservation, often reason of its abandonment, or otherwise the inappropriate functions that determine the loss of the church identity. From this develops the need to recognize these indicators as elements guaranteeing the balance to be maintained between transformation and permanence.

The reuse intervention through these indicators can also represent a tool for revealing the hidden values of the property, revitalizing not only the potential of the building but reactivating dynamics lost over time.

Considering the cultural heritage as the result of the interaction between people and their living context, consequently it is the bearer of both tangible and intangible values (environmental, social/cultural, economic, symbolic, aesthetic, historical, spiritual, etc.) between which symbiotic relationships exist (Fusco Girard 2019). In fact, in turn, the environment shapes people's behavior as the quality of places, infrastructures, human and social capital and institutions affects the entire productivity of an urban system, influencing the quality of life of the people who live there. This circular, symbiotic and reciprocal process between man and the built cultural heritage represents a permanence factor in the transformative dynamics of cultural heritage and landscape, ensuring an evolutionary continuity based on the co-evolution and regeneration of material and immaterial values. The reuse project must also be based on this process, so that the regeneration of disused assets contributes to the reconstruction of symbiotic relationships between people and the environmental and cultural contexts in which they live, preserving and regenerating existing values and generating new ones.

References

ACE (Architects' Council of Europe) (2018) Leeuwarden Declaration. Adaptive reuse of the built heritage: preserving and enhancing the values of our built heritage for future generations. Leeuwarden

Australia ICOMOS (1979) Australia ICOMOS guidelines for the conservation of places of cultural significance

Bianchi A, De Medici S (2023) A sustainable adaptive reuse management model for disused railway cultural heritage to boost local and regional competitiveness. Sustainability 15(6):5127. https://doi.org/10.3390/su15065127

Bosone M, Ciampa F (2021) Human-centred indicators (HCI) to regenerate vulnerable cultural heritage and landscape towards a circular city: from the Bronx (NY) to Ercolano (IT). Sustainability 13:5505. https://doi.org/10.3390/su13105505

Bosone M, Iodice S (2021) Strategie per il riuso adattivo del Monastero di Sant'Agostino a Vicopelago. IN_BO Ric Progett Per Territ Città l'Archit 12:162–187. https://doi.org/10.6092/issn.2036-1602/11278

Bullen PA, Love PED (2011) Adaptive reuse of heritage buildings. Struct Surv 29:411–421. https://doi.org/10.1108/02630801111182439

Caterina G (2013) Conservazione, manutenzione e gestione degli spazi pubblici e dei beni architettonici. In: Castagneto F, Fiore V (eds) Recupero Valorizzazione Manutenzione nei Centri Storici. Un tavolo di confronto interdisciplinare. Lettera 22, Siracusa, pp 14–17

De Medici S, Pinto MR (2012) Public cultural heritage properties enhancement and reuse strategies. TECHNE–J Technol Architect Environ 3(1):140–147. https://doi.org/10.13128/Techne-10839

De Medici S (2021) Building the commons? Feasibility and effectiveness in the shared management of the built heritage. La scuola di Pitagora, Napoli, ISBN 978-88-6542-817-7

Fitch JM (1990) Historic preservation: curatorial management of the built world. University of Virginia Press

Foundation EM (2013) Towards the circular economy: economic and business rationale for accelerated transition. J Ind Ecol

Fusco Girard L (2019) The human-centred city development and the circular regeneration. In: Fusco Girard L, Trillo C, Bosone M (eds) Matera, città del sistema ecologico uomo/società/natura il ruolo della cultura per la rigenerazione del sistema urbano/territoriale. Giannini Publisher, Naples

Gravagnuolo A, Fusco Girard L, Ost C, Saleh R (2017) Evaluation criteria for a circular adaptive reuse of cultural heritage. BDC Boll Cent Calza Bini 17:185–216

Pinto MR (2004) Il riuso edilizio. Procedure, metodi ed esperienze. UTET Università, Torino, IT

Pinto MR, De Medici S, Senia C, Fabbricatti K, De Toro P (2017) Building reuse: multi-criteria assessment for compatible design. Int J Des Sci Technol 22:165–193

Robert P (1991) Adaptations, new uses for old buildings. Princeton Architectural Press

Chapter 61
A Green Technological Rehabilitation of the Built Environment. From Public Residential Estates to Eco-Districts

Lidia Errante

Abstract The environmental quality of the modern city is a central issue in the Italian and international design debate. The pandemic and the perspective of a post-pandemic phase have accelerated the inevitable transformation of the living spaces—indoor and outdoor, urban and domestic—bringing out renewed awareness and new quality requirements. The need to achieve results to limit energy consumption, reduce polluting emissions, promote less land consumption, and conditions of urban resilience is becoming gradually urgent, according to European and national strategic, political and regulatory indications. Space quality requirements, which correspond to different conditions of quality of living, are generally identified in the physical and social accessibility of places and dwellings, in the production and availability of energy from renewable sources, in the availability of green public spaces, and in the opportunity to carry out leisure and sports activities. The paper investigates the transformation of public residential neighbourhoods, highlighting urban and technological design opportunities within the paradigm of eco-district and biophilic urbanism. Two case studies within the INA CASA Plan in Reggio Calabria—Sbarre Inferiori and San Brunello—will be the object of analysis and meta-design transformation scenarios to test with green quality requirements. The scenarios aim to explore microclimatic improvements for the districts, the redefinition of outdoor spaces, the implementations of technologies for clean energy production, and the containment of resources consumption. The object of the contribution goes towards principles of health and well-being of the communities, recognising the urban risk factors implicated in the global pandemic and the need to restore the existing building stock and residential estates. Eventually, the paper suggests a framework of actions, green technologies, and design options to manage those environmental concerns.

Keywords Public residential estates · Eco-districts · Biophilic urbanism · Green technologies · Environmental quality

---

L. Errante (✉)
Mediterranea University of Reggio Calabria, Reggio Calabria, Italy
e-mail: lidia.errante@unirc.it

E. Arbizzani et al. (eds.), *Technological Imagination in the Green and Digital Transition*,
The Urban Book Series, https://doi.org/10.1007/978-3-031-29515-7_61

61.1 Introduction

Cities, their sub-systems and elements, human and non-human actors, are the most complex product of human society, due to the coexistence and interpenetration of multiple spheres in a condition of precarious equilibrium. The social sphere, individual and collective, and the goal of well-being are the reason for technological, scientific, and productive progress as well as the unsustainability of our lifestyles, social injustice and inequality. Today, next to the cultural, scientific and academic debate, public opinion has reached a higher awareness of sustainability issues, recognising how global phenomena can affect their daily lives and vice versa. Such awareness has affected also the use and perception of our living spaces: dwellings, neighbourhoods, public and private spaces. Sustainability has today gained widespread attention thanks to the media and the dramatic events of recent years (from the economic crisis of 2008 to the spread of the Coronavirus in 2020 and the energy crisis of 2022). The city is always the core of this debate as a place where to experiment with new forms of design, production, coexistence, and collaboration. The contribution observes the paradigm shift in the attitude of the citizenry, residents of the residential neighborhood, which no longer corresponds to a passive role of consumer or resident of the working class. In this context, the resident is aware and well happy with his active and proactive role in the policy and decision-making process concerning environmental and urban sustainability, with an increase in direct participation.

Today, the resident is a citizen, is involved and participates in decisions and actions, and can benefit from them. The influence of design for social innovation has played a part in creating socio-technical ecosystems and supporting individual capacity building (cf. Sen). The recognised need to move towards more sustainable lifestyles has opened up new horizons for ecological and green design for the redevelopment and retrofitting of existing buildings, improving energy performance, and environmental well-being, and shared energy production.

This paper aims to outline the salient points of this debate, identifying the challenges posed to contemporary living and cities, focusing on public housing estates. Sustainable development issues and goals are increasingly specific, requiring sophisticated design solutions. The contribution, therefore, reflects on green strategies, processes, and technologies that can positively affect social and environmental well-being, supporting the progress of both spheres.

61.2 Challenges of the 'Modern City'

The contemporary city is often mistreated and scapegoated in the discussion on the unsustainability (physical, social, and environmental) of urban systems. For the same reasons, it is also a privileged object of research applied to its multiple spheres and dimensions falling within the urban and built environment. However, this distinction is not sufficient to frame the daily activities, flows of people, goods, and services which cut across the spaces and functions of the city. Similarly, the sustainability challenges facing the city cannot be resolved exclusively in one dimension or another, as buildings—residential and non-residential—and their urban surroundings must be considered as a *unicum*, held together by multiple cause-effect relationships. Energy consumption, waste production, clean air, green spaces, and social cohesion are just some of the main concerns of contemporary times. These can be addressed through process, design, and technological innovations applied both to the urban environment, mobility, and public spaces and to the built environment, buildings, and housing.

The contribution distinguishes two spheres of environmental concerns: the infrastructural one, related to energy, services, and mobility flows; the construction sector; and the energy cost of buildings in operation. Both have specific issues of energy consumption from non-renewable sources and the production of harmful emissions and different types of waste (UN 2020). In terms of emissions, consumption of resources and production of construction waste, new construction sites have a huge impact on the environment, particularly due to the use of traditional wet construction technologies.

While these structures cannot be reused at the end of their life cycle—due to their many components, such as bricks and plasters—on other hand, they have a wider potential for maintenance, managing to increase the life cycle of the building.

61.2.1 The Post-pandemic Perspective and Beyond: New Scales, Needs and Requirements

The post-pandemic perspective has accelerated both the processes of ecological transition of the built environment and public awareness. During the first lockdown, the decrease in industrial production and the use of fossil fuel transport drastically improved global air quality (WMO 2021). Aerosols from these activities have the most dramatic effect on human health and air quality and are the biggest factor in the amount of PM2.5 in densely populated areas (WMO 2021). These health effects are a major factor in the fatal risk of COVID-19 infection. This environmental and health awareness has motivated different choices related to lifestyles, quality of space, and mobility, resulting in a redefinition of the urban and built environment at the residential and neighbourhood scale. Public spaces, mobility, and the energy grid are pivotal.

The role of urban public space has regained the centre stage in the last 10 years to pursue ecology, cohesion, and democracy (Errante 2019, 2021). Even during the pandemic crisis, public and semi-public spaces such as squares, courtyards, and places of mobility have undergone major perceptual, spatial, and infrastructural transformations. Mainly to make way for pop-up bike lanes and wider walkways to allow social distancing. The EU and individual governments have allocated funds and launched policies to transform road layouts and integrate technology and equipment to encourage individual, non-fossil, and public sustainable transport, which was already a priority for many political agendas (Lozzi et al. 2020). On the other hand, the recent geopolitical situation and the energy crisis are also encouraging containment of consumption and energy self-sufficiency. The production of clean energy and its off-grid accessibility with innovative distribution processes have been experimented with collaborating peer-to-peer micro-grids. The neighbourhood is the place of such cultural and technological innovations, with the active participation of residents. Here, collective self-consumption groups, known as renewable energy communities, are committed to producing and sharing the energy they need. A framework of new sustainable needs and requirements is emerging, with opportunities for technological, design, environmental, economic, and social innovation.

61.2.2 Design Opportunities Towards Eco-Districts

The design opportunities the paper discusses relate to the political role of architectural, technological, and environmental design in responding to the real needs of communities. These are now told through the perspective of sustainability, but they are issues that have always concerned the dynamics of design and refer in particular to access to essential services and their quality. Even today, these quality objectives must be sought in the dimension of the home, the building, and the residential neighbourhood, especially in contexts of physical and social marginality. Here the concept of environmental quality of living spaces refers to two conditions: the physical space, indoor and outdoor, of the home and the public spaces around it; and that of limiting consumption, even before emissions. In contexts where access to energy services is subordinate to the economic capacity of individuals and families, the production of accessible, clean energy and savings from reduced consumption are fundamental objectives to be achieved. Environmental quality can therefore be pursued through a reorganization and re-functionalization of living spaces, integrating energy efficiency and production. The shared management of these environmental technologies drastically affects the economic and social sustainability of the communities that benefit from them.

Biophilic urban and city design. Biophilic design is an approach that holds principles of good design at the building, site, city, and regional scale, including natural elements that humans should have to evolve with (Beatley and McDonald 2021). Beatley and others argue that exposure to green space produces several benefits for the individual mental and physical health. Also, the biophilic design uses the natural asset provided by the context in an environmentally sustainable way by:

- Protecting and enhancing the natural systems and green infrastructures from climatic events, ensuring the provision of essential services;
- Moderate air pollution and balancing microclimate through vegetation that provides cooling, evapotranspiration, shading and reduction in urban flooding.

Green rooftop alone may dramatically contribute to the reduction of urban temperature;
- Allow native biodiversity to adapt and shift as climate changes, according to an extensive and diverse network of parks and green spaces;
- Help cities to become more self-sufficient in times of resource scarcity such as oil, food, or water;
- Encourage a healthier lifestyle, resulting in overall energy saving.

The biophilic approach can be described as an outdoor-oriented design at the neighbourhood scale, with benefits reaching the regional since the individual level (Fig. 61.1).

Green technologies for energy saving and production. The concept of green technologies is closer to the definition of eco-tech or eco-technologies. The green paradigm is declined through wise use of materials from renewable, recyclable, and reusable sources adopting innovative and alternative construction technologies that cannot damage the environment. A building that meets these criteria can be considered green (Fig. 61.2). According to several authors, the criteria of ecological design should be oriented to:

- The correct planning of the intervention and its impact on the social, economic, and environmental spheres; the flexibility of the interventions according to scenario steps; and reuse existing buildings (Bergevoet and van Tuijl 2016);
- Aim at circularity in the production, execution, and management stages (Errante and De Capua 2021);
- The improvement of the overall environmental quality.

In this sense, we can say that the main interventions are related to the improvement of energy performance. Similarly, it has been mentioned how the biophilic approach can intervene on microclimate shading and cooling while reducing energy demand.

Nevertheless, greenery and eco-technologies include energy production and storage systems that can be used to meet these needs. The most common solutions

Fig. 61.1 Overview diagram of possible morphological actions that can be applied, at different scales, by integrating green technologies

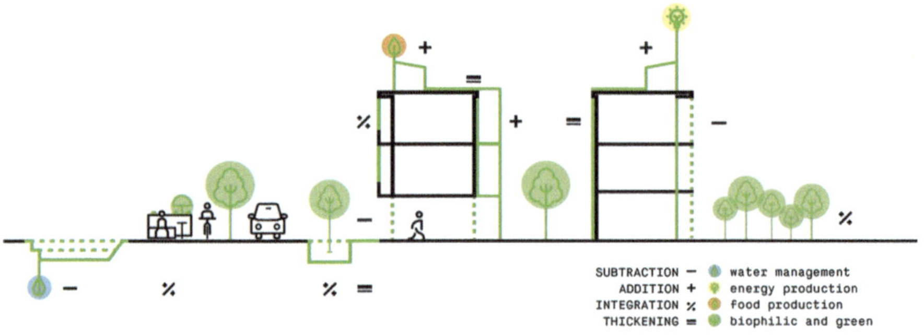

Fig. 61.2 Diagram of the variety of sustainable actions and green technologies that can be adopted at the city block scale to improve energy performance and urban resilience through: heat island effect mitigation systems, clean energy production systems from renewable sources, local production and cultivation, with urban gardens

use solar energy to produce hot water and electricity, but there are also kinetic energy production technologies integrated into the pavement of public spaces.

Technological advances in this regard have enabled the development of new scenarios for off-grid energy production and sharing, such as energy neighbourhoods and communities that, through specific public–private partnerships, provide direct access to energy from renewable sources at a reduced cost or even profit.

61.3 The Research Perspective and the Case Studies

The broader framework of environmental regeneration represents two aspects of technological regeneration for the residential areas capable to address:

- physical and social marginality;
- structural and technological obsolescence of the buildings;
- inaccessibility to essential services for residents.

These contexts are heterogeneous, requiring multi-scalar and multi-criteria methodological approaches. The recent literature on housing shows a hybridization between the urban design and housing dimensions, highlighting their collaboration at the neighbourhood scale. The Public Space-Public Life approach (cf. Gehl) has evolved into the Soft City paradigm (Sim 2019), oriented towards urban density, accessibility and quality of public spaces in neighbourhoods, with an environmental role. UN-Habitat's "Public Space Site-Specific Assessment—Guidelines to Achieve Quality Public Spaces at Neighbourhood Level" (2020) provides an analytical tool to understand the context and act accordingly. The concept of physical and relational proximity has led the neighbourhood to be considered a minimal ecological unit, where accessibility is the result of a space-time-opportunity equation. The topic of "15 minutes city", in Italy well-argued by Manzini (2021) is something already seen and experimented with in the Super-Illas of Barcelona (ES), where urban regeneration, public spaces, mobility, and energy production are part of a wider concept of functional proximity.

Based on these considerations, the research was applied to two case studies of meta-design experimentation, selected in the context of the INA Casa Plan. The cases represent aesthetic, formal, and technological qualities, as a further opportunity to rethink the existing (Di Biagi 2001). The two selected neighbourhoods are located in Reggio Calabria, in the southern and northern suburbs of the city. They present very different spatial characteristics, the common aspects concern:

- the distance from the consolidated historic centre, less than 4 km;
- the coverage ratio and the presence of potential public spaces;
- the wet construction technology, with reinforced concrete framed structure, concrete floors, and brick infill;
- the overall good structural condition;
- the lack of central heating systems and lifts.

At the type-morphological level, the urban layouts are different. Sbarre Inferiori (1959–1964) has 26 buildings of five floors with five aggregate types. The linear aggregations are made of 2–4 modules, of two 75 m^2 units each. The three-lobed typology has hexagonal elements, with 95 m^2 apartments each. The buildings were originally on pilotis, no longer visible due to successive modifications. The residential public spaces are neglected, with no provision for outdoor activities. San Brunello (1950–1954) consists of a long building with four floors that defines an M-shaped layout and a total of 120 dwellings. Within the bends, there are large green spaces, patios, semi-private gardens, and parking spaces for residents. Ten different dwelling dimensions, ranging from 60 to 138 m^2, can be distinguished, all with double overlooking, balconies and loggias, while the ground floor apartments have their entrance.

The research, currently in progress, elaborates design scenarios oriented to the integration of green and eco-technologies applied to the building and the external spaces and three functional operations: thickening, addition, replacement. The design criteria address energy containment and the quality of residential and public spaces in the surroundings. The design of the ground, the hierarchy of pathways, and the leisure activities collaborate in the overall strategy of energy efficiency of the entire neighbourhood. The first phase of the research was conducted and concluded on the case of Sbarre Inferiori, with three scenarios.

1. Functional additions. A buffer space between the building and the environment is designed according to principles of "technological disintegration", DfD and remanufacturing. The use of dry technologies, in steel for the secondary structures and bio X-LAM for the add-on units, allow them to be disassembled—as a whole or in the individual parts of the technological system—replaced—according to those proposed by the functional abacus—and re-assembled into new configurations and different places. The life cycle of the additions is extended through

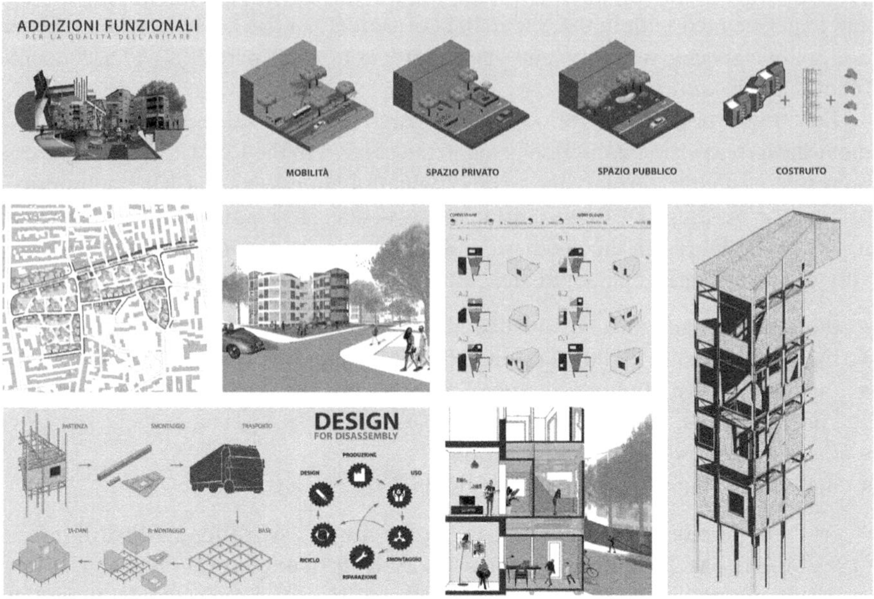

Fig. 61.3 Functional addition connected to the blind front of the building. Together with nature-based solutions, the design scenario bases its strategy on Design for Disassembly. Edited by: A. Quattrone, 2020—M.Sc. thesis in Architecture

the re-aggregation or the disintegration of the elements, to be used as a material bank (Fig. 61.3).

2. The thickening of the envelope reduces energy needs and improve building performance. The double-skin façade includes an elevator integrated with a balcony and ramp system externally accessible by the ground floor by a public portico.
3. The typological redefinition of the dwellings into smaller ones widens the target of residents and integrate air ventilation cavities as a natural cooling system. Aims at the environmental and energy quality of the neighbourhood (Fig. 61.4).

The scenarios involve energy production systems from renewable sources, photovoltaic panels and kinetic paving, green technologies and watersquares in public spaces to enhance climatic and environmental resilience. The scenarios are currently under the examination of the resident community for their feedback.

61.4 Conclusions

The research field of architecture technology is aimed at redeeming its sustainability debt to the natural environment by suggesting green, sustainable solutions to be

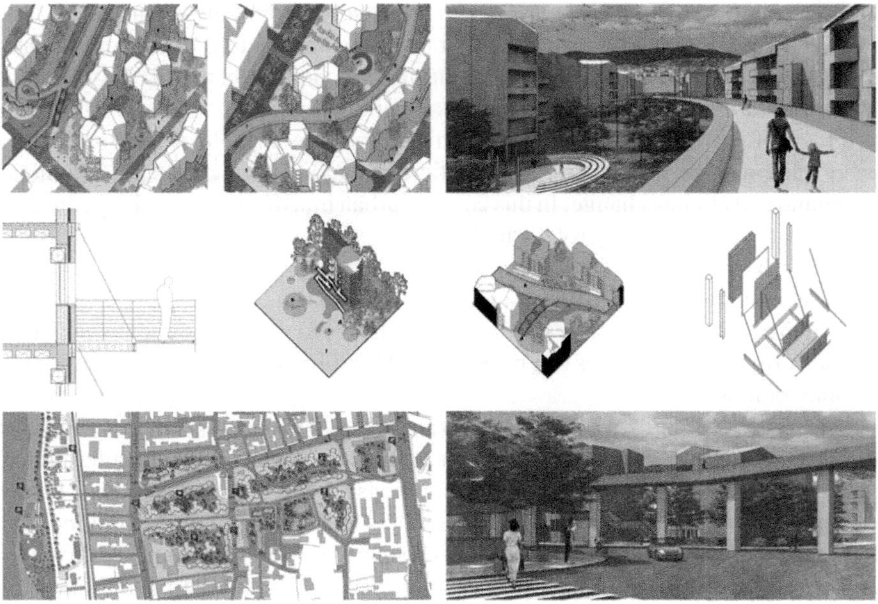

Fig. 61.4 The strategy adopts a general greening of the area, while a twofold solution is used for the buildings: the cladding of the four fronts with external thermal insulation and exposed brick cladding; the insertion of the addition aimed at improving the accessibility of the dwellings through the insertion of appropriate ramps, walkways and a lift body. Edited by: E. Dattoli, 2021—M.Sc. thesis in Architecture

adopted at the scale of the building and the city, to, directly and indirectly, guarantee greater quality and environmental comfort.

Foreseeing energy alternatives in technological design meet environmental, social, and economic sustainability requirements. The emergence of climate change determines new design needs that can only be addressed through an ecological approach that requires a deeper knowledge of the technologies—urban and building—most appropriate to solve local environmental contingencies. Facing these challenges by starting from the first Italian examples of social housing, such as the INA Casa Plan, represents the closing of a circle for research. On the one hand, to "overcome the theories of urbanism of the Modern Movement to define new ways of reading, interpreting and designing the city" (Guidarini 2018). On the other, to restore centrality to Italian excellence in architecture and construction, even in contexts that have remained oblivious because they are outside of authorial production. INA Casa's public residential neighbourhoods allow, more than others, to reflect on the indoor/outdoor, built/empty, private/public space relationships at the local micro-scale, concerning the socio-spatial dynamics and environmental fallout of projects to retrofit the existing and regenerate residential spaces (Ottone and Cocci Grifoni 2017).

The ecosystem approach to urban regeneration and building rehabilitation appears necessary today in order to provide alternative, credible, and sustainable responses to the needs for energy efficiency, health and safety in the built environment. The solutions here described do not represent an exhaustive picture of the overall environmental design panorama but highlight the direction of technological evolution towards solutions that integrate nature as a building material, as a mitigation tool, as a form of production, and for bio-chemical processes capable of constituting real energy alternatives. We are witnessing a time of major social, cultural, geopolitical, economic and climate change. In this context, urban transformation and the construction industry play a driving role, which is as necessary to activate investment as it is to support a necessary ecological transition and urban resilience. For this reason, the premise of the research is to understand these solutions in order to apply them to real case studies through a meta-design approach, thanks to which design, technological and performance alternatives can be developed for discussion with communities and administrations.

The research described so far proposes an approach that can be replicated in other urban and geographical contexts, as it is flexible in capturing the peculiarities of the context, thus proposing appropriate solutions from a bioclimatic point of view. Again, in terms of technological choices, there is an ever more pressing need to formulate nature-based, biophilic, and green solutions that better counterbalance the negative effects of climate change on the quality of life and the natural and built environment.

References

Beatley T, McDonald R (2021) Biophilic cities for an urban century. Why nature is essential for the success of cities. Palgrave Macmillan, Cham

Bergevoet T, Van Tuijl M (2016) The flexible city. Sustainable solutions for a Europe in transition. Nai010 Publishers

Di Biagi P (2001) La grande ricostruzione: il piano Ina-Casa e l'Italia degli anni cinquanta. Donzelli Editore

Errante L (2019) Public space and its challenges. A palimpsest for urban commons. In: Benincasa C, Neri G, Trimarchi M (eds) Art and economics in the city. New cultural maps. Transcript

Errante L (2021) Hybrid communities and resilient places. Sustainability in a post-pandemic perspective. In: Sposito C (ed) Possible and preferable scenarios of a sustainable future, pp 32–45

Errante L, De Capua A (2021) Design for Disassembly and the rehabilitation of public housing stock. A case study. TECHNE J Technol Archit Environ 22:181–191

Guidarini S (2018) New urban housing. L'abitare condiviso in Europa. Skira

Lozzi G, Rodrigues M, Marcucci E, Teoh T, Gatta V, Pacelli V (2020) Research for TRAN committee—structural and cohesion policies. Brussels

Manzini E (2021) Abitare la prossimità. Idee per la città dei 15 minuti. Egea

Ottone F, Cocci Grifoni R (2017) Urban technologies. Built and unbuilt spaces for open spaces configurations. LISt Lab

Sim D (2019) Soft city. Building density for everyday life. Island Press

UN-Habitat—United Nations Human Settlements Programme (2020) Public space site-specific assessment—guidelines to achieve quality public spaces at neighbourhood level
United Nations Environment Programme (2020) Global status report for buildings and construction: towards a zero-emission, efficient and resilient buildings and construction sector. Nairobi
World Meteorological Organization (2021) State of global climate 2021, provisional report

Chapter 62
Adaptive Building Technologies for Building Envelopes Under Climate Change Conditions

Martino Milardi

Abstract Following the widespread recognition of the urgency of environmental and energy issues, cities, now under the influence of the pandemic crisis, are called to cope with them through adaptation strategies to future scenarios that are constantly changing. At the same time, the implementation of adaptive building envelopes seems to be a promising alternative to achieve higher quality levels in the built environment, especially to counter and mitigate climate change, in line with EU directives. Adaptive envelopes can modify physical or chemical characteristics, exploiting environmental stimuli such as solar heat, temperature, airspeed, or atmospheric humidity. In this scenario, the experimental research in progress wants to define a new adaptive model by using innovative materials. It can be applied to curtain wall systems, intended as an element vulnerable to the effects of extreme events in a Mediterranean climate and more stressed by external energy flows. In this work, the author presents some parts of the research results, in which a necessary phase involved the reasoned recognition of adaptive materials for extreme applications or materials that can respond actively to possible external stresses. Research efforts are focused on the choice of the most suitable material to define the levels of environmental adaptability of the model, its constructability, and technological characterization. Finally, the performance verification of the adaptive model will be carried out at the TCLab section of the BFL of the Mediterranean University of Reggio Calabria to develop prototypical lines that can facilitate the new approach to high environmental quality adaptive envelopes.

Keywords Climate change · Building envelope · Adaptive model · Curtain wall · Testing

M. Milardi (✉)
Mediterranea University of Reggio Calabria, Reggio Calabria, Italy
e-mail: mmilardi@unirc.it

E. Arbizzani et al. (eds.), *Technological Imagination in the Green and Digital Transition*, The Urban Book Series, https://doi.org/10.1007/978-3-031-29515-7_62

62.1 Introduction

The current environmental and energy issues affecting the scenario of this time, accentuated by the effects of climate change, such as global warming (IPCC 2021) and the increase in CO_2 emissions, highlight the pressing need to define concrete and measurable responses, moving from a phase considered to be one of transition to one of transformation of the built environment.

This challenge has led to a systematic review of the design modalities, transferring performance issues away from the traditional physical and static properties of building enclosures to a broader discourse in defining how they behave.

For this reason, innovative concepts such as adaptive building façades could play a role in the near future, as their dynamic behavior could optimize the performance of a building in its complex system of internal and external relations (Milardi et al. 2021). In recent years, the climatic conditions of the Mediterranean areas, characterized by islands and heatwaves, extreme rainfall phenomena such as water bombs, micro-typhoons increasingly frequent and known as Medcane, represent a critical point of the building envelope, becoming a specific field of application for research activities.

The contribution refers to ongoing research, financed by funds POR Calabria FESR/FSE 2014–2020, with the ultimate aim of addressing a design methodology related to the concept of adaptability of the building envelope.

The research is in line with the European Directives on energy efficiency 2018/844/EU and is part of the Strategy S3-Region Calabria, related to the development of specific technologies and materials with higher performance and efficient and economically sustainable solutions. Research activities are directed toward the development of simulation scenarios/models and application experimentation, the aim of which is to develop an adaptive model using innovative materials, elements, and components applied to curtain wall systems (Kazmierczak 2010). The latter is vulnerable to the effects of extreme events, such as rising temperatures and precipitation, and more stressed by external energy flows.

This scenario directs experimentation toward realizing efficient and adaptive envelopes through advanced tests conducted at the TCLab Section of the BFL—Building Future Lab, Mediterranea University of Reggio Calabria. It represents a fertile opportunity for research and development of activities and skills to control the overall quality of the building and the urban context of reference.

The laboratory is one of the 40 partners of the project called Metabuilding Labs of Horizon 2020 to create easy access to an extensive network of high-value testing facilities that enable them to develop and test innovative envelope solutions for next-generation buildings.

In this work, the author presents some parts of the research results, on the reasoned recognition of adaptive materials for extreme applications or materials that can actively respond to possible external stresses.

62.2 Adaptive Materials

A review of the state-of-the-art shows how innovative materials, in the construction industry, are those existing materials that, through chemical-physical and electrical changes, acquire adaptive characteristics, offer more advanced performance, and even very different uses from the so-called "starting point" (Buanne 2015).

They belong to two macro-categories: those whose variability is linked to the variability of the shape and configuration, using sensors and handling mechanisms classifiable in pneumatic actuators, hydraulic actuators and drives based on electric motors (Wigginton and Harris 2009), and those whose variability is linked to the intrinsic properties of the material.

The study activities focused on the second macro-category, active materials, which represent the most advanced frontier of innovation.

Within this category, smart and shape-memory materials (SMMs), such as shape-memory alloys (SMAs) and electroactive polymers (EAPs) control and product performance at the level of material shape capable of stretching, bending, or folding, depending on the environmental stimulus, without the need for complex electromechanical systems or energy supply (Elattar 2013).

Thermo-reactive SMMs and SMAs perceive the thermal stimulus and show actuation or some predetermined response, making it possible to adjust different technical parameters such as shape, position, deformation, stiffness, and other static and dynamic characteristics (Wei et al. 1998). An input of thermal energy alters the microstructure through a crystalline phase change, allowing multiple shapes to environmental stimuli (Addington and Schodek 2012), making them promising materials for integration into adaptive façade applications.

The dynamic behavior of electroactive polymers, on the other hand, is triggered when an electric current, the stimulus, flows through the laminate, increasing the electrostatic forces that generate a contraction of the elastomer (Villegas et al. 2020). A typical dielectric electroactive polymer (DEAP)—a type of electronic EAP—consists of a dielectric elastomer membrane placed between two electrodes. When an electric field is applied, the membrane compresses and stretches, causing the material to change shape (Zhao et al. 2016) with the ability to store a large amount of energy.

Based on the analysis done, shape-memory materials, whether alloys or polymers, are identified as adaptive materials that contribute to a more complex and intelligent system. These materials, used for components design for the building envelope, absorb, and dissipate energy, deriving from a rise in external temperature, giving great adaptive response capacity to the entire system.

The direct impact examined in this research concerns the solar radiation reflected directly from the curtain wall surface into the urban microclimate. An adaptive façade improvement could, therefore, reduce the environmental temperature and the impact of the building and, consequently, decrease the magnitude of the phenomenon.

This activity was necessary for knowledge acquisition about material innovation and experimentation and the identification of the most congruent material of the adaptive model.

62.3 Objective and Results

In line with the technical-scientific literature on the difficulties in reconstructing a unique performance profile of the envelopes, the research intends to position itself as an effective tool capable of overcoming the possible criticalities dictated by environmental dynamics. The general objective is to identify operational tools capable of directing interventions on the built environment toward adaptive design with the contributions deriving from technological innovation to respond to the expectations of environmental control.

On the other hand, experimentation with adaptive systems is currently tied to individual cases of architectural design, limiting the large-scale application of innovative envelope technologies that contribute significantly to the control of climate variations. The systems currently on the market are scarcely applicable or used only in buildings with a particular use or size or value that justifies the adoption and thus the cost of such systems (Conato and Frighi 2018).

At the same time, there is a knowledge gap in the literature on future trends and main concepts of adaptive façades. There are knowledge gaps in terms of adaptive façade market share, including the main concepts, the most promising technologies, their categorization, their best use, and the distinction between short-term and long-term structural trends in adaptive façades (Attia et al. 2020).

The research aims at defining an intervention model that can guide design choices toward the elements that make up the building envelope, characterized by an adaptive component.

This study wants to satisfy the need to improve the performance of curtain walls by reducing the impacts of extreme climate change events, through the development of an adaptive model—the result of the research—able to react synergistically to different climatic conditions, ensuring high-quality performance.

Specifically, the following partial and expected results are listed:

- typological classification of curtain walls systems, on construction characteristics, operating principles, and performance requirements to be certified, with respect to critical elements that may be damaged during an extreme event;
- systematization of data and case studies relating to adaptive design through technological elements and systems capable of responding to external environmental conditions;
- elaboration of performance repertoires to guide the design of components with adaptive criteria and reasoned recognition of innovative materials for extreme events;

- identification of the current performance (without the adaptive model) of the curtain wall system with mullions and transoms, which will be chosen for experimentation, by conducting laboratory tests, thanks to the support of the TCLab section of BFL, relating to air permeability and watertightness;
- definition, through a synthesis framework, of the performances to be improved for the adaptive behavior of the façade system, with respect to the increase in temperature and precipitation, and of the materials that will constitute the adaptive model;
- instruction and construction of feasibility checks useful for the development of the prototype to be used for experimentation and testing activities;
- design experimentation through simulation and verification of the adaptive behavior of the model applied on a curtain wall to improve its performance.

The experimentation will focus on the analysis of a type of mullion and transom façade with respect to its adaptive stress behavior with respect to climate change phenomena. The investigations conducted on the study of materials for extreme events, summarized in the section on adaptive materials, allow us to consider the possibility of applying an intelligent material with shape-memory capabilities for the experimental phase. The final phase of the experimental research project will involve the comparison and superimposition of the results obtained from the laboratory tests on a curtain wall system, with and without the adaptive model.

62.4 Methodology

In accordance with the illustrated scenario, the research and its activities are organized in thematic sections to identify advanced technologies and materials for the adaptive management of curtain walls and to arrive at a model idea and the verification of its technical feasibility, which can be implemented through experimentation and evaluation of the results obtained (Grillo and Sansotta 2021).

The first phase concerns the investigation of the state-of-the-art, the European and national regulatory framework and, therefore, the identification of the open problems shaping it. In particular, the focus is on some key assumptions, based on the need for a renewed building design responding to the pressing demands of "new quality." Subsequently, a critical analysis of adaptive components, systems, and materials is useful to highlight the problematic aspects and resolve the criticalities emerging from the conception of the new adaptive model for curtain wall systems.

The experimentation phase is started thanks to the support of the TCLab section (Trombetta and Milardi 2015), carrying out simulations of climatic conditions in an urban environment and their related effects to test the adaptability levels of the model applied on a mullion and transom curtain wall system. The behavior of the curtain wall (mock-up on a 1:1 scale), with and without the adaptive model, in situations of climate change identifies and configures new adaptive scenarios, directing design decisions toward the optimal option for the different reference contexts and relationships.

For this reason, with respect to the climate change phenomena to be examined, the following procedure is identified for the experimentation activities:

- heatwaves: Experimentation is carried out by reproducing a wind flow, through an AAMA/ASTM fan, capable of verifying the performance behavior of the materials of the external façade of the mock-up subjected to a strong pressure and in accordance with ASTM E 283-04 (2012);
- pluvial flooding: The experimentation is carried out by reproducing a constant rain directly on the external façade of the mock-up, through a sprinkler system, with respect to three simulations of the water jet, through a sprinkler network with calibrated nozzles: in the absence of wind, in the presence of wind, and in extreme wind conditions (hurricane power) (Fig. 62.1) and in accordance with ASTM E 331-00 (2009), AAMA 501.1-05-00 (2007) and ASTM E 330-02 (2010).

The experimentation will have double feedback in the progress of the research: one, to test the performance characteristics of the curtain walls according to American standards with respect to extreme events; the other to verify, according to the normative requirements, components of the system, able to guarantee acceptable levels of safety with respect to environmental flows and external stresses, which relate to the building envelope.

Fig. 62.1 Example of hurricane simulation on a curtain wall. *Credit* TCLab

62.5 Conclusions

Research focuses efforts on the assumption that curtain walls guarantee constant performance thresholds calibrated to the average values required by the standards, resulting in envelopes that perform the same even in different contextual conditions.

Therefore, the research, even if still in progress, is oriented toward the design of the optimal combination of the adaptive model, using shape-memory material aiming at the most appropriate adaptive efficiency, calibrated to specific environmental contexts. Incorporating these principles in the prototyping of the adaptive model to be applied to new or existing curtain walls becomes an essential activity for an architecture that must continuously adapt to the effects of climate change.

In this perspective, the prototype to be developed represents the physical element of mediation between the external and internal environment, able to regulate and react to the signals that qualify the external environment, contributing to becoming an element of transformation and control. In this sense, the study and analysis activities of the research cannot be considered exhaustive due to the complexity of the subject and new fields of innovation are still to be investigated, but the experimental approach is functional to define application and simulation potential in the construction sector.

Therefore, it is strategic to pursue the approach based on measurement tests and performance evaluation in a "simulated" regime. The testing process is intended as an adaptive control tool for climate change, based on measurements and performance evaluation concerning specific environmental reference contexts. The coherence of objectives establishes a new mission focused on improving the company's know-how, centered on the realization of innovative products addressing market trends and efficient and adaptive envelopes.

The activities and perspectives of applied research and experimentation can outline concrete solutions on a theoretical and operational level, referring to current and future challenges, moving toward an increasingly controlled vision in the relationship between innovation and architectural design. Work in progress.

References

Addington M, Schodek D (2012) Smart materials and technologies in architecture. Routledge

Attia S, Lioure R, Declaude Q (2020) Future trends and main concepts of adaptive facade systems. Energy Sci Eng 8(9):3255–3272. https://doi.org/10.1002/ese3.725

Buanne M (2015) New materials for smart envelope. TEMA Technol Eng Mater Archit 1(1):178–183

Conato F, Frighi V (2018) Il ruolo dell'innovazione nella definizione di nuovi paradigmi formali in Architettura. TECHNE J Technol Archit Environ 16:105–112. https://doi.org/10.13128/Techne-22965

Elattar SMS (2013) Smart structures and material technologies in architecture applications. Sci Res Essays 8(31):1512–1521. https://doi.org/10.5897/SRE2012.0760

Grillo E, Sansotta S (2021) Experimentation of a new adaptive model for envelope system. In: Sposito C (ed) Possible and preferable scenarios of a sustainable future. https://doi.org/10.19229/978-88-5509-232-6/5102021

IPCC (2021) Climate change 2021. The physical science basis. Summer for policymakers. In: Working group I contribution to the sixth assessment report of the Intergovernmental Panel on Climate Change. Information. https://ipccitalia.cmcc.it/messaggi-chiave-ar6-wg1/

Kazmierczak K (2010) Review of curtain walls, focusing on design problems and solutions, 12–14 Apr 2010. Building Enclosure Science & Technology, Portland

Košir M (2019) Climate adaptability of buildings. Springer International Publishing. ISBN: 978-3-030-18456-8

Milardi M, Mandaglio M, Grillo E, Musarella C, Sansotta S, Tramonti N, Trombetta C (2021) Testing and experimentation on building envelope under climate change. In: Itecons (ed) Proceedings of the CEES 2021, Coimbra, Portugal, 12–15 Oct 2021. ISBN: 978-989-54499-1-0

Trombetta C, Milardi M (2015) BUILDING FUTURE Lab.: a great infrastructure for testing. Energy Procedia 78:657–662. https://doi.org/10.1016/j.egypro.2015.11.053

Villegas JE, Gutierrez JCR, Colorado HA (2020) Active materials for adaptive building envelopes: a review. J Mater Environ Sci 11:988–1009

Wei ZG, Sandstrorӧm R, Miyazaki S (1998) Shape-memory materials and hybrid composites for smart systems: part I Shape-memory materials. J Mater Sci 33(15):3743–3762. https://doi.org/10.1023/A:1004692329247

Wigginton M, Harris J (2009) Intelligent skins. Elsevier Architectural Press, Oxford

Zhao Z, Shuai C, Gao Y, Rustighi E, Xuan Y (2016) An application review of dielectric electroactive polymer actuators in acoustics and vibration control. J Phys Conf Ser 744(1):012162. IOP Publishing. https://doi.org/10.1088/1742-6596/744/1/012162

Chapter 63
The Importance of Testing Activities for a "New" Generation of Building Envelope

Martino Milardi, Evelyn Grillo, and Mariateresa Mandaglio

Abstract The construction sector is considered, directly or indirectly, one of the pillars for the application of technological solutions to rise the quality levels of building envelopes. The need to realize new processes capable of "dynamically" reading the responses of the built systems becomes an essential action to understand how the dynamics of climate change determine and trigger evident effects on the built environment. In this scenario, the contribution describes the experimental research activities on a curved facade—carried out at the TCLab Section of Building Future Lab of the Mediterranean University of Reggio Calabria—to verify its performance responses to extreme events according to specific standardized protocols. Therefore, this study focuses on the building envelope, as the main subsystem through which leakage occurs, not only in terms of thermal and dynamic fluxes, but of air and water permeability. Testing activities, nowadays of fundamental value for climate change phenomena, allow to predict the behavior of the built environment and at the same time to evaluate alternative solution. The research efforts go toward defining a design methodology for a new generation of building envelopes, capable of reacting to different contextual conditions by raising the environmental and performance quality according to adaptive dynamics. From the tests carried out, the results take the form of test protocols, giving real added value to research and implementing applied experimentation actions with highly reliable results.

Keywords Extreme events · Building envelope · Curtain wall · Advanced testing · Performance evaluation

M. Milardi (✉) · E. Grillo · M. Mandaglio
Mediterranea University of Reggio Calabria, Reggio Calabria, Italy
e-mail: mmilardi@unirc.it

E. Grillo
e-mail: evelyn.grillo@unirc.it

M. Mandaglio
e-mail: mariateresa.mandaglio@unirc.it

E. Arbizzani et al. (eds.), *Technological Imagination in the Green and Digital Transition*, The Urban Book Series, https://doi.org/10.1007/978-3-031-29515-7_63

63.1 Introduction

For several years, the construction sector has been called upon to respond to the pressing demand for new energy and environmental qualities, partly because of the acceleration of climate change effects (IPCC 2021). Supporting this scenario, the Global Status Report 2020 for Buildings and Construction, published by the Global Alliance for Buildings and Construction (GlobalABC), notes that the construction sector's share of global energy-related CO_2 emissions rises to 38% (United Nations Environment Programme 2020). Therefore, EU actions and policies will have to contribute to the objectives of the European Green Deal and the Paris Agreement, promoting strategic planning on energy efficiency and climate (Commissione Europea 2019).

Researchers and various players in the construction industry, including companies, businesses, and different stakeholders, want to provide concrete answers to building safer and more efficient environments.

This approach promotes the integration of high-performance components and highly innovative products that facilitate the monitoring and management of environmental and energy performance at the building scale.

Against this, there are new experiments aimed at demonstrating the possibility of equipping buildings with systems that offer "dynamism" useful for the management of flows, like a living organism (Milardi 2016). The complex evolution of the envelope is moving toward its evolved functionalization. The envelope is a three-dimensional closure system composed of several interdependent elements called upon to regulate performance related to the passage of energy and environmental flows.

Therefore, the envelope specializes in the deployment of new control and performance response systems through the interaction between the external world and the internal environmental elements; the maintenance of conditions of well-being and comfort depends on this relationship.

At the same time, it is increasingly evident that climate change requires a substantial modification of building design approaches to make urban systems more adaptive to climate change. In particular, buildings are exposed to increased risks of damage from the expected impacts of climate change, including more frequent high winds, increased heat, particularly in cities (Urban Heat Island effect), floods, and fires that accompany some extreme weather events.

In recent years, climatic conditions in Mediterranean areas have been characterized by rising temperatures, water bombs, and increasingly frequent micro-typhoons known as Medcane (cyclones formed in the Mediterranean) (Mejorin et al. 2018), becoming the new requirements in terms of quality that building envelopes must meet. As multiple factors change, such as the performance conditions of building envelopes and climatic conditions, there is a need for measurable control, at the basis of the "building-context relationship," through new modes of investigation to measure the effects of extreme events on buildings and, where possible, verify the outcomes of their bi-univocal relationship.

The contribution presented here is part of this context and concerns experimental research to verify the performance characteristics of building envelopes in a specific environmental context, developing laboratory experiments, according to standard protocols (UNI/EN, ASTM, AAMA) on a type of transparent façade.

Most of today's curtain wall systems are technologically improved and satisfy higher requirements. However, recent studies indicate that test results from specialized laboratories show that the continuous impact of environmental conditions causes a significant loss of performance (up to 54% due to air infiltration, 18.53% due to wind pressure) over the life of the curtain wall system (Ilter et al. 2015). Mandatory façade performance tests such as air permeability, water impermeability, and wind resistance applied to full-scale curtain wall models can help obtain preliminary information on system performance before the installation process. However, passing these tests does not guarantee that the system's performance in terms of durability will match the test results (Yalaz et al. 2018). Specifically, the case study analyzed is a portion of a curved transparent façade, on a 1:1 scale, made available for experimental research activities by TCLab partner companies, GLASBILT LLC.[1]

The experimentation activities took place at the TCLab Section[2] of the Advanced Testing laboratory facility, the Building Future Lab of the Mediterranean University of Reggio Calabria. The laboratory makes it possible to test new approaches and technical systems to control the overall quality of the building and its urban context. Using the instrumentation and machinery available in the TCLab Section, which reproduces extreme climatic stresses on mock-ups of envelopes, it is possible to study the envelope's performance and measure their resilience characteristics.

63.2 Methodology and Instruments

The approach for the experimental activities lies in the current and fertile scenario of climate change studies in the urban environment. It is an emblematic example of the intellectual and operational challenge posed by today's human fields of action, where technological innovation and experimentation are possible tools for understanding their behavior. Based on current needs, urban growth, and climate change, the development of energy-efficient buildings may become particularly critical for the future (Merlier et al. 2019).

The general objective of the studies is to identify some critical nodes between buildings and their contexts and assess the stresses of climatic phenomena affecting

---

[1] GLASBILT is a custom fabricator of architectural metal, glass and millwork that focuses on the U.S. market with all design and production done in Italy. Online available at: https://www.glasbilt.com/glasbilt.

[2] Large infrastructure of advanced testing, at the Mediterranea University of Reggio Calabria, which bases its activities on a set of standardised and experimental tests (UNI-EN, ASTM, AAMA), having as its object the envelope in its technological and material assets. Online available at: https://www.unirc.it/ricerca/laboratori.php?lab=69.

the building envelopes, recalling the need to verify the technical feasibility of interventions, oriented and supported by experimentation, and validation of the results obtained.

From a methodological point of view, the first part of the research study consists of understanding the evolving climate change scenario in the Mediterranean area (Sannino et al. 2016) due to islands and heatwaves, high-intensity rainfall phenomena such as water bombs, and micro-typhoons. These phenomena can cause direct damage to the envelope of buildings, with particular attention to their façades and the identification of building elements to be reinforced to avoid severe damage to affected properties and the safety of people. The aim is to prevent the "cascading effects," where one defective piece causes other pieces to fail, leading to disproportionate consequences. Studies and regulations prescribe that curtain walls must be air and watertight, prevent the formation of condensation on internal surfaces, and resist wind load and other external forces acting on the building envelope. Therefore, assessment is essential to minimize the risk of undesirable and costly problems during a building's expected lifetime (Gonįalves et al. 2010).

Subsequently, the boundary conditions between the building and its context are recreated in the laboratory to initiate experimentation and identify the behavior of the curved façade in situations of extreme climatic events, in particular heatwaves and rain floods. In this phase, the focus shifts to the performance of the curved façade, with the final objective of developing test protocols, both for subsequent experimental activities and certification in the regulatory framework.

The instrumental research opportunity for the study of environmental phenomena involving the building envelope is provided mainly by two large pieces of equipment in the laboratory: a "test chamber" (Test Lab) and a climatic simulation "Cell" for accelerated indoor and outdoor tests (Test Cell).

Specifically, the Test LAB is a "test chamber" built according to the operational characteristics of the test set. It consists of a steel-framed structure measuring 18 (15 effective) $\times$ 12 $\times$ 2.50 m, where mock-ups of curtain walls (according to UNI definition), windows, and doors (or similar elements) are mounted on a scale of 1:1 and tested according to unified protocols.

The Test Cell is a structure for the thermodynamic characterization of full-scale building envelope systems, usable in a closed or open environment. It consists of three independent units installed on a support platform and managed by a control PC. It calculates the thermal performance of buildings, as indicated by the UNI/TS 11300-1 standard, and is fundamental for the regulatory verification and testing of vertical and horizontal closure components, roofs, doors, windows, etc.

The complex operation of data collection, processing, and systematization led to the drawing up of test reports of the comparison and verification of the experimentation carried out. The following section describes the experimentation activities.

63.3 Experimental Procedure

Field experimentation is commonly interpreted as the study of the behavior of buildings—stressed by the reproduction of climate change phenomena—through experimental activities and tests on full-scale models, called mock-ups.

In this case, the mock-up under examination is a curved curtain wall consisting of an aluminum mullion and transom frame and three openings (one window and two doors), including a balcony, which is not considered for the overall performance of the façade system (Figs. 63.1 and 63.2).

The experimental phase starts with the identification of the urban layout to be simulated in the laboratory: It is assimilated into a building with an external space similar to a courtyard. The test chamber (dimensions 17 × 12 × 4.50 m) is closed on three sides; on the fourth external side, the curved portion of the façade is located, allowing direct interaction with the external climatic conditions and offering the possibility of modeling them to specific requirements. In addition, the test chamber is equipped with three seismic beams for performing displacement and elastic equilibrium tests. Other machines are a big fan that simulates winds of up to 200 km/h,

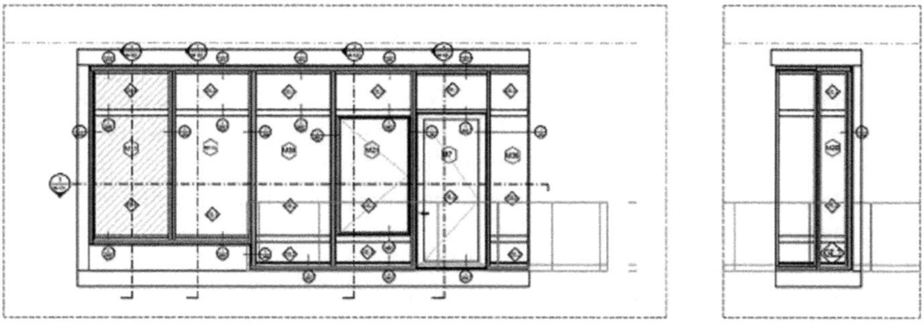

Fig. 63.1 PMU—project mock-up of curved façade, elevation, and sections (Credit Glasbilt)

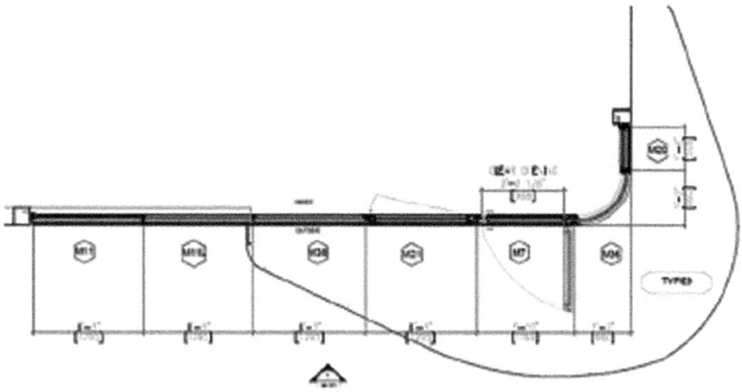

Fig. 63.2 PMU—project mock-up of curved façade, plan (Credit Glasbilt)

a thermal chamber (dimensions 7 × 5 × 1.50 m), and an internationally patented retractable rain simulator (Trombetta and Milardi 2015).

About the phenomena to be analyzed, the following procedure is identified for the experimentation activities:

- Heatwave: The experimentation is carried out by reproducing a wind flow through the AAMA/ASTM fan—simulating the power of a hurricane—to verify the performance behavior of the mock-up's curved external façade subjected to strong pressures;
- Pluvial flooding: The experiment is conducted by reproducing a constant rainfall directly on the mock-up's outer façade. Three simulations of the water jet through a network of sprinklers with calibrated nozzles are carried out: in the absence of wind, in the presence of wind, and under extreme wind conditions (hurricane power).

The objective of the tests is to assess in the laboratory the opening/closing performance, preload, air permeability, water tightness under static pressure, water tightness under dynamic conditions, and structural performance.

For this type of façade, the following test methods are planned and performed:

- Air Leakage Resistance Test (ASTM E283): applying a (negative and positive) pressure differential of 6.24 psf/300 Pa, after determining the loss of the system (Fig. 63.3);
- Water Penetration Resistance Test (ASTM E331): applying a static pressure differential of 10 psf, 480 Pa for a period of 15 min with five gallons of water per square foot per hour (= 3.4 l/min m^2), with a watering device consisting of a square-meshed network of nozzles positioned on a horizontal plane parallel to the plane of the specimen (Fig. 63.4);
- Dynamic Water Resistance Test (AAMA 501.1): applying a dynamic pressure differential of 10 psf for a period of 15 min with five gallons of water per square foot per hour (= 3.4 l/min m^2), with a watering device consisting of a square-meshed network of nozzles positioned on a horizontal plane parallel to the plane of the specimen;
- Uniform Load Deflection Test (ASTM E 330): applying a positive and negative test pressure equal to 50% and 100% of the design wind load, for which measurements and checks are carried out to verify that, under these effects, the mock-up presents an admissible deformation and retains its stability characteristics.

From the experiments carried out, a framework for reading and comparing performance data is constructed to highlight the potential, criticalities, and malfunctions of the technological systems that characterize the typological and geometric layout of the mock-up analyzed.

From the experiments carried out, a framework for reading and comparing performance data is constructed to highlight the potential, criticalities, and malfunctions of the technological systems that characterize the typological and geometric layout of the mock-up analyzed.

Fig. 63.3 Air permeability test (Credit: TCLab)

Fig. 63.4 Watertightness under static pressure (Credit: TCLab)

63.4 Results

The data analyzed do not show negative behavior of the curved façade, for the simulation of the heat island and pluvial flooding in the absence of wind, obtaining a positive and flexible reaction behavior of the system. The performance verification under conditions of pluvial flooding with wind reveals water infiltration, thus requiring attention in the production, installation, and installation of rainwater drainage systems for this type of façade. In extreme wind conditions, on the other hand, the mock-up reacts because the pressure generated by the fan disperses the flow of water on the façade; thus, no water accumulates and avoids infiltration inside the building envelope.

As shown in Table 63.1, the air infiltration test is passed: for fixed windows 0.1 CFM and for opening windows 0.30 CFM.

The optimum results of the structural test are conducted at a negative pressure of 1676 Pa and a positive pressure of 3064 Pa through the use of specific sensors that read the displacements of the façade at particular points (Fig. 63.5). Table 63.2 shows the data of the test performed.

Table 63.3 shows the absolute deformation values for comparison between opening and fixed elements.

From the tests carried out, the results materialized in experimentation protocols that represented a real added value to the research, realizing applied experimentation actions with highly reliable results.

Table 63.1 Results of air infiltration test

Air infiltration	Negative pressure 6.24 psf	Positive pressure 6.24 psf
Fixed	$8.47\ m^3/h < 50\ m^3/h$ 0.028 cfm × $ft^2 < 0.1$ cfm × ft^2	$2.23\ m^3/h < 50\ m^3/h$ 0.008 cfm × $ft^2 < 0.1$ cfm × ft^2
Openable n. 1 window	$0.41\ m^3/h < 6.4\ m^3/h$ 0.03 cfm × $ft^2 < 0.30$ cfm × ft^2	$1.53\ m^3/h < 6.4\ m^3/h$ 0.12 cfm × $ft^2 < 0.30$ cfm × ft^2
Openable n. 2 door	$0.29\ m^3/h < 15.9\ m^3/h$ 0.0095 cfm × $ft^2 < 0.30$ cfm × ft^2	$3.06\ m^3/h < 15.9\ m^3/h$ 0.098 cfm × $ft^2 < 0.30$ cfm × ft^2

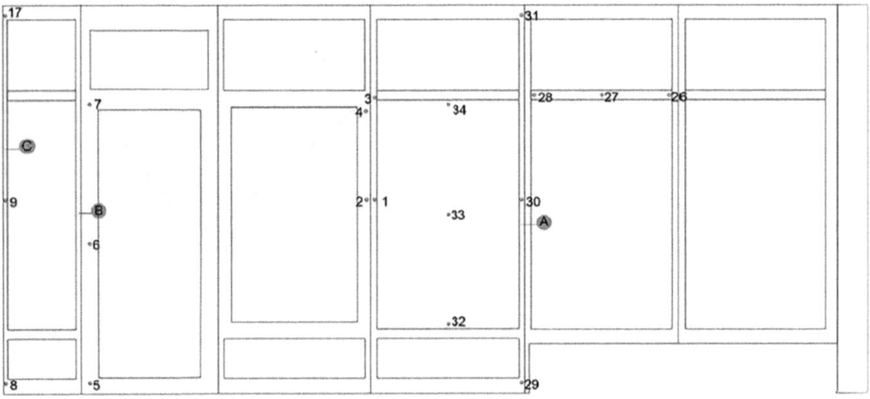

Fig. 63.5 Sensors position inside the chamber (Drawing by TCLab)

Table 63.2 Results of structural performance

Structural performance	Positive pressure 1676 Pa	Negative pressure 3064 Pa
Frontal deflection of upright A (sensors n. 29-30-31)	6.98 mm (limit 18.87)	9.74 mm (limit 18.87)
Frontal deflection of transom (sensors n. 26-27-28)	2.75 mm (limit 6.82)	4.61 mm (limit 6.82)
Frontal deflection of transom (sensors n. 5-6-7)	5.12 mm (limit 14.38)	4.85 mm (limit 14.38)
Frontal deflection of upright C (sensors n. 8-9-17)	1.46 mm (limit 18.76)	7.88 mm (limit 18.76)
Frontal deflection of glass (sensors n. 32-33-34)	10.64 mm (limit 19.05)	10.64 mm (limit 19.05)

Table 63.3 Deformation values

Structural performance	Positive pressure 1676 Pa	Negative pressure 3064 Pa
Sensor n. 1 (opening)	11.07 mm	22.25 mm
Sensor n. 2 (fixed)	10.34 mm	23.87 mm
Sensor n. 3 (opening)	13.61 mm	25.03 mm
Sensor n. 4 (fixed)	14.25 mm	27.09 mm

63.5 Conclusion

The research focused its efforts on the assumption that curtain walls guarantee constant performance thresholds calibrated to the average values required by the standard, resulting in envelopes that perform the same even under different contextual conditions.

The results and evaluations of this experimental research on a full-scale mullion and transom curtain wall mock-up, even if the system passed the performance tests, show how deficiencies or failures can occur during their service life, especially in the face of sudden climate changes. From this point of view, the testing process is understood as an adaptive control tool for climate change, based on measurements and performance assessments against specific environmental reference contexts. That is, the more traditional aspect of modeling and simulation has to be extended, with testing protocols requiring new methodological approaches and infrastructures capable of responding innovatively to market trends and stringent industry regulations (Milardi 2021).

Although the main objectives of the experiment are achieved, it can be considered a work in progress, as it opens up promising fields of investigation. It would be interesting to record the dynamic trend of the thermal values on the mock-up, either with a thermometer or a thermal imaging camera at regular intervals to verify the heat propagation.

References

Commissione Europea (2019) Comunicazione della commissione al parlamento europeo, al consiglio, al comitato economico e sociale europeo e al comitato delle regioni Il Green Deal europeo. https://ec.europa.eu/info/strategy/priorities-2019-2024/european-green-deal_it

Gonįalves MD, Jutras R, Velji M (2010) Evaluating the field performance of windows and curtain walls of large buildings. In: Symposium on building envelope technology, pp 57–65

Ilter E, Tavil A, Celik OC (2015) Full-scale performance testing and evaluation of unitized curtain walls. J Facade Des Eng 3(1):39–47. https://doi.org/10.3233/FDE-150028

IPCC (2021) Climate change 2021. The physical science basis. Summer for policymakers. Working Group I contribution to the sixth assessment report of the Intergovernmental Panel on Climate Change. https://ipccitalia.cmcc.it/messaggi-chiave-ar6-wg1/

Mejorin A, Trabucco D, Stelzer I, Nakada R, Rooprai MS (2018) Cyclone-glazing and façade resilience for the Asia-Pacific region: market and code study. CTBUH J 2:42–47. https://www.trosifol.com/fileadmin/user_upload/TROSIFOL/support/research_testing/CTBUH_CycloneResistantFacades.pdf

Merlier L, Frayssinet L, Johannes K, Kuznik F (2019) On the impact of local microclimate on building performance simulation. Part I: prediction of building external conditions. Build Simul 12(5):735–746. http://doi.org/10.1007/s12273-019-0507-7

Milardi M (2016) Un laboratorio di ricerca applicata per i processi di innovazione tecnologica in edilizia/An applied research laboratory for technological innovation processes in building. Techne 11:113. https://doi.org/10.13128/Techne-18409

Milardi M (2021) Test, simulazioni, protocolli, attività on site e in laboratorio. In: Bologna R, Losasso M, Mussinelli E, Tucci F (eds) Dai distretti urbani agli eco-distretti. Metodologie di conoscenza, programmi strategici, progetti pilota per l'adattamento climatico. Maggioli Editore, Rimini. ISBN: 9788891643209

Sannino G, Pisacane G, Struglia MV, Carillo A (2016) +2 °C: quali rischi per l'area mediterranea? I cambiamenti climatici globali e i loro effetti regionali. Energia, Ambiente e Innovazione 1:56–63. https://doi.org/10.12910/EAI2016-008

Trombetta C, Milardi M (2015) BUILDING FUTURE Lab: a great infrastructure for testing. Energy Procedia 78:657–662. https://doi.org/10.1016/j.egypro.2015.11.053

United Nations Environment Programme—UNEP (2020) 2020 global status report for buildings and construction: towards a zero-emission, efficient and resilient buildings and construction sector. https://globalabc.org/resources/publications/2020-global-status-report-buildings-and-construction

Yalaz ET, Tavil AU, Celik OC (2018) Lifetime performance evaluation of stick and panel curtain wall systems by full-scale testing. Constr Build Mater 170:254–271. https://doi.org/10.1016/j.conbuildmat.2018.03.061

Chapter 64
Data Visualization and Web-Based Mapping for SGDs and Adaptation to Climate Change in the Urban Environment

Maria Canepa, Adriano Magliocco, and Nicola Pisani

Abstract To address Sustainable Development Goals (SDGs) and face climate change effects, it is necessary to adopt multidisciplinary methodologies and strategies for risk prevention and mitigation of the impact in urban contexts. These phenomena represent a risk for cultural heritage conservation, with negative consequences for local economies. To move from the analysis of climate impacts to adaptation measures and governance tools, it is necessary to deal with the different characteristics of the urban context in its physical, historical, cultural, and socio-economic components. The paper focuses on the collaboration between UNIGE Architecture and Design Department (DAD), and Colouree S.r.l. that has developed an analytical platform that uses artificial intelligence, geo-referenced data, and automated analysis to define the characteristics of the urban context. The aim of the research is the identification of parameters and solutions to respond to the effects of climate change in the urban environment, considering risk levels and context settlement; alongside the climatic skills, also the architects' skills in environmental technologies, urban landscape, and cultural heritage have been given relevance. DAD aims to capitalize on the previous and ongoing experiences of Colouree, offering scientific and methodological support, to reach the definition of a detailed settlement analysis, providing indications on the risks associated with the main predictable effects (extreme weather events, heat island effect, water availability). The expected results will define a methodological structure to create a sensitivity mapping to meteorological phenomena, based on the data support from Colouree towards the carrying capacity of the urban fabric, making information more accessible thanks to the data

M. Canepa (✉) · A. Magliocco
University of Genoa, Genoa, Italy
e-mail: maria.canepa@unige.it

A. Magliocco
e-mail: adriano.magliocco@unige.it

N. Pisani
Colouree S.r.l., Genoa, Italy
e-mail: n.pisani@colouree.com

E. Arbizzani et al. (eds.), *Technological Imagination in the Green and Digital Transition*, The Urban Book Series, https://doi.org/10.1007/978-3-031-29515-7_64

visualization and web-based mapping, including, among the stakeholders, not only experts but also professionals and citizens.

Keywords Smart city · Resilience · Data-driven design · Key performance indicators · Climate change

64.1 Introduction

The 2030 Agenda for Sustainable Development, adopted by all United Nations Member States in 2015, provides a shared blueprint for peace and prosperity for people and the planet, for the present and next generations (United Nations 2015). The Agenda defines the 17 Sustainable Development Goals (SDGs), which are an urgent call for action by all countries—developed and developing—in a global partnership (https://sdgs.un.org/goals). These objectives aim to safeguard the health of people and the environment, to cope with climate change, to achieve an effective ecological transition towards renewable sources, and to make cities more resilient. The research presents in this paper will refer to Goal 11 "Make cities and Human settlements inclusive, safe, resilient, and sustainable" and Goal 13 "Take urgent action to combat climate change and its impacts".

Climate change is one of the most demanding challenges that the human being, and consequently the urban environment, must face nowadays. Global Warming caused a rise in temperatures of the planet by 0.98 °C and the trend observed since 2000 suggests that, in the absence of interventions, it could reach + 1.5 °C between 2030 and 2050 (Intergovernmental Panel on Climate Change 2022).

Plant and animal species move unpredictably from one ecosystem to another, creating incalculable damage to biodiversity around the world, even if ecosystems have demonstrated to be extremely resilient (Côté and Darling 2010). The framework of climatic weather phenomena that determine significant impacts in the cities is complex and concerns different natural matrices, due to the transformations that have taken place in urban areas, causing different behaviour from normal dynamics.

Therefore, in urban areas, the consequences of climate change are particularly dramatic, in relation to the strong impermeability of the soils, with consequent phenomena of flooding, or overheating conditions due to the Urban Heat Island effect (UHI) (Watson and Adams 2011).

The warming climate, combined with the UHI effect, will increase air pollution in cities (Akhtar and Palagiano 2018). It is very important to define trends and forecasts that address extreme events, both in terms of safety and risk, adaptation, and defence.

Cities should be prepared to anticipate these changes to fulfil a more sustainable future and be more resilient to climate-rated disasters, also reducing greenhouse gases (GHG) emissions (Rosenzweig et al. 2018).

The use of Key Performance Indicators (KPI) and the monitoring through them of the progress in achieving the SDGs are a way to develop strategies for cities management and to monitor their present conditions (Chan and Chan 2004). Thus, the municipalities must equip themselves with tools capable of translating the information collected through the monitoring of the KPIs through digital tools, such as document reports and interactive dashboards.

The collaboration between Ecosystemics Research Group, which operates within the University of Genoa (UNIGE), Architecture and Design Department (DAD), and Colouree S.r.l., a company that has developed an analytical platform that uses artificial intelligence, geo-referenced data, and automated analysis to define the characteristics of urban context, focuses on the development of a shared methodology for organize KPIs to define trends to help municipality to fulfil targeted SDGs and improve resilience measures. This partnership is supported by the research carried out on the Operative National Programme (PON) Research and Innovation 2014–2020 (currently in progress), which provides for a collaboration between researchers and companies.

64.2 Key Performance Indicators for SDGs and Climate Change

64.2.1 Key Performance Indicators: Role and Selection

To better understand and monitor the progress in achieving the SDGs considering the conditions of each individual country, region, or metropolitan city, it is necessary to identify comparable and universally shared evaluation parameters. KPIs can define in a realistic and measurable way essential factors related to a specific object, such as climate change reduction, or communities' resilience achievement. KPIs are an important instrument also for set environmental policies, especially for public bodies, that can define realistic goals for the fulfilment of SDGs and to understand in which sectors they are most lacking (Schokker et al. 2022).

Accountability and monitoring for city governments is a central concern in local and multilevel governance strategies (Hughes et al. 2020). Compared to monitoring on a national scale, those on a local scale can be extremely significant in mapping the actual differences found on a regional and metropolitan scale. However, it is necessary to select sets of indicators with adequate units of measurement and refer to databases that offer values on a small scale, updated and updatable in future (Arup 2014). To identify a shared statistical information framework as a tool for monitoring and evaluating progress towards the objectives of the Agenda, the United Nations Statistical Commission has set up the Inter Agency Expert Group on SDG which has defined a set of over 200 indicators. The Italian National Institute of Statistic (ISTAT) is involved in the production of statistical measures for monitoring progress towards the Sustainable Development Goals (Rapporto Sdgs 2021).

The measures consider the indicators defined by the Expert Group together with some specific national context data, also deriving from the Equitable and Sustainable Well-being framework (Bes) (Il Benessere Equo e Sostenibile in Italia 2021). The

Bes project was born in 2010 to measure fair and sustainable well-being, with the aim of evaluating the progress of society not only from an economic point of view, but also from a social and environmental point of view (https://www.istat.it).

64.2.2 SDGs Monitoring Experience

The collaborations between DAD and Colouree, started on March 2022, focusing on the DataLab project (Guidelines for Integrated Agenda Monitoring Sustainable Underground) promoted by Metropolitan City of Genoa and Metropolitan City of Milano. DataLab stems from the experience gained within the project Decimetre, an online consultation platform, which allows dialogue between public administration bodies and provides an environment for the analysis of territorial data decision support. Starting from the objectives and targets defined by the Sustainable Metropolitan Agenda, DataLab was imagined as a dashboard for data analysis and monitoring, multitenant, open, and interoperable, which can be shared with other Metropolitan cities, and with European metropolitan areas (access online on https://sdgcittametropolitana.mi.it/).

Colouree offers Artificial Intelligence (AI), Data and Location Intelligence solutions in the Smart City and Real Estate, oriented to monitoring, decision support, and based user engagement on the ability to analyse the interactions between people, activities, and the built space. The DataLab activities consist in the diagnosis and setting of Data Architecture starting from the defined objectives and targets in the Sustainable Metropolitan Agenda and Italian Alliance for Sustainable Development (ASVIS) and ISTAT indicators; data-driven storytelling definitions based on needs and strategic scenarios of the administration; design of the technological platform for indicators monitoring and viewing; implementation of indicators and dashboard (Fig. 64.1).

64.2.3 Data-Driven Storytelling as a Strategy for Sustainable Development

Data-driven storytelling gives the possibility to turn raw data into more easy-to-read and understandable concepts that help users to have a wider view. This approach is very relevant considering KPIs: they could give a very detailed kind of information, especially in relation to SDGs fulfilment and to each goal. Therefore, SDGs are correlated to KPIs that could describe different scenarios and being associated to define a trend. Data-driven storytelling allows to communicate sustainability into its complexity, trough data visualization, infographic elements thank also to dashboard architectures (Ren 2019).

Fig. 64.1 SDGs dashboard for metropolitan city of Milan—copyrights: Colouree S.r.l

DAD, according to Colouree, provides data-driven storytelling methodology, starting from the selection of strategic indicators for the metropolitan city, considering the development lines, the leading sectors (based for example on SWOT analyses provides by municipalities) it is possible to define trends that group proven indicators with different macro-objectives. The municipalities have identified the most relevant indicators for the specific territories among the ISTAT and Bes SDGs indicators. The reference data have been identified for each indicator. Starting from these selected indicators, some reference scenarios have been identified, defining specific trends. The defined trends, according with Municipalities, are "Resilient Communities", "Inclusiveness and Empowerment", and "Smart Connections". "Resilient Communities" trend, for example, helps in understanding, through the selected indicators, how the Metropolitan City of Milan and the communities that inhabit it can respond in time to perturbative events, learning to be more flexible and active in dealing with difficulties, making cities and human settlements more inclusive, safe, sustainable, and self-sufficient. The table below summarizes the objectives and indicators (Table 64.1).

The same approach is applied for "Inclusiveness and empowerment" trend, that allows to understand through the selected indicators how the Metropolitan City of Milan contributes to promoting the coexistence and enhancement of differences, the quality of life and connections, guaranteeing equal opportunities to its citizens, and for "Smart Connections", that shows how the communities and the inhabitant are able to use connections (tangible and intangible networks) to improve synergies, communications, mobility, exchange and access to information (Figs. 64.2 and 64.3).

Table 64.1 "Resilient communities" trend

SDGs goal	Objectives	KPIs
2. Zero hunger	2.3 Double the productivity and income of small-scale food producers	Added value per hectare of agricultural area used
	2.4 Ensure sustainable food production and resilient agricultural practices	Agricultural area utilized (UAA) invested in organic crops
6. Clean water and sanitation	6.6 Protect and restore water-related ecosystems	Efficiency of drinking water distribution networks
7. Affordable and clean energy	7.3 Double the overall rate of improvement in energy efficiency	Total electricity consumption (GWh) required from the distribution networks per 10,000 inhabitants
11. Sustainable cities and communities	11.7 Provide access to safe and inclusive green and public spaces	Availability of urban green public spaces
12. Responsible production and consumption	12.4 Handle chemicals and waste responsibly	Selected collection of urban waste
13. Climate action	13.1 Strengthen resilience and adaptability to climate-related disasters	Population exposed to flood risk
	13.2 Integrate climate change measures into policies and planning	Emissions of tons of CO_2 eq per capita
		Population exposed to landslide risk

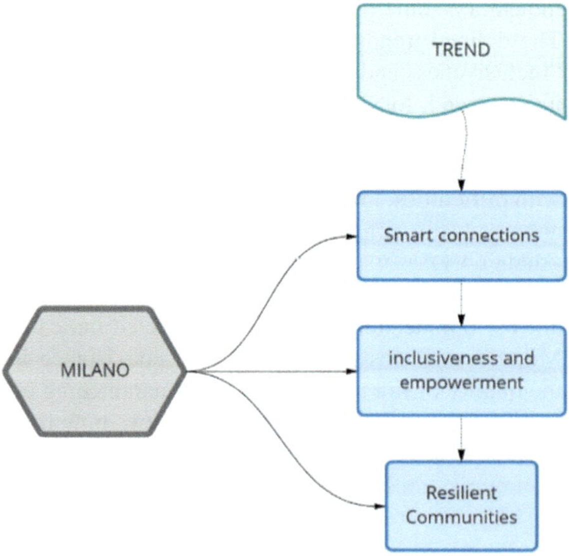

Fig. 64.2 SDGs trends for metropolitan city of Milan—copyrights: authors

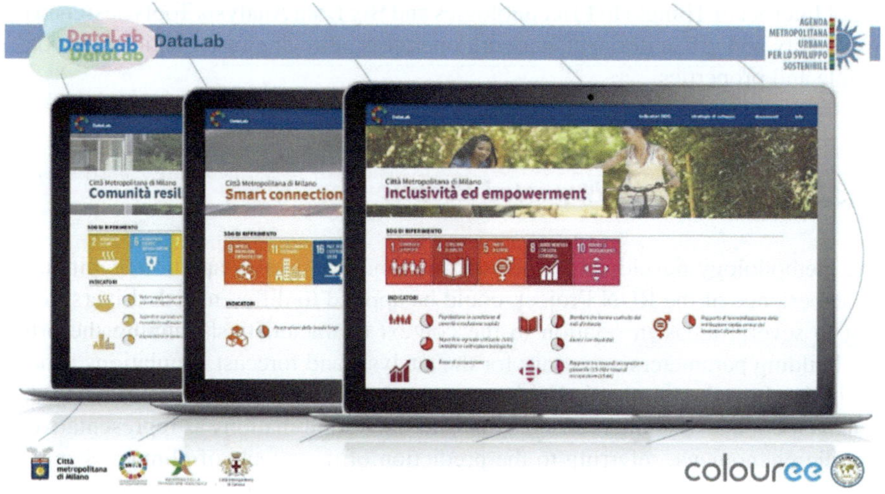

Fig. 64.3 Trends dashboard visualization for metropolitan city of Milan—copyrights: Colouree S.r.l

64.3 Parameters and Solutions to Respond to the Effects of Climate Change

64.3.1 Monitoring Extreme Events in Urban Areas

The project developed during the research and the collaboration with Colouree on KPIs (as requested by RTDa PON agreement) aims also to develop an integrated model for the analysis of urban space, adaptable to different contexts, and the identification of adaptation strategies and solutions to climate change, with the objectives of reducing risk and mitigating residual impacts. To move from the analysis of climate drivers and related impacts to adaptation measures and related governance tools capable of making them operational, it is necessary to deal with the different characteristics of the local context in its physical, spatial, historical, cultural, and socio-economic components (Quijano et al. 2022). Events can lead to a loss of both architectural and landscape values—as in the case of events in a coastal environment—and affect the most fragile social groups (e.g., the elderly)—as happens, for example, in relation to the effects of the UHI, as well as with the loss of private and public assets (as in the case of flooding). This step implies the use of parameters for evaluating the sensitivity of the urban space, at different scales (KPI) (Amoah 2021).

DAD and Colouree will collaborate on the Project "Urban resilience: nowcasting of flood risk with IoT sensors and open data—RUN", promoted by University of Genoa, Colouree and Iride Enia (IREN), an Italian joint stock company, operating as a multiservice. The project involves the development and demonstration of a nowcasting service of the risk of flooding in the presence of intense rains and making use of Internet of Things (IoT) technologies and Big Data Analysis Tools designed for Smart City and urban drainage network operators, allowing faster actions to protect people and properties.

64.3.2 Outlook and Expected Results

The methodology developed to face urban climate change adaptation, capitalizing the experience of the RUN Project, could be applied to different risk drivers, referring to several areas in relation to the context characteristics, defining the urban and building parameters necessary for the analysis and forecast simulations. One of the in-depth analyses intended to be carry out regards the possibility of using data management and visualization systems, and their combinations as representative of specific phenomena, referring to the prediction of the effects of climate change in urban space. Predictive software, for example, starting from climatic data, allow us to anticipate changes in different environmental parameters (atmospheric temperature, relative humidity, speed of air flows, radiant temperatures, thermal comfort parameters) without the need for direct detection, starting from data collected in adjacent areas. It is therefore possible to create scenarios relating to microclimatic (local) variations according to the different characteristics of the urban fabric: height of buildings, settlement density, presence of vegetation, distribution, and shape of buildings, etc. Similarly, it is possible to make forecasts relating to the effects of rainy events on the ability of urban areas to manage water volumes in the disposal process (in relation to soil permeability, capacity of canalizations, etc.). Climate events can have serious consequences on people's safety and cause loss of assets if their effects are not correctly foreseen, therefore mitigating and adaptation solutions are put in place.

64.4 Conclusions

Data visualization and web-based mapping are always more relevant to address SDGs and face climate change effects, offering to municipality the opportunity to use dynamics tools to support governance decision process. Also, it could be possible to monitor the parameters on a reduced scale, operating a downscaling compared to the data already available, allowing to take more effective measures, linked to the specific territories. The limits of these tools can be conditioned by the scarce availability of data, or by the lack of continuity over the years. To achieve more consistent results, it's important to adopt multidisciplinary methodologies and strategies for risk prevention and mitigation of the effects in urban contexts, implementing dashboards with real-time data and specific KPIs for sustainability objectives that are intended to reach.

Although the collaboration between DAD and Colouree is in the initial phase, the first data relating to data-driven storytelling have been published in the first draft of the dashboards, constantly updated.

Considering the ongoing "RUN" Project, the forecasting simulation processes, however, require processing time, as well as expert personnel capable of managing the data and interpreting the results. This specific PON research, among the various objectives, aims to verify the possibility of interactions between meteorological conditions and the urban fabric characteristics, categorizing them in cases, to understand, almost in real time, as the climatic conditions change, the level of risk for people and goods. Although the simulations can be carried out at different dimensional scales, the effects of climate change, in fact, retain a certain uncertainty that should be managed with immediate interventions. Subsequently, it would be possible to identify a set of intervention strategies, which can modify the characteristics of the elements constituting the urban fabric (soil permeability, use of the land, constraints on activities at risk, adoption of Nature Based Solutions, etc.). In certain cases, behavioural strategies can instead be implemented such as those that are currently triggered according to the level of alert that local administrations set and which, due to the complexity of the territory, are often excessively precautionary if applied throughout the territory, with the risk of being underestimated by the population.

Acknowledgements Research carried out in the context of: Operational Programme National (PON) "Research and Innovation" 2014–2020, Green, Researcher Type A (RTDa PON), UNIGE, DAD.

References

Akhtar R, Palagiano C (2018) Climate change and air pollution the impact on human health in developed and developing countries. Springer International Publishing, Cham

Amoah J et al (2021) In: Dima A (ed) Resilience and economic intelligence through digitalization and big data analytics. Sciendo, Warsaw

Arup (2014) City resilience index: research report, vol 2. Fieldwork data analysis. Ove Arup & Partners International Limited. https://www.arup.com/-/media/arup/files/pdf-downloads/cri_research_report_vol2_comp.pdf

Chan AP, Chan AP (2004) Key performance indicators for measuring construction success. Benchmarking Int J. http://doi.org/10.1108/14635770410532624

Côté I, Darling E (2010) Rethinking ecosystem resilience in the face of climate change. PLoS Biol 8(7):e1000438. https://doi.org/10.1371/journal.pbio.1000438

Hughes S, Sarah G, Tozer L (2020) Accountability and data-driven urban climate governance. Nat Clim Chang 10(12):1085–1090. https://doi.org/10.1038/s41558-020-00953-z

Il Benessere Equo e Sostenibile in Italia, Bes 2021 (2021) Technical report. ISTAT. https://www.istat.it/it/archivio/269316

Intergovernmental Panel on Climate Change (2022) Climate change 2022 impacts, adaptation and vulnerability. IPCC, Switzerland

Quijano A et al (2022) Towards sustainable and smart cities: replicable and KPI-driven evaluation framework. Buildings (Basel) 12(2):23. http://doi.org/10.3390/buildings12020233

Rapporto Sdgs (2021) Informazioni Statistiche per l'agenda 2030 in Italia. Technical report. https://www.istat.it/it/files
Ren D (2019) Visualization authoring for data-driven storytelling. eScholarship, University of California. https://escholarship.org/uc/item/1ks25661
Rosenzweig C et al (eds) (2018) Climate changes and cities. Cambridge University Press, New York
Schokker J, Kamilaris A, Karatsiolis S (2022) A review on key performance indicators for climate change, progress in IS. In: Wohlgemuth V, Naumann S, Behrens G, Arndt H-K (eds) Advances and new trends in environmental informatics. Springer, Berlin, pp 273–292
United Nations (2015) Transforming our world: the 2030 agenda for sustainable development. A/RES/70/1
Watson D, Adams M (2011) Design for flooding architecture, landscape, and urban design for resilience to flooding and climate change. Wiley, Hoboken, N.J.

Chapter 65
Fog Water Harvesting Through Smart Façade for a Climate Resilient Built Environment

Maria Giovanna Di Bitonto, Alara Kutlu, and Alessandra Zanelli

Abstract Water emergency is one of the terrible effects of climate change; it is defined as the Blue gold of twenty-first century. In this scenario, fog stands as a potential alternative water resource. Many territories are affected by fog phenomenon; here fog collectors have been developed to extract water from humid mass of air. The aim of this paper is to explore the application of this technology in building sector. The Large Fog Collector is the device commonly used for these projects; it is a textile structure, composed of a mesh, two poles and cables. The exploitation of conventional water resources implies a massive distribution system with significant energy consumption and costs. Otherwise, fog harvesting is a passive system; it relieves the stress upon freshwater resources. Nowadays, fog collectors are low tech devices, and fog harvesting projects are commonly developed in arid areas for agricultural and reforestation purposes. Nevertheless, taking advantage of the vertical development of the device, this textile structure shall be integrated in façade, to promote resilient constructions and make buildings water self-sufficient. The paper explores the design criteria for the development of a novel concept of smart water collecting façade. It can promote also shading effect, reducing the use of cooling system, energy demand, so lowering the ecological footprint. Depending on fog Liquid Water Content, the collected water can be used for the irrigation of green roofs, gardens or in an optimal scenario also for domestic use. The analysis of local weather data is crucial to extend the territories where this system can be applied; but, more important, the improvement of the device's technology is essential to implement it in new application fields.

Keywords Fog harvesting · Water self-sufficiency · Smart façade · Textile architecture · Resilient construction

M. G. Di Bitonto (✉) · A. Kutlu · A. Zanelli
Polytechnic University of Milan, Milan, Italy
e-mail: mariagiovanna.dibitonto@polimi.it

A. Kutlu
e-mail: alara.kutlu@mail.polimi.it

A. Zanelli
e-mail: alessandra.zanelli@polimi.it

E. Arbizzani et al. (eds.), *Technological Imagination in the Green and Digital Transition*, The Urban Book Series, https://doi.org/10.1007/978-3-031-29515-7_65

65.1 Introduction

Water scarcity is the consequence of many interrelated factors. First, the changing of human habits is enhancing climate change, which provokes an alteration on water cycle, so the spread of desertification and unpredictable rain events. In addition, the shift to a more meat base diet is modifying land use, therefore local ecosystems. Furthermore, trade globalization and increasing industrial production contribute to an economic shift towards more resource-intensive consumption patterns. Moreover, during last centuries earth faced a great grown in global population. The presented factors reflect their footprint on global freshwater consumption and availability; water is becoming a scarce resource. In fact, freshwater for agricultural, industrial and domestic use has increased nearly six-fold since 1900 (Ritchie and Roser 2017). The global use of water is divided in three sectors that are Agriculture, Industry and Municipality. In this scenario of water demand increase and supply decrease, fog stands as an alternative source of water. Fog water has been documented to be an efficient source in many territories that are affected by fog phenomenon (Klemm et al. 2012). Fog harvesting projects are usually developed in rural areas for agricultural purpose; instead, fog collectors' potentiality should be explored to integrate water supply also in industrial and domestic sector. The aim is identifying chances of improvement and possibility of generating a positive impact on urban water resource management. For the mentioned purpose, the device's structure should be elaborated, to develop it for urban environment and integrate it in building design. At first, the fog harvesting project requirements should be analysed, and therefore, the design criteria should be individuated. Fog water by itself probably cannot fulfil the water demand, but it could release the stress upon conventional overexploited freshwater resources.

65.2 Water Scarcity

Even though water covers almost 3/4 of the earth's surface, the 97.4% is saline water, while the amount of freshwater available for human consumption corresponds just to 2.5% (Gleick 1993). Moreover, water is not equally present around the planet; this is due to natural climatic conditions, and to overexploitation for anthropic use (Pereira et al. 2009). However, it is not even equally distributed; in fact, in many cases the hydric issues are caused by bad water management. The main sources of freshwater are surface waters and underground basins. To use this water, it should be extracted and, after being treated, it requires a distribution system, which functioning depends on energy supply. Furthermore, the consumption is not uniform; water availability directly determines per capita consumption. In fact, although WSI index states about 100 m^3 of water per capita a year (m^3/c/year) as optimum standard of living (Falkenmark 1986), Greece's water withdrawal per capita in 2018 was 962 m^3/c/year, while

Israel's was 132.7 m^3/c/year (OECD 2022). Moreover, the sources are often contaminated (Vörösmarty et al. 2005). Although water is considered a renewable resource, the consumption is not adequate. In addition to the current water constraint, the world population is predicted to grow; for this reason water demand is forecasted to rise by 55% according to UNESCO (United Nations World Water Assessment Programme Secretariat (WWAP) 2016).

65.3 Fog Harvesting

Fog is a meteorological phenomenon that consists of water droplets suspended in the air. From a physical point of view, when a saturated mass of air, reaches its dew point, fog is generated. Fog can be classified on its forming process; it can be by radiative heat loss, by the mixing of two parcels of saturated air initially at two different temperatures, and by an ascend and resultant cooling of an air parcel (Roach 1994). Many territories worldwide are affected by the fog phenomenon; many of them are in arid areas, while others in territories that will face the hydric crisis in the upcoming years. Since ancient time, in arid fog oasis, the populations adapted to those extreme conditions; here some sort of fog collector have been developed. Some of fog harvesting projects have been mapped, in Fig. 65.1 (Klemm et al. 2012). Usually, those projects are realized in undeveloped areas of rural environment for agricultural purpose (Morichi et al. 2018). Nevertheless, some urban environments are affected by fog phenomenon, and they should take advantage of this alternative resource.

65.3.1 Fog Collector

Fog water is obtained through the fog collector; it is a passive system, made of textile structure. Those collectors must be orientated towards the main wind direction, to optimize the collection; in fact, wind is one of the determining factors for the process.

Therefore, when fog is pushed by the wind towards the mesh, the water droplets get deposited on the mesh's filaments and when they reach a certain weight and dimension they flow down, where they are collected by a gutter and then stored in tanks (de Dios Rivera 2011). During last century many types of fog collector have been developed, they are generally classified on the type of structure, which can be two or three dimensional. The model most used worldwide, is the Large Fog Collector (LFC); it's a bidimensional structure (Holmes et al. 2015), designed by Schemenauer and Joe (1989). It is composed by a mesh, two poles and tensors; its collection panel is of 40 m^2, made by a double layer of Raschel mesh. Generally, for these experiments the Standard Fog Collector (SFC) is used, it consists of a rigid metal frame that measures 1 m $\times$ 1 m; this frame is supported by two poles at 2 m above ground level, to reach stronger winds (Schemenauer and Cereceda 1994).

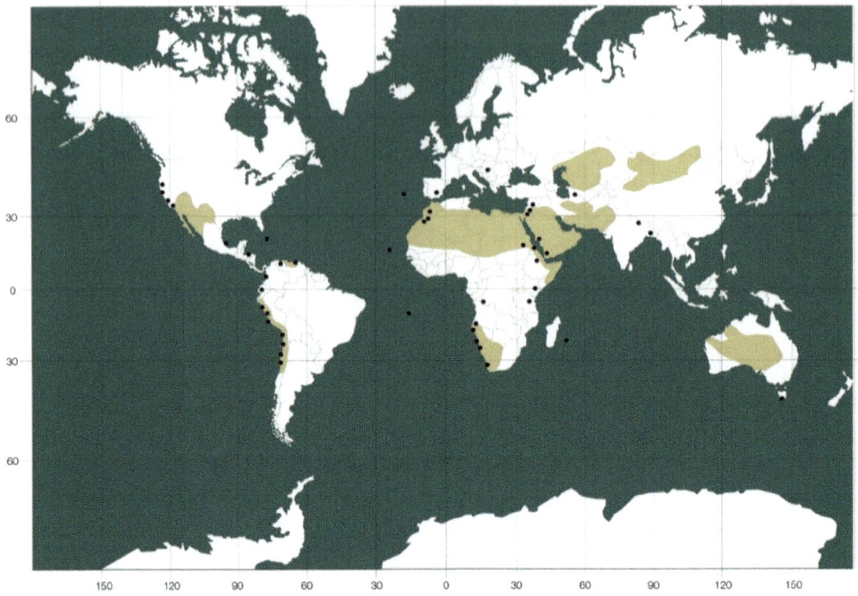

Fig. 65.1 Fog harvesting projects map, elaboration of the authors based on Klemm et al. (2012) and Schemenauer and Cereceda (1994)

The three-dimensional collectors are still not widely explored. Moreover, a three-dimensional device can provide more collection rate for the same soil occupation, but the additional mesh layers permit to collect the fog droplets that already passed through the first mesh. The three-dimensional structures can differ in shape; they can be cylindrical as in Warka Water project, or parallelepiped as in Nieblagua's case (Fig. 65.2).

65.4 Novel Fog Harvesting System

65.4.1 Fog Harvesting Project

In order to develop a fog harvesting project, firstly the climatic conditions of the site should be analysed; in fact, wind speed, wind direction, mean humidity, temperature and Liquid Water Content (LWC) (Holmes et al. 2015) are determining factors for the fog collector disposition and dimensioning. In fact, the collectors should be oriented perpendicularly towards the main wind direction, and generally more wind speed imply more water collection (Schemenauer and Cereceda 1994). Moreover, due to the vertical development and the lightness of the traditional fog collector, the possibility to integrate this device into a smart membrane façade is under consider

Fig. 65.2 Nieblagua structure—three-dimensional parallelepiped fog collector. *Source* Photo by the Author

(Caldas et al. 2018). To do so, the structure of the fog collector must be conceived to integrate it on a building. Furthermore, the novel fog collector should be dimensioned referring to the water demand of such building. The first step to take, to analyse the fog water potentiality of the selected area, is the development of a test campaign. At least a Standard Fog Collector (SFC) (Schemenauer and Cereceda 1994) should be installed for a period, preferably a year or more, to have a mean collection for each season. Several for harvesting projects have been developed which resulted in different collection rates, for example in Cerro Moreno the amount is 8.26 $l/m^2/d$, while in Falda Verde is 1.43 $l/m^2/d$, the amounts refer to the year average (Larrain et al. 2002; Bitonto et al. 2020). Once, the collection property has been stated, depending on the water demand of the building, the smart façade should be dimensioned. The main wind direction could vary during the year, for this reason the façade should be adaptable to any condition. Therefore, flexibility should be one of the main aspect of the smart membrane façade development. Depending on the quantity of water collected on each season, it can be used for domestic purposes, when fog events are abundant, or for watering a green roof or garden in other seasons. The quantity of fog water collection varies from site to site, depends also on the season, and on the structure and mesh used for the fog collector.

65.4.2 Application Field: Built Environment

The construction sector is one of the most impacting for the environment, and its footprint is progressively increasing in carbon dioxide emissions, causing rise in air temperatures, and consuming a large amount of water (European Parliament, Council of the European Union 2010). For this reason, the European commission has stated some guidelines to reach the goal of Near Zero Energy Buildings (NZEB) by 2020 (European Parliament, Council of the European Union 2010). Moreover, in 2015, also the United Nations stated 17 Sustainable Development Goals (SDGs) (General Assembly 2015). Some of these goals can be reached through the application of a novel smart envelope on buildings. The proposed experimental envelope is composed by a double façade, where the first layer is the main closure of the building, while the external one is the smart mesh. This multipurpose textile façade is a passive and 'Km 0' system, because it doesn't require energy to collect or distribute water, and is located where the users are, unlike the traditional hydric systems. Furthermore, since the external skin is composed by a mesh, it can provide shading. In fact, shading a window during summer can reduce the demand of cooling system, which implies notable energy consumption, and related emissions. If the envelope is combined also with vegetation, the façade can also provide purification of the air, given both by the filter mesh and the absorption properties of the plants. It is important to underline that, depending on the location's characteristics, fog is not always present along the year. The proposed device refers to the Mediterranean condition, where water can be collected mostly during winter nights and early mornings, while the shading effect is required just during summer days (Fig. 65.3); moreover, each part of the envelope is more suitable for specific functions (Fig. 65.4).

Nevertheless, this device is a model that can be applied and adapted to any fog oasis referring to external conditions and users' needs. In particular, the design methodology of this smart façade is based both on the requirements of the fog collector previously stated, together with requirements of building envelopes. Therefore, it

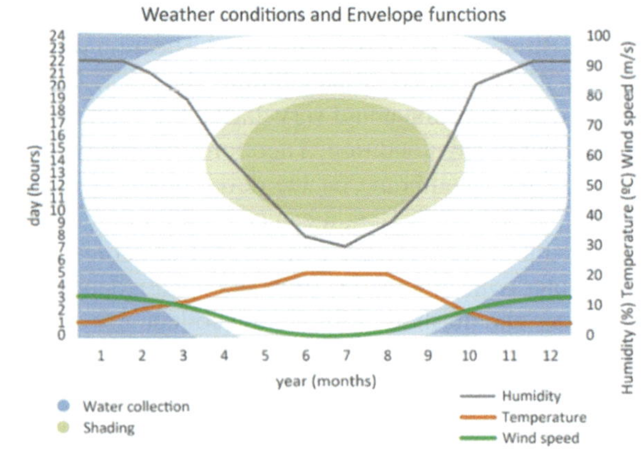

Fig. 65.3 Weather conditions and envelope functions—relation between the weather conditions and the respective envelope functions along the day in a typical year. *Source* Graphic by the Author

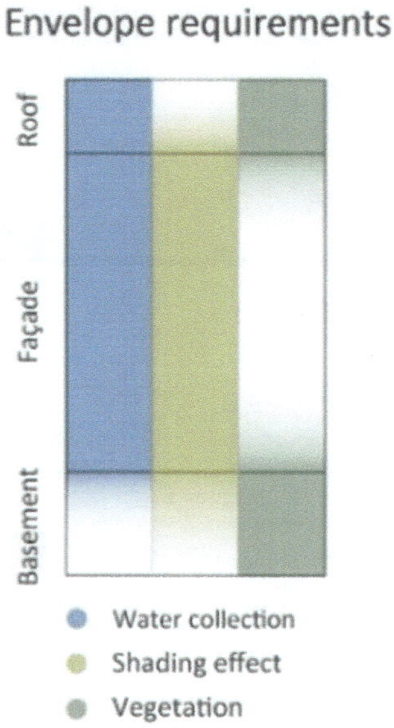

Fig. 65.4 Envelope requirements—water collection, shading and vegetation are arranged on a specific part of the envelope, due to users' comfort but also to functional aspects. *Source* Graphic by the Author

should be composed by modules that can be oriented towards winds direction, for fog harvesting, and towards sun radius for shading (Fig. 65.5).

Today, there are many examples of architectural projects that have worked with adaptive facade systems regarding daylight conditions by implementation of smart systems (Al-Obaidi et al. 2017; Pirouz et al. 2020). Depending on the amount and use of water, the size and placement of the storage must be planned; it can be on the basement or on the roof. In case the amount of water collected is only sufficient for irrigation purposes such as gardening, it can be an autonomous system. Thus, when the amount collected is sufficient for domestic use, it can be connected to the plumbing distribution for utilizing on WC flushes and washing machines after filtration and disinfection processes (Fig. 65.6).

As regarding the maintenance, weakest point of the LFC is the mesh, in fact it often gets broken because of the high velocity of the wind, generally more than 17 m/s (Holmes et al. 2015). Smaller size for the mesh can reduce the risk of damage as stated by Holmes (Holmes et al. 2015). In the urban environment, the wind speed is usually lower, and the proposed façade is composed of reduced fog collecting modules. Therefore, the maintenance of the proposed façade only regards the gutter cleaning; the lifespan of the fog collector depends on the one of the selected mesh. Since a new application field has been proposed, the potentiality of a fog harvesting façade should be explored with some test campaign both in the field, in a fog oasis, and in the lab, in order to understand the behaviour of a mesh placed in front a solid in varying wind velocities and façade dimensions. In order to develop this kind of

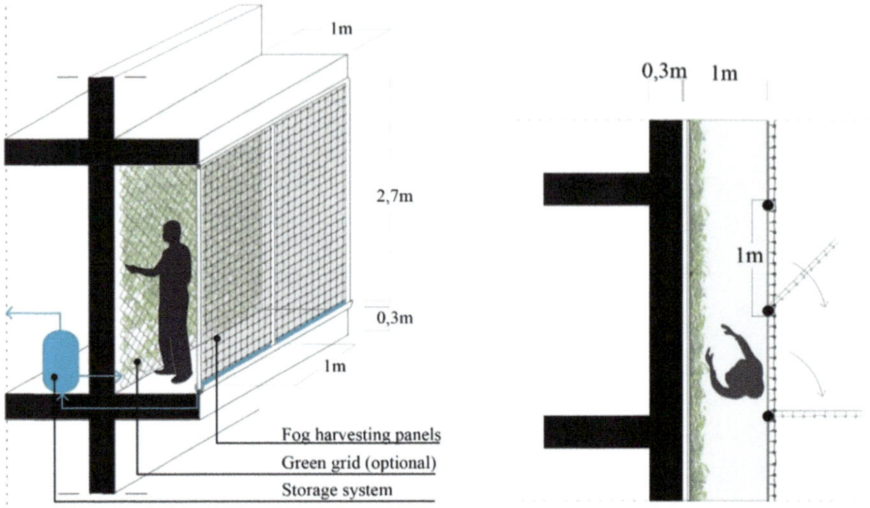

Fig. 65.5 Façade concept—example of fog harvesting project development in building sector. *Source* Graphic by the Author

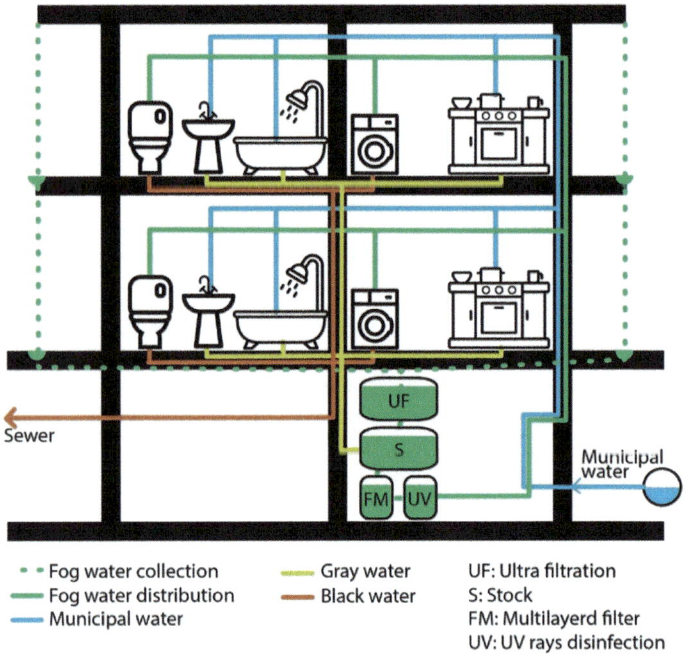

Fig. 65.6 Plumbing system of fog water for domestic use. *Source* Graphic by the Author

façade a parametric design is required to shape it, considering all the variables, so the specific characteristics of the location, the building and the water collection purpose.

65.5 Conclusions

Considered the actual circumstances of traditional resources crisis, due to inadequate management, and the sustainability goals state for many sectors, including construction, a novel architectonical device is essential to enhance the building performance. The smart textile façade proposed can make the building water self-sufficient lowering the ecological footprint and energy demand. The design depends on many factors that refers to the local climatic conditions, to the building features on which it can be attached and on the water demand and use. In order develop the design some parametric tools are needed together with an experimental campaign. In fact, the efficiency of fog collection has been proven in many areas worldwide; however, the innovative solution must be verified and adjusted on the requirement. Extending the system to a bigger scale, the smart textile can cover billboards, construction sites or can be used as a fence in metropolitan cities, collecting water for public use. Depending on the application site, even the fog water collected may not be sufficient for the complete domestic use along the year, it can be stored and used as a support supply for gardening.

References

Al-Obaidi KM, Ismail MA, Hussein H, Rahman AMA (2017) Biomimetic building skins: an adaptive approach. Renew Sustain Energy Rev 79:1472–1491

Caldas L, Andaloro A, Calafiore G, Munechika K, Cabrini S (2018) Water harvesting from fog using building envelopes: part I. Water Environ J 32(4):493–499

de Dios Rivera J (2011) Aerodynamic collection efficiency of fog water collectors. Atmos Res 102(3):335–342

Di Bitonto MG, Angelotti A, Zanelli A (2020) Fog and dew harvesting: Italy and Chile in comparison

European Parliament, Council of the European Union (2010) Directive 2010/31/EU of the European Parliament and of the Council of 19 May 2010 on the energy performance of buildings OJ L 153, 18.6.2010, pp 13–35

Falkenmark M (1986) Fresh water: time for a modified approach. Ambio 192–200

General Assembly (2015) Resolution adopted by the General Assembly on 11 September 2015. A/RES/69/315, 15 Sept 2015. United Nations, New York

Gleick PH (1993) Water in crisis. Pacific Institute for Studies in Development, Environment & Security. Stockholm Environment Institute, Oxford University Press, 473 p, 9, 1051-0761

Holmes R, de Dios Rivera J, de la Jara E (2015) Large fog collectors: new strategies for collection efficiency and structural response to wind pressure. Atmos Res 151:236–249

Klemm O, Schemenauer RS, Lummerich A, Cereceda P, Marzol V, Corell D et al (2012) Fog as a fresh-water resource: overview and perspectives. Ambio 41(3):221–234

Larrain H, Velásquez F, Cereceda P, Espejo R, Pinto R, Osses P, Schemenauer RS (2002) Fog measurements at the site "Falda Verde" north of Chañaral compared with other fog stations of Chile. Atmos Res 64(1–4):273–284
Morichi G, Calixto LB, Zanelli A (2018) Novel applications for fog water harvesting. J Geosci Environ Prot 6(3):26–36
OECD (2022) Water withdrawals (indicator). http://doi.org/10.1787/17729979-en. Accessed on 10 Mar 2022
Pereira LS, Cordery I, Iacovides I (2009) Coping with water scarcity: addressing the challenges. Springer Science & Business Media, Berlin
Pirouz B, Turco M, Palermo SA (2020) A novel idea for improving the efficiency of green walls in urban environment (an innovative design and technique). Water 12(12):3524
Ritchie H, Roser M (2017) Water use and stress. Our world in data
Roach WT (1994) Back to basics: fog: part 1-definitions and basic physics. WEATHER-LONDON 49:411
Schemenauer RS, Cereceda P (1994) A proposed standard fog collector for use in high-elevation regions. J Appl Meteorol Climatol 33(11):1313–1322
Schemenauer RS, Joe PI (1989) The collection efficiency of a massive fog collector. Atmos Res 24(1–4):53–69
United Nations World Water Assessment Programme Secretariat (WWAP) (2016) Water and jobs
Vörösmarty CJ, Douglas EM, Green PA, Revenga C (2005) Geospatial indicators of emerging water stress: an application to Africa. Ambio 34:230–236

Chapter 66
Building Façade Retrofit: A Comparison Between Current Methodologies and Innovative Membranes Strategies for Overcoming the Existing Retrofit Constraints

Giulia Procaccini and Carol Monticelli

Abstract The constant expansion of the cities outside their borders, together with the rapid growth of new technologies and the environmental impact of the building sector, make existing buildings quickly obsolete, both in terms of their functions and their performances. Achieving the goal for greenhouse gas reduction by 2030 implies the necessity to improve the energy performances of the building stock and, for doing so, to overcome the existing constraints that very often prevent builders, tenants and residents from undergoing a renovation process. Given also that the building renovation contributes in the up-cycle strategy of the building stock, avoiding the production of unnecessary waste caused by demolition processes, innovative fast and average costly solutions must be shaped in order to encourage building façade renovation processes at different scales of interventions. Considering that membranes present some inherent properties (such as lightness, thinness, fast assembly, etc.) that make them suitable for both temporary and permanent façade renovations and valuable for overcoming current retrofit constraints, their investigation is of primary interest in order to promote and achieve an extensive building façade renovation. Starting by the investigation of Textile-based Façade Retrofit Solutions (TFRS), this analysis aims at comparing current methodologies with innovative membranes retrofit strategies, in order to evaluate the effectiveness and advantages of textile-based products in overcoming existing constraints to façade retrofit. The main goal of the analysis is to present innovative membrane existing solutions for making building façades resilient and adaptable to the several requirements expected from time to time. The research highlights future developments for TFRS with regard to both temporary and permanent solutions through their employment over existing façades.

G. Procaccini (✉) · C. Monticelli
Polytechnic University of Milan, Milan, Italy
e-mail: giulia.procaccini@polimi.it

C. Monticelli
e-mail: carol.monticelli@polimi.it

E. Arbizzani et al. (eds.), *Technological Imagination in the Green and Digital Transition*, The Urban Book Series, https://doi.org/10.1007/978-3-031-29515-7_66

Keywords Textile façade · Envelope retrofit · Resilient buildings · Innovative strategies

66.1 Introduction

Cities have always accommodated humans' needs and their necessity of permanent dwellings. Actively facing new challenges over time, they have always reflected either the stability of the society or the changes of time.

Nowadays, cities are asked to be resilient more than ever: given that 'resilience' means the ability of an entity to be able to deal and to overcome a traumatic event or a period of difficulty, cities are called to accommodate and to react to extreme conditions such as the COVID-19 pandemic, the climate change and the severe globalization.

Consequently, with the current fast times, temporary constructions are re-gaining popularity as a practical solution for accommodating the recurring changes dictated by the time. The accelerated speed, due and thanks to which the global society is nowadays moving faster and further, is contributing in making cities quickly obsolete.

Additionally, the continuous birth of new technologies is a double-edged sword, affecting on one hand the progress of the society and their cities but intensifying, on the other hand, the process and the speed of the changes, with a consequent mutual impact on climate change, contributing in parallel to its acceleration and to its slowdown.

Cities and buildings obsolescence is reflected by their need to be more resilient and open to changes, both in terms of aesthetic appearance and technical performances.

The climate change has led to the necessity to diminish the greenhouse gas emissions of the building stock, which is responsible for approximately 40% of the energy consumptions, representing the largest energy consumer and one of the most significant CO_2 emission sources in Europe, with a share higher than a third of the total EU emissions (European Commission 2021).

The strategy of the renovation has been continuously spreading in the last years, and new incentives have been launched with the exact purpose of increasing this good practice that contributes in the up-cycle of the building stock and avoids the production of unnecessary waste caused by demolition processes. With the aim to look for fast and average costly solutions in order to encourage building façade renovation processes at different scales of interventions, this paper analyzes innovative solutions on the market for making building façades resilient and adaptable through the application of membranes over façades.

Among the spectrum of lightweight materials, membranes present some inherent properties [such as thinness, easy transportability, fast assembly, etc. (Chilton 2010; Pohl and Pohl 2010)] that make them suitable for both temporary and permanent façade renovations and valuable for overcoming current retrofit constraints. Consequently, taking advantage of membranes potentialities, this analysis aims at comparing current methodologies with innovative membranes retrofit strategies, in order to evaluate the effectiveness and advantages of textile-based products in overcoming existing constraints to façade retrofit and achieving extensive building façade renovations.

66.2 Methodology

Starting from the assumption that architectural textiles present some inherent properties suitable for retrofit applications, the study focuses on a comparative analysis between current Façade Retrofit Measures (FRMs) and Innovative Membrane Strategies for identifying the current façade retrofit constraints that could be overcome by the application of textile-based solutions.

In order to do so, a qualitative and comparative analysis based on the data acquired through the state of the art has been carried out. The methodology applied for the study consists of two parallel analyses and a sequential investigation aimed at comparing the acquired data: The attention has been drawn, on one hand, on current façade retrofit methodologies and their constraints and, on the other hand, on the analysis of textile potentialities for façade retrofit applications. Successively, the investigation of some textile-based façade retrofit best practices and innovative research contributes in supporting the discussion about the advantages of TFRS.

66.3 Review of the Current Retrofit Methodologies and Their Constraints

The retrofit practice is globally increasing its popularity due to the necessity to diminish building energy consumptions and to adapt existing buildings to enhanced performances. Aesthetic and functional reasons usually drive the renovation of a building, due to the aging of the components and the lowering of the performances.

Building façade retrofit is an effective measure to reduce global energy consumptions, considering that only façades account for 20–30% of the total (Dall'O' et al. 2012), being characterized by large thermal transmittance. Therefore, the aim of building façades retrofit is to reduce the use of air-conditioning and heating systems in existing buildings.

The extensive review of Façade Retrofit Measures (FRMs) provided by Sarihi et al. (2021) divides them between Energy Conservation Measures (ECMs), Energy Modulation Measures (EMMs) and Combined Measures (CoMs) (Fig. 66.1), analyzing their effectiveness in different climatic conditions.

The most common FRMs are insulation and shading, respectively, representing the most effective solutions in Heating and Cooling Dominated climates. The main difference between ECMs and EMMs consists in the extension of the measure over

Façade Retrofit Measures (FRMs)		
Energy Conservation Measures (ECMs)	Energy Modulation Measures (EMMs)	Combination of Measures (CoMs)
• Insulation - Exterior Insulation - Interior Insulation **• Window Improvement** **• Airtightness** **• Window to Wall Ratio**	**• Double Skin Glazed Façade** **• Opaque Ventilated Façade** **• Façade Finish Coating** **• Shading** **• Green Façade** **• Phase Change Material (PCM) Façade**	**• ECMs + EMMs**

Fig. 66.1 FRMs classification according to Sarihi et al. (2021)

time: While ECMs are permanently applied to building façades, EMMs measures modulate energy consumptions only in specific periods.

Given that the thermal transmittance of the building envelope represents the main cause of the high energy consumptions, ECMs are being extensively applied in order to stabilize the internal temperature both in summer and winter, minimizing the use of technological appliances for improving interior thermal conditions. The EMMs instead aim at controlling the solar properties by the application of solar thermal-driven heating and cooling technologies. The combination of ECMs and EMMs aims to fully exploit their benefits for reducing the heating and cooling energy demand.

However, each of these strategies implies not only advantages but also various disadvantages (Corrêa et al. 2020) that often prevent tenants and builders from undergoing building renovation processes. The main constraints limiting the extensively application of the practice have been summarized in Fig. 66.2, associating FRMs with each related constraint.

	Insulation ETI (External Thermal Insulation)	Insulation | ITI (Internal Thermal Insulation)	Insulation | CTI (Cavity Thermal Insulation)	Window Properties	Window to Wall Ratio	Double Skin Glazed Façade	Opaque Ventilated Façade	Green Façade	Finish Coating	Phase Change Material (PCM)	Shading Device
Durability issues related to system coating	•••	•	•	•		••••	•	•	•••	••	••
Finishing execution complexity due to architectural constraints	•••		••	••	••	••	••	•••	••	••	•
Vulnerability to mechanical stresses	•••	••	•	••	•••	••	••	•••	•••	•••	•••
Additional weights to the structure	•	•	•	•	•••	•••	••	•••	•	••	•••
Exposed execution to weather conditions	•••			•••	•••	•••	•••	•••	•••	•••	•••
Leak of protection of the façade from outside actions		•••	•••								••
Poor reduction of thermal inertia		••	••		••						•••
Increase of moisture-related problems		•••	••		••			••			
Presence of thermal bridges		•••	••		••		•				
Fire spreading risk			•••			•••	•				
High skilled labour required			•••	••	•	•		•••		••	
Aesthetical modification of the building appearance	•••			•	•••	•••	•	•••	•••	••	•••
High cost investment	••	•	•••	•••	•••	•••	••	•••	•••	•••	••
Disturbance for the occupants during the execution	••	•••	••	••	•••	••	••	••	••	••	••

Fig. 66.2 FRMs and related constraints

On top of the detailed analysis of each single practice, it should be clarified that most presumably there are some parameters, such as (i) the disturbance for the occupants during the execution of the process, (ii) the high-cost investment requested for the retrofit practice and (iii) the limited durability of the strategy in an optic of LCA, that limit the spread of this practice, representing the crucial barrier that must be overcome for increasing the number and possibilities of application of FRMs.

66.4 Membranes for Façade Retrofit Strategies

Starting from the most diffused constraints, it follows the clear necessity to exceed them in order to enhance the building façade retrofit practice.

Textile materials are nowadays re-gaining popularity in architecture thanks to their intrinsic characteristics (Chilton 2010; Pohl and Pohl 2010) and the renovated awareness about environmental and sustainability topics (Mendonça 2010; Monticelli and Zanelli 2021; Sandin and Peters 2018). As highlighted by Monticelli (2015a):

> the success keys for the textile envelopes in architecture are (1) the reduction of weight and stiff parts of the building elements coupled with ensuring high performances […] and (2) the minimisation of the time of installation and maintenance though allowing an easy replacement of the building elements […].

Indeed, with high performance textiles available on the market, new functions and different architectural applications arise. Membrane applications in façades are mainly recorded as cladding systems for commercial buildings or, sometimes, as the second layer of a double skin (existing) façade. Lately, a new diversified range of products has entered the construction market for being applied either in retrofitting practices of existing buildings, either as sunscreens, or backlit surfaces (Monticelli 2015a).

The main characteristics that make textiles attractive as envelope materials are their lightness, thinness and flexibility, together with the light transportability and the increased durability. Additionally, they present pleasant aesthetic qualities and visual properties given by their translucency and the possibility of sun and light control. They can also comply with the mechanical and thermal requirements of a building façade, being able to face daily and seasonally temperature differences (Mendonça 2010; Paech 2016). Given that textile structures have the advantage to be easy dismantled, it is therefore possible to foresee a second (and even a third) life for these type of structures (Monticelli and Zanelli 2021; Sandin and Peters 2018). The efficient use of the materials and the reduced environmental impact contribute to the diffusion of membranes and their spread especially for temporary structures, taking advantage of the minimization time of installation and maintenance. Starting from the most common characteristics that define and differentiate these materials, it is possible to foresee an extensive use of textiles in the retrofit practice with different applications.

66.4.1 Membranes in Façade Retrofit Applications

Although nowadays the use of membranes in façade retrofit practice is still limited and it is majorly employed for the purpose of an aesthetic retrofit with the aim to give a new iconic appearance to a building, there are few cases in which it has been used as a sun-shading device or with the aim to improve the energetic efficiency of the building envelope (Mendonça 2010).

The application of textiles and membranes for the purpose of the façade retrofit lays its ground in the exploitation of the intrinsic properties of the materials and the great freedom they leave: Being very lightweight, their application can be taken in consideration even in a second moment, without interfering on the existing structure with additional and excessive weights.

As testified by the Case Studies in Fig. 66.3, the transparency or translucency given by the material used allows for the control of the natural light penetrating inside the building without obstructing the inside-out view. The creation of a textile second-skin has the aim to protect the building components from the atmospheric aging and, defining a buffer zone, allows for natural ventilation. Textiles and membranes can also be employed as exterior sun-shading devices, modulating the sunlight penetration inside the building (Fig. 66.4a) or obstructing it partially or completely through textile curtains, representing at the same time a useful and decorative element of the façade (Fig. 66.4b). A more common application of the material can be detected in Fig. 66.5, where textiles have, respectively, been used as a cladding system for giving a new appearance to the building or as an additional envelope with the aim to define an iconic façade.

Fig. 66.3 **a** Westraven Office Complex—Utrecht, Netherlands (Credit: Cepezed - picture: Joannes Linders); **b** ETFE Façade Unilever Building—Hamburg, Germany (Credit: Vector Foiltec)

Fig. 66.4 **a** King Fahad National Library—Riyadh, Saudi Arabia (Credit: Gerber Architekten - picture: Christian Richters); **b** Aichinger House—Kronstorf, Austria (Credit: Hertl Architekten - picture: Kurt Hörbst)

Fig. 66.5 **a** McDonald's – Legnano, Italy (Credit: Canobbio Textile Engineering); **b** Gotha Cosmetics Headquarter - Lallio, Italy (Credit: iarchitects - picture: Claudia Calegari)

66.5 Discussion

The use of textiles in façades for the purpose of the retrofit is increasing its popularity thanks to the different FRMs it matches, especially the ones referring to the EMMs (Fig. 66.6). Some recent studies have been conducted for simulating the application of a textile solution onto an existing façade for improving its energy performances.

The tensile second-skin simulation refurbishment (Ciampi et al. 2021) highlighted a yearly heating and cooling energy demand reduction of about 9.8%, while the semi-opaque ventilated façade solution (Cortiços 2020) testified around 23% in heating and 39% in cooling savings in inland areas, with a reduction of 6.45% of CO_2 emissions and around €13.5/m^2 on the heating and €23.5/m^2 on the cooling spending.

Within the spectrum of EMMs, these last two research testify a potential application of TFRS with respect to the first and second strategies. Although there are no studies investigating the energetic behavior of textile sun-shading systems, the

Textile Façade Retrofit Measures (TFRMs)		
Energy Conservation Measures (ECMs)	**Energy Modulation Measures (EMMs)**	**Combination of Measures (CoMs)**
• **Insulation** - Exterior Insulation - Interior Insulation • **Window Improvement** • **Airtightness** • **Window to Wall Ratio**	• Double Skin Glazed Façade • Opaque Ventilated Façade • Façade Finish Coating • Shading • **Green Façade** • **Phase Change Material (PCM) Façade**	• **ECMs + EMMs**

Fig. 66.6 TFRMs (in green) within the more general spectrum of FRMs edited by Sarihi et al. (2021)

numerous applications that can be found and the spread of the practice serve for valuing this measure too.

Given that the analysis of the current FRMs led to the understanding of three main constrains that limit the practice of the retrofit, it is presumable that the application of TFRS could overcome them thanks to their intrinsic characteristics:

- In terms of disturbance for the occupants, it is possible to claim that a TFRS could diminish it given the reduced amount of time for the installation of the system (Paech 2016; Gezer and Aksu 2021);
- The cost investment could be lowered by the use of a minor amount of material (Cortiços 2020; Beccarelli and Chilton 2013);
- The advantage in terms of estimated duration of the practice can be foreseen in an optic of LCA: The use of TFRS could be conceived both for permanent and temporary applications, therefore unveiling new temporary retrofit strategies able to comply with the specific requirements of the time without implying the use of excessive materials, whereas forecasting the re-use of the system (Sandin and Peters 2018; Monticelli and Zanelli 2018).

66.6 Conclusion

Currently, innovative Façade Retrofit Strategies are being investigated with the aim to disclose easier, faster and lower-cost solutions. TFRS perfectly fit in this scenario, envisaging both permanent and temporary solutions.

The paper has presented some TFRS already accomplished, analyzing their applications and advantages in comparison with current methodologies and the main façade retrofit constraints that limit the spread of the practice. The qualitative analysis has been based on the data acquired through the Literature Review and the investigation of some Best Practices. The outcome highlighted the potentialities of TFRS in overcoming some of the current retrofit constraints, such as the disturbance for the occupants and the cost investment, consequently framing new textile-based products development lines.

Further LCA, Life Cost Analysis and Energy Simulations will be run in order to quantitatively value the thesis. Nevertheless, it is already possible to envisage a wider spread of TFRS by taking advantage of textile intrinsic properties for overcoming current retrofit constraints.

References

Beccarelli P, Chilton J (2013) Advantages of lightweight tensioned coated fabrics and foils façades for the building sector. In: Bletzinger K-U, Kröplin B, Oñate E (eds) VI international conference on textile composites and inflatable structures. Structural membranes

Chilton J (2010) Tensile structures—textiles for architecture and design. In: Pohl G (ed) Textiles, polymers and composites for buildings. Elsevier Ltd., Cambridge, pp 229–257. http://doi.org/10.1533/9780845699994.2.229

Ciampi G, Spanodimitriou Y, Scorpio M, Rosato A, Sibilio S (2021) Energy performance of PVC-coated polyester fabric as novel material for the building envelope: model validation and a refurbishment case study. J Build Eng 41. http://doi.org/10.1016/j.jobe.2021.102437

Corrêa D, Flores-Colen I, Silvestre JD, Pedroso M, Santos RA (2020) Old buildings' façades: fieldwork and discussion of thermal retrofitting strategies in a Mediterranean climate. Designs 4:45. https://doi.org/10.3390/designs4040045

Cortiços ND (2020) Improving residential building efficiency with membranes over façades: the Mediterranean context. J Build Eng 32. http://doi.org/10.1016/j.jobe.2020.101421

Dall'O' G, Galante A, Pasetti G (2012) A methodology for evaluating the potential energy savings of retrofitting residential building stocks. J Sustain Cities Soc 4:12–21. http://doi.org/10.1016/j.scs.2012.01.004

Gezer H, Aksu GA (2021) Assessment of textile architecture from a sustainability perspective. Architectural sciences and technology, chap 6

Mendonça P (2010) Low-span lightweight membranes in housing—environmental and structural potentialities. In: Cruz (ed) Structures and Architecture. Taylor & Francis Group, London

Monticelli C (2015a) Membrane claddings in architecture. In: De Llorens JI (ed) Fabric structures in architecture. Elsevier Ltd, Cambridge, pp 449–479. http://doi.org/10.1016/B978-1-78242-233-4.00014-0

Monticelli C (2015b) Steel and textile solar shading. Gerber Architekten, King Fahad National Library, Riyadh, Saudi Arabia, 2004–2013. In: Costruzioni Metalliche – nov. dic. 15

Monticelli C, Zanelli A (2018) Le membrane strutturali in architettura: una soluzione eco-efficiente per il futuro? Techne 16

Monticelli C, Zanelli A (2021) Material saving and building component efficiency as main eco-design principles for membrane architecture: case studies of ETFE enclosures. J Archit Eng Des Manag 17. http://doi.org/10.1080/17452007.2020.1768507

Paech C (2016) Structural membranes used in modern building façades. In: International symposium on "Novel structural skins—improving sustainability and efficiency through new structural textile materials and designs", pp 61–70. Procedia Eng 155. http://doi.org/10.1016/j.proeng.2016.08.007

Pohl J, Pohl G (2010) The role, properties and applications of textile materials in sustainable buildings. In: Pohl G (ed) Textiles, polymers and composites for buildings. Elsevier Ltd., Cambridge, pp 258–289. http://doi.org/10.1533/9780845699994.2.258

Sandin G, Peters GM (2018) Environmental impact of textile reuse and recycling—a review. J Clean Prod 184:353–365. https://doi.org/10.1016/j.jclepro.2018.02.266

Sarihi S, Saradj FM, Faizi M (2021) A critical review of façade retrofit measures for minimizing heating and cooling demand in existing buildings. J Sustain Cities Soc 64. http://doi.org/10.1016/j.scs.2020.102525

Chapter 67
Technologies and Solutions for Collaborative Processes in Mutating Cities

Daniele Fanzini, Irina Rotaru, and Nour Zreika

Abstract The city, a place of contemporary living par excellence, challenges the planner by making it necessary to adapt progressively quicker to changes and to overcome the traditional design approach linked to the modern idea of the industrial city. Indeed, living in non-stationary contexts, the complexity of problems nowadays requires a new planning endeavor capable of testing future solutions 'in the field' rather than 'on paper,' involving citizens, but also continuously adapting processes to achieve the expected results. The proposed contribution aims to document possible ways to trigger virtuous urban renewal processes, sustainably activating tangible and intangible resources. The topic will be investigated from the point of view of the triad: 'project, technology, and digital solutions,' adopting a social perspective. The latter ensures the active involvement of citizens in strategic decisions, increasing their awareness and civic sense, but also supporting the proposition of evolving planning scenarios in order to develop solutions that will be concrete, achievable, and resilient. The core element concerns the way in which it is possible to promote the creation of an extended social mind through which collective behavioral change can be fostered. In some cases, digital technologies prove to be the effective 'expert instrument,' also for understanding the planned intervention, opening the design process for different stakeholders not necessarily familiar with technical conventions. According to Floridi, digital transformation 'disconnects and reconnects specific processes,' and the project represents the most powerful innovation element to promote the ecological transition. These dynamics will be explored through the analysis of some research and project activities that directly involved the authors of this article.

D. Fanzini (✉) · N. Zreika
Polytechnic University of Milan, Milan, Italy
e-mail: daniele.fanzini@polimi.it

N. Zreika
e-mail: nour.zreika@polimi.it

I. Rotaru
Saint Germain-en-Laye, France
e-mail: irina.rotaru@saintgermainenlaye.fr

E. Arbizzani et al. (eds.), *Technological Imagination in the Green and Digital Transition*, The Urban Book Series, https://doi.org/10.1007/978-3-031-29515-7_67

Keywords Mutation cities · Collaborative processes · Phygital ecosystem · Urban change management · Innovative solutions

67.1 Introduction

Cities are going through an inevitable and necessary mutation. The concept of the mutating city, or rather the city as a living organism evolving over time with respect to the circumstances and needs of its inhabitants, requires tailored approaches to the management of this transformation. Design experience and the use of enabling technologies can support the implementation of adaptable systems of governance and participative project activities which exploit the potential of the widespread social network to support the central decision-making processes (Celaschi et al. 2020).

The practice of participatory planning began with the American civil rights movement in the 1950s and later developed with important figures such as Jane Jacobs and Jane Addams in the USA and Charles Rowley in Manchester, whose studies paved the way for 'advocacy planning.' This term, invented by Paul Davidoff, can be defined as a theory of pluralistic and inclusive planning through which the interests of various groups within society can be represented. In her idea of the city, Jacobs includes the concept of social capital deriving from the importance of networks of people. This is also true for Jan Gehl, another important European scholar, who brings the methods of the social sciences into the toolbox of architects and urban planners, to construct a new relationship between physical space and social space, starting from the careful and accurate observation of reality (Ermacora and Bullivant 2016). Gehl considers that cities are for people, and urban diversity can be interpreted as a resource that originates in the past when cities were still designed around people. With his work, Gehl creates a new relationship between physical space and social life; collecting data on people's ways of life brings their needs into urban design, thus breaking the rigidity of traditional planning practices.

The process of deregulation that first began in Great Britain with the Thatcherian policies and gradually spread to other countries, including Italy, created a gap that in the 1990s was partially filled by local activism, allowing the co-creation of services or measures of responsible subsidiarity.

In the new millennium, the growth of environmental awareness will further strengthen the need for more conscious and participatory management of the common good. With explicit reference to urban issues, Goal 11 of the UN 2030 Agenda sets the objective for making human settlements inclusive, safe, enduring, and sustainable, also through the adoption of integrated policies and plans aimed at inclusion. By 2030, the UN Agenda calls for the strengthening of more inclusive and sustainable forms of urbanization and the capacity to participatively, integrally, and sustainably—in a word, 'openly'—plan and manage all human settlements in all countries that are participative, integrated, and sustainable—in a word, 'open.'

The expression 'open government' refers to a way of exercising power, both at central and local level, based on models, tools, and technologies that allow public administrations to be transparent in order to promote effective actions and ensure control over their work. Open Government is widely diffused in Anglo-Saxon countries thanks to the propulsive push of Barack Obama's Open Government Directive. The model is also spreading in Europe thanks to the Recommendation by the Committee of Ministers of the Council of Europe on 'Citizen Participation in Public Life at the local level,' which encourages the concrete opening of the bodies and institutions of European nations to the new information and communication technologies, making them concrete tools for change with a view to transparency and dialogue with citizens.

When adopting Open Government, administrations open up to direct and participatory confrontation with private individuals, focusing decision-making processes on the actual needs and requirements of local communities. Fundamental elements of the open logic are the centrality of the citizen, participatory and collaborative administration, as well as transparency of data and information through new digital technologies. In this sense, the ICT could represent the enabling elements that make the process of reconfiguration of models, tools, and technologies within administrations effectively sustainable, both from the technical-operational point of view and from the investment point of view.

67.2 Digital Tools for Collaboration and Public Involvement

The critical success of collaboration, of which co-creation is the highest level, is based on sharing information, developing honesty with the (potential) client or community, and expressing sincere interest. Co-creation with stakeholders can involve either simple on/off decision-making consultation on previously developed solutions or full partnership involvement. Several types of relationships can arise in the co-creation process and can be traced to four basic relationships: the expert club; the general public; the coalition of parties; the communities of like-minded people (Malakhatka et al. 2021). This categorization refers to the level of openness of the community and the ability of the public administration to capitalize on these energies through appropriate legal, administrative, and technical-operational tools. Thus, a fundamental problem of the practices is related to the forms of public involvement (Ostanel 2017). New formulas to overcome these limitations emerge from crowdsourcing approaches, co-creative digital media, and neogeography tools.

In some contexts (e.g., Germany, France, the Netherlands), methodologies of massive co-design have been developed (Meroni et al. 2018) by using specific computerized solutions to exploit diffused intelligence in creative processes. Interactive systems are capable of facilitating public participation, anticipating problems, and allowing creative interaction between experts, the public, and stakeholders in various capacities. These technologies represent a possible solution to the problem highlighted above if intermediate forms between professional and non-professional operators, and professional operators of different disciplines are provided. The objective is to use data as a resource for the empowerment of communities and as a means to test the effectiveness of solutions. Therefore, the project supports the democratic use of data as a resource to create new values and meanings of action (sense making). Participatory design could also enable social innovation and transform data into something that can be understood and utilized by broader communities (data sensification) (Morelli et al. 2018).

Based on the analysis of the information tools available online, the following types of participatory design emerge:

- Platforms that use the power of design and art to increase citizen engagement (i.e., http://welcometocup.org).
- Think tanks to support the public in defining environmental, economic and social policies (i.e., https://cles.org.uk/about/cles/).
- Tools for community planning (i.e., https://www.canr.msu.edu/michigan_citizen_planner/).
- Solutions for the facilitation of community use of social media as a design tool (i.e., https://ecosistemaurbano.com).
- Solutions for urban planning (i.e., https://minstad.goteborg.se/minstad/index.do).

The daily use of these tools has highlighted the importance of understanding, and boosting the ecosystem of co-creation as the comprehensive organizational, institutional, and cultural setting in which social innovation is embedded (Eckhardt et al. 2021; Steen 2013). The disconnection between various players, and between individual and collective dimensions, leads to the weakening of the ecosystem of co-creation and causes disengagement and loss of motivation. The issue here is how to encourage the direct assumption of responsibility of the parties involved through behavioral dynamics and cognitive skills to clarify why action is important for the community (Kaletka et al. 2017). Obviously, this is not enough; it is also necessary to simplify access to opportunities for action (what concretely makes virtuous behavior feasible) through motivational contributions, and individual and community incentives.

67.3 The Motivational Platform of the Open-Air Urban Market of Saint-Germain-en-Leye

The heritage protected city center of SGL is an exception in Ile de France, being a lively area animated by around 800 commercial and service activities, attracting people from an estensive surrounding area. However, the need to improve the attractiveness of the city center and implicitly to offer new updated opportunities to businesses in the area was felt more and more intensively. In response to this, over the past seven years, several attempts were made to promote simple digital tools for increasing the visibility and selling possibilities of local shops.

Thanks to the EIT-UM funding, a team made up of researchers from the Politecnico di Milano, the local association of retailers (CAP SGL) and the Municipality of Saint-Germain-en-Laye developed an active mobility and community friendly e-commerce platform (Fig. 67.1) was developed to support both the revival of lively people friendly public spaces and the economic relaunch of the city center. Furthermore, it represented an essential component of the people centered approach of the Safely Connected project, backing the sense of community and the creation of a solidarity network necessary especially during difficult times of crisis.

Going far beyond the simple idea of e-marketplace, this tailor-made innovative digital tool resembles a series of advanced services (for local entrepreneurs and their clients) enabling the optimization of business models and the creation of new business opportunities. Through the limitation of necessary travel and a shift toward the use of active modes of transport, by enabling the sustainable sharing of deliveries for both clients and shops (Fig. 67.2), this digital tool also contributes to the general improvement of the public health of local people, directly related to the public space quality and use, and in this specific COVID-19 context, to curb the number of potentially unsafe interactions.

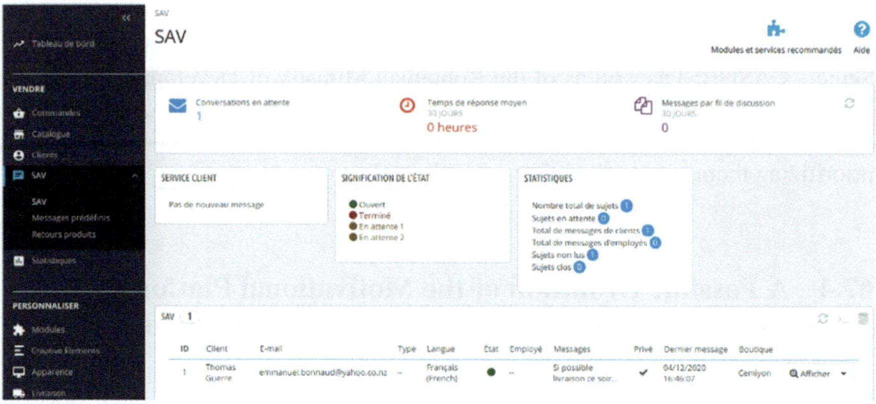

Fig. 67.1 SGL platform

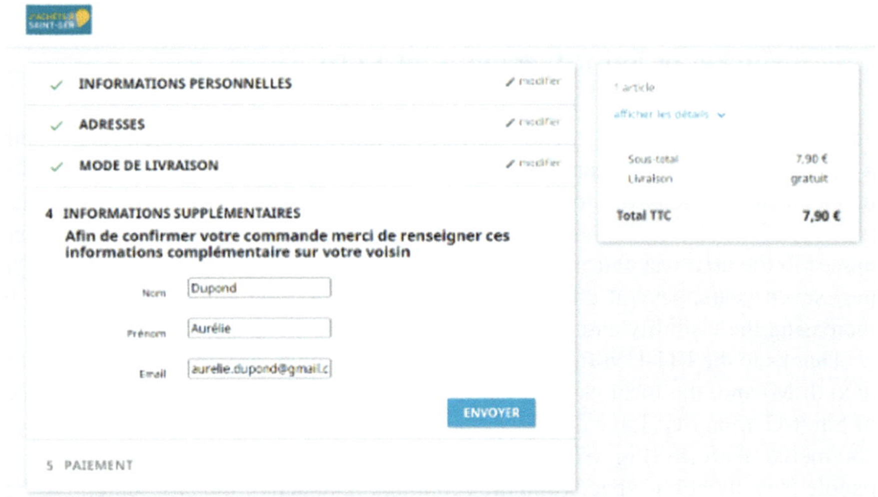

Fig. 67.2 SGL platform: customers can choose to indicate a neighbor for the delivery of their order

During the implementation phase, a new series of services have been added through the organization of an urban logistics platform at the periphery of the city center in cooperation with Urby (a branch of the Post Office). Among these services are the last mile cargo bike shared deliveries for shops, delocalised storage, package collection, and the extension of the area for sustainable deliveries to clients over a radius of 15 km around SGL. The innovative digital tool has been conceived so as to be easily adaptable for a variety of contexts and also to allow further evolutions depending on the advancement of needs, businesses, and technologies. Tested in the city center of SGL, it can easily be scaled-up and replicated for other territories in France or abroad. Once validated, the plan is to promote and replicate it in a first phase with the help of the Ile de France Region and of the Community of cities that SGL is part of (Communauté d'Agglomération Saint Germain Boucles de Seine—CASGBS) as well as of the Romanian Ministry of Development and other foreign cities that have already expressed interest in such a facility. Additionally, it can be introduced to and made available for the members of the EIT UM City Club, prioritizing them among the other possible replicators.

67.4 A Possible Evolution of the Motivational Platform: The Proposal for the Historical 'Borgo di Camugnano'

Camugnano is a small town located in the Bolognese Apennines rich in history and nature. As happens for many similar realities, the desire to enhance and promote one's territory collides with the lack of economic resources. For this reason, the current municipal administration has decided to participate in the tender launched by the Ministry of Culture (MiC) for the implementation of the Borghi Plan envisaged by the PNRR and to announce a collection of ideas by requesting expressions of interest. The Create regional cluster dedicated to Cultural and Creative Industries, in collaboration with e-Making srl, has presented its proposal, centered on the use of enabling technologies, to promote the involvement and creativity of the inhabitants. The proposal aims at creating a physical and digital ecosystem (phygital) of the Borgo di Camugnano as a tool to manage in an integrated way the different lines of action proposed by the call for expressions of interest by the Municipality of Camugnano in the framework of the Ministerial Call ("Riattivazione dei Borghi Storici").

The phygital ecosystem is conceived as a space in which the ideal and the virtual city (digital model) overlap to enable through applications and interactive tools such as iPads, tablets, digital polling, QR code, Augmented Reality (AR), or Virtual Realty (VR), services, products, and suggestions. These can enrich the fruitful online experience, increasing the involvement of the public in a pre- and post-visit phase. Very different actions, ranging from the enhancement of cultural heritage to the rationalization of services, the promotion of tourism and territorial cooperation were proposed, therefore requiring an integrated platform for their effective management.

For its practical implementation, the proposal requires the following key steps:

- Identification of the urban perimeter chosen for experimentation (a limited but significant portion of the territory).
- Identification of the subjects to be involved in the co-design/co-management of the phygital model (recruitment of the practitioners).
- Digitalization of the selected urban context through the Matterport Technology (https://matterport.com) for digitally capturing the perimeter of the urban space (Fig. 67.3).
- Once the digital model has been obtained, construction of the data box for carrying out the activities foreseen by the expression of interest, including those for promotion of the territory, will be realized. By means of tags, the data box will be able to activate interactive environments within the virtual model to exchange information with the users of the platform, thus enriching the virtual space with useful contents (Fig. 67.4).
- Realization of concrete artifacts and services to improve the urban environment and create the interconnections with the digital model. Some QR codes will be placed at the points of interest around the city to provide access to the same contents of the virtual urban model.
- Definition of a communication strategy will be conveyed through existing channels and the new tool itself, which for the public will be a quick way to get in touch with the territory. The instrument could also be a 'landing' base for visitors from which they use the various links, such as the App store and other functions, for sharing contents and co-creation of visitor experiences (Fig. 67.5).
- Testing and collection of monitoring data for the use of technologies.

Fig. 67.3 Example of digital urban space realized by the partner e-Making srl through Matterport technology

Fig. 67.4 Example of data box realized by the partner e-Making srl

The set of activities represents a first action plan which, if funding is granted, will be tested by the regional Create Cluster to promote the area through the contribution of cultural and creative industries.

This social-virtual world provides knowledge of possible destinations, as well as a better understanding of them by optimizing the services and promoting experiential forms of use and stimulating projects by testing their effectiveness in advance.

The creation of a phygital ecosystem of a portion of the Borgo di Camugnano, and its related applications, could represent the necessary step for the future construction of a real metaverse supported by a spatial geolocation, capable of opening up much wider and unforeseen opportunities. In fact, the extremely fast and inexpensive cultural metabolism of virtual worlds makes it possible to overcome the space–time

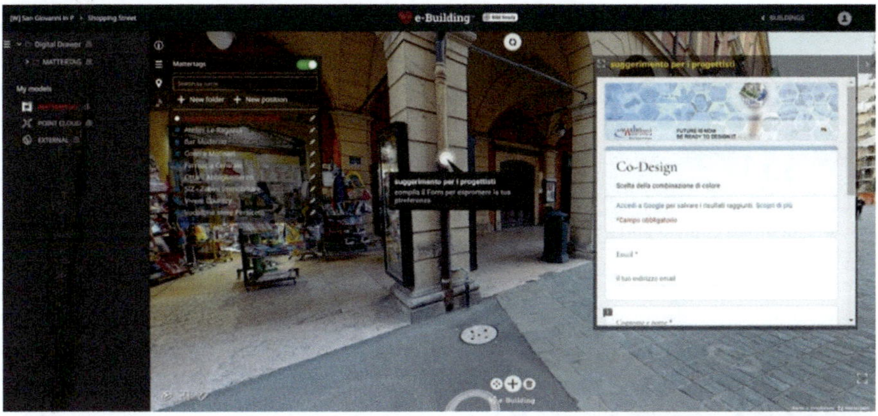

Fig. 67.5 Example of online toolbox for sharing contents. In this case, the sharing box concerns an activity for improving the quality of the public space through co-design activity

limitations of the real world and offer to large numbers of websites. This does not mean abandoning initiatives in the field to dedicate exclusively to the communicative and social formulas of the web, but to encourage them to grow together by making one an extension of the other, through the collaborative involvement of people (Fig. 67.6).

67.5 Conclusion

Starting from the concepts of 'Mutant City' and the need to develop new tools to govern urban transformation, the contribution has analyzed the way in which information technologies can enable new and more effective forms of participatory planning based on the assumption of responsibility by the subjects involved. The practical application of these methodologies was illustrated through two research projects that involved the authors of this contribution. The first research project involved the creation of an online platform for the optimization of the logistics of the open-air urban market of SGL that provides rewarding solutions for virtuous behaviors. The second project foresees the further development of the system for transforming it into a phygital urban space. The latter research, currently under development, is supported by Cluster Create, a non-profit organization set up by the Emilia Romagna Region, which brings together companies, research centers and training institutions to strengthen the regional production system by focusing on the integration of technology, creativity, and culture.

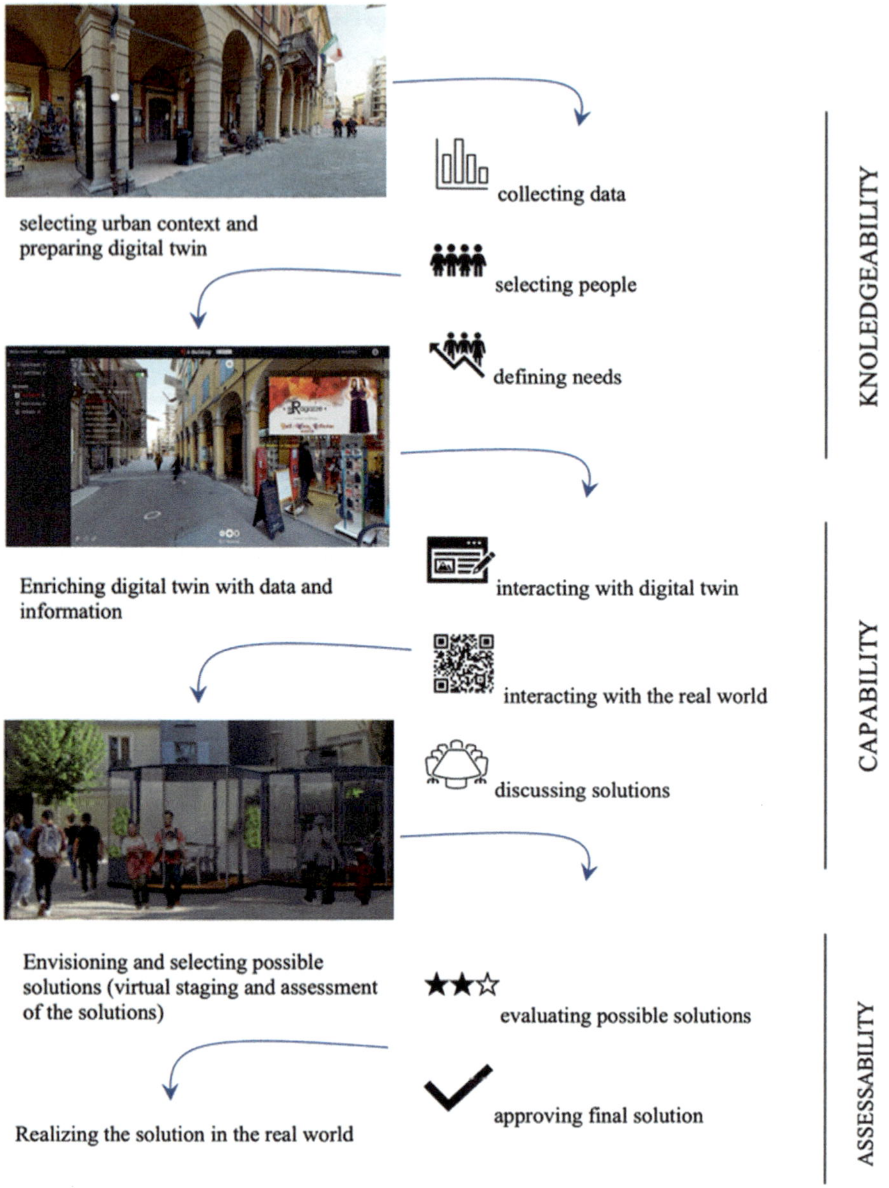

Fig. 67.6 Workflow of the process

References

Celaschi F, Fanzini D, Formia E (2020) Enabling technologies for continuous and interdependent design. In: Lauria M, Mussinelli E, Tucci F (eds) Producing project. Maggioli Editore, Santarcangelo di Romagna

Eckhardt J, Kaletka C, Kruger D, Maldonado-Marsical K, Schulz AC (2021) Ecosystem of co-creation, frontiers. Sociology 6. Retrieved from https://doi.org/10.3389/fsoc.2021.642289. Accessed 19 Mar 2022

Ermacora T, Bullivant L (2016) Recoded city: co-creating urban futures. Routledge, London

Kaletka C, Markmann M, Pelka B (2017) Peeling the onion. An exploration of the layers of social innovation ecosystems. Modelling a context sensitive perspective on driving and hindering factors for social innovation. Eur Public Soc Innov Rev 1(2). Retrieved from https://pub.sinnergiak.org/esir/article/view/42. Accessed 19 Mar 2022

Malakhatka E, Sopjani L, Lundqvist P (2021) Co-creating service concepts for the built environment based on the end-user's daily activities analysis: KTH live-in-lab explorative case study. Sustainability 13:1942. Retrieved from https://doi.org/10.3390/su13041942. Accessed 19 Mar 2022

Meroni A, Selloni D, Rossi M (2018) Massive co-design. A proposal for a collaborative design framework. Franco Angeli, Milano

Morelli N, de Götzen A, Mulder I, Concilio G (2018) Editorial: designing with data, democratisation through data. In: Storni C, Leahy K, McMahon M, Lloyd P, Bohemia E (eds) Design as a catalyst for change—DRS international conference 2018, 25–28 June, Limerick, Ireland

Ostanel E (2017) Spazi fuori dal comune: Rigenerare, includere, innovare. Franco Angeli

Steen M (2013) Co-design as a process of joint inquiry and imagination. Des Issue 29(2). Retrieved from https://www.researchgate.net/publication/249335514_Co-Design_as_a_Process_of_Joint_Inquiry_and_Imagination. Accessed 19 Mar 2022

Chapter 68
New Perspectives for the Building Heritage in Depopulated Areas: A Methodological Approach for Evaluating Sustainable Reuse and Upcycling Strategies

Antonello Monsù Scolaro, Stefania De Medici, Salvatore Giuffrida, Maria Rosa Trovato, Cheren Cappello, Ludovica Nasca, and Fuat Emre Kaya

Abstract The building reuse can reduce both consumption of non-renewable resources and production of construction and demolition waste, preserving the architectural and constructive culture. The progressive depopulation of the European inner areas is an opportunity to discuss the potential of reuse and sustainable adaptation of extensive heritage sites to cope with abandonment processes. The study of depopulation processes, as well as the investigation of case studies, allows to analyze the main strategies implemented to regenerate and repopulate abandoned inner areas, to highlight successful approaches and intervention criteria. In this scenario, "smart shrinkage" emerges as a powerful strategy to systemize resources and values embedded in the territories. On the basis of the economic-territorial interpretation of the performance decay process of buildings and settlement systems, developed by the research group of the Universities of Sassari and Catania, the paper proposes a multi-scale

A. M. Scolaro (✉) · C. Cappello · F. E. Kaya
University of Sassari, Sassari, Italy
e-mail: amscolaro@uniss.it

C. Cappello
e-mail: c.cappello@studenti.uniss.it

F. E. Kaya
e-mail: f.kaya@studenti.uniss.it

S. De Medici · S. Giuffrida · M. R. Trovato · L. Nasca
University of Catania, Catania, Italy
e-mail: stefania.demedici@unict.it

S. Giuffrida
e-mail: salvatore.giuffrida@unict.it

M. R. Trovato
e-mail: mariarosa.trovato@unict.it

L. Nasca
e-mail: ludovica.nasca95@gmail.com

E. Arbizzani et al. (eds.), *Technological Imagination in the Green and Digital Transition*, The Urban Book Series, https://doi.org/10.1007/978-3-031-29515-7_68

methodological approach for the evaluation of enhancement strategies and technological upcycling. The research links building performance with the urban and territorial values, integrating the Performance-Based Building Design in an axiological approach based on the solidarity between functions and values, referring to the economic category of human and urban capital. The model is tailored to the characteristics of Sardinia, the Italian region with the strongest population shrinkage in inner areas. The result is an analysis-evaluation-programming model, based on an iterative process of information/decision-making, allowing to steer intervention strategies toward a balance between the rehabilitation of the built environment and the enhancement of cultural and environmental resources, offering new opportunities for socioeconomic development.

Keywords Performance-based building design · Sustainable building reuse · Multicriteria decision making aid · Axiological approach · Life cycle cost

68.1 Introduction

For several years, the theme of inner areas has been at the center of the debate about the strategies of the European Community for sustainable development. These areas generally devoted to agriculture, widespread in most of the European territory, have been suffering from a depopulation process for many years in favor of urban areas, risking being profoundly marginalized due to the lack of essential services and job opportunities. The strategies of the revitalization of the abandoned areas, which have traditionally ensured the vitality consolidating the flows of wealth coming from the earth in the city forms (Giuffrida et al. 2021), must now be rethought in light of a radical modification of the development models and lifestyles subordinated to the well-being and inhabitant's essential needs.

To this end, the European Commission has launched a study about "shrinking and lonely places" and how these areas can turn their condition into an opportunity (JRC European Commission 2021). For the lonely places characterized by a high level of depopulation and aging, as well as a high level of loss of employment, the European Commission suggests lines of development oriented to: the silver economy (the service sector addressed to over 65); the digital transition; the enhancement of the territory, and the integration of young people and women into the labor market.

These objectives also coincide with a focus on circular development models to improve the life cycle of goods/products/services and in particular the built environment (Foster 2020). The available large building stock due to depopulation and abandonment offers environmental opportunities thanks to incorporated energies and material. Furthermore, the analysis of the interactions between natural and human systems at local and regional scales seems more efficient in site-specific sustainability, through a multi-scale approach (Wilbanks 2015). Therefore, after analyzing the causes of depopulation, as well as European good practices recognized by international awards, this paper proposes a trans-disciplinary methodological approach as a decision support tool for the enhancement of inner areas.

68.2 Depopulation and Abandonment

The World's population has always met chaotic growth rates, never uniform (Vallin 2002); however, the balance between survival and reproduction begins to falter during the eighteenth century, passing from the high levels of birth and death rates of the traditional demographic regime to low levels of the modern period. From the second half of the eighteenth century, advances in agriculture, the containment of infectious diseases, and industrial and educational development led to a strong demographic growth that generated strong migration flows and colonial expansions, until the early 20s of the twentieth century when population growth slowed down to zero in the years of the Second World War. The birth rate rise again between the post-war period and the 1960s, then it reached the so-called growth 0, which is a downward balance between the born and the dead. This profound modification of the demographic mechanisms was accompanied by the rise of industrial and urban societies, contributing both to the social changes linked to the new condition of women and to the modification of the structure of families, thus favoring new dynamics and structures of the population.

Today, however, the inner areas suffer from both a profound "demographic malaise" accompanied by an aging population and low birth rate, economic, social, cultural, and psychological malaise, in particular due to the shortage or lack of work that determines the abandonment and the progressive depopulation of towns up to the risk of disappearance (Golini et al. 2020).

Eurostat's long-term population projections show the EU-27 population increasing from 447 million in 2019, peaking at 449 million in 2026. Inhabitants will gradually decline to 441 million in 2050 and 416 million in 2100 (Eurostat 2019), in different ways in different territories.

In the last decades, the EU funded projects and research to investigate the processes and dynamics of depopulation and abandonment of rural and inner areas, to define more adequate strategies and tackle the resulting environmental, economic, and social consequences. Some studies show that depopulation is not a linear process, but rather influenced by several interconnected factors, dependent on aging, population loss, and urban agglomeration effects. Most European countries have experienced a so-called urban–rural polarization due to the lower population growth in rural areas compared to urban areas where the population is more concentrated (ESPON 2020).

In some Central and Eastern European countries, like Poland, Slovakia, and the Czech Republic, the demographic decline in rural areas is offset by migration from urban areas. In the Western Balkans and post-socialist countries, the demographic decline in rural areas has been caused by distance from socioeconomic hubs or services, geomorphology and soil conditions, and climate. In southern Europe, the depopulation phenomena have increased around the 1950s, especially in the rural areas with the labor loss from the primary sector in favor of the industrial sector located near the more densely urbanized areas. The geographical or economic circumstances such as increased hydrogeological vulnerability, the globalization of markets and loss of competitiveness of production capacity, the inadequate governance and supporting infrastructures/services, the strong economic polarity of coastal areas especially in Mediterranean areas are other factors determining the depopulation phenomena. These factors were also relevant in parts of eastern regions of Germany, or eastern mountainous areas of Austria and in northwestern Europe, Scotland, Ireland, and France.

In Italy, the inner areas cover about three-fifths of the territory, and they are characterized by a low settlement density and peripheral municipalities of high hills or mountains, where the economy is traditionally based on agriculture, the depopulation has resulted in severe environmental imbalances. However, these areas are rich in environmental and cultural resources on which a possible sustainable regeneration is based through the construction of networks of goods, activities, services, and infrastructures.

Depopulation makes clear the need to interpret the expectations and hopes of local communities, along with the residual and potential value of places—economic, historical, and cultural—to define a new framework of meanings and functions in an entirely new dimension of sustainability, in which resources regain value and relational meaning.

68.3 Research and Good Practices

Since the mid-2000s, the research has been funded in Europe to investigate the phenomena of demographic shrinkage, policies, and governance methods of the territories. The study, "Shrinking Regions: a Paradigm Shift in Demography and Territorial Development" (2008) produced by a team of universities at the request of the European Parliament's Committee on Regional Development, emphasizes the need to tackle the demographic decline according to a multi-scale approach. The exchange of multidisciplinary scientific knowledge to define more effective regeneration strategies in the cities with demographic decline is the main idea of the project "Cities Regrowing Smaller—Fostering Knowledge on Regeneration Strategies in Shrinking Cities across Europe" (2008–2013). The project "Shrink Smart—Governance of Shrinkage within a European Context" (2009–2012), identifies and analyzes the main challenges caused by shrinkage for urban and regional development and defines the possible sustainable and transferable policy recommendations with practitioners and policymakers. Other research activities investigate the recurring factors to tackle the depopulation, as labor market; service provision, and housing issues in the North Sea Region (DC NOISE, 2008–2013); education and lifelong learning, health care and social services as well as tradition and innovative economy (DART, 2009–2012); transnational strategies in the field of public infrastructure and services (ADAPT2DC, 2011–2014). In addition, the Shrinking

Cities research network (SCiRN) has advanced the concept of adaptation (the "Cities Regrowing Smaller" project) and Hollander and Németh (2011) speak of "smart decline," approaches consistent with current visions of circularity and enhancement of existing resources-economic, social, cultural. Demographic contraction offers new opportunities for public services and governance to become modern and innovative, place-based and pro-active (Haase et al. 2016), and inland areas could grow smaller and greener again, better suited to the needs of inhabitants.

To discretize the variety of possible scales and approaches, several good practices have been analyzed from which to select possible recurring factors and parameters on which to base a methodological proposal to support the decisions to be adopted in the regeneration and enhancement processes of depopulated areas.

The premise of each good practice is the preliminary research for the evident and potential values of the territories and their subsequent systemization according to local needs. The policies based on sustainable inter-ministerial and inter-sectoral cooperation, and the financial instruments require the use of EU funds for development and territorial cohesion, while the aim is to better understand and enhance the resources and diversities of non-homogeneous regions by encouraging the interchange of experiences. The analysis of good practices (Fig. 68.1) highlighted some recurring key issues, which were taken as a base to develop the methodological proposal to support decisions. This cannot be separated from a local-based approach, the integration of context-aware and expert knowledge, to define new development guidelines oriented to environmental, social, and economic sustainability combined with local opportunities and resources.

This, furthermore, involves the regeneration of settlement systems, infrastructures, and economic organizations as well as the improvement of territorial cohesion policies to cope depopulation (Humer 2018) and improve the quality of life of residents.

68.4 A Methodological Approach for Smart Shrinking and Enhancement of Environment Built in Inner Areas in Sardinia

Sardinia has always been sparsely populated and rich in large uncultivated lands (Day 1973). The demographic growth has always followed innovations: in the administrative, education, health system, and in the mining and manufacturing industry in the mid-nineteenth century, in all production sectors with the 1962 Rebirth Plan. The fragility of the regional economy and territorial fragmentation (377 municipalities of which 31 are at risk of disappearing) are today the main cause of depopulation. The investments in the redevelopment of historic centers and the improvement of municipal solid waste management at the end of the 1990s did not strengthen the economic-productive system. Since the 2000s, the Regional Development Plans

GOOD PRACTICE	PARTNERS	DESCRIPTION	KEY ISSUES
RURITAGE (2018-2022) Source: www.ruritage.eu	14 EU countries, Iceland, Norway, Turkey and Colombia	Knowledge Facilitator Partners develops recommendations and tools within six Systemic Innovation Areas: Pilgrimage, Local Food, Migration, Art & Festival, Resilience, Landscape. Successful strategies are based on Role Models, best practices where rural areas have been regenerated thanks to cultural and natural heritage. Stakeholders (selected Replicators) understand, learn, tailor and replicate such experiences.	• Heritage-led regeneration strategies • Simulation of regeneration scenarios • Involvement of stakeholders • Improvement of knowledge, skills and capacity building
LEADER TRANSNATIONAL CULTURE (2014-2020) Source: https://fr.unesco.org/creativity/policy-monitoring-platform/leader-transnational-culture	Austria, EU	Financial support for projects that aim to activate development processes in rural regions, taking art, culture and creativity as drivers. The focus areas are transformation of professional domains, networking for socio-cultural integration, active participation of people by creative and artistic means, building a new image of the countryside.	• Art and culture as a transformative force • Involvement of local cultural players • Development of new professional activities in the field of art and culture • Transnational cooperation
SHANNON ARTS DEVELOPMENT (2015-2021) Source: https://clarearts.ie	Ireland	Training programmes for local businesses willing to start up or convert their activities in the arts and culture sector. The funding enabled the implementation of a reuse project of an abandoned commercial space, refurbished with recycled materials by a local artist. The space has been reused for project-related activities.	• Art as a regeneration strategy • Involvement of local artists and entrepreneurs • Development of new professional activities in the field of art • Building reuse and recycling of materials
INTESI Alpine Space Project (2015-2018) Source: https://www.alpine-space.eu/projects/intesi/en/home	European Union	Development of Services of general interest (SGI) and the related models of integration in 5 Alpine countries (Italy, Switzerland, Austria, France, and Slovenia), exchange of best practices, implementation of a collaborative platform of pilote activities.	• Strategic reflection to services of general interest • Involvement of local stakeholders • Transnational cooperation
LAVRAR O MAR (2016-2022) Source: https://www.lavraromar.pt/en/	Portugal	Enhancement of the cultural identity and creation of a new brand to increase the attractiveness of the coastal area in low season periods, definition of a comprehensive artistic and cultural programme.	• Art and culture as a driving force for attractivene: • Involvement of local cultural players • Building a brand identity
REGIONALE (2019) Source: https://plovdiv2019.eu /en/platform/fuse/122-regionale/1443-regionale	Bulgaria	Project about a book cataloguing cultural, tourist and social undertakings from Bulgaria's Southern Central Economic Region. Knowledge as a strategy for conservation and enhancement.	• Enhancement of local cultural and artistic resources • Support tools for sustainable tourism • Improvement of the links between generations • Passing down of experience and memory
ARCTIC ARTS SUMMIT (2017 and 2022) Source: https://arcticartssummit.ca	Finland	Organization of summits to strengthen Northern arts and culture and to develop circumpolar cooperation through collaboration in the arts and creative industries.	• Arts and creative industries as a driving force for development • Involvement of local cultural players • Transnational cooperation
THIRD PLACES (2019-2023) Source: https://www.mkw.nrw/kultur/foerderungen/dritte-orte	Germany	Creation and development of cultural centres in rural regions, comparison and evaluation of regional cultural policies, design and implementation of hubs for services of general interest.	• Culture as a driving force for development • Strategic reflection to services of general interest • Involvement of local cultural players • Building reuse
MY GRANDFATHER'S HOUSE (2012-2017) Source: https://www.interreg-central.eu/Content.Node/-MY-GRANDFATHER-S-HOUSE-.html	Hungary	Summer school of master apprentice training for students interested in folk architecture, to introduce young people to the values of traditional architecture, construction techniques and craftsmanship.	• Building rehabilitation • Knowledge of traditional house building techniques as preservation strategy • Investment in youth training

Fig. 68.1 Good practices and key issues

(PSR) have aimed at supporting the recovery of inner areas by strengthening agricultural production activities, public transport, viability, and basic services. The implementation of these plans was based on the division of the territory into four zones: urban poles and rural areas (urbanized, significantly or predominantly rural), distinguished in blue and orange based on population density (Fig. 68.2). However, by superimposing the conditions of demographic malaise on rural areas, a less homogeneous situation emerges, resolved by the subsequent subdivision into Optimal Territorial Areas (ATO) and by Local Actions Groups (GAL), supra-municipal territorial groupings with legal form, capable of reflecting the historical-identity, geographical, and demographics of the territory.

The current territorial support and enhancement interventions are based on the Regional Strategy for Inner Areas (SRAI), which aims to increase the offer of public services, the employment levels and the enhancement of territorial capital in line with Europe 2020 and the SNAI. The effectiveness of this approach requires the systematization of local resources—physical, economic, and social—to be enhanced and reconstituted in aggregative and settlement forms on which economic ethics

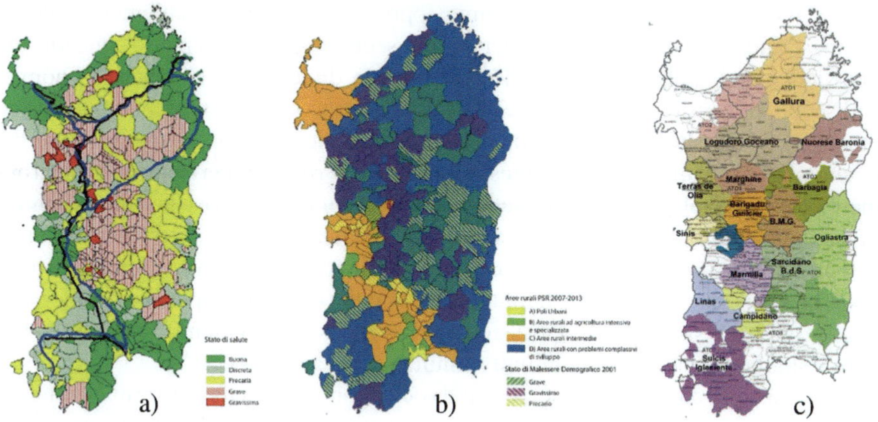

Fig. 68.2 **a** State of demographic malaise (Stato di malessere demografico, SMD); **b** superimposed SMD/PSR territorialization; **c** local action groups (Gruppi di Azione Locale, GAL), LEADER areas, various sources Sardinia Region

governs over the economic-estimative evaluation. On the other hand, abandoned settlements and buildings incorporate energy and materials, consumed resources (often primary), contributing to the depletion of the natural environment (Ecorys 2014). This is the intersection between the tools of the technological project, which investigates functional patterns and performance clusters to maximize efficiency and effectiveness (in particular environmental), and the economic-estimative discipline (in particular the evaluation of the possible outcomes of the project).

The analysis of good practices made it possible to identify key issues reorganized according to the operational peculiarities of the Sardinia region, articulated in a methodological proposal useful for defining enhancement strategies for areas in the process of depopulation (Fig. 68.3). Territorial development strategies should be based on a systemic and multi-scale interpretation of tangible and intangible factors. Such analysis aims to provide a picture of the multiple and sometimes competing needs of settled communities and ongoing transformation dynamics.

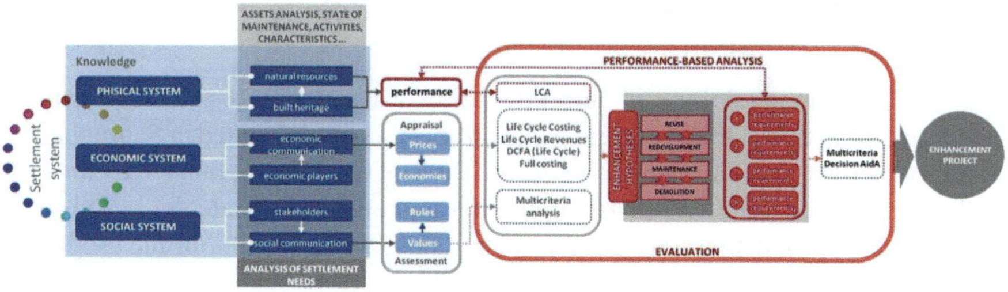

Fig. 68.3 Flow chart of the methodological approach

The methodology involves a systemic approach to the urban settlement scale, to compare and evaluate alternative intervention scenarios. The first phase concerns the knowledge and analysis of the settlement system, divided into physical, economic, and social sub-systems (knowledge), through specific investigation methods for the observed systems (assets analysis). The expected result is both the definition of the potential of the built heritage, expressed in terms of residual performance (environmental and technological) and the settlement needs of the area based on the dynamics of socioeconomic transformation (settlement needs). The analysis of the economic and production system (using the methodology life cycle cost and revenues) defines both the economic value of the built heritage and the economic operators of the area; finally, the analysis of the social system makes it possible to identify the values and norms shared by the local community (multicriteria analysis).

In the second phase, the settlement and socioeconomic needs (performance-based analysis) are identified according to the consistency of the available built heritage, so the method allows to choose the type of intervention from the comparison of project alternatives on a municipal scale, based on multiple factors and available economic and technological parameters. Subsequently, through an LCA based approach, to limit the environmental impacts related to the intervention, the design alternative derives from the comparison between technological requirements associated with the new settlement needs and residual performance of the built heritage. Finally, through Multicriteria Decision Aids, among the possible alternatives, only those associated with limited transformations of the physical system will be compared, thus with a lower associated environmental impact. Physically, the strategy involves alternatives such as the reuse and requalification of abandoned buildings; the requalification and maintenance of those whose intended use is confirmed; or the eventual demolition and recovery of the incorporated materials in an urban mining perspective (Koutamanis et al. 2018). In this scenario, the evaluation defines the individual preference functions (axiological profiles) and the contents of social value (ethical profile) through which the best strategy among those internally coherent on a planning (technological) level is defined by progressive adaptation between constraints and opportunities, thus maximizing the cost-effectiveness and fairness of the revitalization process.

68.5 Validation of the Developed Methodology

The developed methodology is currently being validated on a network of inland Sardinian minor cities. In the present paper an analysis implemented in Samugheo, an urban center of late medieval origin in the province of Oristano, populated by just over 3000 inhabitants, is outlined. In the case of Samugheo, action strategies based on the introduction of productive activities that benefit from the local cultural and productive tradition and rely on the use of digital technologies have been defined. The study led to the setting of new functions for municipally owned buildings, which provide the role of hubs for services targeting residents and visitors and house collective activities. Furthermore, a network of abandoned buildings available to accommodate productive activities and residences has been identified (Fig. 68.4).

The choice of building reuse has given priority to buildings that still maintain their original architectural characters, helping to preserve local identity through conservation, performance improvement, and maintenance actions. The rehabilitation of such buildings is performed by largely reusing materials and components that can no longer be available on the market, with a process of urban mining involving buildings that are in a very advanced state of decay or ruins, which are not considered to be worth (for cultural reasons) and cost-effective (for market reasons) to preserve.

Fig. 68.4 Samugheo (Oristano). Mapping the state of preservation of buildings in the city center

68.6 Conclusions

The variety and complexity of the dynamics at the base of the depopulation and progressive abandonment of inner areas requires integrated, multi-scale interventions, capable of putting values and resources into a system, to be rediscovered, still incorporated in these territories. The definition of new forms of socioeconomic development; new ways of living; and unprecedented levels of well-being for the resident communities, together with the maintenance/strengthening of basic and infrastructural services, are essential objectives to tackle depopulation or encourage repopulation. The proposed methodology represents a decision support tool that systematizes the set of values—environmental, economic, social, and cultural—through the enhancement of the built heritage as an essential premise for sustainability, to define new design approaches toward a smart shrinking. The incremental (relative to factors and parameters to be considered) and multi-scale (building-settlement-territory) character of the methodology requires further adaptations and optimizations to be also adequate for other territorial realities while maintaining its validity.

References

Day J (1973) Villaggi abbandonati in Sardegna dal trecento al settecento: inventario. Ed. du CNRS, Paris

Ecorys (2014) Resource efficiency in the building sector. Final report. https://ec.europa.eu/environment/eussd/pdf/Resource%20efficiency%20in%20the%20building%20sector.pdf

ESPON (2020) Fighting rural depopulation in Southern Europe. https://www.espon.eu/sites/default/files/attachments/ESPON%20Policy%20Brief%20on%20Shrinking%20Rural%20Regions.pdf

Eurostat (2019) Population projections in the EU. https://ec.europa.eu/eurostat/statistics-explained/index.php?oldid=497115#Population_projections

Foster G (2020) Circular economy strategies for adaptive reuse of cultural heritage buildings to reduce environmental impacts. Resour Conserv Recycl 152:1–14. https://doi.org/10.1016/j.resconrec.2019.104507

Giuffrida S, Trovato MR, Strigari A, Napoli G (2021) "Houses for one euro" and the territory. Some estimation issues for the "geographic debt" reduction. Smart Innov Syst Technol 178:1043–1052. http://doi.org/10.1007/978-3-030-48279-498

Golini A, Mussino A, Savioli M (2020) Il malessere demografico in Italia una ricerca sui comuni italiani. Il Mulino, Bologna

Haase A, Bernt M, Großmann K, Mykhnenko V, Rink D (2016) Varieties of shrinkage in European cities. Eur Urban Reg Stud 23(1):86–102. https://doi.org/10.1177/0969776413481985

Hollander JB, Németh J (2011) The bounds of smart decline: a foundational theory for planning shrinking cities. Hous Policy Debate 21(3). http://doi.org/10.1080/10511482.2011.585164

Humer A (2018) Strategic spatial planning in shrinking regions. In: Hospers G, Syssner J (eds) Dealing with urban and rural shrinkage: formal and informal strategies, Chap 5. Lit Verlag, Zurich, pp 73–87

JRC European Commission (2021) From lonely places to places of opportunities. EC report

Koutamanis A, van Reijn B, van Bueren E (2018) Urban mining and buildings: a review of possibilities and limitations. Resour Conserv Recycl 138:32–39. https://doi.org/10.1016/j.resconrec.2018.06.024

Vallin J (2002) The end of the demographic transition: relief or concern? Popul Dev Rev 28(1):105–120. https://doi.org/10.1111/j.1728-4457.2002.00105.x
Wilbanks TJ (2015) Putting “Place” in a multiscale context: perspectives from the sustainability sciences. Environ Sci Policy 53:70–79. https://doi.org/10.1016/j.envsci.2015.04.009

Chapter 69
Climate Adaptation in Urban Regeneration: A Cross-Scale Digital Design Workflow

Michele Morganti and Diletta Ricci

Abstract Urban vulnerability has many facets. Among these, urban texture and plot pattern, building massing and density, greatly affect the microclimate. Thence, redefining urban regeneration design criteria for climate neutrality is crucial, including environmental factors in the design process at different scales. In the light of climate change, despite this urgent call, adaptive design approaches useful to assess trade-offs between urban regeneration scenarios and microclimate quality are lacking. This paper introduces a novel digital design workflow that integrates climate quality and associated indicators in urban and building design, adopting a cross-scale approach. The main goal is to increase the resilience of the built environment in the foresight of future scenarios, by promoting climate-sensitive design solutions. Environmental performances were analysed using digital tools and implemented in a design workflow, allowing urban microclimate analysis. Performance metrics were calculated using Urban Weather Generator and Energy Plus. With the former tool a climate performance comparative study has been run in different scenarios, by varying morphological parameters and computing the intensity of the Urban Heat Island. While, Energy Plus was used to simulate the impact of building form and UHI on building energy demand, highlighting the interdependence of different design scales and addressing optimal building performance. The results provide additional levels of knowledge, both in terms of analysis and design scenario evaluation: urban metrics and climate impacts, building form and envelope design, adaptation solutions. This workflow is tested and a scenario suitability for the Mediterranean city is shown, exploiting the research-by-design transformations of 22@ Innovation District of Barcelona. The paper highlights the correlation between microclimate

M. Morganti (✉) · D. Ricci
SOS Urban Lab, DICEA Department, Sapienza University of Rome, Rome, Italy
e-mail: michele.morganti@uniroma1.it

D. Ricci
e-mail: d.ricci@tudelft.nl

Present Address:
D. Ricci
Delft University of Technology, Delft, Netherlands

E. Arbizzani et al. (eds.), *Technological Imagination in the Green and Digital Transition*, The Urban Book Series, https://doi.org/10.1007/978-3-031-29515-7_69

and design solutions and lays the foundations for a climate/design cross-talk to help policymakers and practitioners achieve urban climate adaptation goals.

Keywords Adaptive design · Urban microclimate · Climate change · Urban vulnerability

69.1 Introduction

Dense cities expose people to different kinds of climate vulnerabilities: extreme events, concentration of inhabitants in risk-prone areas, inadequate buildings and many others.

This is why research attention on spatial configuration and physical features of cities has recently risen in various disciplines, including geography, urban ecology, urban and environmental design, building design, urban climatology and building physics (Erell 2012). As a reaction to the demanding need to mitigate climate change and the ecological impact of the built environment, in order to adapt to inevitable consequences and promote health and well-being, research efforts have focused on the unintended interaction among cities' physical characteristics, microclimate and energy balance with a diagnostic or design perspective (Lenzholer 2015; Stewart and Oke 2012). This led to a number of relevant changes in the design discourse and structure—both in term of process and method—and contributed to highlight the key elements for the decisions about the main enforcement actions to be included in urban regeneration design process (Morganti and Rogora 2021).

However, this subject is still fragmented: studies hardly reach comprehensive outcomes due to the complexity of the above-mentioned interaction in the built environment and to the lack of skilled scholars and professionals. By consequence, practical application in urban regeneration process remains limited.

The present study proposes and discusses a novel digital design workflow (Fig. 69.1) that integrates urban climate quality and associated indicators in urban and architectural design, adopting a cross-scale approach. The novelty of the study lies in permitting architects to analyse environmental performance and to take evidence-based urban and building design choices. The main goal is to help architects and urban designers to easily control climate and energy parameters, impacts and associated urban vulnerability trough well-known digital tools.

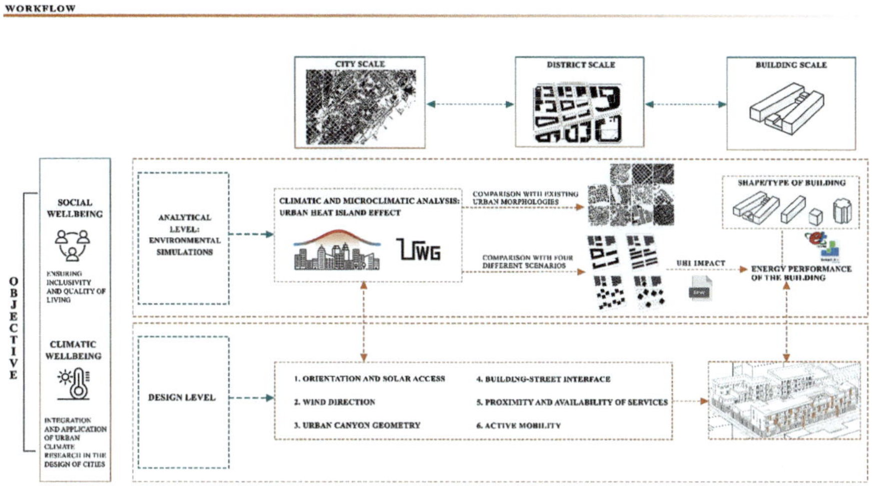

Fig. 69.1 Visual representation of the workflow

69.2 Materials and Method

69.2.1 Analytical and Design Approach

The design process has been supported by the workflow in Fig. 69.2. It is articulated by running both the analytical process of environmental simulations and the design process. The analyses were executed through digital tools. The main objective was to validate the current state of the art about the impact of certain specific parameters on the urban microclimate performance, crossing different scales: the neighbourhood, the district, the island and the building. The impact of urban morphology at neighbourhood level on urban microclimate and UHI was investigated, by focusing on the influence of building typology and form to heating and cooling demand in the Mediterranean climate.

69.2.2 Case Study

A regeneration project of an urban area of 22@ Innovation district of Poblenou in Barcelona has been used as case study to test the novel digital design workflow (Fig. 69.2). The project area is about 8 hectares and currently characterized by a limited number of low-energy efficiency housings, scattered across industrial buildings. The reference scenario is tested by exploiting the research-by-design transformations of the above-mentioned project area. Four different scenarios are presented to compare in detail at both neighbourhood and building scale.

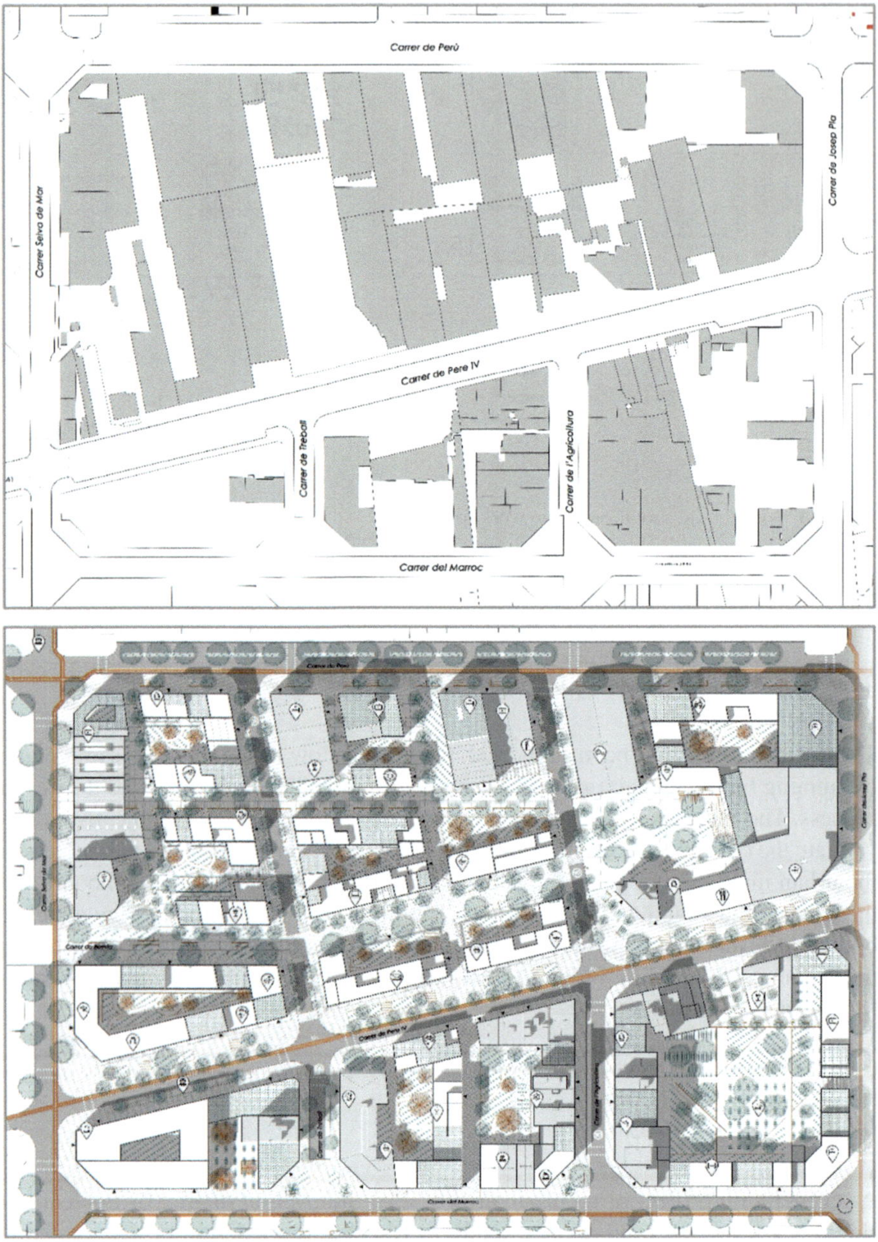

Fig. 69.2 Current state (top) and masterplan (bottom) of the project area

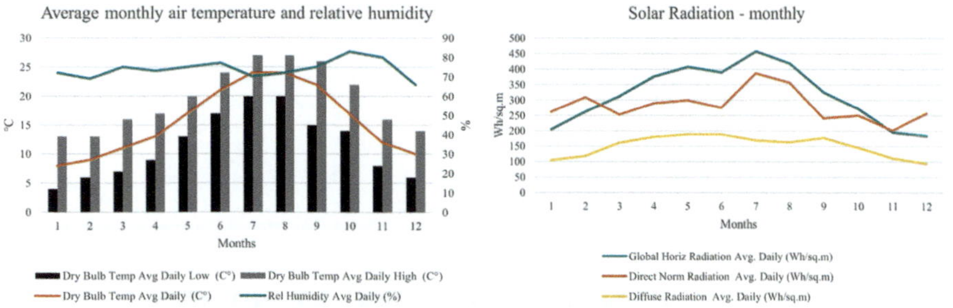

Fig. 69.3 Climatic conditions of Barcelona (based on Barcelona 081,810 IWEC EPW file). Average monthly air temperature and relative humidity values (left); monthly average daily Solar Radiation values (right)

69.2.3 Climate and Microclimate Evaluation

These sections describe how climate and microclimate were studied and the metrics used to compare different scenarios. The study mainly covers the summer period, during which the urban fabric's morphological characteristics influence the urban heat island most significantly. A preliminary analysis was carried out on the Mediterranean climate of Barcelona, characterised by hot, dry summers (Csa Koppen-Geiger climate classification), shown in Fig. 69.3. The analysed climate data from Barcelona El Prat airport will later be considered as rural station data for the urban heat island analysis. The typical climate presents relatively high average outdoor temperatures throughout the year (average annual temperature 16°), and a rather high average annual relative humidity of 73%, due to the proximity of the sea.

69.2.4 UHI District Scale

The effect that the regenerative design of the case study's urban area has on the urban heat island was assessed using Urban Weather Generator (Bueno et al. 2012). The UWG algorithm evaluates the difference in temperatures between a rural context and the urban canopy layer; the UWG calculation tool has already been validated for Barcelona's climate in previous researches (Salvati et al., 2019). The 'rural' EPW climate weather file, which provides the meteorological inputs, was taken from EnergyPlus weather data (Barcelona 081810 IWEC), and includes climate data from Barcelona El Prat airport (Elev. 4 m, 41.33 °N, 2.1 °E). The 3D model of the project was used to provide the morphological input parameters (Fig. 69.4). Table 69.1 lists the input variables that were customised, while others, such as the traffic-sensitive heat flux, kept their default values.

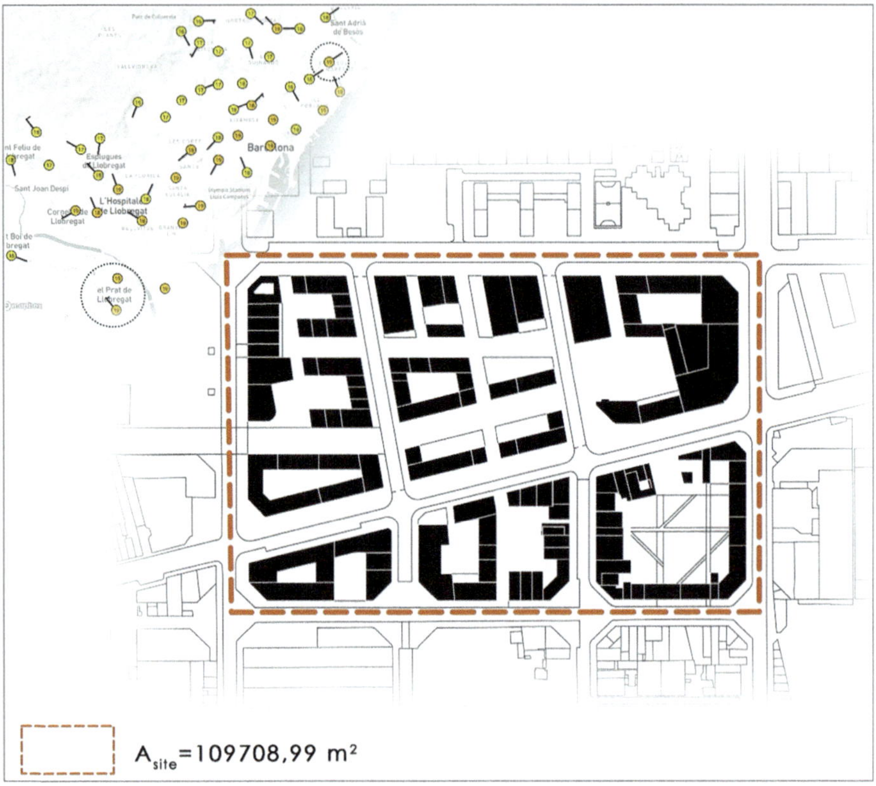

Fig. 69.4 Analysed scenario—district scale

Table 69.1 Customized input UWG parameters

Variable	UWG parameter	Definition	Value
Urban morphology	Site coverage ratio [ρ_{urb}]	Ratio of building footprint to the site area [–]	0.35
	Façade to site ratio [VH_{urb}]	Ratio of the vertical external surface area to the site area [–]	0.64
	Average building height [H_{bld}]	Average building height normalized by building footprint [m]	12.8
Vegetation cover	Urban area veg. coverage	Ratio of vegetation coverage in the urban area to the site area [%]	0.26
Surface albedo	Road albedo	Ratio of reflected radiation from surface to incident radiation upon it [–]	0.15

69.2.5 UHI Island Scale—Comparative Analysis

The UHI effect has been evaluated to compare four different scenarios of an island in the project area. The three alternative scenarios, against which the SA_Courtyard has been cross-referenced, were created, by varying the morphology of the urban fabric, and by fixing the built-up cubature at about 112.000 m^3 (as in the reference block), as shown in Fig. 69.5. The single island was assumed to repeat homogeneously over an area of 1 km × 1 km, and we focused on the impact that the variation of morphological parameters alone (Site coverage, Façade to site ratio, Average building height) can have on the UHI in the Mediterranean context. (Salvati et al., 2019) This comparison was central to validate how different urban forms display different climatic performances, and to prove that some morphological parameters have a more negative effect on the UHI than others.

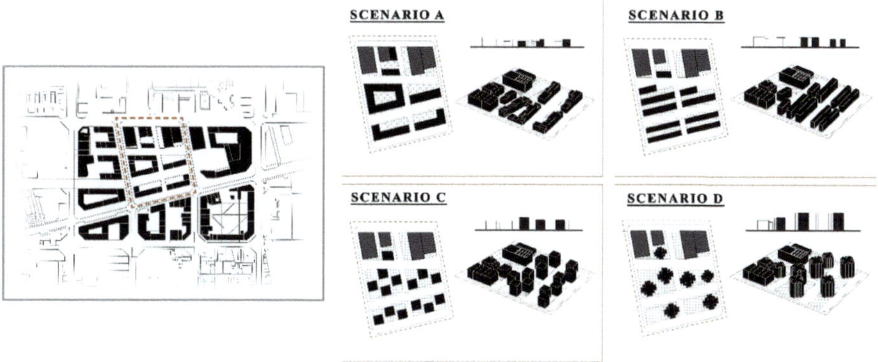

Morphological Parameter	SA_Courtyard	SB_slab	SC_block	SD_tower
Site Coverage Ratio [ρ]	0,31	0,29	0,25	0,21
Facade to site ratio [VH_{urb}]	0,66	0,93	0,95	1,03
Average building height [H_{bld}]	12,4m	15,2m	17,7m	20,8m

HORIZONTAL DENSITY ← → VERTICAL DENSITY

Fig. 69.5 Framework of the city block (top left); visual representation of four different alternative urban fabric (top right); values of the morphological parameters used for calculating the UHI (bottom)

69.2.6 Energy Demand—Building Scale

Moving to the building scale, a courtyard residential building from the reference model was studied and designed in further detail. A comparative analysis of the energy demand for heating and cooling was carried out using EnergyPlus software. Three alternative building types for the courtyard one, based on existing building plans (linear block, low-rise block, high-rise tower) has been computed (Fig. 69.6), highlighting the interdependence of different design scales and addressing optimal building performances. The typologies have been put in their urban context of the four scenarios used for the UHI assessment, by inserting, as input climate data for each scenario modified running UWG from the rural one of Barcelona El Prat airport. The study has a comparative purpose, and it does not constitute an absolute assessment of energy demand, since some factors that have been kept by default or not taken into account. Table 69.2 shows the input values: in modelling the scenarios, the shape (compactness) and window-to-wall ratio parameters were varied, while keeping constant the envelope performance, the activity and the HVAC templates set to the standard values for residential buildings by the *Código Técnico de la Edificación* (Ministerio de Fomento 2017).

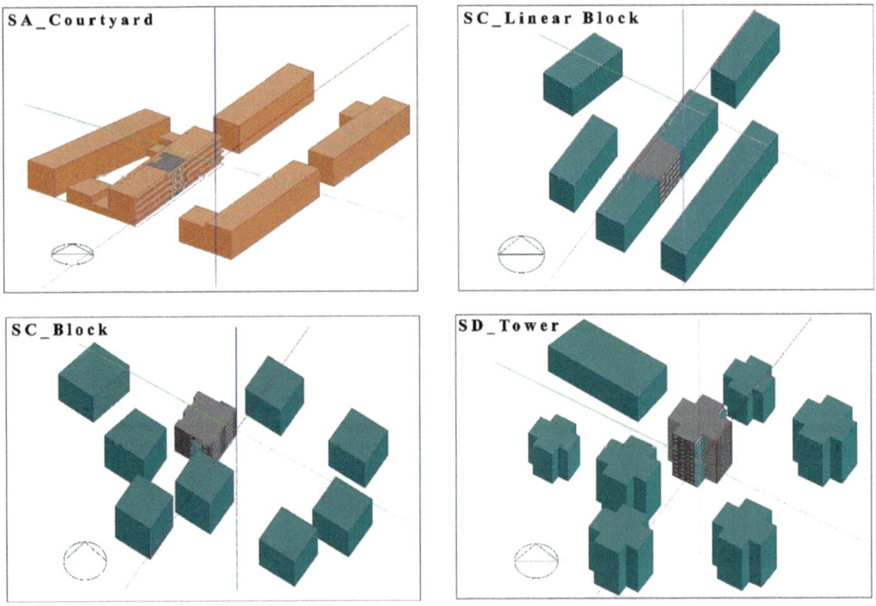

Fig. 69.6 Visual representation of the four scenarios modelled in DesignBuilder

Table 69.2 Building scale analysis input parameter

	Variable parameters		Fixed parameters	
Test cases	Compactness Index	Window-to-wall ratio (%)	U-CLT structural wall (W/m^{2}K)	U-ventilated roof (W/m^{2}K)
SA_COURTYARD	0.66	30.27	0.28	0.25
SB_SLAB	0.72	31.28		
SC_BLOCK	0.74	21.82		
SD_TOWER	0.64	42.86		

69.3 Results and Discussion

The new urban layout has been planned with mixed uses, large amounts of public open spaces and permeable surfaces. This ensures outdoor and indoor comfort and the availability of spaces that inhabitants can use as climate shelters to cope with the increasingly uncomfortable and risky conditions caused by climate change (Taleghani, 2018). The block becomes very permeable to pedestrians and large bicycles. Besides, pedestrian paths are designed to encourage active mobility and the strategic location of services, so as to have attractive, safe and always active streets in the neighbourhood.

69.3.1 UHI—District Scale

The average values of summer and winter UHI, for the transformed project area, are quite low, respectively, 0.9 for the average summer UHI (month of July) and 1.0 average winter UHI (month of January). The summer UHI has been further investigated, since for the Mediterranean climate the UHI in the warm months, as demonstrated in other studies (Natanian and Auer 2020), has more evident effects, including on the energy performance of the building. In July, the hottest month in the city of Barcelona, a maximum average UHI value of 1.4 is reached at midnight. These values were benchmarked by running UWG with input parameters given by the average values of 10 existing urban fabrics taken from the study by (Salvati et al., 2019). The baseline model has well higher average summer and winter UHI values than the new urban settlement, 2.2 summer UHI and 1.5 winter UHI, respectively, as shown in Fig. 69.7.

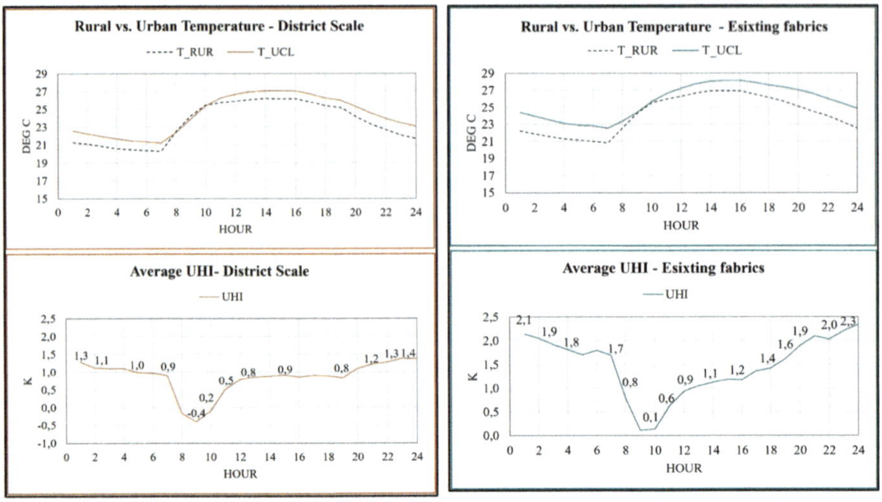

Fig. 69.7 UWG output graphs July UHI analysis: transformed urban area at district scale (top) and baseline model of existing urban fabrics (bottom)

69.3.2 UHI—Island Scale

With the input parameters of Fig. 69.5, we obtain the summer and winter average UHI values reported in Table 69.3. The values refer to low ranges of UHI having all low compactness index fabrics, it is noticeable how, in particular for summer, UHI confirms what has been already proved in other studies: the increase of vertical density (more façade surfaces) in Mediterranean climate contexts, due to the phenomenon of multiple reflections of radiation between surfaces in Urban Canyons, impacts on UHI via the causation of temperatures higher than those in other scenarios, where vertical density is lower.

To verify this trend, the UHI values of the four scenarios were also compared to those provided by the graphical tools created for different mean heights in the study by Salvati et al. 2019. In these diagrams, the UHI values are only reported based on the variation of the morphological facings (ρ_{urb}, VH_{urb}, H_{bld}). In the UHI values provided by the diagrams, there is a more direct and linear correspondence between

Table 69.3 Average UHI values in the four different scenarios

Scenarios	UHI winter UWG	UHI summer UWG	UHI winter graphic	UHI summer graphic
SA_PROJECT	1.00	0.8	0.8	0.8
SB_SLAB	0.9	1.1	1.2	0.8
SC_BLOCK	1.0	1.0	1.5	0.8
SD_TOWER	0.9	1.3	1.6	1.0

the increase of vertical density and summer UHI, and the same output as in the UWG simulation is not available, as the diagrams were created neglecting the parameters of vegetation and tree coverage, albedo of the surfaces. In both evaluations, the less dense urban plot with tower buildings is the one with the highest UHI values, as it has more façade area. On the other side, the project case study with courtyard building typology has the lowest values in both seasons.

69.3.3 Energy Demand

Figure 69.8 shows the results of the analysis of the heating and cooling demand of the buildings of the various case studies, compared with the respective compactness coefficients calculated as

$$Rc = Se\frac{Se}{Sg}4836\frac{Vt^{\frac{2}{3}}}{Sg} \tag{69.1}$$

which refers to the equivalent surface of a sphere with the same volume as the building (Serra and Coch 1995) and the window-to-wall ratio. Since the intention was to focus mainly on the effect of building morphology urban fabric on the building's energy performance, the presence of any shading devices on the windows was not taken

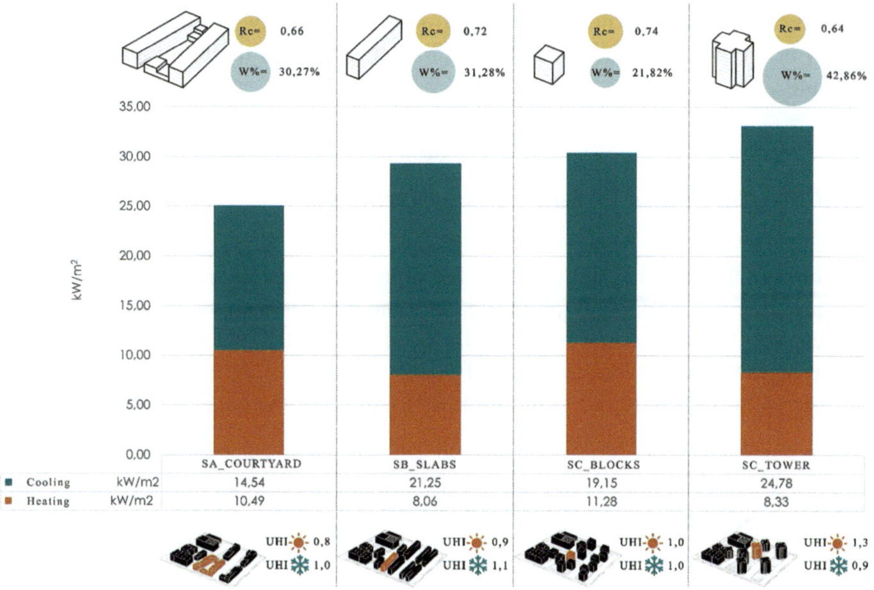

Fig. 69.8 Annual heating and cooling demand (kW/m^2), compactness index, window-to-wall ratio (%), and the average values of summer and winter UHI for the four different scenarios

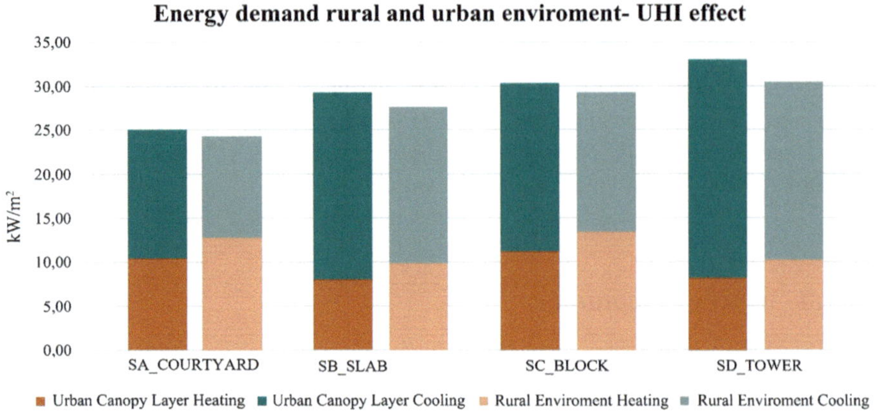

Fig. 69.9 Comparison between annual cooling and heating demand of the four scenarios with and without the effect of the UHI

into account, as this would have considerably modified the results. The courtyard type clearly has the lowest overall annual demand, which result is obtained despite the high values of heating demand with respect to the other buildings, thanks to the lower cooling demand in the summer months. It is evident that there is a more direct relationship between energy performance and the amount of glazed surfaces than between the former and of the compactness index, as already shown by Premrov et al. 2016. The tower typology has the highest cooling demand due to the heat entering from the large number of windows, considered without shading, which in return in the winter months provide high solar gains and therefore a lower heating demand.

69.3.4 Rural and Urban Energy Demand

By running the four simulations with the climate data of Barcelona airport as well, it is possible to see in the graph in Fig. 69.9 the gap between the rural and the urban context which is influenced by the UHI. It can be seen that the benefits in winter due to the higher external temperatures of the rural contexts are in any case lower than the energy surplus required in the summer months for the cooling system.

69.4 Conclusions

Through the workflow described above, an urban regeneration project was developed in the Mediterranean urban climatic context, which, in parallel, acted as a case study for several climatic and microclimatic analyses, focusing in particular on the effects of the urban heat island phenomenon, calculated with the validated UWG tool. By carrying out comparative analyses with other scenarios, different in morphology from the reference one, the influence trend of some morphological variables on the built environment's energy performance has been verified at district, island and building scale. The results suggested that, during the summer period, urban layout with courtyard buildings of low average height is to be preferred over other types of fabrics, which have higher 'vertical density' and contribute to the increase of temperatures in cities, particularly at night. Turning to the building scale, it has been further verified that regenerative design must have a holistic and cross-scale approach given the interdependence of the effects that these different levels have on climatic well-being.

The importance of this type of study is related to the fact that in the preliminary stages of planning and design, the choices that most affect the quality of the space and environmental and the energy performances of buildings are concentrated. For instance, those about the shape of buildings and of urban layout are increasingly difficult to modify as design advances. The study contributes to foster the integration of scientific knowledge on urban climatology and sustainability of urban systems into the planning and design practices for densification and/or regeneration of existing urban areas. Through the use of available digital tools for climate and microclimate assessments, it allows for an integrated, cross-scale control of the design process. This workflow can be considered reliable for a pre-design phase, while for more detailed analyses, it needs further integrations and other tools. A limitation also lies in the reliance on different digital tools that require parallel 3D model creation on different software, without being able to use a single digital graphic interface during the overall analysis development.

References

Bruno B, Norford L, Hidalgo J, Pigeon G (2013) The urban weather generator. J Build Perform Simul 6(4):269–281. https://doi.org/10.1080/19401493.2012.718797

Erell E (2012) The application of urban climate research in the design of cities. In: Advances in building energy research. Vol. 3, pp 95–121. CRC Press. https://doi.org/10.3763/aber.2008.0204

Ministerio de Fomento (2017) Código Técnico de la Edificación de España - Documento Básico HE Ahorro de energía. Madrid: Gobierno de Espana. https://www.codigotecnico.org/images/stories/pdf/ahorroEnergia/DBHE.pdf

Morganti M, Rogora A (2021) Cross-scale adaptive design research: a framework for fragile buildings, urban spaces and neighbourhoods. In: Design and construction. Tradition and innovation in the practice of architecture, pp 1158–1168. Monfalcone (Gorizia): EdicomEdizioni

Naboni E, Havinga LC (2019) Regenerative design in digital practice: a handbook for the built environment. Bolzano, IT: Eurac.

Natanian J, Auer T (2020) Beyond nearly zero energy urban design: a holistic microclimatic energy and environmental quality evaluation workflow. Sustain Cities Soc. 56. https://doi.org/10.1016/j.scs.2020.102094

Lenzholer S (2015) Weather in the city: how design shapes the urban climate. Rotterdam: Nai Uitgevers Pub

Premrov M, Žegarac Leskovar V, Mihalič K (2016) Influence of the building shape on the energy performance of timber-glass buildings in different climatic conditions. Energy 108:201–211. https://doi.org/10.1016/j.energy.2015.05.027
Salvati A, Monti P, Coch Roura H, Cecere C (2019) Climatic performance of urban textures: analysis tools for a Mediterranean urban context. Energy Build 185:162–179. https://doi.org/10.1016/j.enbuild.2018.12.024
Serra Florensa R, Coch Roura H (1995) Arquitectura y energía natural. Universitat Politècnica de Catalunya. Iniciativa Digital Politecnica, Barcelona, ES
Stewart ID, Oke TR (2012) Local climate zones for urban temperature studies. Bull Am Meteor Soc 93(12):1879–1900. https://doi.org/10.1175/BAMS-D-11-00019.1
Taleghani M (2018) Outdoor thermal comfort by different heat mitigation strategies-a review. In: renewable and sustainable energy reviews (vol 81, pp 2011–2018). Elsevier Ltd. https://doi.org/10.1016/j.rser.2017.06.010

Chapter 70
Adaptive “Velari”

Alberto Raimondi and Laura Rosini

Abstract As it is known, the global phenomenon of rising temperatures causes uncomfortable and often harmful conditions for human beings living in moderate-climate zones, such as the Mediterranean area, especially in the hottest periods. Examinations of metropolitan cities can witness that high temperatures generate Urban Heat Island (UHI), due to population, buildings, vehicles and human activities in general. With the increase of rising temperatures in the latest decades, people living in big cities have gotten used to tackling heat discomfort with electricity charged cooling systems. As a result, the energy consumption for air-conditioning causes UHIs’ effects to further grow. It is scientifically confirmed that the behavioral habit of relying on artificially generated cold whenever temperatures rise will eventually make the climate crisis more problematic in the near future. Energy communities are used to producing, storing and consuming energy on site; therefore, power sources must be in close proximity to users. Albeit neglected in the Modern Era, the most proximate and sustainable energy supply is directly available to us: sunlight. The origin of hot temperatures, discomfort and energy waste is, indeed, the most exploitable power generator men can access to. In Southern Europe or Middle East cities, the use of veils as urban-scale shading devices is part of the consolidated tradition; a well-known example can be found in the Spanish city of Sevilla, where textile curtains named “Sevillans” are stretched between buildings. At the present time, we’re witnessing that the climate mitigation action of shading systems can be pursued in combination with energy production, with the development of membrane integrated flexible photovoltaic cells (PV). Masdar City in the United Arab Emirates, designed by the Foster Studio, or the Solar trees of the German pavilion at EXPO 2015 in Milan and the Promenade of the EXPO 2021 in Dubai are some innovative yet relevant cases. The use of PV cells for sun-shielding purposes is optimal to respond to a double-sided problem with a single object. Manufacturing an *adaptive velario* using composite fibers (i-Mesh), could both allow us to design the shape and modulate the density of integrated PV cells as needed. **Method**: To identify the best position for the adaptive tensile canopies, it is necessary to superimpose different site-specific data: temperatures in the urban area, in particular close to buildings; surfaces

A. Raimondi (✉) · L. Rosini
University of Roma Tre, Rome, Italy
e-mail: alberto.raimondi@uniroma3.it

E. Arbizzani et al. (eds.), *Technological Imagination in the Green and Digital Transition*, The Urban Book Series, https://doi.org/10.1007/978-3-031-29515-7_70

that receive most of the daytime radiation; sunlight and ventilation. To develop the most suitable solutions to many environmental scenarios, three-dimensional simulations performed with virtual models must be used both at urban (Envimet) and at building scale (in-Sight). **Expected results**: An algorithm capable of determining the "Velari" best position and the proper shading/density factor. A model, applied to a case study in Rome, to serve an evaluation of the benefits of this technology in terms of decreasing surface temperatures of external horizontal and vertical surfaces of buildings and streets.

Keywords Urban shading · Parametric textile · Heat islands effect · BIM model

70.1 Introduction

The average planet temperature is rising, and cities' are following at a doubled rate due to the phenomenon of urban heat island (UHI). At the latest United Nations Conference (COP 26) for climate change, that took place in Glasgow, the United Nations Environment Programme (UNEP[1]) has presented a manual designed to guide governments in the prevention and mitigation of the negative effects of UHIs. Among all the strategies displayed throughout the handbook, the ones providing "public shading structures" are considered the easiest and most effective ones to be implemented and therefore achieve the highest benefits.[2]

The research is part of the largest ongoing research on the city's potential transformation, carried out by the UR of Roma Tre University in Rome. In particular, the study observes the proximity of buildings in order to identify the actions that ideally take action on climate change effects in urban areas. Roma Tre's research group is operating in the context of a bigger project, conducted by six universities and research centers: *PRIN 2017—TECH-START—key enabling TECHnologies and Smart environmenT in the Age of gReen economy. convergent innovations in the open space/building system for climaTe mitigation.*

70.2 State of the Art

Protection from solar radiation is a widely developed topic in scientific literature, both regarding buildings and external spaces. From traditional applications, as the case of Sevillian tents spreading between buildings of the historic city center, along with contemporary projects, such as the city of Masdar City in the Arab Emirates designed by the studio Foster + Partners, or more experimental structures, as the

[1] United Nations Environment Programme (UNEP) has been the global authority that sets the environmental agenda.

[2] Beating the Heat: A Sustainable Cooling Handbook for Cities, p 56.

Fig. 70.1 Shading systems, Dubai Expo 2020 Promenade—January 2022 (*Source* author's photos)

Solar tree of the German pavilion at EXPO 2015 in Milan or the Promenade of EXPO 2020 in Dubai, the most intuitive ways for human-beings to tolerate heat have always involved shading (Fig. 70.1).

Along with different material canopies, planting in urban environments has been yet another solution of sure validity to screen street surfaces and building stories; however, implanting tall trees remains impossible in many built areas, causing exposure to heavy radiation loads for every irradiated element; this is why, especially in warmer countries during summertime, a large amount of heat gets accumulated inside and outside buildings, leading to air-conditioning exploitation. Studies on the efficacy of urban veils (Garcia-Nevado et al. 2020) state that up to 6 °C of temperature reduction can be obtained with shading systems. The degree reduction span is mainly dependent on which orientation does the investigated road spread over; along with this, differences in results are due to the percentage of perforation of the fabric (Solar Absorbance) and the reflectance power of the material (Solar Reflectance). Integration of thin-film photovoltaic technology in membranes, albeit being an ongoing scientific and industrial development, is the key concept for the proposed study: a drapery with embedded solar cells can perform the double function of shading and producing energy (Xiang et al. 2021).

Therefore, we've set up a case study that pivots around photovoltaic textiles that are composed of natural and polymeric fibers I-Mesh (Cesario 2017). We've worked on a Building Information Modeling (BIM) based mock-up. This model allowed us to observe how the imagined solar veils impacted on the urban context of Testaccio district in Rome, and consequently estimate how to tailor their characteristics to different needs.

The adaptive velarios were, indeed, conceived as adjustable elements whose texture can be adjusted to a range of spot-specific demands, insisting on both SA and SR factors.

70.2.1 Targets

- Verify the effectiveness of the shading solution on city-streets tracks
- Evaluate the best shape and position of the textile canopy in order to lower surface temperature of buildings
- Identify the most suitable solar-cells-weaving for the veils surface (that directly affects the Solar Absorbance Factor) as a function of the radiation intensity.

70.2.2 Case Study

The Testaccio district, in Rome, is the urban context for this study, and it is characterized by a homogeneous configuration of regular routes and squared block buildings. It has been affected by numerous redevelopment projects in recent years, and this, for us, makes it a setting of favor for innovative environmental technologies research.

The application of an urban-scale active shading system, has been set on a selected quadrant, that's bounded by Via Luigi Vanvitelli to the north, Via Nicola Zabaglia to the west, Via Galvani to the south and Via Marmorata to the west. Within this portion of the city, two roads have been chosen on the basis of their features; in fact, the aim was to examine how the shading solution responded to different conditions and, thereby, how to modify targeted parameters in order to enhance its performance. As a starting point, we set three default solar curtains of different geometries and imposed three incremental values for percentage of absorption of the solar flow of the fabric, so that the shielding could be boosted if required.

70.3 Methodology

70.3.1 Equipment

The study was carried out simulating the radiation conditions of the site surfaces with a three-dimensional model of the selected urban area, in such a way as to quantify the accumulation of thermal energy in the outer walls of buildings. The support used is Autodesk Revit 2022 software, which is based on BIM. This program offers the possibility to perform solar radiation analysis thanks to a plug-in: Autodesk Insight. In order to run the process, the first data to be provided are type of analysis, period of the year and time interval of observation. Furthermore, geolocation settings allow to simulate a realistic scenario.

70.3.2 Site

The two roads selected for the analysis of solar radiation have opposite orientations; they meet at right angles and the roads' transversal span and extension are similar; these latter features have less impact on the results, enabling a deeper focus on the characteristics of the surrounding (i.e. proximity to urban voids or buildings, presence planting and green areas or dense built-up area…). As follows, we collected the foregoing information about Aldo Manuzio and via Mastro Giorgio (Table 70.1).

For each route, the study was set up choosing three road sections, corresponding to three building plots with different boundary conditions; from these different points of observation, it was possible to identify two categories of varying properties:

- Architectural characteristics
- Environmental conditions.

As for the architectural factors, the following variations could be pointed out:

- Height of buildings
- Morphology of facades
- Position and amount of openings
- State of preservation of masonry.

The environmental factors are.

- Orientation of irradiated ground areas and external surfaces of the structures
- Path and height of the sun.

Since we can't operate on the environmental traits, it became necessary to lay out a modulation of the solar blinds instead. Choosing three fixed standard shapes and three progressive shading factors, we configured types as the baseline of our

Table 70.1 Analyzed streets characteristics

	Orientation	Section (m)	Spread length (m)
Via Aldo Manuzio	SW–NE	13.5	260
Via Mastro Giorgio	SE–NW	12.5	275

Table 70.2 Geometry of solar veils and solar absorption factor of their fabric

Geometry	Quadrangular	Trapezoid	Triangular
Typology	1	2	3
Average area (sqm)	160	120	100
SA factor (%)	30	30	30
Additional SA factor (%)	60; 90	60; 90	60; 90

study. To respond to the architectural and environmental factors previously listed, the default geometries are: quadrangular, trapezoidal, triangular (Table 70.2).

The ability of the fabrics to filter the incident solar flow (Absorbance Factor) has been set at an initial minimum value of 30%. The adequacy of this basic value has first been verified as the initial step of the analysis and design process.

If the results are not satisfactory in the starting examination, this may lead to an increase in the Absorbance Factor (AF). The other percentages of radiation absorption were set at 60 and 90%, so as to estimate a noticeable increase in shading performances by doubling and thirding the original SA factor of the fabric.

An additional variation of the standard solutions could be set on the reciprocal position of the hanging supports of the sheeting; based on the aesthetics and configuration of facing buildings, one portion of the same road can be suitable for top level anchorages, either aligned or unaligned, or for a height-varying hanging deployment between its two sides.

The following unfolded workflow is based on a cumulative radiation analysis, which indicates the total energy load insisting on an area in a set period of the year, and it's expressed in kWh/m^2. In this case, the results show a single day cumulative analysis.

The date refers to the first day of June (1 June), which has been chosen as the starting point for investigating the whole summer temporal span in a later development of our research.

70.3.3 Analytical model in Autodesk Revit.

Navigating Revit's modeling interface, an Analyze section can be found between the other BIM disciplines. The installation of Insight plug-in provides Solar Analysis and Light Analysis tools (Fig. 70.2). This paper deploys the workflow for the solar analysis, as the main focus of the investigation is to quantify the solar energy radiation on buildings and street pavements. The available solar analysis types are Cumulative Insolation, Peak Insolation and Average Insolation, and they determine the energy loads stored inside the elements of the model. This can be built simply using default Autodesk Revit families for walls and roofs. In fact, there is no need to customize the elements with physical information as the solar radiation analysis does not take into account material features.

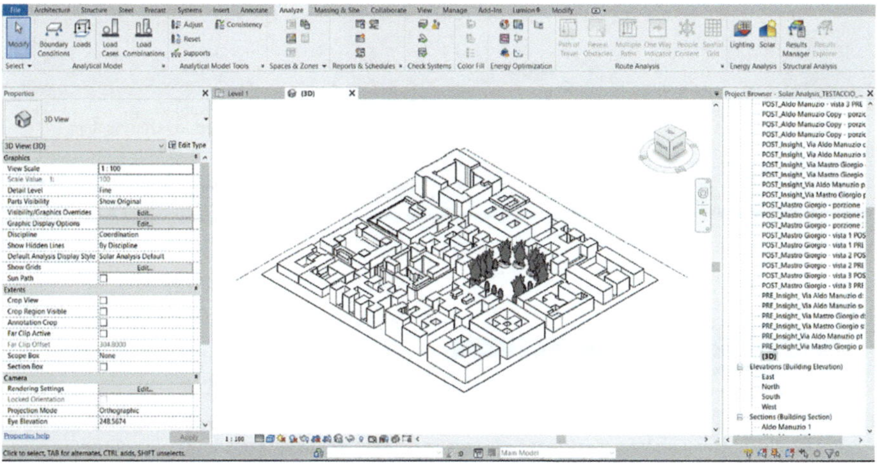

Fig. 70.2 Interface and analysis tools (*Source* Autodesk Revit)

The following method is the one adopted to develop the Adaptive Velario research; thus, the steps are illustrated providing illustrative information relative to the context of this study.

The initial key data to be included in the model are.

- Geolocation of the site: Testaccio, Rome
- Type of study: one day solar study
- Date: 1 June (Fig. 70.3).

The next step is to generate a duplicate of the model (named energy model) from a three-dimensional view, as the Insight plug-in does not run in bi-dimensional views. Therefore, starting the analytical process requires to provide the software with a few further inputs:

- Study Type: custom
- Surfaces: user selection
- Type and units: cumulative insolation—kWh/m^2 (Fig. 70.4).

We can now proceed by selecting wall and roof surfaces in the model. The surfaces in question are.

- Ground surface of the road
- Vertical surfaces of facades that define the road (Fig. 70.5).

By updating the analysis window, the cumulative insolation loads are shown both in kW/h and kWh/mq units.

Aiming to compare a lifelike scenario and a virtual simulation of how the designed velarios would perform in reality, it is necessary to target the right options for shading system modeling. Solar curtain modeling requires several consecutive steps to create recognizable three-dimensional elements, which the software can run the simulation

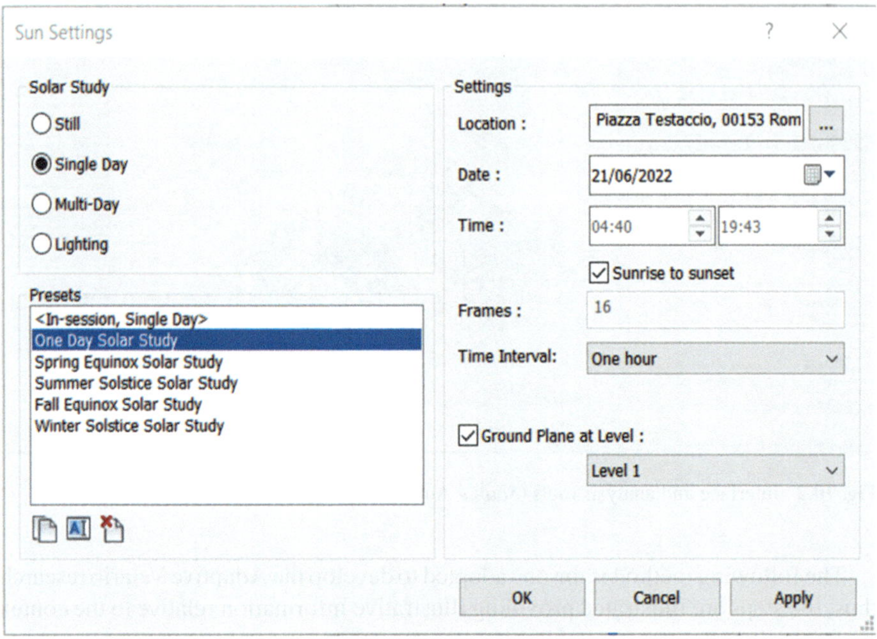

Fig. 70.3 Sun settings (*Source* Autodesk Revit)

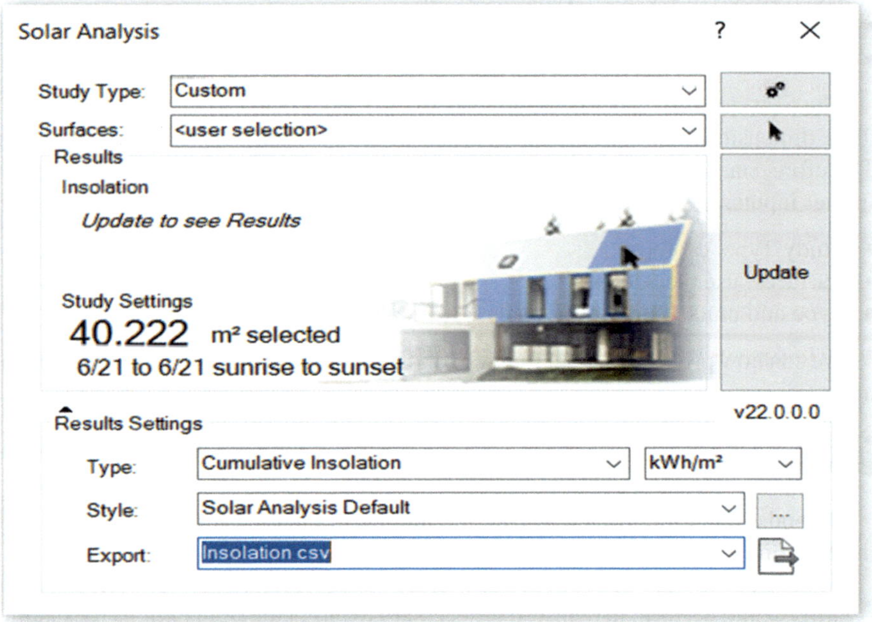

Fig. 70.4 Solar analysis settings (Source: Autodesk Revit)

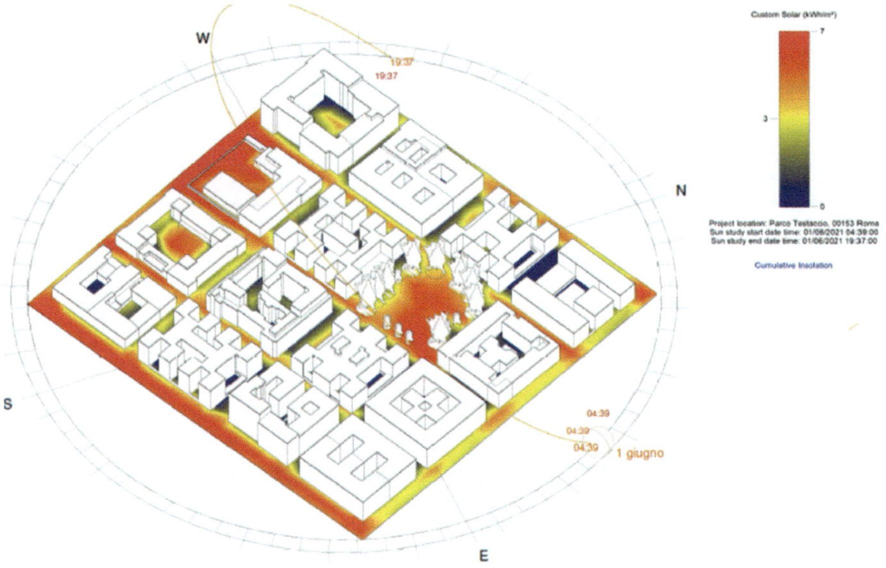

Fig. 70.5 Solar analysis output at ground level (*Source* Autodesk Revit)

with. The program does not provide a special category for shading systems. In contrast to what previously said about the irrelevance of assign material characteristics to wall and roof families, the only way to make the solar fabric interact with the analysis tool is editing the element's material parameters.

The surface type in question is provided in the basic Revit material library: Analytical Surfaces—Shades.

The following steps are.

- Roofing elements modeling using a basic *Roof* family (Fig. 70.6)
- Determination of geometry to be chosen between the three standard ones
- Editing type properties of the element: thickness and material
- Type duplication: three analytical roofing types must be created
- Setting the *Absorptance* parameter of the three types between 30, 60 and 90% (Fig. 70.7).

70.4 Solar Analysis Development

At first, the solar analysis has been performed on the model to produce results that simulate the starting condition of the examined roads (Fig. 70.8).

After the creation of solar veils with *Absorptance Factor* (AF) of 30% and the setup of different position hanging supports according to the morphology of the buildings, the analysis process has been repeated.

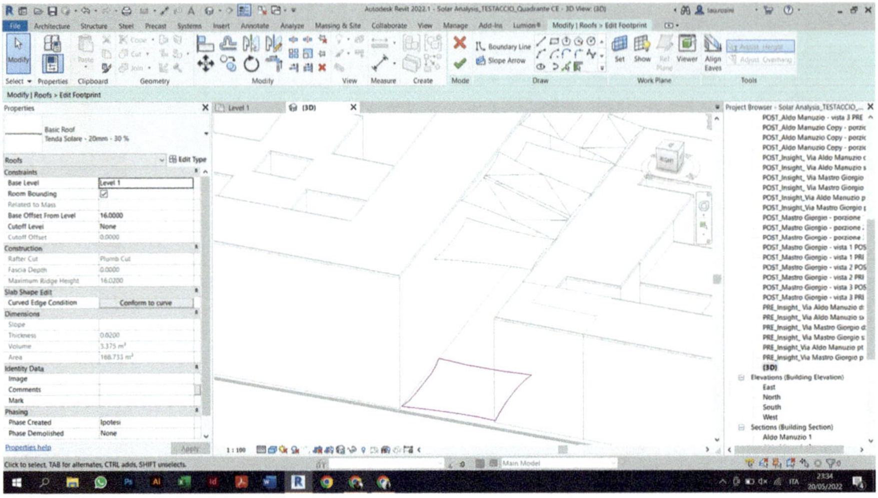

Fig. 70.6 Shading elements design: *Roof* tool (*Source* Autodesk Revit)

We have therefore evaluated the compliance of the basic solution of the fabric with a 30% AF factor to the shading needs of the different cases (Fig. 70.9). In the eventuality of unsatisfactory solutions, the AF factor of the veil could be gradually increased and the analysis of the sector run as many times as required (Table 70.3).

70.5 Solar Analysis Results

The comparison between the before and after scenarios is now summarized by observing results from quadrant 2B and 2E (Fig. 70.10). Via Aldo Manuzio spreads from West to East (to simplify), which means that the facades facing North are sun-lit during the first hours of the day, when temperatures are lower than the rest of the daytime. At the same time, facades facing South get insulated from midday, when they get affected the most by solar energy loads, to sunset; however, during the latter hours of the day, these facades are shaded by the facing buildings (Fig. 70.11).

The observation of this phenomenon brought us to the conclusion that a simple Absorption factor of 30% could be sufficient for this case. In terms of energy loads, the difference between the before and after results is mostly valuable for the ground surface. The road can be affected by a solar energy of 1 to 5 kWh/m^2. The simulation after installing the veil provided a reduction of up to 1 kWh/m^2 (Fig. 70.12) (Table 70.4).

Via Mastro Giorgio spreads from South to North which means that it gets irradiated during the middle part of the day, when sun height is at its highest and so do temperature (Fig. 70.13).

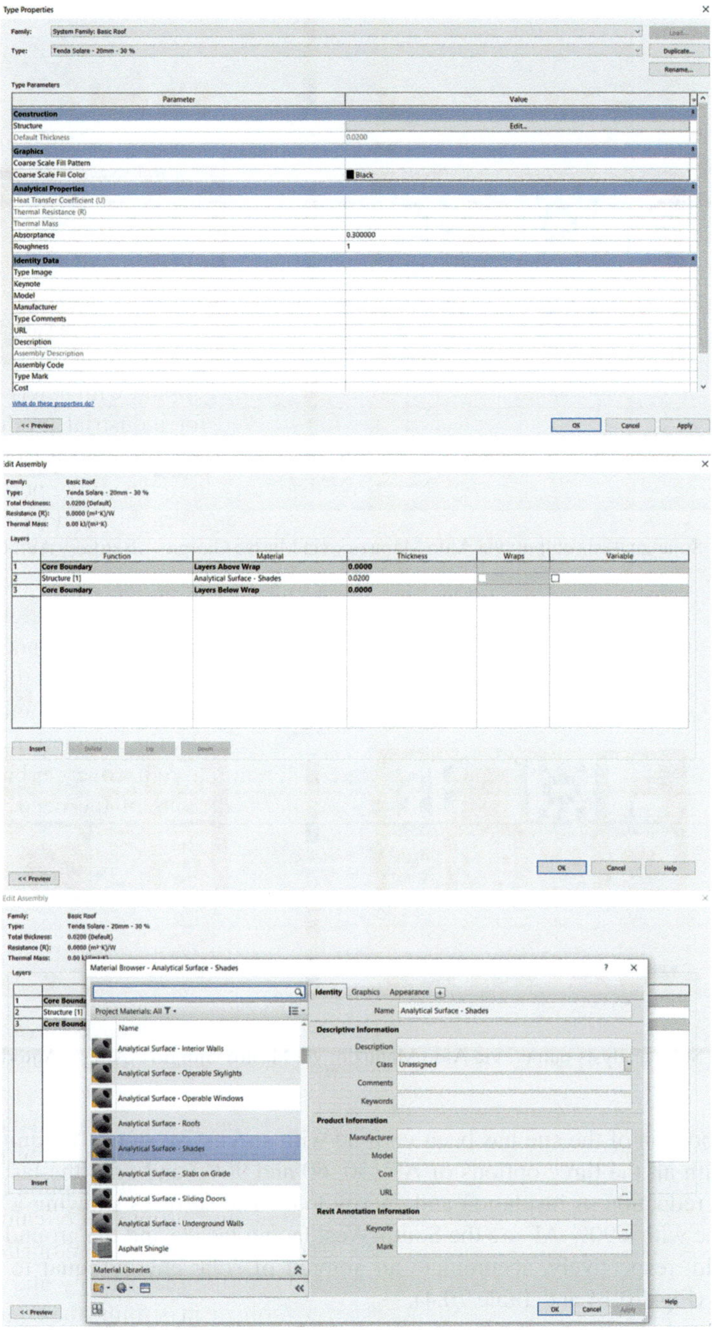

Fig. 70.7 Shading elements material parameters editing steps (*Source* Autodesk Revit)

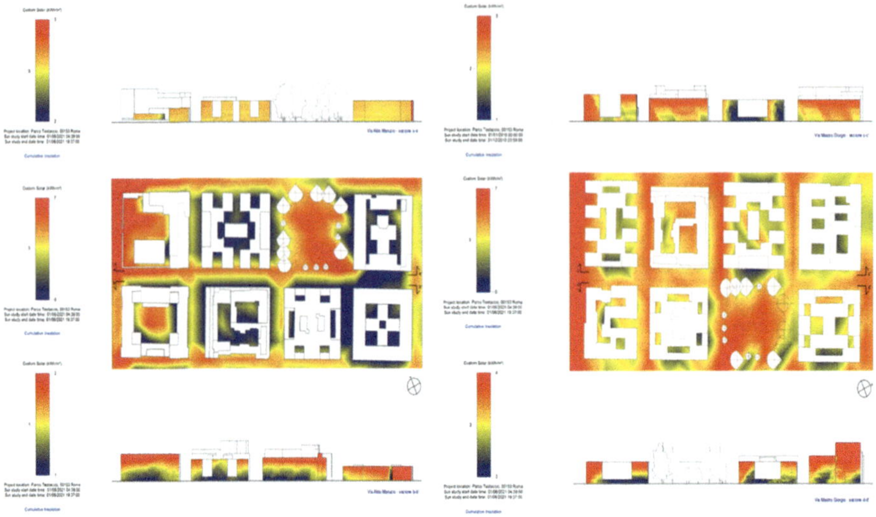

Fig. 70.8 Solar analysis output: via Aldo Manuzio; via Mastro Giorgio—Before—Autodesk Revit

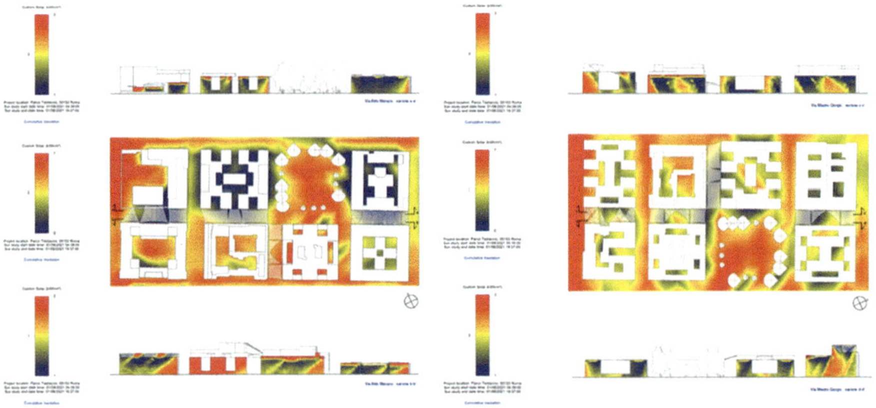

Fig. 70.9 Solar analysis output: via Aldo Manuzio; via Mastro Giorgio—After—Autodesk Revit

This portion of the site has been verified with successive steps. It required to be tested with all the three options of AF (30, 60 and 90%) and only the last one got a decent reduction of insulation and energy load. The surfaces showing a valuable difference with a 90% AF are the South-West facing facade and the ground surface, that could, respectively, accumulate an amount of solar energy equal to 4 and 5 kWh/m^2 (Fig. 70.14 and Table 70.4).

Table 70.3 Adaptive Velari parameters description

Via Aldo Manuzio (SW–NE)	Portion 1 section C	Portion 2 section B	Portion 3 section A
Facing buildings height	Different	Equal	Slightly different
Solar veils typologies	1 + 2 + 3	2 + 3	2 + 3
Anchoring position	uneven	even	uneven
AF (%)	30	30	30
Via Mastro Giorgio (SE–NW)	*Portion 1 section F*	*Portion 2 section E*	*Portion 3 section D*
Facing buildings height	Different	Equal	Equal
Solar veils typologies	1 + 3	1 + 3	1 + 3
Anchoring position	Uneven	Even	Even
AF (%)	90	90	90

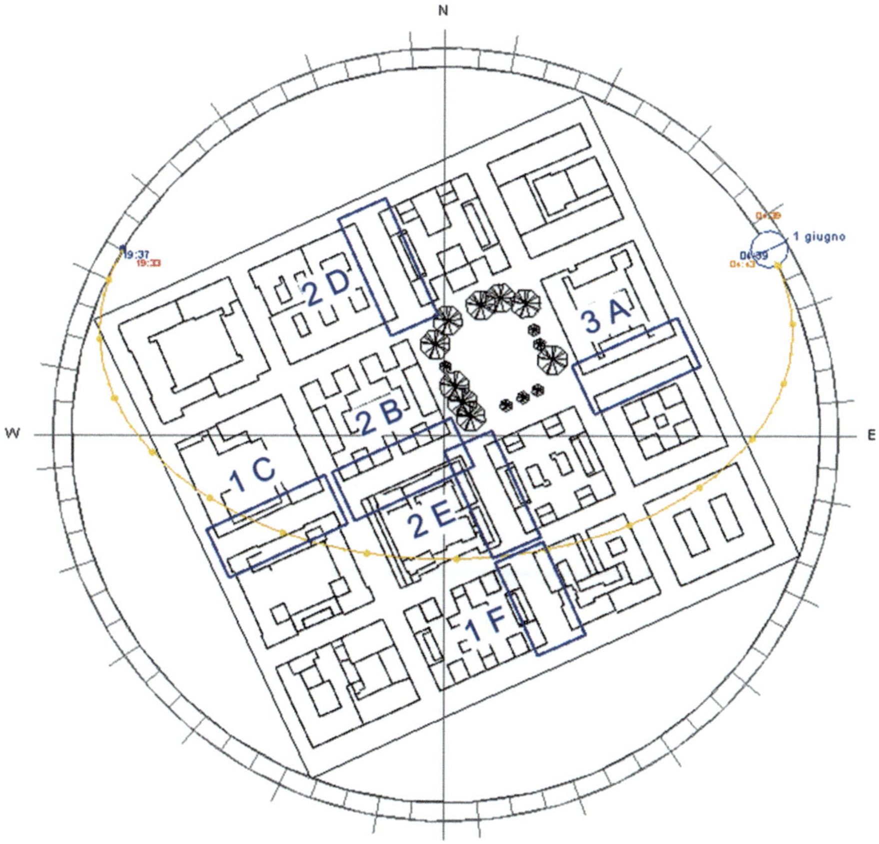

Fig. 70.10 Analyzed sectors. The coding indicates portions and sections along the streets' extension—Autodesk Revit

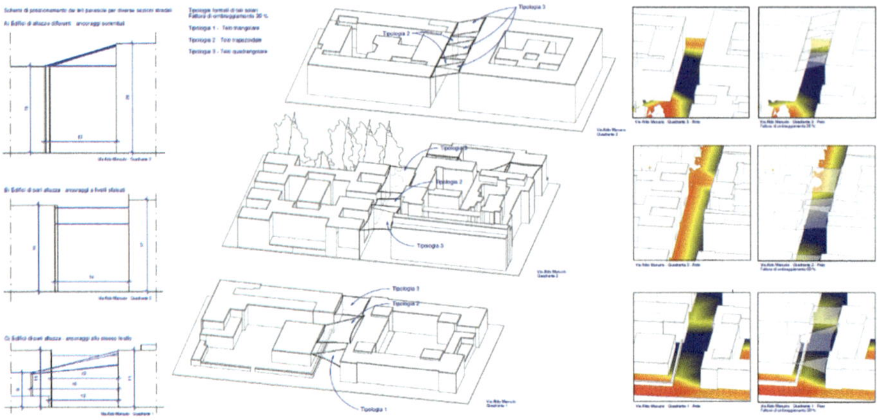

Fig. 70.11 Analysis output comparison: via Aldo Manuzio—Autodesk Revit

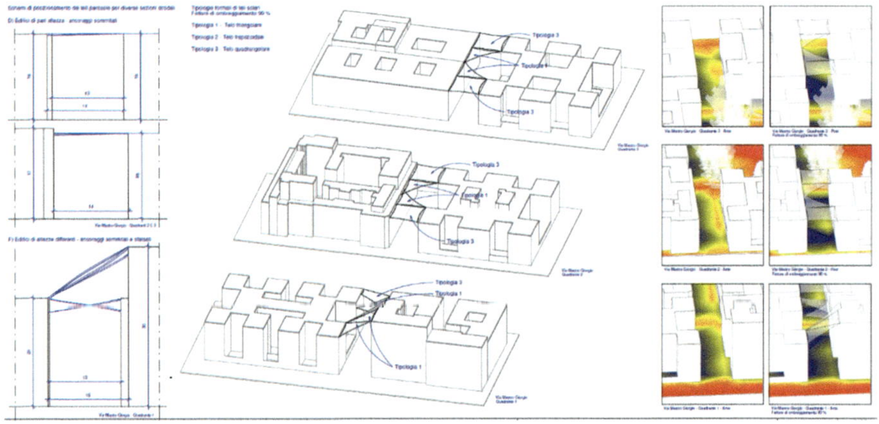

Fig. 70.12 Analysis output comparison: via Mastro Giorgio—Autodesk Revit

Table 70.4 Results details

Via Aldo Manuzio (SW–NE)—2B	Before: irradiation spread	Before: Energy load (kWh/m^2)	After: Energy load (kWh/m^2)
SE facing facade	Even	> 3	≤ 2
NW facing facade	Uneven	≤ 2	≤ 2
Ground surface	Uneven	≤ 5	~ 1
SE facing facade	Even	> 3	≤ 2
Via Mastro Giorgio (SE–NW)—2E	*Before: irradiation spread*	*Before: Energy load* kWh/m^2	*After: Energy load* (kWh/m^2)
NE facing facade	Slightly uneven	≤ 3	~ 1
SW facing facade	Uneven	≤ 4	~ 2.5
Ground surface	Even	> 5	~ 1

Fig. 70.13 2B quadrant—via Aldo Manuzio: results comparison—Autodesk Revit

Fig. 70.14 2E quadrant—via Mastro Giorgio: results comparison—Autodesk Revit

70.6 Conclusion and Future Perspective

In this initial phase of the research, we had the validation that systems such as velars, mounted at the top of buildings are able to produce a significant reduction of temperatures on the surfaces of urban canyons. The reduction of the stored heat load varies depending on the orientation of the road and how the sun radiates on the outer walls of the buildings; as a consequence of environmental effects on street surface and on facades, the type of fabric can be chosen from three Solar Absorbance values: 30, 60 and 90%.

Future developments will consist in studying a further modulation of fabric texture, in order to obtain point-directed shading, insisting where it's strongly needed and a looser tissue density where it is not; this facilitates ventilation and light permeability, along with inhabitants' well-being. Moreover, the research aims to outline the embedding of thin film PV cells into the velarios for in-site electricity production.

References

Akbari H, Cartalis C, Kolokotsa D, Muscio A, Pisello AL, Rossi F, Santamouris M, Synnef A, Wong NH, Zinzi M (2015) Local climate change and urban heat island mitigation techniques—the state of the art. J Civ Eng Manag 22(1):1–16

Cesario E (2017) I-MESH for FACADES - energy efficiency and environmental sustainability of textile devices by analysis and numerical simulation. PhD thesis school of advanced studies, Università di Camerino, Ciclo XXX

Garcia-Nevado E, Beckers B, Coch H (2020) Assessing the cooling effect of urban textile shading devices through time-lapse thermography. Sustain Cities Society. Vol 63, Elsevier

Grifoni RC, Tascini S, Cesario E, Marchesani GE (2017) Cool façade optimization: a new parametric methodology for the urban heat island phenomenon (UHI). In: 2017 IEEE international conference on environment and electrical engineering and 2017 IEEE industrial and commercial power systems Europe (EEEIC/I & CPS Europe), pp 1–5. https://doi.org/10.1109/EEEIC.2017.7977677

Saneinejad S, Moonen P, Carmeliet J (2014) Comparative assessment of various heat island mitigation measures. Build Environ 73:162–170

United Nations Environment Programme (2021). Beating the heat: a sustainable cooling handbook for cities. https://www.unep.org/resources/report/beating-heat-sustainable-cooling-handbook-cities

Xiang S, Zhang N, Fan X (2021) From fiber to fabric: progress towards photovoltaic energy textile. Adv Fiber Mater 3:76–106. https://doi.org/10.1007/s42765-020-00062-8

Chapter 71
Temporary Climate Change Adaptation: 5 Measures for Outdoor Spaces of the Mid-Adriatic City

Timothy Daniel Brownlee

Abstract The paper aims to identify a set of systemic measures that, through the temporary use of devices, space configurators, and installations, is capable of responding promptly to the vulnerability factors of a given outdoor space, flanking adaptation plans which require time to be assimilated into ordinary territory management tools. Based on the INTERREG Italy-Croatia Joint_SECAP project data of 9 target areas located on the two sides of the Adriatic Sea the document refers to a reference framework on risks and vulnerabilities of urban coastal areas and recurring climatic events. From a case studies analysis, built on climate hazards and on outdoor space configurations, the paper extracts replicability features and attempts to propose feasible models based on reversible and reconfigurable matrices that can be exported into contexts with similar characteristics. It follows that a technological design sensitivity capable of enhancing elastic spatial setups must be consolidated in order to address the needs of a specific outdoor space, during a climatic—or non-climatic—event, assuming that the city is a continuously evolving organism, with an in-grown ability to accommodate the variability of events.

Keywords Temporary · Adaptation · Outdoor space · Adriatic · Technological setup

71.1 Introduction

Outdoor urban spaces denote increasing climate change vulnerability in particular to flooding and urban heat islands (Matos Silva 2019; Graaf-van 2021) so much so that they could potentially be uncomfortable, unusable, or dangerous for people. The chronicity of climate extreme episodes has progressively required adaptation strategies capable of taking into account localized characterizations according to the identification of site-specific risks and the formulation of possible climatic scenarios based on the knowledge of the territory, on its socioeconomic and geographical context

T. D. Brownlee (✉)
University of Camerino, Camerino, Italy
e-mail: timothy.brownlee@unicam.it

E. Arbizzani et al. (eds.), *Technological Imagination in the Green and Digital Transition*, The Urban Book Series, https://doi.org/10.1007/978-3-031-29515-7_71

(Hagenlocher et al. 2018). It is no coincidence that many recent Euro-partnership climate change policies are precisely aimed at the coastal territories of the Mediterranean defined as a "hot-spots," one of the most sensitive regions to global warming (Guida 2021). The European Union has promoted the member states' actions in a progressively more structured way, coordinating the sharing of experiences and the development of climate-adaptive actions through a series of specific initiatives (from the 2008 Climate and Energy Package and Directives 2009/28/EC, 2010/31/EU, and 2012/27/EU to the 2021 EU Strategy on Adaptation to Climate Change and Next Generation EU). One of these initiatives, the Covenant of Mayors (CoM), launched by the European Commission in 2008, expanded in 2015 and since then evolved into the Covenant of Mayors for Climate and Energy, brings together in a network the cities that intend to launch a coordinated set of initiatives. The signatories undertake to exceed the European targets for reducing greenhouse gas emissions and increasing the resilience level of their territories.

The public bodies that adhere to the CoM are required to edit a Sustainable Energy and Climate Action Plan (SECAP), drawn up with the participation of civil society and accompanied by monitoring and verification tools. The plan defines a set of comprehensive measures that a municipality intends to implement on the basis of an assessment of risks and vulnerabilities induced by climate change. In this context, the Interreg Italy-Croatia Joint_SECAP (Joint Strategies for Climate Change Adaptation in Coastal Areas) project, conducted between 2019 and 2021, stems from the desire to bring together neighboring territories and build a common methodology to define joint SECAPs focused on sharing knowledge on climate change adaptation and mitigation measures for coastal areas of the Adriatic (Brownlee et al. 2022). The partnership was formed by a network of eight Italian and Croatian partners who have identified at least one target area in contiguous municipalities, on which they carried out the experimentation, in collaboration with local institutions. The partnership shared basic knowledge regarding adaptation strategies to climate change, also through a specific Web platform[1] that acts as a tool for the development of climate scenarios, necessary, and preparatory for the development of the joint actions. The main objective of the project was to improve climate change monitoring and plan adaptation measures to tackle specific climate change effects in the area of cooperation. The project proved to be innovative due to its ability to work on a scale larger than the single district, putting the neighboring territories' vulnerabilities into a system, and identifying common actions.

71.2 Objective

The study starts from the observation that the measures proposed by the climate change adaptation plans such as SECAPs require application processes capable of allowing their assimilation into the ordinary management tools of the territory and

[1] The following Web site is the project's web platform, https://joint-secap.unicam.it.

require variable times for their complete integration, often several years (medium-long time). The issues that may limit the effective implementation of these actions within the territorial management tools or outdoor urban space projects are not only linked to financial or governance matters; nor do they purposely refer to the absence of adequately trained human resources (Messori et al. 2020; Pascali and Bagaini 2021): but often the cause should be sought precisely from the lack of integration between plans that operate at different levels (Pietrapertosa et al. 2019).

In this interval, however, some urban contexts and in particular outdoor spaces could use measures capable of providing timely solutions, immediately applicable, in order to maintain their qualitative connotations of livability. Implementing and developing practices in medium/short times and addressing issues that concern the connection between strategic-programmatic planning and operational-constructive aspects (Angelucci and Di Sivo 2018) could be considered as a first and immediate tactical phase, in which actions could then translate it into ordinary programming with medium-long applicability, supporting some aspects, and facilitating the transition to such measures.

In this sense, the Covid-19 crisis offered several perspectives that picture issues related to urban space use in relation to the need to intervene promptly: an unexpected series of experiences and countless bottom-up practices that emerged highlighted that adaptation also depends on the community's dynamic features to respond to contingencies (McCullough 2020). In fact, communities have often shown their ability to generate transformations even before the necessary decisions were taken in order to convert spaces originally conceived for different purposes into legitimate and consolidated infrastructures (Manzini 2021). Many outdoor spaces in cities around the world have suddenly found a new connotation, animated and put into practice by the will of the local communities. One of the pillars at the base of climate resilience is precisely the transformative capacity: the ability to create an "enabling environment, strengthen the skills of key players, and identify and implement catalytic interventions," to be characterized through the key factor of time (Graaf-van 2021). The literature agrees in defining the urgent need to identify adaptive urban design tools, based on the idea of working on an open project, capable of changing over time, and also capable of creating an added value to the spaces we inhabit, through the variability of "elastic spatial arrangements" (Manigrasso 2019).

The capacity of bottom-up transformations, therefore based on the need to optimize procedures and finalize objectives in a rational way, in a certain sense can be considered as a reference of the technological process, and allow us to understand how some tools, including applicative-constructive ones, managed by associations and local communities, albeit rudimentary, can significantly contribute to increasing the level of urban space resilience.

71.3 Methodology and Results

The study is based on the data collected during the Joint_SECAP project in the target areas analyzed by the project partners. In particular, the risk and vulnerability assessment makes it possible to have a detailed picture of the risks associated with climate change. Through these data, it is possible to obtain general information concerning risks and vulnerabilities of the mid-Adriatic coastal areas and to draw up a list of the most recurring vulnerabilities, and consequently of the most widespread impacts that develop during specific climatic events. Table 71.1 reports in detail the link that is established between vulnerabilities, climatic impact, and the risks for the population, for urban structures, for energy production, and for transportation (Table 71.1). As is known, the recognized definition of adaptation to climate change refers to the ability of a measure to adapt certain contextual conditions to the current or expected climate, and to its effects in the future or during specific climatic events. In human systems, adaptation seeks to moderate or avoid harm and, when possible, to exploit any beneficial opportunities introduced by the measures themselves (IPCC 2018). Normally, these measures can limit the risks through the reduction of vulnerability factors, the propensity or predisposition of a territory to be affected, and in some cases decrease exposure factors, the presence of people, livelihoods, ecosystems, etc., that could be adversely affected. Vulnerability can be reduced by decreasing the sensitivity or increasing the capacity of the systems (Fig. 71.1).

With respect to the topics dealt with, it is urgent to find an applicative characterization for some measures which, in order to be effective, must work on the superimposition of the project's scales, as a matter of connection in the physical and temporal dimensions.

And if adaptation to climate change works precisely on these multiscalar factors, we can see a gap between the macro and the meso-micro-level, between the—programmed strategy and the—activated implementation, in which plans and strategies often risk being described through an incomplete vision of the transformation process underway, of the actions and objectives achieved (Rossi 2019). The study led to the identification of possible meta-design actions aimed at decreasing several vulnerability factors of the mid-Adriatic city and reducing the impact deriving from climate change. The possible actions are based on the physical characteristics of outdoor urban space, in a densely urbanized context and with sealed pavings. The analysis of the vulnerabilities and climatic impact that affect urban spaces leads to the selection of some light transformation actions of the urban environment, which can be implemented through temporary stratification.

The study identified a series of international case studies, relevant since they were designed to facilitate the use of outdoor spaces, sometimes born as a local community answer during the pandemic lockdown intervals.

The case studies used by the research, here only briefly reported, have not been designed as climate change adaptation measures, but they are analyzed with this objective, on the basis of a hidden potential that can be deduced from their characteristics, among all the ability to effectively *stratify* the hosting urban spaces. The

Table 71.1 Most common vulnerability factors (sensitivity) of the mid-Adriatic coastal areas (data from the JointSECAP project, Italian Adriatic coastal area, re-elaborated by the author)

Vulnerability	Related hazards family	Related impact family	Related risk family
Outdated drainage system	Extreme precipitation events Hailstorms	Flooding (river, ground, and coastal) Excessive runoff Landslides and instability phenomena	Damage to people and urban structures, activities, energy production, and transportation
Road network prone to flooding			
Maintenance level of building stock			
Inadequate maintenance of green and river banks			
Water bodies canalization			
High soil sealing level			
Water storage capacity			
Proximity between activities and river banks			
Low, sandy highly urbanized coast prone to erosion	Rise in water level	Flooding (coastal) Coastal erosion	Damage to agriculture, people, activities
Geomorphological aspects regarding the shore			
Shore line modification			
Outdoor surface prone to overheating	Consecutive dry days Dry period with high temperatures Mean of annual precipitation	Droughts Increase of fires	Risks related to water scarcity and fire (to activities, people, health, and transportation)
Inadequate or absent irrigation systems			
Areas with fire risk			
Water storage capacity			
Losses in water supply			
Irrigation capacity			
Increase of water consumption during tourism periods			

(continued)

Table 71.1 (continued)

Vulnerability	Related hazards family	Related impact family	Related risk family
Distance of water in case of fire			
Water scarcity	Extreme heat Higher average temperature Mean of annual precipitation	Alteration of ecosystems	Damage to agriculture tourism, health
Outdoor surface prone to overheating			
Inadequate or absent irrigation systems			
Water storage capacity			

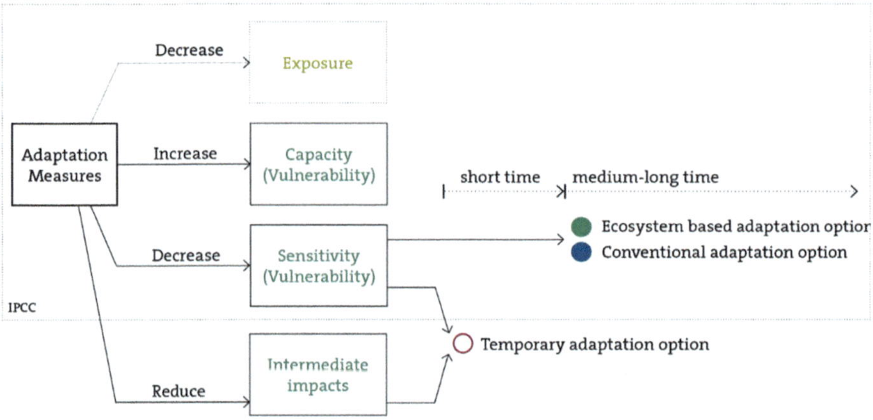

Fig. 71.1 How temporary adaptation measures can intervene in reducing the effect of climate impact and decreasing vulnerability climate impact and decreasing vulnerability

analyzed examples propose interventions that can be rapidly implemented, through reversible stratification actions and reconfigurable matrices, suggesting models that can be exported to contexts with similar characteristics.

The case studies, interpreted in consideration of their transferable potential, give the opportunity to understand that through light construction features it is possible to improve the safety conditions of the urban space. The article summarizes five possible actions (Fig. 71.2), please consider that the mentioned case study for each action is only one of the possible ones.

For a complete analysis of the case studies referring to each action refer to future discussions.

1. *Raise.* Vulnerability factors: water storage capacity, outdated drainage system, road network prone to flooding. Related impact factors combined with extreme

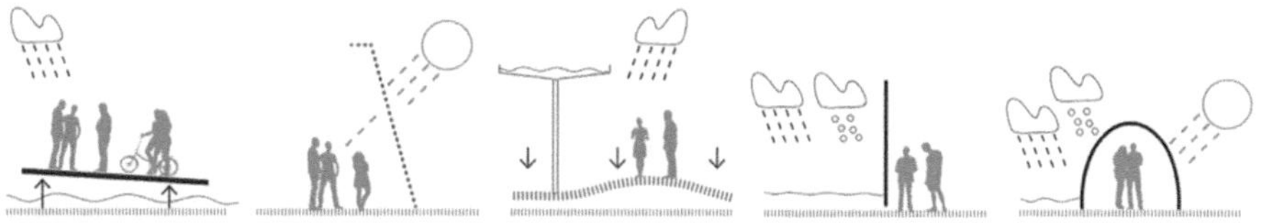

Fig. 71.2 Five possible meta-actions for temporary adaptation

precipitation: flooding, excessive runoff. Possible case study, used as a reference for the constructive implementation transfer: AAIM Architecture + Urban Matters, Urban Bloom, Shanghai, 2019, temporary setup based on the use of a new raised surface of the open space.

2. *Screen*. Vulnerability factors: outdoor surface prone to overheating, water scarcity. Related impact factors: extreme heat, hailstorm. Possible case study, used as a reference for the constructive implementation transfer: Andrés Jaque Arquitectos, Escaravox, Madrid, 2012, structure with variable configurations for temporary cultural events.
3. *Sponge*. Vulnerability factors: water storage capacity, outdated drainage system, road network prone to flooding, high soil sealing level. Related impact factors combined with extreme precipitation: flooding, excessive runoff. Possible case study, used as a reference for the constructive implementation transfer: Shma, Come on/calm on, Bangkok, 2021.
4. *Separate*. Vulnerability factors: road network prone to flooding, shoreline modification. Related impact factors combined with hailstorms and precipitation: flooding. Possible case study, used as a reference for the constructive implementation transfer: Colab-19, SCA, Taller Architects, Instalación Activación Vertical, Bogotà, 2020.
5. *Shelter*. Vulnerability factors: water storage capacity, outdoor surface prone to overheating. Related impact factors: extreme heat, higher average temperature. Possible case study, used as a reference for the constructive implementation transfer: Breath Earth Collective, Airship 01 mobile forest, Roma, Milano, Padova, 2016, pavilion based on air quality improvement.

The methodological approach is illustrated through a case study (Fig. 71.3) referred to the meta-action n° 1 *raise*, explained through a diagram that highlights the main aspects concerning the extrapolation and re-interpretation of several technological-constructive solutions. The research, for the sake of synthesis, is mentioned here only for one of the meta-actions but is based on the application of this methodological approach to multiple case studies for each one of the five described temporary measures. The meta-actions thus described can be added together.

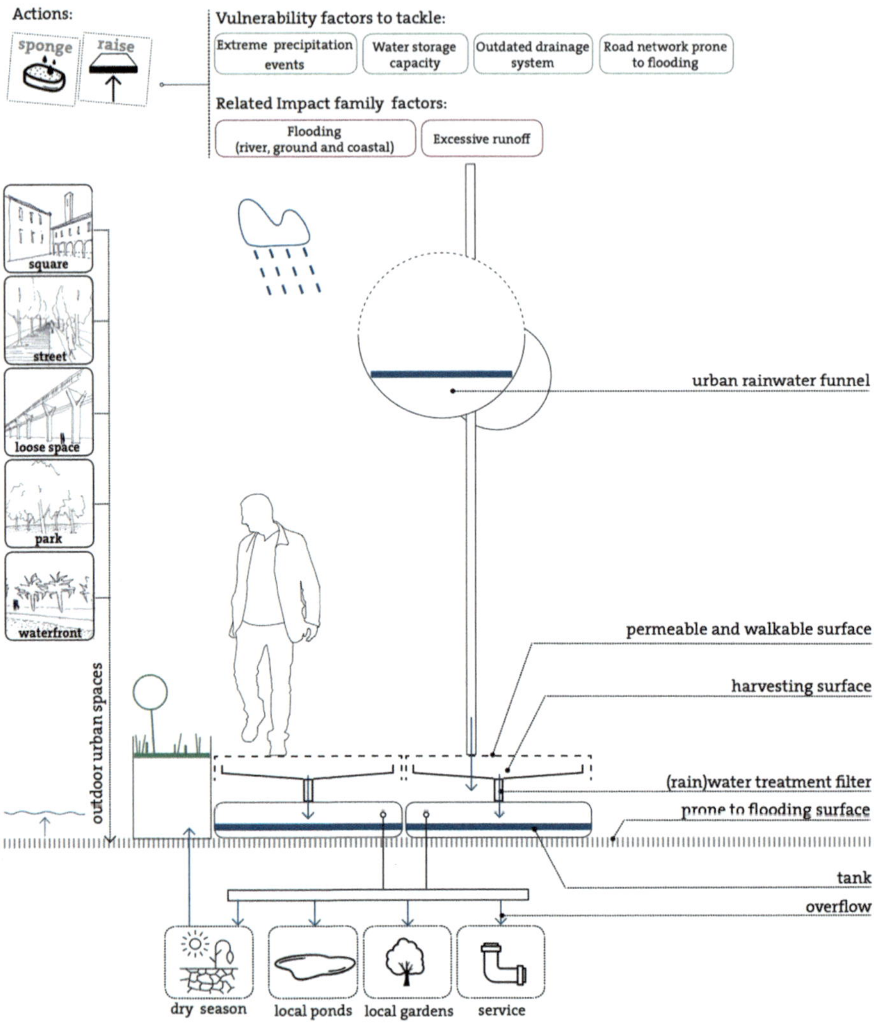

Fig. 71.3 Extrapolation and re-interpretation of the case study's technological-constructive solutions

71.4 Conclusions

The focus introduced by the paper provides a contribution that refers to a broad and trans-disciplinary topic such as adaptation to climate change that needs to be investigated in all the facets of its complexity. The contribution briefly illustrates one of the five temporary adaptation meta-actions and provides a first set of possibilities for provisional intervention aimed at the enhancement and recovery of outdoor urban spaces of the mid-Adriatic Italian cities, vulnerable to climate change, as a measure to respond quickly and not necessarily in a systemic and permanent way to climate events. The mentioned example illustrates how light and temporary layering of open spaces can also be implemented through modular and easily arrangeable solutions. In this sense, some operations aimed at *securing* outdoor urban spaces during climatic events, being quite light, could be easily implemented by local community services. The measures described provide a possible partial response to the lack of integration between the macro and the meso-micro-level, between the—programmed strategy and the—activated implementation level, pending more effective measures that will be implemented over a longer period of time. The topic suggests that the identification of possible transformations that lead to a climate-adaptive built-in environment requires a series of technological and experimental steps to make them possible. The subject is treated only in its general character and requires future investigations and verifications, to be developed through specific projects located on real outdoor urban sites.

References

Angelucci F, Di Sivo M (2018) Il macro requisito dell'accessibilità integrata per la healthy city. In: Angelucci F (ed) Smartness e healthness per la transizione verso la resilienza, Milano, Francoangeli srl, 185–204

Brownlee TD, Camaioni C, Magaudda S, Mugnoz S, Pellegrino P (2022) The INTERREG Italy-Croatia Joint_SECAP project: a collaborative approach for adaptation planning. Sustainability 14(1):404. https://doi.org/10.3390/su14010404

De Graaf-van Dinther R (2021) The five pillars of climate resilience. In De Graaf-van Dinther R (eds) Climate resilient urban areas, Governance, design and development in coastal delta cities. Switzerland, Palgrave Macmillan, Springer Nature

De Pascali P, Bagaini A (2021) L'affermazione del piano d'azione per L'energia Sostenibile (PAES) in Italia. Limiti e tentativi di integrazione con la pianificazione locale. ASUR, 131. https://doi.org/10.3280/ASUR2021-131-S1004

Guida C (2021) Cambiamento climatico, Rischi. Napoli, FedOA—Federico II University Press, Mediterraneo

Hagenlocher M, Schneiderbauer S, Sebesvari Z, Bertram M, Renner K, Renaud F, Wiley H, Zebisch M (2018) Climate risk assessment for ecosystem-based adaptation. A guidebook for planners and practitioners. Bonn, Deutsche Gesellschaft für Internationale Zusammenarbeit (GIZ) GmbH. http://collections.unu.edu/view/UNU:6522 (Accessed March 2022)

IPCC (2018) Annex I: Glossary. In: Matthews JBR (ed) Global warming of 1.5 °C. An IPCC special report on the impacts of global warming of 1.5 °C above pre-industrial levels and related global greenhouse gas emission pathways, in the context of strengthening the global response to the threat of climate change, sustainable development, and efforts to eradicate poverty. https://www.ipcc.ch/sr15/chapter/glossary/. (Accessed 20 March 2022)

Manigrasso M (2019) *La città adattiva, il grado zero dell'urbana design.* Macerata, Quodlibet Studio

Manzini E (2021) *Abitare la prossimità, idee per la città dei 15 minuti*, E.G.E.A. Milano, Spa

Matos Silva M (2019) Public spaces for water: a design notebook. CRC Press, London

McCullough M (2020) Downtime on the microgrid, architecture, electricity, and smart city Islands. The MIT Press, Cambridge

Messori G, Brocchieri F, Morello E, Ozgen S, Caserini S (2020) A climate mitigation action index at the local scale: methodology and case study. J Environ Manag, pp 260, 110024. ISSN 0301-4797. https://doi.org/10.1016/j.jenvman.2019.110024
Pietrapertosa F, Salvia M, De Gregorio Hurtado S, D'Alonzo V, Church JM, Geneletti D, Musco F, Reckien D (2019) Urban climate change mitigation and adaptation planning: are Italian cities ready? Cities 91, 93–105. ISSN 0264–2751. https://doi.org/10.1016/j.cities.2018.11.009
Rossi GE (2019) Adattamento urbano, strategie e progetto; il divario fra le politiche e la loro implementazione. Agathòn 6, pp 46–57

Chapter 72
A Serious Game Proposal for Exploring and Designing Urban Sustainability

Manuela Romano and Alessandro Rogora

Abstract This contribution reports on part of the research carried out in the development of a digital serious game to explore and design urban sustainability. The work investigates the application of role-playing logics in urban transformation planning processes. Digital technologies support the participatory process and allow users to learn, interact and discuss the effectiveness of solutions when co-designing quality conditions for the everyday living environment and the sustainable development of their area. The method intends to develop procedures and simulations that aim to clarify and verify the results that transformations can have on the area's ecological footprint. During the game, players gain knowledge and awareness of the individual behaviour changes and the built environment trasformations, necessary in order to impact their areas as sustainably and as little as possible. Prefiguring resilient and sustainable urban habitats, increasing user awareness of the need to adopt more responsible behaviours, increasing ability of the built environment to meet community needs by practising low environmental impact lifestyles: these key aspects are observing at all phases of the design process. This paper describes the progress of research carried out on the construction of the structure and rules of the game, developed prior to testing the method in the context of the municipality of Rescaldina, in the Milan metropolitan area.

Keywords Sustainable communities · Serious game · Digital tools · Urban game · Urban sustainability

M. Romano (✉) · A. Rogora
Polytechnic University of Milan, Milan, Italy
e-mail: manuela.romano@polimi.it; manuela.romano@giallo.it

A. Rogora
e-mail: alessandro.rogora@polimi.it

E. Arbizzani et al. (eds.), *Technological Imagination in the Green and Digital Transition*, The Urban Book Series, https://doi.org/10.1007/978-3-031-29515-7_72

72.1 Introduction

The technological and digital revolution over the last twenty years has changed many social processes with forms of *gamification* (Deterding et al. 2011) that have changed the habits and behaviours of society in every area of life, such as relationships, work, training and entertainment. Encouraged by common interests or objectives, groups of people have built collaborative enterprises (Barricco 2019) by aggregating skills and experiences that, within digital platforms, have shortened physical and cultural distances and brought about what Pierre Levy has defined as *collective intelligence* (Pierre 1994); namely, the collaborative effort in solving common problems through the sharing of information and results.

This process also applies to design practices, where digital technologies play an increasingly important role in the transformation processes of the built environment (Perriccioli et al. 2020).

Digital tools and interactive technologies are defining new forms of dialogue between project and user. The quality of urban governance appears to be increasing thanks to a greater ability to data collection, orientation and control the impact of project decisions and the involvement of a larger number of stakeholders experimenting with problem-solving-oriented design approaches.

This contribution describes part of a research project carried out on the development of a digital serious game for exploring and designing urban sustainability. The study—carried out by the SEEDItaly research group of the Department of Architecture and Urban Studies at the Polytechnic University of Milan—aims to understand how the serious game aspects and digital technologies can be adopted in the participatory planning processes for territory interventions. The goal is to understand to what extent the logics of role play can support decision-making processes in urban and building transformation initiatives in the prefiguration of sustainable urban habitats. Increasing user awareness of the need to adopt more responsible lifestyles and the ability of the built environment to meet community requirements are observing as fundamental aspects in all phases of the design process.

72.2 The Contribution of Serious Games in the Development of a Sustainable Society

In recent years, in an attempt to develop actions to transition to more sustainable lifestyles and move closer to attaining Sustainable Development Goals (SDGs), the involvement of society in decision-making processes has been considered fundamental.

It has been acknowledged that technological components alone, without a societal assumption of responsibility and the modification of lifestyles practised on the ground, are not sufficient to achieve effective results. In this context, serious games have become increasingly important in communicating and raising awareness of climate change issues (Wu and Lee 2015; Flood et al. 2018), with the goal of engaging users to explore the personal and collective decisions and changes needed to address this global challenge. Most games have been developed to enhance educational experiences, especially in school settings (Neset et al. 2020).

The issues addressed are multiple: pollution, plastic consumption, water resource use, natural disaster management, land resource use, energy and sustainable mobility.[1]

Within digital platforms or with table games, players can immerse themselves in simulations while developing skills and competencies to apply in the real world. Including targets within the game supports social learning (Ensor and Harvey 2015) and an increased awareness of the need to adopt environmentally conscious daily habits.

Recognising the relevance of this issue at the international level, the 'Playing for the Planet Alliance'[2] initiative was launched in 2019, in which many companies operating in the Gaming sector collaborated to develop games with a primary reference to issues related to the 17 SDGs of the 2030 Agenda (Patterson and Sam 2019). The proposed games aim to increase skills on topics such as: sustainable production chains (Anno 1800), biodiversity protection (Away: The Survival Series), renewable energy (Transformers Earth Wars) and building sustainable communities (Minecraft-Climate Hope City). The games demonstrate the technologies and processes needed to solve problems related to pollution and unsustainable resource use and to build a society in harmony with the environment.

In Eco,[3] for example, players are called upon to seek compromises and agreements by developing new laws and economies to manage the resources available to them and avert the destruction of the ecosystem. The game's name was conceived from the relationship between ecology and economics, which must be managed by good governance and appropriate regulations on the use of the resources that the ecosystem provides, without destroying its natural state.

Serious games were theorised in the 1970s (Abt 1970) as games developed for educational purposes that use fun and playful communication to convey educational notions and promote experiential learning. Today, they are increasingly used in various settings (business, education, government) to promote training programmes for individual development and cooperation within work groups by

---

[1] There are online platforms where users can access games geared towards sustainable development education. For example, Games4Sustainability (games4sustainability.org) or GamesForChange (gamesforchange.org), where multiple experiments can be explored.

[2] The Playing for the Planet Alliance initiative was launched at the Climate Summit held in New York in 2019 with the aim of empowering the games industry to support action against the climate crisis. Numerous games were developed through the 'Green Game Jam' 2020 and 2021 competitions (playing4theplanet.org/greengamjam/).

[3] Eco is an online game developed by Strange Lop Games in collaboration with the University of Illinois Urbana-Champaign, with funding from the United States Department of Education (https://play.eco/).

expiring processes. Their main feature is highlighting real problems as interesting and fun, thus enhancing awareness of the actions to be taken towards shared solutions. Knowledge is fostered by exploring reality and carrying out activities that require behavioural changes from actors according to a 'learning by doing' approach. Some studies (Patti 2018; Viola 2011) observe how role-playing games in everyday life can influence emotions, having been deliberately designed by developers with the aim of guiding people's behaviour in and outside of video games.

In academia, the potential of applying role-playing game logic is the subject of several studies (Sousa et al. 2022), aimed at understanding the contributions these tools can make in the development of co-construction processes oriented towards sustainable development goals. For example, some experiments simulate climate negotiations, in which participants can gain insights into the urgency and complexity of environmental action and the influence that policy choices have in the transition to a green economy. Even in urban planning, some studies (Brkovic and Groat 2020, Papathanasiou et al. 2019) note how the use of role-playing can improve knowledge of the urban environment and the exchange of views amongst various stakeholders in the decision-making process for urban habitat transformation. Civic engagement and community participation in urban planning can be supported by providing citizens with the tools and opportunities to address global challenges in solving real, local problems, therefore increasing knowledge on how to develop a more sustainable society.

72.3 TRAceS: A Serious Game that Explores and Designs Sustainable Communities

As part of the research presented in this text, the aim of the game experience in question is to encourage shared and informed planning choices within the regeneration processes of the built environment (Rogora 2022). The game aims to facilitate the planning of urban and building transformations. Digital technologies support the participatory process and enable users to learn, interact and discuss the effectiveness of solutions as they co-design new conditions for the quality of the everyday living environment and sustainable development of their area.

The method aims to involve inhabitants, public administrations, professionals, associations and all those who express a willingness to be involved in their area's transformation processes. Transformation scenarios are constructed within a framework of the resources that are locally available. During the game, players gain knowledge and awareness of the individual behaviour changes and the built environment trasformations necessary in order to impact their areas as sustainably and as little as possible. The method aims to produce simulations that are intended to be a tool for knowledge, comparison and debate when validating the choices to be made, thereby clarifying and verifying the results that transformations can contribute to help determine the area's ecological footprint.

72.3.1 Tools and Methods

The game's universe represents a real context. The municipality of Rescaldina, in the Milan metropolitan area, has been used in the experimentation conducted in the research. The municipal territory has been reconstructed on a digital platform (Fig. 72.1).

The information is organised and processed with the help of GIS software. The initial knowledge phase of the game's universe is aimed at understanding the initial environmental impacts and their main causes, in relation to which the urban sustainability scenarios are designed.

The impact is measured in per capita cost per inhabitant, expressed as the amount of area occupied and tonnes of CO_2 emitted for the production and consumption of goods and services needed to sustain the lifestyles of the community settled in the game area.

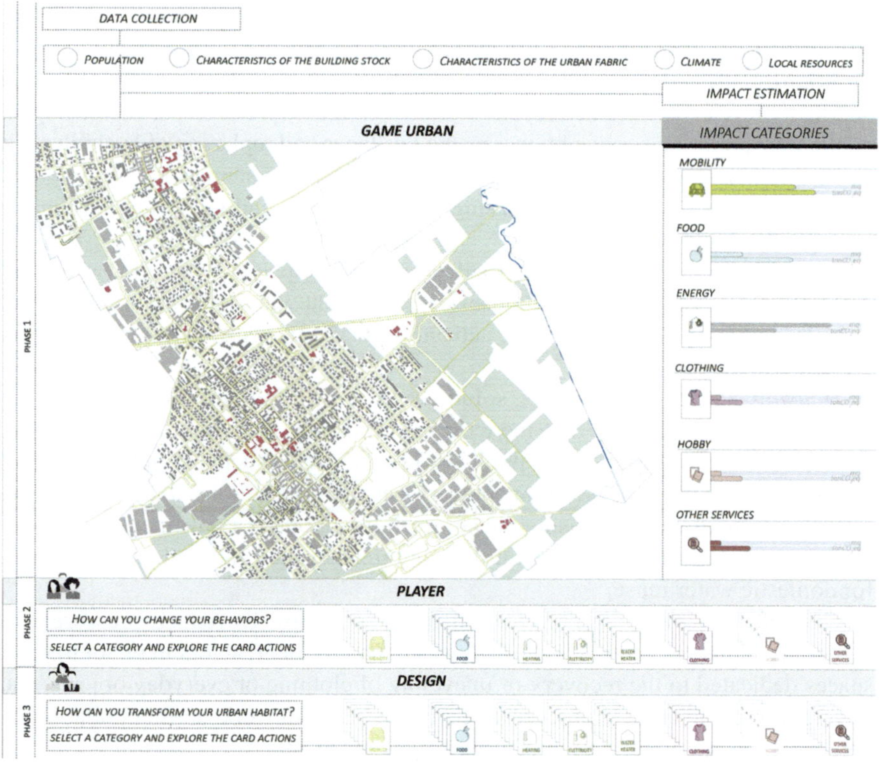

Fig. 72.1 Graphic representation of the game urban

The main data considered include:

- Population and average consumption;
- Characteristics of the urban fabric (morphology, infrastructure, services and their distribution);
- Characteristics of the public and private building stock (typological, constructive and performance related);
- Climate (solar radiation, temperatures, prevailing winds, etc.);
- Local resources (arable land, forests, water resources, renewable energy sources).

Data collection is also aimed at defining transformations that are appropriate to the needs and potential of the context. The possible transformations are described on the playing cards. The transformation actions represent suggestions in the form of best practices, technologies and processes aimed at developing self-sufficient communities (Clementi 2019). The actions are grouped by impact categories related to the main types of expenditure for everyday goods and services that are consumed by the population settled in the game area (mobility, food, housing, clothing, leisure and services). The solutions refer to technical and/or functional choices, briefly described below, that would impact local urban metabolism in the way energy, water, food, transportation, goods or daily services are used, managed and produced.

Mobility. To reduce pressure from the transportation sector, options include building bicycle lanes, and bicycle workshops and express stops, increasing the public transportation service by increasing bus lanes and replacing old vehicles with electric vehicles, replacing owned vehicles with electric systems and installing electric charging stations, reducing parking for private cars and increasing parking for low-speed vehicles.

Food. To reduce impact from food procurement, players can choose to build urban gardens, productive greenhouses or experiment with domestic solutions such as hydroponics or vertical gardens, activate solidarity purchasing groups or market for zero kilometer food.

Housing. Players can opt for solutions to reduce building and urban energy consumption with actions that increase the efficiency of public or private assets, produce energy locally to cover public or private needs, such as lighting and build energy communities. In addition, players can opt for actions that reduce urban water consumption by increasing the permeability of outdoor pavements or building roofs, creating bio-retention and infiltration green spaces and collecting rainwater for domestic water reuse.

Clothing and other everyday products or services. To reduce pressure from secondary industrial sectors and urban waste management, actors may opt to create spaces dedicated to the recovery or upcycling of clothing or everyday objects, such as neighbourhood tailor shops, creative workshops, craft workshops or collaborative enterprises, possibly located in disused spaces in the area.

For each action, transferability is checked during the preliminary stage of preparing the call for proposals. The environmental effect and economic cost required to implement the action are estimated and described. In addition, best practices are suggested to encourage the development of new local entrepreneurial activities related to proposed activity management.

The design process takes place throughout the game, which is organised into three main phases (Fig. 72.2). Once the game urban is constructed, we move to the Player phase, which is preparatory to the Design phase. In the Player phase (Fig. 72.3), with reference to the six impact categories, players are asked to choose from alternative proposed actions to change their daily behaviours and understand the extent to which adopting more responsible lifestyles affects the environmental impact of the game's field. The changes that players can make are contained in the 'actions of behaviour' deck of cards. During the Player phase, the goal is to provide players with the tools to, first, understand the individual contributions they can make in building sustainable communities and second, to understand the context's ability to facilitate the sustainable behaviours suggested by the cards and the intervention needed to ease these transitions.

Information gathering guides choices in the second design phase (Fig. 72.4). Players learn about creating healthy and thriving urban environments and help prefigure local sustainability scenarios by exploring the effectiveness of design solutions contained in the 'project actions' deck of cards. Actions aim to maximise locally available resources and/or mitigate impacts associated with the consumption of everyday goods and services.

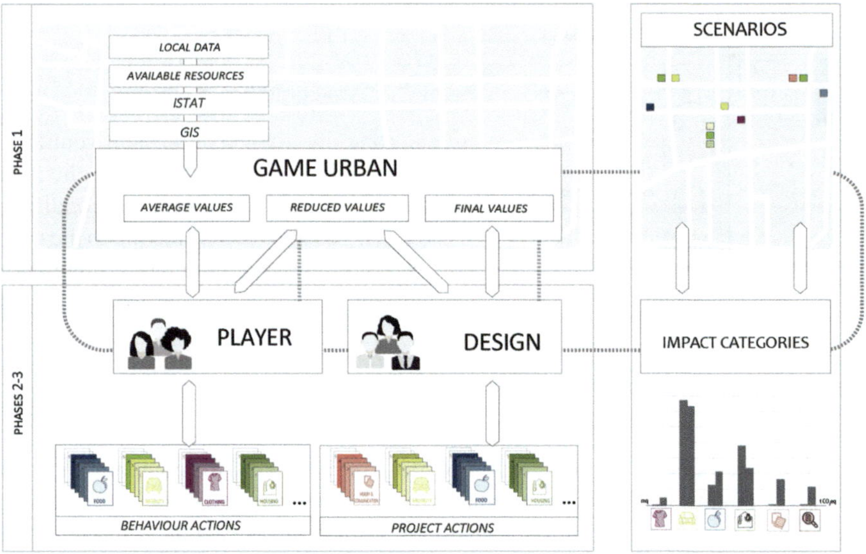

Fig. 72.2 Structure of the TRAceS method

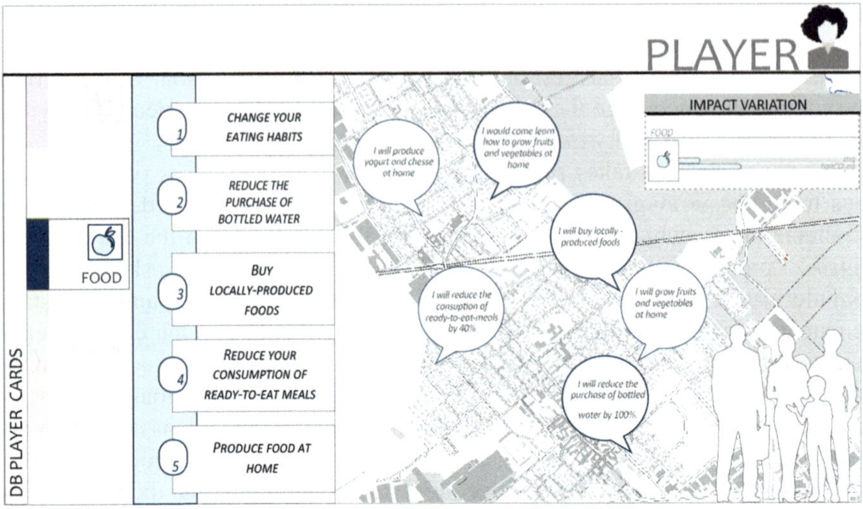

Fig. 72.3 Graphic representation of the player phase in the food category

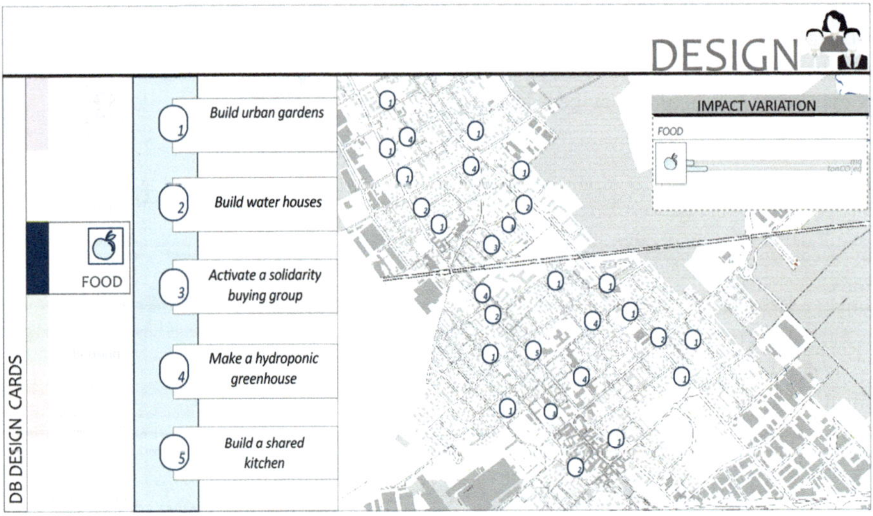

Fig. 72.4 Graphic representation of the design phase in the food category

During the game, in the prefiguration of possible scenarios, the population's ability to accept certain transformative actions is tested. The design solution's effectiveness is evaluated with respect to the propensity of the actors to accept the transformations achieved by implementing a participatory process between citizens, professionals and public administration.

72.4 Conclusions

Digital games represent an emerging field of research (Papathanasiou et al. 2021) and application that enables playful learning about the urban environment and an exchange of views in the planning choices to be made in transformation actions.

The development of TRAceS is intended to represent an innovative approach aimed at participatory planning processes in which involvement from the community in improving the quality of life in their area is supported by tools for knowledge building and maximising locally available resources.

The objective of the game is twofold: to increase knowledge about the difficulties of the climate adaptation challenge and the importance of individual and collective engagement amongst governments, professionals and citizens and to understand the extent to which urban and building transformations can foster more responsible human behaviour.

The method is currently being tested within the research team and in educational workshops. To test its effectiveness and refine its potential, simulations will be conducted in the coming months to test the ability to imitate and represent the results.

The test will be carried out by students under the direction of the research group in order to fine-tune the rules of the game and the calculation procedures for the implementation of the results obtained in preparation for the experiment in the municipality of Rescaldina.

References

Abt C (1970) Serious game. The Viking Press, New York

Barricco (2019) The game, Einaudi

Brkovic DM, Groat NL (2020) The routledge companion to games in architecture and urban planning. Tools for design, teaching and research, Routledge, New York

Clementi M (2019) Progettare l'auto-sostenibilità locale, Edizioni Ambiente, Milano

Deterding S, Dixon D, Khaled R, Nacke L (2011) From game design elements to gamefulness: defining gamification. In: Proceedings of the 15th international academic MindTrek conference: Envisioning future media environments, New York ACM Press, pp 9–15

Ensor J, Harvey B (2015) Social learning and climate change adaptation: evidence for international development practice. Wiley Interdiscip Rev Clim Chang 6:509–522

Flood S, Cradock-Henry NA, Blackett P, Edwards P (2018) Adaptive and interactive climate futures: systematic review of "serious games" for engagement and decision-making. Environ Res Lett

Neset T, Andersoon L, Uhrqvist O, Navarra C (2020) Serious game for climate adaption—assessing the potential and challenged of a digital serious game for urban climate adaption. Sustainability 12:1789. https://doi.org/10.3390/in12051789pp.1-18

Papathanasiou J, Tsaples G, Blouchoutzi A (2021) Urban sustainability. A game-based approach, Springer Cham

Papathanasiou J, Armenia S, Barnabè F, Carlini C, Ciobanu N, Digkoglou P, Jarzabek L, Kulakowska M, Lanzuisi A, Morfoulaki M, Myrovali G, Onofrei C, Pompei A, Scolozzi R, Tsaples G (2019) Game based learning on urban sustainability: the "sustain" project. pp 9240–9246. https://doi.org/10.21125/edulearn.2019.2293.

Patterson T, Barratt S (2019) Playing for the planet—how video games can deliver for people and the environment. UN Environment/GRID-Arendal. Arendal, Norway
Patti I (2018) Serious game design: Storia e teorie sull'esperienza ludica applicata, Franco Angeli, Milano
Perriccioli M, Rigillo M, Russo ES, Tucci F (2020) Design in the digital age. Technology nature culture, Maggioli, Santarcangelo di Romagna
Pierre L (1994) L'intelligenza collettiva. Per un'antropologia del cyberspazio, Milano, Feltrinelli, ISBN 978-88-07-81716-8
Rogora A, Carli P, Trevisan A (2022) New proposals for sustainable design: the imitation game as an experience of shared co-design, In Bernardes O, Amorim V, Moreira AC (2022) Handbook of research on gamification dynamics and user experience design. IGI Global, Pennsylvania
Sousa M, Antunes AP, Pinto N, Zagalo N (2022) Serious games in spatial planning: strengths, limitations, and support frameworks. Int J Serious Game 9:115–133. https://doi.org/10.17083/ijsg.v9i2.510
Viola F (2011) Gamification—I Videogiochi nella Vita Quotidiana
Wu JS, Lee JJ (2015) Climate change games as tools for education and engagement. Nat Clim Chang

Chapter 73
Energy Efficiency Improvement in Industrial Brownfield Heritage Buildings: Case Study of "Beko"

Jelena Pavlović, Ana Šabanović, and Nataša Ćuković-Ignjatović

Abstract Brownfield sites often form on industrial sites of once successful companies dating from the era of industrialization, due to loss of active function and despite their historical significance. Accompanied by urban decline, they contribute to continuous pollution, decrease in economic values, as well as loss of local identity. On the other hand, they represent a reserve of space of great potential in central urban locations. The main purpose of this research is to examine possibilities for improvement by their reuse, while preserving built-in cultural values and acknowledging contemporary requirements. A review of contemporary literature considering the concept of brownfield sites provides a starting theoretical basis for understanding their strengths and potentials, as well as the problems when redeveloping such sites. The subject of the research is exploring strategies for brownfield revitalization while reactivating industrial buildings through adaptive reuse. This includes sustainable solutions in accordance with modern requirements, especially energy efficiency, as one of the main concepts of existing building stock improvement that recognizes importance of responsible energy resources management. The paper includes a case study of the previously devastated brownfield site of "Beko" industrial building, located in the central urban area of Belgrade. Its former state, as well as parts of the documentation for reconstruction and its conversion into a modern business facility "Kalemegdan Business Center," is thoroughly analyzed, emphasizing the positive results of energy efficiency improvements despite the restrictions intended for historic buildings alterations. The aim of the paper is to create a theoretical platform that provides firm arguments in favor of realizing the importance and potentials of industrial brownfield sites revitalization at present, as well as the constraints regarding its practical implementation considering buildings of cultural value.

J. Pavlović (✉) · A. Šabanović · N. Ćuković-Ignjatović
University of Belgrade, Belgrade, Serbia
e-mail: j.pavlovic.phd@edu.arh.bg.ac.rs

A. Šabanović
e-mail: ana.sabanovic@arh.bg.ac.rs

N. Ćuković-Ignjatović
e-mail: natasa@arh.bg.ac.rs

E. Arbizzani et al. (eds.), *Technological Imagination in the Green and Digital Transition*,
The Urban Book Series, https://doi.org/10.1007/978-3-031-29515-7_73

Keywords Sustainability · Brownfield · Industrial heritage · Adaptive reuse · Energy efficiency

73.1 Introduction

After a global shift of paradigm in the industrial processes, once successful companies established during the period of industrialization experienced downfall lacking the resilience to change. It gradually led to the loss of their active function and formation of industrial brownfield sites. With some of the vital functions of the city diminishing, urban decline ensued, deepening the loss of local identity. The very motive for writing this paper is exploring possibilities for reviving industrial brownfield locations by improvements regarding contemporary demands, primarily respecting and preserving their built-in cultural values.

Reinstating an active function of an industrial brownfield is facilitated through the concept of adaptive reuse. Theoretical basis regarding brownfield characteristics, classification, and inventory mainly considers contemporary authors within Serbia and the neighboring region, with respect to the regional specifics on the matter in comparison to developed EU countries.

An overview is given of contemporary strategies for buildings' improvement that is necessary for contemporary purposes, emphasizing energy efficiency as the first and foremost requirement for compliance, with essential legislative framework presented.

The case study presented serves as an example of a building conversion, formerly used for an industry function. After losing its original purpose, it has become part of an industrial brownfield, deteriorating rapidly despite a favorable location within the central region of Belgrade. A previous state analysis and evaluation of industrial brownfield characteristics according to a referent model are included, as well as an assessment of planned and executed works on improving the energy efficiency during the process of adaptive reuse carried out in order to revive the brownfield site.

73.2 Adaptive Reuse and Revitalization of Brownfield Sites of Industrial Heritage

Even regardless of a building's historical significance, treating the existing unused building stock as a resource provides substantial benefits with retaining the built-in materials and embodied energy, thus reducing raw material use, pollution, and waste.

On a larger scale, utilizing inactive central urban capacities contributes to the compactness of the city, optimizing infrastructure and population density, resulting in considerable energy savings, and preventing the urban sprawl at the expense of natural resources.

Adaptive reuse, a contemporary concept of protecting the architectural heritage, represents improvement of the built environment through adaptation for new purposes and requirements, while preserving the identity and intrinsic collective memory. It gains foothold in international conventions regarding industrial heritage, such as the Dublin Principles (International Council on Monuments and Sites (ICOMOS) 2011), and architectural heritage, as in the Leeuwarden Declaration (Architects' Council of Europe (ACE) 2018). It is important for heritage preservation, sustainable urban development, and growing environmental concerns.

In the most developed EU countries, the reuse of urban areas is a priority of spatial development strategies, meeting economic, environmental, and social conditions—the three basic principles of sustainable development. (Danilović et al. 2008) Even though the issue of brownfield sites revitalization has been present globally since 1980s, within the post-socialist countries, it has been considered only since the 2000, due to their transition causing delay (Đukić et al. 2014).

In relation to greenfield investments, brownfield revitalization is a complex process with uncertainties, additional risks, and costs, involving significant private investment and public intervention, while in some cases it is not even possible. However, investing in brownfield sites is a better long-term investment due to increasing economic values of the land in question, less investment in infrastructure, growing employment, and overall activities on the local level. Brownfield sites represent an important reserve of urban space which, if strategically designed in accordance with the actual development capacities, can significantly improve the characteristics of an entire urban area. They are more of an opportunity and a challenge than a problem for the local context.

The ecological load of the brownfield site implies that its decontamination is necessary for the reuse or conversion, unless the pollution is resolved naturally. The need for investments in consolidation of a site can significantly affect its market price, even to its negative value. Liability for environmental pollution can be determined and charged according to the "polluter pays" rule. Since the Republic of Serbia is the inheritor of formerly state-owned companies, a more significant government-funded realization of such consolidations is necessary.

The key problems are stereotypes of unprofitability of investing in brownfield sites, neglect of their spatial potentials as an "inherited burden," lack of development strategies and declarative commitment without practical engagement. Investment models that are supporting greenfield investments and excessive urbanization, with perception through the prism of economic aspects, result in their suburbanization. (Đukić et al. 2014).

73.3 An Overview of Strategies for Buildings' Improvement

Contemporary solutions that can be used to improve existing structures are the same as in new buildings construction, but the implementation is as difficult as the existing conditions and/or conservation requirements are challenging. As shown in Table 73.2, evaluation of the current state is the first step for any strategy, in order to determine the factual state as accurately as possible before selecting applicable measures.

Energy efficiency is one of the basic contemporary requirements, with European legislation on the matter being transmitted into national strategies and regulations of the Republic of Serbia, as part of the Accession to the European Union. The EU Directive 2018/844 (Directive (Eu) 2018) amending Directive 2010/31/EU on the energy performance of buildings and Directive 2012/27/EU on energy efficiency thus were transferred, respectively, into the Law on Planning and Construction (Gazette office of RS 2021), as well as the Rulebook on Energy Efficiency of Buildings (Gazette office of RS 2011), and Rulebook on the Conditions, Content, and Manner of Issuance of Certificates of Energy Performance of Buildings (Gazette office of RS 2018). The European Standard for Conservation of cultural heritage—Guidelines for improving the energy performance of historic buildings (Standard 2016) give further directions regarding acknowledged heritage, although serving only as a recommendation.

The most immediate energy performance upgrades are the enhancements of the envelope with the aim of reducing energy consumption required for heating and cooling. External walls' thermal characteristics are directly advanced by adding thermal insulation. Integrated and interactive façade systems can prevent overheating during the summer and heat loss during the winter but are disputable regarding heritage buildings.

Improving the roof means adding waterproofing and thermal insulation and can include "green roofs." The structures on the ground are amended by performing proper drainage, waterproofing, and thermal insulation. By repairing or replacing doors and windows, their thermal properties can be increased. In the case of buildings with strict conditions for protection of façade elements, these measures are applied accordingly.

Reducing the need for heating, cooling, and ventilation is achieved primarily through the use of passive solar systems, such as: greenhouse, Trombe's wall, or double façade.

The advantages of the latter are reduced energy consumption, comfort, and esthetics, as it is most often used as a design solution when renovating the façades of protected buildings. Passive natural ventilation systems are achieved through appropriate positioning of openings or introducing solutions such as the Venturi effect. When it comes to retrofit, the design of natural lighting is an important component, because most conversions involve changing the elements that affect the entry of daylight, and analysis of daylight is included in all relevant performance evaluations. Further reducing the demand for electricity is accomplished by modifying electrical installations and equipment according to new purposes and requirements of efficiency.

Due to high share in carbon dioxide emissions, the decarbonization of the building stock in the EU will be mandatory by 2050 according to the Directive 2018/844 (Directive (Eu) 2018).

This implies ever stricter requirements for the building sector and would not be possible without the use of renewable resources that should therefore be promoted in Serbia.

All things considered, one should keep in mind the possible integration of individual solutions, which decreases the amount of space and substructure required. Adapting all applied systems after the final phase of execution of works enables fine-tuning and optimization of the whole instead of maximizing the individual elements.

73.4 Case Study of "Beko"—Kalemegdan Business Center

The case study discloses data on the previously abandoned textile industry facility on Vojvode Bojovića Boulevard in Belgrade, as well as the phase of its conversion into the Kalemegdan Business Center office building, which was completed in 2019 with a comprehensive adaptation of the assembly.

73.4.1 Previous State of Industrial Brownfield

The "Beko" (Fig. 73.1) was built in 1931 and is especially important because of its location in the very heart of the city in Dorćol, within the Belgrade Fortress assemblage, which was declared a cultural monument in 1965 and determined to be immovable cultural heritage of exceptional importance for the Republic of Serbia in 1979. It has lost its active function in 2002 and gradually turned into ruin, despite its advantageous position.

The technical and environmental characteristics of the brownfield location of the Belgrade clothing brand "Beko" building were originally analyzed in the research by Krstić-Furundžić et al. (Krstić-Furundžić et al. 2014) among eleven other industrial brownfield locations in Belgrade. An evaluation was performed according to a selected model, which used technical criteria and brownfield factors to identify properties and priorities significant for their treatment, in support to creating a central register for brownfield refurbishment.

The extracted data for the location in question are listed in Table 73.1.

The data show that "Beko" brownfield site was characterized by moderate environmental impact hazards and having substantial economic potential that would be fulfilled through its adaptive reuse. This would result in increased values, improving

Fig. 73.1 "Beko" building before reconstruction: **a** street view (*Author* Đorđe Kojadinović, RAS Srbija), **b–d** interior and structure. (*Source* Remorker Architects)

Table 73.1 Characteristics of the brownfield location "Beko" (*adapted from* Krstić-Furundžić et al. (Krstić-Furundžić et al. 2014))

Technical criteria	Toxicity of contaminated sites	Concentration in traces, toxicity below 100, bioactive below 500
	Site characteristics	Large volume of waste, low concentration of settlements, existing structure not maintained and cannot be maintained, possible displacement
	Human exposure	High possibility of human exposure—residential neighborhood
	Environmental exposure	Possible exposure of protected habitats
Brownfield factors	Potential for site restoration with benefits	High possibility of renewal
	Possibility of creating public benefit	Creation of amenities at locality required by the local community
	New jobs openings	20–50 new job openings
	Increase in tax revenues	Moderate increase based on mean increase in value or job openings
	Location within or near a poor neighborhood	Site far from underprivileged community

the quality of life for the local community, enhancing both social and environmental sustainability.

73.4.2 Improving Energy Efficiency Within Adaptive Reuse

According to the Technical Description of the Architectural Design for building permit (Kopring 2018a), the evaluation of the current state executed by the Faculty

Table 73.2 Observed building improvement strategies

	Improvement strategies	Applied measures
1	Evaluation	Technical expertise
2	Conservation measures	Issued and respected
3	Envelope enhancements	
3.1	External walls	Façade reconstruction, thermal insulation added
3.2	Roof cover	Flat roof, thermal and hydro insulation added
3.3	Ground structures	Thermal, hydro insulation
3.4	Doors and windows	All new, Al, 3-layer glass
4	Passive systems	
4.1	Passive solar systems	No specific system
4.2	Passive natural ventilation	No specific system
5	Electrical equipment	All new installations
6	HVAC	All new, standard
7	Renewable energy sources	No

of Civil Engineering showed that the six floors high structure was practically in a derelict condition, with no roof construction, roof cover, doors, windows, nor internal installation. Respecting former methods of building, the façade was left without any thermal insulation, and the final layer with façade plastic was in very poor condition. The structure expertise pointed out the need for improvement of the load bearing elements, so the design officially included the structural rehabilitation of the building.

Conditions for conservation, maintenance, and use of cultural heritage and heritage that has status of preliminary protection and protection measures for the preparation of the Detailed Regulation Plan (No. 39/10, dated 09.11.2009) were issued by the Institute for the Protection of Cultural Monuments of Serbia, which provided opportunities and conditions for the transformation that the authors respected throughout the design.

All floors have been completely adapted, in terms of function and construction, and instead of an attic, an additional floor has been formed, as shown in Fig. 73.2.

The Energy Efficiency Study (Kopring 2018b) included detailed calculations for the structure, renewed in a combined structural system—reinforced concrete columns and beams, solid façade walls made of bricks of the old format, massive reinforced concrete finned ceilings. Regarding the improvement of the envelope, in accordance with received requirements for conservation that were not restrictive in terms of retaining the existing façade surface, the design included its complete reconstruction with the addition of stone mineral wool thermal insulation layer, plastering and fabrication of façade elements according to the original appearance. The new façade mimics the look of the original but is made in a modern system. The roofs were planned as flat, laying on reinforced concrete construction, with vapor barrier, rock wool thermal insulation, waterproofing membrane, and mechanical protection. The ground floor was also repaired by performing waterproofing, placing XPS boards for

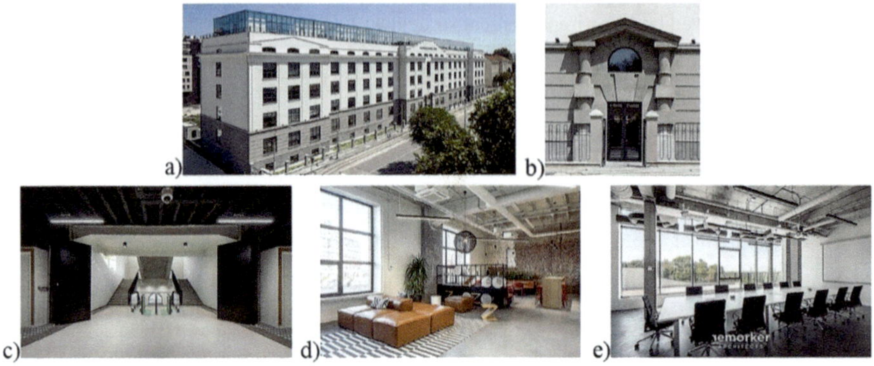

Fig. 73.2 "Kalemegdan Business Center" (ex "Beko") after adaptation: **a** exterior, **b** main entrance, **c–e** interior. (*Source* https://www.gradnja.rs/beko-kalemegdan-business-center-remorker/)

thermal insulation, cement screed, and final floor coverings. Doors and windows of aluminum profiles with thermal break and three-layer low-emission glass package were installed. The document concluded that all individual assemblies of the new thermal envelope met the conditions of the maximum allowable heat transfer and water vapor diffusion.

Existing geometry and position of the building allows for proper insolation, as well as natural ventilation, so no new passive systems were deliberately introduced. Natural lighting was maximized through open space design concept in addition to organizational flexibility. All new electrical and mechanical equipment were installed, ensuring modern-day efficiency. The heating is supplied via district heating system, and renewable sources of energy were not proposed, even though a more substantial utilization of these strategies could have contributed to an even greater advancement.

As the result of improvements, current design puts the building into Energy Performance Class "C," according to national regulations (Fig. 73.3).

73.5 Conclusions

Despite the less strict attitude toward conservation today and existing standards for the application of energy efficiency, industrial brownfield heritage buildings will hardly meet contemporary requirements in terms of environmental certifications that are constantly being improved. Instead of further tightening the rules for values to be achieved, it is more urgent to insist on a greater degree of existing buildings' improvements, through various renovation programs, and with an emphasis on public buildings as representatives of the concept.

The analysis of the "Kalemegdan Business Center" case study undoubtedly confirmed that, despite additional restrictions in the field of architectural heritage

ENERGIJA POTREBNA ZA GREJANJE

TRANSMISIONI GUBICI		**Qt = 466809.01 kWh**
VENTILACIONI GUBICI		**Qv = 753320.53 kWh**
SOLARNI DOBICI	(koristi se)	**Qsol = 155474.2 kWh**
DOBICI OD LJUDI	(koristi se)	**Qp = 80738.62 kWh**
DOBICI OD EL.UREDJAJA	(koristi se)	**Qel = 184334.74 kWh**
ENERGIJA POTREBNA ZA GREJANJE (razlika izmedju gubitaka i dobitaka)		**Qh,nd = 799581.94 kWh**
Energija potrebna za grejanje po m²		**Qh,an = 41.59 kWh/m²a**

Energetski razred

Za usvajanje energetskog razreda koristi se specifična godišnja energija potrebna za grejanje za sisteme koji rade bez prekida

En. razred	Qh.rel = 64.0 %	Qh = 41.59 kWh/m²
A+	<=15	<=10
A	<=25	<=17
B	<=50	<=33
C	**<=100**	**<=65**
D	<=150	<=98
E	<=200	<=130
F	<=250	<=163
G	>250	>163

Na osnovu energije potrebne za grejanje po m², objekat spada u C energetski razred

Energent	**Daljinsko grejanje na fosilna goriva**
Faktor pretvaranja	**1.8**
Primarna energija	**2187946.75 kWh**
Emisija CO_2	**822945.70 kg CO2**

Fig. 73.3 Excerpt from the energy efficiency study—energy performance class (EPC) evaluation (Kopring 2018b)

protection, the potentials for improving the built environment through the revitalization of industrial brownfields by adaptive reuse are substantial. Furthermore, the application of measures for improving energy efficiency realized in this particular case met the set requirements for new construction in Serbia of a minimal "C" Energy Performance Class. This is still far from the coveted concept of energy-neutral buildings but represents a significant achievement since it is not mandatory for buildings of cultural heritage.

Acknowledgements This paper is a part of the research funded by the Ministry of Education, Science and Technological Development of the Republic of Serbia, grant number 451-03-68/2020-14/200090. The research is done under research lab SaRA (Sustainable and Resilient Architecture), University of Belgrade—Faculty of Architecture.

References

Architects' Council of Europe (ACE) (2018) Leeuwarden declaration: adaptive re-use of the built heritage: preserving and enhancing the values of our built heritage for future generations. In: Conference on the adaptive re-use and transition of built heritage. Leeuwarden, Netherlands

Đukić A, Simonović D, Vujičić T (eds) (2014). University of Banja Luka, Faculty of Architecture, Banja Luka

Krstić-Furundžić A, Grbić M, Grahovac D (2014) Identification and evaluation of brownfield sites in the wider area of the city of Belgrade. In: Đukić A, Simonović D, Vujičić T (eds) BrownInfo: toward a methodological framework for brownfield database development. University of Banja Luka, Faculty of Architecture, Banja Luka, pp 315–333

Danilović K, Stojkov B, Zeković S, Gligorijević Ž, Damjanović D (ur.) (2008) Oživljavanje braunfilda u Srbiji—Priručnik za donosioce odluka i profesionalce. PALGO centar, Beograd. (In Serbian)

European Union (2018) Directive (Eu) 2018/844 of the European parliament and of the council of 30 May 2018 amending Directive 2010/31/EU on the energy performance of buildings and Directive 2012/27/EU on energy efficiency, Official Journal of the European Union (19.6.2018, L 156/75-91)

European Standard (2016) FprEN 16883:2016 E, Conservation of cultural heritage—guidelines for improving the energy performance of historic buildings, ICS 91.120.10; 97.195 (equivalent to СРПС ЕН 16883)

International Council on Monuments and Sites (ICOMOS) (2011) The Dublin principles: joint ICOMOS—TICCIH principles for the conservation of industrial heritage sites, structures, areas and landscapes. 17th ICOMOS General Assembly, Paris

Mašinoprojekt Kopring (2018a) Excerpt from the architectural design for the building permit, technical description (p 9–17). Mašinoprojekt Kopring, Beograd. (In Serbian)

Mašinoprojekt Kopring (2018b) Excerpt from the architectural design for the building permit, energy efficiency study. (p 5–29), Mašinoprojekt Kopring, Beograd. (In Serbian)

Official Gazette of RS (2011) Rulebook on energy efficiency of buildings. Official Gazette of RS, No. 61/2011. Official Gazette of RS, Belgrade. (In Serbian)

Official Gazette of RS (2018) Rulebook on the conditions, content, and manner of issuance of certificates of energy performance of buildings. Official Gazette of RS, No. 69/2012 and 44/2018. Official Gazette of RS, Belgrade. (In Serbian)

Official Gazette of RS (2021) Law on planning and construction. Official Gazette of RS, No. 72/2009, 81/2009., 64/2010, 24/2011, 121/2012, 42/2013, 50/2013, 98/2013, 132/2014, 145/2014, 83/2018, 31/2019, 37/2019, 9/2020 and 52/2021. Official Gazette of RS, Belgrade. (In Serbian)

Chapter 74
Industrial Heritage of Belgrade: Brownfield Sites Revitalization Status, Potentials and Opportunities Missed

Jelena Pavlović, Ana Šabanović, and Nataša Ćuković-Ignjatović

Abstract Being created during the period of intensive industrialization, industrial buildings and landscapes carry importance as birth places of rapid technological progress, social and economic changes, which has established their great significance for modern human history and identity. Termination of their active function causes symptoms of decline to appear gradually, and their number decreases as the time passes. They are often endangered regardless of their protection status. Implementation of adaptive reuse principles allows for less strict approach to conservation practice, and its benefits are demonstrated worldwide. Despite that fact, industrial heritage has not been revitalized enough through adequately treated sites in Serbia. That indicates lack of understanding of the value of this cultural and historical heritage, as well as its suitable future purposes, impossible without some form of active dialog between participants in the planning process. The legal preconditions for this collaboration exist in Serbian regulations, and they are examined in the paper, but other potential causes of the lack of consensus that result in failure are also explored. In Belgrade, industrial brownfields occupy attractive locations, often targeted for market-driven redevelopment. For that reason, the paper explores current practice of revitalizing brownfield sites of industrial heritage in Belgrade. It considers the achievements, probable missed opportunities, and remaining potentials where acquired knowledge can be utilized. Key results of the research define critical points in the planning process for the preservation of values despite the modernity of brownfield sites transformations. The purpose of this paper is to help safeguard industrial landscapes of Belgrade and Serbia, and their sustainable conversion, the most adequate for the present moment, as well as to contribute to urban reconstruction of declining landscapes to which these brownfield sites belong.

J. Pavlović (✉) · A. Šabanović · N. Ćuković-Ignjatović
University of Belgrade, Belgrade, Serbia
e-mail: j.pavlovic.phd@edu.arh.bg.ac.rs

A. Šabanović
e-mail: ana.sabanovic@arh.bg.ac.rs

N. Ćuković-Ignjatović
e-mail: natasa@arh.bg.ac.rs

E. Arbizzani et al. (eds.), *Technological Imagination in the Green and Digital Transition*, The Urban Book Series, https://doi.org/10.1007/978-3-031-29515-7_74

Keywords Sustainability · Industrial heritage · Brownfield · Revitalization

74.1 Introduction

The historical phenomenon that began with the industrial revolution has affected the entire world and its consequences continue to this day. It represents a period of accelerated social and economic changes, mechanization of production, and the transition from fragmented manufacturing to purpose-built factories with specialized production. The interconnectedness of the development of cities and industry is unquestionable and complementary. However, the restructuring of the economy and the change in the production processes triggered the industrial decline. In Serbia, overall conditions of socioeconomic transition and the failed processes of privatization of publicly owned enterprises caused the withdrawal of industrial production, thus creating deteriorating industrial brownfields. Numerous examples are situated in today's urban areas, having lost their previous purpose, or adequacy for contemporary requirements. They have great potential for improving the local context in social, economic, and environmental terms through interventions in the form of adaptive reuse and are therefore important for sustainable urban development. Supporting industrial landscapes and their conversion offers significant opportunities for the urban context and preserving cultural identity. Enhancing the vitality of the sites enables urban reconstruction of declining landscapes to which these brownfield sites belong.

Furthermore, it supports the responsible attitude toward resources, including energy and spatial resources, which represents an important aspect of climate change mitigation concept.

The paper is considering characteristics of industrial brownfields and the revitalization process. Due to the particularities, it requires responsible management that includes a participative planning process, clearly defined responsibilities, and targeted investments for each individual case. Various data on industrial heritage sites in Belgrade are analyzed, providing a multi-layered image and identifying potentials for intervention, both in legislation and in practice.

74.1.1 Industrial Architecture

Even the earliest examples of industrial buildings indicate high quality of elements with the design following the functional aspects. Improvement of building materials led to the discovery of newer structural systems that bridged ever-increasing spans and enabled the creation of new spatial effects. Although depending on the function, the artistic aspect cannot be a neglected part of industrial architectural design since it acts psychologically by stimulating esthetic feelings or causing resistance to its evaluation. (Mirković 1964).

Extensive studies have been conducted on industrial heritage buildings of Serbia built before the Second World War, while even significant buildings built afterward are less often subject of research. This is mostly due to them being a priori excluded (Kadijević 2012), or insufficient time has passed for them to be evaluated. In contrast, examples of good practice in neighboring European countries show the introduction of legal protection for historic industrial buildings as soon as signs of decline in their function are noticed.

The role of public participation of the local community is important in that because timely reaction and establishing financial resources can prevent deterioration of industrial sites where the loss of the original purpose is recorded. At the same time, this increases the chances of a higher success rate of reactivation success, which directly corresponds to the fulfillment of the principles of sustainability.

Architectural transformation involving improvements intended for contemporary purposes and requirements, that preserve the cultural feats of built heritage, listed or not, represents the concept of adaptive reuse. Due to its proven potentials, it is recognized within the heritage charters regarding industrial architecture, such as the Dublin Principles (International Council on Monuments and Sites (ICOMOS) 2011), as well as international architectural heritage conventions, such as the Leeuwarden Declaration (Architects' Council of Europe (ACE) 2018). It contributes to heritage preservation, urban sustainability as well as the environment.

74.2 Brownfield Locations and Their Revitalization

Brownfield revival addresses numerous problems regarding the environment. Most importantly, it supports rational use of the finite green areas, as well as decontamination—which is implied in the process of industrial brownfield redevelopment. It also offers significant contribution to socioeconomic improvements, and cultural identity preservation.

Emphasizing the sustainable use of natural and cultural heritage and natural resources, the "Spatial Development Strategy of the Republic of Serbia 2009–2013—2020" (Ministarstvo životne sredine i prostornog planiranja 2009) was adopted in 2009. It recognized the importance of utilizing brownfields and set brownfield recycling as one of fourteen key strategic priorities to help urban renewal and solving numerous economic, social, and environmental problems, preventing excessive urbanization and conversion of fertile agricultural land for construction purposes.

In 2019, a new National Strategy of Sustainable Urban Development of the Republic of Serbia until 2030 (Official Gazette of RS 2019) was adopted, in accordance with the New Urban Agenda (United Nations 2017), with accent on the Goal 11—"Make cities and human settlements inclusive, safe, resilient and sustainable," from the 17 Sustainable Development Goals (United Nations 2015). Recycling the existing architectural heritage by revitalizing brownfield sites represents a vast potential for lessening the stress put on greenfield construction, ultimately attenuating the effects of climate change. The National Strategy served to create a legislative framework for the utilization of existing urban assets with the aid of stakeholders' participation.

74.2.1 Jurisdiction and Funding

An integrated approach to brownfield reactivation brings together all its potentials. Successful regeneration implies active cooperation with a focus on the roles, responsibilities, and limitations of the institutional instruments. Horizontal collaboration represents the cooperation of different sectors, disciplines, and institutions at the same organizational level, while vertical collaboration is between the national, regional, and local levels. (Perić and Furundžić 2014).

Greatest responsibility for the treatment of industrial brownfields in the Republic of Serbia lies with the municipalities in which they are located given their role in urban planning and raising local awareness of the importance of brownfield revitalization. On the other hand, the fact that fiscal and legislative instruments for consolidating brownfield sites are still regulated on the national level can prevent greater implementation.

Insufficient data on brownfield sites, private ownership, or private financial institutions controlling the necessary funds also pose challenges, according to Danilović et al. (2008) Funding is provided mostly from private sources, while local authorities have a wide range of formal and informal incentive instruments—formal instruments being creating strategic priorities, directing investment programs and marketing, and informal such as assistance in pollution detection, land remediation, and consolidation.

The instrument of private–public partnership, which can range from non-profit cooperation to co-financing of the project, is irreplaceable in the case of brownfield site redevelopment in order to balance private and public interest. The basic principles of the collaboration are clear common goals, transparency and public involvement, coordination of processes and properly distributed risks. Since it is vital to gain support of the public that will be affected by the planned revitalization, involving the public in the participatory decision-making process from the very beginning of the project is advised.

74.2.2 Brownfield Types and Their Inventory

Preventing further deterioration of brownfield sites by timely revitalization and increasing revenues is of the utmost importance. In order to be well informed about the economic potential in a particular area, the brownfield sites must be properly assessed.

According to Danilović et al. (2008), 5 types of brownfields can be distinguished according to the exclusivity of the location, investment needs and opportunities for return on investment, as well as contamination. The first type is characterized by excellent location and best effects of private investment. The second and third require greater involvement of public funds due to less attractive locations. The fourth type is characterized by pollution that needs to be remediated before the revitalization can take place. The final, fifth type, does not have any market potential, and in most cases will be demolished and restored to construction land or green areas. A unified manner, used to determine accurate data on abandoned and insufficiently utilized land, would enable a reasoned assessment of the problems and priorities of rehabilitation, as well as comparability of values in different municipalities.

Shortcomings of current databases for brownfields represent practical obstacles for accomplishing such reconstructions. Thus, a methodological framework for creating a register of brownfield sites is necessary for effective urban planning. (Đukić et al. 2014).

74.3 Industrial Brownfield Heritage Location of Belgrade

Data on Belgrade brownfield sites and industrial heritage buildings in this research have been extracted from the databases of the Museum of Science and Technology, primarily the Department for Industrial Heritage, as well as the publication "Industrial Heritage of Belgrade" by the former director of the museum, Kulenović (2009). Information obtained then was updated by researching the cadaster database and conducting field research for selected locations. The research findings are presented in the form of summarized data in Table 74.1, data distribution (Fig. 74.1), and selected features' percentages (Fig. 74.2). A map of Belgrade is provided with positioning of sites as well as their classification according to characteristics of significance for the analysis (Fig. 74.3). The limitation of the research is the lack of updated data on certain buildings, such as structures involved in litigation.

The majority of buildings or their parts are under some level of formal protection, but the fact that Belgrade is the capital city must be considered, having regulatory systems that are generally higher than the national average. The largest number of facilities is in public ownership, with a declining trend due to numerous privatizations and the limited number of brownfields. (Fig. 74.1).

Only about one fifth of the facilities are in adequate condition for their purposes, while the rest of them have an equal share of buildings in bad or very bad condition, regarded as unsafe and requiring urgent intervention, and in good condition, where safety is not questioned, but they require renovation or reconstruction. (Fig. 2a).

Buildings that are still in their primary industry, or out of function, are less in number than the repurposed buildings, while the demolishment is the least present (Fig. 2b).

Table 74.1 Industrial heritage brownfield sites in Belgrade

No	Name of building	Construction year	Main industry	Current function	Ownership status	Formal protection	State of building
01	Brewery I. Bajloni and Sons	1850–1929	F	H	PR	+	3
02	Bell foundry and tower clock factory	1854	M	N	PU	+	3
03	The main railway station Belgrade	1883–1885	TR	C	PU	+	4
04	Belgrade waterworks	1890–1992	W	U	PU	–	3
05	Meteorological observatory	1891	S	U	PU	+	3
06	Belgrade wool factory	1898–1923	TE	N	PR	+	1
07	Royal Serbian privileged sugar factory	1899–1901	F	C	PU	+	2
08	Thermal power plant for electrical companies	1899–1901	E	C	PU	+	3
09	Belgrade cotton factory	Around 1900	TE	N	PR	+	2
10	Steam bakery Soko	Around 1900	F	D	PR	–	0
11	Gliša Josipović wire factory	Around 1900	M	N	PR	+	2
12	Brickyard Polet	Around 1900	I	H	PR	–	1
13	Brickyard Record	Around 1900	I	H	PR	–	1
14	Brickyard Kozara	Around 1900	I	H	PR	–	1
15	Craft and art foundry Skulptura	Around 1900	M	O	PR	+	2
16	K. M. Shonda confectionery factory	1900–1910	F	O	PR	+	4
17	Mill Suvobor	1900–1920	F	U	PU	–	3
18	Mill of the first joint stock bakery association	1901–1902	F	H	PR	+	4
19	Turner and mill M. Simovic	1903	F	N	PR	–	1
20	Seismological institute	1906–1909	S	U	PU	+	4
21	Captaincy Belgrade	1906–1912	TR	U	PU	+	3

(continued)

Table 74.1 (continued)

No	Name of building	Construction year	Main industry	Current function	Ownership status	Formal protection	State of building
22	Captaincy Zemun	1908	TR	H	PU	+	3
23	Old telephone exchange building	1908–1911	CO	C	PU	+	3
24	Reversible water pumping station Boljevci	1910–1911	W	U	PU	–	2
25	Reversible water pumping station Jakovo	1910–1911	W	N	PU	–	2
26	Reversible water pumping station Fenek	1910–1911	W	U	PU	–	5
27	Factory of modern knitted products	1912–1914	TE	O	PR	–	5
28	Nikola Bošković bank warehouse	1920	I	O	PR	+	4
29	Milan Vape paper factory	1921–1924	P	O	PR	+	3
30	Power plant of the municipality of Belgrade	1922–1928	E	C	PU	+	3
31	Ball bearing industry-airplane factory	1923–1924	M	D	PR	+	0
32	Umčari railway station	1924	TR	U	PU	–	2
33	Belgrade railway station furnace	1926	TR	C	PU	+	1
34	Rakovica engine industry	1927–1928	M	N	PU	+	2
35	Institute for production of banknotes and coins	1927–1929	P	U	PU	–	4
36	Hangar of the old Belgrade airport	1927–1931	TR	O	PR	+	4
37	Reversible water pumping station Zidina	1929	W	N	PU	–	2
38	Astronomical observatory	1929–1932	S	U	PU	+	3
39	Mill Dimitrijevic	Around 1930	F	N	LI	–	2
40	Forge shop rad	1930–1931	M	O	PR	+	3
41	Thermal power plant power and light	1930–1932	E	C	PU	+	1
42	BeKo-Belgrade clothing	1931	TE	O	PR	+	3

(continued)

Table 74.1 (continued)

No	Name of building	Construction year	Main industry	Current function	Ownership status	Formal protection	State of building
43	Danube railway station	1931–1935	TR	N	PU	–	2
44	Rev. centrifugal pumping station for water Borča	1934	W	U	PU	–	5
45	Rev. centrifugal pumping station for water Ovča	1934	W	U	PU	–	3
46	Kosutnjak railway station	1934	TR	N	PU	+	2
47	Garage of court compositions Topčider	1934	TR	U	PU	+	2
48	Belgrade main post office	1935–1938	CO	U	PU	+	4
49	Mill Markovic	Around 1935	F	U	PR	–	3
50	Mill Falkenburger	Around 1935	F	U	PR	–	3
51	Mill S. And Z. Maric	Around 1936	F	U	PR	–	3
52	Topčider Dvorska railway station	1936	TR	N	PU	+	3
53	Warehouse of the Port of Belgrade	1936–1939	TR	H	PU	+	3
54	Belgrade publishing and graphic institute	1936–1940	P	O	PR	+	3
55	Airplane factory Ikarus	1938	TR	D	PR	–	3
56	Teleoptics	1938–1939	CO	N	PR	+	2

Main industry	Current function	Formal protection status
CO–Communication	D–Demolished	+ Existent
E–Energy production	N–Not functioning	– Non-existent
F–Food industry	U–Unchanged function	**State of the building**
I–Industrial warehouse	C–Culture	0–Ruined
M–Metal industry	H–Hospitality industry	1–Very bad
P–Paper industry	O–Office and commercial space	2–Bad
S–science institutions	**Ownership status**	3–good
TE–textile industry	PU–public property	4–very good
TR–traffic industry	PR–private property	5–exceptional
W–water supply	LI–in litigation	

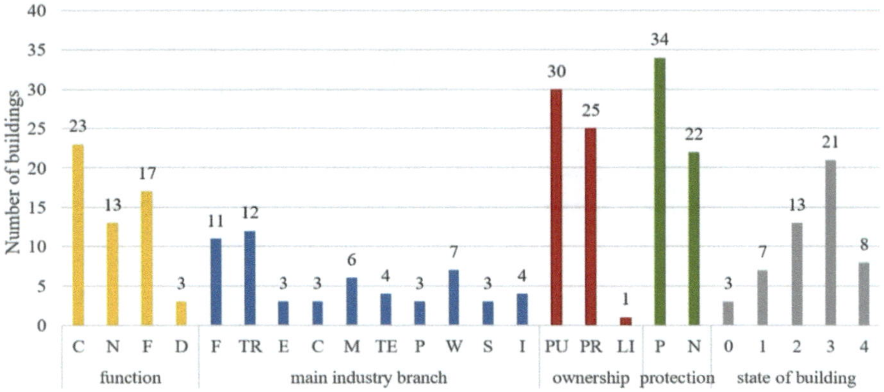

Fig. 74.1 Data distribution chart (abbreviations as per Table 74.1)

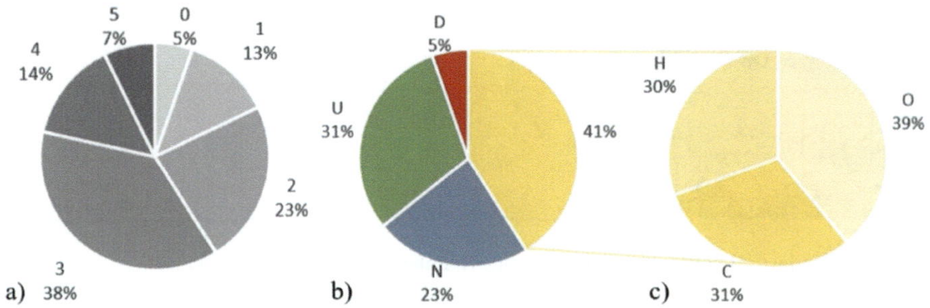

Fig. 74.2 Circular charts for selected characteristics' percentages—**a** state of buildings, **b** current function status, **c** repurposed buildings' function (abbreviations as per Table 74.1)

All buildings that were converted or demolished are in the central zone of the city (Fig. 74.3). Most of them are converted by market-oriented construction into residential, office or commercial space, and privately owned. A smaller part of the conversion was made for public purposes, primarily culture, and all these facilities are in public ownership (Fig. 2c). These buildings represent the few examples of good practice.

From the spatial disposition (Fig. 74.3), it can be observed that the contents occupying most attractive property (central city area) are mostly repurposed or in the phase of conversion. Few buildings are not in function and currently unconverted. These are publicly owned facilities that are not included in detailed regulatory plans that would define their possibility of conversion. The peripheral zone, on the other hand, has most sites out of function.

Regarding the demolished sites, field research revealed that new office and residential buildings were constructed in their place, having a much greater index of occupation.

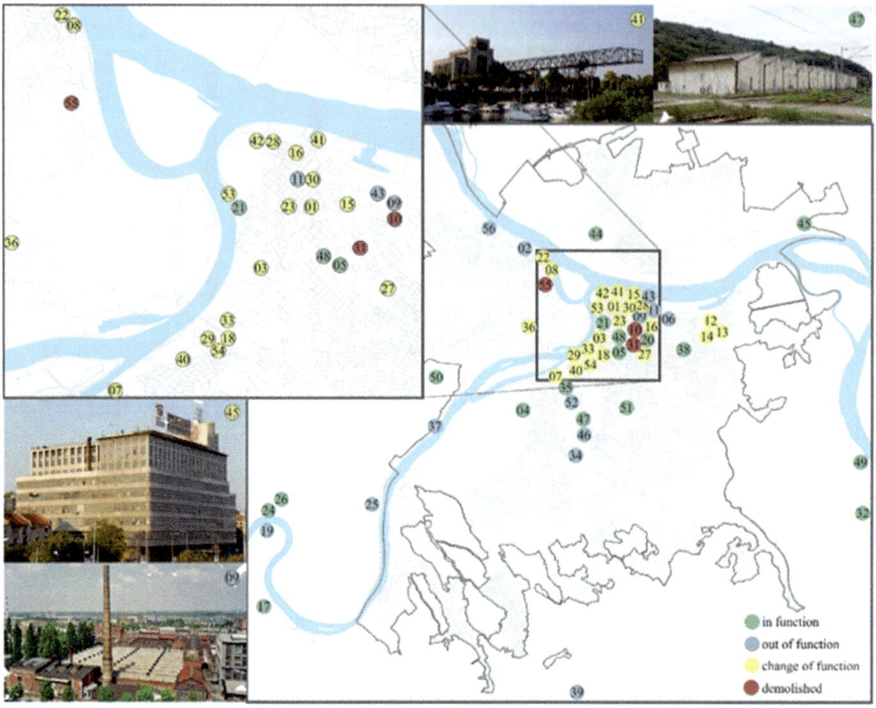

Fig. 74.3 Industrial heritage sites in Belgrade (numbering as per Table 74.1)

74.4 Conclusion

According to the data, most of the conversions of buildings were done for market demands. Only a few good practice examples applied the principles of adaptive reuse for the benefit of all stakeholders, so the opportunity for its significant gains and promotion of the concept were notably missed.

It can be concluded that the conversion has not consistently followed the plans, directives, and strategies that emphasize the significance of public functions and participation. The redevelopment of industrial brownfields in Belgrade is mostly done with the aim of maximizing revenues and profits, not paying attention to the importance of continuity of identity and cultural diversity in given locations, thus allowing damage to existing social structures, which is contrary to the strategies.

The partially incomplete data of the researched relevant sources indicate a discrepancy between plans and on-site circumstances, marking the necessity for a centralized system for planning, and monitoring the condition of industrial heritage. It indicates undefined competency for the application of existing knowledge in the field of industrial brownfields revival, lack of implementing strategies and plans defined, poor cooperation between institutions on different levels and stages of the process. A great responsibility lies with the local governments to continue working on the data collection and synchronization of databases and strategies for reviving the sites in question carrying urban renewal.

At the same time, a potential for further research is highlighted, both thematical and practical, because there is only a finite number of brownfields of industrial heritage that is only decreasing, either by devastation or inadequate conversions that do not meet the full potential of said heritage.

A more substantial implementation of industrial brownfield revitalizations, with respect to sustainable development and growing environmental concerns apart from economical features, would bring benefits to the overall struggle against the consequences of the climate change that were aided by the process of industrialization.

Acknowledgements This paper is a part of the research funded by the Ministry of Education, Science and Technological Development of the Republic of Serbia, grant number 451-03-68/2020-14/200090. The research is done under research lab SaRA (Sustainable and Resilient Architecture), University of Belgrade—Faculty of Architecture.

References

Architects' Council of Europe (ACE) (2018) Leeuwarden declaration: adaptive re-use of the built heritage: preserving and enhancing the values of our built heritage for future generations. In: Conference on the adaptive re-use and transition of built heritage. Leeuwarden

Đukić A, Simonović D, Vujičić T (eds) (2014). University of Banja Luka, Faculty of Architecture, Banja Luka

Danilović K, Stojkov B, Zeković S, Gligorijević Ž, Damjanović D (eds) (2008) Oživljavanje braunfilda u Srbiji—Priručnik za donosioce odluka i profesionalce, PALGO centar, Beograd. (In Serbian)

International Council on Monuments and Sites (ICOMOS) (2011) The Dublin Principles: Joint ICOMOS—TICCIH principles for the conservation of industrial heritage sites, structures, areas and landscapes. 17th ICOMOS General Assembly, Paris

Kadijević A (2012) Industrijska arhitektura Beograda i Srbije: Problemi istraživanja i tumačenja. In: Spasić M (ed) Godišnjak rada Beograda, knjiga LIX. Muzej grada Beograda, Beograd. (In Serbian) (pp. 11–36).

Kulenović R (2009) Industrijsko nasleđe Beograda. Muzej nauke i tehnike, Beograd. (In Serbian)

Ministarstvo životne sredine i prostornog planiranja (2009) Strategija prostornog razvoja Republike Srbije 2009–2013–2020, Republička agencija za prostorno planiranje i Ministarstvo životne sredine i prostornog planiranja, Beograd. (In Serbian)

Mirković K (1964) Industrija—prostorni modulator. In: Minić O (ed) Arhitektura urbanizam, br 26. Savez arhitekata Jugoslavije i Urbanistički savez Jugoslavije, Beograd. (In Serbian), pp 27–29

Official Gazette of RS (2019) Strategy of sustainable urban development of the Republic of Serbia until 2030. Official Gazette of RS, No. 47/2019. Official Gazette of RS, Belgrade. (In Serbian)

Perić A, Furundžić D (2014) institutional framework of brownfield regeneration in Serbia. In: Schrenk M, Popovich V, Zeile P, Elisei P (eds) Proceedings REAL CORP 2014 Tagungsband, (pp 303–307), Vienna, Austria. http://www.corp.at, ISBN: 978-3-9503110-6-8 (CD-ROM); ISBN: 978-3-9503110-7-5

United Nations (2015) Transforming our world: the 2030 agenda for sustainable development, A/RES/70/1. https://sdgs.un.org/2030agenda
United Nations (2017) New Urban Agenda. Habitat III Secretariat, ISBN: 978-92-1-132731-1

Chapter 75
Challenges and Potentials of Green Roof Retrofit: A Case Study

Nikola Miletić, Bojana Zeković, Nataša Ćuković Ignjatović, and Dušan Ignjatović

Abstract Green roofs are becoming common practice in building new public buildings and are considered the roofs for the future since they address the issue of energy and environment simultaneously, providing social, environmental and economic benefits. Despite these benefits, retrofitting an existing building with a green roof is not widely practiced. Undergoing such a project is no small task since it requires a thorough investigation of existing building's constraints, functional, material, and technological to even begin considering design options. Therefore, this process requires specific, case-sensitive approach, especially with the aim of improving the building's energy performance. This paper presents a methodological approach and design proposals of a green roof retrofit project, through a case study of Belgrade's "City Housing" building. This retrofit project presents an interesting research topic since it incorporates three distinct roofs, of all of different types, different ways of accessibility and levels of privacy, varying top-to-bottom from a simple extensive roof through a semi-public semi/intensive roof garden to a ground-level public park with trees and intensive vegetation. Also, since this building provides socially significant services, it is frequently visited by general public which presents a potential for introducing educational and demonstration elements in the retrofit project, not only the functional and technological ones. That way, this project can be a showcase example, promoting greening the roofs of Belgrade's existing public buildings as a way of improving their energy performance.

Keywords Green roof · Building retrofit · Energy efficiency

N. Miletić (✉) · B. Zeković · N. Ć. Ignjatović · D. Ignjatović
University of Belgrade – Faculty of Architecture, Belgrade, Serbia
e-mail: nikola.miletic@arh.bg.ac.rs

B. Zeković
e-mail: bojana@arh.bg.ac.rs

N. Ć. Ignjatović
e-mail: natasa@arh.bg.ac.rs

D. Ignjatović
e-mail: ignjatovic.dusan@arh.bg.ac.rs

E. Arbizzani et al. (eds.), *Technological Imagination in the Green and Digital Transition*, The Urban Book Series, https://doi.org/10.1007/978-3-031-29515-7_75

75.1 Introduction

Energy refurbishment includes wide spectrum of actions among which improving thermal properties of building envelope takes the central spot. Applying green roof systems represents one way of improving energy properties of the roof since there are systems which are light in weight thus allowing for adding enough thermal insulation on the existing roof construction. While the thermal insulation affects thermal properties of the roof, the outer layer of vegetation brings numerous additional benefits, from decreasing heat in summer months, rainwater retention, air filtration, increasing biodiversity to psychological and social effects on their users. Green technology application is therefore being adopted and even incorporated into regulations by many countries around the world to address the issue of energy and environment simultaneously. However, this concept is relatively new for most of the builders, developers, and designers in Serbia. In Belgrade, Serbia's capital, there are not many buildings with green roofs due to lack of knowledge and awareness, financial affordability, lack of statutory mandate, or regulation in building design.

Apart from the excessive energy consumption, air pollution in Serbia has become a serious problem for human health. More parks and gardens are needed to restore balance. As free space for new greenery is limited on the ground, the alternative is to place them on the roofs of the existing buildings. In developed cities, roof areas account for about 40–50% of total impermeable surfaces of urban areas. It is also argued that retrofitting an old building with a green roof in is more cost-effective compared to a new building since old buildings are mostly poorly insulated and use a lot of energy annually. (Shafiquea et al. 2018).

The implementation of green roofs in Belgrade will certainly provide benefits as well as challenges to both public and private sector. The starting premise of this paper is that this new solution would most certainly increase thermal properties of the existing roof due to adding more thermal insulation, but the true question lies in discovering if the chosen green roof system allows for enough thermal insulation to be added, due to weight, building geometry, and existing infrastructure limitations, while providing additional benefits unique to a green roof. This paper aims to discover and address these benefits and challenges based on a case study of one public company administrative building—Belgrade's "City Housing" building.

75.2 Green Roof Retrofit—Pros and Cons

Climate of Serbia can be described as moderate-continental with more or less emphasized local characteristics. This climate is characterized by significant precipitation, and green roofs have proven their efficiency in rainwater retention, which ranges from 55 to 88%. (Shafiquea et al. 2018) Very high temperatures that are characteristic for summer season in Serbia can be effectively alleviated by existence of greenery on the roof tops, thus reducing the effects of urban heat island effect. A study in Toronto,

Canada, found that the heat gain through the green roof was reduced by an average of 70–90% in the summer and heat loss by 10–30% in the winter. (Castleston et al. 2010) It also adds thermal mass to help stabilize internal temperatures year round.

In addition to previously mentioned benefits, green roofs help mitigate air pollution. Two-year study conducted with the aim to determine the effect of extensive green roofs on the surrounding air resulted in conclusion that an extensive green roof absorbs and retains 189 g of CO_2 per m^2 per year. (Getter et al. 2009) Apart from the direct reduction, green roofs also have an indirect positive influence on building's thermal envelope, resulting in less energy spent for heating and less CO_2 released in the atmosphere. (Djordjević et al. 2018).

Green roofs are commonly classified into two major categories—intensive and extensive green roofs (Wilkinson and Dixon 2016). Intensive green roofs are categorized on the basis of substrate thickness (> 30 cm.), with a wide variety of plants/vegetation similar to ground-level landscapes, high water holding capacity, high capital, and maintenance costs and larger weight. Extensive roofs have substrate thickness of 7–10 cm, use mostly sedum as the vegetation layer and typically do not require irrigation. They require less capital and maintenance costs as compared to all other roofs. These roofs are usually very lightweight and useful where there are building weight restrictions.

There are also those green roofs with 15–30 cm substrate thickness and are referred to as semi-intensive and usually have small plants, shrubs, and grass. These roofs require regular maintenance and high capital costs for the better performance.

The major problem with green roofs is that they usually require high level of maintenance which building occupants have to organize by themselves. However, the advantage of public buildings is that they have publicly organized maintenance. In Serbia, availability of green roof manufacturers is also a problem, since the offer is limited, and the price of installation and maintenance is high compared to regular roofs. Green roofs also require highly skilled and experienced workers for installation and maintenance, which Serbia lacks.

75.3 Methodology

A set of requirements was created in order to verify if the building's roofs were suitable to undergo a process of green roof retrofit in terms of the following:

- Evaluating the bearing capacity of the existing building—aimed at identifying the possible solutions due to weight of different types of green roof;
- Analyzing pros and cons of different green roof types to know which one is most suitable to install. This step directly relates to the previous one. For example, installing extensive green roof offers ecological benefits, while intensive green roof can provide more substantial benefits like public spaces and allow more plants species but requires more space and has more weight;

- Examining the accessibility to roof, for construction and post-construction maintenance, including the availability of sufficient roof space;
- Valorization of different scenarios, determining their benefits in terms of socio-esthetic, functional, energy and financial aspect.

To calculate the potential influence of the green roof as well as the properties of the existing roof, a software package available on Serbian market was used—Knauf TERM 2 PRO since it complies with the current law and is widely used by professionals. Software is calculating annual energy need for heating according to the HDD methodology defined in the current regulations in Serbia. (Rulebook on Conditions, Content and Method of Issuing Energy Performance Certificates 2012; Rulebook on Energy Efficiency in Buildings 2011) Constraints coming from this kind of thermal analysis are not taking into account different occupancy regimes and heating regimes, as well as dynamic aspects of heat transfer through building elements as explained in previous research. (Ignjatović et al. 2018).

75.4 The Case Study—Belgrade's "City Housing" Building

The administrative building that is the subject of this research is part of a larger building complex, in a form of the letter "H." Each of its parts represents a separate functional zone, with different heating regimes and heating systems. The subject of this work is improvement of energy performance of one such independent part—office building of the "City Housing" public enterprise that occupies building's south-western part. (see Fig. 75.1) Considering that some parts of the building's envelope were reconstructed in the previous period—building's facade walls and window components, this research and design project is reduced exclusively to interventions on roof planes, specifically:

- Access plateau above the garage of the complex (zone 1—area of 597 m^2);
- Flat roof above the entrance hall (zone 2—area of 160 m^2);
- Flat roof above the first floor of the building (zone 3—area of 646 m^2).

This research has been used in real-case scenario and has been finalized in the form of technical documentation required for the construction.

The existing condition of the analyzed roof areas is rather poor. The pedestrian surface is not even, due to the deterioration of concrete tiles and their joints, which represents a potential risk. Also, concrete tiles are problematic from the aspect of rainwater retention and heat island effect because of concrete's high Solar Reflective Index (SRI). As for the rainwater runoff, since there is only one gutter per roof, inappropriately sized, overflows often occur. Finally, the structure of the roof layers does not have adequate thermal insulation properties, so additional insulation is required.

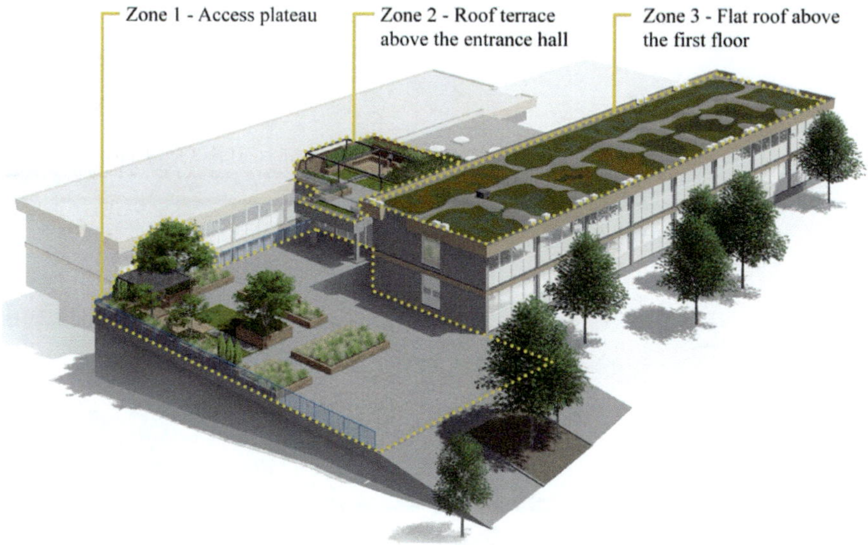

Fig. 75.1 Roof zones included in the intervention on "City Housing" building

75.4.1 Zone 1—Access Plateau Above the Garage

Even though the access plateau is not part of the building's thermal envelope, it is part of the retrofit project because of its functional, esthetic, and environmental potential.

This roof combines several different parts and processing: areas under extensive, semi-intensive and intensive green roof, plateaus made of composite wooden decking covered by a steel rectangular pergola used as the leisure area and paths made out of granite ceramic tiles, providing access for pedestrians and motor vehicles (see Fig. 75.2).

The plant species planned on this roof are various medium-sized deciduous trees, evergreen and deciduous shrubs, and ornamental grasses planted in planters made of light aggregate blocks. Benches are added on some of the planter walls.

As this part of the building is frequently visited by the members of the general public and as it is located next to a university, it possesses great educational and demonstrational potential and is designed to be a "showcase" example—showing how the green roofs can look and what types there are—having extensive, semi-intensive, and intensive roof segments all in one place, how they can be adapted to existing buildings, reduce energy consumption, and have an overall positive effect on the environment. Educational info-graphic board is planned to be placed on the plateau, toward the building entrance, explaining the measures taken in the retrofit project as well as the benefits such an intervention has in regard to building energy consumption, CO_2 emission, rainwater retention, and urban heat island reduction.

Fig. 75.2 Zone 1—access plateau above the garage—existing state (top), top view (bottom left), and rendering (bottom right)

75.4.2 Zone 2—Roof Terrace Above the Entrance Hall

Roof above the entrance hall of the building is retrofitted in the form of a mixed green roof with elements of the roof terrace. This roof area was designed as a kind of "garden" intended for the employees that work in the building, making it a semi-public place suitable for presentations, company meetings, and coffee breaks. It incorporates several different elements: plateaus and paths made of granite ceramic tiles or composite wooden decking with appropriate spaces for sitting and gathering, covered area under a steel pergola serving as a sunshade, and a base for the creeping plants to grow and green segments consisting of both extensive green roof, sedum as ground cover, and medium-sized vegetation that offers privacy and contributes to the atmosphere (see Fig. 75.3).

75.4.3 Zone 3—Flat Roof Above the First Floor

The area above the first floor is envisioned in the form of an extensive green roof. Plants of the succulent species (sedum species) of different colors and types are

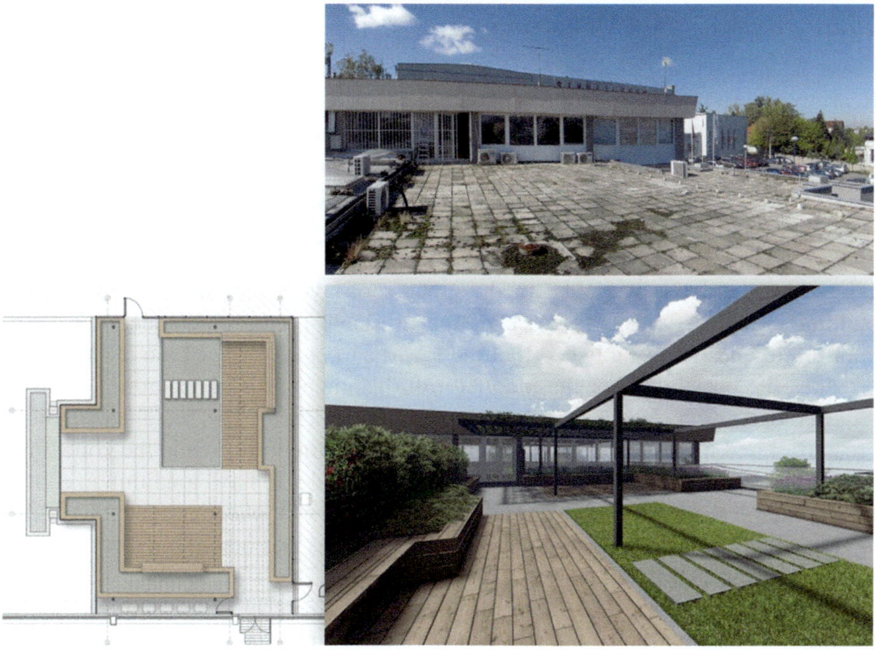

Fig. 75.3 Zone 2—roof terrace above the entrance hall—existing state (top), top view (bottom left), and rendering (bottom right)

planted in organic geometric form on the appropriate sublayers of the extensive roof garden. Gravel is used for filling the drainage edges of the extensive roof, in order to rationalize the solution. For safety reasons, access to upper roof is restricted only to construction and maintenance workers via steel ladder (Fig. 75.4).

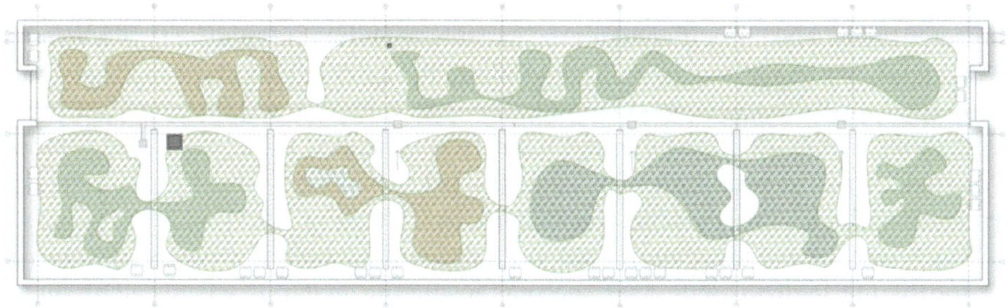

Fig. 75.4 Zone 3—flat roof above the first floor—top view

75.5 Results and Discussion

Bearing capacity of the existing construction is the most important requirement when discussing green roof retrofit project since it determines what type of the green roof and the depth of the substrate is possible to apply on the building in question. If the existing construction is not able to bear the load of the preferred green roof, it is possible to reinforce the construction. However, this is a complex process requiring a lot of additional work and substantial financial investment. Also, such an action can reduce the depth available for the green roof layers depending on the existing building's geometry. The research of the available project documentation of the building in question established that the load-bearing capacity of the existing slabs provides the possibility for the planned refurbishment project according to the set requirements. Zone 1 and zone 2 roofs have a 0.55 kN/m^2 less weight with all the green roof and insulation layers added than the existing roof, while the zone 3 roof weighs 0.6 kN/m^2 more than the existing roof. In order to have medium-sized trees on the zone 1 roof, which require larger amounts of soil, planters made out of light aggregate blocks were placed directly above columns and other structural elements to support their additional weight.

The refurbishment of existing roof by implementing green roof solution requires freeing the construction of the excess layers in order to place the new ones. Some research suggests leaving existing screed to falls made of perlite concrete, since it is a light material with thermal properties better than other materials used for this purpose like cement (Djordjević et al. 2018), but since we cannot be sure what the current state of the materials in existing roof is, it may be best to remove them all, up to the structural slab.

When adding layers of the new roof, it is necessary to calculate the heat transfer coefficient of the newly formed structure of thermal envelope. By current regulations (Rulebook on Energy Efficiency in Buildings 2011), the highest allowed value of the heat transfer coefficient of the refurbished flat roof is Umax = 0.20 W/m^{2}K for the elements above heated areas and Umax = 0.40 W/m^{2}K for those elements above unheated areas. The U coefficient for the existing roofs has the value of 0,545 W/m^{2}K. When applying suggested green roof solution, the calculations show that only 8 cm of thermal insulation is needed for the zone 1 to fulfill the requirements set by the regulations, while 20 cm of thermal insulation is needed for zones 2 and 3 (about a quarter of the thermal envelope), and bring U coefficient to 0.174 W/m^{2}K thus reducing buildings energy needs for heating by 11%. (Fig. 75.5).

Since the current legislation does not take into account the layers of substrate on top of the hydro-isolation as thermal insulating material when calculating U coefficient, it is obsolete to speak about green roof's thermal characteristics without the simulation and calculation procedures which take into account the positive effects of thermal mass in lowering energy need. Researches that take these effects into detail consideration show that energy consumption for cooling can be reduced up to 35% and for heating up to 10%. (Cascone et al. 2018).

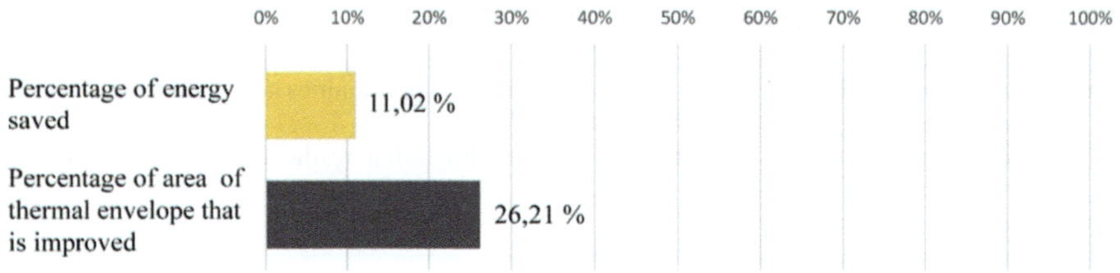

Fig. 75.5 Diagram comparing the percentage of energy saved for heating with thermal envelope area intervened upon

75.6 Conclusions

This model offers a demonstrational know-how for old building refurbishment using green roof implementation. In this approach, energy reduction is not a main motive of building refurbishment, but it is seen as just one of the benefits.

On the urban scale, it is shown how green roofs contribute to reduction of urban heat island effect by reducing solar radiation index, mitigation of air pollution by absorbing CO_2 and better rainwater retention, thus reducing the negative effect our cities have on the environment and presenting an adequate architectural response to our planet's changing climate.

On the building scale, even though the green roof has no positive impact on roof's thermal properties itself (without additional thermal insulation), it can be concluded that green roof presents a god solution for future refurbishment projects since it allows for enough thermal insulation to be added having in mind load-bearing characteristics of the existing structure. Therefore, it makes room for improvement in aspect of building's thermal and structural properties, while improving its user's quality of life from ecological, psychological, and social aspect.

References

Cascone S, Catania F, Gaglian A (2018) A comprehensive study on green roof performance for retrofitting existing buildings. Build Environ 136:227–239

Castleston HF, Stovin V, Beck S, Davison B (2010) Green roofs; building energy savings and the potential for retrofit. Energy Build 42:1582–1591

Djordjević K, Joksimović O, Jovanović-Popović M (2018) Energy consumption and CO_2 emission reductions trough refurbishment of residential buildings' roofs by applying the green roof system—case study. Therm Sci 22:1217–1229

Getter K, Rowe B, Robertson P, Cregg B, Andresen J (2009) Carbon sequestration potential of extensive green roofs. Environ Sci Technol 43(19):7564–7570

Ignjatović D, Rajčić A, Stanković B (2018) Building simulations and modelling: energy. In: Konstantinou T, Ćukovic IN, Zbašnik-Senegačnik M (eds) Energy resources and building performance. TU Delft Open, pp 129–144

Rulebook on conditions, content and method of issuing energy performance certificates (2012) The Official Gazette of Republic of Serbia No. 69/2012

Rulebook on energy efficiency in buildings (2011) The Official Gazette of Republic of Serbia No. 61/2011
Shafiquea M, Kima R, Rafiq M (2018) Green roof benefits, opportunities and challenges—a review. Renew Sustain Energy Rev 90:757–773
Wilkinson SJ, Dixon T (2016) Green roof retrofit: building urban resilience. Wiley-Blackwell

Chapter 76
Designing with Nature Climate-Resilient Cities: A Lesson from Copenhagen

Maicol Negrello

Abstract Climate change is accelerating at a faster rate than previously anticipated, and a significant number of cities remain unprepared for this transition. There is a pressing need to reconsider the approach to the design of public spaces, directing attention towards the development of design concepts that can impart knowledge for adaptation to climate change. Landscape architects, through nature-based solutions, can emerge as key figures capable of regenerating urban spaces. The case study of this research is the city of Copenhagen, which has become the stage of the most innovative experiments to create climate-resilient urban spaces. It is evident that a multidisciplinary and site-specific approach can be the critical components for a successful transition. Such a transition necessitates innovative project management that involves the collaboration of municipalities, private stakeholders, and citizens. Natural-based solutions, through an ecosystem approach, can effectively address the environmental, social, and economic challenges presented by climate change.

Keywords Nature-based solution · Climate change adaptation · Landscape architecture · Urban design · Copenhagen

76.1 Introduction

Climate change is showing that our cities are still inadequate to respond resiliently to hazards (Roggema et al. 2021) such as flash floods or urban heat island, caused by the interaction of climatic factors and anthropic activities within cities (Panno et al. 2017). In the Mediterranean urban contest, rapid floods and heatwaves are among the main threats to human safety and well-being in European cities (Sanesi et al. 2011); specifically, the associated hot temperatures in urbanized settlements are becoming a significant public health challenge (Hajat and Kosatky 2010; Bosch and Ode Sang 2017).

M. Negrello (✉)
Polytechnic University of Turin, Turin, Italy
e-mail: maicol.negrello@polito.it

E. Arbizzani et al. (eds.), *Technological Imagination in the Green and Digital Transition*, The Urban Book Series, https://doi.org/10.1007/978-3-031-29515-7_76

Considering that over 70% of Europe's population lives in cities and is expected to increase (Kabisch and Haase 2011), it is necessary to establish strategies for a healthier and less polluted urban environment for better well-being, by reconsidering the living space not limited to the residential unit, but the whole urban scale. Nature can be a mitigation tool both to increase urban resilience and to reduce psycho-physical problems generated by degraded places, that lack green areas (Bosch and Ode Sang 2017; Vries et al. 2003; Negrello and Ingaramo 2021). This would prevent socioeconomic inequality, offering all direct access to natural spaces and achieving equal resilience.

Studies on the regenerative potential of green space planning show that the use of nature as a design strategy contributes to improving urban conditions (Inostroza 2014) and approaching climate mitigation and adaptation goals (Kabisch et al. 2016). In Europe, urban planners have continuously advocated that nature and climate factors, proximity, and access to natural spaces should be integrated into sustainable urban planning (Fan et al. 2017; Gehl 2013).

76.2 Managing the Climate Transition

Climate impacts are expected to fundamentally change the way we live, and plan and design, our cities, and landscapes (Roggema et al. 2021). From policy to action, the architect plays a key role in the creation and management of adaptation and mitigation projects. In fact, in the last fifteen years, the architect's area of action in responding to climate change has broadened, no longer limited to the power of technology to make buildings more efficient but going on to re-consider the urban fabric as part of the metabolism that needs to be re-designed.

It is imperative to go beyond the logic of the individual building. This approach, although aiming at maximum local energy efficiency, does not contribute to generating an adequate impact on the urban energy transition goals in a decisive way because it tends to consider each building cell as a closed-loop system (Leone and Tersigni 2018); this reflection highlights the need for a shift in scale that involves individual building structures and public space in a comprehensive transition project that has an impact on the urban microclimate and, consequently, on energy consumption, the environment, and the well-being of citizens. A transition model based on this approach requires a professional figure capable of managing a transition that considers built and unbuilt, permeable, and impermeable layers, and natural and man-made surfaces. The role of the architect, and in particular the landscape architect, requires additional knowledge to deal with a new form of design that integrates urban and environmental needs, as they are two sides of the same coin. Today, we can see an evolution of this role which has become the emblem of a multidisciplinary player able to approach different scales of implementation.

At the same time, skills from different disciplinary fields (technology and humanities) and local engagement become indispensable building blocks for redefining an environment capable of influencing the microclimatic behavior of space and that of its inhabitants.

However, the challenge is not only related to the technical–technological aspects of this transition: once the concept of the machine city—whose objective is maximum efficiency, also referring to Le Corbusier's concept of the *machine à habiter*—has been abandoned, there is still the question of creating an environment that can meet the demands of citizens, promoting better well-being through the creation of an urban space designed using biophilic criteria.

The natural element does not only represent a decorative element or an urban planning standard but also a substantial climatic device for both public and built environment, to face the challenges of the new urban climate scenarios, and to create environmental conditions in symbiosis with the inhabitants (Eekelen and Bouw 2020).

The public space adaptation project becomes emblematic of developing site-specific solutions that transform these places into resilient urban systems. However, given the intensification of the impacts of climate change on urban areas, both now and in the future (according to the 2050 and 2100 scenarios studied by the IPCC in its annual reports), adaptation must be carried out in a capillary way, starting with the common urban fabrics, such as squares, small green areas, but also pavements, roofs, and not merely in the current green areas, which are already areas of resilience.

76.3 A Lesson from Copenhagen: Changing to Adapt

Among the European cities that have acted on morphological modification of urban space to increase the resilience of the built fabric, Copenhagen stands for having displayed interesting approaches and rapid transformations. The Danish capital represents a virtuous example of how climate change represents an excellent chance for urban development (Xu et al. 2021). Starting in 2011, the municipality to face the climate crisis has adopted a proactive plan to reduce the impact of future hazards, after the extreme cloudbursts that hit the Greater Copenhagen area on July 2, 2011, causing over DKK 6 billion in damage (Gerdes 2012). This flood has represented how future climate events could impact the city: It has been estimated that the costs of damage will be higher (DKK 18 billion) than DKK 16 billion based on an assessment of rainfall events in recent years (City of Copenhagen 2015).

In particular, the Copenhagen Municipality has launched three main plans:

- *Copenhagen Climate Adaptation Plan (2011)* Long-term and massive actions to strengthen resilience to extreme events by ensuring that stormwater is retained or discharged into the harbor and lakes, consequently reshaping a greener city and new jobs.

- *Cloudburst Management Plan (2012)* Prioritize cloudburst management efforts where the risk is greatest and where there is an opportunity for synergies with urban development in general
- *Climate Change Adaptation and Investment Statement (2015)* It describes the inputs, effectiveness, and challenges of the climate change adaptation plan (Xu et al. 2021). It also identifies the potential for urban space improvements in the districts (City of Copenhagen 2015).

76.4 Nature-Based Solutions as Strategical Action for Adaptation/Mitigation

The plans described are based on simple principles: prevent the damage, minimize the damage, and reduce the city's vulnerability. Following those principles, actions for adaptation and mitigation have been deployed, building on the ecosystem-based approach (EbA), from which the nature-based solutions (NBSs) derive and find bases (Eggermont et al. 2015). According to the EU Commission, NBSs are defined as actions that are inspired, supported by, or copied from nature; those have an enormous potential to be energy and resource-efficient and cope with the challenge of climate change in urban areas. The European Commission Directorate (Commission 2015) has identified solutions for reducing greenhouse gas emissions goals and helping to conserve and expand carbon sinks, which include: ecosystem restoration, greening of gray surfaces, and integrated broad-scale climate change mitigation and adaptation measures (Bosch and Ode Sang 2017). However, it is necessary to emphasize how these strategies have to be adapted to site-specific conditions to have results that provide economic, social, and environmental benefits (Commission 2015). Moreover, the NBSs are ecosystem services that play a critical role in promoting a cultural shift from a resource-intensive growth model toward a more resource-efficient, inclusive, and sustainable one. Those radical innovations involve actors from different sectors, domains, and scale levels in the co-design and co-implementation of solutions (Faivre et al. 2017; Meene et al. 2011), as in the case of Copenhagen, where negotiation and cooperation have been applied among multiple stakeholders (Xu et al. 2021).

76.5 Building the Resiliency: Case Studies in Copenhagen

The Danish capital has defined strategies and areas of urban regeneration, investing in the reconversion of public infrastructures, and ambitious projects of climate adaptation parks to respond to the heavy rainfall. Copenhagen has taken the opportunity to redesign its city considering water as a risk, firstly, but also as a resource and designing new attractive urban spaces. In this section, some case studies of resilient urban development based on NBS are presented, chosen for their esthetic, environmental, and inclusion qualities—values of the New European Bauhaus.

The Copenhagen adaptation project starts from a systemic vision of interventions in the public space. In the first phase, it has been defined an area particularly subject to climate risks: Østerbro. The idea was to create the first climate-resilient neighborhood within which to develop a series of urban redesign interventions to increase the quality of public space as well as ecosystem service. This plan expects to realize green spaces for 50,000 m^2 and slowly disconnect 30% of the stormwater from the underground mixed storm sewer network (City of Copenhagen 2016).

In 2013, the Danish firm Tredje Natur realized the masterplan for the project identifying major development areas characterized by low permeabilization: Skt. Kjelds Plads, Bryggervangen, and Tåsinge Plads.

Skt. Kjelds Plads and Bryggervangen (2016–2019). The projects, realized by SLA, transformed a congested infrastructure (the traffic circle Skt. Kjelds Plads and Bryggevangen road, Fig. 76.1) into a new green space and responsive to different climatic hazards, such as extreme heat waves or flooding. It has been designed with the idea of bringing a new nature into the city to benefit both the environment and the well-being of all citizens. The intervention has provided for the depavimentation of a large area around the traffic circle and along the avenue. The new retention surface hosts an ecosystem of 586 native trees belonging to the local biotype, which absorb CO_2 and pollutants. The Bryggevangen has been transformed into a "green corridor:" the new zig-zag green spots, that occupy part of the packing area, contribute to slowing down cars and increasing the detention area. The plaza of 34,900 m^2 is designed for containing and delaying rainwater in green spaces. As a result of this NBS approach, biodiversity has increased as well as new vibrant places have been built that improve health and quality of life for all the residents.

Tåsinge Plads (2014). The square represents the first climate change-adapted urban space in Copenhagen designed by GHB. Once 1.000 m^2 of flat asphalt has been transformed with a new land morphology: slopes of grass allow the rain collecting is conveyed to areas with moisture-tolerant plants that filter and absorb the rainwater in the storage facility, by reducing the pressure on the urban pipe network. The paths, built on a higher level, allow fruition even when the square is completely flooded. Nowadays, the project delays and infiltrates about 8.000 m^3 of rainwater from the site, making it the flagship of climate change adaptation projects in Copenhagen (Meene et al. 2011).

In Copenhagen, the NBS approach is becoming common in most future projects, such as the Hans Tavsen Park with Korsgade Avenue (Fig. 76.2); it involves the creation of a large green–blue infrastructure, a natural basin that intercepts and drains excess rainwater (18,000 m^3), subsequently purified and conveyed toward the Peblinge Sø waterway. The project (by SLA, Rambøll, and Dreiseitl) has been developed through a co-design process with non-profit associations and residents who will be able to use this space in different climatic conditions.

Fig. 76.1 Traffic circle Skt. Kjelds Plads and Bryggevangen road (Courtesy of SLA)

76.6 Conclusion

The use of NBS for urban design is a necessity for any designer to meet climate adaptation needs, as well as to reduce greenhouse gas emissions and pollutants. The case of Copenhagen represents how the management of natural resources becomes fundamental to designing new resilient public urban spaces. Green is not conceived as a mere necessity prescribed by urban standards, but the functionality of natural elements also responds to the need to create an environment that is resilient to extreme events, as well as providing an eco-systemic service. Outcomes of this approach are a better quality of life and well-being for the communities that live there.

Fig. 76.2 Hans Tavsen Park by SLA, Rambøll, and Dreiseitl (Courtesy of SLA)

The cases show how the approach is necessarily site-specific: the natural elements (from the choice of plants to the materials used) and the solutions adopted must respond to climate, culture, and local uses. It follows that the skills involved in those projects are vast, as demonstrated by the team of designers of the studio SLA and Trejed Nature. In addition to the technical and design aspects, the social and community aspects are fundamental in the design of a project that is also responsive to the needs of neighborhood life. In the climate-resilient neighborhood project, the contribution of co-design has involved the local Østerbro Environmental Center, the municipality, stakeholders, and over 10,000 residents who have developed a greater sense of belonging and community.

These projects have enabled the municipality to save money and have a more resilient city for future climate challenges, demonstrating how NBSs should be considered as key climate devices for both urban and architectural design.

This approach, common in northern countries, can be applied also in other extreme climatic contexts, such as the Arab Emirate, where it has been tested. Some projects, such as Al Fay Park in Abu Dhabi by SLA (Fig. 76.3), have shown that it is possible to reduce the temperature to encourage the creation of public places rich in biodiversity through NBS and material contextualized to the climate zone.

Fig. 76.3 Al Fay Park in Abu Dhabi by SLA (Courtesy of SLA)

References

City of Copenhagen (2015) Climate change adaptation and investment statement (Part I)

City of Copenhagen (2016) Copenhagen's first climate resilient neighbourhood (2nd edn). http://klimakvarter.dk/wp-content/uploads/2015/08/Copenhagens-first-climate-resilient-neighbourhood_WEB_low.pdf

de Vries S, Verheij RA, Groenewegen PP, Spreeuwenberg P (2003) Natural environments—healthy environments? An Exploratory analysis of the relationship between greenspace and health. Environ Plan Econ Space 35(10):1717–1731. https://doi.org/10.1068/a35111

European Commission (2015) Towards an EU research and innovation policy agenda for nature-based solutions and re-naturing cities: final report of the Horizon 2020 expert group on "Nature-based solutions and re-naturing cities": (Full version). Publications Office. https://doi.org/10.2777/479582

Eekelen EV, Bouw M (2020) Building with nature: creating, implementing and upscaling nature-based solutions (1° edizione). Nai Uitgevers Pub

Eggermont H, Balian E, Azevedo JMN, Beumer V, Brodin T, Claudet J, Fady B, Grube M, Keune H, Lamarque P, Reuter K, Smith M, van Ham C, Weisser WW, Le Roux X (2015) Nature-based solutions: new influence for environmental management and research in Europe. GAIA—Ecol Perspect Sci Soc 24(4):243–248. https://doi.org/10.14512/gaia.24.4.9

Faivre N, Fritz M, Freitas T, de Boissezon B, Vandewoestijne S (2017) Nature-based solutions in the EU: innovating with nature to address social, economic and environmental challenges. Environ Res 159:509–518. https://doi.org/10.1016/j.envres.2017.08.032

Fan P, Ouyang Z, Basnou C, Pino J, Park H, Chen J (2017) Nature-based solutions for urban landscapes under post-industrialization and globalization: Barcelona versus Shanghai. Environ Res 156:272–283. https://doi.org/10.1016/j.envres.2017.03.043

Gehl J (2013) Cities for people. Island Press

Gerdes J (2012, ottobre 31) What Copenhagen can teach cities about adapting to climate change. Forbes. https://www.forbes.com/sites/justingerdes/2012/10/31/what-copenhagen-can-teach-cities-about-adapting-to-climate-change/

Hajat S, Kosatky T (2010) Heat-related mortality: a review and exploration of heterogeneity. J Epidemiol Commun Health 64(9):753–760. https://doi.org/10.1136/jech.2009.087999

Inostroza L (2014) Open spaces and urban ecosystem services. cooling effect towards urban planning in South American Cities. Tema. J Land Use Mobility Environ SI, 523–534. https://doi.org/10.6092/1970-9870%2F2541

Kabisch N, Frantzeskaki N, Pauleit S, Naumann S, Davis M, Artmann M, Haase D, Knapp S, Korn H, Stadler J, Zaunberger K, Bonn A (2016) Nature-based solutions to climate change mitigation and adaptation in urban areas: perspectives on indicators, knowledge gaps, barriers, and opportunities for action. Ecol Soc 21(2), art39. https://doi.org/10.5751/ES-08373-210239

Kabisch N, Haase D (2011) Diversifying European agglomerations: evidence of urban population trends for the 21st century: diversifying European agglomerations. Popul Space Place 17(3):236–253. https://doi.org/10.1002/psp.600

Leone MF, Tersigni E (2018) Progetto resiliente e adattamento climatico: Metodologie, soluzioni progettuali e tecnologie digitali. CLEAN edizioni

Negrello M, Ingaramo R (2021) Lo Spazio Del Burn-out. ARDETH 8:18

Panno A, Carrus G, Lafortezza R, Mariani L, Sanesi G (2017) Nature-based solutions to promote human resilience and wellbeing in cities during increasingly hot summers. Environ Res 159:249–256. https://doi.org/10.1016/j.envres.2017.08.016

Roggema R, Tillie N, Hollanders M (2021) Designing the adaptive landscape: leapfrogging stacked vulnerabilities. Land 10(2):158. https://doi.org/10.3390/land10020158

Sanesi G, Gallis C, Kasperidus HD (2011) Urban forests and their ecosystem services in relation to human health. In: Nilsson K, Sangster M, Gallis C, Hartig T, de Vries S, Seeland K, Schipperijn J (A c. Di), Forests, trees and human health (pp 23–40). Springer Netherlands. https://doi.org/10.1007/978-90-481-9806-1_2

Van de Meene SJ, Brown RR, Farrelly MA (2011) Towards understanding governance for sustainable urban water management. Glob Environ Chang 21(3):1117–1127. https://doi.org/10.1016/j.gloenvcha.2011.04.003

van den Bosch M, Ode Sang Å (2017) Urban natural environments as nature-based solutions for improved public health—a systematic review of reviews. Environ Res 158:373–384. https://doi.org/10.1016/j.envres.2017.05.040

Xu H, Liu L, Ding P (2021) Building climate resilient city through multiple scale cooperative planning: experiences from Copenhagen. IOP Conf Series Mater Sci Eng 1203(3):032063. https://doi.org/10.1088/1757-899X/1203/3/032063

Chapter 77
New Urban Centralities: Universities as a Paradigm for a Sustainable City

Camilla Maitan and Emilio Faroldi

Abstract The urgency of facing up to the new challenges posed by climate change has relaunched the debate on the future of the city, a thought which points to public buildings as the way to promote the main processes of urban regeneration, both physical and social. Universities and their spaces are among the permanent architectures of the consolidated fabric. With their educational role par excellence, they are one of the most suitable urban institutions through which to promote an environmentally responsible attitude. The concept of sustainability is now an undisputed value which involves all applied disciplines, but the debate on how architecture should respond to this urgent need is still open. This paper proposes *dialectical ecologism* as an attitude to achieve significant environmental standards not only through the choice of sustainable materials and technologies, but also by rediscovering in public and community spaces, as well as in the character and language of architecture, the potential to make the built environment adaptable and resilient to contemporary challenges. In these terms, an initial theoretical reflection, whose main subject is the university campus and its contribution in terms of ecological impact and wise use of resources, will be followed by a description of the themes that guide sustainable design, which find application in the actions recently conducted on two Milan campuses of the Polytechnic University of Milan.

Keywords University campus · Dialectical ecologism · Urban and social regeneration · Multi-scalar approach · Polytechnic University of Milan

---

C. Maitan (✉) · E. Faroldi
Polytechnic University of Milan, Milan, Italy
e-mail: camilla.maitan@polimi.it

E. Faroldi
e-mail: emilio.faroldi@polimi.it

E. Arbizzani et al. (eds.), *Technological Imagination in the Green and Digital Transition*, The Urban Book Series, https://doi.org/10.1007/978-3-031-29515-7_77

77.1 Introduction

The contemporary debate on the future of cities tries to respond to the need to make the built environment adaptable to the challenges posed by climate change by investing in redevelopment or the addition of new public buildings. Within this scenario, the university campus appears as a suitable urban element to promote a regeneration that must be rooted in a mature awareness of the environmental issue, orienting the design discussion towards issues such as the growing urgency of using effectively the available resources, in a respectful and resilient way.

In recent decades, there has been a growing awareness of the influence that the university campus has on the surrounding area, identifying in the dialectic with the city the tool through which the university, initially a passive holder of culture, can play, now, a primary role as an educational institution for sustainability.

The present contribution intends to bring attention to the potential of this urban complex by adopting the architect's point of view, and guiding towards a multi-scalar analysis that makes explicit how to make the built environment resilient and sustainable.

Therefore, the reflection will be dedicated to the definition of a correct approach, proposing dialectical ecologism as the manifesto of an effective response to the environmental issue that, although not yet structured as an objective judgement tool through which to scientifically evaluate the performance of architecture, it identifies the issues necessary for sustainable construction. In these terms, a brief discussion will be focused on some specific interventions carried out on two Milan campuses of the Polytechnic University of Milan, paradigms of an effective response to spatial and environmental issues.

77.2 Beyond Rhetoric. The Responsibilities of Architecture on the Environmental Issue

For more than half a century, the ecological question has been increasingly important in the development of architectural theories and projects, trying to counteract the negative impact that human artifice has on nature (Ingersoll 2009). Although it was initially raised as an ethical issue, sustainability has now become so widely recognised that it has become a dominant criterion in the evaluation not only of architecture, but also of all other practical disciplines. The need to act in this way is therefore undisputed, but the debate on how architecture can and must respond to sustainability remains open.

There are an increasing number of examples of architects whose projects limit their efforts to adopting new technologies as the only way to achieve the high energy and environmental performance standards required, and even more widespread are the cases in which architects see superficial imitation of nature as their passe-partout for an presumed, but often hypocritical, adherence to the logic of sustainability.

Thinking that an effective response to the ecological question can only be achieved through these attitudes reveals, however, a major shortcoming: architecture has always been a discipline interested in sustainability, whose contribution must primarily be sought in the form, orientation and relationship with the neighbouring context at various scales; nevertheless, there are many contemporary cases in which the identity and specificity of the place, which have always been fundamental to the definition of architecture, take a back seat, leading to a reduction in cultural diversity (Dematteis 2009) and an impoverishment of the built heritage; With such superficial attitudes, there is a risk that architecture will be emptied of its importance, leading architects to hide behind the "rhetoric of environmental sustainability" (Dematteis 2009) instead of encouraging actions aimed at conscious design. The challenge of sustainability to which architecture is called to respond cannot be extinguished just by creating zero-impact buildings, because this is not enough to build a sustainable city; the ecological footprint of a city must be measured through density and compactness, and through an idea of environmental comfort that goes beyond the limits of home spaces to create public spaces in compliance with a sustainable urban model. Architecture has the responsibility to trigger a real paradigm shift, through those criteria of accessibility, permeability and aesthetic durability which find, in form and in dialectic with the context, the main tool through which to become part of the built heritage.

77.3 University Campuses. From Heterotopias to New Urban Centralities

The greatest contribution architecture can make in terms of sustainability is, therefore, the promotion of urban regeneration, through an increase in density and a reduction in the ecological footprint: it seems reasonable, hence, to start again from those nodal elements with a public character and unexpressed potential which, instead of embodying the role of new urban centralities, still remain faithful to their image as Foucauldian heterotopias, that are isolated objects, with a distinct architectural identity and linked to function, which recreate a "space that is other, another real space, so perfect, so meticulous, so well organised" to be inevitably at odds with the city (Foucault 1986); a definition which also includes universities that have historically been established far from cities, due to the desire to recreate isolated and introverted communities of knowledge, in order to prevent their function, so encumbered, from affecting the urban conformation of the city. The strong expansionist growth, however, has led in most cases to the inclusion in the consolidated urban fabric of numerous campuses which, over time, have progressively broken down their boundaries, seeking a dialogue with the neighbouring city, through public spaces which transform the sites of the universities into engines of development and, therefore, into potential new urban centralities.

A similarly attentive approach is increasingly being adopted with newly founded campuses, for which peripheral sites previously used for industrial production—now abandoned—are often chosen and whose spaces are redeveloped to host new knowledge communities, proving to be a valid strategy for redesigning public space and giving rise to new forms of peripheral centralities (Biraghi and Pierini 2016); in these places, borns a hybrid campus model (Postiglione et al. 2016), open to the city, which responds to the needs of the community, no longer composed exclusively of students and teachers, but also of the citizens of the neighbouring urban fabric. Universities, as Bernard Tschumi suggests, become "condensers of the city. Through their programmes, and through their spatial qualities, they accelerate or intensify a cultural and social transformation that is already underway", and the task of architects is to foster, as much as possible, the interaction between campus and city, so as to propose "a model of living that rediscovers urban values at the local scale, that is ecologically aware and conscious of the need for local connections" (Guidarini 2018).

77.4 Dialectical Ecologism: A New Methodological Setting

The choice of the site therefore proves to be an important first decision in terms of sustainability. The desire to work for grafts on existing university campuses and the decision to regenerate large disused industrial spaces to host the new sites reinforces the desire to trigger spontaneous regeneration phenomena, a very effective response if we consider that university campuses are able to recreate a network of relations in the peripheral areas. It is enough to think about this simple assumption to reflect on the fact that the criteria with which the performance of university campuses is currently assessed reveal a general lack of interest in the specific contribution of architecture, looking instead only at contributions in terms of energy and environmental performance—equally important, but not holistic in defining the contribution architecture can make to the construction of a more ecologically responsible city; and it is precisely this first reason that has revealed the need to define a new methodological setting within which to assess the architectural performance of university campuses.

The method presented here, which is still in the process of being developed, is called dialectical ecologism and aims to define the first guidelines which, through a multi-scalar approach, respond to the urgencies of ecology by basing its contribution not only on economic, energy and environmental choices, but also on architectural ones. The challenge is to provide a method capable of guiding towards an effective, rather than efficient, design logic (McDonough and Braungart 2002) by shifting the focus to the achievement of objectives such as creating healthy, sustainable urban habitats with a propositive response to environmental issues. The strategic design framework presented here defines the design themes to be investigated at different scales:

- Relations with the city at the territorial planning scale. Starting from an analysis of the quality of the consolidated urban fabric, it will be possible to identify areas where the abandonment of old industrial sites and peripheral location will be the key elements in deciding where to build new university campuses, which will become spaces serving the neighbouring city, that lacks services and needs regeneration;
- The dialectic between the campus and the neighbouring urban context. The typological analysis of the context and of the scale of the neighbouring urban structures will provide the tools through which to start designing the new university complex, so that it does not become an out-of-scale element. In these terms, there has already been a tendency to abandon the image of the campus based on the hospital citadel model, in favour of the construction of autonomous and identity-driven buildings that at times suggest a "discordant landscape, an assemblage of quality architectural objects on a programmatically open surface" (Biraghi and Pierini 2016). Reflections on issues such as the environment, and thus on the ecological footprint of the building and the ability to promote urban and social regeneration should then follow;
- Quality of the architectural complex. This parameter is measured on the basis of the heterogeneity of the spaces and the amount of green and open spaces present. At this scale, the university complex should be concerned with providing services to the city, such as sports facilities, green areas and recreational spaces so that the campus can become an integral part of the urban fabric. Attention is also paid to the economy, under three headings: the promotion of a circular economy, the search for an economy of means—which is reflected in the choice of local materials and prefabrication techniques—and economy of language—in other words, the definition of buildings without unnecessary superfetations or decorations;
- Choice of materials and technologies. The challenge lies in the ability to use new technologies that contribute effectively to the fight against climate change without degrading the language of architecture. It is therefore a question of building a new architectural artefact which has a language coherent with the tradition of the place, but which at the same time is a synthesis of zeitgeist (spirit of time), making the best use of new technologies.

77.5 The Milanese Paradigms of the Polytechnic Experience

This is the background behind the action taken by the Politecnico di Milano, whose two campuses in the homologous city offer themselves as paradigms in terms of regeneration, valorisation and urban redesign, taking their first steps from profoundly different evolutionary histories and identities (Faroldi 2020).

The Campus in Città Studi, built in what was then the Milanese countryside, has progressively broken down the boundaries of the hospital pavilions, expanding northwards in the second half of the twentieth century with new buildings that went along with Gio Ponti's innovative vision of a school open to cultural exchanges and dialogue with the adjacent context, and welcoming the recent mending action proposed by Renzo Piano's studio and implemented by Ottavio di Blasi and Partners, and the intervention conducted by the Vivi.Polimi group. This last renewal, radical in its paradigm shift but delicate in its terms, has contributed to rediscovering the existing building, consolidating its historical and architectural value and giving back to the city a campus with new values, from which emerges an idea of a campus with greater density, whose mending sees its maximum expression in the quality of the semi-public space, open to the community, in the regeneration of the external spaces, until now scarcely used, and recreating a system of interrelationships between pre-existing and empty spaces able to accommodate the new buildings (Faroldi and Vettori 2021), responding to the growing need for dialectic between the campus and the neighbouring urban context (Figs. 77.1, 77.2 and 77.3). Even the choice of materials and technologies used, which have reduced the language of architecture to its essentiality, have supported the vision of a "new poetics of building" (Piano in Irace 2021), representing a natural extension of open space rather than an antithetical relationship with it (Faroldi and Vettori 2021).

Equally valid is the approach used for the La Masa Campus, located in an area still on the periphery of the city: here the large abandoned industrial buildings, which bear witness to the area's historical and industrial past, have undergone a significant transformation since the second half of the 1990s, when the Politecnico di Milano settled in the area, building teaching spaces and regenerating the existing heritage, thus triggering policies of social, economic and productive change in the neighbouring territory as well (Faroldi 2019). La Masa Campus is in continuous evolution, within a framework that is extremely dynamic and receptive to new stimuli. The last element to be built is the Collina degli Studenti (Figs. 77.2, 77.3 and 77.4), a project created by the Vivi.Polimi group, conceived as an open space for students and teachers to socialise, which takes on the appearance of a university infrastructure, seeing it as an element capable of connecting different urban functions and insisting on its social and educational role in sustainability. In the area in which this project stands, primarily lacking in vegetation and identity, a programmatically complex building has been inserted, whose use of materials recalls the historical and industrial identity of the place, and whose transparencies permit a more direct relationship between the internal, collective space and the green hill outside, also a place of encounter and exchange. From an anonymous contextual space, the new building has been able to transform itself into a centrality for the entire Campus, also contributing in terms of sustainability, fighting to reduce the existing heat island.

77.6 Conclusions

The latest projects on the two Milan campuses support the vision of a dialectical ecologism, in that the Politecnico di Milano's contribution does not end with the creation of new high-performance and/or zero-impact buildings, responding only

Fig. 77.1 Leonardo Campus, Polytechnic University of Milan, photographer Marco Introini

"passively" to urgent environmental demands, but seeks to adopt a truly sustainable approach, capable of becoming a manifesto of the demands of the new generations, thus becoming the bearer of a paradigm shift that sees its highest dimension in its relationship with the city and in the creation of new centralities. Universities can no longer be seen as cumbersome portions of the city, characterised by monofunctional

Fig. 77.2 La Masa Campus, Polytechnic University of Milan, photographer Marco Introini

spaces that blend into the urban fabric: they must also commit themselves to actively contributing to the development of the city. It must be recognised that the main contribution architecture can make in terms of sustainability is to be found in its strong social vocation, and that therein lies the potential for urban regeneration through which it is possible to contribute to the creation of healthy, sustainable habitats.

Fig. 77.3 Bonardi Campus, Politecnico di Milano, photographer Marco Introini

Fig. 77.4 La Masa Campus, Politecnico di Milano, photographer Marco Introini

Universities, hence, as a paradigm of a design education that sees architecture as its noblest tool.

References

Biraghi M, Pierini SO (curated by) (2016) Campus controcampus. Architetture per studiare e ricercare, vol 3. Universitas/Universities. Architecture Schools in the World, Maggioli editore, Sant'Arcangelo di Romagna
Dematteis G (2009) La sostenibilità territoriale dello sviluppo. In: Lotus 140. Sustainability?, pp 84–89
Faroldi E (2019) Università è città. Milano paradigma della cultura politecnica. In: Modulo, vol 421, pp 58–65
Faroldi E (2020) UniverCity. Università come infrastruttura per la rigenerazione urbana. In: The Plan, vol 422, pp 58–61
Faroldi E, Bucci F (2021) Architetture al Politecnico di Milano. Università è città, Silvana Editoriale, Cinisello Balsamo
Faroldi E, Vettori MP (2021) Campus Leonardo. Politecnico di Milano. In: Arketipo, vol 151, pp 70–81
Foucault M (1986) Di altri spazi. Diatrics 16:22–27
Guidarini S (2018) Live-Work Complex Kalkbreit. In: Lotus 165. Politics of the Campus, p 124
Ingersoll R (2009) Questione ecologica in architettura. In: Lotus 140. Sustainability?, pp 36–49
Irace F (2021) Sguardi sull'architettura contemporanea. Interviste di Fulvio Irace, Libri Scheiwiller, Milan
McDonough W, Braungart M (2002) From cradle to cradle. Remaking the way we make things. North Point Press, New York
Postiglione G, Rocca A, Bassanelli M (curated by) (2016) Campus controcampus. Architetture per studiare e ricercare, vol 1. Ultra/beyond Città Studi: projects for Politecnico di Milano, Maggioli editore, Sant'Arcangelo di Romagna

Part V
Session | Health

The topic of people's health is relevant to the full range of potential ushered in by technological innovation and processes of ecological and digital transition, touching on all its component factors. In international conventions, the definition of health has evolved to "a state of complete physical, mental and social wellbeing", and not simply "an absence of illness or infirmity", so that it is now viewed as "a resource for everyday life". Health is a proactive concept whose promotion is not relegated exclusively to the medical sector's ability to meet the population's needs, but rather represents the measures through which "an individual or group must be able to identify and to realize aspirations, to satisfy needs, and to change or cope with the environment" (WHO). Health, the result of a complex system that is both adaptive and dynamic, evolves through interaction between the potential of individuals and the social and environmental determinants to which they are subject.

The session is organized as a discussion of how the environmental determinants of health, along with their 'tangible manifestations', can be characterized and examined within the framework of architectural technology, at the various working scales, as well as through an osmotic dialogue with other disciplines. The challenge is to draw up visions of planning, decision-making, design, and execution that focus on people, foreseeing the short-, medium-, and long-term impacts on their health.

Chapter 78
Environment for Healthy Living

Francesca Giofrè

Abstract How do we build a healthy vision of the future? What interventions should architects promote to support human health and well-being, and from where do we start? The paper discusses the concept of health in a broader vision through international documents, focusing on the area of action for architects, and stressing the crucial role of collaborative, multisectoral, and transdisciplinary approaches to achieving optimal health balance. Moreover, it argues that the various contributions presented at the Conference link them to a unique vision.

Keywords Health · Environment · People · Technology

78.1 Introduction

The topic of people's health is relevant to the full range of potential ushered in by technological innovation and processes of ecological and digital transition, touching on all its component factors. In international conventions, the definition of health has evolved to "a state of complete physical, mental, and social well-being," and not simply "an absence of illness or infirmity," so that it becomes "a resource for everyday life" (WHO 1948).

Health is a proactive concept whose promotion is not relegated exclusively to the medical sector's ability to meet the population's needs, but rather represents the measures through which "an individual or group must be able to identify and to realize aspirations, to satisfy needs, and to change or cope with the environment." Moreover, health is "created and lived by people within the settings of their everyday life; where they learn, work, play, and love" (WHO 1986).

F. Giofrè (✉)
Sapienza University of Rome, Rome, Italy
e-mail: francesca.giofre@uniroma1.it

E. Arbizzani et al. (eds.), *Technological Imagination in the Green and Digital Transition*, The Urban Book Series, https://doi.org/10.1007/978-3-031-29515-7_78

Starting from its definition, health is the result of a complex system that is both adaptive and dynamic, and it evolves through interaction between the potential of individuals and the social and environmental determinants to which they are subjected (Meikirch Model). This interaction must be sufficient to respond satisfactorily to the demands of life, and it varies across individuals and contexts; for these reasons, the state of health is the result of the functioning of the system as a whole (Bircher and Hahn 2016, 2017).

The Meikirch model is an evolution of the one formulated by the northern European School of public health. The concentic model has at is center the individuals with their biological and genetic characteristics, as invariable factors. The focuses are the social determinants of health and the environmental determinants of health, and their "material manifestation," namely the built environment, in the awareness that they affect or influence the demands of life and the individual determinants of health.

Social determinants of health (SDH) include the immediate social surroundings of each individual as well as the larger social context. According to Tarlov's definition, they are "the social characteristics within which living takes place" (Tarlov 1996).

Environmental determinants of health (EDHs) include the whole biosphere, and the environment individuals need for nutrition and recreation, as well as the quality of water, air, and soil. Moreover, they include the place where we are born, live, work, move, and enjoy housing, workplaces, streets, schools, hospitals, open spaces, and more. In short, EDHs constitute the environment where people carry out their daily activities.

SDHs interact strongly with the demands of life and the potential of the individual. Equity and equality, social concerns, working conditions, autonomy, and social participation are all aspects that affect health and longevity. Likewise, EDHs influence living and working conditions and affect quality of life outcomes, and therefore a wide range of health issues, including risks. Both SDHs and EDHs also have a global and local impact on natural resources, catastrophes, population growth, and climate change. For these reasons, the topic of health must be considered as cutting across the other sessions presented at the Conference, underlying every anthropic action. Every design action has in itself the responsibility to create environments that promote the health and well-being of people or reduce the negative effects on their health.

Over the past twenty years, the topic of health has gained increasing relevance with a particular focus on the "city," because of the very large number of inhabitants in urban areas. The COVID-19 pandemic proclaimed by the World Health Organization on 11 March 2020, and still ongoing, impacts intensively built and extended cities, reinforcing the debate on how to rethink and reshape their social, cultural, economic, and environmental sustainability (UNESCO 2020). In fact, continued world urbanization and the population's demographic characteristics have a great impact on the city, putting it under pressure in terms of the consumption of natural resources, organization of physical and virtual infrastructure, demand for public services, and social relations. This scenario produces significant differences among the world's countries and brings negative effects on the health of the city inhabitants, as confirmed by several studies (Rydin et al. 2012).

Among the seventeen Sustainable Development Goals (SDGs) in the 2030 Agenda, the third one, "Ensure healthy lives and promote well-being for all at all ages," and the eleventh, "Make cities and human settlements inclusive, safe, resilient, and sustainable," highlights the importance of planning and managing these processes within the city as well (UN 2015). In detail, the 11th Goal is articulated in seven primary targets that relate to different dimensions and elements, as follows: housing and basic services; transport systems; urbanization; cultural and natural heritage; natural disaster; environmental impact; and green and public spaces. In 2030, the world's countries must reach the targets for developing and investing in safe and affordable housing, basic services, and transport systems; inclusive and sustainable urbanization; protecting and safeguarding cultural and natural heritage; reducing the negative effects of natural disasters, and preventing them; reducing the environmental impact of cities; and guaranteeing universal access to green and public spaces that must be safe and inclusive.

Although the SDGs have a universal character, the world's countries show inequalities in terms of living conditions, access to public services, transport, economy, and more. Consequently, meeting the targets will depend on the different specificities and capabilities of the Global South and the Global North; however, in both cases, there is still a strong need for investments and policies to achieve them.

"Rich and poor people live in very different epidemiological worlds, even within the same city. And such disparity occurs in both high-income and low-income countries" (Rydin et al. 2012). Given the above, it may be stated that each target considers in its achievement the positive impact on the health of the population that lives within the city, which becomes the locus of action, and strategies focused on people's needs and respect for the environment.

Moreover, in keeping with the global policies, the European Healthy Cities Network—launched by the WHO in 1986—plays a crucial role in promoting policies and strategies for urban areas, placing the principle of population health at its core.

The WHO defines the Healthy City as the "one that is continually creating and improving those physical and social environments, and expanding those community resources which enable people to mutually support each other, in performing all the functions of life and developing to their maximum potential" (WHO 1998).

The Healthy City aims to create a health-supportive environment: to achieve a good quality of life, provide basic sanitation and satisfy hygiene needs, and supply access to health care. The documents produced by the European Healthy Cities Network, and in particular the Zagreb Declaration (WHO 2009), are the starting point in defining intervention areas for urban planners, architects, and engineers, as well as for other professionals, because building a Healthy City requires interdisciplinary action.

According to these documents, the attention focuses on three core themes and illustrates their main aims.

- "Caring and supportive environments." A healthy city must be a city for all its citizens, and must be inclusive, supportive, and responsive to the needs and expectations of the various age groups that inhabit the city, such as children, young adolescents, adults, and the elderly. Because of Europe's ageing population, policies and universal action plans that address the health needs of this vulnerable group must be introduced, promoting participation, empowerment, independent living, supportive and safe physical and social environments, and accessible services.
- "Healthy living." A healthy city must provide conditions and opportunities that support healthy lifestyles, enhancing active living, physical activity, and pedestrian mobility, as a strategic plan of urban environment development policies.
- "Healthy urban environment and design." A healthy city must offer a physical and built environment supporting health, recreation and well-being, safety, social interaction, easy mobility, a sense of pride, and cultural identity, and it must meet the needs of all its citizens.

The three themes are strictly interconnected and suggest different kinds of interventions at different scales. Planning and design actions refer to various disciplines and different scales of interventions and, as mentioned above, they are all strictly connected and they all influence one another.

Nevertheless, it is important to consider that these actions and strategies extend their impacts beyond the boundaries of the city, and for this reason, the dichotomy between urban areas, which include the city itself, and rural areas must be overcome, and a multisectoral approach focusing on patterns of territorial continuity must be proposed. The features and development of urban and rural areas are interdependent (Cork.2 Declaration 2016) and linked to one another, and far more than being buildings, streets, open spaces, landscape, and agriculture, they embody a whole and complex system as a living, breathing organism, triggering an osmotic process. The 11th SDG highlights this interrelationship, underlining the need to "support positive economic, social, and environmental links between urban, peri-urban, and rural areas by strengthening national and regional development planning."

The main question to "answer" is how urban designers, architects, engineers, and other researchers and practitioners from different disciplines take account, in the design or research process, of the strategies and solutions for improving health or minimizing health impacts on people, and of what their vision is. How spaces are planned, designed, managed, and dreamed can make a significant difference in the impact on their users' health.

78.2 Vision for a Healthy Living Environment

The "Health | Environment for healthy living" conference session aimed to discuss how the environmental determinants of health, along with their "tangible manifestations," can be characterized and examined within the framework of architectural technology, at the various working scales, as well as through an osmotic dialogue with other disciplines. The challenge was to draw up visions of planning, decision-making, design, and execution that focus on people, foreseeing the short-/medium- and long-term impacts on their health. The tracks identified are public and urban health; inclusive designing; accessible environments; green and open spaces; designing for people with fragility; healthy individual buildings and building types; and the Internet of Things in healthcare and public buildings.

Two video interviews opened the session, held by experts with different expertise and belonging to different generations, who were asked to answer the following question: "What interventions should architects promote to support human health and well-being, and from where do we start?".

Professor Ruzica Bozovic-Stamenovic[1] argues that despite the quick spread and potentiality of information technology, humans do not change, and their sensorial mechanisms remain the same; for that reason, she assumed, building "human well-being" requires starting from recognition of the human sensorial mechanisms, because "people, and not the system, are smart." The Ph.D. student Giorgio Caprari[2] tackles different aspects, one of which being the importance of deep knowledge of the design context, considering it within its spatial and temporal dimension, and of the dialogue between designers and clients, or the local community, in order to meet their needs and reach health equity.

Professor Matteo Vitali[3] was the discussant of the live session, focusing on the possibilities of finding a field of collaboration between research in the field of public health and infectious diseases, and design research, in a transdisciplinary way. He showed the importance of creating disciplines such as environmental hygiene (1854)—taught in the past within the Faculty of Architecture as well—and environmental chemistry (1962), which investigate the environment's impact on human health.

Moreover, he showed several research results on outdoor and indoor living environments. The impacts of these research efforts on how the built environment is made[4] brought up a new perspective for collaboration among those and other disciplines (i.e. technology of architecture, planning, etc.).

---

[1] D.Sc. Deputy Head, Associate Professor, Leader Urbanism Research Cluster, National University of Singapore, College of Design and Engineering, Department of Architecture, 4 Architecture Drive.

[2] Landscape architect, PhD "Sustainable Urban Planning and Inland Areas Development Strategies," University of Camerino (UNICAM-CNR scholarship holder).

[3] Full Professor, Sapienza University of Rome, Head of Environmental Chemistry Laboratory, Department of Public Health and Infectious Diseases.

[4] For the outdoor environment, see: Guidotti, M., Stella, D., Dominici, C., Blasi, G., Owczarek, M., Vitali, M., Protano, C. (2009). Monitoring of traffic-related pollution in a province of central Italy with transplanted lichen Pseudovernia furfuracea. *Bulletin of Environmental Contamination and Toxicology*, 83(6), 852–858.

For the indoor environment, see: Manigrasso, M., Vitali, M., Protano, C., & Avino, P. (2017). Temporal evolution of ultrafine particles and of alveolar deposited surface area from main indoor combustion and non-combustion sources in a model room. *Science of the Total Environment*, 598, 1015–1026.

Fourteen papers have been selected for the conference and are published below. A matching of each paper's keywords shows their interrelated conceptualization: healthy leaving, quality of life, and human well-being. The papers assume that social and environmental determinants impact equity in human health and well-being. The words most frequently quoted by the authors are cohesion policies, social sustainability, accessibility, education, emergency, physical activity, and risk reduction. Each paper tackles different scales of intervention from citywide to neighbourhood and district, from local and natural environment to construction, and they discuss various "spaces" and "places" like housing, schools, healthcare facilities, nursing home, snoezelen rooms, sports facilities, ecclesiastical heritage, and artistic monuments. The presented research investigates the topic of the impact on human health and well-being of different "kinds" of people/users such as communities, children, pedestrians, the elderly, tourists, the homeless, users with intellectual disabilities, and healthcare professionals. Furthermore, two categories of the research method applied within the papers may be identified, as follows: observation method—through tools such as sustainability assessment, transit-oriented development, and evidence-based design—and experimental, simulation methods, using the Internet of Things, Advanced technology, ENVI-met (three-dimensional microclimate model), Environmental Modelling and Simulation, Environmental Sensing, and Advanced prefabrication.

It is interesting to note how research is moving more in the direction of the experimental simulation method through information technologies than towards applying the observation method. The papers offer a valuable contribution to understanding various points of view and identifying the areas of research interested in designing the built environment for healthy living.

78.3 Conclusion

To conclude, designing an environment for healthy living is a challenge that requires skills to make people's health needs the driving principle in the designing process, while remaining fully aware of the context under consideration, in an interdisciplinary way. However, people are not the only focus overcoming the vision of people's health; nowadays the challenge is "One health" as "a collaborative, multisectoral, and transdisciplinary approach—working at local, regional, national, and global levels—in order to achieve optimal health (and well-being) outcomes while recognizing the interconnections between people, animals, plants, and their shared environment" (OHHLEP 2022). This concept is also recalled in the Italian Recovery and Resilience Plan.

In this framework, the discipline of architectural technology plays a strategic role, as a "mediator" in connection with various disciplines such as engineering, the social, economic, and medical sciences, and others. Digital technology, artificial intelligence, and the metaverse will allow a vision of "One health" to be created while not losing sight of the fact that they are tools used by humans and that people have to be involved in the processes of building "one" healthy future.

References

Bircher J, Hahn EG (2016) Applying a complex adaptive system's understanding of health to primary care. F1000Research 5:1672. https://doi.org/10.12688/f1000research.9042.2

Bircher J, Hahn EG (2017) Will the Meikirch model, a new framework for health, induce a paradigm shift in healthcare? Cureus 9(3):e1081. https://doi.org/10.7759/cureus.1081

One Health High-Level Expert Panel (OHHLEP), Adisasmito WB, Almuhairi S, Behravesh CB, Bilivogui P, Bukachi SA et al (2022) One Health: a new definition for a sustainable and healthy future. PLoS Pathog18(6):e1010537. https://doi.org/10.1371/journal.ppat.1010537

Rydin Y, Bleahu A, Davies M, Dávila JD, Friel, De Grandis G, Groce N, Hallal PC, Hamilton I, Howden-Chapman P, Lai KM, Lim CJ, Martins J, Osrin D, Ridley I, Scott I, Taylor M, Wilkinson P, Wilson J (2012) Shaping cities for health: complexity and the planning of urban environments in the 21st century. Lancet (London, England) 379(9831):2079–2108. https://doi.org/10.1016/S0140-6736(12)60435-8

Tarlov AR (1996) Social determinants of health: the sociobiological translation. In: Blane D, Brunner E, Wilkinson R (eds) Health and social organization. Towards a health policy for the 21st century. Routledge, London, pp 71–93. https://doi.org/10.4324/9780203434932

UNESCO (2020) Urban solutions: learning from cities' responses to COVID-19: online meeting report. Retrieved from: https://unesdoc.unesco.org/ark:/48223/pf0000373940 [Online Resource]

United Nation (2015) Resolution adopted by the General Assembly on 25 September 2015. Transforming our world: the 2030 Agenda for Sustainable Development. Retrieved from: https://sdgs.un.org/publications/transforming-our-world-2030-agenda-sustainable-development-17981

World Health Organization (1948) Constitution of the World Health Organization. Retrieved from: https://www.globalhealthrights.org/wp-content/uploads/2013/10/Constitution-of-the-World-Health-Organization-WHO.pdf

World Health Organization (1986) Ottawa charter for health promotion. Retrieved from: https://www.euro.who.int/__data/assets/pdf_file/0004/129532/Ottawa_Charter.pdf

World Health Organization (1998) Health promotion glossary. Retrieved from: https://www.who.int/healthpromotion/about/HPR%20Glossary%201998.pdf

World Health Organization (2009) Retrieved from: https://www.euro.who.int/__data/assets/pdf_file/0015/101076/E92343.pdf

World Health Organization (2010) A conceptual framework for action on the social determinants of health. Retrieved from: https://www.who.int/sdhconference/resources/ConceptualframeworkforactiononSDH_eng.pdf

Chapter 79
New Paradigms for Indoor Healthy Living

Alberto De Capua

Abstract For some time, we have been witnessing a series of alarms that warn us that our wealth is leading to an exaggerated and increasing use of energy and resources, as well as the deep and irreversible transformation of natural systems and social inequality, causing a continuous growth of the overall impact of the human species. "Ecology" and "environment" have become the key words of the third millennium: a media bombardment that has helped to overcome the insurmountable barrier of indifference and insensitivity. Everything that has to do with architectural design, from the choice of materials to the technologies used, has been confronted with the term "sustainability," whose meaning, despite the attempt to place it in a univocal definitional apparatus, always takes on different nuances and meanings. The text defines the complex system of principles that animate architecture today whether those aimed at greater attention to and protection of the health of users and the environment or also concerns social and economic issues when it is proposed as a cultural, social, ecological and economic change necessary to safeguard future generations. Today, we are often witnesses of an inadequacy and poor quality that concerns precisely the aspects of health and safety. The origin of this failure is attributable to attitudes, indifferent to housing needs, easily found in the majority of designers: the environmental scale of the project intended mainly as morpho-typological abstraction, the superficiality in technological choices, and the poor verification of interactions between technical elements and housing needs.

Keywords Health · Indoor air quality · Environmental quality · Sick building syndrome · Housing

A. De Capua (✉)
Mediterranea University of Reggio Calabria, Reggio Calabria, Italy
e-mail: adecapua@unirc.it

E. Arbizzani et al. (eds.), *Technological Imagination in the Green and Digital Transition*, The Urban Book Series, https://doi.org/10.1007/978-3-031-29515-7_79

79.1 Introduction

The control of the environmental quality of an architectural project is one of the most recent and essential requirements of the design activity, whose main purpose is to control those physical parameters closely related to the definition of the microclimate of confined spaces and environmental comfort. It is a matter of identifying parameters and related methodologies able to control the state of the environment, according to needs arising from existential conditions and activities of users with reference to the phenomena of "heat," "cold," "noise," "light," "air purity," defining the structured set of environmental "surroundings."

Each of these parameters can be considered as indicative of a measure of environmental quality and, at the same time, can be considered as an initial element of assessment of the so-called "technological quality," that is, indicator of the effectiveness of the technical-constructive solution.

Controlling the environmental quality of the project, in the awareness of these aspects, is obviously difficult. Also because, to this new need, there are several unfavorable elements: the increase in the technological complexity of the building product; the introduction of new technologies; the lack of knowledge of the behavior of building materials with respect to external actions and over time; the difficult measurability of energy costs, and the polluting effects of building production (Stiglitz 2016).

Today, in addition, it is not possible to give up the idea that the new basic indications for the design and implementation of architecture, in a perspective of circularity (Losasso 2021), should concern the impact of buildings both on the environment and on the consumption of resources, but above all, on the health, comfort, and safety of the occupants, as is increasingly emerging from the scientific and technical debate.

The fundamental performance that users require from built spaces is that of having conditions suitable for living well and being able to carry out their activities well, and indoor pollution is an objective indicator of the general inadequacy of buildings and the discomfort of users. We are, however, often witnesses of inadequacy and poor quality that concerns precisely the aspects of health and safety. The origin of this failure, the demon of development (Faloppa 2016), is attributable to attitudes, indifferent to housing needs, easily found in the majority of designers: the environmental scale of the project intended mainly as morpho-typological abstraction, superficiality in technological choices, and poor verification of interactions between technical elements and housing needs.

Situation further worsened by the advent of industrialization and, within this, the contribution of chemistry, due to which hundreds of new materials have been introduced into the construction field that should have perfected the living and building.

These, really, made more difficult the quality control of the built environment and its compliance with the needs of users, more and more distant from the process for the production of housing.

79.2 A Public Health Problem

For some years now, in the most developed countries, a new issue of primary interest for public health has emerged: the air we breathe inside non-industrial buildings is more polluted than the air outside, and most of the population spends up to 90% of the time of their lives in such places (U.S. Environmental Protection Agency 1989; Lepore et al. 2010).

The house, after decades of building policy oriented toward objectives of "quantity" first, and then "quality," has long since lost that aura of sacredness that has historically distinguished it, becoming, by now, a dangerous place exposed to radiation of natural and artificial origin. Houses, hotels, offices, schools have concentrations of pollutants dangerous to health higher than those detectable outsides. The causes of this situation are attributable to several factors, such as the introduction of new products in the building cycle, the lack of attention to the design of technical solutions, and the decrease in ventilation due to the increased sealing of buildings.

The buildings we inhabit are often polluted, dangerous, degraded, poorly ventilated, and unsuitable for the climate. It seems natural to assume that the quality of the project is inadequate with respect to the complexity of the problems, knowledge, and requirements. Designers and builders have rarely corresponded with a real technical-professional adaptation to technological innovation, to the evolution of the social framework or to the mutation of housing needs (Mercalli 2018).

And just for a few years, there has been insistent talk of building-related illness[1] or sick building syndrome[2] (De Capua 2008, 2019). Houses, schools, hospitals that until a few years ago escaped any control of air quality and healthiness of the environment are now the subject of specific studies.

At present, we cannot say that the problem has been solved, nor that the knowledge acquired is able to describe and manage all the mechanisms through which the phenomenon occurs.

In Italy, the same legislation in force is certainly not suitable to deal with the problem: only sporadically, in fact, the attention of the legislature and the public administration has been addressed to the confined living environments, and when this happened, it was only for specific sectors and pollutants to correct situations

---

[1] The symptomatology is specific and can be traced back to a precise etiological agent: these are almost always microbial agents. Typical examples are legionnaire's disease, a form of acute pneumonia due to a microorganism present in air-conditioning systems or particular allergic reactions. These types of diseases can then be attributed to a specific source of indoor pollution: the contamination of air-conditioning systems in the first place, but also by the action exerted by air humidifiers in a polluted environment (humidifier fever). B.R.I. occurs in a limited number of people living in a given indoor environment.

[2] It has different characteristics from B.R.I. The symptoms are nonspecific and can be grouped into five basic categories. All of these symptoms, unlike B.R.I., regress or disappear as individuals move away from diseased buildings. Since S.B.S. has been observed and described almost exclusively in offices, the regression of symptoms has been, as a rule, observed when returning from work in the evening and during the weekends. There is a wide range of factors potentially associated with the symptoms that determine sbs: chemical, physical, biological, or psychosocial.

now worrying (Ministero della salute 2015). The search for solutions that would allow the exploitation of valuable areas has caused designers to forget the beneficial influence, both psychologically and physically, of air and natural light. The rush to build and the cost of labor has imposed the use of materials that are easier to use, but often unproven.

The spread of "do it yourself" has put in the hands of the common man highly dangerous products, if used without due caution.

All these actions, but above all the combination of them, have had the effect of worsening indoor air quality. It is not just a matter of discovering that indoor pollution situations can occur equal to or greater than those of the outside; but of tracing all the transformations and changes, that technological evolution suggests to us to improve our environment, to the qualitative control of the built environment and its compliance with the needs of users, in the integration of old and new knowledge.

79.3 Envelope and Environmental Quality

The building, according to its morphological, dimensional, organizational, and technical-constructive characteristics, is able to establish a relationship with the external environment, such as to produce variations and significant alterations in the conditions of thermal comfort. The indoor environmental comfort depends not only on the quality, but also on the ways of using building technologies, depending on the complex of environmental variables that act outside the building.

The building envelope is no longer conceivable as a simple barrier, but rather as a selective filter, with the ability to annex and/or reject the effects induced by external environmental conditions.

Research for a low-impact technology, which is able to set the conditions for the construction of spatial and functional arrangements in controlled ecosystem, is achievable only when the operation of technical transformation is not indifferent to the places, but find its declination with respect to the physiographic and bio-ecological characteristics of the same. Therefore, the definition of the conditions of organization and structuring of the building (construction technology, size, shape, configuration, distributional characters) should not be separated from the consideration of the physical-spatial and material conditions of the surrounding environment of reference. The controlled interaction of these two factors (Özdamar and Umaroğullari 2018) can contribute (Brown 2002) to:

- improving the quality of the project, limiting the conflicting conditions of interference with elements or parts of the natural system;
- optimizing the material and technical-constructive characteristics in relation to the needs of protection (from moisture, thermal load, air infiltration), aero-illumination (field of view, air purity, spare parts, etc.);

- encouraging the natural functioning of buildings (passive lighting and natural air-conditioning systems) also through the use of hybrid passive technologies, biocompatible materials, and building components.

Opacity and transparency of the envelope, moreover, determine the onset of different problems, mainly due to the use of different materials, with their own behaviors and characteristics. Opaque walls are more critical in relation to possible indoor pollution, transparent walls have more influence on aspects related to thermohygrometric comfort.

The pollution produced by buildings depends mainly on the fact that the materials used; the substances and the manufactured articles are not sufficiently tested from the point of view of emissions and their effects on the users' health.

The concept of healthiness of a building is very articulated and complex and cannot be defined only in relation to the absence of pathologies and influencing factors, but must meet the needs of comfort and physical, psychological, and social well-being. The building is a container of activities and behaviors, and to design healthy buildings, it is necessary to think about each environment in relation to all factors, including microclimatic conditions, which have some relevance in improving or worsening the indoor air. It is already from the earliest stages of the project, that the location, orientation, size, shape, and envelope of a building, as interacting with the air-conditioning systems and techniques for the containment of energy consumption, must be verified also in reference to their indirect "harmfulness" (Zannoni 2006).

A first definition of the functioning of the technological building system must be based, therefore, on the analysis of the external environment, that is, according to the climatic characteristics at regional (macroclimate) and local scale without neglecting the possible intervention of modification of the latter to obtain a particular microclimate.

The result of this first phase is the definition of the output to be required to the technological system in terms of strategies to control ventilation, summer and winter solar radiation, and natural lighting: a behavior model to be completed with a first performance dimensioning of the various technological subsystems, in relation to the functions attributed to them.

Just as the definition of spaces and the choice of construction, materials have a significant impact on indoor air quality: with a correct distribution, it is possible to isolate and enclose the most polluting activities and make the systems more accessible, in order to facilitate their maintenance; with a conscious choice of materials and construction techniques, it is possible to keep under control the potential level of pollution.

It is, therefore, to set up a rigorous design procedure equipped with verification tools to exclude or minimize errors in air quality. An analytical design process that highlights the requirements and interactions derived from the needs of the inhabitants, especially those requirements related to the healthiness of the building.

A clear understanding of how multiple factors can affect indoor air quality, and how they should be controlled, can enable the designer to develop an effective design for indoor air quality verification. To this end, real theoretical models have been developed by several parties that define the problem from several points of view and lead the factors to be taken into account in three main categories:

- Physical: temperature, relative humidity, ventilation speed and direction, artificial lighting, solar radiation, noise, ions, presence of electromagnetic radiation fibers, and particulate matter.
- Chemical: ETS, formaldehyde, VOCs, biocides, CO_2, CO, NO_2, O_3, SO_2.
- Biological: microorganisms, bacteria, etc.

It is clear that in order to address the problem, a global approach is fundamental. A building is a system and as such should be studied and treated in the totality of its interactions: one should not act on one of its parts without considering the effects on the others, and one should not act on air quality without considering the implications in the management—including energy—of the building, and it is necessary to verify that problems are not solved because of a polluting substance or source by adopting techniques that may have negative effects on the building elements or that in turn contaminate the air. To this aim, it is necessary that for each intervention, the designer manages all aspects of the problem and is able to choose among the possible techniques, those best suited to the building/user system.

Most of these techniques have already been used, experimentally, in the United States, Canada, and Northern Europe and tend to act directly on the polluting sources or treating the polluted air. It is clear that it is preferable to act directly on the source, avoiding those pollutants reach the air, but this is not always possible: the choice of the most suitable techniques is conditioned by technical and economic constraints or guided by strategic considerations of a more general nature, and moreover, it is often not possible to identify the pollutant source; it is therefore more frequent to resort to strategies of ventilation or air treatment.

Poor ventilation of rooms is a major cause of poor air quality problems, and ventilation standards are an important method of control, effective, above all, in removing pollutants released from indoor sources and in removing excess moisture.

If the outdoor air is of acceptable quality, ventilation can effectively reduce pollutant concentrations by diluting and expelling contaminated air with cleaner outdoor air. If the outside air should be polluted before using it, it must be treated. External air exchanges take place through: infiltration and extrafiltration mechanisms; natural ventilation; mechanical ventilation.

79.4 Parameters and Indicators for a New Scenario

In planning a scenario that encompasses the new principles of ecology, we must then consider that the objective is not to control all the variables involved, but to reflect on the main indicators, on which the achievement of sustainable quality may depend, within the more traditional construction processes: investigate new thematic areas, integrate existing ones with new specific requirements, relate the new indicators to the areas of application and the different levels of the project.

Show designers the way to improve the environmental efficiency performance of the building and to promote the use of "clean" technologies.

It is an approach that can realize its effects in the construction phase of the intervention programs and in the operations of verification on the quality of the constructions, in the intent to develop a whole series of collaborative aspects—according to common and shared methodological and operational lines—between structures that operate with technical-building competences also very different from each other.

A demand-performance approach, containing clear and measurable quality levels, possibly without specific constructional references, and giving the designer and builder the responsibility of ensuring that designs and constructions conform to the prescribed requirements.

The difficulties are not few, especially with regard to the need to prepare operators and have adequate equipment to measure performance. The problem remains open.

It is a matter of leaving the current phase and preparing an instrument for evaluating environmental quality, which does not exist today; convincing oneself that there can be no seriously managed legislation, but above all no quality policy, without adequate experimentation, without control of real quality in actual operating conditions, and without adequate technical support information.

Something complex that, among other things, requires:

- to redesign local regulations as many regions are already doing
- the congruity of local regulations with national and community constraints
- the attribution of progressive importance to the norm as a recommendation, seen as an opportunity for operators to grow technically and culturally.

The designer must be aware of the new needs to be met and guide the individual project choices toward the awareness of natural balances, toward efficient transformation interventions, using technologies and tools for bioclimatic control of the system (Paolella 2006).

In these concepts, the term efficiency refers to the measurement of the quantity and frequency of exchanges, at various levels of relationship between physical and biological systems. Inherent in the term are precise indicators:

- contextuality, understood as the ability to recognize oneself in the place by determining and enhancing it, also in terms of resources fed by the system;
- adaptability, understood as the ability of the system to adapt to environmental conditions;
- duration, or the ability to maintain itself over time;
- deconstruction, understood as the system's ability to use, produce and recover resources.

Our approach is structured through this key of interpretation, with the aim of contributing to a significant improvement of housing quality, respecting the receptive limits of ecosystems, the possibility of renewal of natural resources (especially for their conservation), the balance between natural and anthropic systems (eco-sustainable needs).

79.5 Conclusion

The new rules of living quality now inseparably include the updating and cultural actualization of the past concept of healthiness. Combining rehabilitation and redevelopment interventions with the problems of indoor air quality control will be the challenge of the coming years, and this involves a multi-level approach. I believe that the only tool today that can deal with this complexity is the *Building Regulations* on which the different scales of intervention can be brought together and in which multiple disciplines can converge with important repercussions, at a cognitive and operational level, on design activity. Because of its complexity, it can fully address what are the fundamental actions for technological control of indoor air quality:

- awareness of the context in which we intervene;
- the control over materials and products;
- control over microclimatic conditions.

The innovations of a new and current regulation could be:

- The acquisition that indoor air quality is no longer definable through ventilation and carbon dioxide content requirements alone but calls into question the many unexpected sources of pollution: from materials used to microclimatic conditions.
- A set of prescriptions on the basis on requirements to be met, referring to the needs of users, accompanied by systematic verification methods, making mandatory only a limited package of requirements related to mandatory safety and health requirements.
- Become the only instrument for regulating building activity present at the municipal level: this means that (as far as building prescriptions are concerned) it replaces and cancels any other instrument.
- Verification of conformity in place of habitability, no longer a bureaucratic act, but an effective guarantee of performance subject to risk assessment on what the designer declares in a descriptive technical sheet.

References

Brown LR (2002) Eco economy. Editori Riuniti, Roma
De Capua A (2008) Tecnologie per una nuova igiene del costruire. Gangemi editore, Roma
De Capua A (2019) Il Miglioramento della Qualità Ambientale Indoor negli Interventi di Riqualificazione Edilizia, Roma, Legislazione Tecnica
Faloppa F (2016) Alexander Langer, Non per il potere, chiarelettere, Milano
Lepore A, Ubaldi V, Brini S (2010) Inquinamento indoor: aspetti generali e casi studio in Italia, quaderni ISPRA, n. 117/2010
Losasso M (2021) Transizione circolare: scenari per il futuro. In: TECHNE, 22/2021, pp 7–9
Mercalli L (2018) Non c'è più tempo, Einaudi
Ministero della Salute (2015) Direzione generale della comunicazione e dei rapporti europei e internazionali. Vari Reports. www.salute.gov.it

Özdamar M, Umaroğullari F (2018) Thermal comfort and indoor air quality. Int J Sci Res Innov Technol 5(3)
Paolella A (2006) L'ombra dei grattacieli, il prato. Saponara (PD)
Stiglitz JE (2016) Invertire la rotta. Editori Laterza, Bari
U.S. Environmental Protection Agency (1989) Office of Air and Radiation. Report to Congress on Indoor Air Quality, Volume II: Assessment and Control of Indoor Air Pollution, pp I, 4–14, EPA 400-1-89-001C
Zannoni G et al (2006) Gas Radon, tecniche di mitigazione. Edicom, Monfalcone

Chapter 80
Healthy and Empowering Life in Schoolyards. The Case of Dante Alighieri School in Milan

Valentina Dessì, Maria Fianchini, Franca Zuccoli, Raffaella Colombo, and Noemi Morrone

Abstract This paper presents a participatory process aimed at improving outdoor education in a primary school in Milan. The rationale of this work was that the psycho-physical benefits for children from outdoor living could be enhanced through outdoor education. Indeed, open-air environments are fit for supporting learning experiences, bringing out different abilities and improving well-being. Moreover, during the COVID-19 pandemic, schoolyards turned out to be a resource for overcoming physical distancing. Nevertheless, the availability of flexible physical environments and proper equipment for the educational goals is a basic condition for overcoming difficulties in the extensive use of outdoor spaces in schools. The purpose of this work was to support the school in designing new outdoor educational environments with a focused vision on the pedagogical context. Thus, the process was developed by a multidisciplinary team with the involvement of the students and the teaching staff. By the initial analytical stage, site and use conditions as well as emerging needs were enlightened. These outcomes were assumed to develop a design solution both suitable for the innovation goals and attentive to environmental aspects. The proposal was selected for funding by the Municipality and implemented. Finally, a three-year post-occupancy evaluation program started in the earliest stages of use. In conclusion, by the first monitoring activities, it emerged that outdoor educational

---

V. Dessì · M. Fianchini (✉)
Polytechnic University of Milan, Dept. of Architecture and Urban Studies, Milan, Italy
e-mail: maria.fianchini@polimi.it

V. Dessì
e-mail: valentina.dessi@polimi.it

F. Zuccoli
University of Milano-Bicocca, Dept. of Human Sciences for Education "Riccardo Massa", Milan, Italy
e-mail: franca.zuccoli@unimib.it

R. Colombo · N. Morrone
Istituto Comprensivo Rinnovata Pizzigoni, Milan, Italy
e-mail: raffaella.colombo@scuolarinnovata.edu.it

N. Morrone
e-mail: noemi.morrone@scuolarinnovata.edu.it

E. Arbizzani et al. (eds.), *Technological Imagination in the Green and Digital Transition*, The Urban Book Series, https://doi.org/10.1007/978-3-031-29515-7_80

experiences increased and diversified from the past, together with students' perception of opportunities and benefits achieved from more frequent and longer work in external environments, and their expectation of involvement in proposing further implementations.

Keywords Outdoor learning · Children well-being · Healthy life · Environmental education · Monitoring process

80.1 Introduction

Life in the open air is essential for the psycho-physical development of children and teenagers, for the benefits and stimuli deriving from contact with nature and the opportunity to overcome sedentary lifestyles that favor obesity and limitation of social relations (Muñoz 2009; Knight 2013). However, most of their day is spent inside the school building, with limited time in the schoolyards, mostly for recess. Outdoor education favors the development of children's multiple skills; in particular, direct experience, practical exercises and interaction with the surrounding environment promote learning and school performance, as well as well-being and social inclusion. At the same time, they reduce the risk of discomfort and disease (Dessì and Fianchini 2021; Faskunger et al. 2018).

Various national and international experiences focused on the problem and promoted initiatives to encourage outdoor activities in the school's outdoor spaces and enhance them (Broda 2011; Gamson 2010). The new configuration of outdoor school spaces should help teachers overcome operational difficulties and imagine new ways of working (Bellomo 2019; Dessì and Piazza 2020; Gilbertson et al. 2006; Boston Schoolyard Initiative 2013).

In the years of the COVID-19 pandemic, the classrooms arrangements returned to fixed and outdated solutions; thus, schoolyards were rediscovered as resources for overcoming physical distancing. Providing flexible environments and equipment suitable for teaching objectives is a fundamental condition for overcoming cultural resistance and operational difficulties in the extensive use of outdoor spaces in schools. However, it is not always enough to set up the places; it is also necessary to promote awareness-raising paths on the potential that different school contexts can offer and involve the school communities in designing the most suitable solutions.

The case study of the new outdoor learning environments in the Dante primary school in Milan is an example of place adaptation and practice innovation. It has been done through the synergy between different disciplines, the development of design visions anchored to the context but open to the future, the renewal of decision-making and operational processes.

The Dante Primary School is part of the Rinnovata Pizzigoni Comprehensive School, where the Pizzigoni Educational Method is followed. This is an early XX century experimental program aimed at reforming learning methods, for whose application a new school was purposely built, with all classrooms opening onto a large

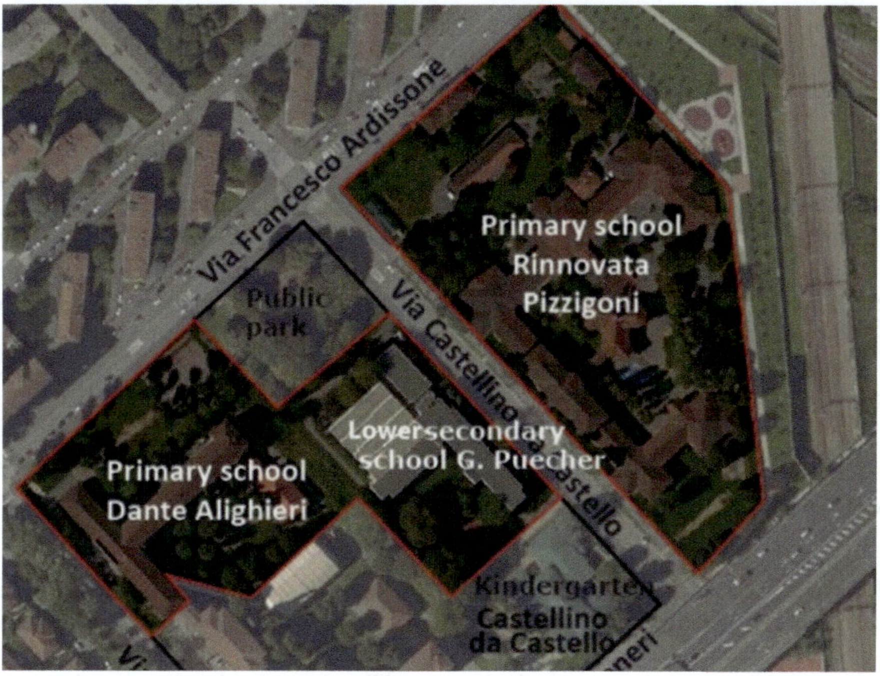

Fig. 80.1 Aerial view of the three schools of the Rinnovata Pizzigoni Comprehensive School (Dessì and Piazza 2020)

well-equipped garden, to offer children various opportunities for direct experiences through observations and practical activities (Pizzigoni 1914).

The Dante School is located on the third floor of a traditional school building, located on the opposite side of the road (Fig. 80.1). Since the method was adopted, the spaces and special equipment available in the Rinnovata Pizzigoni school have been shared with them, causing an increase in use and greater access difficulties. Hence, the need to adapt the external spaces of the Dante school to Pizzigoni's educational approach and thus rebalance the school's resources.

80.2 Process Development: Methodology and Results

The enhancement of the outdoors of the Dante school through new equipment for educational use is the result of a process started without a precise plan, but developed step by step with a multidisciplinary scientific approach and experimental participation practices.

In 2017, as part of the activities of the Ambiente Scuola team of the DAStU dept. of the Politecnico di Milano in collaboration with the Department of Human Sciences for Education "Riccardo Massa" of the University of Milano-Bicocca, a research agreement was stated with the Rinnovata Pizzigoni school. Then, the process started to analyze the site, identify users' needs and expectations (students and teachers) and define project proposals by university students of the School of Architecture (2017–2018).

In July 2020, a competitive call by the Municipality of Milan for the "support of projects aimed at the innovation of learning environments for the first cycle of education schools in Milano area" gave the chance to resume the project and deepen it, also to meet a pedagogical plan. The project proposal got funding, so the new learning environments were built in the early 2021 and occupied in April 2021, while the monitoring process started a month later.

80.2.1 Phase 1. Analytical Activities

During the collaboration with the Rinnovata Pizzigoni Comprehensive School, some activities were carried out on the external spaces of all its schools.

Some of these didactic experiences were aimed at defining guidelines and design proposals for outdoor learning in the Dante and Puecher schools. However, the close connection between all schools often prompted to include the Rinnovata Pizzigoni school in the analysis as well. The works have been developed through a method based on the enhancement of spaces, starting from a careful process of knowledge from different points of view. According to this approach, strategic lines of intervention have been proposed, that must take this into consideration:

Functional aspect. Flows between the school building and the open space, flows among the three schools, the entrances.

The maps in Fig. 80.2 show the external space of the Rinnovata school compared to the other ones. The balance between close and open spaces in this school encourages the use during break time and outdoor learning activities and the pavilion type offers more chances of moving from indoors to outdoors than in the Dante school, that is located on the third floor and consequently with few connections. Space functions are less in this latter and, apart from the vegetable garden, the outdoor space is more suitable for play than for learning. For this reason, flows are mainly from the Dante school to the Rinnovata one, which shares special spaces (a farm, a science pavilion, a greenhouse).

In Dante, once the outdoor experience related to the observation or contact with natural elements is over, the lack of a gathering space for the class is evident.

Physical aspect. The resources, in terms of vegetation, equipment, and the relationship between permeable/waterproofed surface (analysis of materials).

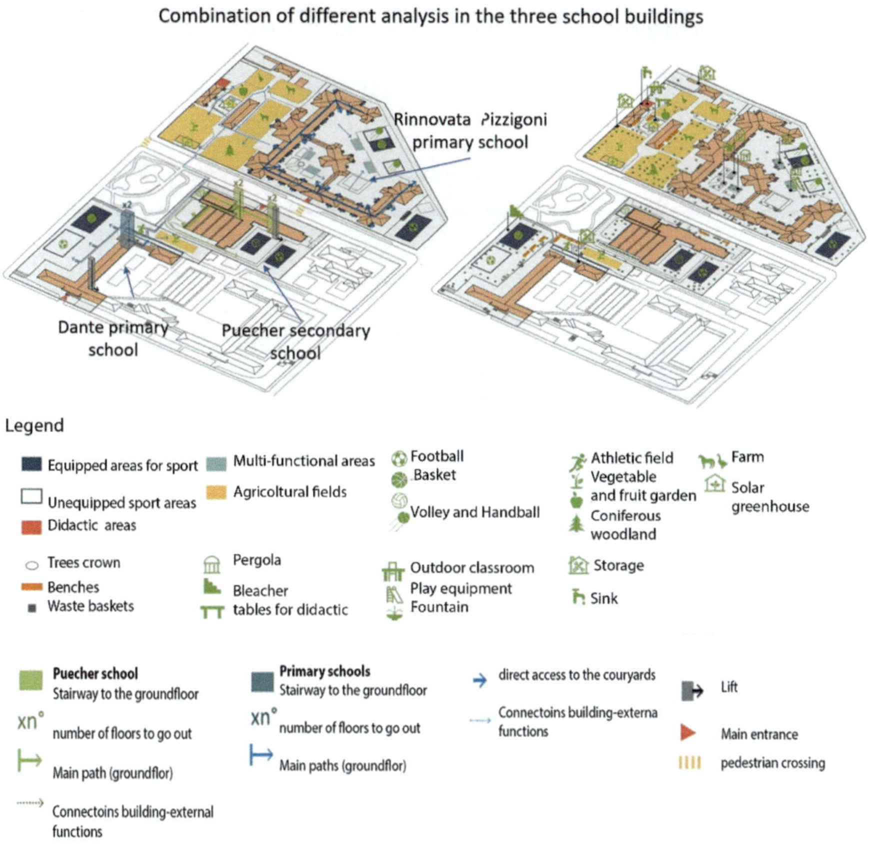

Fig. 80.2 Maps of the School Institute combining different aspects: learning/sport areas with pedestrian flows/accessibility (left) and with equipment (right) (Dessì and Piazza 2020)

In the three schools, there is a high percentage of draining lawn flooring, but also a part of waterproof, in asphalt; the existing concrete or earth sports fields present critical issues both for the type of activity (no suitable paving materials) and in the moments after rainfall or, on the contrary, when it is too dry because of the dust. At the main entrances of the schools, there are concrete tile floors, while the concrete sidewalks run all around the school buildings (Fig. 80.3).

Energy aspect and environmental comfort. Identifying areas with more potential to host/equip specific functions.

It can be done with a series of assessments of the microclimatic conditions that change according to the urban morphology, the orientation, the materials, and the vegetation. The abacus of the vegetation reports, among others, the characteristics of the size and shape, necessary to build the models for the simulation.

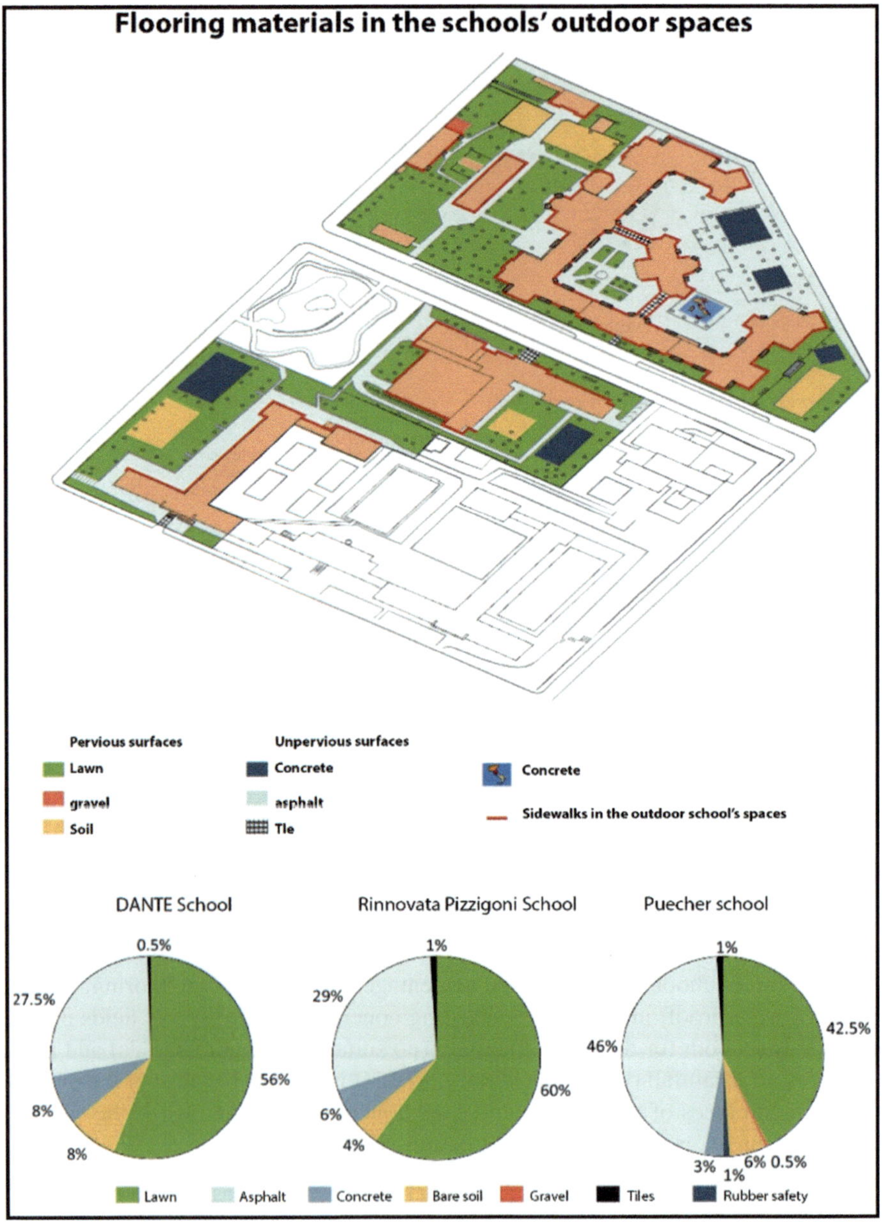

Fig. 80.3 Distribution of pervious and waterproof materials in the three schools. *Source* (Dessì and Piazza 2020)

The first campaign of microclimatic measurements was carried out in spring (March 2017) and made it possible to calibrate the OTC model simulation tool. Through simulations, the thermal comfort conditions were assessed with the UTCI comfort indicator (comfort conditions is in the range 9–26 °C and up to 32 °C with a slight discomfort) in three seasons and three hours/day (Fig. 80.4).

Thermal comfort conditions are strongly conditioned by the component of solar radiation incident on the ground, and by the coating material of the irradiated surfaces. The solar radiation control, the shading, and the materials choice (including vegetation and water) are the main strategies to improve comfort conditions and give a distinctive sign to the project. The maps (Fig. 80.4) show a spring day (1 pm), the summer solstice (2 pm), and the winter solstice (12 noon), i.e., times when the children are at school.

Considering the maps of the UTCI comfort indicator, in the part concerning the area of the Dante Alighieri school, it emerges that in the spring (1 pm), the two most used areas have different behaviors depending on the position with respect to the solar radiation. In spring, at 1 pm, the two sport fields and the area next to the vegetable garden—suitable for placing an outdoor classroom—are mostly sunny and in the

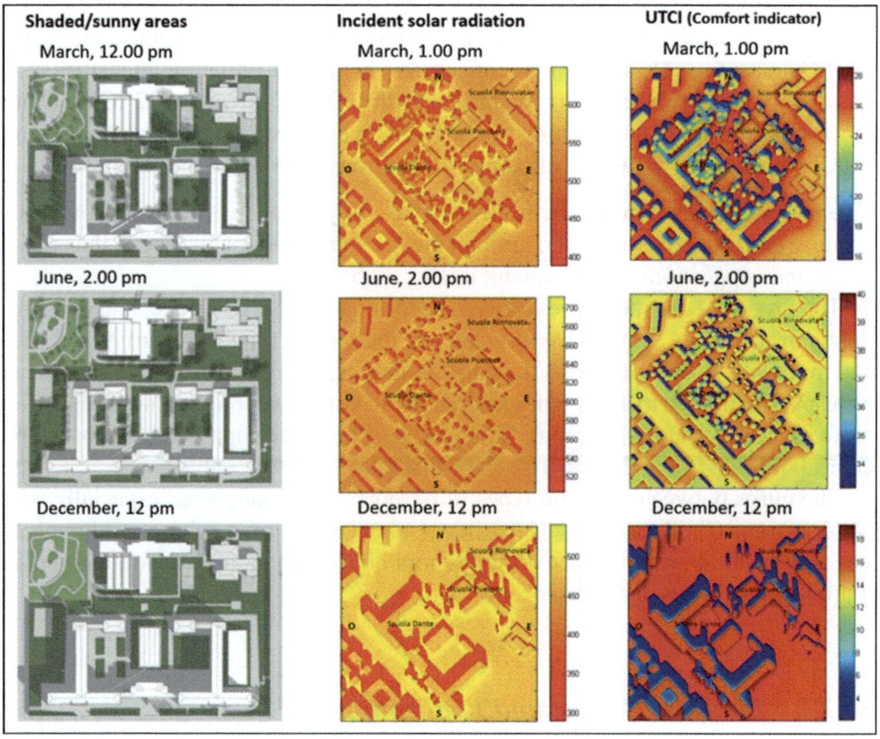

Fig. 80.4 Simulations of the shadows, of the incident radiation, and the comfort conditions in the Dante and Puecher schools (Dessì and Piazza 2020)

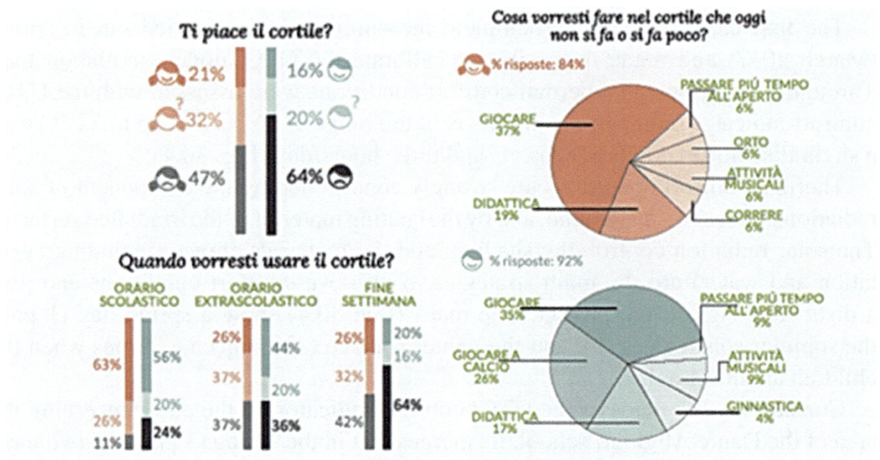

Fig. 80.5 Students' questionnaire, answers on the appreciation of the schoolyard, the time to spend outdoor, and suggestions on possible uses (play, vegetable garden, music, run, learning activities gym) (Dessì and Piazza 2020)

shade only near the trees and result in comfort conditions. In summer, the lack of shaded areas, high air temperature values, and high solar radiation intensities cause unsatisfactory thermal comfort conditions, with UTCI values above 32 °C. In winter, this area is shaded in the morning by the building and trees; the air temperatures and UTCI values are low and out of comfort conditions even when in the sun.

The users' point of view. The needs expressed by users (teachers and students of primary and secondary schools) through the questionnaires on the use of the outdoor space and suggestions for improving it (Fig. 80.5).

The proposed questionnaires differed according to the student age. Answers from pupils of the Dante school have brought out interesting aspects regarding the space and functions. In general, those of the first grades express the need to play and run and have areas equipped to do so. The children of the last classes require more vegetation, more spaces for outdoor learning, and more benches. They appreciate the basketball court, but find critical the concrete floor that makes it little usable.

The curbs between different types of pavements are dangerous especially when children run. Children have diverse ideas about the timing of use. Although the majority favor the use of space during school hours, they are less favorable to using it after school hours and on weekends.

80.2.2 Phase 2. Project and Realization

The analysis results were essential for responding quickly, effectively, and successfully to a municipal tender aimed at funding new learning spaces and plans able to strengthen educational effectiveness and at the same time approach the pandemic limitations.

According to the pedagogical vision of this school, the open-air classroom was intended as a bridge between the primary and the lower secondary schools, to experiment the Pizzigoni's method in an innovative way and to guarantee physical distancing by outdoor activities, while involving the whole school community in its development. Thus, the place was expected to be full of significant and significative elements, to enhance the relationship between learning and experience (Pizzigoni 1921). In fact, outdoor lessons and observation of nature have been hallmarks of this school long before the pandemic emergency made these practices necessary for school life. Specifically, the new space should have had a hybrid and innovative function in which all the pupils could experience new ways of learning through peer education and cooperative learning.

Consistently with the school pedagogical objectives and with the technological-environmental approach, the new project was oriented not only to the search for flexible and adequate solutions for the foreseen use scenarios, but also to the promotion of users' well-being, the implementation and enhancement of the natural resources, the integration of new spaces in the physical and functional system, and the search for solutions with low environmental impact.

The new learning environment was built in a marginal area, on the limit of the courtyard of the Dante school near the connection path to the lower secondary school (Fig. 80.6). It consists of two spaces with different features and functions, available both for parallel activities of different classes, and the joint use of larger and heterogeneous groups, in line with the peer education program. The first space (Fig. 80.7) is about m^2 75, and it is fitted with fixed equipment (a gazebo with a steel frame that supports a packable waterproof roof, wooden flooring and fence) and furnished with tables and stools. The second one (called the "bucolic classroom") (Fig. 80.8) is smaller and more informal; it is bordered by benches made with tree trunks placed under some olive trees, to which are added a dozen seats in wooden blocks. Both are fully integrated into the larger green area of the schoolyard and in connection with the vegetable garden (Fig. 80.9).

Fig. 80.6 Site before the intervention (left) and in the project rendering (right) (elaborated by E. Cusato and A. Esposito)

Fig. 80.7 Use of the open-air classroom with (left) and without (right) tables

Fig. 80.8 Use of the Bucolic classroom during a learning activity (left) and during recess (right)

80.2.3 Phase 3. The Monitoring Process

The Municipality tender required a three-year monitoring program with annual reports.

The proposal assumed a multidisciplinary methodological approach, and a mix of references, both from the post-occupancy evaluation (Way and Bordass 2005) and from the operational and methodological activation of the Student Voice (Cook-Sather 2002, 2009; Flutter et al. 2004). Three objectives were focused: to highlight the educational changes achieved through the of the physical environment modification (Imms et al. 2016); to bring out experiences and encourage comparison between different groups of users, for an expansion of the outdoor learning plan and project; to point out initial critical issues and outline possible interventions.

Fig. 80.9 Partial view of the vegetable garden

In May 2021, the monitoring process was launched through field observations and focus groups with students. It emerged that both spaces had already been used in different ways, especially the main one, whose fittings are very flexible, both for the furniture that can be easily moved by the students (Fig. 80.10) and for the opening roof of the gazebo. The lower secondary students occupy this space in a deeper and more natural way than the primary ones, while standing, moving, sitting on the floor, etc. Conversely, the bucolic classroom is used both for more concentrative activities and during recess. The different but connected use of this pair of spaces typically occurs in peer tutoring activities between secondary and primary classes. The students' most appreciated conditions are: to stay outdoors/in nature; the possibility of looking around, breathing, and seeing schoolmates from other classes; materials other than the internal classrooms and the use of timber.

A second phase was opened in February 2022, when teachers were engaged through two surveys: by electronic questionnaires delivered to all staff and internal interviews to the staff of just the Dante school. Forty teachers from the primary schools and sixteen from the secondary one answered the questionnaires. It resulted that over 70% of them have used the new outdoor learning environments, but only 10% regularly.

As about the activities, the same done inside prevail, followed by those that need larger spaces or in relation with natural elements. It seems significant that better well-being and health conditions than in indoor environments are the main reasons

Fig. 80.10 Pupils put away the stools at the end of their activity

to work outdoors, followed by the greater interest and participation on the part of students, and by the available space and equipment. Conversely, the most highlighted criticality is the disturbance from other classes in the courtyard, followed by feeling cold in autumn.

About half of the teacher sample changed the setting according to the different activities, while a minimal percentage used the coverage opening system to regulate the shading. Finally, almost all the teachers consider these environments as an opportunity to be further enhanced. From the focus with the Dante school teachers more intensive and diversified use of both outdoor environments has emerged, due to the opportunity to experience the school environment differently, in contact with nature, with ample freedom of movement. Again, the most underlined criticalities are the noise from the other classes, the cleaning and furniture keeping, and thermal comfort in the later spring.

80.3 Conclusions

The opportunity of realizing the new outdoor learning environments was offered by the pandemic emergency, which prompted the municipality to invest in innovative fit-out projects to improve school environments and limit risks of contagion. However, a school community already oriented toward an experimental education and open to academic research, together with a base of knowledge previously developed, were the actual conditions to meet this objective in such a short time as the result of careful evaluations from the different points of view.

From the monitoring activities, it emerged that outdoor educational experiences increased and diversified from the past, together with students' perception of opportunities and benefits achieved from more frequent and longer work in external environments, and their expectation of involvement in proposing further implementations.

The process of enhancing the outdoor environments of the Dante school is still in progress, both in terms of the space layout and the operational conditions. Future goals will include favoring a more extended use, to be pursued by both dealing with emerged criticalities and enhancing the evidence of appreciation brought by students and teachers that experienced them with great awareness of the benefits got in terms of greater freedom and psycho-physical well-being.

Acknowledgements We thank Anna Ferri, the principal of the Rinnnovata Pizzigoni Institute for promoting the implementation of the outdoor classroom project in the Dante school and supporting this research work.

References

Bellomo A (2019) Outdoor experience in schoolyards aimed at widening learning opportunities. In: Fianchini M (ed) Renewing middle school facilities. Springer, pp 33–51

Boston Schoolyard Initiative (2013) Schoolyard design guide. Available at http://www.schoolyards.org/pdf/SYDesignGuide.pdf (Accessed 03/2023)

Broda HW (2011) Moving the classroom outdoors. Schoolyard-enhanced learning in action. Ed Stenhouse, USA

Cook-Sather A (2002) Authorizing students' perspective: toward trust, dialoguer, and change in education. Educ Res 31(4):3–14

Cook-Sather A (2009) I am not afraid to listen: prospective teachers learning from students. Theory Into Practice 48(3):176–183

Dessì V, Fianchini M (2021) The schoolyard: a resource for health and educational innovation. Sustain Mediterr Constr 13:160–165

Dessì V, Piazza AI (2020) La scuola è in cortile. Strategie e buoni esempi per valorizzare il cortile scolastico. UNA Press, Pescara. Available at https://urbannarraction.net/ (Accessed 03/2023)

Faskunger J, Szczepanski A, Åkerblom P (2018) Teaching with the sky as a ceiling. Reports from Forum för ämnesdidaktik No 11. Linköping University, Swedish University of Agricultural Sciences and Utenavet

Flutter et al (2004) Consulting pupils: what's in it for schools? Routledge-Farmer, London

Gamson Danks S (2010) Asphalt to ecosystems. Design ideas for schoolyard transformation. Ed. New Village Press, Oakland, CA

Gilbertson K et al (2006) Outdoor education: methods and strategies. Human Kineticx Publisher, Champaign, USA

Imms W, Cleveland B, Fisher K (2016) Learning environments evaluation. Snapshots of emerging issues, methods and knowledge. Sense Publishers, Rotterdam

Knight S (2013) International perspectives on forest school. Sage, Londra

Muñoz SA (2009) Children in the outdoors. A literature review. Sustainable Development Research Centre. Available at https://ltl.org.uk/wp-content/uploads/2019/02/children-in-the-outdoors.pdf (Accessed 03/2023)

Pizzigoni G (1914) Indicazioni per la costruzione della "Rinnovata" rivolte al progettista ingegner Erminio Valverti, s.e., Milano

Pizzigoni G (1921) La scuola rinnovata secondo il metodo sperimentale, Tip. Luigi Luzzatti, Roma
Way M, Bordass B (2005) Making feedback and post-occupancy evaluation routine 2: soft landings—involving design and building teams in improving performance. Build Res Inf 33(4):353–360

Chapter 81
Design for Emergency: Inclusive Housing Solution

Francesca Giglio and Sara Sansotta

Abstract The paper describes a study on the growing emergency of homelessness, of which alarming data are estimated at national and European levels and which the Cohesion policies of the European Community are addressing. Thanks to the recent launch of the Collaboration Platform on Homelessness to stimulate dialog, improve data collection and monitoring and strengthen cooperation between all actors involved in the fight against the phenomenon. The emerging concept of 'Design for Emergency' highlights the historical link between temporary and emergency living regarding the welfare and health implications of the weak. The aim is twofold: to define a theoretical and design model that can be repeated, contributing on the one hand to a process of social reintegration for fragile realities and on the other to the circularity of construction processes and the recovery of resources and components, through innovative housing solutions, with characteristics of modularity, disassembly and dry connections. The results, deriving from a deductive scalar methodological approach, concern: (i) data collection is inherent to the issues addressed, the emergency conditions; (ii) a critical analysis of the data acquired and systematized; (iii) methodological and design experimentation. The research hopes for repeatable results in diverse marginal contexts, respecting the disparate needs not only of the users but of the place where the temporary installation will be needed. This is an aspect in which the intervention of the municipal administrations and all possible stakeholders involved is fundamental and which at the moment may represent a limitation, albeit a surmountable one of the research.

Keywords Cohesion policies · Design for emergency · Homelessness · Quality of life

F. Giglio (✉) · S. Sansotta
Mediterranea University of Reggio Calabria, Reggio Calabria, Italy
e-mail: francesca.giglio@unirc.it

S. Sansotta
e-mail: sara.sansotta@unirc.it

E. Arbizzani et al. (eds.), *Technological Imagination in the Green and Digital Transition*, The Urban Book Series, https://doi.org/10.1007/978-3-031-29515-7_81

81.1 Introduction

Meeting human needs at sustainable levels of energy use is critical to avoiding catastrophic climate change and ensuring the well-being of all people (Vogel et al. 2021). While global trajectories are focused on responding to the environmental emergency through *decarbonization of global energy systems and reducing global energy use* (Grubler et al. 2018; IPCC 2018), social emergency strategies aim to ensure *the well-being and decrease inequalities among various population groups* (Ranmal et al. 2021), whose needs relate primarily to the issue of housing. Indeed, the relationship between health and housing is complex and influenced by many interrelated factors. The relationship between emergency conditions and housing responses is becoming more complex over the years due to the increase in the types of vulnerable groups that no longer only affect developing countries but constantly belong to the everyday emergency life of our urban centers (Kidd et al. 2021).

The social goals of the European Community, as set out in the challenges of the Green Deal, can also be interpreted through a new role for design as an opportunity to propose emerging innovations, even on a small scale, helpful in improving the quality of life and providing answers to the most disadvantaged. Therefore, the consequences of environmental and social emergencies regarding people's health are the background for extensive debates by the European community. The 20 principles of the European pillar on social rights aim at strategies whose goals are inclusiveness and well-being, especially toward population groups with fragility (Cantillon 2019). Responding to these needs through temporary housing models that contribute to giving new dignity to fragile segments of society is a debate that has been ongoing for many years and has led to many experiments.

As a backdrop to this scenario, the emerging concept of 'Design for Emergency' highlights and reinforces the historical link between temporary and emergency housing, in favor of quality of life. The paper proposes an intervention model aimed at the homeless through innovative housing solutions. It is intended to activate social and environmental processes, outlining a social reintegration activity and strategies contributing to the circularity of construction processes and the recovery of resources and components.

From the results of the systemic work of retrieval, classification and interpretation of the themes and case studies, the experimentation of a multifunctional urban micro-architecture was carried out in response to the housing demand of the same. The study also hopes for repeatable results in diverse marginal contexts, respecting the disparate needs of users and places and contributing to the integrated programs of municipal administrations through the involvement of all possible stakeholders.

81.2 The Homeless Condition Between Design for Emergency and Unpleasant Design

Billions of people worldwide are still deprived of basic needs, and current pathways to sufficient needs satisfaction seem to involve highly unsustainable levels of resource use (O'Neill et al. 2018). Carbon or environmental intensities, understood as measures of unsustainability, increase inequalities as an outcome of urbanization processes, showing asymmetrical relationships between 'developed' and 'less developed' countries (Greiner and McGee 2020).

For almost a decade, the 'European Observatory on Homelessness' has promoted the production of an increasing number of policy analyses on existing homelessness strategies in member states (Benjaminsen and Dyb 2008; Baptista 2009; Wygnańska 2009; Hansen 2010; Houard 2011; Hermans 2012; Sahlin 2015). The focus of the European Pillar on Social Rights is on the principles '16—Health care' and '19—Housing and assistance for the homeless' (Commission 2019). Regarding the latter, the European Commission and the European Federation of National Organizations (FEANTSA) signed and launched the *European Platform on Homelessness.* The Platform's launch is the beginning of a process to establish common understanding and commitment and ensure concrete progress in Member States' social cohesion policies in the fight against homelessness.

For the commitment to respond concretely to this emergency, the report 'Technological innovation for humanitarian aid and assistance' (EPRS 2019) highlights how technological innovations are recognized as being capable of playing a crucial role in addressing humanitarian challenges. Through human-centered solutions, shaped to provide users with shelter-related solutions and aimed at a broader social inclusion strategy, the ultimate goal is to improve the health of the most vulnerable. This contribution is part of this challenge that looks at an increasingly extended, varied, and hybrid emergency condition.

The focus on health and well-being for the weakest groups has been accentuated in some ways in the fight against the COVID-19 emergency, through the development of 'Design for Emergency', an initiative of the Centre for Design (CAMD) at Northeastern University (Boston, USA) as an open platform, created to gather the needs of different social groups and to respond through innovative solutions that connect, inform and support communities.

While, on the one hand, the community is engaged in designing projects and strategies that aim at the concepts of equity, inclusiveness, and welfare of the weak, on the other hand, urban planning seems to increase social stratification. This concept reinforces hierarchies and transforms the landscape into a battleground, blaming the disenfranchised, masquerading as design. In fact, for some years now, the trend of so-called *Hostile Architecture* or *Unpleasant Design* has been emerging, i.e., a particular urban design that, behind the apparent functionality or aesthetics, has an exact goal, namely to exercise a kind of social control over public space and discourage particular behavior considered 'anti-social' (Savičić and Savić 2013).

Hostile Architecture reveals itself on several levels, spreading compulsively in several 'developed' countries such as the United States, Europe and Italy included, identifying itself not as a product of accident or carelessness but as a thought process.

Savić and Savicic (2013) extensively discuss the reasons for these interventions, pointing out that the administrations most inclined to use this type of design are those of the richest and largest cities, which have to deal with a high number of people. Defensive urban interventions have the advantage of eliminating the need for surveillance and human intervention. Still, they actually solve some social problems by simply moving them elsewhere and trying to hide them, to make them invisible.

Finally, in addition to the political and social issues, one of the main problems with these interventions is that they are definitive: they do not allow for negotiation, limit activities and deny interactions.

81.3 Inclusive Housing Solution: A Research Path

The reflection, and thus the study, is translated as a (counter) proposal to the model of Hostile Architecture and in line with 'Design for Emergency', with a long-term research goal as that of resolving the distance between a state of affairs and a series of contemporary pressures of use that tend to define public spaces in cities in an impulsive manner. This strategy aims to connect shelter environments and common areas without feeling the need to classify spaces but to define them by collective social use. The experimental project (Fig. 81.1), Structured on a theoretical intervention model, was based on research conducted by the authors at the Department Architecture and Territory, Mediterranea University of Reggio Calabria. In its twofold morphological configuration, the experimentation is intended to represent a strategy, through human-centered processes, oriented at developing intervention models that can fit harmoniously into the social inclusion projects of public administrations. This, through a structured planning process, so that the location of the shelters is strategic for social re-inclusion aimed at improving the quality of life of the homeless through a technological system that combines dual functions.

The innovative aspect of the proposal does not lie in the use of 'new' materials and technologies but in the promotion of a more feasible social inclusion between homeless municipalities and citizens by identifying useful solutions to address health inequalities related to housing. The research path was articulated following a deductive, systematic and scalar methodological process, structuring a thematic frame of reference, moving from the definition of the macro-theme of 'Design for Emergency' to the construction of the variables that define an emergency condition through which to respond effectively and rapidly to it. For this phase, a theoretical reference model was developed for the social inclusion and innovation of vulnerable groups, the homeless. Specifically, the research is structured in three phases:

- preparation of data on the issues addressed and emergency conditions, with case studies and types of intervention plans. European strategic approaches to social

Fig. 81.1 Experimental Project: Home or Shelter? A filter between man and word

inclusion of homeless people show an increasing tendency among member states, '*to move toward the development of more comprehensive and integrated strategies or at least approaches*' and their monitoring;

- critical analysis of acquired and systematized data. The analysis of the case studies on the most common homeless shelters, together with the statistics (Figs. 81.2 and 81.3), led to a design reflection on the space and configuration of possible shelters, closely linked to the urban design of cities, as a theoretical assumption of reference for the research;
- methodological and design experimentation. The experimentation is presented as an opportunity for research activity, still in progress, whose first results have concerned the development of an urban micro-system, a micro-architecture developed as a transitional, multifunctional housing system, which in the nighttime

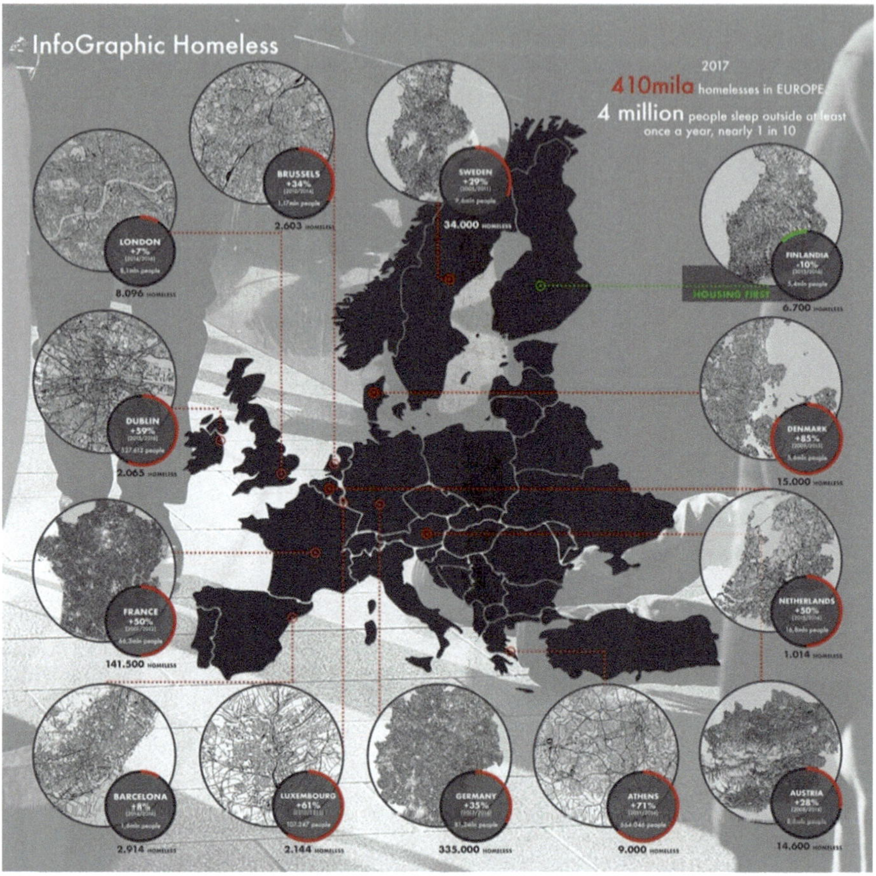

Fig. 81.2 InfoGraphic Statistics Homeless in Europe

configuration, the most critical, is configured as a shelter (specifically for homeless people) and in the daytime configuration, is configured as a system of shaded urban seating, supplemented by small micro-activities such as book sharing.

In this sense, the research aims to outline the close relationship between well-being and housing by setting up a technological system that can meet the basic needs of vulnerable groups and by being part of a real social inclusion plan. This is done through what is called *Housing Led*, i.e., those strategies that promote forms of residency combined with assistance and support services. The project has thus been translated as a theoretical foundation, proposing a way out of the superficiality with which forms and images without substance are habitually elaborated, with outcomes unrelated to the real issues of the housing emergency (Agnoletto 2008).

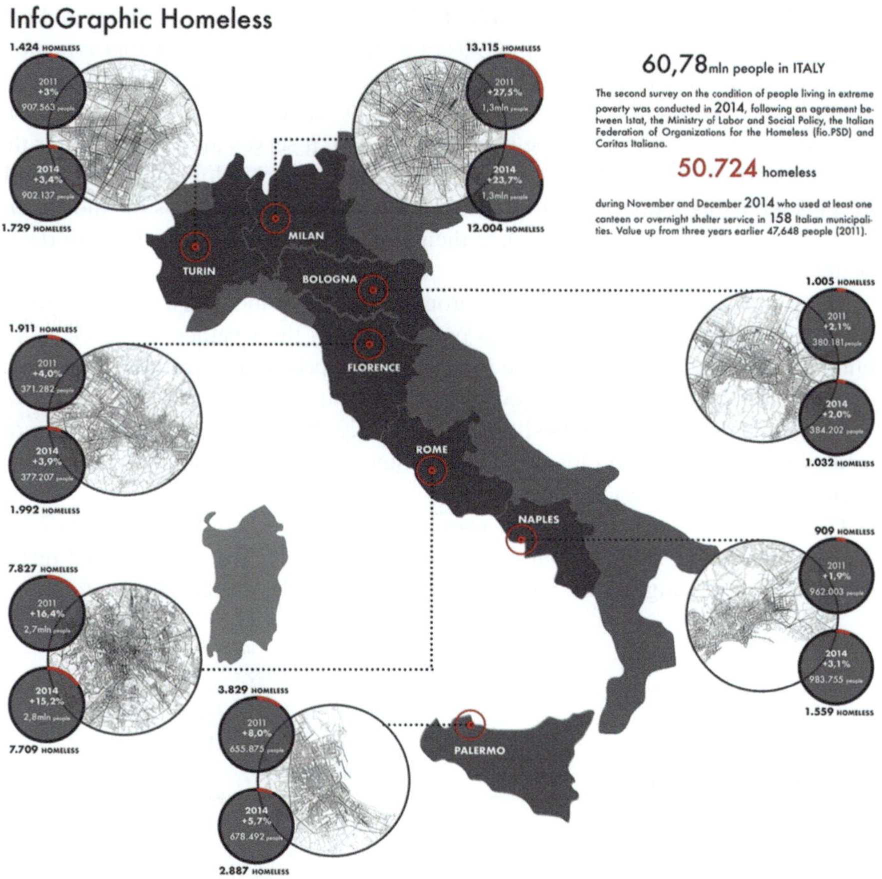

Fig. 81.3 InfoGraphic Statistics Homeless in Italy

81.4 The Design Experimentation

The composition of the shelter is based on a modular grid applied to the architectural and compositional concept of Japanese vernacular architecture, the tatami, and in particular to the tearoom (minimal living space), resulting in the 'Tatami Shelter'.

The 'Tatami Shelter' shares Perriand's (1949) observation that the constructive and compositional value of the Tatami lies in its ease of assembly and disassembly, and according to De Lucchi (2017), Japanese architecture refers to the micro dimension. He explains that it is an intrinsic characteristic of the Japanese; thanks to the Tatami, they can sleep on the table on which they eat and eat on the bed on which they sleep; it is natural to think that space is as compressed as possible.

The shelter consists of four and a half tatami mats (four equal modules and one half-module), each measuring 180 × 90 cm with an overall height of approximately 2.6 m, as in traditional Japanese culture, with a 45 cm cantilever (Antonini et al. 2020). In the daytime phase (Fig. 81.4), the cantilever serves as a seat for the users of public spaces. In the nighttime phase (Fig. 81.5), it serves as a container for the activities carried out by the homeless; finally, it is a technological solution as it prevents the rising humidity. By installing a track system that allows the movement of three components, the seats change their configuration to become a comfortable shelter.

The micro-architecture is designed through a glulam frame with a curtain of wood panels and compact polycarbonate to protect it from the weather. The roof is not reminiscent of the archetypal house but follows the principle of vertical closures, a glulam frame with a slatted plywood panel for shading, and compact polycarbonate.

The infill frames also act as sliding frames (Fig. 81.6). They are equipped with handles that facilitate the closing of the shelter and lock the modules themselves when they reach the preset configuration.

There are four main activities that a homeless person can carry out within its 7sqm space: *Access*, *Rest*, *Refreshment* and *Storage* (Fig. 81.7). Two steps, stored inside a module, can be dragged via a rail system to allow the homeless person to access

Fig. 81.4 Tatami Shelter—day time configuration

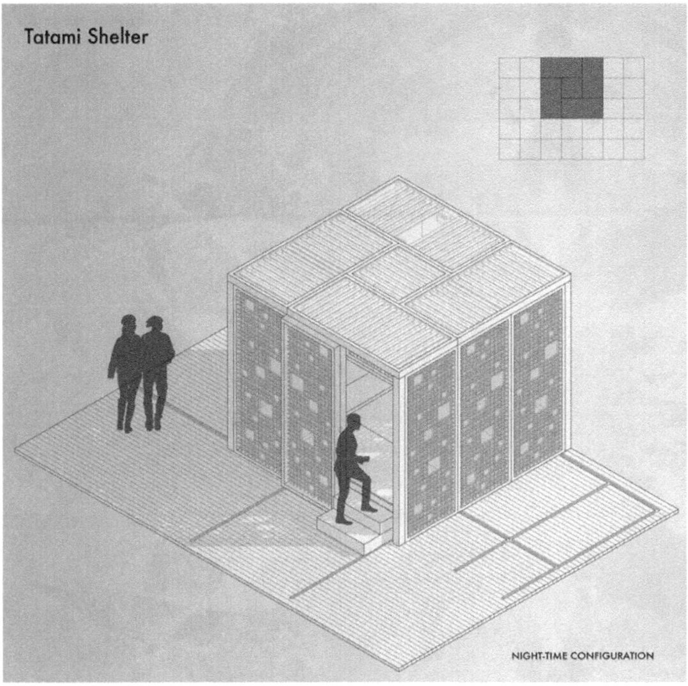

Fig. 81.5 Tatami Shelter—night time configuration

it. Refreshment is provided by a hydraulic piston underneath the 90 × 90 cm half-module, which can be raised to become a table; the chairs are obtained by tilting part of the tatami, reclining at 90°. Among the modules designed, the homeless person can store his personal belongings, thanks to the possibility of lifting them with a hinge system.

Finally, 'Tatami Shelter' is equipped with a special toilet module intended for the exclusive use of the homeless and equipped wall including a washbasin under which is a rotating toilet. The plant engineering part is off grid: water is collected from the roofs of the modules and stored on-site to reuse for the toilet.

In its double configuration, the urban micro-architecture assumes the location in public spaces equipped with drinking water useful for integrating the supply of the same, inserting itself within the renewable energy systems to which the urban areas in which it will be installed are predisposed, guaranteeing adequate standards of use.

Although to be considered a future development, the maintenance aspect is considered for the location of the project proposal of the integrated social cohesion plans. The use of the shelter finds the solution of cleaning and maintenance by the users, not only for the micro-architecture but also for the park where the shelter will be installed.

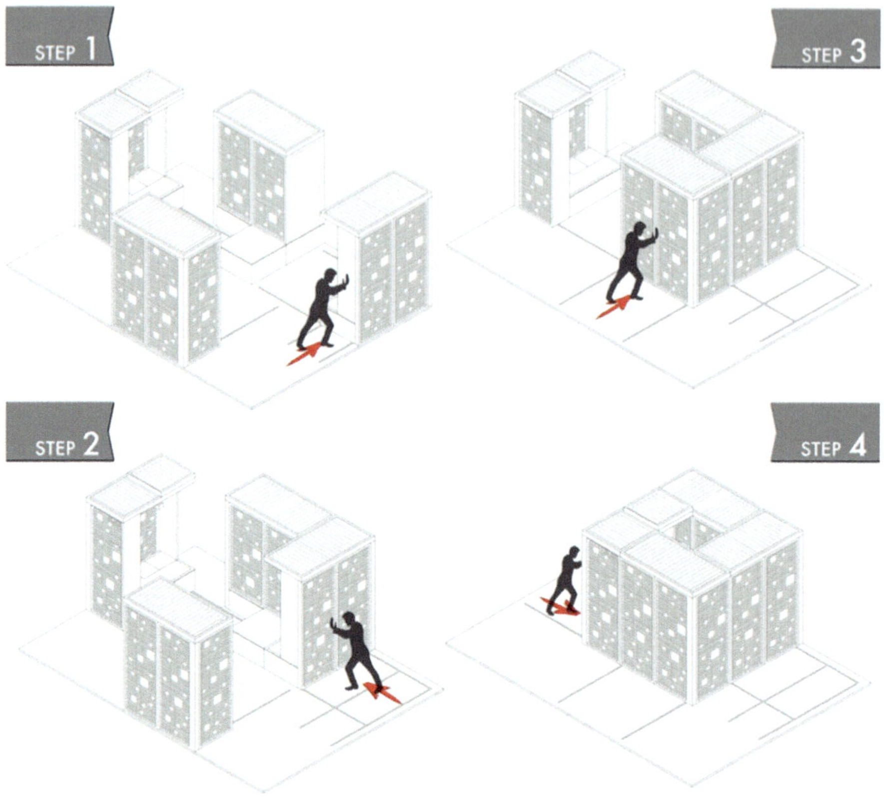

Fig. 81.6 From Bench to Shelter, sliding frames

81.5 Conclusions and Future Developments

The twofold challenge to emergency conditions, both environmental and social, has as a common goal the rethinking of housing models that, regarding temporariness, can become an opportunity for new projects, useful to improve the quality of life and provide answers to the most disadvantaged groups. The need to intervene in and for emergency conditions is a driving principle behind the goals of constructive and productive innovation, intending to use material and immaterial resources through socially and materially sustainable models for the contexts of intervention. This scenario aims to identify the relationships between health and housing for the main mechanisms of housing accessibility and housing conditions.

On a theoretical level and an applicative one, the contribution provides ideas to feed the debate on the emerging condition of 'Design for Emergency' and the social cohesion policies in place in this particular historical moment. An aspect that must

Fig. 81.7 Main activities: access, rest, refreshment and storage

be read about the need/opportunity to propose new temporary housing models of an emergency nature that can guarantee the dignity and well-being of those who use them and at the same time represent new design, construction and realization parameters.

A small model of intervention, therefore, that combines cultural identities and housing models to improve the quality of life of specific vulnerable groups, interpreting reversibility as a paradigm in the relationship with the context, in urban connections, and in the revitalization of brownfields.

The replicable theoretical model of the research allows for possible future developments in other contexts where similar problems need immediate and operational responses on the territory. The desire to involve public administrations and stakeholders is connected to this perspective, with a view to the participation and involvement of all the actors belonging to the chain, which guarantees, through a bottom-up approach, in the most complete and broadest sense of the term, an opportunity to improve the quality of life, respecting people and places.

References

Agnoletto M (2008) Smallness. Area n. 98
Antonini E, Boeri A, Giglio F (2020) Emergency driven innovation. Low tech buildings and circular design. Springer
Baptista I (2009) The drafting of the Portuguese homeless strategy: an insight into the process from a governance-oriented perspective. Eur J Homelessness
Benjaminsen L, Dyb E (2008) The effectiveness of homeless policies—variations among the Scandinavian countries. Eur J Homelessness 2(1):45–67
Cantillon B (2019) The European pillar of social rights. Towards a European Social Union, p 54
De Lucchi M (2017) Microarchitecture and minimal habitat. Area 151:4–11
EPRS (European Parliament Research Service) (2019) Technological innovation for humanitarian aid and assistance. https://www.europarl.europa.eu/RegData/etudes/STUD/2019/634411/EPRS_STU(2019)634411_EN.pdf. Accessed 1 Mar 2021
European Commission (2019) The European pillar of social rights in 20 principles
Greiner PT, McGee JA (2020) The asymmetry of economic growth and the carbon intensity of well-being. Environ Sociol 6(1):95–106
Grubler A, Wilson C, Bento N, Boza-Kiss B, Krey V, McCollum, DL, Rao ND, Riahi K, Rogelj J, De Stercke S, Cullen J, Frank S, Fricko O, Guo F, Gidden M, Havlík P, Huppmann D, Kiesewetter G, Rafaj P, Schoepp W, Valin H (2018) A low energy demand scenario for meeting the 1.5 C target and sustainable development goals without negative emission technologies. Nat Energy 3(6):515–527
Hansen FK (2010) The homelessness strategy in Denmark. Eur J Homelessness 4:113–125
Hermans K (2012) The Dutch strategy to combat homelessness: from ambition to window dressing? Eur J Homelessness 6(2):101–118
Houard N (2011) The French homelessness strategy: reforming temporary accommodation, and access to housing to deliver 'Housing First': continuum or clean break. Eur J Homelessness 5(2)
IPCC (2018) Global warming of 1.5 °C: an IPCC special report on the impacts of global warming of 1.5 °C above pre-industrial levels and related global greenhouse gas emission pathways, in the context of strengthening the global response to the threat of climate change, sustainable development, and efforts to eradicate poverty. Intergovernmental Panel on Climate Change
Kidd SA, Greco S, McKenzie K (2021) Global climate implications for homelessness: a scoping review. J Urban Health 98(3):385–393
O'Neill DW, Fanning AL, Lamb WF, Steinberger JK (2018) A good life for all within planetary boundaries. Nat Sustain 1(2):88–95
Perriand C (1949) Au Japon. L'Architecture d'Aujourd'hui
Ranmal R, Tinson A, Marshall L (2021) How do health inequalities intersect with housing and homelessness. Eur J Homelessness 15(3)
Sahlin I (2015) Searching for a homeless strategy in Sweden. Eur J Homelessness 9(2)
Savičić G, Savić S (eds) (2013) Unpleasant design. GLORIA
Vogel J, Steinberger JK, O'Neill DW, Lamb WF, Krishnakumar J (2021) Socio-economic conditions for satisfying human needs at low energy use: an international analysis of social provisioning. Glob Environ Change 69:102287
Wygnańska J (2009) The development of a national homeless strategy in Poland: achievements and challenges. Eur J Homelessness 3:201–214

Chapter 82
Environmental Sensing and Simulation for Healthy Districts: A Comparison Between Field Measurements and CFD Model

Matteo Giovanardi, Matteo Trane, and Riccardo Pollo

Abstract Atmospheric Particulate Matter (PM) is considered among the main risk factors for cardiovascular, respiratory, and carcinogenic diseases. Besides, heat waves accounted for 68% of natural hazard-related deaths in Europe between 1980 and 2017 and many climate models project a global rise in climate hazards. Environmental Monitoring (EM) is a key resource to control health determinants, addressing threats arising from unhealthy external conditions. Forecasting models may need data coming from pervasive distributed sensor networks and computational simulations. Moreover, district-scale Environmental Sensing (ES) and Environmental Modelling Simulation (EMS) may identify criticalities and specific strategies to mitigate climate risk affecting physical health. This paper compares the output from ES, by field measurements during a "climate walk" joined by more than 60 people, with EMS, by a Computational Fluid Dynamic software (CFD). The assessment has been performed on a real urban district. For on-site measurements, data were acquired by low-cost IoT-based sensors developed by the authors. For simulations, we used ENVI-met, a prognostic non-hydrostatic CFD. Potential Air Temperature and PM 10-2.5 concentration parameters have been measured and simulated on a specific winter day. Results are presented and discussed through a visualisation matrix making the comparison direct. The analysis of the results pointed out the role of ES and EMS for high-resolution scenarios assessment. Although real-time monitoring needs extensive infrastructure at the urban scale, the use of low-cost sensors and a citizen science approach could provide precise input data to support even more accurate models, towards a healthy district site-specific design perspective. This may finally contribute to achieving the Sustainable Development Goal 11.6, aiming at reducing the adverse environmental impact of cities, thus paying particular attention to air quality.

M. Giovanardi (✉) · M. Trane · R. Pollo
Polytechnic University of Turin, Turin, Italy
e-mail: matteo.giovanardi@polito.it

M. Trane
e-mail: matteo.trane@polito.it

R. Pollo
e-mail: riccardo.pollo@polito.it

E. Arbizzani et al. (eds.), *Technological Imagination in the Green and Digital Transition*, The Urban Book Series, https://doi.org/10.1007/978-3-031-29515-7_82

Keywords Environmental sensing · Environmental modelling and simulation · Healthy district · ENVI-met · IoT

82.1 Introduction

By aiming at making cities and human settlements inclusive, safe and resilient, the Sustainable Development Goal 11 points out to reduce the adverse per capita environmental impact of cities, including by paying special attention to air quality (target 11.6). Despite some progress achieved in reducing exposure in certain countries, the global health burden of ambient fine Particulate Matter (PM) is still increasing annually (Southerland et al. 2022). Air pollution causes a wide range of adverse health effects, even at the lowest observable concentrations (Strak et al. 2021). Alongside this, heat waves accounted for 68% of natural hazard-related deaths in Europe between 1980 and 2017 and many climate models project a global rise in climate hazards (Woetzel et al. 2020). Emissions by anthropogenic sources are the main factors in the processes causing air pollution and heat waves in cities. Even though some of these processes regard planetary-scale climatic phenomena, planning at regional and local scale has to respond to imminent challenges due to global warming and threats arising for human health. For instance, the role of greenery has been largely discussed as a pollution mitigating element (Rui et al. 2019). The urban fabric can allow natural ventilation or obstruct the wind flows, influencing PM concentrations and temperature cool down.

Urban surface materials are determinant in lowering the air temperatures, thus improving comfort for people in public spaces.

Given the complexity of these multi-scale issues, Environmental Monitoring (EM) is a key resource for health determinant control. EM asks for data that may come from Environmental Sensing (ES), by a distributed pervasive sensor network monitoring several Environmental Parameters (EPs), and Environmental Modelling and Simulation (EMS), by advanced computational tools forecasting patterns based on given boundary conditions and site-specific features. This research combines both approaches, supported by a Citizen-Science (CS) experience, for EM purposes. The second section of this paper describes the Materials and Methods applied and introduces the case study.

The third section illustrates the Results obtained both from the on-field measurement and a Computational Fluid Dynamics (CFD) simulation, while the fourth one discusses them by a visualisation matrix. Finally, the Conclusions present the advantages coming from the combination of ES and EMS for scenarios assessment, towards a healthy district site-specific design perspective.

82.2 Materials and Methods

The research was carried out in Turin, a city in north-western Italy, surrounded by the western Alpine (Cfa climate according to Köppen–Geiger classification). The case study is within Regio Parco district (45°04′ N 7°42′ E), a peripheral area in the north-eastern part of the city, located near the Po River and some main green infrastructures.

The area of analysis extends for approximately 640,000 m^2 (800 m × 800 m).

82.2.1 A Citizen-Science Experience for Environmental Sensing

ES can be defined as the process by which acquiring real-time data on several EPs through a distributed pervasive sensor network. ES systems can range from dynamic (mobile) to purely static deployments and can monitor different built environment parameters, to improve process efficiency, ensure optimal environmental conditions, highlight patterns, detect anomalies, or avoid stress conditions. ES allows for major knowledge on dynamic phenomena as it is enabled by an Internet of Things (IoT) virtual infrastructure, consisting of a network of interconnected objects based on standard communication protocols (Giovanardi et al. 2021).

In the context of an innovative teaching experience, 62 students were part of the on-field measurement campaign. This CS approach led us to acquire real-time data on several EPs at the same time: air temperature (AT), relative humidity (RH), PM 2.5-10 concentration, and air pressure. The on-field measurement campaign was carried out by using IoT-based devices (Fig. 82.1). Although in a prototypal status, these devices were successfully used in previous research, after its calibration and validation (Montrucchio et al. 2020). The device consists of four PM sensors using laser scattering technology, a DHT22 sensor for air temperature and relative humidity, a barometric sensor for air pressure, and real-time clock for temporal data synchronisation. It also incorporates a Raspberry Pi simple-board computer, and its micro-SD card stores data by a Python script. The external case was 3D printed and it measures 44 mm × 36 mm × 12 mm.

The low-cost device (total cost around 40 euros) is powered by portable batteries.

The campaign was organised in five different paths, namely Climate Walks (CWs) *A-B-C-D-E* (Fig. 82.2), and took place on 29 November 2021 from 1.45 to 3.45 P.M. approximately. Each CW consisted of some stop points, where students were given pre-printed surveys too, to fill in with the time of arrival at and departure from each stop (approximately 15 min per stop), data about traffic (number of cars passing by per road lane), and personal feelings about environmental quality. At the end of the walks, surveys were collected and data on road traffic was used as input to model the pollutant sources with reliable site-specific values. The paths were also recorded and geo-referenced by using the application Open GPX tracker, to match the data

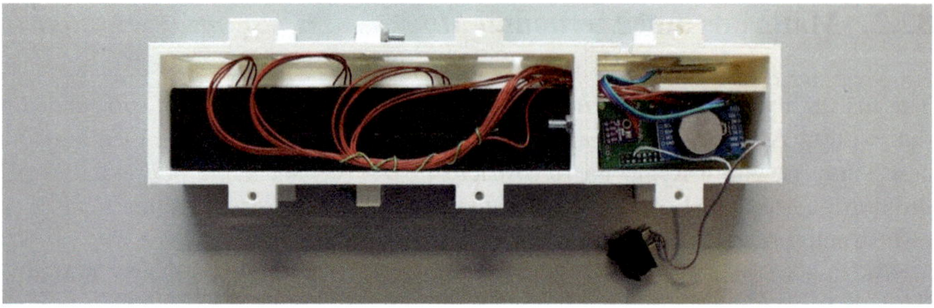

Fig. 82.1 IoT-based device for monitoring air quality, temperature, humidity, and pressure

acquired from the devices with the time interval spent in each stop and the traffic data coming from surveys.

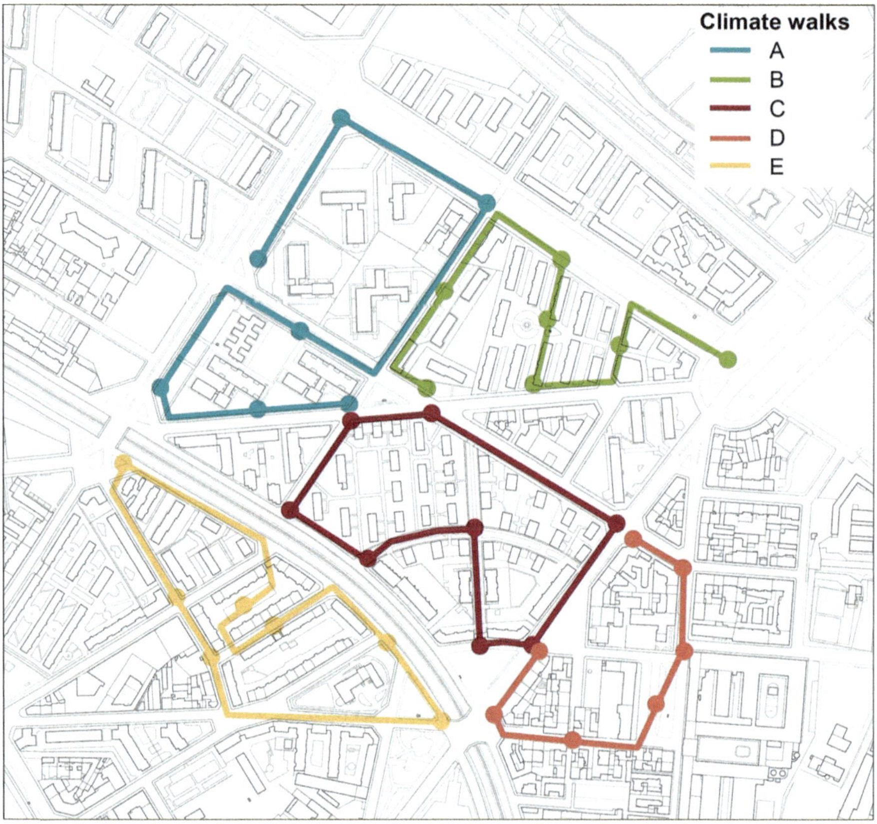

Fig. 82.2 Climate walks

82.2.2 CFD for Environmental Modelling and Simulation

For EMS purposes, we used the software ENVI-met[1] version 5.0.2, a holistic three-dimensional non-hydrostatic CFD for the simulation of surface-plant-air interactions in urban environments. The district area was modelled with a 5 × 5 × 2 m grid cell resolution (xyz), where buildings, horizontal and vertical greenery, roads and pavements, natural surfaces, and pollutant sources were digitised. As for the meteorological input data, we used the ones provided by ARPA Piedmont[2] on 29 November 2021 (Table 82.1). Data were acquired from the nearest urban meteorological station, 4 km far as the crow flies from our campaign's start meeting point (Torino Grassi station). The simulation time was set to run for 48 h. We considered the second 24 h results, which are more accurate as ENVI-met requires some spin-up time. Although doubling the simulation timing, this could turn out into more accurate results, especially in the afternoon and evening hours (Middel et al. 2014). For the purposes of this research, we mainly focused on Potential Air Temperature (PAT) (°C) and PM 2.5-10 concentration (μg/m^3). PAT and PM were evaluated at 2 m height from the soil.

As for the pollution sources modelling, in absence of detailed data, punctual emissions due to heating from buildings were not considered. However, this approximation does not invalidate the results: as reported by ARPA Piemonte (2019), the main source of PM in Turin is actually linked to the traffic (Fig. 82.3). Thus, linear traffic sources were sized by combining on-site traffic measurements, carried out during the CWs, and the Traffic Tool in the Database Manager.

Specifically, it calculates the emission profiles per linear source type by providing an equivalent hourly flow rate profile after injecting a type-day total car volume (Veh/h). Its calculations are based on standard emission rate (HBEFA 2022). PM 2.5 was calculated as a fraction of PM 10 according to Schafer et al. (2021) (36% out of PM 10 in inner roads, 53% in roads at urban fringe and suburban roads) (Fig. 82.4). For roads with no traffic measures, we used data provided by a regional report, providing traffic volumes per hour on a standard day in November (Regione Piemonte 2017). In total, we created 11 linear emission profiles (Table 82.2). The estimation of the urban bus rate over the total traffic volume was carried out by considering the number of bus lanes crossing the roads,[3] number of passages throughout the day according to specific hour intervals[4] and real-time data on bus lines.[5]

Background levels were set (6 μg/m^3 for PM 2.5, 10 μg/m^3 for PM 10) according to the lowest most recurrent values acquired by the sensors.

[1] Developed by M. Bruse (ENVI-met GmbH, Essen, Germany).

[2] Regional Agency for the Environmental Protection: http://www.arpa.piemonte.it/.

[3] https://www.gtt.to.it/cms/risorse/urbana/mappa/mapparete.pdf.

[4] https://www.gtt.to.it/cms/risorse/urbana/intervalli_sito.pdf.

[5] https://www.gtt.to.it/cms/percorari/urbano?view=linee&bacino=U.

Table 82.1 Meteorological data on 29 November, 2021 by Torino Grassi ARPA station

Hour	Air temperature (°C)	Relative humidity (%)	Wind speed (m/s)	Wind direction (deg.)	Global radiation (W/m^2)
00:00	1.6	80	0.6	227	–
01:00	1.4	82	0.7	161	–
02:00	0.3	87	0.6	316	–
03:00	– 0.1	88	1.1	89	–
04:00	– 0.2	81	0.4	326	–
05:00	– 0.1	80	1.5	276	–
06:00	0.4	74	1.3	242	–
07:00	0.8	71	1.2	224	–
08:00	1.6	68	1	221	43
09:00	3.1	73	0.5	94	253
10:00	7.1	62	0.9	163	372
11:00	9.5	38	1.4	126	427
12:00	11	22	0.6	120	411
13:00	10.9	17	0.9	106	351
14:00	11.3	18	1.4	88	275
15:00	10.6	17	3.4	317	208
16:00	9.3	15	4.4	328	–
17:00	7.9	17	2.8	323	–
18:00	7.4	21	2.4	247	–
19:00	6.9	23	3.4	290	–
20:00	6.6	24	2.9	263	–
21:00	5.3	25	1.4	100	–
22:00	5	30	1.2	174	–
23:00	4.4	39	1.4	150	–

82.3 Results

82.3.1 Results from ES Campaign

About 120′000 PM data were collected during the CWs. Data coming from four sensors within the devices were averaged to obtain a single PM 2.5 and PM10 data, and the results refer to a time period of ~ 90 min grouped in 10-min steps. The PM 10 average data varies between 6 and 13 μg/m^3, while PM 2.5 ranges between 4 and 9 μg/m^3, accounting for about 60% of the PM 10 share. As shown in Fig. 82.5, the variance of the average data is minimal, with the exception of CW *E*. A more in-depth analysis for each walk was carried out to deepen the correlation with endogenous factors. For example, in CW *C*, a higher level of PM is recorded at the road junction

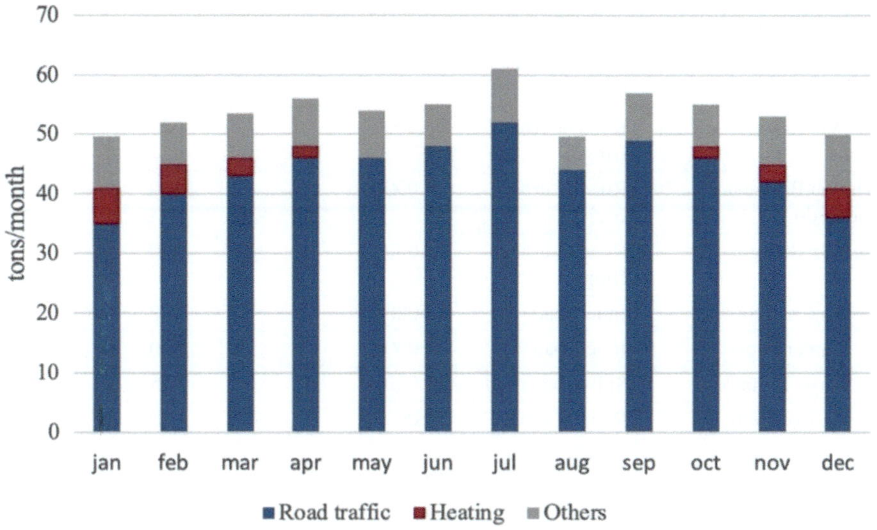

Fig. 82.3 PM 10 emission profile in Turin. Based on ARPA 2019

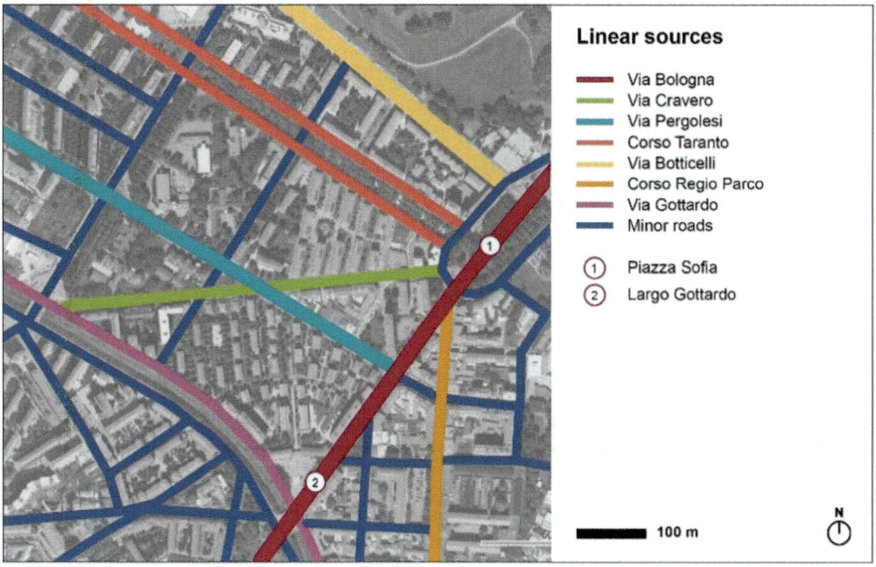

Fig. 82.4 Road linear sources

Table 82.2 Traffic volumes/road/day

Road	Road type on ENVI-met	Traffic volume (Veh/d)	Public transport (Veh/d)
1a. Via Bologna—P.zza Sofia	Road at urban fringe	23,200	620
1b. Via Bologna—L. Gottardo	Road at urban fringe	21,000	400
1c. Via Bologna	Road at urban fringe	21,000	540
2. Via Cravero	Inner road	5000	100
3. Via Pergolesi	Inner road	5000	100
4. Corso Taranto	Sub-urban road	14,000	115
5. Via Botticelli	Road at urban fringe	21,000	170
6. Corso Regio Parco	Sub-urban road	17,000	–
7. Via Gottardo	Inner road	5000	140
8a. Minor roads (1 lane)	Inner road	2100	–
8b. Minor roads (2 lanes)	Inner road	2100	–

between Via Bologna and Via Pergolesi, while in CW *E*, higher pollution levels are monitored at the intersection of via Maddalene, via Sempione, and via Bologna. The PM values were usually higher at main street intersections. As for AT and RH, the values recorded are partially higher than those officially monitored. More precisely, between 2 and 3 P.M., 11 °C (AT) and 17% (RH) was recorded by ARPA, compared to 15 °C and 20% respectively acquired by the devices on average.

82.3.2 Results from EMS

PM 2.5 and PM 10 concentration peaks were present at 7:00 A.M., while at 2:00 pm and 3:00 P.M., slightly lower values resulted from EMS compared to the ones acquired from the devices (Fig. 82.6). However, the major criticalities were present in correspondence with the main traffic sources, namely Corso Taranto, Via Botticelli, and Via Bologna. In the first, pollutants were prevented from removal by the building curtain in the north direction (considering the prevalent direction of the wind on that day), while in the second and the third, the traffic volume was much higher than in all other roads. One can still highlight how pollutants are generally lower in inner areas, where the traffic is generally lower or absent and the amount of greenery is higher.

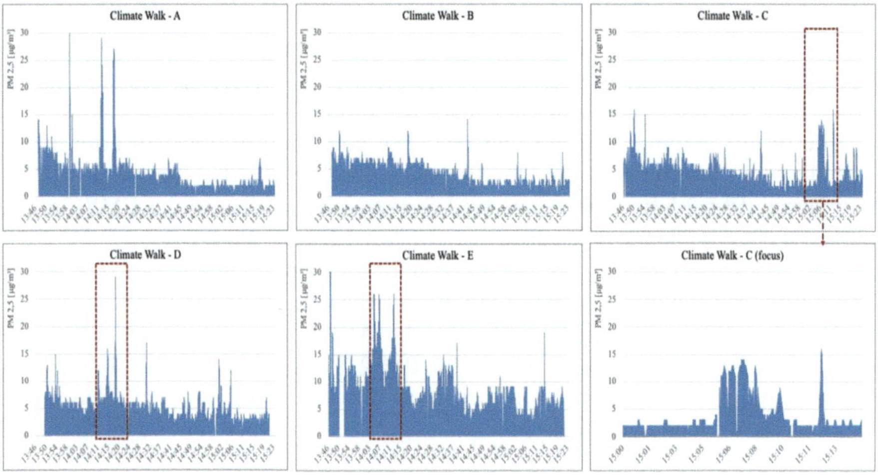

Fig. 82.5 PM2.5 values from different climate walks and a focus on path C. In the red boxes time-steps at crossroads with Via Bologna

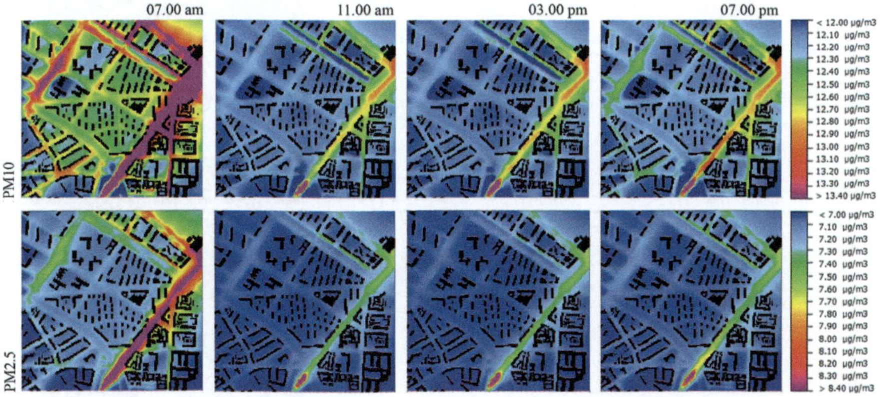

Fig. 82.6 PM10 and PM2.5 concentration at 7:00 A.M., 11:00 A.M., 3:00 P.M., and 7:00 P.M.

82.4 Discussion

Constant development in computational abilities has been allowing more advanced approaches for microclimate analysis and modelling, emphasising its high capacity of solving complex phenomena and nonlinearity of urban climate systems (Liu et al. 2020). Indeed, EMS makes it possible to analyse EPs over a relatively wide area, also predicting the microclimate conditions under different planning scenarios (Bartesaghi Koc et al. 2018). While data coming from sensors point out values that are valid for a certain path (if they are dynamic sensors, as in our case) or a single point in the space (if they are static ones), relative to a specific narrow time, EMS offers a more comprehensive overview on several EPs, describing trend and pattern throughout a type-day with a higher space resolution.

The results coming from ES and EMS are compared in a visualisation matrix including the data coming from the official meteorological urban station (Fig. 82.7).

Average PM 2.5 and 10 values by ES were very similar to the EMS outputs, although in certain areas, the values acquired by ES were slightly higher. However, in both cases, we can still assume a certain correspondence between peak values and traffic, especially at the intersections between Via Bologna and minor roads. Daily urban average values provided by the ARPA station were actually higher (28 μg/m^3 for PM 10 and 19 μg/m^3 for PM 2.5), but they do not allow for any high-resolution information and further in-depth considerations. Specifically, acquisition by the devices could be considered more accurate as they capture traffic-related

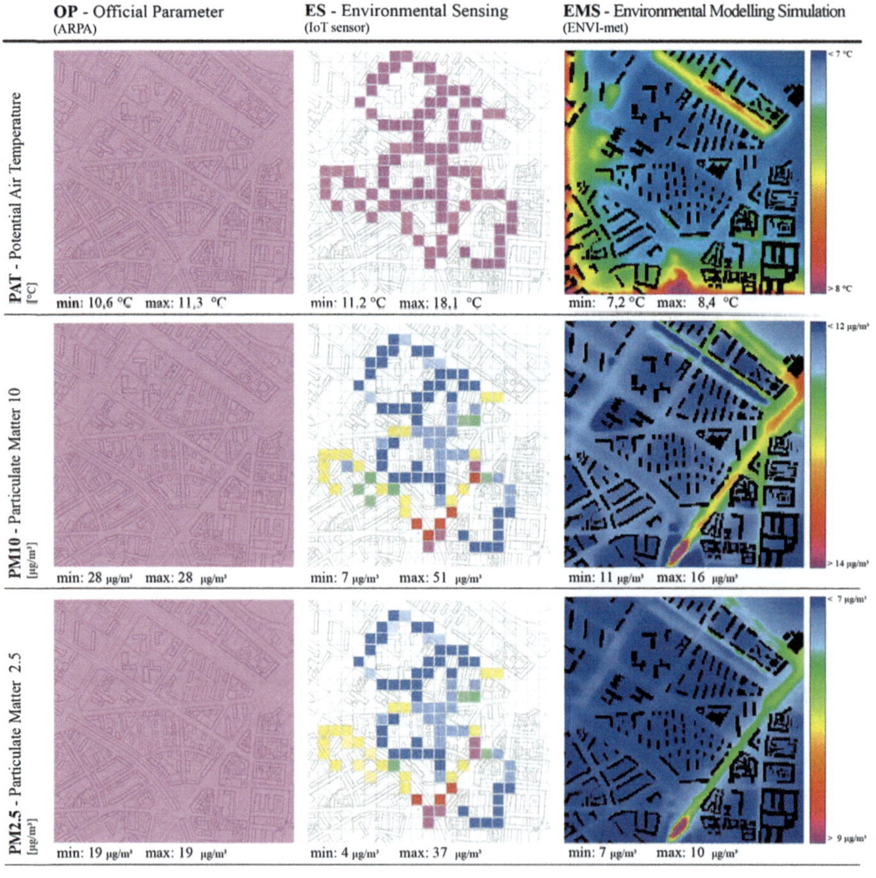

Fig. 82.7 Visualisation matrix at 3:00 P.M

instant conditions, while the simulation outputs rather highlight a certain trend, as they are based on traffic volume approximations. In these terms, sensors highlighted a higher PM concentration in some crossroads.

The research is burdened by several constraints. The simulation timing didn't allow us to model the area with a greater resolution. Although the main pollution sources are related to traffic, modelling other sources could have affected the total PM concentrations and the distribution pattern, as they finally account for ~ 15% of the total PM in November (Fig. 82.2). Besides, we could not force the wind speed and direction (apart from injecting initial values), as this would have required 30-min interval data. This may have affected the PM distribution and concentration, especially if we consider that the simulation day was characterised by highly variable wind speed and direction. As for the ES, we had to clean data, as some outliers were present. Finally, although necessary for data sampling and processing information, data acquired were overabundant, thus hard to manage.

82.5 Conclusions

The aim of the paper was to compare the results of an on-field measurement campaign with modelling and simulating, towards a site-specific assessment of the environmental quality in a real district. The originality of this research lays on performing an environmental assessment by combining ES and EMS, in mutual support for a comprehensive overview on several EPs. Both approaches "fed" from a CS experience, which is also meant to have a major role in sensibilising people towards more pro-environmental consciousness. The findings may encourage the extension of a network of sensors for a more accurate analysis of the urban environment conditions over time and space. This is especially true if we imagined a distributed sensor network for EPs and traffic monitoring, to be spread all over the city in parallel with respect to the official meteorological stations, supported by a proper IoT infrastructure. Indeed, these only provide hour data on a few EPs and a daily average values on PM concentration, which may actually strongly vary from one point of the city to another and cannot highlight any site-specific distribution pattern.

On the other hand, the on-site survey and acquired data were crucial for EMS. In this perspective, EMS could count on real-world high-resolution data, which may turn into a robust environmental time-series and "labelled" environmental conditions (i.e. hot summer day, rainy autumn day, dry spring day, etc.) for scenario assessment, design, and validation. This may finally lead to more and more accurate models, depending on-site-specific boundary conditions forcing. The finding may encourage expanding EMS to the whole city too, by discretizing the urban area to optimise the computational timing but still providing a space–time resolution allowing micro-urban scale analysis. Apart from the limitations described, EMS, supported by ES, plays a major role in knowing, thus representing input data to manage urban environmental conditions that may threaten health. Combining the approaches would finally lead to setting digital worlds for real cities, with a deeply site-specific perspective at the district scale, where the effects of policies, personal choices and habits, projects, anthropogenic processes, patterns of use are much more immediately visible and correlatable to human health and well-being.

References

ARPA Piemonte (2019) Inquinamento da particolato PM10: il riscaldamento domestico. http://www.arpa.piemonte.it/news/inquinamento-da-particolato-pm10-il-riscaldamento-domestico. Accessed 8 Mar 2022

Bartesaghi Koc C, Osmond P, Peters A (2018) Evaluating the cooling effects of green infrastructure: a systematic review of methods, indicators and data sources. Solar Energy 166:486–508

Giovanardi M, Trane M, Pollo R (2021) IoT in building process: a literature review. J Civ Eng Archit 15(9):475–487

HBEFA (2022) Documentation of update. https://www.hbefa.net/e/documents/HBEFA42_Update_Documentation.pdf. Accessed 2 Mar 2022

Middel A, Hab K, Brazel AJ, Martin CA, Guhathakurta S (2014) Impact of urban from and design on mid-afternoon microclimate in Phoenix Local Climates Zones. Landsc Urban Plan 122:16–28

Liu D, Songshu H, Jiaping L (2020) Contrasting the performance capabilities of urban radiation field between three microclimate simulation tools. Build Environ 175(13)

Montrucchio B, Giusto E, Ghazi Vakili M, Quer S, Ferrero R (2020) A densely-deployed, high sampling rate, open-source air pollution monitoring WSN. IEEE Trans Veh Technol 69(12)

Regione Piemonte (2017) Report 2017 sulla mobilità sostenibile in Piemonte. https://www.regione.piemonte.it/web/sites/default/files/media/documenti/2019-04/report_2017_sulla_mobilita_veicolare_in_piemonte_completo.pdf. Accessed 1 Mar 2022

Rui L, Buccolieri R, Gao Z, Ding W, Shen J (2019) The impact of green space layouts on microclimate and air quality in residential districts of Nanjing, China. Forests 9(4):224

Schafer M, Salari HE, Köckler H, Xuan TN (2021) Assessing local heat stress and air quality with the use of remote sensing and pedestrian perception in urban microclimate simulations. Sci Total Environ 794

Southerland V, Brauer M, Mohegh A, Hammer MS, Van Donkelaar A, Martin RV, Apte JS, Anenberg SC (2022) Global urban temporal trends in fine particulate matter (PM2.5) and attributable health burdens: estimates from global datasets. Lancet Planet Health 6:139–146

Strak M, Weinmayr G, Rodopoulou S et al (2021) Long term exposure to low level air pollution and mortality in eight European cohorts within the ELAPSE project: pooled analysis. BMJ Clin Res 374(1904)

Woetzel J, Pinner D, Samandari H et al (2020) Climate risk and response. Physical hazards and socioeconomic impacts. McKinsey Global Institute. https://www.mckinsey.com/~/media/mckinsey/business%20functions/sustainability/our%20insights/climate%20risk%20and%20response%20physical%20hazards%20and%20socioeconomic%20impacts/mgi-climate-risk-and-response-full-report-vf.pdf. Accessed 8 Mar 2022

Chapter 83
A Synthesis Paradigm as a Way of Bringing Back to Life the Artistic Monuments Inspired by the Motives of the People's Liberation Struggle and Revolution of Yugoslavia

Meri Batakoja and Tihana Hrastar

Abstract The *artistic monuments inspired by the motives of the People's Liberation Struggle and Revolution of Yugoslavia* are stratified phenomena constructed through a multitude of meanings. They are an indispensable part of important movements of architectural thought: the Yugoslavian avant-garde and the synthesis of all arts as its distinctive operational mode; the process of how the modern non-figurative means became an official language to Yugoslavian socialist ideology as opposed to the soc-realist figuration of the Eastern Bloc; the popularization of the Yugoslavian heritage taking place in the last decade as a unique enclave of the world scene of modern architecture, etc. This paper aims to argument their conception and meaning as a complex synthesis of art, architecture, technology and nature, and at the same time to open a potential health paradigm positioning them in the context of open-air culture, new ritual practices of social memory and new museum-destination types of contemporary art forms. In terms of methodology, this study puts together architectural scholars who actively work on the problem of the synthesis of the arts, offers a detailed review of existing scientific knowledge on artistic monuments in order to provide a new conceptual frame, an instructive yet speculative material that can alter the perception away from the old ideological stand into a new and healthy synthesized experience. Apart from the theorization of this synthesis paradigm as a way of bringing back to life the *artistic monuments inspired by the motives of the People's Liberation Struggle and Revolution of Yugoslavia*, this paper provides criteria for this new conceptual frame and showcases them upon selected case study.

The authors have contributed equally to this work.

The complexity of the title "Artistic Monuments Inspired by the Motives of the People's Liberation Struggle and Revolution of Yugoslavia" is taken from the then prevailing Yugoslavian nomenclature of monuments.

M. Batakoja (✉)
Ss. Cyril and Methodius University, Skopje, North Macedonia
e-mail: batakoja.meri@arh.ukim.edu.mk

T. Hrastar
University of Zagreb, Zagreb, Croatia
e-mail: thrastar@arhitekt.hr

E. Arbizzani et al. (eds.), *Technological Imagination in the Green and Digital Transition*, The Urban Book Series, https://doi.org/10.1007/978-3-031-29515-7_83

Keywords Synthesis of the arts · Advanced technology · Nature · Yugoslavian artistic monuments · Health

83.1 Introduction | Previous Research

This paper uses all previous research on the subject of Artistic Monuments, from academic context (dissertations, papers) to popular dissemination (artistic approaches, digital databases and exhibitions)[1] in order to extract a new theoretical approach and to propose a synthesis paradigm to add to Monuments' prospective survival.

What does a complex title *Artistic Monuments Inspired by the Motives of the People's Liberation Struggle and Revolution of Yugoslavia* signify in the first place? The People's Liberation Struggle was waged as a war of the partisans for freedom from the fascist occupation which resulted in the forming of the new Yugoslavian state in 1945. In parallel, the Socialist Revolution of the people further transformed Yugoslavia as a distinct socialist country. The ideology of Yugoslavia was thus based upon two ground-breaking events—People's Liberation Struggle (as a finished process) and the Revolution (as an ever-going process). Due to this self-consciousness, the Communist Party never treated those processes independently while establishing its ideological structures. Even formally, and put in the simplest of terms, the Artistic Monuments consisted of horizontally manipulated form that signified the victims of the fascist terror and a dominating vertical that symbolized the victory and the revolution.[2] These two elements were often placed next to each other in a monumental location, an elementary nature by itself that added upon this effect of sublime encounter with a monolith of above-average dimensions placed in the middle of wildlife (Tepina 1961).

The Artistic Monuments nurtured rituals of practicing collective memory. Even the dates of inaugurations of the monuments followed strict policy of remembrance, affirming important events of uprising and resistance of the individual Republic, but also of Yugoslavia as a whole. Monuments' immediate surroundings were equipped with functional elements[3] that supported established protocols for mass gatherings utilizing a directed play with numerous actors and a convincing ceremonial scenography. Beyond their usage on important dates, the Artistic Monuments were used as a secular space for "common folk celebrations" in the form of organized excursions dedicated to quality time for socialization in open-air and nature, with accompanying cultural and educational activities. Once the focus in the 1980s shifted to

[1] The topic has gained a well-deserved place in the popular culture on the regional and world scene, and its place in the canon of modern architecture was confirmed in the MoMA's exhibition "Toward a Concrete Utopia: Architecture in Yugoslavia, 1948–1980" of 2019.

[2] The Artistic Monuments exemplify the process of how the modern non-figurative means became an official language to Yugoslavian socialist ideology as opposed to the soc-realist figuration of the Eastern Bloc.

[3] Commemorative plateau, space for wreaths, torches, flagpoles, etc.

the younger generation, the Artistic Monuments began to further participate in this popular cultural milieu, by hosting youth festivals, car rallies, maritime regattas or rock-and-roll concerts (Horvatinčić 2017). Such performing rituals sustained the group identity and started to signify a distinct mass culture of "yugoslavianism". As such, Artistic Monuments provide a fascinating example of an adaptable and durable ideological construct that extends from the sublime to the everyday life.

According to the typology of monuments of the Socialist Era made by Horvatincic (2017),[4] the Artistic Monuments that are the subject of this research are a synonym of Monuments Based upon Innovative Structural Principles. The technological innovation refers to the complex extension of the monument's utilitarian dimension that crossed the category of architectural sculpture toward the category of public museum, cultural and education center, etc. The technological innovation also stands for the complex sculptural solutions that sought constructively demanding structures and use of innovative delicate materials. But the building process of those complex structures in previously untouched natural sites was a technological challenge in itself.

This organic fusion of the Monument with its natural surrounding formed the foundations for a theory of "placing of the monument" which combines techniques of stage design, control of the mise-en-scène and the morphology of nature, all resulting in a dramatic corporeal experience (Bogdanovic 1961).

The Artistic Monuments represented a common problem for architecture, art, urban design, horticulture and politics and demanded a synthesized approach based on creative collaborations. Therefore, this synthesis paradigm is probably the Monuments' most important and distinctive mark.

83.2 A Synthesis Paradigm in the Context of Health

This paper introduces a comparison of the Yugoslavian Artistic Monuments with the artworks produced by the "new art tendencies" of the period between 1960 and 1990, which expands the current theory of the Monuments and brings forth an addition to their existing synthesis paradigm.

The new art tendencies, represented by the **Minimal** and **Land Art Movements**, are based upon similar conceptual grounds as the Artistic Monuments. Formally, they aim at using the mere physical facts of scale structure and materiality of the artwork to provoke strong encounter through a real bodily experience (including the mind as the most immaterial part of the body). They are often colossal in magnitude (measuring kilometers and being visible on satellite photos) and are inseparable from the ambient they are placed or exhibited in.

[4] It is a single comprehensive survey of the monuments from Socialist Era on the territory of Croatia.

Fig. 83.1 Petrova Gora Monument—original state from the 1980s. *Source* Copyright 2022 by CCN-images/Turistkomerc Archive

But what is less known about new art tendencies is that both Minimal and Land Art left us with alternative museal[5] types situated outside the crowdedness of cities, in a form of decentralized, dispersed, faraway "museums" that promote a richer and healthier real life experience (Batakoja 2015).

> "When we have to spend hours traveling in order to see something, we do so only when we have a special interest, a special reason to go. We are more willing to spend the time that is necessary to experience the full potential of the works we have gone to see, and we are ready for a different approach to the art. The visit becomes a real life experience and, by spending perhaps a full day on site, we try to understand, in the maximum way possible, what the artists are doing. This is a completely different approach to that of the museum in the city ..." (Papadakis 1991)

In a similar way that the Land Art allowed us to understand the fragile beauty of nature as an early chapter in our evolving notion of the environment, the Artistic Monuments could also unveil an eco-paradigm in the museum context. If the traditional museum paradigm consists of a building, collections, experts and the public, the eco museum paradigm consists of territory, heritage, memory and the community (Babić 2009). This paper applies this eco-paradigm in the context of Artistic Monuments, and repositions the focus from the monument as an isolated and standalone building to the totality of the territory it belongs to. The aesthetic heritage then becomes just one layer in the heritage palimpsest of the given territory. Also, the anti-fascist ideology, reviewed and accepted as authentic by the experts, becomes just one memory layer in the totality of memories produced by every visitor and the community. Tomislav Šola (Babić 2009) says that to follow an eco-paradigm in the museum context is to follow a philosophy, a mental and social behavior, transformed into a professional methodology.

[5] Museum-like.

Let us consolidate a possible philosophy: Artistic Monuments should be rethought as **museum-sites** far away from the cities, reachable by prolonged **pilgrimage through the landscape** and therefore valued in the context of environmental issues. This new representation of them, as dispersed, faraway "museums," with their awareness of the "**sense of the place**," enables healthier and spiritually rich experience by exploiting the territory they are placed in. Additionally, the Artistic Monuments through the "**delicacy of scale structure and material**" are individually capable as the "new art tendencies" of providing authentic aesthetic encounter through a real bodily experience. By putting aside the deeply rooted ideological connotations (one of the main reasons for their neglected state), they should focus toward new "**performing rituals**", even those not yet discovered. Therefore, by using the totality of available natural and cultural heritage, the Monuments could potentially grow into a once again relevant and healthy building type (Fig. 83.1).[6]

83.3 Case Study of Petrova Gora

The Monument to the Uprising of the People of Kordun and Banija (1970–1981), simply known as Petrova Gora Monument, was conceived as a multifunctional monument building that hosted a Museum of Revolution, consequently exceeding the definition of traditional monuments. It formed a focal point of a larger memorial area with over 70 dispersed historical objects and sites in the midst of preserved nature.[7]

The authentic historical territory was equipped for organized political, cultural and educational visits and recreation so it could be included "in the continuity of contemporary life" (Pavlović 1975). Therefore, the case study of Petrova Gora is elaborated not only as an isolated building, but also through the totality of the territory it belongs to, and illustrates how "the synthesis paradigm" was actually a part of its original concept.

The monumentality of the carefully picked position and its formal appearance (size, form, materials) came as a result of strong intentions for this monument to become a landmark—a representation of Petrova Gora and its fighters, as well as a symbol of the "entire area" and the "entire Revolution" (Knežević 1981). The monument is located on the highest summit in Petrova Gora range, making it visible from great walking distance and via satellite images as well (Fig. 83.2).

During the "pilgrimage" through the wooded terrain, the observer can hardly grasp its actual size of 37 m, which can only be felt at its immediate proximity. This introduces a play with the observer's perception of scale and invites a bodily experience of the monument. Monument's verticality and organic curves underline the characteristics of its natural environment and the topography of the hill, while the all-encompassing metal cladding dematerializes its form and reflects the light,

[6] The impact on health is highly multifaceted, and it extends its exact quantitative aspects connected to the physical wellbeing.

[7] Petrova Gora Biljeg is protected as "significant landscape" since 1969.

Fig. 83.2 Petrova Gora Monument—aerial photography of the current state. *Source* Photo by Ivica Pezić, printed with permission (up left); photo by Mladen Ivanović, printed with permission (up right); photo by Alan Čaplar adopted from Planinarski izazov za ožujak (2021). https://gojzeki.com/izazov-za-ozujak/ (down)

sky and the surrounding trees. These effects point out to the delicacy of the material that embraces engagement with natural elements such as fog or snow, forming new visual frames in different seasons and periods of day (Hrastar 2020).

Carefully designed objects in its micro context directly contribute to the monument's experience, with a series of axially aligned sequences. The vertical of the monument has its counterpoint in an ascending horizontal axis of a ceremonial path, accompanied with ancillary spaces (service and catering). These secondary objects are located partially underground and covered with grass roof, blending in with the environment (Fig. 83.3).

Fig. 83.3 Experiencing Petrova Gora Monument. *Source* Photos by Ivica Pezić, printed with permission (up); photo by Mladen Ivanović, printed with permission (down left); photo by Marko Mihaljević, printed with permission (down right)

This kind of synthesis of art, architecture and nature required intense collaboration between sculptor Vojin Bakić and architect Berislav Šerbetić, who sought the optimal relation between the utilitarian function and the form (with all its symbolic character).[8] The authors achieved a corresponding interior negative of the outer shell that allowed the sculpture to "live outside and inside" (Spomenik Revoluciji na Petrovoj gori 1981). The complex structural solution with a curved cavity provokes spiral movement that culminates in the open observation point at the top of the monument, thus accentuating the element of time and making this event extremely contemplative and corporeal (Fig. 83.4). This ambitious but never completely finished

[8] In order to achieve a balanced correlation with the nature a landscape architect Dragutin Kiš was included.

project[9] provided the necessary infrastructure (water, electricity, pedestrian paths, roads, parking spaces, etc.) that enabled the monument to be accessible both by foot and by motor vehicles, and allowed for a wide range of new activities to take place today. Thus, regardless of its current neglected state, the fascination with its form fused with untouched nature attracts experts, scientists, artists and common citizens, forming new secular rituals of practicing collective memory that go far beyond the memorial's initial use.

The monument was the subject of artistic interventions such as the one of Tei Pelant's light painting in 2021, which shows the monument's potential for becoming part of extended in situ art narratives.[10] Due to low level of light pollution, the site is used for various astronomy activities (stargazing, astrophotography), which resulted in the area receiving International Dark Sky Park certification in 2019. Petrova Gora is easily accessible from Croatia's three large towns (Zagreb, Sisak and Karlovac), making it attractive for one-day leisure activities, such as hiking, biking, archery, horse riding, paint ball or visit to Petrovac ornithological park, including organized events such as seasonal mushroom picking, chestnut harvest, movie nights, quad or running races (e.g., Stairs Trail Petrova Gora 2021). Cultural and educational activities are combined with recreational ones on the main Hiking trail of Petrova Gora and two additional educational trails, which pass by several cultural historical sites scattered in the monument's close proximity.

Regardless of its radically reduced number of visitors compared to the Yugoslavian period, the monument's timeless characteristics and today's activities point out to its secularization[11] and to the potential for achieving cultural sustainability. Through its size, symbolic expansion of the surrounding nature and the reduction of means of expression (elementary form and uniform material), this monument opens room for individual interpretation and subjective emotional impact, hence strikingly resembling the narrative of new art tendencies. Its conceptualization as a monument with utilitarian function through the synthesis of the art, architecture and nature opens the possibilities for a variety of future uses, as well as for conceptual repositioning towards a new kind of a "decentralized" and "living" museum site. Thus, combining its past connotations and present values, Petrova Gora illustrates the synthesis paradigm in the context of health, an integrated approach that will unite all of its potentials in a single platform of national care (Fig. 83.5).

---

[9] Next to budget issues, project's incompletion is attributed to the first signs of failure of the utopian self-governing socialism and political disintegration of the Yugoslavian society.

[10] Projects such as "Yesterday tomorrow" in 2012, video installations from David Maljković (Scene for New Heritage, 2004), or photographs of Roman Bezjak and Jan Kempenaers turn the public attention towards rethinking and preserving the heritage of socialist modernism. The site has also been used for filming series and music videos.

[11] The only persisting memorial activity left is the wreath-laying ceremony on the May 14, that marks the anniversary of the Partisan resistance fighters breaking the siege at Biljeg.

Fig. 83.4 Petrova Gora Monument as a sculpture. *Source* Family Archive of Vojin Bakić (up left); copyright 2022 by Nenad Gattin, printed with permission by Photoarchives Nenad Gattin—Institute of Art History Zagreb (up right); photos by Tihana Hrastar (down)

83.4 Conclusion

Although the new national narratives followed by the bloody dissolution of Yugoslavia made the Artistic Monuments obsolete as a common ideological ground, this paper introduces a new paradigm built upon novel perceptions of their shared values and prospects for sustainability. Reviewed through the extended synthesis paradigm in the context of health, every Artistic Monument could unveil a different philosophy for sustainability and healthy living, based upon an authentic eco museal experience that integrates the cultural and the natural, the tangible and the intangible as well as the expert's positions and the common citizen's unfolding narratives. Some of the criteria of this conceptual re-framing are a conscious exploitation of the **destination far away from the cities' cultural centers, a pilgrimage through the**

Fig. 83.5 Petrova Gora Monument's secular rituals. *Source* Photo by Ivica Pezić, printed with permission (up); adopted from Stars Trail Petrova Gora (2021). https://kaportal.net.hr/skriveni-raj-nadohvat-ruke/4131454/ (down left); copyright 2022 by Tei Pelant, printed with permission (down center); photo by Ivan Čujić for the bend "Cura i dečko", printed with permission (down right)

landscape with its totality of heritage, and a real (bodily) experience of nature juxtaposed to the colossal magnitude and the delicacy of the monuments' materiality. The Petrova Gora Monument includes all those criteria and exemplifies the capacity of Artistic Monument to go far beyond its initial use and to **continually mobilize new secular rituals of performing collective memory** by **attracting the creative and recreational industries** (art, architecture, entertainment, media, sports, etc.). This case study reveals that our relationship with the Artistic Monuments should be treated as fluid and ever evolving, based upon our responsibility toward the environment as an integrated, synthesized category. In the midst of the pandemic, this insight into the Artistic Monuments, further acknowledged a lack of systemic understanding of health in relation to architectural and urban design and a lack of secure spatial patterns for sustaining togetherness, such as planned and programmed retreats in nature and open-air activities. By reviving the forgotten philosophy of the humanist ideal of "leisure" as the total experience of arts and nature, such sites as the Yugoslavian Artistic Monuments can be reimagined from the ideologically obsolete toward relevant synthesis paradigm in the context of health.

References

Babić D (2009) Experiences and (hidden) values of ecomuseums. Ethnol Res 1(14):237–252
Batakoja M (2015) Conceptual interaction between art content and museum architecture in art museum design. Unpublished Dissertation, University of Zagreb
Bogdanovic B (1961) O postavljanju spomenika—Odlomci. Arhitektura Urbanizam 10:26–33, 47–48
Horvatinčić S (2017) Memorials from the socialist era in Croatia—typology model. Unpublished Dissertation, University of Zadar
Hrastar T (2020) Models of integrating architecture, design and art—collaborations in Croatian architecture from 1951 to 2016. Unpublished dissertation, University of Zagreb
Knežević S (1981) Projekt spomenika na Petrovoj gori. In: Mohorovičić A (ed) Projekt spomenika na Petrovoj gori. Zavod za arhitekturu Arhitektonskog fakulteta Sveučilišta u Zagrebu, Zagreb, pp 10–17
Papadakis A (ed) (1991) New museology—museums and alternative exhibition spaces [Special issue]. Art & Design (AD), pp 51–62
Pavlović B (1975) Memorijalni park NOB Petrova Gora. Arhitektura 19(155):24–29
Spomenik Revoluciji na Petrovoj gori (1981) Arhitektura 34(176–177):4–6
Tepina M (1961) Spomenici naše epohe. Arhitektura Urbanizam 10:6, 46

Chapter 84
Social Sustainability and Inclusive Environments in Neighbourhood Sustainability Assessment Tools

Rosaria Revellini

Abstract Neighbourhood Sustainability Assessment (NSA) tools are voluntary rating systems for certifying sustainable neighbourhoods in case of new constructions or urban renewals. They consist of categories and indicators to value specific performances. Their purpose is to objectify planned interventions assigning a final score which identifies the overall performance of the district in terms of sustainability. However, is it possible to affirm that these systems actually contribute to the improvement of inclusiveness and healthy living in the neighbourhoods? This question arises as a reflection on the two main issues that contemporary cities have to face urgently which are urbanization and ageing population, focusing attention on developed countries. In this regard, "new" urban spaces are called to achieve inclusion and healthy living for all the people and the neighbourhood represents the right scale for reasoning about. The present study investigates some of the most commonly used neighbourhood scale tools (BREEAM Communities, GBC Italia, DGNB Districts, Living Community Challenge, EcoDistricts) looking at how these systems can help to create more inclusive districts. In particular, the analysis aims to understand how much the social pillar of sustainability affects on urban wellbeing. In fact, there is the evidence that in most NSA tools environmental dimension shall prevails on the others. Through a review of each protocol's "social" categories and of the recent literature on these topics, the study wants to underline criticalities and potentialities of NSA systems and tries to understand in which way a new protocol should act in order to help municipalities, planners and stakeholders in designing inclusive and accessible environments for all.

Keywords Neighbourhood sustainability assessment · Urban wellbeing · Social sustainability

---

R. Revellini (✉)
IUAV University of Venice, Venice, Italy
e-mail: rosaria.revellini@iuav.it

E. Arbizzani et al. (eds.), *Technological Imagination in the Green and Digital Transition*,
The Urban Book Series, https://doi.org/10.1007/978-3-031-29515-7_84

84.1 Introduction

According to United Nations (2019a), currently there are four demographic megatrends: population growth, international migration, population ageing and urbanization. These trends follow differing geographies with disparities between developed countries and developing ones. Furthermore, they all affect the sustainable development of nations.

With a focus on developed countries, it is possible to see that the number of over 65 persons is growing exponentially. For the first time in the history, in 2018, the elderly cohort has exceeded the one of under 5 children. This trend is expected to go faster during the next years—despite the Sars-CoV-2 pandemic—and over 65 will exceed the 15–24 group too by 2050 (UN 2019a).

Meanwhile, it is estimated that about 68% of the world's population will live in urban contexts by the same date (UN 2019b). Therefore, it is crucial to understand the future aspects related to urbanization to ensure a sustainable development of cities and implement the goals of the *2030 Agenda for Sustainable Development* (UN 2015). Specifically, the goal n. 11 *Make cities and human settlements inclusive, safe, resilient and sustainable* is closely related to the development of cities that recognize the centrality of people in transformation processes by providing equal opportunities for all. In this way, the *New Urban Agenda* (UN-Habitat 2017; UN-Habitat 2020) acts as an accelerator to achieve this goal.

In this scenario, population ageing together with migratory pressure influence changes in the urban environment. Specifically, population ageing can be considered as an opportunity to rethink cities as physically and socially inclusive environments, which means suitable places for all people and all ages.

This paper recognises social sustainability as the key to a broader understanding of the concept of inclusivity. Through the literature review on the topic, it aims to understand how this issue is addressed within the most known Neighbourhood Sustainability Assessment tools, with a view to the drafting of a new protocol for the evaluation of social sustainability at neighbourhood scale.[1]

84.2 The Social Dimension of Sustainability

Sustainability is a very complex concept in which different "dimensions" are intersected: environmental, political, regulatory, economic, social, cultural one. In 1994, Elkington coined the term *triple bottom line* by which sustainability has to be considered through the 3 "P": people, planet and profit. It implies an approach that promotes economic growth while minimizing environmental impacts and ensuring social inclusion.

[1] It is the protocol developed within the Ph.D. research carried out by the author at Università Iuav di Venezia. In particular, it refers to the assessment of age-friendly neighborhoods.

Sustainability is achieved when there is a balance between its three dimensions, which are dependent on one another (Colantonio 2009). Nonetheless, unlike the economic and environmental dimensions, the social one has always disregarded in policies and practices, probably because of its "immaterial" nature.

It is not easy to find a unique definition of social sustainability (SS) because this concept includes multiple facets. The definition by Polese and Stern (2000: 15–16) seems to be one of the most interesting, by which SS is the «development (and/or growth) that is compatible with harmonious evolution of civil society, fostering an environment conducive to the compatible cohabitation of culturally and socially diverse groups, while at the same time encouraging social integration, with improvements in the quality of life for all segments of the population».

Starting from this statement, the main aspects of SS are: social equity, social cohesion and participation, social exclusion, environmental justice, security, urban livability, quality of life (Colantonio 2009; Shirazi and Keivani 2019). As Colantonio (2009) says, it is possible to distinguish "soft" components (intangible, as social cohesion) from "hard" ones (tangible, as facilities presence).

The urban environment is the "cradle" of SS. The physical characteristics of the city have considerable influence on the components of the SS and vice versa (Bramley et al. 2006). For this reason, it seems necessary to measure "sociality" in urban context, in order to improve both spatiality and policies when necessary. However, its "intangible" nature and the lack of a unique definition make the evaluation difficult to achieve (Colantonio 2009), as it is possible to see below.

84.3 Neighbourhood Sustainability Assessment Tools

Sustainability assessment tools are voluntary systems edited by no-profit organization to certify specific performances of the "object" to be assessed. The first examples were born in the 90s with the aim of controlling and limiting buildings energy consumption. In these cases, the number of environmental and economic criteria predominates over the social ones. This trend is still present in the most recent tools, confirming «the fundamental misunderstanding according to which sustainability is mainly intended in environmental terms, despite its strongly anthropocentric nature» (Acierno and Attaianese 2018: 267).

Neighbourhood Sustainability Assessment (NSA) tools were born in the early 2000s. During these years, in fact, cities, neighbourhoods and public spaces have been subjects of interest of sustainable studies since they can play a key role in sustainable development processes (Sharifi et al. 2021).

NSA tools are used both in the case of new constructions and urban renewals. They consist of categories, indicators and benchmarks to evaluate specific performances. Their purpose is to objectify planned interventions assigning a final score which identifies the overall performance of the district in terms of sustainability (Boyle et al. 2018). The assessment process is led by independent third parties and it has a cost. These characteristics are seen as the main critical aspects in fact they constitute a limit in the dissemination of these systems above all in developing countries.[2]

The most used NSA tools worldwide are: BREEAM Communities (UK), LEED Neighbourhood Development (USA) and GBC Quartieri (Italy), ITACA Scala Urbana (Italy), DGNB Districts (Germany), Living Community Challenge (USA), CASBEE for Urban Development (Japan), Green Star Communities (Australia), EcoDistricts (USA), HQE2R (France).[3]

They can be divided in "spin-off"—the most of them—which are the ones derived from building-scale systems, and "others"—as EcoDisctricts and HQE2R—which instead have been specifically created for urban-scale interventions (Sharifi and Murayama 2012).

This study aims to investigate the social dimension in five of these tools,[4] in order to understand which are the actual limitations regarding SS and how these systems can help municipalities and planners to create more inclusive districts.

84.4 BREEAM Communities

The *Building Research Establishment Environmental Assessment Method* (BREEAM) was born in UK in 1990 and it is the first sustainable assessment tool at building-scale worldwide. In 2008 the urban-scale version was published with the name of *BREEAM Communities.*

This tool is organized in three steps (establishing the principles; determining the layout; designing the details) and six categories (*Governance*; *Social and economic wellbeing*; *Resources and energy*; *Land use and ecology*; *Transport and movement*; *Innovation*). The 2012 version—the most update one—consists of a total of 40 individual assessment issues. The certification is obtained when at least 30% score is reached.

The subcategory *Social wellbeing*[5] is the 17.1% of the total and it aims «to ensure a socially cohesive community» (BRE 2017: 15). Here the social theme is expressly stated thanks to its 9 criteria, which are listed in Table 84.1.

---

[2] Especially in Italy, the district-scale tools are rarely used because they should be provided by public administrations (as promoters).

[3] These tools are globally widespread, but they are built based on priorities and regulations of the countries in which they are developed.

[4] Protocols have been selected based on opensource availability and international scientific relevance.

[5] The main category is *Social and economic wellbeing.*

Table 84.1 Social wellbeing criteria in BREEAM communities

ID	Criterion	Aim in brief	Weighting (%)
SE02	Demographic needs and priorities	To ensure that design is based upon the local demographic trends and priorities	2.7
SE05	Housing provision	To ensure appropriate housing provision for all within the development	2.7
SE06	Delivery of services, facilities and amenities	To ensure essential facilities are provided and that they are located within a reasonable and safe walking distance	2.7
SE07	Public realm	To encourage social interaction by creating comfortable and vibrant spaces in the public realm	2.7
SE09	Utilities	To provide easy access to site service and communications infrastructure	0.9
SE11	Green infrastructure	To ensure access to high-quality space in the natural environment or urban green infrastructure for all	1.8
SE12	Local parking	To ensure parking is appropriate for the expected users and well-integrated into the development	0.9
SE14	Local vernacular	To ensure that the development relates to the local character whilst reinforcing its own identity	0.9
SE15	Inclusive design	To create an inclusive community by enhancing accessibility for as many current and future residents as possible	1.8

84.4.1 EcoDistricts

EcoDistricts was born in Portland, Oregon (USA) in the first decade of the 2000s to promote health and justice in the cities. Its tool—named *EcoDistricts* too—is designed exclusively for neighbourhood scale. The protocol is «a rigorous, sustainable urban development framework for achieving people-centred, economically vibrant neighbourhood and district-scale sustainability» (EcoDistricts 2018: 7).

EcoDistricts has three key-elements: three *Imperatives* (*Equity*; *Resilience*; *Protection*), six *Priorities* (*Place*; *Prosperity*; *Health and Wellbeing*; *Connectivity*; *Living infrastructure*; *Resource regeneration*) and three implementation phases. If all the requirements are achieved, it is possible to have the final certification.

Table 84.2 Most relevant social indicators in EcoDistricts

Priority	Objective categories	Aim in brief
Place (*create inclusive and vibrant communities*)	Engagement and inclusion	Civic engagement is strong and processes are inclusive and representative. Sharing programmes are robust
	Culture and identity	Historic and culturally significant places are preserved and celebrated. Participation in cultural events is high
	Public spaces	Public spaces are accessible to all. They are high quality, engaging and active
Health and wellbeing (*nurture people's health and happiness*)	Active living	Access to recreation facilities and services is improved. Walkability is enhanced
	Health	Health outcomes and life expectancy are more equitable. Affordable, high-quality health care is accessible. Toxic environments are remediated and regenerated
	Safety	Public safety is enhanced. The built environment is designed for public safety
	Food systems	Healthy and affordable fresh food is accessible Food production in the district is encouraged

Among the six *Priorities*, only *Resource regeneration* concerns strictly the environmental field, in fact social aspects are more integrated than in other tools. Some of the most relevant social indicators, according to the author, are listed in Table 84.2.

84.4.2 DGNB Districts

The *German Sustainable Building Council* (DGNB) was born in 2007 in Germany to promote sustainability in the building sector. After its version at building-scale, in 2012 the NSA tool was published with the name *DGNB Districts*.

In the last version (ed. 2020), there are 5 thematic areas (*Environmental quality*; *Economic quality*; *Sociocultural and functional quality*; *Technical quality*; *Process quality*), all of them with the same weighting (20%), and 31 criteria. It is possible to achieve the certification with a minimum of 50% score.

Table 84.3 Main social criteria in DGNB Districts

ID	Criterion	Aim in brief	Weighting (%)
SOC1.6	Open space	To satisfy the need for recreation and interaction by providing high-quality open spaces within walking distance	3.5
SOC2.1	Barrier-free design	To make the entire environment accessible to everyone and without restrictions on its use	2.6
SOC3.1	Urban design	The objective is to contribute cultural identity by establishing and maintaining consistent urban structure as part of the city as a whole	2.6
SOC3.2	Social and functional mix	To make the district adaptable to social change and ensure a socio-functional mix	3.5
SOC3.3	Social and commercial infrastructure	To ensure close, easily accessible and commercial infrastructure, creating social acceptance of the district	2.6
PRO1.7	Participation	To involve all those affected by the planning at an early stage	3.3

In this protocol, «people's health and happiness should be a focal point when making design and construction decisions» (DGNB 2020: 7). The main social criteria are listed in Table 84.3.

84.4.3 Living Community Challenge

In 2014, the *International Living Future Institute* of Seattle (USA) created *Living Community Challenge* (LCC), a district-scale assessment tool, after the building-scale version.

It has a different structure from the other tools because it is not prescriptive, but it allows the analysis of urban areas to understand the potential of the place in improving citizens experience. For this reason, there are no benchmarks, and this represents its most important limit in application.

It has 7 categories (called *Petals*: *Place*; *Water*; *Energy*; *Health and happiness*; *Materials*; *Equity*; *Beauty*) and a total of 20 imperatives (ILFI 2017). The certification is possible only if all the imperatives are checked. The main social imperatives are listed in Table 84.4.

Table 84.4 Main social imperatives in LCC (ed. 2019)

ID	Criterion	Aim in brief
04	Human-powered living	To create walkable, pedestrian-oriented communities
08	Healthy neighbourhood design	To promote and optimize the health and wellbeing of its residents
14	Human scale and human places	To create human-scaled rather than automobile-scaled places
15	Universal access to nature and place	All primary transportation, roads and non-building infrastructure must be equally accessible to all people
16	Universal access to community services	To have basic community services and amenities that support the health, dignity and rights of all people
19	Beauty and spirit	To have public art and design features in urban spaces intended solely for human delight
20	Inspiration and education	To ensure participation through education of the community

84.4.4 GBC Italia Quartieri

The last analysed tool is *GBC Italia Quartieri*, made by *Green Building Council Italia* in 2015. It is the Italian version of the *LEED Neighbourhood Development* (USA).

It has three main assessment categories (*Site Location and Connections*; *Neighbourhood Planning and Organization*; *Sustainable Infrastructure and Buildings*) and two optional ones (*Design Innovation*; *Regional Priority*). There are 42 credits and 12 required prerequisites. The certification is available only with the minimum score of 40 points (GBC Italia 2015).

Compared to the other tools, *GBC Italia Quartieri* does not contribute much to social sustainability aspects. According to the author, the main social credits are related to "spatial quality" as showed in Table 84.5.

84.5 Discussion and Conclusions

The carried-out analysis shows that the environmental aspects are the most considered, followed by the economic ones and finally by the social ones. In particular, regarding SS, the most common criteria refer to its "hard" part rather than the "soft" one. In fact, there are no methods to evaluate it objectively considering its partly intangible nature.

Table 84.5 Main social credits in GBC Italia Quartieri

ID	Credit	Aim in brief	Max pt (out of 100)
OPQ1	Streets friability for pedestrians	To promote efficient transport and walking	9
OPQ3	Mixed use neighbourhoods	To group and make accessible different uses in central areas of the neighbourhood	4
OPQ6	Connected and open communities	To promote projects that have high levels of internal connection and are well connected to the city	2
OPQ9	Access to public spaces	To improve citizens social life by offering them a variety of open spaces	1
OPQ10	Access to recreational activities	To improve citizens social life by offering them a variety of recreational activities	1
OPQ11	Universal accessibility	To allow all citizens to participate more easily in community life	1
OPQ12	Involvement and openness to the community	To promote awareness of community needs by activating participation	2

Looking at the various criteria in the previous tables, it is possible to affirm that: (1) the different tools are not comparable with each other because they differ from the criteria and weights assigned[6]; (2) the "spatial" criteria (e.g. public spaces, accessibility, mixitè) are much greater in numbers than those concerning social activities and sense of community (such as participation, involvement, equity). EcoDistricts and LCC are the only tools that work more in this direction, even if their weakness, as mentioned, consists in the fact that they do not provide for an objective measurement through benchmarks.

However, trying to evaluate the SS as a whole could be helpful in order to achieve inclusivity for all people in urban areas. This is one of the most important issues of the *New Urban Agenda*, which in fact promotes actions for inclusive cities and human settlements right through participation, civic engagement, sense of belonging, social and intergenerational interaction (UN-Habitat 2020).

[6] For example, in the case of "public space" criterion (BREEAM—S07 Public realm; EcoDistricts—Public spaces; DGNB Districts—SOC 1.6 Open space; LCC—10 Human scale and human places; GBC Italia Quartieri—OPQ9 Access to public spaces) the aims are similar but the weighting differs from a tool to the other.

Fig. 84.1 Monday farmer market in the Santa Marta neighbourhood in Venice stimulates social interaction in an area with high number of elderly residents. Rosaria Revellini (2021)

In the ongoing Ph.D. research, the author is elaborating a protocol focusing both on tangible and intangible aspects of the SS. Starting from the carried-out analysis on the existing NSA tools (briefly reported in this paper), the new tool aims to reconsider the social dimension of sustainability as the key element for urban regeneration processes in developed areas. In fact, it seems necessary considering the two trends of urbanization and population ageing.

Specifically, the new criteria will all have the same weighting for two reasons: each aspect of SS is equally important and avoiding subjectiveness of the assessment process. The purpose is to have a tool that can be easily used by municipalities, planners and stakeholders in designing inclusive and accessible environments for all (Figs. 84.1 and 84.2).

Fig. 84.2 "Afternoon" bench where elderly people in Santa Marta neighbourhood in Venice are usually seated, favoured by the tree shade. Rosaria Revellini (2019)

References

Acierno A, Attaianese E (2018) Fattore umano e sicurezza nei protocolli di certificazione a scala di quartiere. BDC. Bollettino Del Centro Calza Bini 18(2):267–284. https://doi.org/10.6092/2284-4732/6241

Boyle L, Michell K, Viruly F (2018) A critique of the application of neighborhood sustainability assessment tools in urban regeneration. Sustainability 10.https://doi.org/10.3390/su10041005

Bramley G, Dempsey N, Power S, Brown C (2006) What is 'social sustainability', and how do our existing urban forms perform in nurturing it. http://www.city-form.org/uk/pdfs/Pubs_Bramley etal06.pdf. Accessed 15 May 2022

Building Research Establishment—BRE (2017) BREEAM communities. Technical Manual, SD 202—1.2:2012

Colantonio A (2009) Social sustainability: a review and critique of traditional versus emerging themes and assessment methods. In: Horner M, Price A, Bebbington J, Emmanuel R (eds) Sue-Mot conference 2009: second international conference on whole life urban sustainability and its assessment: conference proceedings, pp 865–885

DGNB GmbH (2020) DGNB system. Districts criteria set, version 2020. Stuttgart

EcoDistricts (2018) EcoDistricts protocol. The standard for urban and community development, version 1.3. EcoDistricts, Portland

Green Building Council Italia—GBC Italia (2015) Manuale GBC Quartieri. Per progettare, realizzare e riqualificare aree e quartieri sostenibili

International Living Future Institute—ILFI (2017) Living community challenge. A visionary path to a regenerative future. Seattle

Polese M, Stren R (ed) (2000) The social sustainability of cities: diversity and the management of change. University of Toronto Press

Sharifi A, Murayama A (2012) A critical review of seven selected neighborhood sustainability assessment tools. Environ Impact Assess Rev 38:73–87. https://doi.org/10.1016/j.eiar.2012.06.006
Sharifi A, Dawodu A, Cheshmehzangi A (2021) Limitations in assessment methodologies of neighborhood sustainability assessment tools: a literature review. Sustain Cities Soc 67. https://doi.org/10.1016/j.scs.2021.102739
Shirazi MR, Keivani R (ed) (2019) Urban social sustainability. Theory, policy and practice. Routledge, Oxon; New York
United Nations (2015) Transforming our world: 2030 agenda for sustainable development (A/RES/70/1)
United Nations, Department of Economic and Social Affairs, Population Division (2019a) World population prospects 2019a: highlights (ST/ESA/SER.A/423)
United Nations, Department of Economic and Social Affairs, Population Division (2019b) World urbanization prospects: the 2018 revision (ST/ESA/SER.A/420). United Nations, New York
United Nations, Human Settlements Programme, UN-Habitat (2017) New urban agenda (A/RES/71/256*)
United Nations, Human Settlements Programme, UN-Habitat (2020) The new urban agenda illustrated (HS/035/20E)

Chapter 85
Inclusive Neighborhoods in a Healthy City: Walkability Assessment and Guidance in Rome

Mohamed Eledeisy

Abstract With the increasingly global and European attention toward healthy inclusive cities, the focus on pedestrian-friendly environments, as a tool to encourage and support healthy lifestyles for people of all social groups and ages, continues to rise. This article aims to assess "walkability" as one of the main conditions of a built environment that enhances "healthy living", a core theme of the Zagreb Declaration for Healthy Cities. Transit-Oriented Development Standard (TOD Standard) is used as a tool to evaluate the walkability level in San Giovanni area in Rome, Italy. Through urban plans and measurements of the pedestrian realms, the research evaluates the state of walkability through a metric scoring method of the walkways' segments. The analysis demonstrates the percentages of all-accessible walkway segments and crosswalks that are safe in all directions; segments with visually active frontages; physically permeable frontages; and segments that incorporate adequate shade or shelter. The results show the pedestrian realm's level of safety, completeness, accessibility to all; its activeness and vibrance; and its level of comfort. The conclusions provide guidance for areas of intervention to make walking accessible for everyone and support decision-making processes to develop inclusive neighborhoods, as a part of the future policies for equitable access and mobility in a healthy city.

Keywords Healthy living · Neighborhood accessibility · Pedestrian-friendly · Transit-oriented development

85.1 Introduction

The WHO European Healthy Cities' movement aims to make a difference in health and well-being and to improve equity through action on underlying urban factors (WHO European Healthy Cities Network 2014). According to the Zagreb Declaration, a healthy city offers a physical and built environment that enables and supports health, recreation and well-being, safety, social interaction, accessibility,

M. Eledeisy (✉)
Department of Architecture and Design, Sapienza University of Rome, Rome, Italy
e-mail: mohamed.eledeisy@uniroma1.it

E. Arbizzani et al. (eds.), *Technological Imagination in the Green and Digital Transition*, The Urban Book Series, https://doi.org/10.1007/978-3-031-29515-7_85

and mobility to the needs of all its citizens (World Health Organization 2009). Physical activity and walkability have been linked to positive health outcomes and well-being in all age groups, especially among older adults (Cauwenberg et al. 2016; Wang and Yang 2019). In Rome, the aging index is 166.2% and the demographic dependence ratio is 54.7% where 22% of the population is 65+ years old, and 5.2% of the Italian population is with severe health limitations or a disability (Roma Capitale 2018; Istituto nazionale di statistica 2019). This puts emphasis on the importance of developing pedestrian-friendly neighborhoods, where all have the right to access, to reach close and distant destinations easily and affordably, and to live a good life free of dependence on cars.

85.2 Methodology

This paper aims to investigate walkability as a tool to increase accessibility and inclusiveness on a neighborhood scale, with consequent benefits to support healthy lifestyles in Rome, Italy. The first step was to choose a case study that respected the following criteria: densely populated residential area, representative of the Roman population demography, representative of a consolidated urban morphology with well-defined building typologies, and presence of a high-quality transit station (i.e., metro station connected to more than one line).[1] Accordingly, San Giovanni (SG) in Municipality area VII was chosen. It is the densest neighborhood in Rome with 20,910 residents/km^2, 26.1% of its population over the age 65 years old, 13.9% over 74, and 11.1% under the age of 15. SG's buildings (developed between 1920 and 1980s) which can be classified into four types: Palazzo (4–8 story apartment block, most common type), Palazzini (3–4 story apartment building), Villino (two-story building), and Container (non-residential, single-purpose building style). It is considered to have few public spaces, and the street typologies are high volume traffic artery, two-lane road, and local single-lane road (Shin et al. 2017).

The second step was the definition of the study area with the selection of the assessment tool. Urban plans and G-Earth imaging of the area were revised. City's urban department website and other public administration's resources were reviewed for zoning and strategic plans (Roma Capitale 2022). Three field visits were performed for comparison and calibration between the plans and the neighborhood and to assess the behavioral patterns. Parallelly, the choice of an assessment tool was based on the following criteria: based on health and inclusiveness principles, well-established (i.e., officially recognized standard), applicable to urban areas and to existing built environments, and based on numerical metrics that can be independently and easily observed and verified.

[1] A mobility system that is safe, healthy, and sustainable encourages two things: the adoption of soft travel methods or non-motorized means and the switching between transport methods. For this, the vicinity to a transit station represented a selection criterion.

Transit-Oriented Development (TOD) Standard, defined by the Institute for Transportation and Development Policy (ITDP), was chosen. It is used to score products of urban development according to their adherence to a set of principles, with easy walking and cycling, and near-excellent transit service to the rest of the city (ITDP (Institute for Transportation and Development Policy 2017). It represents a recognition system based on a 100-point score, distributed between eight principles. For this study, principle 1 (walk) is used to provide a reference with clear definitions and a rapid assessment of the pedestrian realm. Its objectives and measurement methods are applied through 15 points. It motivates three key implementation objectives:

- Objective A: The pedestrian realm is safe, complete, and accessible to all.
- Objective B: The pedestrian realm is active and vibrant.
- Objective C: The pedestrian realm is temperate and comfortable.

It is recommended that the walking distance between the transit station and the building entrance in the project farthest from the station should be 500 m (10-min walk). This was applied to SG's metro station, overlapping it with the area within the jurisdictions of Municipality VII to define the case study area (Fig. 85.1). Between January and March 2022, six visits were performed to assess the urban properties of the pedestrian realm, measure the metrics, apply scoring, and perform interviews.

Objective A: The pedestrian realm is safe, complete, and accessible to all

The most basic feature of urban walkability is the existence of a complete and continuous walkway network including safe crossings. It must be accessible to

Fig. 85.1 Study area in San Giovanni, Rome, Italy

all persons, including older people and people with disabilities. This objective is measured by the following metrics:

1.A.1. Walkways (3 points). This metric shows the percentage of segments with safe, complete all-accessible walkways, where a block's walkway is measured as a segment between two adjacent intersections in the pedestrian network and can be of the following types:

(a) Dedicated sidewalks protected from vehicular traffic by a curb or other adequate device.
(b) Shared streets designed for sharing between pedestrians, cyclists, and vehicles (speed limit).
(c) Pedestrian paths or pedestrian–cyclist shared paths.

The percentage is calculated by dividing qualifying walkway segments measured by the total segments abutting the block, where qualifying walkways are:

(a) Designed for easy pedestrian access to all abutting buildings on the block frontage.
(b) Unobstructed and barrier-free for people with disabilities, including wheelchair users and people with low vision.
(c) Receive street lighting at night (adequate for pedestrian safety and security).

Although almost all segments are considered safe and secure due to the adequate night lighting, no walkway segment can be considered qualifying due to the lack of the first two conditions in all the study areas, i.e., a complete absence of accessibility (to all) to almost all buildings on the block frontages, and they are obstructed for users of walking or carrying aids and severe long-term limitations due to a health problem.[2] None of the segments were found free or near free of architectural barriers, especially steps at entrances of apartment blocks and stores. The walkways' condition is poor with inadequate surface materials, irregular surfaces, and unsafe changes in elevation, incompliant gaps, and grates, causing tripping hazards and difficulties for pedestrians, especially those with vision and mobility disabilities (Fig. 85.2). For this reason, 1.A.1. walkways obtained 0 out of 3 points (walkway network that is complete is less than 80%).

1.A.2. Crosswalks (3 points). This metric shows the percentage of intersections with safe, all-accessible crosswalks in all directions (compliant with the below):

(a) Barrier-free for people with disabilities, including wheelchair users and people with low vision.
(b) Minimum 2 m width and are demarcated.
(c) Feature all-accessible refuge islands if crossing more than two traffic lanes.
(d) Receive adequate street lighting at night.

---

[2] According to the data from the Istat survey "Aspects of daily life", 6.2% of females and 4.1% of males in Lazio region have severe long-term limitations in usual activities due to a health problem according to the Global activity limitation indicator (Gali).

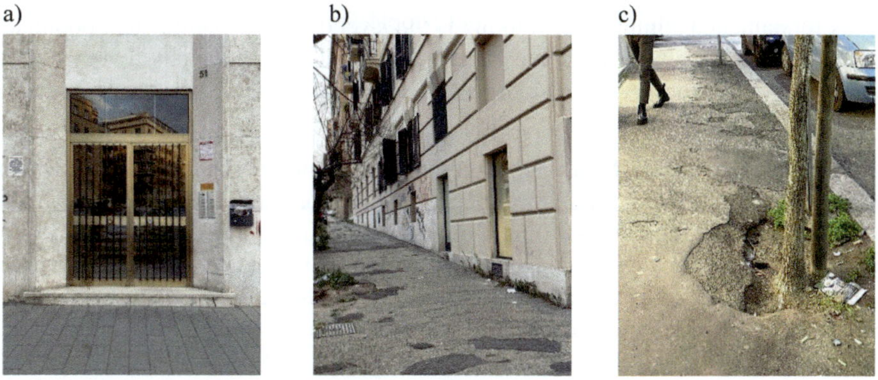

Fig. 85.2 Walkways in the case study. **a** An architectural barrier at building entrance in Via Urbino. **b** Irregular surfaces in Via Fregene. **c** Incompliant gaps in Via Veio

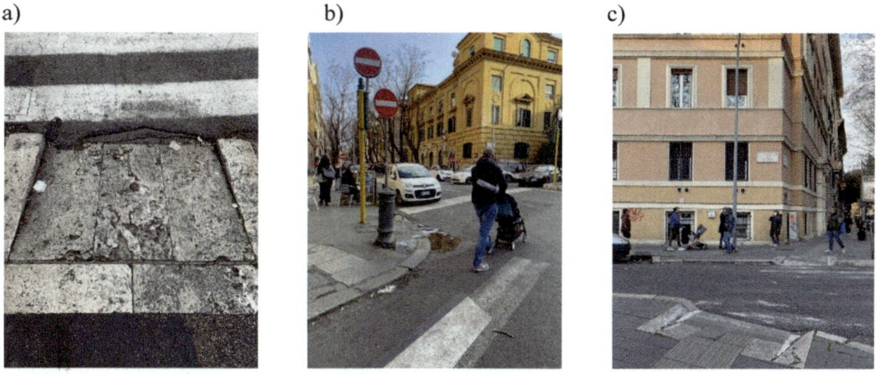

Fig. 85.3 Crosswalks in the case study. **a** Cracks and holes on the ramp in Via Appia Nuova. **b** Inadequate slope and change in level in Via Magnagrecia. **c** Faded crosswalk paint in Via Ardea

Eighteen intersections (67%) were found compliant (i.e., safe and all accessible) of 27 intersections. Therefore, 1.A.2. crosswalks obtained 0 out of 3 points (less than 80% complete crosswalks). The assessment was based on local and international standards (Il presidente della repubblica 1996; The Department of Economic and Social Affairs 2013).

Even though multiple intersections were, presumably, compliant when realized, the maintenance is low, presenting cracks and holes on the surfaces, irregular ramps, inadequate slopes and changes in level, and inadequately painted crosswalks (Fig. 85.3). This was less evident in the main streets,[3] especially Via La Spezia and Via Taranto, with significantly higher safety and all-accessible crosswalks.

Objective B: The pedestrian realm is active and vibrant.

---

[3] Streets with a right of way from building line to building line of 20 m or more.

A fundamental stimulant of walking is the level of activity. Animated, populated sidewalks make walking attractive, secure, and productive, especially with the presence of commercial services and ground floor activities like storefronts, restaurants, etc. This objective is measured by the following metrics:

1.B.1. Visually active frontage (6 points). This metric shows a percentage of walkway segments with a visual connection to interior building activity. A segment is active if 20% or more of its ground floor frontage abutting public walkways is visually penetrable with a visual connection to interior building activity. This comprises partially transparent materials and windows (even with operable curtains or shutters) at any point between ground level and 2.5 m. Therefore, all 73 segments studied were found active. This is due to the significant commercial presence. Even, walkways with no commercial activity (e.g., segments in Via Isernia, Via Amiterno, or Via Ardea) have frontages that are more than 20% active (i.e., with openings). Except for Via La Spezia, visual activity is higher on main streets such as Via Magnagrecia, Via Aosta, and Via Appia Nuova, which was considered 100% active on both sides. Therefore, this metric obtained all 6 points.

1.B.2. Physically permeable frontage (2 points). This metric represents the average number of shops, building entrances, and other pedestrian access per 100 m (m) of block frontage. To quantify this metric, the total length of block frontage that abuts public walkways (12,106 m) is divided by 100 m. Then, the number of entrances along walkways (744) is divided by the previous measure. The average number of entrances per 100 m of block frontage is 6.15 entrances. Therefore, this metric obtained 2 out of 2 points (100 m frontages with more than 5 entrances).

Objective C: The pedestrian realm is temperate and comfortable.

This objective is measured by one metric:

1.C.1. Shade and shelter (1 point). This metric reflects the percentage of walkway segments that incorporate adequate shade or shelter amenities, including trees, building elements (e.g., arcades, awnings, cast shadows), freestanding structures (e.g., public transport shelters), and vertical wind and solar screens (e.g., walls and lattices). Fifty-six segments (77% of the total) were incorporating climate-adequate shade and shelter elements. The shade or shelter was mainly provided by trees and buildings (compact urban fabric cast shadows). The field visits were in January–March, where the solar angles are low and the deciduous trees' impact is different from summer; therefore, measurements were supported by a conceptual solar analysis (LoD1) and interviews for the residents' feedback. The street with the least shade and shelter was Via La Spezia, especially the northern sidewalks. This metric obtained 1 out of 1 point (75%+ adequate).

Table 85.1 TOD (walk) score in San Giovanni

Metric		Case study points	Maximum score
1.A.1	Walkways	0	3
1.A.2	Crosswalks	0	3
1.B.1	Visually active frontage	6	6
1.B.2	Physically permeable frontage	2	2
1.C.1	Shade and shelter	1	1
Total		9	15
Total %		**60% Bronze-standard**	

85.3 Results of the Scoring and Assessment

San Giovanni study area was assessed according to TOD principle 1 (Walk) scoring (Table 85.1). No points were obtained for the completeness and safety of walkways and road-crossings, indicating that the pedestrian realm is not safe and accessible to all, especially for users of walking or carrying aids. The pedestrian realm, however, is active and vibrant; building frontages are highly visually active and physically permeable, where both metrics scored the maximum points. The assessment of the walkway's temperateness and comfort was also evaluated positively, indicating adequate shade and shelter in the neighborhood. The final score is 9 out of 15 points (60%), corresponding to a Bronze-standard TOD (if, hypothetically, the other seven principles were to obtain the same score), indicating an area that satisfies most of the objectives of best practice. The assessment and recommendations of each metric are described in Table 85.2.

85.4 Conclusion

This study has provided a methodological assessment of the walkability, as a tool to enhance "healthy living" and increase accessibility and inclusiveness. The case study was found active and vibrant, due to the mixed-use and commercial presence on the ground floor. This encouraged foot traffic, enhanced commercial activity due to the exposure of retail and services. It also reinforced a sense of security due to the visual interior–exterior interactions, through passive and informal observation and surveillance. The work highlighted that the study area has multiple challenges regarding accessibility and safety of the pedestrian realm, creating difficulty for aging inhabitants, an issue for the inclusion of people with vision and mobility disabilities, and those with limitations due to a health problem. This calls for the activation of programs for the walkway network maintenance, and the architectural barrier removal, which on one hand, is already on the agenda of the local administration and, on the other hand, requires major efforts to be achieved. Additionally, there is a

Table 85.2 Evaluation and recommendations of the walkability metrics

Objective		Performance	Recommendations and comments
1.A.1	Walkways	0% very weak	Major requalification interventions are needed to reach complete, all-accessible walkways, especially for people with disabilities and people with severe long-term limitations in usual activities due to a health problem
1.A.2	Crosswalks	0% very weak	With minimal requalification and maintenance interventions, it is possible to adapt the crosswalks to local standards, reaching a full score
1.B.1	Visually active frontage	100% excellent	Highly present in the whole study area. Highest score is reached in the main streets with commercial activity on ground floor
1.B.2	Physically permeable frontage	100% excellent	Highly present in the study area. Highest score is reached in the main streets with commercial activity on ground floor
1.C.1	Shade and shelter	100% excellent	Highly present in the study area due to the compact urban morphology and the presence of trees in most streets. The main streets, however, are the ones lacking adequate shading and shelter, especially during the hot season

lack of amenities and streetscape-enhancing elements. Future projects should ensure the availability of universal design furniture. It is necessary to provide action plans and design solutions that aspire to reach the aims of a healthy city, a city for all citizens, inclusive, supportive, sensitive, and responsive to their diverse needs and expectations.

85.5 Limitations and Future Work

TOD is a simple scoring system that can be independently and reasonably applied for a prompt evaluation. It was designed by ITDP, operating in urban contexts that are different from Rome/Italy. However, it has a margin of adaptability in the evaluation of some aspects (for example, it indicates to use local or international regulations' references). This could be seen both as a limit or a strength based on the work scope. It is possible to furtherly define its application method(s) according to the work objectives, especially for research purposes. Future work could perform detailed climatic simulations to achieve a developed evaluation of comfort (1.C.1). It could also measure the other TOD principles for a comprehensive assessment of the study area.

Acknowledgements The author wishes to thank Carlo Cecere, Helena Coch, Carlos Alonso, Judit López, and Michele Morganti.

References

Cauwenberg JV, Holle VV, De Bourdeaudhuij I, Van Dyck D, Deforche B (2016) Neighborhood walkability and health outcomes among older adults: the mediating role of physical activity. Health Place 37:16. https://doi.org/10.1016/j.healthplace.2015.11.003

Il presidente della repubblica (1996) Decreto del Presidente della Repubblica 24 luglio 1996, n. 503. https://www.gazzettaufficiale.it/eli/id/1996/09/27/096G0512/sg

Istituto nazionale di statistica (2019) Conoscere il mondo della disabilità: persone, relazioni e istituzioni. https://www.istat.it/it/files/2019/12/Disabilit%C3%A0-1.pdf

ITDP (Institute for Transportation and Development Policy) (2017) The TOD Standard v3.0. ITDP. Institute for Transportation and Development Policy, New York

Roma Capitale (2018) La popolazione di roma struttura e dinamica demografica anno 2017. https://www.comune.roma.it/web-resources/cms/documents/La_popolazione_a_Roma_2017.pdf

Roma Capitale (2022) Municipi di Roma Capitale. http://www.urbanistica.comune.roma.it/mappa-municipi.html

Shin B, Galvin M, Laude R, Hussam S (2017) San Giovanni living next to a transit corridor. Cornell University

The Department of Economic and Social Affairs (2013) Mainstreaming disability in the post-2015 development agenda. United Nations. https://www.un.org/disabilities/documents/accessibility_and_development.pdf

Wang H, Yang Y (2019) Neighbourhood walkability: a review and bibliometric analysis. Cities 93:43. https://doi.org/10.1016/j.cities.2019.04.015

WHO European Healthy Cities Network (2014) Healthy cities. Promoting health and equity—evidence for local policy and practice. WHO Regional Office for Europe, Copenhagen

World Health Organization (2009) Zagreb Declaration for healthy cities. WHO Regional Office for Europe, Denmark

Chapter 86
Tools and Strategies for Health Promotion in Urban Context: Technology and Innovation for Enhancing Parish Ecclesiastical Heritage Through Sport and Inclusion

Francesca Daprà, Davide Allegri, and Erica Isa Mosca

Abstract The relationship between the built environment and health is an increasingly important issue in the planning and regeneration of the contemporary city. The contribution reflects on the impact of sport and social inclusion on the population's health and well-being, moving from the results of a research experience. The project involves the parish ecclesiastical heritage. It proposes methods committed to its regeneration and innovation, aiming to enhance the *oratorio* sports facilities in a multi-generation, inclusive, and health education perspective. The definition of a multidisciplinary and analytical tool is based on a set of qualitative and quantitative criteria, for assessing the structures in different aspects, to reconsider the pre-existing sport facilities, and suggesting strategies for the renovation and innovation of their spaces and services. The application of the tool to the parish facilities brings extensive reflections on the importance of promotion of physical activity and of the creation of accessible social environments, suggesting strategies for more liveable and healthy community spaces; moreover, it contributes to the definition of systemic strategies and scientific tools for the enhancing of built heritage in the urban context.

Keywords Ecclesiastical Heritage · Social Inclusion · Healthy Communities · Physical Activity · Sport Facilities

---

F. Daprà (✉) · D. Allegri · E. I. Mosca
Polytechnic University of Milan, Milan, Italy
e-mail: francesca.dapra@polimi.it

D. Allegri
e-mail: davide.allegri@polimi.it

E. I. Mosca
e-mail: ericaisa.mosca@polimi.it

E. Arbizzani et al. (eds.), *Technological Imagination in the Green and Digital Transition*, The Urban Book Series, https://doi.org/10.1007/978-3-031-29515-7_86

86.1 Introduction

The current research considers the importance of the relationship between the built environment and health promotion, as a key factor for urban development and regeneration. The study reflects on the impact of sport and social inclusion on the population's health and well-being, moving from the results of an experience—the Research Project "SPèS—Sport è Società", funded by PoliSocial Award 2019.

By collocating sport as a driver for social inclusion, the research analyzed the parish ecclesiastical heritage—best described by the Italian word *oratorio*—highly consolidated and diffused in the urban context. Such a system demonstrated its power and potential as a collective infrastructure to be re-discovered and integrated with the proximity service system, especially concerning sport and health promotion. Despite its great potential, there is no systematic strategy for *oratorio* facilities planning and renovation: often, the initiative for the redevelopment and innovation of their structures, services, and activities takes place in a non-structured and informal way.

The project proposes methods committed to the regeneration and innovation of such heritage in the city, aiming at enhancing the parish sports facilities in a multi-generation, inclusive, and health education perspective.

86.2 Built Heritage and Health Promotion in Urban Regeneration Processes

The contribution deals with the topic of built heritage within the processes of urban and social regeneration, concerning redevelopment practices able of promoting health and well-being in the urban environment.

During the last years, more and more studies highlighted that physical inactivity impacts people's health (WHO 2014) and that sports and exercise can lead to the prevention of different diseases, promoting physical and mental well-being for various ages (WHO 2018). Therefore, strategic actions aimed at transforming urban settlements and programmatic actions focused on Public Health and urban quality promotion are needed (Capolongo et al. 2018). In this regard, initiatives aimed at promoting physical and sport activity are constantly increasing (EU Physical Activity 2008; WHO 2010).

In this panorama, the collective structures devoted to sports in the urban context are passing through processes of transformation, due to the renovated needs of the social system and the changing in urban structure and livability (Vettori 2020), even more, after the pandemic period which raised the attention for health and accessibility to urban services (Capolongo et al. 2020). For these reasons, the transformation of existing facilities becomes a relevant issue, considering the non-consumption of soil trends and the urgency of improving the resilience of existing artifacts (Allegri and Vettori 2018).

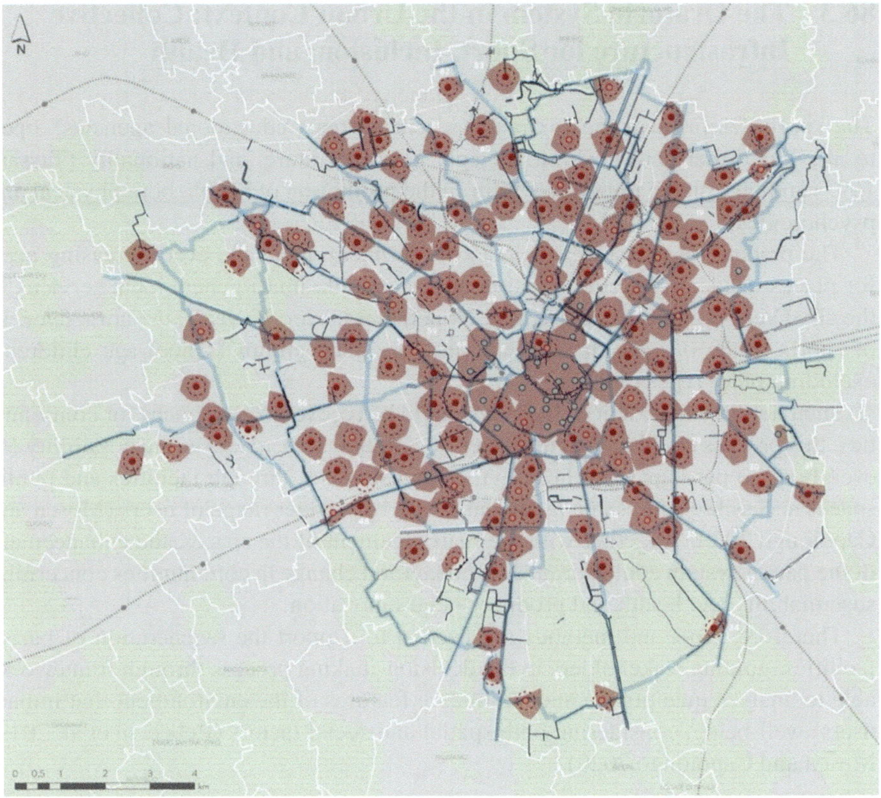

Fig. 86.1 Map of the *oratorio* system and its accessibility in the city of Milan (by foot, by public transport, and soft mobility). Map developed during Project SPèS, Marika Fior, 2021

The emerging concepts of the "city of proximity" lead to the reconsideration of systems under-used and under-valorized (Duany and Steuteville 2021), reflecting the importance of the transition from the centrality of "hard" sport infrastructure to "soft" ones, capillary diffused and present in public spaces where physical activity, leisure, and socialization become more horizontal and inclusive (Faroldi 2020; Vettori and Cognigni 2020) (Fig. 86.1).

86.3 The Oratorio System in the Urban Context: Collective Infrastructure for Sport, Inclusion, and Health

The sports facilities in the *oratorio* present as real "educational agencies" open to all, where there are no limits of age, gender, culture, and nationality (Tassani 1997) and they are an important factor in the promotion of health, both physical and psychological.

The practice of grassroots sport in the *oratorio* meets the ever-increasing need for places of identity and integration and accessible multi-generational services in the city (Vettori and Cognigni 2020). Nonetheless, over the years, the attractiveness of sports in the *oratorio* has been limited to some groups (school-age children), excluding other types of users.

Such heritage represents a large system to be recovered as services for communities: this process could lead to increased social inclusion, meant as the possibility for everybody to participate in the activities and services, through facilities and public spaces accessible and usable in an equal way by a wider range of users (Mosca and Capolongo 2020). Together with the rearrangement of the spaces, the enhancement of the parish system could stimulate a behavioral change in communities concerning sustainability and health and produce social innovation.

Therefore, tools and methods are needed to support the regeneration of parish facilities, guiding stakeholders in the decision-making process through frameworks able to analyze quantitative and qualitative features of the environment that impact users' well-being, considering both spatial and social factors (Rebecchi et al. 2019; Mosca and Capolongo 2020).

86.4 Technology and Innovation for the Parish System: A Methodological Setting

The research set as the main objective the definition of a methodology to "enhance" and innovate the assets of the *oratorio* sports facilities in a multi-generational, inclusive, and educational perspective on health. The project developed a *Strategic Design Framework* (SDF) as the final output: a multidisciplinary analysis tool capable of identifying the current and potential characteristics of the structures, concerning their urban and social contexts.

The articulation of the SDF derives from the objective of the research and from the multidisciplinary competencies of the research group (urban analysis, multi-criteria analysis, social impact evaluation, and strategic design). The methodology was set to achieve an organic and synthetic analysis of three scales through four specific tools: (I) Macro-analysis of the urban context (urban); (II) Micro-analysis of the facilities (architectural); (III) Analysis of social context through a Social Impact Evaluation (social); and (IV) Tool Report.

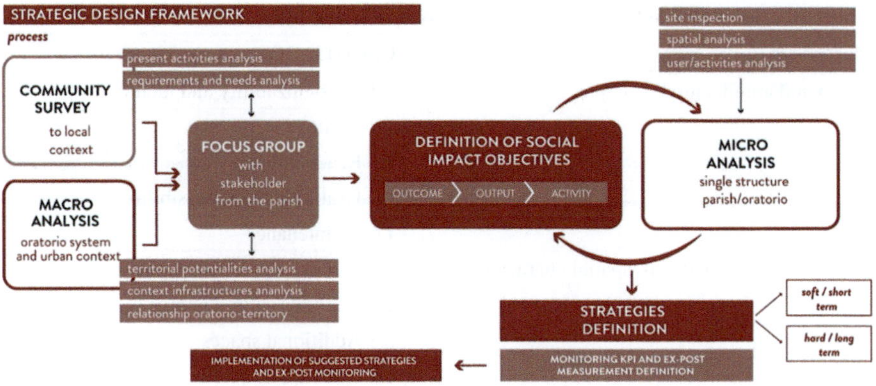

Fig. 86.2 Scheme of the elements of the *Strategic Design Framework* and its process of application. Scheme developed during Project SPèS, Francesca Daprà, 2021

The process of application of the tools was guided by the necessity of relating and putting in sequence the actual state of the context and the community (through community survey and macro-analysis) and the issues and needs that emerged from the dialogues (Focus Group) with the analysis of the building considered and its physical potentialities (micro-analysis). The matching of these data enables the definition of social impact objectives, from which the final strategies suggested derive (Fig. 86.2).

The application of this tool allows the definition of strategies both for the physical and organizational regeneration and social reactivation of the areas considered. The SDF has been applied simultaneously to five pilot cases in the city of Milan, for a few months, involving both remote actions and data analysis, both on-site analysis, data collection and activities. The surveys to the communities involved more than 850 people (in a range distributed from 100 answers per parish up to 250 in other cases), while the Focus Group held with the communities up to 12 people per parish. Following is a description of the four tools composing the SDF.

86.4.1 Macro-analysis of the Urban Context

The first element of the tool consists of an analysis of the relationship between the parish system and urban context, carried out through GIS tools. The objective is the definition of relevant elements for the *oratorio*-city relationship and their systemic reading, to evaluate possible synergies with local contexts, and the definition of key elements for their regeneration. The analysis maps specific indicators from the urban context, such as the system of city sport services and the parish one, other social and aggregative services, environmental and infrastructural factors.

Table 86.1 Structures and criterion of the tool for the microanalysis

Macro-area	Criteria
Urban and architectural character	1.1 Recognizability and identity
	1.2 Context
	1.3 Reachability
	1.4 Usability and accessibility
	1.5 Maintenance
Technological–functional–spatial character	2.1 Indoor and outdoor sport spaces
	2.2 General spatial features
	2.3 Additional spaces
	2.4 Sport space quality
Organizational– management character	3.1 Users' characteristics
	3.2 Sport spaces' use

86.4.2 Micro-analysis of the Structures

The second element considers the building scale through a panel of quantitative and qualitative indicators for its evaluation. The objective is a detailed analysis of the physical space of the parish hall and its activities—focusing on sport—to define physical and organizational–managerial regeneration strategies for the regeneration of the oratorio. The tool considers three macro-areas with related criteria (see Table 86.1).

86.4.3 Analysis of Social Context and Social Impact Evaluation

The third element considers the social framework and defines the social impact objectives for the territorial context (both parish and neighborhood/district). The methodology set a *Community-Based Survey*: a participation tool composed of a "Community Questionnaire" and "Focus Group". The objective is the involvement of the community and the collection of the needs and wishes of the parish and the local context, to propose strategies that improve the social impact of the parish facilities.

The Focus Group, conducted by a group of experts, consists of a meeting with some parish community members and stakeholders, representing different categories of the parish population and responsible. The Community Questionnaire is administered to a broader spectrum of users, to obtain a heterogeneous panel of responses, intercepting both those attending the parish and inhabitants of the neighborhoods of reference.

86.4.4 Tool Report

The final report intends to highlight all the outcomes from urban and social analyses and inspections. In addition, the report returns some strategic elements that indicate some medium or long-term regeneration actions for the assets considered. This document is conceived as an address to the projects that local contexts will be able to implement, providing scientific support, and some guidelines for the definition and planning of interventions (Fig. 86.3).

86.5 Results and Strategies for Health Promotion in the Urban Context

Through the application of the tools, the project results highlighted some common risks and potential of the parish system concerning the topics considered: the main issues can be described as follows.

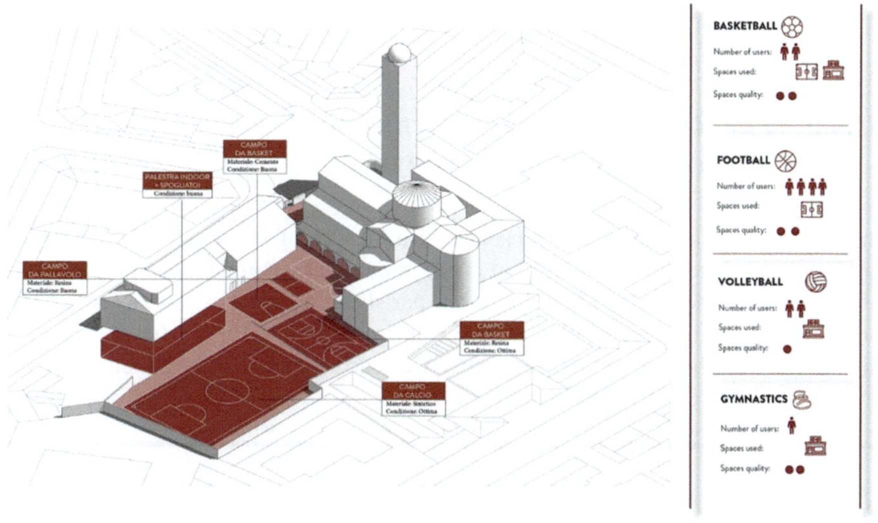

Fig. 86.3 Extract from the report that describes the spatial characters and the sport offer of the parish *San Luigi Gonzaga* in Milan. Scheme developed during Project SPèS, Francesca Daprà, 2021

86.5.1 The Potentiality of Increasing Users from a Multi-generational Perspective

The *oratorio* developed an offer that intercepts a narrow range of sport users both in terms of age (range 6–18) and proximity to the structure. The strategies proposed are oriented toward a larger involvement of different age groups, favoring generational exchange phenomena and meeting the potential demand for sport and aggregative activity.

86.5.2 The Possibility of Improving Social Inclusion Through Sport

The sporting activity represents an important moment of cultural integration in the parishes: the tool suggests the involvement of various fragile categories, lacking specific services for their needs in the analyzed context. In this way, the parish would provide social services within the neighborhood and contribute to improving the living conditions of the citizens.

86.5.3 The Rehabilitation of Structures in Terms of Accessibility and Usability

In most of the structures, the presence of a variety of spaces—both indoor and outdoor—for different sports offered is accompanied by maintenance problems and obsolescence. The sports areas, in some parishes, are poorly visible, accessible, or enhanced, while they could represent an important landmark for the community. Moreover, they often lack support spaces and services, leading them to be less competitive than other services in the district.

86.5.4 The Enhancement of Physical Activity Through Facility Accessibility

The localization of these structures, often accessible and well connected with public transport, concurs to promote soft mobility and physical activity to reach them. The revitalization of the spaces' facility, together with the empowerment of the system of soft mobility and safe pedestrian routes in the proximity of the structures, would support a correct and healthy lifestyle also for other people.

86.5.5 The Enhancement of Outdoor Spaces as "Public Spaces"

In most cases, the structures have important areas of outdoor relevance, which are not always exploited to their full potential, and barely accessible and visible from the districts. Such spaces represent a huge potential for the proximity context, to promote physical activity and social meetings.

86.5.6 The Exploitation of the Heritage in a Systemic and Synergic Way

The main activities in the parishes are held in the afternoon and night, leaving free spaces during the mornings. Innovation in the management system and the proposal of activities for different users, as well as the creation of new synergies with the city (e.g., with schools in the neighborhood), would represent an opportunity to implement the space use, and consequently, its maintenance and profitability.

86.6 Conclusions

The increase in demand for grassroots sport, urban health conditions, as well as the pandemic effects reinvigorated the importance of proximity services and healthy and active lifestyles. In this sense, studying the *oratorio* in this delicate period for cities was an opportunity to intercept social and urban issues and systemically review the phenomena.

As demonstrated by the study, sport in the parishes can represent a powerful driver for health promotion and social inclusion, improving well-being, intercultural relations, and the overall cultural and social level.

The unique and "tailor-made" qualitative and quantitative analysis tool developed for ecclesiastical structures represents a contribution to scientific research and the practices of analysis and evaluation of built heritage. The application of the *Strategic Design Framework*—applied to the Milanese territory—could be replied in the rest of the national territory, and on other properties, of similar characteristics, which need decision-support guidance for the facilities' regeneration.

The planned redevelopment of the enormous built heritage—linked directly or indirectly to sports activities—is now an unavoidable action, acquiring relevance in the Italian context, and becoming a strategic asset of the post-pandemic recovery, also in relation to exceptional financing instruments recently implemented by nations (such as Italian PNRR). Such trends require innovative tools and technologies, capable of governing the complexity of these transformations and integrating social, architectural, health, and economic values and needs.

References

Allegri D, Vettori MP (2018) Complex sports infrastructure and urban resilience: technologies and paradigms. TECHNE J Technol Archit Environ 15:165–174. https://doi.org/10.13128/Techne-22124

Capolongo S, Rebecchi A, Dettori M, Appolloni L, Azara A, Buffoli M et al (2018) Healthy design and urban planning strategies, actions, and policy to achieve salutogenic cities. Int J Environ Res Public Health 15(12):2698. https://doi.org/10.3390/ijerph15122698

Capolongo S, Rebecchi A, Buffoli M, Appolloni L, Signorelli C, Fara GM, D'Alessandro D (2020) COVID-19 and cities: from urban health strategies to the pandemic challenges. A decalogue of public health opportunities. Acta Biomed 91(2):13–22. https://doi.org/10.23750/abm.v91i2.9515

Duany A, Steuteville R (2021) Defining the 15-minute city. Public square: A CUN J

EU Physical Activity (2008) Guidelines recommended policy actions in support of health-enhancing physical activity. EU Working Group "Sport & Health". https://ec.europa.eu/assets/eac/sport/library/policy_documents/eu-physical-activity-guidelines-2008_en.pdf

Faroldi E (2020) Sport infrastructure. Origin, evolution, transformation. In: Faroldi E (ed) Sport architecture. Design construction management of sport infrastructure. Lettera Ventidue, Siracusa, pp 117–133

Mosca EI, Capolongo S (2020) A universal design-based framework to assess usability and inclusion of buildings. In: Gervasi O et al (ed) Computational science and its applications—ICCSA 2020. Springer, Cham, pp 316–331. https://doi.org/10.1007/978-3-030-58814-4_22

Rebecchi A, Buffoli M, Dettori M, Appolloni L, Azara A, Castiglia P et al (2019) Walkable environments and healthy urban moves: urban context features assessment framework experienced in Milan. Sustainability 11(10):2778. https://doi.org/10.3390/su11102778

Tassani G (1997) L'oratorio. In Isnenghi M (ed) I luoghi della memoria. Strutture ed eventi dell'Italia unita. Laterza, Roma-Bari, pp 136–172

Vettori MP (2020) Sport and public space. The role of sport infrastructure in the evolution of the city. In: Faroldi E (ed) Sport architecture. Design construction management of sport infrastructure. Lettera Ventidue, Siracusa, pp 185–197

Vettori MP, Cognigni M (2020) Space, sport, society. The practice of sport in the design of contemporary public space. TECHNE J Technol Archit Environ 19:142–152. https://doi.org/10.13128/techne-7832

WHO World Health Organization (2010) Global recommendations on physical activity for health. https://www.who.int/dietphysicalactivity/global-PA-recs-2010.pdf

WHO World Health Organization (2014) Global status report on non-communicable diseases. WHO Press, Ginevra. https://apps.who.int/iris/handle/10665/148114

WHO World Health Organization (2018) Global action plan on physical activity 2018–2030. More active people for a healthier world. https://apps.who.int/iris/bitstream/handle/10665/272722/9789241514187-eng.pdf

Chapter 87
Nursing Homes During COVID-19 Pandemic—A Systematic Literature Review for COVID-19 Proof Architecture Design Strategies

Silvia Mangili, Tianzhi Sun, and Alexander Achille Johnson

Abstract The immense impact of the coronavirus disease 2019 (COVID-19) pandemic on older adults living in nursing homes (NH) and other long-term care facilities, who at baseline are at increased risk of infection due to fragility, cognitive impairments, and complex comorbidities, has renewed the attention of researchers to the unmet needs of this population. It is well known that the built environment can significantly influence human health, a reality which is often overlooked in the setting of NHs. Recognizing how qualities of the NH built environment can influence resident outcomes, particularly in the context of the COVID-19 pandemic, can provide architects and medical professionals implementable strategies. As such, we conducted a systematic literature review from May to November 2021 to identify components of the NH built environment and their potential impacts on the health and well-being of NH residents during the COVID-19 pandemic. Relevant articles were identified with a search of Scopus, Web of Science, and PubMed scientific databases, as well as a search of gray literature. The initial search resulted 481 articles, though after the application of eligibility criteria and full-text screening, 17 articles remained for inclusion. From these, a total of 24 built environment features were identified, divided across four domain levels of NHs: Overall Facility, Building, Service Space, and Residential Room. These features were differentially linked to improved facility infection control, decreased COVID-19 incidence and mortality from COVID-19, better air quality, and enhanced resident health, quality of life, and socialization. This research defines a set of design/architecture strategies that NHs

S. Mangili (✉) · T. Sun
Department of Architecture Built Environment and Construction Engineering (DABC), Polytechnic University of Milan, Milan, Italy
e-mail: silvia.mangili@polimi.it

T. Sun
e-mail: tianzhi.sun@polimi.it

A. A. Johnson
Vagelos College of Physicians and Surgeons, Columbia University, New York 10032, USA
e-mail: aaj2152@cumc.columbia.edu

E. Arbizzani et al. (eds.), *Technological Imagination in the Green and Digital Transition*, The Urban Book Series, https://doi.org/10.1007/978-3-031-29515-7_87

may implement to improve COVID-19-related outcomes as well as the overall health and quality of life of their residents. Additional research utilizing primary data and testing these identified interventions is needed to provide stronger evidence-based suggestions.

Keywords Elderly · Built environment · Nursing homes · COVID-19 · Evidence-based design

87.1 Introduction

The global population is aging. By 2050, the number of people aged 65 years or older is predicted to double while also occupying a greater percentage of the overall population, increasing to 16% from 9% in 2020 (United Nations Department of Economic and Social Affairs 2020). The Italian context is expected to see an even more dramatic change over this same time frame, with the proportion of those age 65 years or older projected to increase from 23% in 2020 to 35% (Instituto Nazionale di Statistica 2021). As such, it is vital to recognize the increasing demand that global populations will have for appropriate facilities to care for the elderly, as well as the critical role that these facilities play in health care and social infrastructures.

Throughout the course of the coronavirus disease 2019 (COVID-19) pandemic, its disproportionate toll on older adults, in terms of both infection and mortality rates, has been well documented. The effects of COVID-19 were compounded in long-term care facilities, places comprised primarily of older adults and previously known to be vulnerable to respiratory disease outbreaks (McMichael et al. 2020). By May 2020, 37–66% of COVID-19-related deaths in European countries were attributed solely to residents of long-term care facilities (ECDC 2021).

Emerging research investigating the characteristics of long-term care facilities that may have contributed to these COVID-19 death rates has primarily focused on characteristics of residents, characteristics of the surrounding community, management strategies, or performance metrics (Zhu et al. 2022). Less attention has been paid to the built environment of these facilities and how their design features may have impacted resident outcomes during the COVID-19 pandemic. These outcomes extend beyond COVID-19 incidence and mortality, instead including resident physical and mental health, socialization, and quality of life, all of which may be impacted by the built environment (Evans 2003).

The role of the built environment on improving health and well-being outcomes in health care (Ulrich et al. 2008) and broader urban environments (Rao et al. 2007) is well known, with growing evidence of this relationship emerging in the wake of the COVID-19 pandemic (Capolongo et al. 2020a; Capolongo et al. 2020b). These discussions have been comparatively quiet in the context of long-term care facilities (Parker et al. 2004) with the COVID-19 pandemic revealing how their built environments represent yet another inadequacy in this sector (Fulmer et al. 2020).

Search Topic	Search Terms
Nursing homes	"Nursing home" OR "Healthcare facilities" OR "Long-term care"
Built environment	"Built environment" OR "Physical environment" OR "Indoor quality"
Older adults	"Older adults" OR "Senior" OR "Elderly"
COVID-19	"COVID" OR "COVID-19" OR "Sars-Cov-2"

Fig. 87.1 Search topics and terms

This systematic review, focusing on 24-h residential long-term care facilities typically referred to as nursing homes (NH), aims to identify components of the NH built environment and their corresponding impacts on the health and well-being of NH residents during the COVID-19 pandemic. In doing so, we aim to synthesize evidence-based design interventions that have emerged throughout the pandemic, providing both NH design recommendations to prepare for the future of elderly adult care and avenues for future research in the field of NH evidence-based design.

87.2 Methods

To accomplish the aims of the research, a systematic literature review of existing articles related to NH design was completed (Pati and Lorusso 2018). This was conducted from May to Nov 2021 using the scientific databases Scopus, Web of Science, and PubMed, as well as gray literature. Key search terms were identified within the following thematic categories: nursing homes, built environment, older adults, and COVID-19 (Fig. 87.1).

87.3 Results

The initial literature search conducted on scientific databases identified 456 articles with an additional 25 articles identified from gray literature databases, for a total of 481 articles. Following the removal of duplicates and initial screening of these articles, 41 articles remained for full-text evaluation. Of these, 22 were excluded based on their content being outside the scope of this review, leaving 17 eligible articles for data extraction and analysis in this study. The literature search and selection process is summarized in Fig. 87.2.

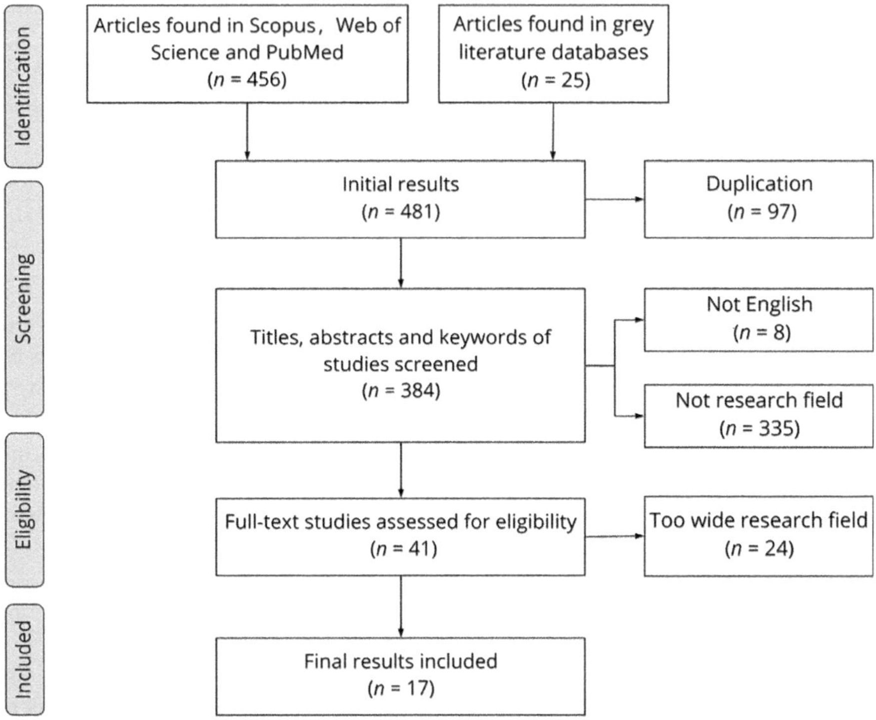

Fig. 87.2 Flow diagram of the literature search and selection process

87.4 Characteristics of Included Studies

The general characteristics of the 17 included articles are detailed in Fig. 87.3. The article types were classified as either research, review, or theoretical studies. Four articles collected primary data in the form of a case study, questionnaires, or interviews, though secondary data sources (databases, guidelines, previously published articles, etc.) were the predominately utilized data source.

A total of 24 elements of the built environment of NHs were extracted from the 17 included articles. These were classified into four domain levels of NHs: Overall Facility, Building, Service Space, and Residential Room. The corresponding impacts of each built environment feature on outcomes related to the health and well-being of older adults in NHs are identified and organized in Fig. 87.4. Seven outcomes were identified and are listed here from most to least discussed: infection control, quality of life, COVID-19 incidence, COVID-19 mortality, overall health, socialization, and air quality improvement.

N°	Author	Title	Year	Source Name	Country	Article Type	Data Source
1	Abrams H.R. et al.	Characteristics of US Nursing Homes with COVID-19 Cases	2020	Journal of the American Geriatrics Society	USA	Research	Secondary
2	Anderson D.C. et al.	Nursing Home Design and COVID-19: Balancing Infection Control, Quality of Life, and Resilience	2020	Journal of the American Medical Directors Association	USA	Review	Secondary
3	Brown K.A. et al.	Association between Nursing Home Crowding and COVID-19 Infection and Mortality in Ontario, Canada	2020	JAMA Internal Medicine	Canada	Research	Secondary
4	Cazzoletti L. et al.	Risk Factors Associated with Nursing Home COVID-19 Outbreaks: A Retrospective Cohort Study	2021	International Journal of Environmental Research & Public Health	Italy	Research	Primary
5	Chow L.	Care homes and COVID-19 in Hong Kong: how the lessons from SARS were used to good effect	2021	Age and Ageing	Hong Kong	Research	Secondary
6	Dosa D. et al.	Long-Term Care Facilities and the Coronavirus Epidemic: Practical Guidelines for a Population at Highest Risk	2020	Journal of the American Medical Directors Association	USA	Editorial	Secondary
7	Estabrooks C.A. et al.	Restoring trust: COVID-19 and the future of long-term care in Canada	2020	Facets	Canada	Review	Secondary
8	Harrington C. et al.	Nurse Staffing and Coronavirus Infections in California Nursing Homes	2020	Policy, Politics, & Nursing Practice	USA	Research	Secondary
9	He M. et al.	Is There a Link between Nursing Home Reported Quality and COVID-19 Cases? Evidence from California Skilled Nursing Facilities	2020	Journal of the American Medical Directors Association	USA	Research	Secondary
10	Khairat S. et al.	US Nursing Home Quality Ratings Associated with COVID-19 Cases and Deaths	2021	Journal of the American Medical Directors Association	USA	Research	Secondary
11	Konetzka R.T. et al.	A systematic review of long-term care facility characteristics associated with COVID-19 outcomes	2021	Journal of the American Geriatrics Society	USA	Review	Secondary
12	Liu M., et al.	COVID-19 in long-term care homes in Ontario and British Columbia	2020	Canadian Medical Association Journal	Canada	Research	Secondary
13	Olson N.L. and Albensi B.C.	Dementia-Friendly 'Design': Impact on COVID-19 Death Rates in Long-Term Care Facilities around the World	2021	Journal of Alzheimer's Disease	Canada	Review	Secondary
14	Spaetgens B. et al.	The Post-Acute and Long-Term Care Crisis in the Aftermath of COVID-19: A Dutch Perspective	2020	Journal of the American Medical Directors Association	Nether-lands	Editorial	Secondary
15	Verdoorn B.P. et al.	Design and Implementation of a Skilled Nursing Facility COVID-19 Unit	2021	Journal of the American Medical Directors Association	USA	Research	Primary
16	Wang Z.	Use the Environment to Prevent and Control COVID-19 in Senior-Living Facilities: An Analysis of the Guidelines Used in China	2021	Health Environments Research and Design Journal	China	Research	Primary
17	Zimmerman S. et al.	Nontraditional Small House Nursing Homes Have Fewer COVID-19 Cases and Deaths	2021	Journal of the American Medical Directors Association	USA	Research	Primary

Fig. 87.3 Characteristics of included studies

87.5 Overall Facility Level

The Overall Facility Level represents the broadest domain analyzed and refers to aspects of the built environment related to the entirety of the nursing home site. There are four specific features organized at this level and include: (i) small NH size, (ii) crowding index, (iii) integration with health/social services, and (iv) proximity to home community. NH size (i) was discussed in eight articles and was suggested or found to be related to decreased incidence of COVID-19 infections, decreased mortality due to COVID-19, and increased quality of life for NH residents (Cazzoletti et al. 2021) also evaluated the role of NH size on rates of COVID-19 incidence

Domain	Feature of the Built Environment	Outcome on Older Adults
Overall Facility	Small nursing home[1,2,7-11,17*]	COVID-19 mortality[9,11,17]; COVID-19 incidence[2,8-11,17]; Quality of life[2,17]
	Crowding index[3]	COVID-19 mortality; COVID-19 incidence[3]
	Proximity to and integration with health and social services[2]	Overall health[2]
	Proximity to resident home community[2]	Quality of life; Socialization[2]
Building	Dedicated resident, visitor, and staff access areas[2,16]	Infection control[2,16]
	Large internal circulation space[2,7,16]	Infection control[2,16]
	Alcohol hand sanitizer widely available in residential and public spaces[6,14]	Infection control[6]
	Dedicated and separated infectious unit[2,7,15,16]	Infection control[2,16]
	Outdoor areas and spaces for social interaction/exercise[2,13,16]	Infection control[2,16]; Socialization[2,16]; Overall health[13]; Quality of life[13]
	Natural and mechanical ventilation[2,13,16]	Air quality improvement[13]; Infection control[2,13,16]
	Homelike environment[2]	Quality of life[2]
	Telemedicine/telecommunication capabilities[7]	Quality of life[7,15,16]
	Adequate natural light[2,13]	Overall health[13]; Quality of life[2,13]
	Wayfinding and orientation[13]	Overall health; Quality of life[13]
Service Room	Reception room for visitor screening[16]	Infection control[16]
	Compartmentalized staff hygienic areas[2,15]	Infection control[2]
	Decentralized care stations[2,15]	Infection control[2]
	Designated space for contaminated waste[16]	Infection control[16]
	Compartmentalizable common spaces[2,16]	Infection control[2,16]
	Onsite medical clinic[16]	Infection control; Overall health; Quality of life[16]
Residential Room	Private rooms and bathrooms[2,5,12,17]	Infection control[5]; Quality of life[2]
	Adequate transitional spaces[2]	Quality of life[2]
	Wide and accessible walking area[2]	Quality of life; Infection control; Socialization[2]
	Outdoor views[2,5]	Quality of life[2]; Socialization[2,5]

Fig. 87.4 Relationships between elements of the built environment and outcomes for older adults across organizational levels

but found no significant association. The second feature is (ii) crowding index, a quantitative metric relating the number of NH residents to the number of bedrooms and bathrooms within a NH site.

Higher crowding index was associated with both increased incidence of COVID-19 infection and mortality due to COVID-19. The final two features, (iii) integration with health/social services as well as (iv) proximity to a resident's home community, are related to the geographic location of NHs. These features are suggested, respectively, to be related to enhanced overall health and enriched quality of life/socialization for residents.

87.6 Building Level

The next domain of the NH built environment is the Building Level and includes ten specific features. Elements within this level are related to the organization and structure of the individual buildings that comprise NHs. The first of these features is (i) dedicated resident, visitor, and staff access areas with its potential impact focused on improved infection control in NHs by reducing non-essential interactions. The second, third, and fourth features at this level all similarly are suggested to support infection control in NHs. These features are: (ii) a large internal circulation space, referring to large enough corridors to provide for social distancing and one-way flows; (iii) alcohol hand sanitizer widely available in residential and public spaces; and (iv) dedicated and separated infectious units within NHs. While also being linked to improved infection control, the fifth feature, (v) outdoor areas and spaces for social interaction/exercise, is suggested to promote resident socialization, overall health, and quality of life, and the sixth feature, (vi) adequate natural and mechanical ventilation, can improve overall NH building air quality. The final four features within the Building Level are: (vii) telemedicine/telecommunication capabilities, (viii) a homelike environment, (ix) wayfinding and orientation, and (x) adequate natural light. Through different mechanisms, these all are suggested to improve the quality of life for NH residents, with features (ix) and (x) having an additional benefit for resident overall health.

87.7 Service Room Level

There are six features of the built environment organized within the Service Room Level. These are (i) reception room for visitor screening, (ii) compartmentalized staff hygienic areas, (iii) decentralized care stations, (iv) designated spaces for contaminated waste, (v) compartmentalizable common spaces, and (vi) an onsite medical clinic. All factors within this domain are suggested to positively impact infection control in NHs. An onsite medical clinic, which may double as a geriatric care site for the surrounding community, additionally expands overall access to medical care for residents, thus improving health and quality of life.

87.8 Residential Room Level

The fourth and final domain is the Residential Room Level, with four associated built environment features. All four features at this level are suggested to improve resident quality of life. The first of these is (i) adequate transitional spaces, referring to spaces such as porches and corridor alcoves where residents can sit and receive visual and cognitive stimuli. The second feature, (ii) private rooms and bathrooms, supports infection control as well as quality of life. The third feature, (iii) wide and accessible walking areas, and fourth feature, (iv) outdoor views, both support resident socialization, while (iv) also provides elements of infection control.

87.9 Discussion

Our population is aging and will be increasing shuttled to a systemically and extensively flawed system of NHs, spaces that function both as institutional care centers as well as homes. This sector requires sweeping reform across several domains, one of which is the built environment. As places of healing and as residential spaces for the elderly, it is critical that these facilities can be designed with the specific needs and vulnerabilities of their residents in mind. This systematic review identified 17 articles that discussed 24 features of the NH built environment, providing variable targets in four different domains (Overall Facility, Building, Service Space, and Residential Room) to improve outcomes during the contemporary COVID-19 situation as well as in the future. These features were most frequently implicated in improving general infection control in NHs, which helps combat infectious disease outbreaks, improve resident morbidity and mortality, and reduce healthcare costs. Quality of life was the second most frequently discussed outcome which, through its impact on well-being, stress, and resilience, can improve longevity (MacLeod et al. 2016). Improved socialization, only discussed in three of the articles, is relatively underrepresented as an outcome and target for the built environment. Social isolation, a significant existing challenge facing older adults, was amplified during the COVID-19 pandemic with the potential to increase the risk of premature death, mental health disorders, cardiovascular disease, and dementia in this population (Centers for Disease Control 2021). Designers should follow the guiding principle of creating a more homelike environment for the residents of these NHs, following or expanding upon established nontraditional designs for NHs. From this principle of creating an ideal home environment for older adults, many of the desired design features highlighted in this review would follow, including private rooms and bathrooms, outdoor areas, outdoor views, natural light, ventilation, large circulation spaces, adequate way finding, and telecommunication, all within a non-crowded building with < 20 residents that is close to and integrated within their home community.

Designers can then incorporate features specific to the healing environment of the NH, making sure that they are compartmentalizable and decentralized, including staff hygienic areas, common spaces, contaminated waste areas, visitor screening areas, and infectious units.

87.10 Conclusions

This research represents a starting point in defining a set of design/architecture strategies that NHs may implement to improve COVID-19-related outcomes as well as the overall health and quality of life of their residents. Creating smaller nursing homes integrated with the surrounding community that are designed to be home-like with single rooms/bathrooms, adequate outdoor space/views, and telecommunication capabilities emerged as features with significant positive impact on residents. Additional research utilizing primary data and testing these identified interventions is needed to provide stronger evidence-based suggestions in this sector. This research must be applied within the larger sociologic and political situations that shape outcomes in NHs, yet nonetheless provides insight into the benefit of interdisciplinary research and collaboration between healthcare designers and healthcare providers, while also arming these actors with new strategies to support and not to forget the elderly. Though this review fills a knowledge gap by synthesizing published information about the NH built environment during COVID-19, results of this review are limited by its brief and early time frame that may have missed relevant research articles still underway.

References

Capolongo S, Gola M, Brambilla A, Morganti A, Mosca EI, Barach P (2020a) COVID-19 and healthcare facilities: a decalogue of design strategies for resilient hospitals. Acta Biomed Atenei Parmensis 91(9-S):50–60. https://doi.org/10.23750/abm.v91i9-S.10117

Capolongo S, Rebecchi, Buffoli M, Appolloni L, Signorelli C, Fara GM, D'Alessandro D (2020b) COVID-19 and cities: from urban health strategies to the pandemic challenge. A decalogue of public health opportunities. Acta Biomed Atenei Parmensis 91(2):13–22. https://doi.org/10.23750/abm.v91i2.9615

Cazzoletti L, Zanolin ME, Tocco Tussardi I, Alemayohu MA, Zanetel E, Visentin D, Fabbri L, Giordani M, Ruscitti G, Benetollo PP, Tardivo S, Torri E (2021, Dec 14) Risk factors associated with nursing home COVID-19 outbreaks: a retrospective cohort study. Int J Environ Res Public Health 18:8434. Int J Environ Res Public Health 18(24):13175. https://doi.org/10.3390/ijerph182413175. Erratum for: Int J Environ Res Public Health 18(16), 2021 Aug 10. PMID: 34949038; PMCID: PMC8701741

Centers for Disease Control (2021) Loneliness and social isolation linked to serious health conditions. https://www.cdc.gov/aging/publications/features/lonely-older-adults.html

European Centre for Disease Prevention and Control (ECDC) (2021) Surveillance of COVID-19 in long-term care facilities in the EU/EEA. European Centre for Disease Prevention and Control. https://www.ecdc.europa.eu/en/publications-data/surveillance-COVID-19-long-term-care-facilities-EU-EEA

Evans GW (2003) The built environment and mental health. J Urban Health: Bull New York Acad Med 80(4):536–555. https://doi.org/10.1093/jurban/jtg063

Fulmer TT, Koller CF, Rowe JW (2020) Reimagining nursing homes in the wake of COVID-19. NAM Perspect 2020. https://doi.org/10.31478/202009a

Instituto Nazionale di Statistica (2021) Households and population projections—base 1/1/2020. https://www.istat.it/en/archivio/263998

MacLeod S, Musich S, Hawkins K, Alsgaard K, Wicker ER (2016) The impact of resilience among older adults. Geriatr Nurs 37(4):266–272. https://doi.org/10.1016/j.gerinurse.2016.02.014
McMichael TM, Currie DW, Clark S, Pogosjans S, Kay M, Schwartz NG, Lewis J, Baer A, Kawakami V, Lukoff MD, Ferro J, Brostrom-Smith C, Rea TD, Sayre MR, Riedo FX, Russell D, Hiatt B, Montgomery P, Rao AK, Duchin JS (2020) Epidemiology of Covid-19 in a long-term care Facility in King County, Washington. N Engl J Med 382(21):2005–2011. https://doi.org/10.1056/NEJMoa2005412
Parker C, Barnes S, Mckee K, Morgan K, Torrington J, Tregenza P (2004) Quality of life and building design in residential and nursing homes for older people. Ageing Soc 24(6):941–962. https://doi.org/10.1017/S0144686X04002387
Pati D, Lorusso LN (2018) How to write a systematic review of the literature. HERD: Health Environ Res Des J 11(1):15–30. https://doi.org/10.1177/1937586717747384
Rao M, Prasad S, Adshead F, Tissera H (2007) The built environment and health. The Lancet 370(9593):1111–1113. https://doi.org/10.1016/S0140-6736(07)61260-4
Ulrich RS, Zimring C, Zhu X, DuBose J, Seo HB, Choi YS, Quan X, Joseph A (2008) A review of the research literature on evidence-based healthcare design. HERD 1(3):61–125. https://doi.org/10.1177/193758670800100306. PMID: 21161908
United Nations Department of Economic and Social Affairs, Population Division (2020) World population ageing 2020 highlights: living arrangements of older persons. https://www.un.org/development/desa/pd/sites/www.un.org.development.desa.pd/files/undesa_pd-2020_world_population_ageing_highlights.pdf
Zhu X, Lee H, Sang H, Muller J, Yang H, Lee C, Ory M (2022) Nursing home design and COVID-19: implications for guidelines and regulation. J Am Med Dir Assoc 23(2):272-279.e1. https://doi.org/10.1016/j.jamda.2021.12.026

Chapter 88
A New Generation of Territorial Healthcare Infrastructures After COVID-19. The Transition to Community Homes and Community Hospitals into the Framework of the Italian Recovery Plan

Andrea Brambilla, Erica Brusamolin, Stefano Arruzzoli, and Stefano Capolongo

Abstract COVID-19 disrupted existing processes and accelerated the rethinking of healthcare spaces, functions, and model of care, stressing the ineffectiveness of the territorial health network in the Italian National Health System (NHS). Within the framework of European Recovery Plan (Next Generation EU), Italy's Piano Nazionale di Ripresa e Resilienza (PNRR) allocated €15.63 Bn in the Mission 6 "Health" to strengthen proximity networks, facilities, and telemedicine for territorial healthcare. Aware of the importance that the physical built environment plays in the process of care delivery and health promotion and prevention, €3 Bn has been allocated to the planning, design, and construction of two new low-care typologies in a vision of person-centered healthcare: the Community Home (Casa della Comunità-CdC), and the Community Hospital (Ospedale di Comunità-OdC). It has been estimated that 795 new CdCs and 381 new OdCs will completed before 2026 as novel buildings or renovation of existing healthcare facilities. Although in European context several best practices are present in terms of integration of healthcare architectures into the urban context (Spanish Health Centers or Swedish Primary Care Centers), the Italian experience is generally outdated, with some regional exceptions; there is the need to understand the architectural characteristics of such new typologies. Therefore, the aim of the paper is to shed light on the spatial, functional, technological, and organizational needs and requirements of CdC and OdCs and to

A. Brambilla (✉) · E. Brusamolin · S. Arruzzoli · S. Capolongo
Polytechnic University of Milan, Milan, Italy
e-mail: andrea1.brambilla@polimi.it

E. Brusamolin
e-mail: erica.brusamolin@polimi.it

S. Arruzzoli
e-mail: stefano.arruzoli@polimi.it

S. Capolongo
e-mail: stefano.capolongo@polimi.it

E. Arbizzani et al. (eds.), *Technological Imagination in the Green and Digital Transition*, The Urban Book Series, https://doi.org/10.1007/978-3-031-29515-7_88

map the different regional requirements in a systematic and structured framework. The methods adopted in the study include a review of national and regional guidelines, data collection from National agency for regional health services (AGENAS) databases, and comparison matrix development of the different requirements in Italian regions. The results will highlight technological and architectural implications of territorial health centers implementation.

Keywords Territorial health center · Recovery plan · Healthcare facilities · COVID-19 · Healthcare infrastructure

88.1 Introduction

Territorial care services are a fundamental part of the whole healthcare network of assistance, prevention, and health promotion with specific qualities of healthcare facilities (Brambilla et al. 2021a, b; Capolongo et al. 2015; Wilhelm and Battisto 2020). The literature has shown that health systems with "strong" primary and proximity care services have also better population health outcomes (Rifkin 2018), defined by the WHO as a fundamental human right (World Health Organization 2020; Capolongo et al. 2021; Starfield et al. 2005). In fact, in 2019, the latter included primary care among the most relevant humanitarian threats to be solved with the new 5-year strategic plan to achieve the universal health coverage, protection from health emergencies and more health accessibility.

The global SARS-CoV-2 pandemic has underlined the crucial role of community health, especially in Italy (Vinceti et al. 2021; AGENAS 2021) where interactions between the different levels of healthcare systems, excessive waiting times in the provision of health services, and their unequal territorial distribution were particularly problematic (Filippini and Vinceti 2021). (AGENAS 2021). The Italian peninsula currently lacks proximity structures and telemedicine for territorial healthcare assistance to decrease inequalities in healthcare access and improve healthcare overall services. Community Homes ("Case della Comunità"—CdC) and Community Hospitals ("Ospedali della Comunità"—OdC) are among the responses to the problem that the Italian government has decided to implement through the recovery and resilience plan (Gola et al. 2020). These new territorial structural typologies (Fig. 88.1) aim to provide more sustainable, uniform, inclusive, and equitable healthcare by fitting into the existing National Health System with two different objectives reflecting the following definitions:

- Community Homes are multi-purpose facilities that are easily recognizable and reachable and that will allow patients to get in touch with all the health, social, and health services offered by the NHS on a territorial scale. These facilities promote an integrated and multidisciplinary intervention model by providing needs assessment, orientation to services, integrated planning, and management of home care,

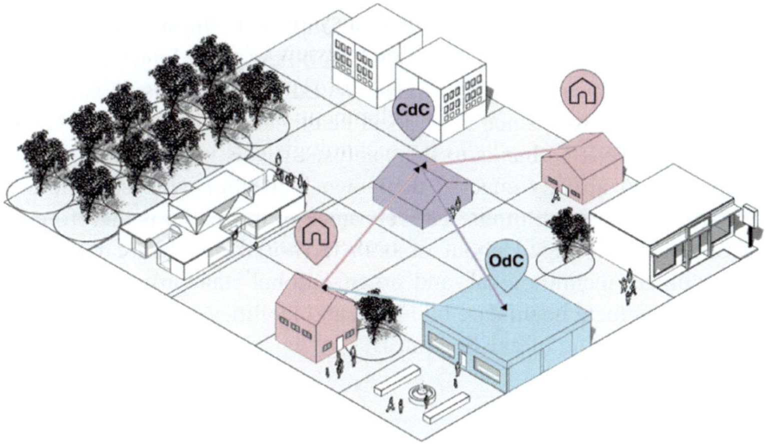

Fig. 88.1 New territorial healthcare system in Italy. Personal elaboration by the authors, 2021

ensuring continuity of care, targeted assistance to chronically ill and frail patients (DM71).

- As defined by Ministerial Decree 2 April 2015 n. 70, Community Hospitals are territorial intermediate healthcare facilities for short-term stays that aim to improve the quality and appropriateness of care while avoiding unnecessary hospitalizations and improper access to hospitals. They are for patients who require low-intensity clinical interventions and continuous nursing care and supervision, which cannot be provided at home. Furthermore, these two new facilities will be geared toward social inclusion and well-being, sustainability and climate resilience, energy and plant efficiency, and the safety and security of users and all citizens, in accordance with the main international guidelines (2030 Agenda for Sustainable Development and its 17 Sustainable Development Goals, SDGs; the European Green Deal, the Urban Health Rome Declaration).

88.2 Objectives of the Study

Despite the National Healthcare System has already introduced local healthcare buildings and facilities, each Italian region applied them in various ways and no systematic frameworks are available for mapping and comparison; in fact, regions have defined very different accreditation standards from each other. Italian Government provided a document (Dossier n° 144 of the Italian Chamber of Deputies, 2021) which lists the regional legislations about local healthcare facilities, but it is missing a model that compares and collects the accreditation standards required by each Italian region.

The lack of data collection and awareness on such healthcare facilities is also part of the research gap formulation (Banchieri 2021).

Therefore, the objective of this study is to outline the framework about regional accreditation standards for local healthcare facilities, providing the general understanding of the technical and performance-based requirements of such healthcare infrastructures. Moreover, since these local healthcare facilities have been recently introduced in the PNRR, thanks to this analysis, it was possible to find out which Italian regions had already enacted their own legislation with specific accreditation standards for these buildings. The relevance of this paper is therefore to reflect the generally acknowledged about of both territorial healthcare facilities needs, spatial, functional, technological, and organizational standards that can be useful for designing new local healthcare buildings and facilities as well as for guiding the remodeling/restructuring of existing ones.

88.3 Methods

The research grounds on the PNRR document and the release process of Ministry of Health Decree D.M.71 on "Standards and Models for the Territorial Healthcare Development in the NHS" which provides some general guidelines for the new local healthcare facilities defined inside the national PNRR document and have been anticipated by several draft documents that have been used as the starting point of the study. Even if they cannot be considered official accreditation standards, they contains very useful information about activities (as mandatory and recommended functional areas), technological (as plants and medical equipment), and organizational features (as staff type).

From the methodological perspective, it was important to understand the definition of accreditation standards for buildings and facilities provided by the Italian national legislations: Legislative Decree DL 502/92 defines 3 different kinds of them, as minimum performances that facilities and building have to guarantee. As reported in DL n. 502/92—in both the initial draft, and the following updates—and in D.P.R 42/1997, accreditation is a right that must be provided to all facilities that can reach the minimum required performances. Then, it was necessary to discover which national and regional laws define accreditation standards for the existing local healthcare buildings and facilities in Italy. For this step of the research, 10 national documents have been studied, among them, it was Dossier n° 144 of the Italian Chamber of Deputies "Case della salute ed Ospedali di comunità: i presidi delle cure intermedie. Mappatura sul territorio e normativa nazionale e regionale", which provide a list of the regional legislations about local healthcare facilities, especially referred to "Healthcare Homes (Case della Salute-CdS)" and to OdCs. Since CdCs have been recently introduced in Italy, both national and regional accreditation standards, which are currently in development.

Anyway, CdCs can be considered as an evolution of CdSs (Banchieri 2021), so regional accreditation standards for CdSs have been examined for this analysis. In facts, in Italy, the existing local healthcare facilities are CdSs and OdCs, and regions which introduced them have already provided to define their own accreditation standards.

From the list of regional legislations reported in Dossier n° 144 of the Italian Chamber of Deputies, 62 documents about local healthcare buildings and facilities have been studied, looking for which of them provide accreditation standards for these kind of facilities.

Qualitative and quantitative differences emerged regarding accreditation standards provided by each region (Fig. 88.2): in fact, through researching keyword such as "performance", "accreditation", and "standard", it has been discovered that just 7 regions out of 20 provided accreditation standards for CdSs, as well as 7 regions out of 20 for OdCs. The full list of regulation analyzed provided in Fig. 88.1 has been considered for this study; each document has been read and reviewed by the authors, and the specific requirements for OdC and CdC provided by the different region have been highlighted and collected in a systematic and structured way.

Indeed, a comparative matrix has been chosen as the methodologic strategy for collecting and effectively categorizing the accreditation standards found. Four macro-categories for the analysis of regional standards have been used as well as sub-categories, as reported below:

(I) *Catchment Area—Standard*, about the number of users of the single facility. PNRR gives specific information about the catchment area:

- type "Hub" 1/100.000 residents;
- type "Spoke" divided into 3 sub-categories: (Metropolitan Area: 1/30–35.000 residents; Urban/Sub-urban Area: 1/20–25.000 residents; Rural Area: 1/10–15.000 residents).

National Level(n=10)
Documento AGENAS: Modelli e standard per lo sviluppo dell'Assistenza Territoriale nel SSN; D.L. 502/1992; D.P.R 42/1997; Dossier n° 144 della Camera dei Deputati "Case della salute ed Ospedali di comunità: i presidi delle cure intermedie. Mappatura sul territorio e normativa nazionale e regionale" e Documento PNRR; Regolamento 2021/1058 del parlamento europeo e del consiglio del 24 giugno 2021 relativo al Fondo europeo di sviluppo regionale e al Fondo di coesione; Associazione "Prima la Comunità", in collaborazione con CERGAS-SDA Università Bocconi di Milano e Laboratorio Management e Sanità della Scuola Superiore Sant'Anna di Pisa: "La Casa della Comunità_PROGETTO COMPLETO", 27 Aprile 2021; Decreto Ministeriale 2 aprile 2015 n. 70; Intesa in sede di Conferenza Stato-Regioni, 20 Febbraio 2020; ALLEGATO B alla Dgrn. 2271 del 10 dicembre 2013; Schema di Intesa alla Conferenza Stato-Regioni sui requisiti strutturali, tecnologici ed organizzativi minimi dell'Ospedale di Comunità, Allegato A;

Regional level (n=62)
Abruzzo Region: *DGR 591/P_11 dic 2008; DCA 20_8 mar 2016; DGR 616_26 sett 2016;*
Basilicata Region: *LR 2_12 gen 2017;*
Calabria Region: *DPGR 185_4 dic 2012; DCA 81_22 luglio 2016;*
Campania Region: *DCA 83_31 ott 2019;*
Emilia-Romagna Region: *DGR 327_23 feb 2004; DGR 514 _20 apr 2009; DGR 291_8 febb 2010; DGR 221_6 mar 2015;*
Friuli Venezia Giulia Region: *LR 13_1995; LR 17 _ 2014; Linee guida strutture intermedie;*
Lazio Region : *DCA 90_ 2010; DCA 8_ 10 feb 2011; DCA 228_22 giu 2017; DCA 51_8 apr 2020;*
Liguria Region: *DGR 1367/1368_16 nov 2007; DAL 22_30 set 2009; DAL 21_ 5 dic 2017;*
Lombardia Region: *LR 23_ 08 nov 2015; DGR 3383_10 apr 2015; DGR 6551_ 4 mag 2017; DGR n° XI_31 lug 2019; DGR 5373_11 ott 2021; DGR 5723_15 dic 2021;*
Marche Region: DGR 735_20 mag 2013; DGR 452_14 apr 2014; DGR 139_22 feb 2016; RR 1_1 feb 2018; DGR 1572_16 dic 2019;
Molise Region: DCA 26_19 mag 2016; DCA 27_19 mag 2016; DCA 18_28 feb 2017; DCA 21_2 mar 2018; DCA 65_26 ott 2020;
Puglia Region: DGR 333_10 marzo 2020;
Sardegna Region: DGR 32/10_4 giu 2008; DGR 42/3_20 ott 2011; LR 23_17 nov 2014; DGR 60/1_2 dic 2015; DGR 17/14 _ 4 apr 2017; LR 24_11 set 2020;
Sicilia Region: DA 723 _ 10 mar 2010; GURS n. 15 _ 10 apr 2015;
Toscana Region: *DGR 117 _ 16 feb 2015; DGRT 679 _ 12 lug 2016; DGRT 909 _ 7 ago 2017; DGR 770_22 giu 2020; DGR 818 _ 29 giu 2020;*
Veneto Region: DGRV 2481 _ 6 ago 2004; DGR 84_16 gen 2007; DGR 2067_ 03 lug 2007; DGR 2718_ 24 dic 2012; DGR 751_2015; LR 48_28 dic 2018; DGR 1107_2020;

Fig. 88.2 List of documents analyzed both at national and regional level

(II) *Structural standards* collect minimum spatial and dimensional requirements provided by each regional legislation. This class is divided into 3 sub-categories: minimum dimensions of the building; list of the functional area (which reports information about typology, minimum quantity, and dimensions), single units (which provide indications about typology of the single locals, minimum quantity, and dimensions).

(III) *Technological standards* structured in to 2 sub-categories: typology of plants and minimum performance required.

(IV) *Organizational standards* which provide information about healthcare, socio-healthcare, nursing and technicians staff, reporting typology, and quantity defined by each Region.

88.4 Results

Community Homes. It was possible to find available documents about CdSs and CdCs for 7 Italian regions: Calabria, Emilia-Romagna, Lazio, Marche, Molise, Sicilia e Toscana. Results are reported in the Fig. 88.3. The presence or absence of specific data in each document is reported with a binary system of yes(Y) or no(N) items (Fig. 88.3).

Community Hospital. It was possible to find the legislation containing the accreditation requirements for Community Hospitals in 7 regions: Abruzzo, Emilia-Romagna, Lazio, Lombardy, Marche, Tuscany, and Veneto. For convenience in consulting the results, the contents are summarized in Fig. 88.4. The presence or absence of specific data in each document is reported with a binary system of yes(Y) or no(N) items (Fig. 88.4).

88.5 Discussion

This research is the first step of a bigger project of the National Agency for Regional Healthcare System (AGENAS), which main objective is to define metadesign guidelines for the new local healthcare facilities, using actual regional accreditation standards as background. So, this research project helped AGENAS to map the state of the art and collect in a single document all the design standards defined by each region, useful for eliciting national metadesign guidelines. Moreover, there are strict time limits for deploying this project at the national level: in facts, PNRR established the intent of the Italian Government to realize at least 1350 CdCs by the year 2026. For this reason, mapping accreditation standards at the regional level is fundamental to speed up the definition of national guidelines for designing these facilities. Moreover, another positive aspect is that the existing local healthcare facilities can be considered as very useful case studies for designing the new ones. But, on the other hand, the existing framework is confused, as the identity of these facilities,

CATEGORIES			DETAILS	REGIONS						
				CALABRIA	EMILIA-ROMAGNA	LAZIO	MARCHE	MOLISE	SICILIA	TOSCANA
CATCHMENT AREA - STANDARD				**HUB** 1 / 100,000 inhabitants	**HUB** 1 / 100,000 inhabitants	**HUB** 1 / 100,000 inhabitants	**HUB** 1 / 100,000 inhabitants	**HUB** 1 / 100,000 inhabitants	**HUB** 1 / 100,000 inhabitants	**HUB** 1 / 100,000 inhabitants
STRUCTURAL STANDARDS	DIMENSIONS OF THE STRUCTURE			**SMALL, MEDIUM, LARGE**	**SMALL** (Min: 600 m2, Max: 1400 m2) **MEDIUM** (Min: 2500 sq m, Max: 3900 sq m) **LARGE** (Min: 5900 m2, Max: 6200 m2)	BASIC COMPLEX	**A** typology **B** typology	BASIC COMPLEX	Local Assistance Medical Unit	BASIC STANDARD COMPLEX
STRUCTURAL STANDARDS	ENVIRONMENTAL UNITS (ROOMS AND SPACES)	FUNCTIONAL AREA	NAME	5 Functional Areas grouped in 22 specific functional areas	SMALL: reception and health services, MEDIUM and LARGE: SMALL's services + socio-sanitary services	**BASIC**: 3 functional areas (Outpatient Area, Diagnostic Service and Reception Area). **COMPLEX**: BASE's services + socio-sanitary services, intermediate care and complex outpatient care.	**Type A**: 9 Functional Areas (reception; waiting area; booking center; general practitioner; specialist medicine activities; Home Care; ER; pediatrician; continuity care services). **Type B**: type A services + echography; vaccinations; certifications; outpatient nursing clinic; outpatient observation-therapies; obstetrics; distribution of prosthetic aids; home care coordination.	**BASIC**: 9 functional areas (access point; booking point; voluntary work areas and mutual aid; general practitioner; pediatrician; specialist medicine activities; outpatient nursing services; first level diagnostics) **COMPLEX**: BASE's services + nursing inpatient unit; complex outpatient area; anti-violence centre; pharmacy; ER.	**9 Functional Areas:** access point + reception; specialist medicine activities; diagnostics; vaccinations; screening; primary care unit with general practitioner and pediatrician; urgent care services; center for drug addicts; ER.	**MANDATORY:** Reception activities; booking point; general practitioner; nursing area; first level diagnostics; telemedicine. **OPTIONAL:** care programs for patients with chronic diseases; pediatrician; sampling point; vaccinations.
STRUCTURAL STANDARDS	ENVIRONMENTAL UNITS (ROOMS AND SPACES)	FUNCTIONAL AREA	N°	Y	Y	N	N	N	N	N
STRUCTURAL STANDARDS	ENVIRONMENTAL UNITS (ROOMS AND SPACES)	FUNCTIONAL AREA	m²	Y	Y	N	N	N	N	N
STRUCTURAL STANDARDS	MINIMUM DIMENSIONS AND PARTICULARITIES	SINGLE UNIT	NAME	Operating rooms and auxiliary rooms	N	Operating rooms and auxiliary rooms	N	N	N	N
STRUCTURAL STANDARDS	MINIMUM DIMENSIONS AND PARTICULARITIES	SINGLE UNIT	N°	Y	N	N	N	N	N	N
STRUCTURAL STANDARDS	MINIMUM DIMENSIONS AND PARTICULARITIES	SINGLE UNIT	m²	Y	N	Y	N	N	N	N
TECHNOLOGICAL STANDARDS	SYSTEM STANDARDS (INSTALLATIONS + SPECIAL INSTALLATIONS)	REQUIREMENTS		Min and max temperature in clinics and in the diagnostic area (20-26 ° C). Percentage of relative humidity (40-60%). Lux of the lighting system (300-500 lux) Air changes (5 v / h).	N	Temperature range, Humidity to be maintained in sterilization environments, Air exchange and filtering.	N	N	N	**BASIC:** digital and shared medical history, staff network, telemedicine. **STANDARD**: BASE + spirometer, electrocardiograph, projection technology. **COMPLEX**: BASE + STANDARD + ultrasound, retinograph, interactive totems.
TECHNOLOGICAL STANDARDS	SYSTEM STANDARDS (INSTALLATIONS + SPECIAL INSTALLATIONS)	TYPOLOGIES		Air condition, lighting system, Fire detection, Sanitary water, Access control, Radiation detection.	N	BASIC: Air condition, climatizzazione, ventilation, telephone and fax line equipment, internet connection, private information system and workstations equipped with PC. COMPLEX: BASE's services + socio-sanitary services, intermediate care and complex outpatient care.	N	N	N	N
ORGANIZATIONAL STANDARDS	HEALTHCARE PERSONNEL	TYPOLOGY		17 professional figures: health personnel, technicians, nursing staff, socio-medical staff, administrative staff.	Medical director/manager nursing staff administrative staff Auxiliary with executive tasks Social worker District doctor	6 professional figures: Director/health manager; nursing staff; administrative staff; auxiliary with executive tasks; social worker; district doctors.	N	N	N	**Tipologia BASE**: 6 general practitioners; nurses; office staff; booking point staff; social workers. **Tipologia STANDARD e COMPLESSA**: BASE typology + two additional GPs, with the addition of specialists and administrative staff.
ORGANIZATIONAL STANDARDS	HEALTHCARE PERSONNEL	N°		Y	N	N	N	N	N	N

Fig. 88.3 Matrix of Community Home characteristics in different Italian region

and every regional legislation is very different among each other. Starting from this study, preliminary challenges for defining CdC and Odc are reported below.

88.5.1 The Challenge for Defining Community Homes (CdC)

PNRR has allocated EUR 2 billion for the realization of 1288 CdCs by the first half of the 2026, local facilities which provide primary healthcare and social healthcare services.

CdCs are divided in to 2 different kinds, related to the complexity of healthcare and social healthcare services provided: The most advanced ones are called "Hub" by the PNRR document, and their catchment area is 1 every 100.000 residents; but, there are even smaller CdCs, called "Spoke", which have 3 different catchment areas: Metropolitan Area: 1 every 30–35.000 residents; Urban/Sub-urban Area: 1 every 20–25.000 residents; Rural Area: 1 every 10–15.000 residents. This regional requirement is in line with the national suggestions.

CATEGORIES			DETAILS	ABRUZZO	EMILIA-ROMAGNA	LAZIO	LOMBARDIA	MARCHE	TOSCANA	VENETO
				REGIONS						
CATCHMENT AREA - STANDARD				N	N	N	N	N	N	1,5 beds per inhabitant
STRUCTURAL REQUIREMENTS	DIMENSIONS OF THE STRUCTURE			15 - 20 beds	15 - 20 beds	min 10 - max 40 beds (40/45 sqm per guest)	Basic Level Advanced Level (max 20 beds)	Type C	Setting 1 LOW CARE (minimum 8 beds), Setting 2 INTERMEDIATE HEALTH RESIDENTIALITY minimum 8 beds and maximum 20), Setting 3 INTERMEDIATE CARE RESIDENTIALITY	min 24-25 beds
STRUCTURAL REQUIREMENTS	ENVIRONMENTAL UNITS (ROOMS AND SPACES)	FUNCTIONAL AREA	NAME	N	Refer to the accreditation requirements for the hospitalization area provided for the Internal Medicine facilities	N	**5 Functional Areas:** reception area; hospitalization area; area at the service of residential care and patient mobilization; area for health activities; area intended for support services.	Functional Areas of Type B + **Social and Health Services**, for instance: RSA; Psychiatric residential center; Residential center for the disabled; Psychiatric day center; Disabled day center; Elderly day center	N	N
			N°	N	N	N	N	N	N	N
			m²	N	N	N	N	N	N	N
STRUCTURAL REQUIREMENTS	MINIMUM DIMENSIONS AND PARTICULARITIES	SINGLE UNIT	NAME	**17 Environmental Units**: hospital rooms (from 1 to 4 beds), work rooms for staff, toilets, storage spaces (dirty, clean, for equipment and for storing medicines) and refreshment and relaxation rooms (rooms living room and kitchen). "	Operating rooms and auxiliary rooms only for the inpatient area	**31 Environmental Units**, grouped into: kitchen service, laundry, sterilization, disinfection, hospitalization area and rehabilitation area	**5 Environmental Units**: Reception area; Hospital area; Area at the service of residential care and patient mobilization. Area for healthcare activities, Area for support services.	N	**12 Environmental Units**, relating to areas of hospitalization, health services and support services	**3 Environmental Units**, in the hospitalization area: double rooms with bathroom, rooms for early patient mobilization and living / dining area.
			N°	Y	Y	N	Y	N	Y	N
			m²	Y	N	Y	Y	N	Y	N
TECHNOLOGICAL REQUIREMENTS	SYSTEM REQUIREMENTS (INSTALLATIONS + SPECIAL INSTALLATIONS)	REQUIREMENTS		N	N	Temperature between 20-27 °C, Relative humidity 40-60%, Replacement of 15 v / h Air filtration.	N	N	Temperature between 20 C° and 26 C°	Internet access; standard mobility aids; tele-consultation instrumentation
		TYPOLOGIES		Emergency lighting system; Power installation in the rooms; Call system with acoustic and light signaling; Centralized system for medical gases with empty and oxygen outlets.	N	Air conditioning system; emergency lighting system; compressed air system.	The technological equipment, between the basic and advanced types of structures, differs in the presence of EcoFAST only in the Advanced model.	N	The following systems must be inserted inside the rooms: air conditioning system; emergency lighting system; 3 electrical sockets per bed; oxygen supply system and vacuum sockets for each bed; call system with acoustic and light signaling.	N
ORGANIZATIONAL REQUIREMENTS	HEALTHCARE PERSONNEL	TYPOLOGY		**6 professional figures**: General Practitioner; Continuity of care doctor; Free Choice Pediatrician; Specialized doctor; Nurse; Healthcare Manager of the Structure.	N	**6 professional figures**: Director / health manager; nursing staff; administrative staff; Auxiliary with executive duties; Social worker; District doctor.	**9 professional figures**: Clinical manager; Case Manager; Doctor of general medicine; Free Choice Pediatrician; Psychologist; Physiotherapist; Social worker; Nurse; OSS.	N	**Setting 1:** Internal Medicine Doctor / Geriatrician / Physiotherapy or equivalent; Nurse; OSS; Rehabilitation staff; **Setting 2:** Internal Medicine and / or other discipline in the medical area; Nurse; OSS; Rehabilitation staff;	**10 professional figures**: Nursing coordinator; Shift nurse; Nursing daytime case manager; Social Health Operator; Physiokinesitherapist; Physiokinesitherapist TPE; Palliativist; Physiatrist; Geriatrician; Psychologist;
		N°		N	N	N	N	N	Y	Y

Fig. 88.4 Matrix of Community Hospitals characteristics in different Italian region

Generally speaking, these facilities provide services as: medical check-up; reception; services for continuity of care; general medicine; polyclinics; first aid; first-line diagnostics services; sampling point; rehabilitation; homecare services; primary care pediatricians; vaccination center; screening; social healthcare services; telemedicine.

The functional areas have different spatial units as: hall; sanitary facilities both for users and for staff; archives; offices; locker rooms for staff; relaxation room; clinic rooms; scan room; x-ray room; rehabilitation gym.

Since the list of function is very variable in the national territory and reflects the peculiarity of each regional system, the maximum level of flexibility in national standards as well as in the design or refurbishment of such infrastructures should be guaranteed (Brambilla et al. 2021b).

Finally, these facilities will have conventional systems and equipment such as air conditioning system and emergency lightings but some region provide information about the installation of Internet of things sensors.

Finally, only 57% of the regional legislation studied provides information about types of staff for CdCs or CdSs: general doctor; primary care paediatricians; specialist doctor; nursing staff; social healthcare staff; administrative personnel; healthcare professionals.

88.5.2 The Challenge for Defining Community Hospitals (OdC)

OdCs are facilities which provide intermediate healthcare level services, and PNRR has allocated EUR 1 billion for designing 381 OdCs by the first half of the 2026.

They can host a variable number of beds, which is indicated at national level of 20 or 40 but highlighted in regional documentation from 8 to 25. This discrepancies should be carefully considered in the strategic planning and design. Only 29% of the regional legislations of this study provides information about functional areas as: reception; hospitalization (max 4 beds per room) and rehabilitation; healthcare services; support services; morgue area. On the contrary, specific attention has been dedicated to single environmental unit as data have been provided by 86% of the regional legislation studied: single room (9–12 m^2); multi-occupancy room (16–30 m^2); sanitary facilities both for users and for staff; archives; waiting room; relaxation room; rehabilitation gym; offices; morgue; staff rooms.

These facilities will have special and conventional systems and equipment as reported by 71% of the regions: air conditioning system; emergency lighting; system for medical gas administration, and Internet connection, configuring the OdC as a functional hospital ward.

Finally, in each OdCs, it is recommended to have a multidisciplinary staff, composed by: clinical manager; general doctor; specialist doctor; nursing case manager; primary care paediatricians; psychologist; physical therapist; social worker; nurse; just one region provided quantitative information about staff workforce needs.

88.6 Conclusion

88.6.1 Final Remarks and Outlooks

The analysis and comparison of regional regulations have brought out evident differences in the quantity and type of indications provided. The result obtained confirms the initial hypothesis about the needs to map and compare Italian region guidelines and position this study as a meaningful starting point to further extrapolate the common and virtuous factors to be considered in the national guidelines. In fact, each regional contribution can offer insights for the development of macro-project about national meta-design guidelines definition as much more comprehensive, effective, and high quality as possible: a contribution that remains significant despite each region is unique and influenced by its own territorial, socio-economic, and epidemiological characteristics.

88.6.2 Research Limitations and Further Development

The novelty and time dependence of this issue represent the main limitation of the research. In fact, at the time of data collection, no region had yet decided on accreditation regulations for Community Homes, and less than half had provided indications regarding the similar type of "Health Homes" and Community Hospitals.

In addition, the matrices could be implemented with other types of information useful at the operational level for the implementation of these facilities, such as the time and cost of implementation, information which, however, currently no region has provided.

Starting from these limitations, the systematic collection of data presented is the starting point for several further research developments. Among these, the tables could become an operational tool used by decision-makers to sort, monitor, and map what has been decided within each regional regulation. In addition, if the cataloging method of each region was shared at national level, this would encourage synergy, communication, and virtuous competition between the different territorial healthcare systems and infrastructures.

Acknowledgements The authors would like to acknowledge the full working group of Politecnico di Milano (S. Capolongo, M. Buffoli, M. Gola et al.) and AGENAS (D. Mantoan, A. Fortino, A. Borghini, A. Mutti, R. Bucci et al.) that is participating to the wider research cooperation project on metadesign guidelines definition for territorial healthcare centers in Italy (2021–2022).

References

AGENAS (2021) Piano Nazionale di Ripresa e Resilienza. Missione Salute (No. II–45). AGENAS Agenzia nazionale per i servizi sanitari regionali. https://www.agenas.gov.it/images/agenas/monitor/Agenas_Monitor_45_DOPPIE_Web.pdf. Accessed 14 March 2022

Banchieri G (2021) Case di Comunità. Facile a dirsi...più difficile a farsi. Ecco una road map per farle sul serio. Quotidiano Sanità. https://www.quotidianosanita.it/studi-e-analisi/articolo.php?articolo_id=101059

Brambilla A, Lindahl G, Dell'Ovo M, Capolongo S (2021a) Validation of a multiple criteria tool for healthcare facilities quality evaluation. Facilities 39(5/6):434–447. https://doi.org/10.1108/F-06-2020-0070

Brambilla A, Sun T, Elshazly W, Ghazy A, Barach P, Lindahl G, Capolongo S (2021b) Flexibility during the COVID-19 pandemic response: healthcare facility assessment tools for resilient evaluation. Int J Environ Res Public Health 18(21):11478. https://doi.org/10.3390/ijerph182111478

Capolongo S, Mauri M, Peretti G, Pollo R, Tognolo C (2015) Facilities for Territorial Medicine: the experiences of Piedmont and Lombardy Regions. TECHNE—J Technol Archit Environ, 230–236 pp. https://doi.org/10.13128/TECHNE-16128

Capolongo S, Brambilla A, Girardi A, Signorelli C (2021) Validation checklist for massive vaccination centers. Annali Di Igiene: Medicina Preventiva E Di Comunita. https://doi.org/10.7416/ai.2021.2460

Filippini T, Vinceti SR (2021) Italian National Recovery and Resilience Plan: a healthcare renaissance after the COVID-19 crisis recovery and resilience beyond COVID-19. Acta Biomed Atenei Parmensis 92(S6):e2021463. https://doi.org/10.23750/abm.v92iS6.12339

Gola M, Brambilla A, Barach P, Signorelli C, Capolongo S (2020) Educational challenges in healthcare design: training multidisciplinary professionals for future hospitals and healthcare. Ann Ig 32(5):549–566. https://doi.org/10.7416/ai.2020.2375

Rifkin SB (2018) Alma Ata after 40 years: primary health care and health for all—from consensus to complexity. BMJ Glob Health 3(Suppl 3):e001188. https://doi.org/10.1136/bmjgh-2018-001188

Starfield B, Shi L, Macinko J (2005) Contribution of primary care to health systems and health. Milbank Q 83(3):457–502. https://doi.org/10.1111/j.1468-0009.2005.00409.x

Vinceti M, Filippini T, Rothman KJ, Di Federico S, Orsini N (2021) SARS-CoV-2 infection incidence during the first and second COVID-19 waves in Italy. Environ Res 197:111097. https://doi.org/10.1016/j.envres.2021.111097

Wilhelm JJ, Battisto D (eds) (2020) Architecture and health: guiding principles for practice. Routledge, Taylor & Francis Group, New York

World Health Organization (2020) Basic documents, 49th edn. World Health Organization, Geneva. https://apps.who.int/iris/handle/10665/339554. Accessed 13 March 2022

Chapter 89
Wood Snoezelen. Multisensory Wooden Environments for the Care and Rehabilitation of People with Severe and Very Severe Cognitive Disabilities

Agata Tonetti and Massimo Rossetti

Abstract The paper wants to present the progress of a research project focused on the study and design of a Snoezelen room made with wooden components. "Snoezelen" refers to closed environments capable of stimulating the senses in people with severe intellectual disabilities and non-self-sufficient people. Within these environments, multisensory stimulation takes place through various equipment and instruments such as optical fibers, water-powered light columns, systems for the reproduction of sounds or vibrations, and materials with different surface treatments. Many studies and researches have demonstrated the effectiveness of multisensory stimulation generated within a Snoezelen environment on children with autism spectrum disorder, and how this has led to beneficial effects such as the reduction in aggressive and/or self-injurious behaviors. The increase and diffusion of these environments, especially in schools, could lead to both an increase in the number of students who can undertake a rehabilitation process, and an increase in the sensory-perceptive aids available to the school. The rehabilitation and multisensory aspect of the Snoezelen methodology is, in this context, emphasized by the use of wood, thanks to its beneficial properties in terms of healthiness, comfort, and influences on psychological aspects. The goal of the research is the design of a wooden Snoezelen environment, the drafting of guidelines for its implementation, and an evaluation schedule for its use. In particular, the environment will be made by applying construction techniques in wood as a material for cladding, finishing, and components of the interior environment.

Keywords Wooden constructions · Intellectual disabilities · Snoezelen · Healthy indoor environments · Advanced prefabrication

A. Tonetti (✉) · M. Rossetti
IUAV University of Venice, Venice, Italy
e-mail: atonetti@iuav.it

M. Rossetti
e-mail: rossetti@iuav.it

E. Arbizzani et al. (eds.), *Technological Imagination in the Green and Digital Transition*,
The Urban Book Series, https://doi.org/10.1007/978-3-031-29515-7_89

89.1 Introduction

The Italian demographic scenario shows a constant decrease of new births against progressive aging of the population (Fig. 89.1), with important consequences on the social, health, economic, and productive structure of the country. Moreover, thanks to an increased diagnostic capacity, statistics show that these two age groups represent the segment of the population most exposed to disability. In 2019, there were about 3,150,000 disabled people, 5.2% of the population, of whom almost 1.5 million are over 75 (Blangiardo 2021). The number of disabled children is not insignificant, with a constant percentage increase over the years of 6% (Fig. 89.2). In the 2019/2020 school year, there were 283,856 disabled scholars, 3.5% of those enrolled (Istat 2020). In this scenario, a response is needed in the architectural sphere through the design of facilities that can meet the needs of the future population by designing residential, care, assistance, and rehabilitation facilities for disabled people. These environments include Snoezelen rooms or multisensory rooms that provide non-pharmacological therapy for people with severe and/or very severe multi-disabilities. Recent research has observed how multisensory stimulation generated within a Snoezelen environment on people with severe and/or very severe mental retardant has led to beneficial effects such as, for example, a reduction in aggressive and/or self-harming behavior (Lancioni et al. 2004).

An increase in the number of such environments would therefore lead to an increase in the number of people who could undertake rehabilitation. The design of care and rehabilitation environments requires an appropriate and careful choice of technologies and building products; in this context, recent studies (Demattè et al. 2018) have shown that wood environments can benefit people, thanks to the specific characteristics of the raw material; similarly, other studies have highlighted the potential of wood as an ideal material for the design of environments for the accommodation of dependent people (Camerin et al. 2020); furthermore, the effectiveness of wood in environments designed for people with autism spectrum disorder has been demonstrated (Venturini 2010).

The use of wood as a building material is therefore shown to be optimal for the creation of Snoezelen environments.

89.2 Organization and Research Methodology

The main objective of the research project "wood Snoezelen. Multisensory wooden environments for the care and rehabilitation of people with severe and very severe cognitive disabilities", which was developed by Iuav University of Venice, is to contribute to research aimed at assisting people with severe and/or very severe intellectual disabilities by designing a Snoezelen environment. The project involves other Veneto territorial realities, namely (1) "La Nostra Famiglia", based in Conegliano Veneto and Treviso, (2) I.S.R.A.A., based in Treviso, (3) "Progetto Legno Veneto"

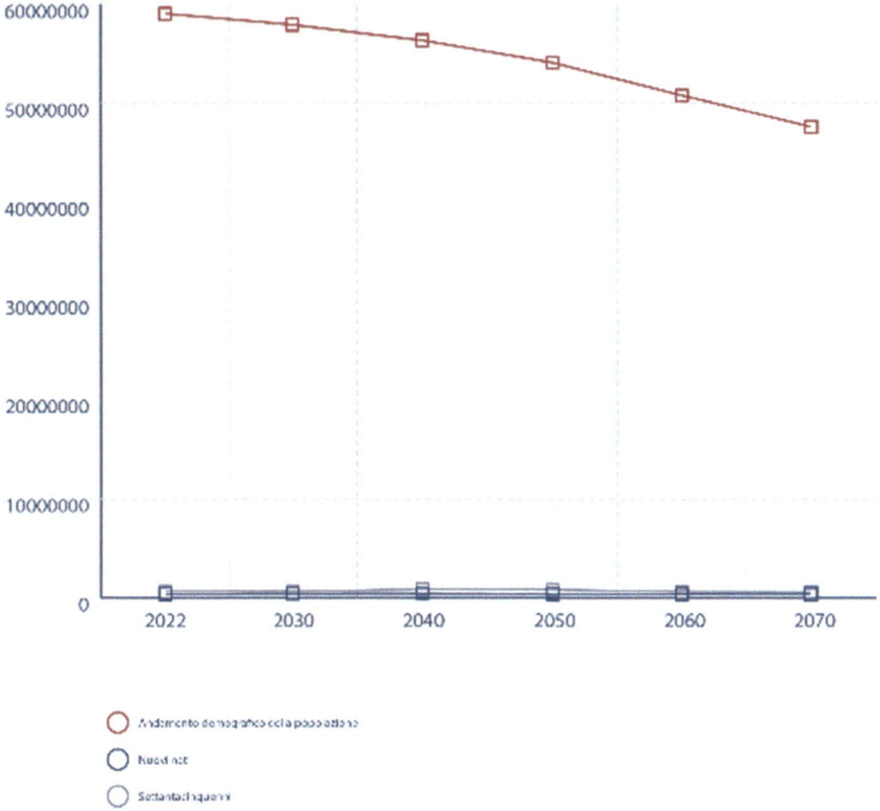

Fig. 89.1 Population projection: years 2022–2070. *Source* Istat (2022)—Graphic elaboration by the authors

Consortium, the Regional Innovative Network Foresta Oro Veneto and the company Bozza S.r.l. The research project, which is under development, is developed through 6 phases (1) reconstruction of the state of the art related to intellectual disabilities, wooden environments, and Snoezelen rooms; (2) identification of the characteristics of a Snoezelen room, with particular reference to the use of wood as the prevalent material for its construction in an existing space; (3) identification of an unused room of a Veneto School Institute for the construction of the Snoezelen environment; (4) design of a prefabricated Snoezelen room with wooden structure; (5) creation of guidelines for the design and use of the Snoezelen room and an evaluation system of the Snoezelen approach; (6) construction of the Snoezelen room inside the previously identified school.

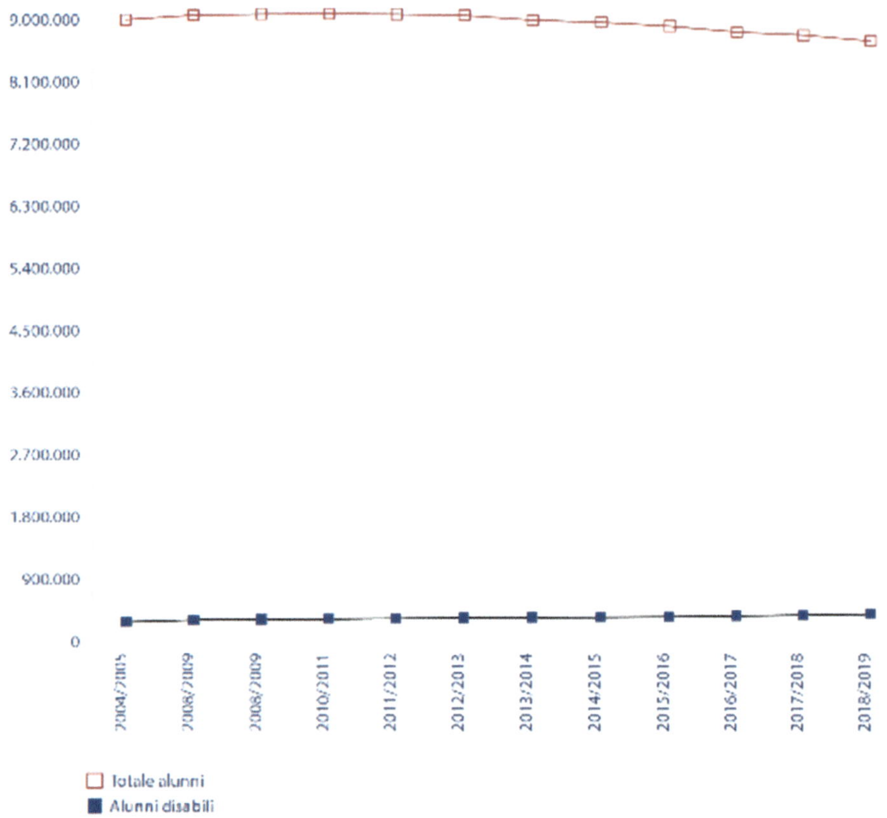

Fig. 89.2 Scholars with disabilities, the school year 2018/2019. *Source* Istat (2020)—Graphic elaboration by the authors

89.3 Snoezelen

The Snoezelen approach is a non-pharmacological therapy aimed mainly at severely and/or very severely disabled people, during which selective sensory stimulation is carried out, under the guidance of an expert operator, in one or more sensory domains, according to the needs of the individual. This approach is carried out in special adaptive spaces, called Snoezelen rooms, equipped with sensory instruments in which the amount, arrangement, and intensity of stimulation are controlled and modulated by the operator. The Snoezelen approach aims to find a balance between relaxation and activity within the multisensory environment. The combination of these two contrasting activities constitutes the term Snoezelen, which derives from the contraction of the Dutch verbs *snufelen* (to sniff) and *doezelen* (to doze), meaning, respectively, to explore and to rest. Numerous studies have demonstrated the effectiveness of the Snoezelen approach, particularly concerning dementia (Strøm et. al. 2016), severe and complex disabilities (Glenn et. al. 1996), autism spectrum disorder (Germeau 1998), and special educational needs (Carter et. al. 2012). The Snoezelen approach originated in the Netherlands in the 1970s as a result of experimentation carried out by Ad Verheul and other Dutch occupational therapists at the De Hartenberg Center, the Haarendael Board, and the Piusoord Board. The first permanent Snoezelen Room was built in 1982 at the De Hartenberg Center; it was a significant event as it was the beginning of the dissemination and testing of the Snoezelen approach. Snoezelen rooms usually consist of two separate rooms: (1) the *white room*, the relaxation room equipped with bubble tubes, waterbed, soft seating, fiber optics, and projectors and (2) the *dark room*, within which the senses of touch, hearing, and proprioception are developed through different equipment such as tactile surfaces, sound and light projectors, and balance tools (Fig. 89.3).

The Snoezelen method has spread mainly in Holland, Denmark, Canada, the United States, and Australia, where there are large Snoezelen room facilities made up of several rooms, such as, for example, the Lacey A. Collier Snoezelen Complex on the campus of the Escambia Westgate School in Pensacola, Florida. In Italy, this approach is still little known; in Italy, currently, there are only approximately 102 Snoezelen rooms in different types of structures such as RSAs, Day Care Centers, and Schools. Snoezelen rooms are unevenly distributed throughout Italy: mainly in Tuscany (18), Lombardy (17), and Veneto (15) followed by Piedmont (9) and Sicily (8), while the other Italian regions have less than 5, except for Basilicata which does not have any (Fig. 89.4.). The majority of Italian Snoezelen consists of a single room in which there is the coexistence of relaxation and exploration equipment.

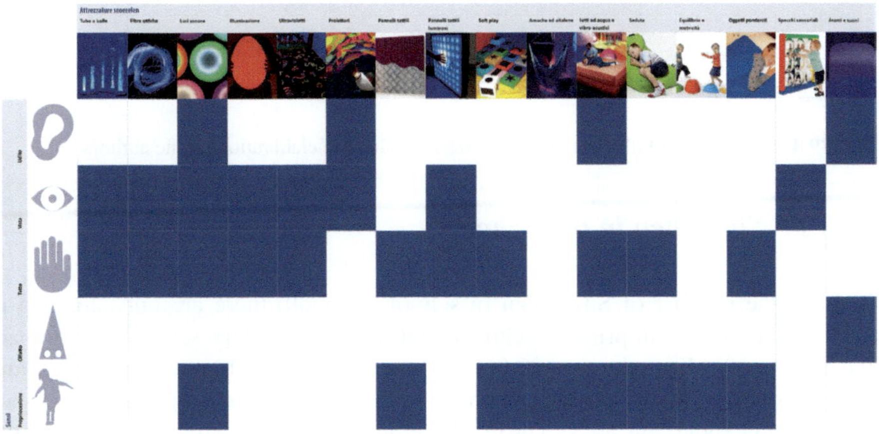

Fig. 89.3 Relationship between Snoezelen equipment and use of the senses—Graphic elaboration by the authors

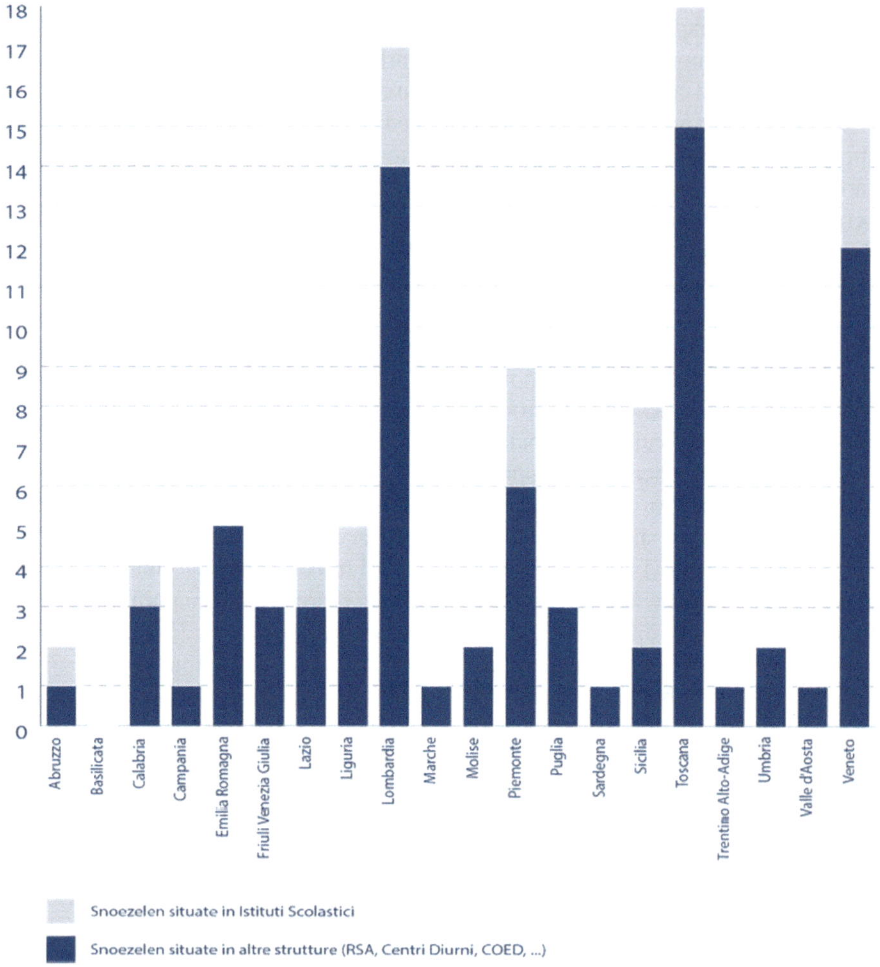

Fig. 89.4 Snoezelen in Italy, last update 02/02/22—Graphic elaboration by the authors

89.3.1 Snoezelen in the Schools

In Italy, the number of Snoezelen in schools is small; there are only 26: two in nursery schools, 23 in primary schools, and one in secondary schools. If compare the number of Italian schools—22,863 nursery schools, 16,840 primary schools, and 8048 secondary schools—with the number of Snoezelen, it is clear that the difference is quite remarkable (Fig. 89.5). The first Snoezelen located in an Italian school is at the Istituto Comprensivo "De Amicis" Primary School in Eraclea, Veneto, which opened in 2012 (Fig. 89.6). Specifically, there are only three Snoezelen rooms in schools in Veneto: in addition to Eraclea, the other two are the Collodi Primary School in Ceggia, which opened in 2014, and the XI Aprile 1848 Preschool in Castelnuovo del Garda,

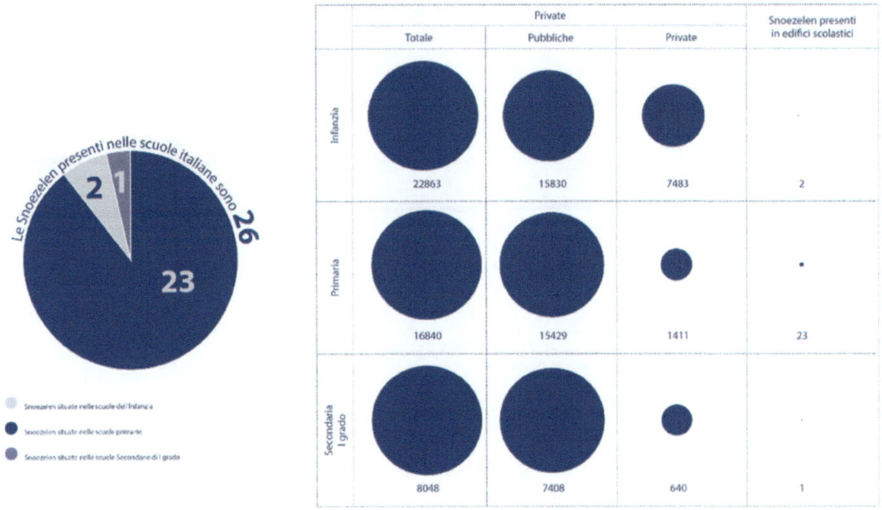

Fig. 89.5 Snoezelen about schools by grade and level, last update 02/02/22—Graphic elaboration by the authors

which opened in 2021 (Fig. 89.7). It is thought that the low number of Snoezelen in Italian educational institutions is caused by almost no knowledge of this approach, as evidenced by various newspaper headings and municipal administration websites that bear the words "the first Snoezelen room in a school" (GenoaMunicipality 2018) probably ignoring the existence of other rooms. Almost all Snoezelen rooms in Italian schools have been located in classrooms not used by the school, placing the usual catalog equipment inside them without thinking about possible solutions of architectural integration between the internal envelope and the equipment. Most of these are in a single, small room, where there is a coexistence of the relaxation area and the activity area.

The Snoezelen rooms analyzed have in common what can be considered the basic equipment: bubble tubes, fiber optics, projectors, soft seating, ball pool, and soft play. One example is the case of the Istituto Comprensivo Alessio Narbone in Caltagirone, which has become a spokesman for the Snoezelen approach in schools throughout Italy by organizing the National Network of Snoezelen Schools.

The institute has a Snoezelen space of 150 square meters consisting of a sensory corridor, a white room, and a workshop for motor and manipulative activities. The Snoezelen space, inaugurated in 2016, is open to the territory and used by 10 other schools for a total of almost 80 students divided into 17 preschool students, 30 primary students, 18 secondary students, and 12 secondary students (Sestina 2020).

Fig. 89.6 Overview of Snoezelen in Italian schools and the year of the opening, last update 02/02/22—Graphic elaboration by the authors

Fig. 89.7 Snoezelen in Veneto, last update 02/02/22—Graphic elaboration by the authors

89.4 Project

The therapeutic quality of Snoezelen environments is significant and, if installed in schools, can become part of the long process of school inclusion as well as compensating for the scarce presence of systems for sensory disabilities in schools.

This research aims to design Snoezelen environments in educational institutions through the architectural integration between the internal envelope and Snoezelen equipment. The integration will be accomplished through the design of prefabricated wooden modules. Recent projects have shown how wood lends to the making

Fig. 89.8 Creative Crews, Classroom Makeover for The Blind, Pattaya, Thailand, 2018, credits: Mana and Friends, Ekkachan Eiamananwattana, Jirakit Panomphongphaisarn

of therapeutic-sensorial environments. There are two exemplary cases of wooden multisensory spaces in school buildings: (1) Noverca house by the architecture studio Atelier JQTS, which built a prefabricated wooden pavilion housing a multisensory space for disabled pupils at the Maria Veleda school; (2) classroom makeover for the blind by the architecture studio creative crews (Fig. 89.8), which designed a wooden multisensory room in line with the Pre-Braille Curricula for kindergarten children.

Wood can also be used to design sensory systems, of which "Dear Disaster" and "Sensorium" exhibition are two emblematic cases. "Dear Disaster" is a cabinet designed by Jenny Ekdhal to help victims of natural disasters recover from their traumatic experiences, while "Sensorium" is a temporary installation designed by Les M where the parametric wooden floor is able to move according to the pressures caused by the human body.

89.4.1 Advance Modular Wooden Components

The considered examples were fundamental for the design—which is still in progress—of the advanced modular wooden components, such as false wall, false ceilings, containing elements, and raised floor systems. The modules are designed to create a neutral room that can be customized as needed by the operator to adapt the room to the patient's therapeutic needs. Specifically, the false walls feature a micro-perforated, acoustic, and backlit acoustic panel in wood (Fig. 89.9). The backlighting and acoustics enable the operator to choose the combination of sounds and colors useful for individual therapy. In addition, the holes allow the installation of additional modules consisting of different surface finishes—smooth, rough, movable shingle, and wood skin—which the operator can choose according to tactile stimulation. The raised floor (Fig. 89.10), in addition to offering optimal acoustic performance and the passage of equipment, allows for the integration of Snoezelen equipment such as bubble tubes.

The finishing layer also consists of different modules with multiple textures—bright and rough surfaces—and various sections that offer tactile, visual, and proprioceptive stimulation. Containing elements (Fig. 89.11) can also be placed on the floor to act as containers for the vibroacoustic waterbed and ball pool. The acoustic false ceiling (Fig. 89.12) allows for the integration of the lighting system, spotlights, hammocks, and ultraviolet. The components, dry-assembled, can generate different spatial configurations resulting in three predefined Snoezelen rooms having three different sizes: S (25 sqm), M (50 sqm), and L (100 sqm). The setup of these

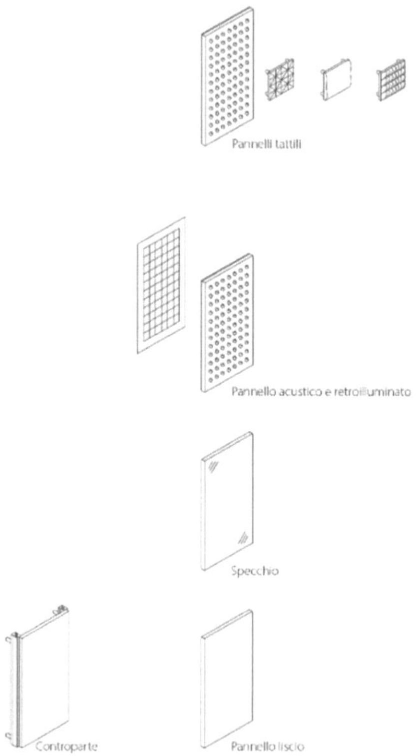

Fig. 89.9 False wall modules—Graphic elaboration by the authors

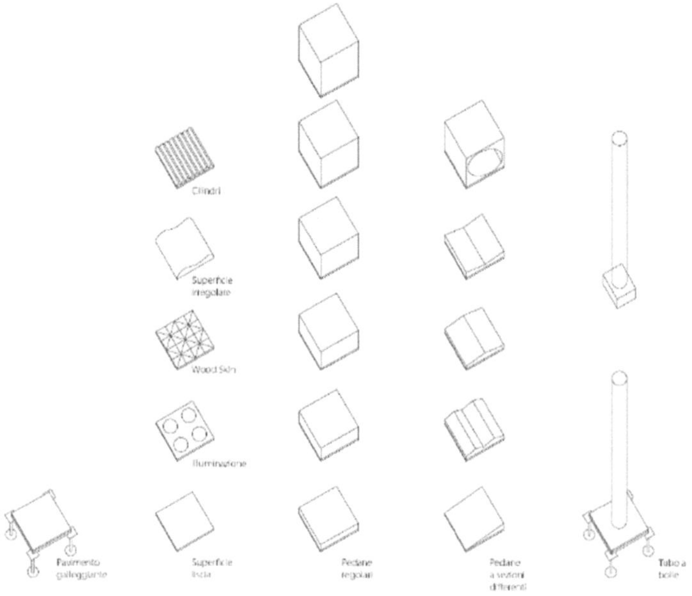

Fig. 89.10 Raised floor systems modules—Graphic elaboration by the authors

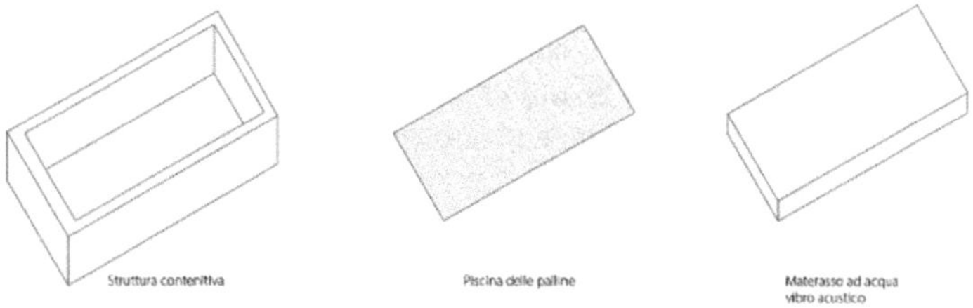

Fig. 89.11 Containing elements modules—Graphic elaboration by the authors

rooms and the modular components, illustrated in the guidelines, will be used by future stakeholders and designers as a basic tool for the making of the desired Snoezelen room.

89.5 Conclusions

The research project aims to contribute to studies on environments for the rehabilitation of people with cognitive disabilities, in particular concerning the potential provided by the advanced prefabrication of wooden construction elements, components, and furniture. This objective will be achieved through the design and implementation of a Snoezelen room with wooden components in a school in Veneto,

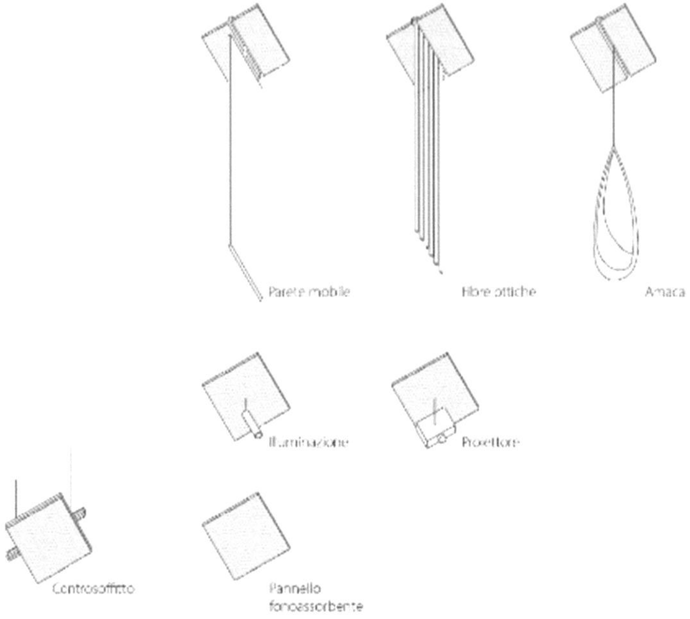

Fig. 89.12 False ceiling modules—Graphic elaboration by the authors

which will be able to open to the surrounding area. Moreover, guidelines for the design of Snoezelen rooms will be drawn up, so that other institutes can also make use of this type of inclusive structure and an assessment system to check the patient's therapeutic progress.

References

Blangiardo GC (2021) Audizione dell'Istat presso il Comitato Tecnico Scientifico dell'Osservatorio Nazionale sulla condizione delle persone con disabilità. https://www.osservatoriodisabilita.gov.it/it/notizie/audizione-dell-istat-presso-il-comitato-tecnico-scientifico-dell-osservatorio/

Camerin F, Incelli F, Rossetti M (2020) Confrontarsi con il tempo. Unità abitative temporanee in legno per anziani non autosufficienti. Techne 20:282–291. ISSN online: 2239-0243

Carter M, Stephenson J (2012) The use of multi-sensory environments in schools servicing children with severe disabilities. J Dev Phys Disabil 24(1):95–109. https://doi.org/10.1007/s10882-011-9257-x

Demattè ML, Zucco G, Roncato S, Gatto P, Paulon E, Cavalli R, Zanetti M (2018) New insights into the psychological dimension of wood–human interaction. Eur J Wood Wood Prod. 76:1093–1100.https://doi.org/10.1007/s00107-018-1315-y,CorpusID:21662687

GenoaMunicipality (2018) A Genova la prima snoezelen room in una scuola, perché nessuno resti indietro, Comune di Genova. http://www.comune.genova.it/content/genova-la-prima-snoezelen-room-una-scuola-39perchegrave-nessuno-resti-indietro39. Accessed 10 Dec 2022

Germeau I (1998) *Une approche particulie're dans la relation personne autiste/e'ducateur en milieu institutionnel: Le Snoezelen.* Faculte´ de Psychologie et des Sciences de L'E´ ducation, Universite de Lie'ge
Glenn S, Cunningham C, Shorrock S (1996) Social interaction in multi-sensory environments. Routledge
Istat (2022) Previsioni della popolazione—Anni 2020–2070: Indicatori demografici. http://dati.istat.it/Index.aspx?QueryId=19668. Accessed 15 Mar 2022
Istat (2020) L'inclusione scolastica degli alunni con disabilità. https://www.istat.it/it/files/2020/12/Report-alunni-con-disabilit%C3%A0.pdf
Lancioni et al (2004) Effects of Snoezelen room, activities of daily living skills training, and vocational skills training on aggression and self-injury by adults with mental retardation and mental illness. Res Dev Disabil 25(3):285–293. https://doi.org/10.1016/j.ridd.2003.08.003
Sestina, G., (2020) Progetto Stimolazione multisensoriale in ambiente Snoezelen, ISTITUTO COMPRENSIVO STATALE "ALESSIO NARBONE" A.S. 2019/2020 AMBITO. https://www.alessionarbone.edu.it/index.php/area-snoezelen/24-monitoraggio-di-esiti-e-processi.
Strøm SB, Ytrehus S, Grov EK (2016) Sensory stimulation for persons with dementia: a review of the literature. J Clin Nurs 25(13–14):1805–1834. https://doi.org/10.1111/jocn.13169
Venturini V (2010) Residenze temporanee presso il Castello di Granarola (PU). Spazi interni ed esterni a misura di bambino speciale. In: Francesca Giofré (a cura di), *Autismo protezione sociale e architettura* (133–150). Aliena Editrice

Chapter 90
The Proximity of Urban Green Spaces as Urban Health Strategy to Promote Active, Inclusive and Salutogenic Cities

Maddalena Buffoli and Andrea Rebecchi

Abstract Urban Green Spaces (UGS) have several positive effects on Public Health, environmental quality, and cities' resilience to climate change; UGS are crucial in urban regeneration actions and urban health purposes. Moreover, to better define the UGS' health impacts, it is important to define and guarantee UGS' proximity, accessibility, and quality. Aim of the research is a quali-quantitative assessment of the UGS in Italian metropolitan cities, taking Milan, Turin, Florence, and Bologna as preliminary case studies. One of the 1st phases was to draw up dynamic and descriptive GIS-based maps of the relationships between density of population and of urban fabric, UGS' availability, and their accessibility. Only the areas with a size greater than 15,000 square meters were considered; three buffer zones of proximity were defined: 250, 500, and 750 m. By combining the UGS' availability with the population's density, it was possible to quantify the citizens included in the three buffer zones. From the 1st analysis, it is observed that about 90% of the population is served by a quality green area within a buffer area of 750 m; 78% by the buffer zone of 500 m; 49% by the buffer zone of 250 m. Both the elaborated maps and graphs obtained show how population is not equally served by close and accessible UGS. Their geo-localization it's a preliminary quantitative step (process started in Italy with the introduction of regulations like green areas' census, mapping, maintenance legislation, and strategic plans), but it's even more crucial to evaluate the UGS' quality in terms of accessibility, safety and security features, provision of services and paths.

Keywords Urban green spaces availability · Urban green spaces accessibility · Urban green spaces proximity · GIS-based approach · Urban public health perspective · Healthy cities

M. Buffoli · A. Rebecchi (✉)
Design & Health Lab, Department of Architecture, Built Environment and Construction Engineering (DABC), Polytechnic University of Milan, Milan, Italy
e-mail: andrea.rebecchi@polimi.it

M. Buffoli
e-mail: maddalena.buffoli@polimi.it

E. Arbizzani et al. (eds.), *Technological Imagination in the Green and Digital Transition*, The Urban Book Series, https://doi.org/10.1007/978-3-031-29515-7_90

90.1 Theoretical Scenario

The recent COVID-19 sanitary and social emergency are a crucial demonstration of the dual effects between the ongoing urbanization phenomena that represent the contemporary cities' capacity to be a place of economic, social, and cultural opportunities, and—at the same time—a situation of multiple environmental risk factors for Public Health, well-being, and welfare (Capolongo et al. 2020). The fast change in lifestyles in the period of physical and social distancing (lockdown) which has interested several World and European Countries, its emphasizing and making urgent the cities' transformation into resilient ecosystems, capable to promote healthy lifestyles (Capolongo et al. 2020) and Public Health, and preventing the spread of the infectious diseases, of today and tomorrow (Lee 2020).

In particular, the urbanization phenomena taking place all over the World, with particular reference to the developed countries and emerging economies, imply a series of reflections on population density as a vehicle for social inequalities and Public Health emergencies (Pisano, et al. 2020). In fact, the population living in urban areas today represents 54%, and forecasts indicate that it will increase up to 70% by 2050 (Talukder 2015).

The recent COVID-19 sanitary and social emergency, moreover, have renewed the demands of a social, environmental, and digital needs to those already in development (Azzoppardi-Muscat et al. 2020),the link between the morpho-typological and functional features of the urban contexts, and the Public Health conditions, had risen up the attention to the Urban Public Health scenario (Capolongo et al. 2018), which shifts the focus of the "concept of health" from a medical model, merely focused on the individual care, to a social model, in which health is the result of various socio-economic, cultural, and environmental factors (Moscato and Poscia 2015).

Among the health determinants, environmental and behavioral risk factors, such as sedentary lifestyle or physical inactivity, represent a key role within the global health balance, with key reference to the risk of onset by the non-communicable diseases (NCDs). In other words, health is no longer an aspect of life exclusively rely on the health protection and promotion, but an individual and collective condition, strongly influenced by the environmental context and by the strategies implemented by local Governments.

The main *environmental risk factors* that affect our cities are urban heat island effect; air, noise, and soil pollution; vehicular traffic; and urban safety and security.

The possible *health-related outcomes* are non-communicable diseases (NCDs) like cardio-respiratory diseases and sedentary, cause of diabetes; and mental health disorders like stress condition, ansia, sleeping disorders, and cognitive development.

The main *urban health strategies* to improve and promote Urban Public Health, and to renovate cities according to the urban resilience to climate change are increasing green and blue infrastructures; promoting biodiversity protection; increase the quantity and quality of outdoor public spaces; reducing vehicular traffic; promoting walkable environment; improving social and functional mix of the ground floor of the city, looking for the city of proximity; promoting the accessibility of the city with the universal design approach; realizing free outdoor places to physical activity like playground; and considering the rooftop of the buildings like a common public floor for the communities, and make it livable or greener.

During the spent three/four waves of the COVID-19 sanitary crisis, urban systems were stressed. During this period, the resilience of our cities and population was tested, highlighting the criticalities in both sanitary systems and urban planning. Urban green spaces (UGS) have proved essential role, not only as a safe recreational space but also as an irreplaceable tool to improve Urban, Public, and Mental Health (World Health Organization 2016), due to its formidable restorative capabilities (Capolongo et al. 2015). Unfortunately, the heterogeneous distribution of UGS inside of urban environments (Gianfredi et al. 2021), and the disparity in quality of such spaces, led to some exclusion phenomena, further underlining their relevance in egalitarian and democratic cities (D'Alessandro et al. 2017).

90.2 Research Purpose and Method

Aim of the research is a quali-quantitative assessment of the UGS in several Italian metropolitan cities, taking *Milan, Turin, Florence, and Bologna* as preliminary case studies. One of the 1st phases was to draw up dynamic and descriptive GIS-based maps of the relationships between density of population and of urban fabric (Rydin et al. 2012), UGS' availability and accessibility, and city users' perception in terms of recreational opportunities and healthy lifestyles' promotion propensity.

Geographic information systems (GIS) are a computerized information system that allows the acquisition, recording, analysis, display, return, sharing, and presentation of information deriving from geographic data. The benefits obtained from its application and use are the ability to organize and archive a large amount of geographic data (Zhang and Yu 2020) and to conduct operations that are directly useful for decision-making, within a georeferenced environment (Irvine et al. 2012). GIS is usually conceptualized as a collection of georeferenced layers, which can be classified according to two different types of data: raster and vector. The main difference between the two data series is in the number of properties it can collect. While raster data are able to describe only a single property, vector data can collect a large number of information and therefore of attributes (Oppio et al. 2016). GIS-based software therefore offers the possibility of supporting territorial analyzes allowing a dynamic graphical rendering capable of associating data of a different nature with a direct display of the results obtained in the area under consideration assessment.

The methodological scheme developed and subsequently applied to evaluate the availability of UGS in the major Italian metropolitan cities is presented below. The proposed analyzes exploit the potential of GIS through the use and subsequent processing of vector data. The approach was divided into three phases.

The first phase of data collection consists of three steps: Definition of the decision-making context: It allows to identify the limits of the analysis and to identify the territorial context being assessed; Source map: It consists of the collection of georeferenced data using existing databases; Screening phase: The data previously collected is selected in order to process those deemed suitable for the purposes of the evaluation and research.

The second phase of data processing was developed through spatial operations: definition of buffer areas from the geometries selected in the data collection phase.

The third phase of data analysis consists in reading the partial and total results obtained through the spatial elaborations developed in the previous phase and considering the selected data.

The previously described methodological approach made it possible to evaluate the availability of green areas in metropolitan cities, the most affected in the national context by the COVID-19 pandemic currently underway, which made it possible to highlight the existing vulnerabilities and inequalities, as well as accelerate the most radical dynamics of change.

90.3 Findings

The municipalities of *Bologna, Florence, Milan, and Turin* were selected in the regional and municipal areas and presented comparable open-source GIS data available. One of the first phases of the research, preliminary to the further assessment phases, was to draw up—referring to the GIS-based software QGIS—dynamic and descriptive maps of the relationships between density of population, density of the urban fabric, UGS (in all its forms) availability and accessibility, and city users' perception of those UGS and infrastructures existing in the context of Milano city.

Only the significant green areas for the research purposes were identified and selected; therefore, poor-quality green areas and those insignificant have been eliminated, as they are not accessible, private, and not suitable for recreational / sporting activities, or not capable of creating opportunities in terms of healthy lifestyles' promotion.

Among the green areas previously identified, it was decided to carry out a further selection that allowed to analyze only the areas with a size greater than 15,000 m2, as they were considered the most representative for their relevant size and sufficient for outdoor physical activity promotion.

Subsequently, three buffer zones were defined, respectively, of 250, 500, and 750 m, which allow to evaluate the distance from the green areas (Figs. 90.1, 90.2, 90.3 and 90.4). These distances can be easily reached on foot from 3 to 10 min and are significant for assessing the quality of the available green areas into the neighborhoods.

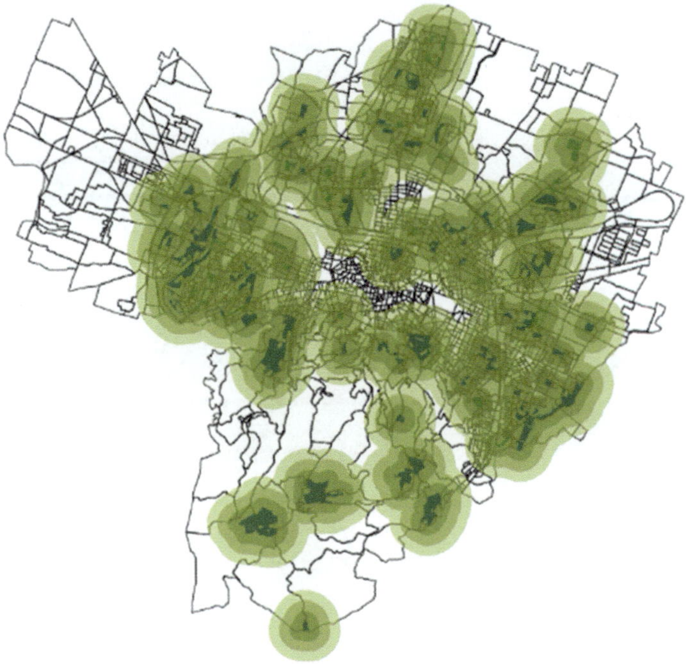

Fig. 90.1 Bologna: buffer for analysis and definition of the distance from UGS with an area greater than 15,000 m^2

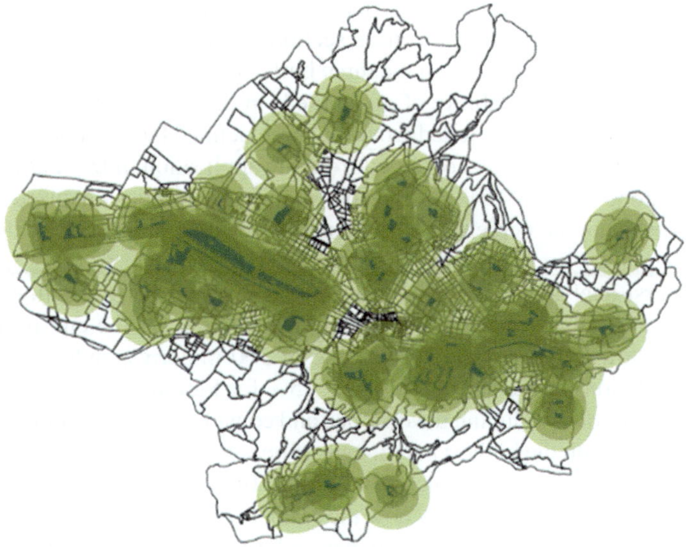

Fig. 90.2 Florence: buffer for analysis and definition of the distance from UGS with an area greater than 15,000 m^2

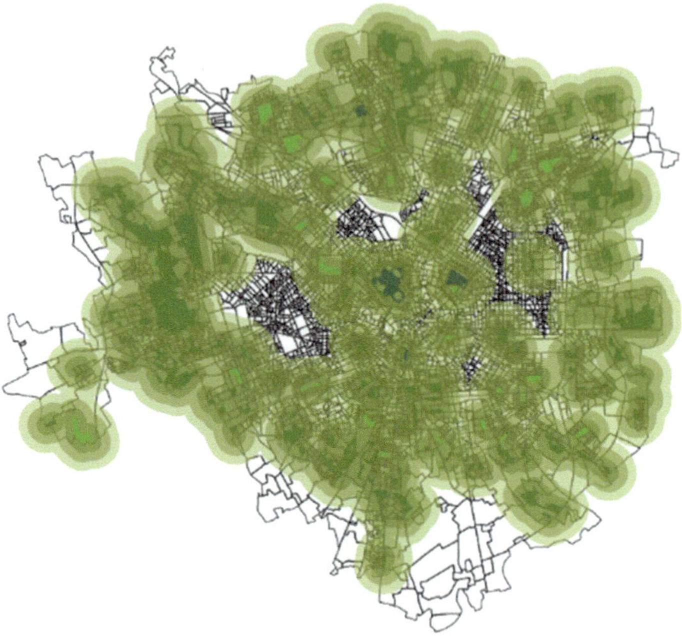

Fig. 90.3 Milan: buffer for analysis and definition of the distance from UGS with an area greater than 15,000 m^2

By combining the elaborations carried out with the information relating to the resident population in the several census sections (ISTAT 2011), it was possible to quantify the population included in the three buffer zones (250, 500, and 750 m) and therefore evaluate the availability of green areas in the cities of Milan, considering the served population VS excluded from the three defined buffer zones.

The graphs show the results obtained from the application of the methodological process developed (Fig. 90.5). For the four metropolitan cities analyzed, it is thus possible to evaluate the percentage of population residing in a radius of distance of 250 m, 500 m, and 750 m from the selected green areas.

From a first analysis, it is evident that about 90% of the population is served by a quality green area within a 750 m buffer area, only for the city of Turin, the percentage is lower (81%). The buffer zone of 500 m corresponds to 83% of the resident population for the city of Bologna, 75% for Florence, 78% for Milan, and 65% for Turin. Further reducing the distance and considering the buffer of 250 m, we note that the city with the highest percentage of population served is Bologna (61%), followed by Milan (49%), Florence (44%), and Turin (37%). Both the elaborated maps and the obtained graphs show how the resident population in urban areas is not equally served by quality green areas, in particular the central areas which appear to be those with a higher population density and high land consumption.

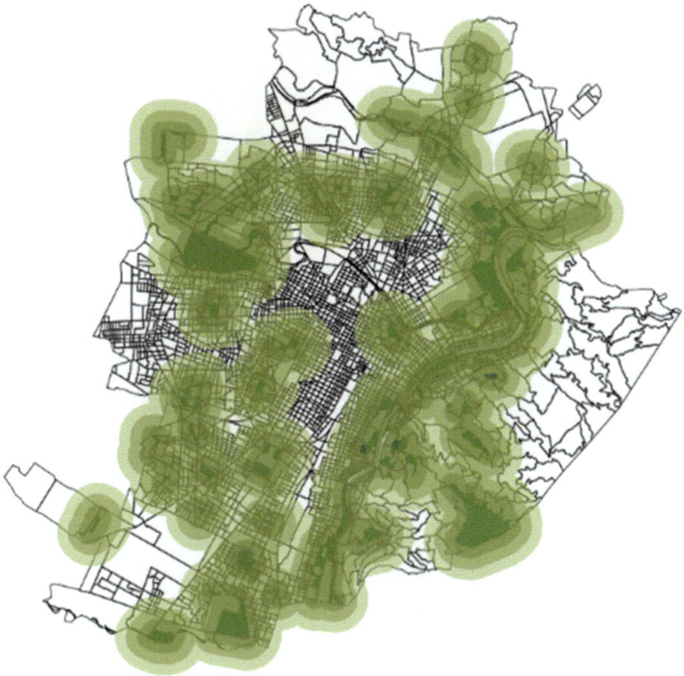

Fig. 90.4 Turin: buffer for analysis and definition of the distance from UGS with an area greater than 15,000 m^2

The proximity to UGS positively influences the well-being of people, the livability of the neighborhoods, and the quality of the public space, bringing environmental, social and economic benefits, as well as effects on health. A densely built area with a reduced availability of green areas on the contrary has negative effects on the urban microclimate, on pollution and consequently on all the previously defined dimensions.

Both the elaborated maps and the graphs obtained show how the resident population in urban areas is not equally served by quality green areas, in particular the central areas are those with a higher population density, with a high consumption of land and less availability of UGS.

90.4 Conclusions and Research Outlooks

The conducted research is the basis for developing further methods, tools, and indicators to design and assess the environmental quality of the public spaces (Capolongo et al. 2020). It is necessary and crucial to evaluate the public space with qualitative and quantitative standard, highlighting the urban context and public spaces features

Fig. 90.5 Comparison of the quantification of the population included in the three buffer zones

evidence-based oriented to promote healthy lifestyles. Making cities more walkable and cyclable means to improve their physical factors—as network of public spaces, land use mix, street design, etc.—to create a more convenient, safe, comfortable, and attractive places.

About the outcomes, some benefits are known and certain—like the environmental ones—and published lots of time in the scientific journals that investigate topics like climate change, urban resilience, environmental issues, and Public Health. At the same time, research projects that quantify, at the statistical level, the direct health benefits, are missing (Bedimo-Rung and Gustat 2006). Furthermore, in all urban projects, part of the economic investment should be dedicated to post-intervention forecasting researches (Knobel and Dadvand 2020), aimed to define health evidence-based data and direct correlation with population health status.

The Sustainable Development Goals (SDGs) from the 2030 Agenda for Sustainable Development by United Nations represent the roadmap that we should follow. The number 3 *"good health and well-being"* is strongly connected with the number 10 "social inequalities reduction" and social inclusion; the number 11 *"sustainable cities and communities"* and the number 13 *"climate actions for climate resilience"* and environmental sustainability.

A multidisciplinary approach is needed, putting together different stakeholders, like Designers (Urban Planners and Architects), Policy and Decision Makers, Public Health experts and City Users, Citizen and Associations (Lenzi et al. 2020), to enhance a direct interaction with decision-maker (Gianfredi et al. 2019),economist, that evaluates the project feasibility, with a cost–benefit approach; and finally, environmental monitoring agencies.

References

Azzoppardi-Muscat N, Brambilla A, Caracci F, Capolongo S (2020) Synergies in design and health. The role of architects and urban health planners in tackling key contemporary public health challenges. Acta Biomed 91(Supplement 3):9–20

Bedimo-Rung A, Gustat J (2006) Development of a direct observation instrument to measure environmental characteristics of parks for physical activity. Human Kintetc J

Capolongo S, Buffoli M, Oppio A (2015) How to assess the effects of urban plans on environment and health. Territorio 73:145–151

Capolongo S, Rebecchi A, Dettori M, Appolloni L, Azara A, Buffoli M, Capasso L, Casuccio A, Oliveri Conti G, D'Amico A et al (2018) Healthy design and urban planning strategies, actions, and policy to achieve salutogenic cities. Int J Environ Res Public Health 15:2698

Capolongo S, Buffoli M, Brambilla A, Rebecchi A (2020) Healthy urban planning and design strategies to improve urban quality and attractiveness of places. TECHNE 19:271–279. https://doi.org/10.13128/techne-7837

Capolongo S, Buffoli M, Mosca EI, Galeone D, D'Elia R, Rebecchi A (2020) Public health aspects' assessment tool for urban projects, according to the urban health approach. Springer, Research for Development, 2020, pp 325–335. https://doi.org/10.1007/978-3-030-33256-3_30

Capolongo S, Rebecchi A, Buffoli M, Appolloni L, Signorelli C, Fara GM (2020) D'Alessandro D. COVID-19 and cities: From urban health strategies to the pandemic challenge. a decalogue of

public health opportunities. Acta Biomedica 91(2):13–22. https://doi.org/10.23750/abm.v91i2.9515
D'Alessandro D et al (2017) Public health and urban planning: a powerful alliance to be enhanced in Italy. Ann Ig 29:452–463
Gianfredi V, Buffoli M, Rebecchi A, Croci R, Oradini-Alacreu A, Stirparo G, Marino A, Odone A, Capolongo S, Signorelli C (2021) Association between urban greenspace and health: a systematic review of literature. Int J Environm Res Public Health 18:5137. https://doi.org/10.3390/ijerph18105137
Gianfredi V, Balzarini F, Gola M, Mangano S, Carpagnano LF, Colucci ME, Gentile L, Piscitelli A, Quattrone F, Scuri S, Mantovani LG, Auxilia F, Castaldi S, Capolongo S, Pelissero G, Odone A, Signorelli C (2019) Leadership in public health: opportunities for young generations within scientific associations and the experience of the academy of young leaders. Front Public Health 7:378
Irvine K, Warber S, Devine-Wright P, Gaston K (2012) Understanding urban green space as a health resource: a qualitative comparison of visit motivation and derived effects among park users in Sheffield, UK. Int J Environ Res Public Health 10(1):417–442
Knobel P, Dadvand P (2020) Development of the urban green space quality assessment tool (RECITAL). J Urban Forest Urban Greening 1–8
Lee VJ et al (2020) Epidemic preparedness in urban settings: new challenges and opportunities. Lancet Infect Dis
Lenzi A, Capolongo S, Ricciardi G, Signorelli C, Napier D, Rebecchi A, Spinato C (2020) New competences to manage urban health: health city manager core curriculum. Acta Biomed 91:21–28. https://doi.org/10.23750/abm.v91i3-S.9430
Moscato U, Poscia A (2015) Urban public health. In: Boccia S, Villari P, Ricciardi W (eds) A systematic review of key issues in public health. Springer, Geneva, Switzerland
Oppio A, Bottero M, Giordano G, Arcidiacono A (2016) A multi-me- thodological evaluation approach for assessing the impact of neighbourhood quality on public health. Epidemiol Prev 40:249–256
Pisano GP et al (2020) Lessons from Italy's response to coronavirus. Harvard Bus Rev
Rydin Y, Bleahu A, Davis M, Dávila JD, Friel S, de Grandis G, Groce N, Hallal PC, Hamilton I, Howden-Chapman P et al (2012) Shaping cities for health: complexity and the planning of urban environments in the 21st century. Lancet 379:2079–2108
Talukder RS et al (2015) Urban Health in the post-2015 Agenda. Lancet 385:769
World Health Organization (WHO) (2016) Health as the pulse of the new urban Agenda. United Nations Conference on Housing and Sustainable Urban Development. Quito, Ottobre 2016. Available on-line: http://apps.who.int/iris/bitstream/handle/10665/250367/9789241511445-eng.pdf;jsessionid=07F882D99F1E1AF399B57D5546EEB2BB?sequence=1
Zhang J, Yu Z (2020) Links between space and public health: a bibliometric review of global research trends and future prospects from 1901 to 2019. Environ Res Lett 15

Chapter 91
Environmental Attributes for Healthcare Professional's Well-Being

Zakia Hammouni and Walter Wittich

Abstract The COVID-19 pandemic has been stressful for everyone and even more so for hard-working healthcare professionals. Contemporary hospitals are now endowed with environmental attributes that contribute to achieving well-being within their environment. However, these attributes tend to be focused on the patient and their experience. This paper examines these issues and describes the attributes of the physical environment that support healthcare professional's well-being. Within a constructivist approach, the study was conducted in two care units in a mega hospital in Canada, before the arrival of the COVID-19 pandemic. Data collection includes a spatial evaluation of these care units, healthcare professionals' spatial behavior, and 44 semi-structured interviews with various healthcare professionals, completed by the mental images. Thematic analysis and triangulation of the data set were conducted. Key attributes identified as promoting healthcare professionals' well-being include light-color in care units, corridors and public areas of the hospital, and the cleanliness and art elements. Furthermore, panoramic views from the staff lounge, corridors, or elevator lobbies provide access to daylighting. This study highlights the importance of providing healthcare professionals break areas that allow them to find respite, particularly during periods of extreme stress such as COVID-19 pandemic.

Keywords Hospital · Well-being · Healthcare professionals · Environmental attributes

---

Z. Hammouni (✉) · W. Wittich
CRIR (Centre for Interdisciplinary Rehabilitation Research of Greater Montréal), Université de Montréal, Montréal, Canada
e-mail: zakia.hammouni@umontreal.ca

Z. Hammouni
Université McGill, Montréal, Canada

Université du Québec à Trois-Rivière, Trois-Rivière, Canada

E. Arbizzani et al. (eds.), *Technological Imagination in the Green and Digital Transition*, The Urban Book Series, https://doi.org/10.1007/978-3-031-29515-7_91

91.1 Introduction

The work of healthcare professionals in normal times is complex (Canouï et al. 2015). With the advent of COVID-19 as a world-wide pandemic declared in 2020, new measures against this global epidemic require healthcare professionals to work harder and longer than before, due to the high number of patients affected by this disease. With all the fatigue that healthcare professionals faced during this pandemic, policymakers, managers and designers need to pay more attention to healthcare professionals' well-being.

In recent years, models have been developed with attributes of the hospital environment targeting different users of the hospital (patients or healthcare professionals) and describing attributes of the physical environment that influence the outcomes of health and well-being (Ulrich et al. 2010; Huisman et al. 2012; Zhang et al. 2019).

Given the complexity of the hospital environment, these models include many attributes, and yet, these are not exhaustive.

In office work environments, by comparison, there is ample research that has focused on employee well-being to improve business performance. More and more companies are incorporating design elements to achieve the goal of improving employee performance by promoting employee well-being.

In work environments, on the one hand, WELL certification (created by the International WELL Building Institute), attempts to improve occupants' health, comfort, and well-being through environmental attributes (Lambert 2019). On the other hand, the biophilic design approach attempts to reconnect individuals to nature by introducing design elements that are inspired by nature (Akrami and Habibi 2018).

91.2 Towards Less Stressful, User-Centered Contemporary Hospitals

Healthcare professionals face a multitude and a diverse amount of stress-related. Literature reveals that in addition to other stressors, the physical environment is a stressor that negatively impacts healthcare professionals, their well-being, and their performance and quality of care, in addition to its impact on patient and visitor recovery and well-being (Canouï et al. 2015; Hammouni 2021).

The physical environment for healthcare professionals is not limited to functional aspects, even though these are more than essential to ensure the quality of care, but also to limit stress among these professionals. Considering the lived experience of these healthcare professionals, some attributes of the environment were identified to improve the quality of their work life by considering the hospital not just as a "healing machine", but as a living environment with all that it represents for its users.

There is already evidence that built environment can have a positive or a negative impact on hospital users. However, several aspects remain to be developed based on healthcare professionals' perceptions (Hammouni 2021). The focus of the existing literature is on the patients' perspective and not on the healthcare professionals' perspective. Some studies describe how some accommodations and features of working conditions improve staff working conditions. However, there is a lack of evidence on factors related to the physical environment and healthcare professionals' well-being (Huisman et al. 2012).

In this perspective, this paper focuses on a contemporary hospital with an interest in the lived experience of healthcare professionals and the environmental attributes that promote their well-being.

91.3 Methodology and Setting

Within a constructivist approach, the study was conducted in two care units in a large hospital in a major urban center in Canada. Data collection was completed before the COVID-19 pandemic—from August to November 2018.

91.3.1 Data Collection and Data Analysis

A spatial evaluation of the two care units was conducted; understand the interaction of health professionals in their work environment and to understand their spatial behavior (Ziesel 2006) during the 3 shifts of a 24-h workday (day, evening, and night shifts). The observations of both the physical setting and the healthcare professionals' behavior were conducted within 8 weeks (270 h). Subsequently, 44 semi-structured interviews (Gauthier 2009) were conducted with various healthcare professionals working in the two care units studied (physicians, physiotherapists, occupational therapists, nutritionist, pharmacists, beneficiary attendants, nurses).

During the observation periods, we invited and scheduled appointments with health care professionals to participate in the interviews. The questions asked were for example: 1. For your well-being at work, how do you find the design of this hospital and your care unit? 2. What attributes are/would be favorable for your well-being in this hospital? 3. Which attributes in the care unit/hospital generate stress during your work shift?

These interviews were completed by asking participants to create a mental map. Healthcare professionals were asked to draw environmental' attributes whether they feel important as supportive or negative for them in the care unit or the whole hospital. These mental images validated the salient components of the hospital environment as perceived by the healthcare professionals.

Using a comparative and interpretive approach for data analysis, an inductive approach was used and that included emergent themes and categories (Gauthier 2009). Highlighted categories of themes were subsequently compared with existing data in the scientific literature. Emergent environmental attributes for healthcare well-being are listed in the Table 91.1.

Table 91.1 The well-being environmental attributes studied

Well-being attributes for healthcare professionals	
Spatial pattern	Quality and accessibility of break and lounge areas Proximity to gardens/terraces
Art	
The views and access to nature	
Air conditioning	Temperature • Temperature control • Air quality
Daylighting	Daylight in resting areas Daylight in workspaces Light control: existence of blinds—convenience of handling
Artificial lighting	Quantity and quality of lighting Control, Absence of glare
Acoustic comfort	Surface finish (noise absorbing vs noise reflecting (ceiling, walls, floor)) Equipment noise: monitors, paging, alarms, call systems, cards, doors Conversation noise Acoustic walls Music
Spatial orientation and wayfinding	Colors Landmarks Clear signage
Other well-being attributes	Communications technology Internet access Furniture in break area, furniture finish, furniture design

91.4 The Hospital Case Study

This mega hospital opened in September 2017 in a major metropolitan city. It is composed of several buildings connected by corridors and/or footbridges. The inpatient departments (care units) are in the same building stacked from the 8th to the 20th floor.

At each floor, circulation spaces (elevators, large corridor, and staircases), 2 family lounges and a bathroom, connect two care units (south and north one). Within a linear layout, two long corridors (79 or 90-m length) run along each care unit. Four other corridors are perpendicular to the 2 long ones and 3 of them serve the nursing workstations. Peripheral spaces include 36 single patient rooms, and the central area in the care unit is dedicated to staff spaces (Fig. 91.1).

Interior attributes include daylighting in the circulation spaces, waiting areas, and patient's room. These spaces allow panoramic views, particularly, in the patient room.

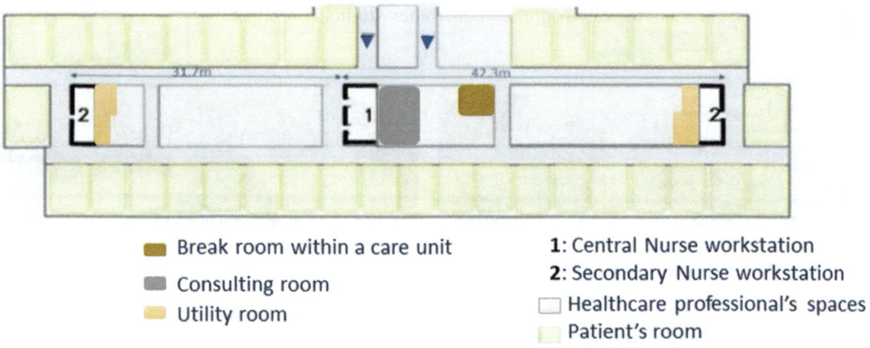

Fig. 91.1 South care unit in hospital case study. *Source* modified from Hammouni (2020)

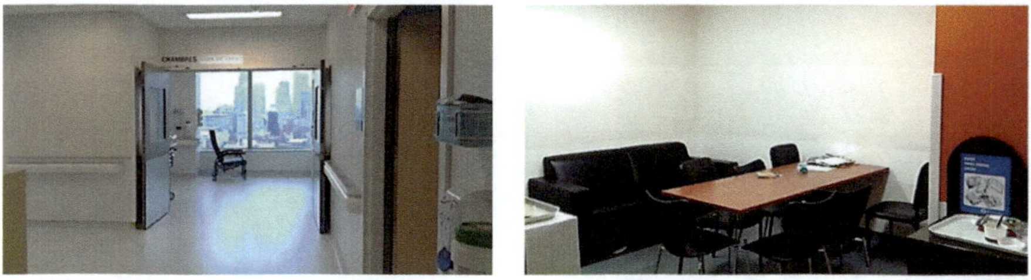

Fig. 91.2 Left: View from care unit corridors. Right: Break room for staff within the south care unit. *Source* Hammouni (2021)

This room (20 m^2) includes large windows, a warm ambience, furniture for patients, and their family members and facilities for healthcare professionals.

The break room (($\pm$) 12,7 m^2) dedicated to the healthcare professionals is not provided with windows and daylighting with and include few furniture (Fig. 91.2).

The staff lounge is a quiet place in the hospital and offers beautiful panoramic views of city. It is dedicated solely to the staff and includes a variety of furniture. In the peripheral zone, extendable chairs are oriented toward the windows allowing panoramic views. A system of blinds allows a control of daylighting and the sun's glare.

A second are, in front of the bay window overlooking a small terrace, contains computers which can be used by the staff during their break. The other space of the lounge is dedicated to allow socialization among the staff, by integrating a variety of furniture (tables, seats, kitchenette area, sink) (Fig. 91.3).

The cafeteria located on the second floor of the care units building includes service space connected to a large room, a small dining rooms and a terrace (total capacity over 600 seats). It is arranged in two floors including these rooms and the agora located on the ground floor (Fig. 91.4). It offers views to the outside through curtained walls that let in natural light throughout the space and contains artworks as in all public spaces of the hospital.

Fig. 91.3 Staff lounge outside. *Source* Hammouni (2021)

Fig. 91.4 Agora/cafeteria. *Source* Hammouni (2021)

91.5 Results and Discussion

According to healthcare professionals' perception of the physical environment in the hospital, despite their positive evaluation of this environment, some attributes are not ideal for their well-being. Break areas are not very popular because of their distance from the care unit or lack of space, with the cafeteria located on the second floor in same building, and the staff lounge located on the 15th floor in another building connected to care units one. This lounge has access to one care unit in 15th floor, but it is used on a limited basis to reduce the number of people coming and going.

However, the observations we made revealed results that do not systematically align with the narratives of the professionals that we interviewed (Gauthier 2009). The use of the hospital cafeteria studied is a representative example of this dissociation of results. In our observations, we noted that this space is occupied continuously, specifically during the lunch hour period which runs from 11:00am to 1:30 pm. The cafeteria is very dense, and the seats are all occupied during this time interval. We also noted a reverberation of noise due mainly to conversations. In the interviews, the vast majority of healthcare professionals stressed that they do not use this space because they cannot always find available seats, and the noise in the cafeteria does not bring them well-being during their breaks.

The spaces dedicated to healthcare professionals do not have windows and do not offer views of the outside in the care unit, except for the trainee rooms in the north care unit. The colors and windows in the hallways are viewed positively by the overall healthcare professionals and are positive elements for healthcare professionals' well-being in this hospital. The views from the workstations are also an element of well-being at work that was noted.

The healthcare professionals expressed their expectations of well-being in their work environment in the care unit. They emphasize the importance of providing them a larger break room within the care unit to allow them to recover from their fatigue and stress, even if these breaks are short. The current surface of this room does not allow to offer well-being and relaxation for the healthcare professionals.

In relation to the existing physical setting, a break room promotes socialization between different types of professionals. Social relationships that promote teamwork in a physical setting are the healthcare professional's well-being expectations. But decentralized nurse workstations, which are functionally distant [08], do not promote socialization among healthcare professionals.

Access to daylighting and outdoor views is an important attribute of well-being during the professionals' break periods (Zhang et al. 2019; Lambert 2019; Akrami and Habibi 2018; Hammouni 2021; Ziesel 2006; Gauthier 2009; Dalke et al. 2006; Alimoglu and Donmez 2005). Alimoglu and Donmez (2005) had pointed out that exposure to natural light for at least 3 h for healthcare professionals can limit burnout (Dalke et al. 2006; Alimoglu and Donmez 2005). Thus, daylight positively affects the quality-of-care services.

The analyzes show that daylight improves the relationship between professionals and the patient and his family. Healthcare professionals are positively affected by their exposure to daylight during their shift in the hospital as the previous studies have pointed out (Marberry 2006). When exposed to daylight, these professionals experience less fatigue and are more motivated in their work (Marberry 2006).

Noise is a negative attributor for the well-being of healthcare professionals and can generate stress. It is one of the attributes of the physical environment that needs to be further improved in the hospital work environment, particularly in the nursing stations and the consultant's room as they are open and very dense, especially during shift change periods.

Among the organizational elements, participants highlighted several factors that affect their well-being at work. The element most frequently mentioned by most healthcare professionals was the possibility of having a flexible schedule that would allow healthcare professionals to have the choice of changing their schedule by period and not staying on the same schedule (for example: a night shift only or having the obligation to work only between 8:00 am and 5:00 pm).

On the other hand, these healthcare professionals expressed a need for adequate staffing levels in relation to the workload. The lack of nursing and orderly staff generates stress among healthcare professionals and leads them to perform additional tasks. This agrees with (Canouï et al. 2015) who pointed out that the lack of nursing staff is a stressor. Increasing the number of nurses and orderlies and a professional-to-patient ratio that needs to be reduced could work to limit stress in the care unit workplace.

Another well-being attribute that was highly cited by the interviewed healthcare professionals was the management of patient records, because healthcare professionals spend a lot of avoidable time searching for displaced paper patient records. These records, which are stored in a cart at each nursing station, are often moved from their place.

One of the insights developed in this study pointed to issues related to the values attributed to healthcare professionals. It is obvious that healthcare professionals work for patients' well-being, but they also need to benefit good conditions in their workplace. This issue emerges from the healthcare professionals' narratives concerning their quality of life at work and their break areas and its impact on the perceived quality of care they provide to the patient.

Lastly, a physical environment in the hospital that considers the needs of different hospital users can therefore improve well-being (Bates 2018) and thus allows for healthcare professional satisfaction and staff retention (Hammouni 2021).

91.6 Conclusion

This study considered the lived experience of various healthcare professionals in a complex work environment of the hospital setting. It generated reflections on the important environmental attributes that can ensure their well-being while ensuring the patient's well-being. It is not enough to design hospitals that are esthetically more pleasing or more sophisticated in terms of introducing advanced technologies than existing traditional hospitals. The analysis reveals the importance of creating hospitals that help patients to recover by ensuring their well-being, but also that help healthcare professionals to perform better in their work by ensuring their well-being as well. This new crisis has demonstrated the importance of providing hospitals with sufficient break areas relative to the number of professionals working in an institution to allow them to quickly recover from the extreme stress they face even during short break periods.

This study is an exploratory one and therefore must be viewed with an appropriate understanding of these limitations. As a qualitative study, the sample size was intentionally small. Future studies could consider expanding the sample for generalization, dig deeper into the attributes that specifically promote well-being, as well as more explanatory assessments of the lived experience of healthcare professionals on the wards based on the results of this study.

Acknowledgements We acknowledge the participants in this study and Dr Tiiu Poldma for her significant support.

References

Akrami B, Habibi S (2018) Applicable biophilic principles on hospitals retrofitting. The case study of turkish public hospitals. In: ANFA 2018 conference, Danemark

Alimoglu MK, Donmez L (2005) Daylight exposure and the other predictors of burnout among nurses in a University Hospital. Inter J of Nurs Stud 42:549–555

Bates V (2018) Humanizing healthcare environments: architecture, art and design in modern hospitals. Des Health 2:5–19

Canouï P, Mauranges A, Florentin A (2015) Le burn out à l'hôpital: le syndrome d'épuisement professionnel des soignants. Issy-les-Moulineaux

Dalke H, Little J, Niemann E, Camgoz N, Steadman G, Hill S, Stott L (2006) Colour and lighting in hospital design. Optics Laser Tech 38:343–365

Gauthier B (2009) Recherche sociale: de la problématique à la collecte des données. Paris, France: PUF

Hammouni Z (2020) The contemporary hospital design: a contextual study of the lived experience of healthcare professionals and quality of care. In: Proceedings of the 6th International Conference on Design4Health 1: 179–188

Hammouni Z (2021) L'expérience vécue des professionnels de la santé dans un grand centre hospitalier universitaire à Montréal : un regard contextuel, Université de Montréal

Huisman E, Morales E, Van Hoof J, Kort HSM (2012) Healing environment: a review of the impact of physical environmental factors on users. J Buil and Env 58:70–80

Lambert C (2019) Répercussions de la certification WELL sur la performance environnementale du cycle de vie d'un bâtiment, Université de Sherbrooke

Marberry SO (2006) Improving healthcare with better building design. Health Administration Press, Chicago, IL

Ulrich RS, Berry LL, Quan X, Parish JT (2010) A conceptual framework for the domain of evidence-based design. HERD 4:95–114

Zhang Y, Tzortzopoulos P, Kagioglou M (2019) Healing built environment effects on health outcomes: environment–occupant–health framework. Build Res Inf 47(6):747–766

Ziesel (2006) Inquiry by design. Environment/Behavior/Neuroscience in Architecture, Interiors, Landscape and Planning. WW Norton & Co

GPSR Compliance

The European Union's (EU) General Product Safety Regulation (GPSR) is a set of rules that requires consumer products to be safe and our obligations to ensure this.

If you have any concerns about our products, you can contact us on ProductSafety@springernature.com

In case Publisher is established outside the EU, the EU authorized representative is:

Springer Nature Customer Service Center GmbH
Europaplatz 3
69115 Heidelberg, Germany

Batch number: 10852827

Printed by Printforce, the Netherlands